Handbook of Primate Behavioral Management

T0225345

Handbook of Primate Behavioral Management

Edited by
Steven J. Schapiro

CRC Press
Taylor & Francis Group
Boca Raton London New York

CRC Press is an imprint of the
Taylor & Francis Group, an **informa** business

CRC Press
Taylor & Francis Group
6000 Broken Sound Parkway NW, Suite 300
Boca Raton, FL 33487-2742

First issued in paperback 2020

ISBN 13: 978-0-367-57367-6 (pbk)
ISBN 13: 978-1-4987-3195-9 (hbk)

Library of Congress Cataloging-in-Publication Data

Names: Schapiro, Steven Jay, editor.
Title: Handbook of primate behavioral management / editor, Steven J. Schapiro.
Description: Boca Raton : Taylor & Francis, 2017. | Includes bibliographical references and index.
Identifiers: LCCN 2016057593 | ISBN 9781498731959 (hardback : alk. paper)
Subjects: | MESH: Primates | Behavior, Animal | Behavior Control--methods | Animal Husbandry--methods | Animals, Laboratory
Classification: LCC QL737.P9 | NLM QY 60.P7 | DDC 599.815--dc23
LC record available at https://lccn.loc.gov/2016057593

**Visit the Taylor & Francis Web site at
http://www.taylorandfrancis.com**

**and the CRC Press Web site at
http://www.crcpress.com**

Contents

Part I
The Basics

Chapter 1

Steven J. Schapiro

Chapter 2

**Kate C. Baker, Mollie A. Bloomsmith, Kristine Coleman, Carolyn M. Crockett,
Julie Worlein, Corrine K. Lutz, Brenda McCowan, Peter Pierre, and Jim Weed**

Chapter 3

Jann Hau and Kathryn Bayne

Chapter 4

**Tammie L. Bettinger, Katherine A. Leighty, Rachel B. Daneault,
Elizabeth A. Richards, and Joseph T. Bielitzki**

Part II
Content Areas with Behavioral Management Implications

Chapter 5

John P. Capitanio

Chapter 6

**Melinda A. Novak, Amanda F. Hamel, Amy M. Ryan, Mark T. Menard, and
Jerrold S. Meyer**

Chapter 7

Kristine Coleman

Chapter 16
Collaborative Research and Behavioral Management...243

Steven J. Schapiro, Sarah F. Brosnan, William D. Hopkins, Andrew Whiten, Rachel Kendal, Chet C. Sherwood, and Susan P. Lambeth

Chapter 17
Pairing Strategies for Cynomolgus Macaques..255

Keely McGrew

Chapter 18
Managing a Behavioral Management Program..265

Susan P. Lambeth and Steven J. Schapiro

**Part IV
Genera-Specific Behavioral Management**

Chapter 19
Behavioral Management of *Macaca* Species (except *Macaca fascicularis*) 279

Daniel Gottlieb, Kristine Coleman, and Kamm Prongay

Chapter 20
Behavioral Management of Long-Tailed Macaques (*Macaca fascicularis*)..................................305

Paul Honess

Chapter 21
Behavioral Management of *Chlorocebus* spp...339

Matthew J. Jorgensen

Chapter 22
Behavioral Management of *Papio* spp..367

Corrine K. Lutz and C. Heath Nevill

Chapter 23
Behavioral Management of *Pan* spp...385

Lisa Reamer, Rachel Haller, Susan P. Lambeth, and Steven J. Schapiro

Chapter 24
Behavioral Management of Neotropical Primates: *Aotus*, *Callithrix*, and *Saimiri*.....................409

Lawrence Williams and Corinna N. Ross

Preface

Nonhuman primates (NHPs) live in captive settings for a variety of reasons. No matter what the reason, it is imperative that these behaviorally, physiologically, and socially sophisticated animals receive absolutely the best care possible while living in captivity. The *Handbook of Primate Behavioral Management* aims to provide the reader with a wealth of information relevant to the behaviotral management of captive NHPs. The *Handbook* contains genera-specific chapters on the behavioral management of macaques, African green monkeys, baboons, chimpanzees, Neotropical monkeys, and prosimians. As importantly, the *Handbook* also contains sections/chapters focusing on the basic science content and implementation programs that are driving ongoing refinements in applied primate behavioral management. Every chapter in the *Handbook* is written with the intent of providing readers with information that will be useful in their own behavioral management programs.

This handbook was developed in tandem with the Primate Behavioral Management Conference, as part of our continuing quest to provide captive NHPs with the best physical and social environments possible. Many of the *Handbook*'s content, implementation, and products/services chapters have evolved from presentations given at the first two Primate Behavioral Management Conferences. The inspiration for both the book and the conference came from our desire to provide those responsible for the captive management of NHPs with access to the most meaningful applications of NHP behavioral research. We felt that it would be extremely valuable for those managing primates to understand how the results of biobehavioral assays (Capitanio, this volume), temperament assessments (Coleman, this volume), studies of depressive behavior (Shively, this volume), and/or social network analyses (McCowan and Beisner, this volume) could be applied in their NHP situations.

I have had the privilege of working with some really forward-thinking scientists in my career. All have helped me along the way and certainly deserve some of the credit if you think this handbook is valuable to you, and perhaps more importantly, if you think the handbook is valuable for your primates. G. Mitchell, PhD and Bernadette Marriott, PhD taught me how to be a primatologist. Michale E. Keeling, DVM and Christian R. Abee, DVM allowed me to become a behavioral manager for NHPs. Mollie A. Bloomsmith, PhD was a critical mentor for many years. Jann Hau, MD and William D. Hopkins, PhD have provided numerous opportunities for me to facilitate collaborative science and to expand my horizons. Susan P. Lambeth has been an inspirational colleague for almost 30 years.

NHPs are special animals that attract and require special people to work with them. None of the information included in this handbook would have been attainable without the efforts of many groups of skilled, dedicated, and caring researchers, technicians, students, caregivers, and veterinarians. This handbook is intended to provide you, the reader, with insight and guidance related to behavioral management; to acknowledge the commitment and advancements made by those who currently work with NHPs; and most importantly, to *honor the contributions made by the primates themselves* to the welfare of humans.

Steven J. Schapiro
The University of Texas MD Anderson Cancer Center

Editor

Steven J. Schapiro is an associate professor of comparative medicine in the Department of Veterinary Sciences at the Michale E. Keeling Center for Comparative Medicine and Research of The University of Texas MD Anderson Cancer Center. Dr. Schapiro earned his PhD from the University of California at Davis in 1985 after receiving his BA in behavioral biology from Johns Hopkins University. He completed a postdoctoral research fellowship at the Caribbean Primate Research Center of the University of Puerto Rico.

In 1989, he joined the Department of Veterinary Sciences at MD Anderson's Keeling Center and has been there ever since. In 2009, Dr. Schapiro was designated an honorary professor in the Department of Experimental Medicine at the University of Copenhagen, Denmark.

He is a founding faculty member of both the Primate Training and Enrichment Workshop and the Primate Behavioral Management Conference, educational programs conducted at the Keeling Center that have reached over 800 individuals from primate facilities around the globe.

Dr. Schapiro has coauthored approximately 170 peer-reviewed papers and book chapters examining various aspects of nonhuman primate behavior, management, and research.

Dr. Schapiro has participated in international meetings and courses on primatology and laboratory animal science in North America, Europe, Asia, and Africa. He has edited the three volumes of the third edition of the *Handbook of Laboratory Animal Science* along with Jann Hau. He has also coedited one issue of the *ILAR Journal*. He is a member of a number of primatology and animal behavior societies and is currently the treasurer and vice president for membership of the International Primatological Society. He is also a past president, former treasurer, and former meeting coordinator of the American Society of Primatologists, as well as an honorary member of the Association of Primate Veterinarians. Dr. Schapiro is an advisor or consultant for a number of primate facilities that produce, manage, and conduct research with nonhuman primates in the United States and abroad.

Contributors

Kate C. Baker
Tulane National Primate Research Center
Tulane University Covington, Louisiana

Kathryn Bayne
AAALAC International
Frederick, Maryland

Brianne Beisner
California National Primate Research
    Center
University of California, Davis
Davis, California

Tammie L. Bettinger
Disney's Animal Kingdom
Lake Buena Vista, Florida

Joseph T. Bielitzki
Consultant
Orlando, Florida

Mollie A. Bloomsmith
Yerkes National Primate Research Center
Emory University
Atlanta, Georgia

Sarah F. Brosnan
Michale E. Keeling Center for Comparative
    Medicine and Research
The University of Texas MD Anderson Cancer
    Center
Bastrop, Texas
and
Language Research Center
Georgia State University
Atlanta, Georgia

Nancy G. Caine
Department of Psychology
California State University San Marcos
San Marcos, California

John P. Capitanio
California National Primate Research Center
University of California, Davis
Davis, California

Kristine Coleman
Oregon National Primate Research Center
Oregon Health Sciences University Beaverton,
    Oregon

Carolyn M. Crockett
Washington National Primate Research Center
    University of Washington
Seattle, Washington

Rachel B. Daneault
Disney's Animal Kingdom
Lake Buena Vista, Florida

Meg H. Dye
Duke Lemur Center
Durham, North Carolina

Luis Fernandez
Bioculture (Mauritius) Ltd.
Riviere Des Anguilles, Mauritius

Daniel Gottlieb
Oregon National Primate Research Center
Beaverton, Oregon

Melanie L. Graham
Department of Surgery
University of Minnesota
Saint Paul, Minnesota

Mary-Ann Griffiths
Bioculture (Mauritius) Ltd.
Riviere Des Anguilles, Mauritius

Rachel Haller
Michale E. Keeling Center for Comparative
    Medicine and Research
The University of Texas MD Anderson Cancer
    Center
Bastrop, Texas

Amanda F. Hamel
Department of Brain and Psychological
    Sciences
University of Massachusetts, Amherst
Amherst, Massachusetts

Jann Hau
Department of Experimental Medicine
University of Copenhagen
Copenhagen, Denmark

Paul Honess
Bioculture (Mauritius) Ltd.
Riviere Des Anguilles, Mauritius

William D. Hopkins
Neuroscience Institute and Language Research
  Center
Georgia State University
Atlanta, Georgia
and
Michale E. Keeling Center for Comparative
  Medicine and Research
The University of Texas MD Anderson Cancer
  Center
Bastrop, Texas

Lydia M. Hopper
Lester E. Fisher Center for the Study and
  Conservation of Apes
Lincoln Park Zoo
Chicago, Illinois

Eric Hutchinson
Department of Molecular and Comparative
  Pathobiology
Johns Hopkins University Medical School
Baltimore, Maryland

Matthew J. Jorgensen
Department of Pathology
Wake Forest School of Medicine
Winston-Salem, North Carolina

Rachel Kendal
Department of Anthropology
Durham University
Durham, United Kingdom

Susan P. Lambeth
Michale E. Keeling Center for Comparative
  Medicine and Research
The University of Texas MD Anderson Cancer
  Center
Bastrop, Texas

Robert D. Latzman
Department of Psychology
Georgia State University
Atlanta, Georgia

Katherine A. Leighty
Disney's Animal Kingdom
Lake Buena Vista, Florida

Corrine K. Lutz
Southwest National Primate Research
  Center
Texas Biomedical Research Institute
San Antonio, Texas

Elizabeth R. Magden
Michale E. Keeling Center for Comparative
  Medicine and Research
The University of Texas MD Anderson Cancer
  Center
Bastrop, Texas

Brenda McCowan
California National Primate Research
  Center
University of California, Davis
Davis, California

Keely McGrew
Charles River Laboratories
Houston, Texas

Mark T. Menard
Department of Brain and Psychological
  Sciences
University of Massachusetts, Amherst
Amherst, Massachusetts

Jerrold S. Meyer
Department of Brain and Psychological
  Sciences
University of Massachusetts, Amherst
Amherst, Massachusetts

C. Heath Nevill
Southwest National Primate Research
  Center
Texas Biomedical Research Institute
San Antonio, Texas

Melinda A. Novak
Department of Brain and Psychological
    Sciences
University of Massachusetts, Amherst
Amherst, Massachusetts

Peter Pierre
Wisconsin National Primate Research Center
University of Wisconsin
Madison, Wisconsin

Kamm Prongay
Oregon National Primate Research Center
Oregon Health Sciences University
Beaverton, Oregon

Lisa Reamer
Michale E. Keeling Center for Comparative
    Medicine and Research
The University of Texas MD Anderson Cancer
    Center
Bastrop, Texas

Elizabeth A. Richards
Disney's Animal Kingdom
Lake Buena Vista, Florida

Corinna N. Ross
Department of Arts & Sciences
Texas A&M University San Antonio
San Antonio, Texas

Amy M. Ryan
Neuroscience and Behavior Graduate
    Program
University of Massachusetts, Amherst
Amherst, Massachusetts

Steven J. Schapiro
Michale E. Keeling Center for Comparative
    Medicine and Research
The University of Texas MD Anderson
    Cancer Center
Bastrop, Texas
and
Department of Experimental Medicine
University of Copenhagen
Copenhagen, Denmark

Carrie L. Schultz
LabDiet
St. Louis, Missouri

Chet C. Sherwood
Department of Anthropology
George Washington University
Washington, DC

Carol A. Shively
Department of Pathology
Wake Forest School of Medicine
Winston-Salem, North Carolina

Jim Weed
Centers for Disease Control and
    Prevention
Atlanta, Georgia

Andrew Whiten
School of Psychology & Neuroscience
University of St. Andrews
St. Andrews, Scotland

Lawrence Williams
Michale E. Keeling Center for Comparative
    Medicine and Research
The University of Texas MD Anderson
    Cancer Center
Bastrop, Texas

Teresa Woodger
Lomir Biomedical Inc.
Notre-Dame-de-l'Île-Perrot, Quebec,
    Canada

Julie Worlein
Washington National Primate Research
    Center
University of Washington
Seattle, Washington

The Basics

CHAPTER 1

Introduction to the *Handbook of Primate Behavioral Management*

Steven J. Schapiro
The University of Texas MD Anderson Cancer Center and University of Copenhagen

CONTENTS

Welcome to the *Handbook of Primate Behavioral Management* (*HPBM*). This handbook contains 29 chapters divided into six parts, all of which focus on aspects of primate behavioral management. The overall goal of the *HPBM* is to provide those responsible for the development and/or implementation of behavioral management programs for nonhuman primates (NHPs) with a plethora of information, guidance, and data that will allow them to do everything within their power to guarantee that their animals are living in the best conditions possible. A more specific goal involves the presentation of the science of behavioral management, so that behavioral managers can base their decisions on relevant empirical evidence. If the data show that the subadult male offspring of high-ranking females cause social instability in large groups of rhesus macaques living in field cages (McCowan and Beisner 2017), then this information should be acted upon to prevent instability in large rhesus groups.

While the *HPBM* does include an entire part (Part IV; 7 chapters) on the behavioral management of specific taxonomic groups (macaques, chimpanzees, prosimians, etc.), I did not want this volume to simply be a collection of genera-specific or species-specific behavioral management plans. In my opinion, the value of presenting such plans is infinitely enhanced by the inclusion of the scientific underpinnings that drive the development and implementation of those plans. Hence, Parts II and III ("content" and "implementation," described in more detail later) precede the discussion of behavioral management plans by taxonomic groups. Similarly, I wanted to take this opportunity to present their work to those who design, produce, and supply products, equipment, techniques, and services that facilitate behavioral management efforts (Part V).

Most of the chapter authors in this handbook work with nonhuman primates in "research" settings; so the information contained in their chapters is likely to be most relevant for primates living in such settings. However, many of the guidelines and recommendations contained in this volume will also be valuable to those managing nonhuman primates in other settings, including zoological parks and sanctuaries.

The *HPBM* coevolved with the Primate Behavioral Management Conference (PBMC) that we conduct at the Michale E. Keeling Center for Comparative Medicine and Research of The University of Texas MD Anderson Cancer Center in Bastrop, TX. Both the handbook and the conference arose from the ever-increasing importance of primate-focused, *theoretical* investigations to the *applied* behavioral management of nonhuman primates. Capitanio and colleagues (Capitanio et al. 2011, 2017; Gottlieb and Capitanio 2013; Gottlieb et al. 2013; Capitanio 2017) were conducting biobehavioral assays to study temperament, and the results of these investigations had important implications for pairing, grouping, and training rhesus macaques (Capitanio et al. 2017). Similarly, McCowan and colleagues (McCowan et al. 2008, 2011, 2016; Beisner et al. 2011; Beisner and McCowan 2013; McCowan and Beisner 2017) were performing social network analyses, the results of which had important practical applications to prevent deleterious aggression in large groups of rhesus monkeys. This handbook, and the PBMC, are filled with similar examples of science resulting in, and driving, applied behavioral management programs and decisions.

The scientific methods and findings included in Part II (Content Areas with Behavioral Management Implications) of the *HPBM* are concisely described and rigorous. It takes many hours of sophisticated observations and analysis to even partially understand the theory and meaning of biobehavioral assays (BBA; Capitanio 2017); stress (Novak et al. 2017), temperament (Coleman 2017), and depression (Shively 2017) assessments; antipredator behavior (Caine 2017); and social network analysis (McCowan and Beisner 2017). Those reading this handbook (and those attending the PBMC) are unlikely to have the time, skills, and/or resources to apply the relevant methods, conduct appropriate observations, and perform the statistical analyses. However, almost all of those reading this handbook should be able to apply specific findings from each of these research programs to enhance the welfare of the primates in their care. Therefore, most of the chapters are written in such a way that important assessment techniques and/or findings from each investigational approach (that should be applicable at any primate facility) have been identified.

Part I of the *HPBM* includes four chapters by Schapiro, Baker and colleagues, Hau and Bayne, and Bettinger and colleagues, respectively. Schapiro's chapter, which you are currently reading, establishes the motivation for the handbook and describes the other chapters herein. The chapter by Baker and colleagues beautifully describes the Behavioral Management Consortium (BMC) of the National Primate Research Centers in the United States. The BMC is in the process of attempting to apply standard definitions and descriptions to many aspects (behaviors, abnormalities, therapies, etc.) of primate behavioral management, applicable not only at the National Primate Research Centers themselves, but should also be pertinent at most, if not all, primate facilities (e.g., academic laboratories, contract research organizations, zoological parks, sanctuaries, breeding "farms"). This chapter has much to offer those responsible for primate behavioral management. In Chapter 3, Hau and Bayne provide a brief description of the regulations and guidelines around the globe that are relevant when keeping primates in captivity, especially in research settings. Their chapter emphasizes those portions of *The Guide* (NRC 2011) and the European Directive (2010) that have the strongest implications for primate behavioral management. Part I closes with a general discussion by Bettinger and colleagues of the effects of the obvious, and not-so-obvious, aspects of the environment on primates in captivity, helping to establish a foundation for information contained in some of the later chapters.

Part II of the *HPBM* includes seven chapters by individuals/groups considered to be "content experts" in their respective fields. As already mentioned, the focus of the chapters in this part is the *application of the theoretical findings* from the work described to aspects of behavioral management. All of the chapters in Part II are written in a more or less similar format, with a description of the techniques used to study the "content" first, followed by descriptions of findings that are particularly relevant to address behavioral management questions, and concluding with recommendations and guidelines concerning the ways in which the methods and/or data from the content area can be *directly applied* to enhance behavioral management.

In Chapter 5, Capitanio describes his BBA program and the ways in which the data gathered when rhesus monkeys are 3 months old can provide valuable insight and guidance for multiple behavioral management decisions during the course of the animals' lifetime. In many ways, the results of Capitanio's work provided the stimulus for both the *HPBM* and the PBMC. Novak and colleagues, renowned experts in the study of abnormal behaviors in NHPs, discuss, in Chapter 6, abnormal behaviors and conditions in captive NHPs. As hair loss and self-injurious behavior can be extremely problematic for those working with NHPs, their empirical work should be applicable in many captive settings. Chapter 7, written by Coleman, describes her work on primate temperament and provides a set of simple, yet highly valuable, assessments that can be performed to help minimize incompatibility when forming pairs of macaques. These same simple temperament assessments can also be used to address questions related to trainability and a variety of other issues that inform behavioral management decisions. Shively, in Chapter 8, focuses on the behavior and physiology of depressed cynomolgus macaques. Her multifaceted research program has addressed many issues related to depression and its effects on scientific data, and she provides invaluable guidance for dealing with NHPs exhibiting depressed behaviors/postures. In Chapter 9, Caine describes primate antipredator behaviors and how these natural behavior patterns can impact behavioral management decisions. Responses to humans, appropriate sleeping sites, and the aftermath of simulated predatory events all must be considered if one is to achieve optimal behavioral management. Hopkins and Latzman, in Chapter 10, describe how noninvasive behavioral research procedures can positively affect the welfare of captive primates, especially chimpanzees. They strongly emphasize the need to continue to obtain valuable data from captive chimpanzees to benefit humans, captive chimpanzees, and even endangered wild chimpanzees. Part II closes with Chapter 11, which presents a discussion on social network analysis by McCowan and Beisner. This chapter contains a fair bit of "math," which may put some readers off, but I strongly encourage you to read this chapter. As mentioned previously, and as will be discussed by Bloomsmith (2017) in the final chapter of the handbook, the types of "deep" analyses that are possible with network analysis are among the most important tools for the continuing evolution of primate behavioral management strategies and programs.

Part III of the *HPBM* includes seven chapters by individuals/groups considered to be "implementation experts" in their respective fields. The focus of these chapters is the application of specific techniques to behavioral management. Some of the techniques are experimental, some are procedural, and some emphasize communication and organizational strategies. However, all of the chapters include specific and useful guidance that can be used to enhance the behavioral management of captive primates.

In Chapter 12, Graham describes her inspiring work that incorporates positive reinforcement training (PRT) techniques into diabetes-related research. This chapter clearly demonstrates how refinements associated with the application of behavioral management techniques can significantly enhance the scientific research endeavor. Magden, in Chapter 13, discusses additional benefits associated with PRT, demonstrating how the application of these techniques can significantly enhance the health management of captive NHPs. Primates that voluntarily participate in their own health care are easier to treat, can be treated using additional modalities, and are likely to be healthier. In Chapter 14, Hutchinson discusses *telos*, the "primateness of the primate," as well as the value of effective communication to the ultimate success of behavioral management programs. While he emphasizes communication among veterinarians and behavioral managers, effective communication involving all parties interested in the welfare of captive primates is essential. Hopper (Chapter 15) provides an insightful discussion of the ways in which findings related to social learning can be utilized to facilitate behavioral management. Understanding how socially housed primates learn from one another, and specifically, who learns from whom, can be extremely useful when making socialization-related decisions. Schapiro and colleagues, in Chapter 16, expand some of the ideas presented by Hopkins and Latzman (2017), specifically describing how collaborative research projects, especially those involving noninvasive and stimulating "cognitive" tasks, can positively affect the NHPs that participate. In Chapter 17, McGrew describes her techniques for pairing macaques, an issue that is of extremely high interest to those who manage primates in captivity. This chapter is based on the establishment of an extremely large number of successful pairs. Part III closes with Chapter 18, by Lambeth and Schapiro, which provides some straightforward guidance for building and supporting functional behavioral management programs.

Part IV of the *HPBM* is comprised of seven chapters describing behavioral management programs for different taxonomic groups of primates. Most of the chapters follow a similar format, beginning with a discussion of the natural behaviors of animals and ending with "expert recommendations" that you should be able to incorporate into your own behavioral management program. As Bloomsmith (2017) emphasizes in the final chapter of the handbook, *understanding the natural behavior of the primates under our care is absolutely essential for the design and implementation of the highest quality behavioral management programs.* The authors of these chapters have quite a bit of relevant experience, and so reading their work should provide you with useful ideas concerning strategies to implement as well as those to avoid.

Part IV begins with a fantastic chapter (Chapter 19) by Gottlieb and colleagues on the behavioral management of most macaque species. This chapter only casually deals with *Macaca fascicularis*, as Honess, in Chapter 20, deals specifically with this frequently utilized NHP. Those who manage macaques will learn a great deal from these two chapters. Jorgensen's chapter (Chapter 21) is next, with a take-home message that *Chlorocebus* are not *Macaca*. Chapter 22, by Lutz and Nevill, describes behavioral management strategies for baboons, a large, potentially destructive, but ultimately manageable genus. Reamer and colleagues (Chapter 23) then discuss behavioral management strategies for *Pan* species (common and pygmy chimpanzees), another large, potentially destructive genus. The fact that *Pan* are both extremely intelligent and quite amenable to positive reinforcement training make them an especially important group for identifying techniques to take behavioral management to the next level. Behavioral management strategies for three genera (*Aotus, Callithrix, Saimiri*) of Neotropical primates are discussed in Chapter 24 by Williams and Ross. Clearly, behavioral management programs for small, New World monkeys should differ from programs for large, Old World monkeys. Similarly, behavioral management strategies for prosimians differ from those for simian primates, the implications of which are discussed in some detail by Dye in the closing chapter (Chapter 25) of this part.

Part V of the *HPBM* contains three short chapters by entities/companies that design, produce, and supply items that facilitate the performance of behavioral management activities with captive NHPs. Woodger, in Chapter 26, describes the way that jackets can be used to help collect reliable and valid data, while allowing NHPs to express as much species-typical behavior as possible. Such jackets, in combination with new, noninvasive data collection procedures, continue to have a major impact on many aspects of primate research. Nutritional contributions to behavioral management programs are discussed in Chapter 27 by Schultz. Primarily of value to the environmental enrichment components of behavioral management, feed manufacturers have made considerable strides toward supplying foodstuffs that are both nutritionally complete and stimulate foraging behaviors. The final chapter (Chapter 28) in this part, by Fernandez and colleagues, describes efforts made by breeding farms to "condition" the animals that they supply for research, so that the animals are less fearful and, most importantly, are likely to provide reliable and valid data when they participate in scientific research projects.

The handbook comes to a close with an excellent final chapter (*Part VI*, Chapter 29) by Bloomsmith that summarizes and integrates the previous chapters within the context of the present and future of primate behavioral management. Going forward, she provides recommendations that are likely to significantly enhance the utility of the study and application of the science of primate behavioral management. As with all of the chapters in this handbook, there is much of value to be gleaned from this chapter.

There are just three more things to quickly mention before you venture off into the important chapters of the *HPBM*.

1. You will encounter some common themes, techniques, and terminology as you read the chapters in this handbook. These include, in no particular order: temperament/personality, the human intruder test, *telos*, behavioral assessments, functional simulations, functionally appropriate captive

environments, personalized treatments, prevention vs. cure, refinements, research-management synergisms, PRT, enhancing the definition of animal models, minimizing confounds, and better scientific data. Please pay attention to these when you encounter them.

2. In a similar vein, you will read about similar things that are, at times, labeled in slightly different ways. I tried to establish a consistent framework across chapters, but sometimes it did not make sense to change what the authors had written. Keep your eyes open for similarities and differences in terms used. For instance, the definitions for abnormal behaviors or types of environmental enrichment may differ slightly across chapters, but there are really more similarities than differences in these cases.

3. And finally, do not forget that the goal of the *HPBM* is to provide you with research-derived information that you can use to benefit the nonhuman primates that you care for. Behavioral management is good for the primates and good for science, contributing to the optimization of the welfare of the animals and the reliability and validity of the data. Behavioral management is constantly evolving; we will always be assessing and improving behavioral management strategies; we can always learn more and we can always do more.

REFERENCES

Beisner, B.A., M.E. Jackson, A. Cameron, and B. McCowan. 2011. Effects of natal male alliances on aggression and power dynamics in rhesus macaques. *American Journal of Primatology* 73:790–801.

Beisner, B.A. and B. McCowan. 2013. Policing in nonhuman primates: Partial interventions serve a prosocial conflict management function in rhesus macaques. *PLoS One* 8 (10):e77369.

Bloomsmith, M.A. 2017. Behavioral management of laboratory primates: Principles and projections, Chapter 29. In: Schapiro, S.J. (ed.) *Handbook of Primate Behavioral Management*, 497–513. CRC Press, Boca Raton, FL.

Caine, N.G. 2017. Anti-predator behavior: Its expression and consequences in captive primates, Chapter 9. In: Schapiro, S.J. (ed.) *Handbook of Primate Behavioral Management*, 127–138. CRC Press, Boca Raton, FL.

Capitanio, J.P. 2017. Variation in biobehavioral organization, Chapter 5. In: Schapiro, S.J. (ed.) *Handbook of Primate Behavioral Management*, 55–73. CRC Press, Boca Raton, FL.

Capitanio, J.P., S.A. Blozis, J. Snarr, A. Steward, and B.J. McCowan. 2017. Do "birds of a feather flock together" or do "opposites attract"? Behavioral responses and temperament predict success in pairings of rhesus monkeys in a laboratory setting. *American Journal of Primatology* 79(1):1–11.

Capitanio, J.P., S.P. Mendoza, and S.W. Cole. 2011. Nervous temperament in infant monkeys is associated with reduced sensitivity of leukocytes to cortisol's influence on trafficking. *Brain, Behavior, and Immunity* 25:151–159.

Coleman, K. 2017. Individual differences in temperment and behavioral management, Chapter 7. In: Schapiro, S.J. (ed.) *Handbook of Primate Behavioral Management*, 95–113. CRC Press, Boca Raton, FL.

EU Directive. 2010. Directive 2010/63/EU of the European Parliament and of the Council of 22 September 2010 on the protection of animals used for scientific purposes. *Official Journal of the European Union* L 276/33 (October 20, 2010).

Gottlieb, D.H. and J.P. Capitanio. 2013. Latent variables affecting behavioral response to the human intruder test in infant rhesus macaques (*Macaca mulatta*). *American Journal of Primatology* 75:314–323.

Gottlieb, D.H., J.P. Capitanio, and B. McCowan. 2013. Risk factors for stereotypic behavior and self-biting in rhesus macaques (*Macaca mulatta*): Animal's history, current environment, and personality. *American Journal of Primatology* 75:995–1008.

Hopkins, W.D. and R.D. Latzman. 2017. Future research with captive chimpanzees in the USA: Integrating scientific programs with behavioral management, Chapter 10. In: Schapiro, S.J. (ed.) *Handbook of Primate Behavioral Management*, 139–155. CRC Press, Boca Raton, FL.

McCowan, B., K. Anderson, A. Heagarty, and A. Cameron. 2008. Utility of social network analysis for primate behavioral management and well-being. *Applied Animal Behaviour Science* 109:396–405.

McCowan, B. and B. Beisner. 2017. Utility of systems network analysis for understanding complexity in primate behavioral management, Chapter 11. In: Schapiro, S.J. (ed.) *Handbook of Primate Behavioral Management*, 157–183. CRC Press, Boca Raton, FL.

McCowan, B., B. Beisner, E. Bliss-Moreau, J. Vandeleest, J. Jin, D. Hannibal, and F. Hsieh. 2016. Connections matter: Social networks and lifespan health in primate translational models. *Frontiers in Psychology* 7:433. doi:10.3389/fpsyg.2016.00433.

McCowan, B., B.A. Beisner, J.P. Capitanio, M.E. Jackson, A.N. Cameron, S. Seil, E.R. Atwill, and H. Fushing. 2011. Network stability is a balancing act of personality, power, and conflict dynamics in rhesus macaque societies. *PLoS One* 6(8):e22350.

National Research Council. 2011. *Guide for the Care and Use of Laboratory Animals*, 8th Edition. National Academies Press, Washington, DC.

Novak, M.A., A.F. Hamel, A.M. Ryan, M.T. Menard, and J.S. Meyer. 2017. The role of stress in abnormal behavior and other abnormal conditions such as hair loss, Chapter 6. In: Schapiro, S.J. (ed.) *Handbook of Primate Behavioral Management*, 75–94. CRC Press, Boca Raton, FL.

Shively, C.A. 2017. Depression in captive nonhuman primates: Theoretical underpinnings, methods, and application to behavioral management, Chapter 8. In: Schapiro, S.J. (ed.) *Handbook of Primate Behavioral Management*, 115–125. CRC Press, Boca Raton, FL.

CHAPTER 2

The Behavioral Management Consortium
A Partnership for Promoting Consensus and Best Practices

Kate C. Baker
Tulane National Primate Research Center

Mollie A. Bloomsmith
Yerkes National Primate Research Center

Kristine Coleman
Oregon National Primate Research Center

Carolyn M. Crockett and Julie Worlein
Washington National Primate Research Center

Corrine K. Lutz
Texas Biomedical Research Institute

Brenda McCowan
California National Primate Research Center

Peter Pierre
Wisconsin National Primate Research Center

Jim Weed
Centers for Disease Control and Prevention

CONTENTS

The seven National Primate Research Centers (NPRCs) form a network of institutions serving as a national scientific resource for research using nonhuman primates (NHPs). Each center provides animals, expertise, and specialized facilities and equipment to scientists conducting research aiming to advance human health. In the early 2000s, the coordinators of NHP behavioral management programs from the NPRCs and the National Institutes of Health (NIH) began conferring, sharing information and strategies, collaborating on workshops and symposia at scientific meetings, and coauthoring publications (e.g., Weed et al. 2003; Baker et al. 2007). During this same time frame, NIH was encouraging an increased level of collaboration across the NPRCs with the aim of strengthening communications, leveraging system-wide resources, and facilitating the sharing of information and best practices across institutions (see nprcresearch.org). Established in partnership with the NIH's National Center for Research Resources (now the Office of Research Infrastructure Programs), the NPRC Consortium consists of working groups in a number of areas, including behavioral management.

The Behavioral Management Consortium (BMC) was established and held its inaugural meeting in April of 2007. The BMC has since held annual face-to-face meetings, as well as monthly web conferences. As a formalized working group, the BMC has three main goals: (1) to build evidence-based consensus on behavioral management, enrichment, and the promotion of psychological well-being; (2) to develop plans for scientific collaborations and resource sharing; and (3) to serve as a resource for behavioral management best practices and recommendations for primate facilities outside of the consortium.

A significant focus of the BMC has been to establish common procedures and tools to facilitate cross-facility communication and collaboration. Over the decades, as each facility's behavioral management program evolved, commonalities in aims and program components had not been mirrored in the use of terminology, scoring systems, and documentation of program implementation. The BMC group quickly realized that the lack of common language due to this convergent evolution hampered our ability to address our goals. Therefore, one of our top priorities became the development of consensus-based terminology and coding methodology; capitalizing on the variety of species (10 or more taxa), behavioral management techniques, and large sample sizes of NHPs (over 20,000) across facilities. We are now able to build databases and support prospective studies to identify best practices in behavioral management.

DEVELOPING AND USING COMMON MEASUREMENT TOOLS

The BMC employs a consistent methodology for developing common tools for describing and documenting behaviors, such as scoring systems, ethograms, and record-keeping. The first step in the process is to share existing methodologies, procedures, and tools utilized across facilities. A designated BMC member compares the various NPRC tools to identify areas of congruence, as well as incongruence, that need to be reconciled, and adds each facility's unique elements to form an aggregate tool. As a group, the BMC evaluates and aligns incongruities to permit comparability across the centers. Common categories can be combined, while fine detail may be retained at individual facilities, as long as the broader categories remain congruent. For example, on the common documentation of social introductions (see below), rearing background is recorded as one of five categories (see Table 2.1). Some facilities make additional distinctions but retain the ability to combine categories to conform to the five BMC categories. The remaining task is to systematically examine unique elements in order to decide which are relevant and feasible to implement across all facilities. Facilities also retain unique elements, of course, for the continued collection of their internal metrics and to permit continued intrafacility retrospective assessments. For example, prior to standardization, some facilities had been categorizing social introductions as successful or unsuccessful according to whether the new pair or group remained together for 1 month, as opposed to the 2-week

Table 2.1 BMC Documentation of the Introduction of Pairs or Small Groups of Caged NHPs

Scope: Indoor caging: pairs, trios, and quads

- Species/strain.
- Sex.
- Date of birth.
- Body weight at the time of introduction.
- Rearing: five categories: mother-reared, mother-only, nursery/peer-reared, nursery/singly housed, mixed (at least 2 months in more than one category), or unknown. (For cercopithecines, rearing categories pertain to the first 6 months of life, but this time span would need to be adjusted for some other taxa.)
- Identified/monitored/treated at the time of introduction for SIB? Yes/no and indicate severity using BMC Self-Injurious Behavior scale.
- Identified/monitored/treated at time of introduction for other abnormal behavior? Specify the type of abnormal behavior using BMC Abnormal Behavior Ethogram.
- Class of introduction (i.e., age and sex class).
- Introduced in neutral caging (i.e., enclosure not previously occupied by any pair or group member)?
- Introduction in single-sex room?
- Prior social contact in the last year (e.g., previously co-housed in same breeding group)?
- Intended group size (pair, trio, quad).
- Start date of introduction.
- Initial panel type. Differentiate as clear, fingertip (mesh or small-hole panel), protected contact (defined as a partition through which more than just the fingertip can pass through), or none (introduction began with full contact).
- Dates and type of panel changes (as many columns as needed).
- Wounds during protected contact phase? Date and description of wounds during protected contact phase.
- Date of initial full contact.
- Wounds in first 14 days after initiation of full contact? Date and description.
- Success (Using BMC Common Definition of a Successful Social Introduction)? Facilities with different preexisting definitions of success that they want to retain can use BMC criteria plus their own in the document.
- Wounds 14–30 days? Date and description.
- Later wounds? Date and severity.
- Date(s) of changes in social status after introductory period (termination of social housing, temporary separation, etc.).
- Reason(s) for changes, e.g., research requirement (assignment or protocol-related), incompatibility (aggression, wounding, food monopolization, fear), clinical issues, reassignment of one individual to a different research project, pair moved to different social group, sold, or euthanized.
- 1-year postintroduction weight.
- Weight at final separation.

time point employed in the BMC's definition. These facilities have added a 2-week time point to enable cross-facility comparisons, while continuing to also document success at 1 month, as part of their ongoing internal retrospective assessments. Nonetheless, increasing the congruence across the centers may complicate some future intrafacility retrospective assessments and must be carefully weighed against the benefits of aligning tools across all centers for the management of over 20,000 NHPs.

The purpose of this chapter is to outline seven of the common tools developed by the BMC:

1. Common definition of a successful pair or small-group introduction
2. Documentation of the introduction of pairs or small groups of caged NHPs
3. Factors Leading to the use of single housing

4. Abnormal behavior ethogram
5. Self-injurious behavior scale
6. Alopecia scoring scale
7. NHP transfer form

For each tool, we outline the purpose and importance; how we implement it; and illustrate the significance of having common definitions, tools, and measurement scales for improving practices and facilitating cross-facility communication and collaboration. Tools are provided in Tables 2.1 through 2.6. Finally, we generate recommendations for the adoption of such tools outside of the NPRC system.

DEFINITION OF A SUCCESSFUL PAIR OR SMALL-GROUP INTRODUCTION

The BMC developed a common definition to apply when determining whether a social introduction of individual NHPs should be considered "successful": *A successful social introduction is defined as one in which the animals can be maintained together for a minimum of 2 weeks after the final step in the introduction process has been completed.*

Having a common definition to establish when an introduction of a pair of NHPs is successful allows us to "speak the same language" when it comes to comparing and analyzing many aspects of social introductions. Given that there is considerable variation in the methods used to implement social introductions (see Truelove et al. 2017 for a discussion), a common definition needs to be applied to determine whether one particular method is more effective than others. Similarly, to compare introductions across different species of NHPs, or across populations with differing characteristics (e.g., age, sex, early social history), a common definition must be employed. As a model for such comparisons, one retrospective examination of different pairing methodologies used with rhesus macaques at four facilities was completed utilizing the 2 weeks common definition for pairing success (Baker et al. 2014). Using this common definition permitted us to analyze more than 4300 pairing attempts and conclude that no single method resulted in a significantly higher success rate than any of the others for all rhesus macaques. However, the outcomes of isosexual pairing of males varied more between facilities than outcomes of female introductions, suggesting that they are more influenced by methodological variation. Even the slightest variation in encoding introductions as successful or unsuccessful can eliminate any possibility of making comparisons across facilities. A common definition for pairing success is necessary, but not sufficient, for making robust comparisons, since there are any number of complicating facility-specific factors that influence the success of socializations (e.g., personnel experience, differences in caging, or selection criteria of potential partners, as well as tolerance of aggression during introductions). Even within a single institution, using a common definition consistently across time will enable the detection of trends when comparing the use of different types of caging (e.g., dividers, novel or familiar caging), in different locations, or across species or age/sex classes; the kind of information that can lead to program improvements. Such evaluations can also be used to illustrate the effectiveness of a pairing program during assessments by other groups (e.g., Institutional Animal Care and Use Committees, USDA, AAALAC International).

DOCUMENTATION OF THE INTRODUCTION OF PAIRS
OR SMALL GROUPS OF CAGED NHPs

Detailed documentation of individual introductions is crucial for refining social introduction methodology. The BMC Documentation for Pair or Small-Group Introductions (see Table 2.1) includes variables that we recommend be recorded when pairs or groups are formed. While this information is intended to be used when pairs or groups of three or four NHPs are introduced to one another in indoor caging, it could be adapted for other situations. The variables enumerated contain characteristics of the individuals to be introduced (e.g., sex, age, weight, early rearing, behavioral history), features of the introduction process (e.g., types of panels used, whether the room contains animals of only one sex), whether wounding occurred over the course of the introduction, and the final outcome. In addition, the documentation captures longer-term information on these pairs and groups, including the later separations of the animals, both temporary and permanent, and the reasons for these separations.

The purpose of this documentation is to assist in standardizing information obtained when introductions are conducted so that comparable databases can be developed and combined, and these variables can be analyzed retrospectively, not only within, but across facilities as well. Some of the variables in the database are known to influence the outcome of introductions (e.g., sex, weight), whereas others are commonly thought to be influential, but have not yet been thoroughly evaluated (e.g., neutral caging, single-sex room). The information obtained through this documentation will provide data to allow these kinds of comparisons. The database will improve our understanding of social introductions and enable us to take multifactorial approaches to analysis. The complexity of various introduction processes (Truelove et al. 2017), as well as the individual differences brought to the process by the animals and humans involved, is best addressed using data collected across facilities at which differences in some practices are common.

FACTORS LEADING TO USE OF SINGLE HOUSING

There is universal agreement that meeting social needs is the foundation of welfare for NHPs. It is incumbent on facilities to optimize their use of social housing for species that live socially in the wild. Social housing programs are dynamic, particularly in a laboratory environment. The social housing status of an individual NHP can vary over time for a number of reasons among group, pair, or single housing, depending on its status as a breeding or research subject, its needs for clinical treatments, or the type of research project to which it is assigned.

Exceptions to the Animal Welfare Act's requirement for social housing involve several conditions or circumstances. Single housing is permissible only when an animal is deemed vicious or overly aggressive such that compatible partners cannot be identified, is debilitated, has or is suspected of having a contagious disease, or is exempted because of its health condition or scientific requirements approved by the local Institutional Animal Care and Use Committee (USDA 2013, pp. 100–101). In practice, there are numerous other situations responsible for an animal's housing situation, some ephemeral and some relatively long-lasting. Maximizing an animal's lifetime social housing is aided by constant awareness of the reason for the use of any single housing and swift attention to any issues that can be addressed. The BMC recommends that facilities document and regularly review not only *exemptions* from social housing (for the reasons covered in the Animal Welfare Act) but also reasons that are not covered in the regulation. We have developed a categorization system outlining 12 different factors that relate to housing NHPs singly (see Table 2.2).

The development of this system was geared toward allowing facilities to identify the circumstances associated with single housing and to quantify their impact on their social housing programs. As a simple example, quantifying the number of animals that are singly housed at any particular time because appropriate caging is not available can determine the need for additional caging purchases and guide decisions regarding the number of cages to be ordered. For other factors, changes over time in the number of animals housed alone permits prompt detection of functional problems. For example, large facilities will often have several animals in the queue for social introduction as processes are undertaken for identifying potential partners and coordinating with other stakeholders. By promptly detecting any increase in the length of the queue over time, programs can identify

Table 2.2   BMC Categorization of Factors Leading to the Use of Single Housing

Animal Welfare Act exemptions

Scientific: Scientific exemption approved by the Institutional Animal Care and Use Committee.

Veterinary: Approved veterinary exemption. Includes quarantine.

Behavioral: Animal has intrinsic behavioral problems that preclude social housing (e.g., hyperaggressivity).

Reasons for single housing: practical issues not addressed in the AWA

No potential partner: Odd numbers of animals in a study or within the treatment groups that cannot be intermingled; only potential partners are the opposite sex, and breeding cannot be permitted.

No potential partner (viral): Nonexperimental viral status [e.g., specific pathogen free (status)].

No compatible partner: Tried and failed with all available partners.

Space: No space in the room where the animal must remain due to project assignment, status, or room dimensions.

Caging: No caging permitting social housing is available.

Investigator information required: Information needed from Principal Investigator, such as treatment group assignment, current phase of project, and approval of proposed pairs.

Experimental results required: Waiting for experimental results (e.g., experimental viral status).

Nonexperimental viral results required: Animals cannot be paired until results concerning viral status of naturally occurring pathogens are received.

Clinical procedures required: Waiting for clinical procedure that will allow pairing (e.g., canine dulling, vasectomy).

Moves: Waiting for appropriate caging to be moved into place and animals to be relocated so that introductions can occur.

Behavioral management queue: Animals are in the queue for future social introduction, and behavioral management staff is preparing to make queries that require responses or actions of individuals outside of the unit (e.g., reading scientific protocol, preparing questions for Principal Investigator, determining potential partners, preparing move requests), or the introduction is scheduled but has not yet commenced.

Timing: Potential partners will not be able to remain together for a long enough period of time for a net gain to animal well-being to be likely, given transient introduction stress and possible separation stress; animals will be sold within a short period of time, and potential pair will imminently be assigned to research projects that do not permit social housing.

needs for reprioritization of staff efforts. Also, the increasing numbers of potential partners that cannot be introduced because they have yet to be moved into place suggest the need for coordination and planning with animal care personnel responsible for animal moves. Not only is it critical to detect and address bottlenecks, but it is also important to be able to articulate and quantify the effects of behavioral management decision-making to *not* socially house certain individuals. For instance, an introduction may not be pursured when the expected tenure of social housing for the primates is very short and unlikely to impart net benefits to the animals' well-being. The number of nonexempted singly housed NHPs at a facility does not necessarily reflect the success of the social housing program or the amount of effort exerted to maximize the use of social housing. The BMC recommends the use of this categorization process not only to strengthen social housing programs but to allow objective evaluation by other groups as well (e.g., Institutional Animal Care and Use Committees, USDA, AAALAC International).

ABNORMAL BEHAVIOR ETHOGRAM

Behaviors are considered to be abnormal in captive NHPs if they differ from wild NHPs, either in kind (i.e., typically not observed in the wild) or in degree [i.e., the behavior occurs at levels significantly different (higher or lower) than what is observed in wild populations (Erwin and Deni 1979)]. The presence of abnormal behavior in a captive population of NHPs may indicate a welfare issue, either past or present (Mason 1991), and therefore needs to be addressed when it is observed.

Prevalence of abnormal behavior can vary greatly across facilities. For example, in rhesus macaques, pacing has typically been reported to be the most common abnormal behavior, but its prevalence can vary from 23% to 87% a cross laboratory populations (Lutz et al. 2003; Vandeleest et al. 2011; Pomerantz et al. 2012). The variance in the reported prevalence of pacing may be due to differences in husbandry, social setting, enclosure type and size, and research practices, but additional variation may be due to the ways in which the behaviors are defined or recorded. Consistent assessment of prevalence rates is necessary to identify risk factors for abnormal behavior, an important step in ascertaining appropriate prevention or treatment methods. If facilities maintaining captive NHPs differ in how they define and identify abnormal behavior in their populations, comparisons of prevalence rates and risk factors become less generalizable. Therefore, it is important to have a common language across facilities to make direct comparisons possible. To better facilitate this process, the BMC developed the Abnormal Behavior Ethogram, which describes and defines abnormal behaviors that occur in captive NHP populations (see Table 2.3).

Table 2.3   BMC Abnormal Behavior Ethogram

Bizarre posture: Holding a seemingly uncomfortable or contorted position.

Bob: Rapid and repetitive[a] up and down motion of the body on flexed limbs; animal does not leave the cage surface.

Bounce: Repetitively[a] using one's hind legs or all four limbs to push oneself off the cage surface.

Coprophagy: Ingesting or manipulating feces in the mouth.

Feces paint: Smearing and/or rubbing feces on a surface.

Flip: Repeated forward or backward somersaults, may utilize the cage sides or ceiling.

Floating limb: An arm or leg rises into the air and may or may not contact the body (e.g., gently stroking the body). The action appears to be nonvolitional; the animal may interact with the limb as if it is not part of the body. This behavior may be associated with SIB, such as self-biting or self-hitting.

Food smear: Spreading of chewed food on a surface with the mouth; food is often licked off surface.

Hair pluck: Removal of hair from one's own body by pulling with teeth or hands, often seen with a quick jerking motion.

Head banging: Repetitively[a] and forcefully hitting the head against an object or surface.

Head toss: Repetitively[a] moving head side to side, or in a circular manner.

Pace: Repetitive[a] locomotion following the same path; for example, walking back and forth on the ground, around the enclosure, or back and forth across bars.

Periorbital contact (*saluting, eye poke*): Animal holding hand, digit, and/or object against/near one's eyebrow or eye.

Regurgitate: Backward flow of already swallowed food; the material may be retained in the mouth or deposited on a surface and reingested.

Repetitive licking: Prolonged or excessive contact of the tongue with a surface or object for no apparent reason.

Rock: Any repetitive motion of the body from a stationary position. Animal remains sitting or standing, while the upper torso sways back and forth.

Self-bite: Closing the teeth rapidly and with force on oneself.

Self-clasp: Clutching one's own body with hands or feet.

Self-injure: Any behavior by the animal that causes physical trauma to itself such as bruising, lesions, lacerations, or punctures.

Self-oral: Sucking a part of one's own body.

Self-slap: Forcibly striking oneself with the hands or feet.

Spin: Repetitive[a] circling of body around a pivot point.

Urophagy: Licking or ingesting urine.

Withdrawn: Slumped or hunched body posture, often accompanied by dull eyes, and relatively unresponsive to environmental stimuli to which other monkeys are or typically would be attending.

Other stereotypical locomotion: Idiosyncratic repetitive[a] whole body movements, particular to an individual; does not meet criteria for other behaviors defined above.

[a] Repetitive = a minimum of two to three times, depending on the current facility criteria.

Although the behaviors identified in this ethogram are focused on those that are seen in macaques, it can be adapted to other species as well, by adding behaviors that are idiosyncratic to those species. With the use of this ethogram, abnormal behavior can be more consistently defined across facilities, allowing investigators to better compare prevalence rates and potential risk factors for the development of these behaviors. This information will allow for a better understanding of cross-facility variation and guide the development of refinements in prevention, assessment, and treatment of abnormal behavior at all facilities.

SELF-INJURIOUS BEHAVIOR SCALE

In some captive NHPs, self-directed behaviors that cause pain or damage to tissue can occur and are behavioral problems of great concern to all stakeholders involved in the animals' care. The phrase self-injurious behavior (SIB) includes activities such as self-biting, head banging, injurious hair plucking, and self-injury which can result in tissue trauma (for a review, see Novak 2003). This pathology can be found in a range of facility types. It has been described in a number of species living in zoos (e.g., Novak et al. 2006; Hosey and Skyner 2007), and numerous surveys have revealed that 10%–12% of singly housed laboratory-housed macaques have a history of performing SIB (e.g., Bayne et al. 1992; Bellanca and Crockett 2002; Lutz et al. 2003; Rommeck et al. 2009). However, there is a lack of uniformity with respect to defining and categorizing this behavior.

When the BMC was first established, some centers defined SIB as behavior that *causes* injury to the animal, while the others included behaviors that *could* cause or progress to injury. This one-word distinction may not seem significant, yet it can have huge implications. For example, the centers that use the more inclusive definition will likely have many more incidences of SIB than those that use the more conservative term, making cross-center comparisons difficult. Thus, the BMC decided to standardize the definition of SIB, as well as assessment of the severity of the self-injurious events in our colonies. SIB that results in tissue damage presents across a spectrum of severities, ranging from minor surface abrasions or small lacerations to deep muscle injuries requiring surgery. All of these presentations have different welfare implications for the individual. To better understand the etiology, the possible progression of the behavior, and the development of effective treatments, it is important to accurately assess each SIB episode.

Our first step in developing this scoring system was to establish a uniform definition. At the time we started this process, a third of the participants defined SIB as a self-directed behavior that *resulted* in tissue damage or broken skin, and the other two-thirds defined it as self-directed behavior that *could result* in pain or tissue damage. After a great deal of discussion, we decided to use the more inclusive definition and incorporate behaviors that could result in tissue damage in the definition. This definition is now utilized by all members of the BMC. Our next step was to develop common methods for categorizing the events. As with the definition, we first examined the ways in which we were categorizing SIB at our various centers. Several centers already used some sort of severity scale, in which the wound was given a score. For the most part, the scores were relatively consistent across the centers that used a severity scale; bruising was generally classified as "mild," while deep lacerations were categorized as "severe." We were therefore able to use these existing scores as our starting point.

The BMC Self-Injurious Behavior Scale contains five categories (see Table 2.4). Two of these categories cover noninjurious SIB incidents (i.e., those that do not cause wounding). These categories are broken down by behavior; one category involves SIB as a result of self-biting, and the other covers behaviors other than biting (e.g., head banging).

The other three SIB categories pertain to events that result in wounding, and are based on the severity of the wound (i.e., mild, moderate, severe). The "mild" category includes superficial wounds that do not require veterinary treatment (other than possibly the provision of psychotropic

Table 2.4 BMC Self-Injurious Behavior Scale

Score	Wound	Observations	Notes
Noninjurious; self-biting	No wound	Self-biting behavior observed or reported. No visible wound.	In all cases, behavior must be observed by an appropriately trained individual.
Noninjurious forceful self-directed behavior; nonbiting	No wound	Potentially injurious behavior (other than self-biting) observed or reported. No visible wound. Could include head banging.	In all cases, behavior must be observed by an appropriately trained individual.
SIB 1	Mild	Superficial wounds that do not require medical treatment. May include superficial abrasions, pinpoint lesions, small puncture wounds, bruising, and calloues. Does not include lacerations. Does not require medical treatment.	Provision of psychotropic drugs does not constitute medical treatment for the purposes of this classification.
SIB 2	Moderate	Surface wounds such as lacerations and puncture wounds. Requires assessment for possible veterinary treatment (including sedation for assessment, or minor treatment such as wound cleaning, pain medication, antibiotics; does not include major medical procedures such as suturing, amputation, or surgery).	In all cases in which there is a wound, injury must be assessed to be self-inflicted after ruling out other possible causes.
SIB 3	Severe	Deep or subcutaneous wounds, such as large lacerations or deep puncture wounds that require major medical treatment. Treatment may include suturing, amputation, surgery, or other major medical procedure under sedation.	In all cases in which there is a wound, injury must be assessed to be self-inflicted after ruling out other possible causes.

medications). Wounds that qualify for this category might include superficial abrasions, pinpoint lesions, small puncture wounds, bruising, and callouses. The "moderate" category includes surface wounds, such as lacerations and puncture wounds, that require assessment for possible veterinary treatment (including sedation for assessment, or minor treatment, such as wound cleaning, pain medication, and antibiotics). Finally, SIB resulting in deep or subcutaneous wounds, such as large lacerations or deep puncture wounds that require major medical treatment, would be classified as "severe." Treatment for injuries in this category might include suturing, amputation, and other surgeries or major medical procedures. In order to be considered SIB, the injury must be assessed to be self-inflicted after ruling out other possible causes, including injury due to caging, enclosure features, or another animal. Even singly housed NHPs may be bitten by a neighbor, for instance, if they reach their fingers outside of their cages.

One source of variation remains, despite the adoption of a common scoring system. The Self-Injurious Behavior Scale relies on the clinical approach taken to treat any resulting wounds as treatment for SIB may vary across individual veterinarians and across the centers. For example, veterinary staff at some facilities may be more likely to examine animals under sedation, while others may rely more on cage-side observation, choices that affect how we score the wound. We recognize this potential challenge to our codification; however, the severity categories are broad enough to encompass wounds for which treatment is likely to be somewhat uniform across facilities. For example, a deep laceration is likely to be sutured at any facility, as part of the basic standard of care.

It should be noted that this scoring system evaluates the *event* and not the *patient*. Classifying the patient would need to take into account frequency as well as severity. While we have not tackled this categorization as a group, several individual centers classify animals in this way, often in an effort to determine treatment plans and/or as part of a humane endpoint policy. For example, at one facility, individual patients' SIB is classified as either mild (infrequent episodes of SIB with no or minor wounds), moderate (at least one episode of minor wounding every 1–2 weeks or episodes of moderate wounding), or severe (minor or moderate wounding episodes occurring every 1–2 days or severe wounding episodes). Animals are then treated based on these classifications. This shared assessment tool has allowed cross-center assessments permitting us to better understand the phenomenon of SIB (Bloomsmith et al. 2015).

ALOPECIA SCORING SCALE

The BMC has also developed a standardized method for scoring alopecia, or hair loss. Alopecia is a ubiquitous phenomenon among captive NHPs, whether they are housed in zoos, laboratories, or breeding facilities; recent studies at large NHP facilities have shown that up to 87% of the population may show some degree of alopecia at any given time (Lutz et al. 2013; Novak et al. 2014). While the exact welfare implications of alopecia are not clear, its high prevalence is of concern to behavioral managers, veterinarians, and regulatory agencies. Several factors have been implicated as contributing to alopecia, including species, compromised immune function, dermatological pathologies, hair plucking, and environmental factors (Novak and Meyer 2009). However, these findings are often not replicated, highlighting the need to examine alopecia across facilities (Luchins et al. 2011; Coleman et al. 2017). A critical first step toward cross-center comparisons is developing a common tool for measuring hair loss (Crockett et al. 2009).

There are different methods by which cross-center comparisons can be standardized. One multicenter approach involves having a centralized scoring facility. For example, in one large study involving four NPRCs, all participants photographed subjects and sent the photos to a single center for analysis (e.g., Novak et al. 2014). In this particular case, Image J software was utilized (Novak et al. 2014). While accurate, assessing alopecia from photographs takes a great deal of time, specialized software, and anesthetic events. Furthermore, although this approach may be feasible for funded research projects, it may not be practical for day-to-day observations of animals. Therefore, the BMC created a method by which animals could be scored during cage-side observations. Our goal was to develop a system that both provided useful information and was simple to implement in large facilities.

The first step in the BMC scoring system is to estimate the amount of the subject's body surface affected by alopecia, using the "Rule of Nines," a tool developed by the medical community to estimate the extent of burns [i.e., Wallace Rule of Nines (Hettiaratchy and Papino 2004)]. This tool breaks the body into 11 "parts," each of which represents ~9% of the body. Figure 2.1 shows the distribution for NHPs. The Rule of Nines is meant to provide a relatively simple and quick way to assess the total amount of the body affected by a particular issue (e.g., burns in humans or alopecia in monkeys). For example, suppose that a monkey presents with a large patch of alopecia covering the entire lower part of his left leg and a large patch covering about half of his back (see Figure 2.2). If one were just looking at the monkey, it might be difficult to estimate the amount of hair missing; indeed, the large patch on his back might make it seem as though almost half of his body is affected. Using the Rule of Nines, however, one would calculate alopecia on the upper part of the left leg as ~4%–5% of the body surface (i.e., half of the 9% allocated to the upper leg) and the area covering the back as ~9% of the body surface, for a total of about 13%–14% of the animal's body affected. The percent of the body affected by alopecia is then categorized using a six-point scale (Table 2.5).

The BMC Alopecia Scoring system includes several assumptions to facilitate consistency in its use. One assumption involves which body parts should be included in the analysis. The chest and stomach of rhesus and other macaques naturally have very fine hair, and it can be difficult to determine whether or not the hair is actually missing, particularly when assessing the animal from a photograph. Thus, we assume that these areas are "haired" unless it is obvious that hair is missing (e.g., the observer is assessing alopecia on a sedated animal and thus can feel the hair). Areas clearly

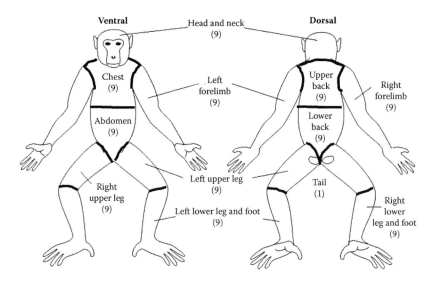

Figure 2.1 Schematic for the "Rule of Nines" modified for use with NHPs. The body surface is divided into 11 parts, each of which represents approximately 9% of total surface area. The tail comprises the remaining 1%.

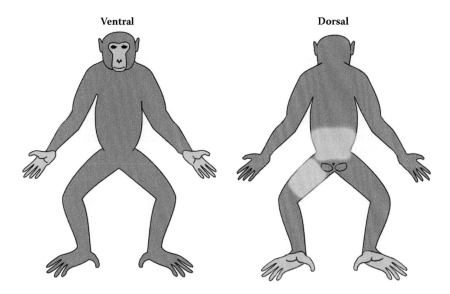

Figure 2.2 Example of how the "Rule of Nines" can be used to estimate the amount of alopecia on a monkey. Using the "Rule of Nines," one would estimate that approximately 13%–15% of the monkey's body is affected by alopecia [the entire lower back (9%) and half of the upper leg (4%–5%)].

shaved for medical or experimental procedures are also assumed to be haired since hair loss was the result of human intervention. Areas of regrowth (determined by the presence of shorter hair), on the other hand, are counted as "haired."

While assessing the percent of body affected by alopecia is useful, this number may not represent the whole picture. Along with the overall amount of hair loss, the pattern of alopecia (e.g., small patches of localized hair loss compared to generalized hair loss) might provide better insight into the nature of the condition (Coleman et al. 2017). For example, Kramer et al. (2011) compared rhesus

Table 2.5   BMC Alopecia Scoring Scale

Score	Area Affected by Alopecia (%)
0	No alopecia
1	1–10
2	>10–25
3	>25–50
4	>50–75
5	>75–100

macaques with alopecia localized at distal forearms to those with generalized alopecia. Animals with localized alopecia had less inflammation than those with generalized alopecia, suggesting a psychogenic etiology for animals with small patches of hair loss (Kramer et al. 2011). Indeed, several centers do include patterns (e.g., thin, patchy) in alopecia assessments at their facilities. At the time this chapter is being written, the pattern component of the BMC Alopecia Scoring Scale is still undergoing refinement.

Unlike the SIB Scoring System, the BMC Alopecia Scoring Scale has not yet been adopted by all NPRCs, pending interobserver reliability testing both within and between all facilities. However, one NPRC has trained over 35 behavioral and clinical staff to assess alopecia using this scoring system, with a reliability of over 85% agreement, demonstrating its feasibility as a standardized tool (Heagerty et al. 2015). We are confident that it will greatly aid data sharing across the centers.

ANIMAL TRANSFER FORM

Individual NHPs may be transferred between facilities when they are purchased for use in research or breeding programs, or when investigators move their laboratories and study subjects between facilities. The transportation and relocation of NHPs between facilities can lead to changes in behavior, body weight, clinical chemistry values, fecal cortisol levels, and immunological measures (Honess et al. 2004; Watson et al. 2005; Schapiro et al. 2012), and some of these changes are likely to be indicative of stress. The goal of using the Animal Transfer Form is to have a simple and consistent manner for sharing information related to the behavioral management of transported NHPs so that staff in the receiving facility can be prepared to best meet the needs of the animals when they arrive. This facilitates continuity of care for the animals (for example, they can have familiar enrichment waiting for them on their arrival) and may help to mitigate some of the stress associated with the relocation, particularly when former social partners can be reunited. In addition, the receiving staff will be aware in advance of behavioral problems that may be expressed if the animal experiences stress during transportation, which can be useful for assessing how the animals are adjusting to their new facility. The *Guide* recommended that information regarding rearing and housing histories, as well as behavioral profiles, be provided when animals are transferred between institutions (NRC 2011).

The Animal Transfer Form is a simple document that allows for qualitative notes concerning the individual animal's social history, enrichment, caging, social group characteristics, behaviors of concern and any treatments provided for that problem, and behaviors for which the individual has been trained. The Animal Transfer Form is intended to improve the welfare of NHPs when they are moved between two facilities by increasing the behavioral information available to the receiving institution. Some of the information included on the form may also be available in the animals' general records, but we have found it useful to have all of this readily accessible for behavioral managers at the receiving institution (Table 2.6).

Table 2.6   BMC Animal Transfer Form

Behavioral information for animals transferring between facilities

This information will be completed for each NHP being moved between facilities. The purpose is to inform the receiving institution about the behavior of each arriving primate, so they can be prepared to best accommodate the needs of that animal.

Animal ID:_____   Species:_____ Sex:_____   Date of birth (or facility of origin if birth date unknown): _____

Who completed form:_____   Facility:_____

INFORMATION	RESPONSE
Standard enrichment provided:	
Type and size of current caging:	
Social history (social background including early rearing, later social housing, and any known social problems):	
Current social status (include partner and type of access to partner such as full contact and protected contact):	
Training history (what behaviors animal has been trained for, what training techniques were used, etc.):	
Behavior(s) of concern (e.g., SIB with and without wounding, hair plucking, pacing, unusual aggression or fear):	
Current treatment (beyond standard enrichment):	
Outcome of treatment:	
Other comments:	

CONCLUSION

We invite the behavioral management community to examine our common tools and consider their adoption at other facilities. Collaborating on these tools has enabled us to benefit from group expertise, and the tools described in this chapter represent the combined thinking of experts in behavioral management. This may be particularly valuable to newer or smaller departments that may not employ behavioral management scientists. Adopting these tools will also open up additional opportunities for cross-facility collaboration and aggregating data across a wider variety of facilities, species, and contexts in which NHPs are housed.

We also promote the establishment of detailed record-keeping concerning program implementation. Documenting behavioral management activities and keeping records for immediate practical reasons present excellent opportunities for generating data for retrospective assessments to guide future program modifications. By recording activities in a consistent, systematic, and detailed way, with an eye toward unanswered questions, a facility can make increasingly objective refinements. Coupled with efforts across facilities, we can all continue to move behavioral management programs forward in providing the best care possible to laboratory NHPs.

ACKNOWLEDGMENTS

Members of the BMC are supported by the National Center for Research Resources and the Office of Research Infrastructure Programs (ORIP) of the National Institutes of Health through Grant Numbers P51 OD01107 (California National Primate Research Center), P51 OD011092 (Oregon National Primate Research Center), P51 OD011133 (Southwest National Primate Research Center), P51 OD011104 (Tulane National Primate Research Center), P51 OD010425 (Washington National Primate Research Center), P51O D011106 (Wisconsin National Primate Research Center), and P51 OD011132 (Yerkes National Primate Research Center). We would like to acknowledge the behavioral management teams at all centers.

REFERENCES

BMC Collaborations Using Common Tools

Baker, K. C., K. Coleman, M. A. Bloomsmith, B. McCowan, and M. A. Truelove. 2014. Pairing rhesus macaques (*Macaca mulatta*): Methodology and outcomes at four National Primate Research Centers. *American Journal of Primatology* 76(Supplement 1):65.

Baker, K. C, J. L. Weed, C. M. Crockett, and M. A. Bloomsmith. 2007. Survey of environmental enhancement programs for laboratory primates. *American Journal of Primatology* 69:377–394.

Bloomsmith, M., K. Baker, R. Bellanca, K. Coleman, M. Fahey, C. Lutz, A. Martin, B. McCowan, J. Perlman, and J. Worlein. 2015. An analysis of self-injurious behavior in a large sample of captive primates. *American Journal of Primatology* 77:96.

Coleman, K., C. K. Lutz, J. Worlein, D. Gottlieb, E. Peterson, G. Lee, N. D. Robertson, K. Rosenberg, M. T. Menard, and M. A. Novak. 2017. The correlation between alopecia and temperament in rhesus macaques (*Macaca mulatta*) at four primate facilities. *American Journal of Primatology*. doi:10.1002/ajp.22504.

Crockett, C. M., K. C. Baker, C. K. Lutz, K. Coleman, M. A. Fahey, M. A. Bloomsmith, B. J. McCowan, J. Sullivan, and J. L. Weed. 2009. Developing a reliable laboratory primate alopecia scoring system for interfacility collaboration and on-line training. *American Journal of Primatology* 71(Supplement 1):73.

Heagerty, A., R. Wales, A. Daws, D. H. Gottlieb, A. Maier, K. Andrews, K. Prongay, K. Rosenberg, M. T. Menard, J. S. Meyers, K. Coleman, and M. A. Novak. 2015. Effect of sunlight exposure on hair loss in captive rhesus macaques (*Macaca mulatta*). *American Journal of Primatology* 77:67–68.

Weed, J. L., K. C. Baker, and C. M. Crockett. 2003. Managing behavioral health and environmental enrichment of laboratory primates. *American Journal of Primatology* 60:34.

Other References Cited

Bayne, K. A. L., S. Dexter, and S. J. Suomi. 1992. A preliminary survey of the incidence of abnormal behavior in rhesus macaques (*Macaca mulatta*) relative to housing condition. *Laboratory Animals* 21:38–46.

Bellanca, R. U. and C. M. Crockett. 2002. Factors predicting increased incidence of abnormal behavior in male pigtailed macaques. *American Journal of Primatology* 58:57–69.

Erwin, J. and R. Deni. 1979. Strangers in a strange land: Abnormal behaviors or abnormal environments? In *Captivity and Behavior: Primates in Breeding Colonies, Laboratories, and Zoos*, eds Erwin, J., T. L. Maple, and G. Mitchell, 1–28. New York: Van Nostrand Reinhold.

Hettiaratchy, S. and R. Papini. 2004. Initial management of a major burn: II—A ssessment and resuscitation. *British Medical Journal* 329:101–103.

Honess, P. E., P. J. Johnson, and S. E. Wolfensohn. 2004. A study of behavioural responses of non-human primates to air transport and re-housing. *Laboratory Animals* 38:119–132.

Hosey, G. R. and L. J. Skyner. 2007. Self-injurious behavior in zoo primates. *International Journal of Primatology* 28:1431–1437.

Kramer, J. A., K. G. Mansfield, J. H. Simmons, and J. A. Bernstein. 2011. Psychogenic alopecia in rhesus macaques presenting as focally extensive alopecia of the distal limb. *Comparative Medicine* 61:263–268.

Luchins, K. R., K. C. Baker, M. G. Gilbert, J. L. Blanchard, D. X. Liu, L. Myers, and R. P. Bohm. 2011. Application of the traditional veterinary species alopecia diagnostic evaluation to laboratory rhesus macaques (*Macaca mulatta*). *Journal of the American Association for Laboratory Animal Science* 50:926–938.

Lutz, C. K., K. Coleman, J. Worlein, and M. A. Novak. 2013. Hair loss and hair-pulling in rhesus macaques (*Macaca mulatta*). *Journal of the American Association for Laboratory Animal Science* 52:454–457.

Lutz, C., A. Well, and M. A. Novak. 2003. Stereotypic and self-injurious behavior in rhesus macaques: A survey and retrospective analysis of environment and early experience. *American Journal of Primatology* 60:1–15.

Mason, G. J. 1991. Stereotypies: A critical review. *Animal Behaviour* 41:1015–1037.

National Research Council (Institute for Laboratory Animal Research). 2011. *Guide for the Care and Use of Laboratory Animals*, 8th Edition. Washington, DC: National Academies Press.

Novak, M. A. 2003. Self-injurious behavior in rhesus monkeys: New insights into its etiology, physiology and treatment. *American Journal of Primatology* 59:3–19.

Novak, M. A., A. F. Hamel, K. Coleman, C. K. Lutz, J. Worlein, M. Menard, A. Ryan, K. Rosenberg, and J. S. Meyer. 2014. Hair loss and hypothalamic-pituitary-adrenocortical axis activity in captive rhesus macaques (*Macaca mulatta*). *Journal of the American Association for Laboratory Animal Science* 53:261–266.

Novak, M. A. and J. S. Meyer. 2009. Alopecia: Possible causes and treatments, particularly in captive nonhuman primates. *Comparative Medicine* 59:18–26.

Novak, M. A., S. T. Tiefenbacher, C. Lutz, and J. S. Meyer. 2006. Deprived environments and stereotypies: Insights from primatology. In *Stereotypic Animal Behaviour: Fundamentals and Applications to Welfare*, eds Mason, G. and J. Rushen, 153–189. Wallingford, UK: CABI.

Pomerantz, O., A. Paukner, and J. Terkel. 2012. Some stereotypic behaviors in rhesus macaques (*Macaca mulatta*) are correlated with both perseveration and the ability to cope with acute stressors. *Behavioural Brain Research* 230:274–280.

Rommeck, I., K. Anderson, A. Heagerty, A. Cameron, and B. J. McCowan. 2009. Risk factors and remediation of self-injurious and self-abuse behavior in rhesus macaques. *Applied Animal Behaviour Science* 12:61–72.

Schapiro, S. J., S. P. Lambeth, K. R. Jacobsen, L. E. Williams, B. N. Nehete, and P. N. Nehete. 2012. Physiological and welfare consequences of transport, relocation, and acclimatization of chimpanzees (*Pan troglodytes*). *Applied Animal Behaviour Science* 137:183–193.

Truelove, M. A., A. L. Martin, J. E. Perlman, J. S. Wood, and M. A. Bloomsmith. 2017. Pair housing of macaques: A review of partner selection, introduction techniques, monitoring for compatibility, and methods for long-term maintenance of pairs. *American Journal of Primatology* 79:e22485.

U.S. Department of Agriculture. Animal Welfare Act and Animal Welfare Regulations. 2013. Section 3.81—Environmental enhancement to promote psychological well-being. Animal Welfare Act and Animal Welfare Regulations ("Blue Book").

Vandeleest, J. J., B. McCowan, and J. P. Capitanio. 2011. Early rearing interacts with temperament and housing to influence the risk for motor stereotypy in rhesus monkeys (*Macaca mulatta*). *Applied Animal Behaviour Science* 132:81–89.

Watson, S. L., J. G. McCoy, R. C. Staviskiy, T. F. Greer, and D. Hanbury. 2005. Cortisol response to relocation stress in Garnett's bushbaby (*Otolemur garnettii*). *Journal of the American Association for Laboratory Animal Science* 44(3):22–24.

CHAPTER 3

Rules, Regulations, Guidelines, and Directives

Jann Hau
University of Copenhagen

Kathryn Bayne
AAALAC International

CONTENTS

INTRODUCTION

This handbook discusses behavioral management of nonhuman primates (NHPs) within the existing framework of the many regulations and guidelines that provide the present day's concept of minimum acceptable standards for the housing and care of NHPs. The word "minimum" is important as is the realization that regulations and guidelines have evolved markedly during the past 30 years with respect to, for example, minimum space requirements, social housing, and husbandry. Hopefully, guidelines will continue to evolve and books like this one will help pave the way toward further refinement of guidelines for the care and use of NHPs in biomedical research. This is an evolutionary process, which is never completed, as captive care techniques will continue to be refined and improved, leading to future modifications of current regulations and guidelines.

During the last century or so, research involving NHP models has resulted in many significant advancements in human medicine, including polio vaccines, life-support systems for premature babies, kidney dialysis, antirejection drugs in xenotransplantation, surgical treatment of eye diseases, new drugs and treatments for neurological diseases and asthma, as well as new techniques in stroke rehabilitation therapy (NC3R 2006).

Nonhuman primates are still, to this day, important models for humans in biomedical research and preclinical testing of new drug candidates. The most common areas of research using NHP models are microbiology, neuroscience, and biochemistry/chemistry, with most primate research being conducted in North America, Europe, and Japan (Carlsson et al. 2004). In Europe, 67% of NHPs participate in safety and efficacy regulatory testing, 15% participate in research and development of medical products and devices, and 15% participate in basic research (Burm et al. 2014). Biomedical research using NHPs is strictly regulated in those parts of the world where it is permitted.

However, conducting research projects that involve NHPs is increasingly difficult because of changes in public sentiment, active lobbying of special interest groups, and a seemingly gradual decline in public acceptance of science. Examples include several initiatives such as "Stop Primate Experimentation at Cambridge," which achieved its goals within 1 year, and its successor, SPEAK, which forced contractors to stop building laboratory animal facilities at Oxford. In the United States, many airlines have effectively ceased NHP transport by air within the country, and the majority of airlines in Europe have taken similar steps, with Air France as a notable exception.

Changes in legislation and government policies are also restricting NHP research. Examples of this include (1) the decision by the National Institutes of Health (NIH) of the United States to phase out support for invasive research involving chimpanzee models (National Institutes of Health 2013), effectively ending biomedical (but not behavioral) chimpanzee research worldwide, and (2) the 2010 revised European Directive (EU Directive 2010/63/EU), which not only bans the use of chimpanzees but also prohibits, in principle, the use of other NHPs for biomedical research, albeit with very important exceptions for certain types of investigations.

This development contrasts somewhat with the need for NHPs in biomedical research, as reflected in a report (EU Scientific Committee on Health and Environmental Risks SCHER 2009) from the EU's Scientific Committee on Health and Environmental Risks (SCHER), which concludes (section 3.1.6) that "from a scientific point of view, the use of NHPs, at the present time, is essential for scientific progress in a number of important areas of disease biology research and in safety testing:

- Development of pharmaceuticals, in particular safety testing, to assess potential toxicity in animals to identify unacceptable adverse reactions in humans. For specific pharmaceuticals, including antibodies, NHPs may represent the most relevant animal model for specific aspects of toxicity testing because of their close similarity to humans.
- Understanding the pathophysiology of infectious diseases such as HIV/AIDS, where the NHP is the only susceptible species and therefore the only useful animal model to study the disease, and to develop safe and effective vaccines and therapies. Learning how complex brains of primates, humans included, are structured and function. Again, NHPs are the best model due to their close similarity to humans with regard to brain complexity and function. In addition, NHPs are the best model for some human brain conditions and have been critical in developing and testing novel and current treatments.
- Developing and testing xenotransplantation methodologies."

The independent UK Weatherall report (2006) also concluded that there is a strong scientific case for continuing studies involving NHPs for carefully selected research problems in many of the areas studied, at least for the foreseeable future. Scientific analyses of public opinion seem to indicate an understanding and acceptance of the participation of NHPs in vital research, and support their continued participation in research. Three out of four medical and veterinary students from Sweden and Kenya in the late 1990s (currently practicing physicians and veterinarians) understood the need for NHPs in medical research and found it morally acceptable (Hagelin et al. 2000).

Ironically, perhaps especially in Europe, biomedical research involving NHPs may be winding down more quickly than anticipated as a consequence of the new Directive. There are an increasing number of examples of scientists closing down projects, which were fully licensed by local authorities, due to court cases and the influence of politicians (Suran and Wolinsky 2009). As a consequence of the changing research climate, European scientists are increasingly moving their NHP research to Asia (Abbott 2014) and, to a lesser extent, the United States.

There are well-recognized ethical challenges associated with housing NHPs in captivity, which, at the moment, do not apply to the captive housing of many other species. In contrast to the majority of laboratory animals, NHPs are basically wild animals, and although they may have been housed in captivity for a few generations, they remain undomesticated. Unlike most other species used in research, primates typically have a long life span, often spend years in captivity, and are frequently reused in several independent studies during the course of their lives (Boccia et al. 1995; van Vlissingen 1997; Carlsson et al. 2004). The captive production and housing of these intelligent and sentient animals, including that which takes place in wildlife parks, zoos, and sanctuaries, are associated with close contact with humans and some degree of deprivation, compared with a life in undisturbed freedom. Thus, captive NHPs are prevented from fully exhibiting their behavioral repertoire and prevented (and protected) from interacting freely with animals of other species. Taking control of the lives of NHPs in captivity, therefore, poses serious moral issues and obligations for those responsible for their housing, care, and management.

Fortunately, increasing knowledge and understanding, from a variety of sources (captive studies, field studies, etc.), of NHPs' use of space and environmental features make it possible to design captive environments that meet the animals' fundamental biological and behavioral needs. Empirical studies of behavior and space use within captive environments can provide important information addressing the animals' requirements and preferences (Mench and Mason 1997). Measuring the ways in which animals utilize their available space is a common method to determine both positive and negative aspects of captive environments. When combined with preference testing, in which subjects are given opportunities to choose between two or more specific environmental features (Dawkins 1983, 1990; Bayne et al. 1992; Fraser et al. 1993), useful information relevant to the design of an appropriate captive environment may be obtained. Space use is influenced not only by features and preferences related to aspects of the physical environment but also by social and biological factors. Analyses of the ways in which animals choose to use the many different features of their captive environment, including social partners, will provide important information on the appropriateness of that environment and will help establish guidelines for the design of future environments.

Early reports on the influence of cage size on primate behavior (Bayne and McCully 1989) made the case that considerations of only the volume of space available to the animal were inadequate. Rather, the *quality* of the space provided to the animal (i.e., the complexity of the space, such as cage furnishings, environmental enrichment, etc.) better defined the necessary "size" of the cage and was more focused on the desired outcome of improved animal welfare (Bayne and Turner 2013). This notion has been reinforced in the most recent *Guide for the Care and Use of Laboratory Animals* (*Guide*, NRC 2011), which notes that "space quality affects its usability" and that cages that are "complex and environmentally enriched may increase activity and facilitate the expression of species-specific behaviors," implying that the expression of the typical behavioral repertoire is positively correlated with the animal's welfare. This theme is further developed by the authors of the *Guide* when they recommend that (1) cage height should take into account an animal's typical posture and allow arboreal species to stand or perch while keeping their body and tail above the cage floor; (2) at a minimum, animals should have enough space to express their natural postures and postural adjustments without touching enclosure walls or ceiling; (3) adolescent animals, which may be more physically active, may require more cage space; and (4) socially housed animals should have sufficient space and structural complexity in their cages to allow them to escape from aggression or hide from other animals. As Hansen and Baumans (2007) have noted, cage space is often determined by the number of grams or kilograms of animals; however older, heavier animals may be less active than younger, lighter animals. They suggest, therefore, that a suitable approach to determining minimum allowable cage space may be to base minimums on the average adult weights for the species, thereby providing younger, lighter animals relatively more room for their activities. For example, primates with long tails, but that are not necessarily "heavy" (e.g., female *Macaca fascicularis*), should be provided sufficient cage height such that the tail is above the cage floor when the monkey is perching (NRC 2011), and primates with large body frames, but that are again not necessarily heavy (e.g., female *Cercocebus atys*), might require a larger cage space than the minimum recommended, if only body weights were considered. Thus, new approaches are being introduced to continuously improve the captive environment for the animals, as exemplified by Ross et al. (2009, 2011), who introduced the "electivity index" as an additional tool to assess the degree of preference for structural features and vertical tiers within and between different sizes and types of enclosures.

The increased understanding of the behavioral needs of NHPs is reflected in modern guidelines that advocate group housing in large enclosures. These enclosures must supply the NHPs with ample foraging opportunities; furniture and fixtures that allow the animals to make full use of the space and to flee vertically when frightened; and safe havens to escape unwanted attention from dominant animals. Older guidelines and regulations were based on what has been termed a "prescriptive engineering" approach, while newer guidelines take a "performance" approach, defining the desired outcome while acknowledging that multiple methods may be used to achieve that outcome. The performance approach relies on sound professional judgment and thus the competence and organization of those managing the NHP care and use program (Bayne 1998).

The moral imperative to provide captive NHPs with a sufficient quality of life increases significantly when the NHPs housed in captivity are used for research associated with pain, distress, suffering, or lasting harm. In spite of many years' of striving to maximally implement the Three Rs, the performance of some *vital* research associated with negative effects on the welfare of the animals has proven unavoidable. The absence of benefits to the individual animals associated with these research projects, combined with the risks of negative effects on their well-being, requires that critical harm–benefit analyses of research project proposals be conducted by relevant oversight bodies (World Organisation for Animal Health (OIE) 2012; AAALAC [Association for Assessment and Accreditation of Laboratory Care] International FAQ, http://www.aaalac.org/accreditation/faq_landing.cfm#B3; the *Guide* (NRC 2011)). However, oversight bodies may be permissive, especially if research project proposals incorporate due consideration to implementation of the Three Rs. It has been argued that a risk threshold should be defined for justification of research with NHPs, comparable to the way in which risk thresholds are defined for vulnerable human subjects who cannot provide informed consent (Ferdowsian and Fuentes 2014). NHPs are often reused in multiple protocols, and in order to minimize the impact on the individual research animal, The EU Directive includes requirements that (1) cumulative severity analyses be performed, in which the entire experience of research animals is considered; and (2) retrospective analyses be performed after a project is finished to analyze the success of the project and its impact on the animals. The lifetime experiences of the animals are a challenge to address qualitatively, as well as quantitatively, but publications are beginning to appear in the literature addressing this complex issue (Honess and Wolfensohn 2010; Wolfensohn et al. 2015). Legislation requires that every NHP must have an individual health record containing all medical information, and Hau and Schapiro (2004) have advocated that NHP health records should include a psychosocial profile, including information on the animal's housing history, social partners, dominance rank, compatibility with other animals, etc. When relevant, the file should include the animal's responses to training and human contact and all of the experimental procedures in which the animal has participated during its lifetime.

Considering the ethical issues inherently connected with the involvement of NHPs in research projects, their complex physiological, psychological, and behavioral needs, and the management requirements in a research setting, primate caging or housing systems (and the regulations/guidelines related to such apparatus) should be designed carefully and include advice from specialists in primate behavioral management while ensuring the safety of the personnel working with the animals (NRC 2003).

LEGAL REQUIREMENTS AND GUIDELINES FOR THE CARE AND USE OF NONHUMAN PRIMATES IN RESEARCH

Housing and working with NHPs in research environments is a heavily regulated endeavor. This chapter focuses on legislation and guidelines that provide specific directions of relevance to the behavioral management of NHPs. There are excellent recent, in-depth descriptions of laws, regulations, and policies regulating NHP research, including transportation, and health and safety (Bayne and Morris 2012). Particularly in Europe, there are a number of guidelines and legal regulations that provide the framework for the ways in which laboratory NHPs should be housed and cared for. Some of these include important advice on occupational health and safety issues, quarantine issues, transportation issues, and biosecurity. These are all naturally very relevant aspects of animal care and use programs for NHPs presenting a number of unique, species-specific challenges. However, because the present chapter is focused on the implications of the regulations/guidelines on the behavioral management and welfare of NHPs participating in research, readers interested in broader issues related to safety, quarantine, biosecurity, etc., are referred to the several excellent sources that provide basic information on ways to address these issues appropriately in an NHP care and use program (e.g., Bayne and Morris 2012).

The present chapter reports on five highly relevant sets of regulations/guidelines: European, American, Japanese, Australian, and Canadian, some in considerable detail (European and American) and some in less detail (Japanese, Australian, and Canadian). In addition, international societal guidelines, such as those of the Association of Primate Veterinarians (APV) and the International Primatological Society (IPS), which provide very relevant guidance, will be discussed when relevant.

Guidelines that recommend social housing for NHPs are not new. Indeed, in 1987, the NIH of the United States developed an intramural "Nonhuman Primate Management Plan" that stated that social housing be considered "an appropriate means of providing enrichment...." (Bayne 2014). As our knowledge concerning appropriate methods to provide social housing for NHPs increased, the philosophical tone in the *Guide* became more strongly biased toward social housing. The seventh edition of the *Guide* (NRC 1996) stated, "It is desirable that social animals be housed in groups.... When it is appropriate and compatible with the protocol, social animals should be housed in physical contact with conspecifics." This edition of the *Guide* (NRC 1996) further stated, "Appropriate social interactions among members of the same species (conspecifics) are essential to normal development and well-being....," noting that social housing might buffer the effects of a stressful situation, reduce behavioral abnormality, increase opportunities for exercise, and expand species-typical behavior and cognitive stimulation.

However, the most important legislation and legislation-enforced guidelines to the laboratory animal community are the European Convention ETS 123 A (1986), the European Directive (EU Directive 2010/63/EU), incorporating most of the guidelines from the European Convention ETS 123 Appendix A, and the American *Guide* (NRC 2011). The global impact of the *Guide* is reflected with it having been translated into Chinese, Japanese, Portuguese, and Thai, with additional translations (e.g., Spanish) underway. There is fairly good agreement between these three documents with regard to performance requirements, such as social housing of all NHP species. Social housing is considered by AAALAC International as the "default method of housing unless otherwise justified based on social incompatibility resulting from inappropriate behavior, veterinary concerns regarding animal well-being, or scientific necessity approved by the Institutional Animal Care and Use Committee (IACUC) (or comparable oversight body). When necessary, single housing of social animals should be limited to the minimum period necessary, and where possible, visual, auditory, olfactory, and, depending on the species, protected tactile contact with compatible conspecifics should be provided. In the absence of other animals, additional enrichment should be offered, such as safe and positive interaction with the animal care staff, as appropriate to the species of concern; periodic release into larger enclosures; supplemental enrichment items; and/or the addition of a companion animal in the room or housing area. The institution's policy and exceptions for single housing should be reviewed on a regular basis and approved by the IACUC (or comparable oversight body) and/or veterinarian" (http://www.aaalac.org/accreditation/positionstatements.cfm#social and http://www.aaalac.org/accreditation/faq_landing.cfm#C6).

The US Department of Agriculture (USDA) regulates the use of NHPs in research in the United States. Institutions using primates for research and testing must comply with the standards for their care as set forth in the Animal Welfare Regulations (AWRs). The AWRs (USDA 1991) include a section (3.81) pertaining to environmental enrichment that endorses social housing and states, "The environment enhancement plan must include specific provisions to address the social needs of NHPs of species known to exist in social groups in nature." In those limited cases where a primate must be individually housed (e.g., behavioral incompatibility, infectious disease state), the AWRs state, "Individually housed nonhuman primates must be able to see and hear nonhuman primates of their own or compatible species unless the attending veterinarian determines that it would endanger their health, safety, or well-being." The AWRs also address nonsocial enrichments by requiring that "the physical environment in the primary enclosures must be enriched by providing means of expressing noninjurious species-typical activities. Species differences should be considered when determining the type or methods of enrichment."

The Office of Laboratory Animal Welfare (OLAW), which oversees the care and use of research animals in US Public Health Service (PHS)-funded research (e.g., the NIH), also considers housing of primates in social groups or in pairs as the default setting and states that clear medical or scientific justification is required for any other method of housing: "A primate may be exempted from the environmental enrichment plan by the institutional veterinarian for health reasons but unless the exemption is permanent it must be reviewed every 30 days. The IACUC may exempt a primate from environmental enrichment or social housing but this must be based on valid scientific justification." (OLAW: http://grants.nih.gov/grants/olaw/NHP_Enrichment_transcript.pdf). In addition, OLAW sponsored the development and publication of a series of six booklets "that serve as an introduction to the subject of environmental enrichment for primates housed in a diversity of conditions" to assist institutions in meeting the recommendations of the *Guide* (http://grants.nih.gov/grants/olaw/request_publications.htm).

The APV also emphasizes the importance of social housing and has issued Socialization Guidelines for Nonhuman Primates in Biomedical Research (http://www.primatevets.org/Content/files/Public/education/APV%20Social%20Housing%20Guidelines%20final.pdf). The APV states that scientists, laboratory animal veterinarians, animal caregivers, and IACUCs/ethical review committees must work together to fully implement regulatory expectations to provide the most appropriate environment for captive NHPs. On their website, the APV has additional relevant guidelines, including *Guidelines for Use of Fluid Regulation in Biomedical Research*, *Laparoscopic Reproductive Manipulation of Female Nonhuman Primates Guidelines*, *Social Housing Guidelines*, *Food Restriction Guidelines*, *Jacket Use Guidelines*, and *Cranial Implant Care Guidelines*.

Both the American (USDA AWRs) and European (the Directive and the European Convention) legislations stipulate minimum acceptable measurements ($l \times w \times h$) for enclosures used to house different NHP species as does the *Guide* (NRC 2011). Further, in their *Animal Care Policy Manual*, the USDA defines the additional spatial requirements necessary for those species considered to be brachiating NHPs. It is important to emphasize that the guidelines provide *minimum* enclosure measurements, and in many circumstances (but not all), it is beneficial to the animals concerned to be provided with more than the absolute minimum amount of space. The minimum cage sizes in these guidelines differ somewhat between Europe and America and, the American guidelines, in particular, remain modest considering the naturally high levels of locomotor behavior and the often arboreal nature of many of the primates they are intended to accommodate. In practice, in recent years, significant progress has been made in many animal care and use programs; more and more NHPs are currently housed in large group enclosures with access to outdoor facilities, increasing the quality of life (welfare) for the animals (Baker et al. 2007; Baker 2016). These improvements have occurred in conjunction with the modernization of husbandry techniques, including the introduction of positive reinforcement training (Reinhardt 1991; Schapiro et al. 2001) procedures to allow the animals to voluntarily participate in necessary activities. Voluntary participation is an example of a refinement that significantly decreases the stress associated with capture, restraint, local transport, and other simple procedures involved in the daily management of NHPs (Lambeth et al. 2006).

The Directive's Annex III lists fairly detailed requirements for the care and accommodation of laboratory animals, with species-specific guidelines for NHPs. These include sections on health, breeding and separation from the mother, enrichment, and handling, including training of the animals. Supplementing these general considerations for NHPs, Annex III also contains additional detailed guidelines on the environment and its control, health, housing enrichment and care, training of personnel, and transport for marmosets and tamarins, squirrel monkeys, macaques, vervets, and baboons.

A number of additional guidelines of relevance to the present discussion (in addition to the detailed legislation in the Directive, which have been transposed into national legislation in the individual EU member states) and the American guidelines (ILAR *Guide*), in practice, must be adhered to in North American NHP care and use programs as well as all other programs that seek AAALAC International accreditation or receive PHS funds.

In Japan, animal experimentation is regulated via (1) a number of laws, (2) amendments to these laws, and (3) ministerial guidelines. The Primate Research Institute, Kyoto University (2010) has issued very detailed guidelines for the care and use of NHPs, including sections on facility design and equipment. These guidelines, which are well known and used by Japanese scientists, are in agreement with the European and American guidelines and provide useful information, often in greater detail than the Western guidelines. They emphasize the importance of providing a captive environment in which the animals can perform their species-specific behavioral patterns at an optimal level, as determined by each individual's physiological, ecological, and behavioral characteristics, within a range that does not interfere with the objectives and methods of the research. In sections on environmental enrichment, the guidelines state that if not all aspects can be improved adequately, due to experimental and environmental limitations, improvement of selected aspects may be able to compensate for the loss of other aspects. For example, if there are necessary limitations to the social environment, efforts must be made to enrich the physical environment and increase human contact. This is in agreement with the *Guide*, which emphasizes the importance of supplemental enrichment to compensate for situations in which an animal has to be singly housed for a certain period of its life. The Kyoto guidelines advocate that the animals' (functional) living space should be as large as possible and should include novelty, manipulatable tools for foraging, and a suitable social environment. Interestingly, they advocate that if an adequate conspecific social environment cannot be provided for research reasons, then positive relationships with humans will, to some extent, compensate for the lack of conspecific social interaction.

The IPS has prepared detailed guidelines (McCann et al. 2007) that recognize the need for internationally accepted standards for primate acquisition, care, and experimentation that are attainable worldwide, regardless of legal, cultural, or economic backgrounds. The IPS document is in agreement with the regulatory guidelines in Europe and North America, described earlier, and includes informative sections on transportation, breeding, weaning, and rearing, as well as codes of practice for housing and environmental enrichment, levels of training for primate care staff, and health care. The IPS guidelines are currently available in English, French, Spanish, Chinese, and Japanese.

Combined, all of the different guidelines on NHP care and use provide detailed information concerning virtually all topics associated with the housing and care of NHPs. Through the increasing number of AAALAC International-accredited NHP facilities in Asia that require compliance with the *Guide*, most NHP research facilities will be motivated to follow the international guidelines.

The European Directive's minimum sizes for caging are expensive for research facilities to comply with. In Europe, compliance is obviously a must, but NHP facilities outside Europe (e.g., China and Africa) often seek to comply with the European standards as well. Compliance with the European Directive by programs outside of Europe facilitates collaboration with European scientists in industry and academia, who do not wish to be associated with NHP research at institutions with lower than European standards.

It is well recognized that an increasing proportion of global NHP research and testing is being carried out in Asia, especially in China, where the local guidelines for care and use of NHPs were historically less well delineated than those in Europe, the United States, or Japan. Recently, however, the Chinese government has developed animal welfare and ethics standards for traditional laboratory animal species, with an anticipated release date of 2016/2017. These standards include sections on the provision of environmental enrichment, the Three Rs, and training of staff with regard to correct deportment in handling animals. In a parallel effort, the Chinese government has been in the process of developing *Regulations for the Administration of Affairs Concerning Experimental Animals*. In addition, a domestic accreditation system has been implemented that is based on national standards that encompass many of the same approaches as the *Guide*.

To a lesser extent, NHP research is also being performed in South America and Africa. Although these continents lack supranational legislation or guidelines setting out rules for NHP research, some countries have implemented legislation that incorporates the Three Rs (Bayne et al. 2015), but information specific to NHPs has not been included in legislative frameworks. An exception to this is the National Museums of Kenya's Institute of Primate Research (IPR) located in the Nairobi suburb of Karen. It is perhaps the best known primate research center in Africa. Primate models, in particular vervet monkeys and baboons, are used in studies of tropical infectious diseases, human reproductive disorders, and conservation strategies. Many high-quality, peer-reviewed scientific articles have been published from IPR during its 50 years of existence, including studies of stress susceptibility of NHPs captured from the wild (Uno et al. 1989; Suleman et al. 2000, 2004; Kagira et al. 2007) and methods to increase the positive outcome of relocation of endangered species (Moinde et al. 2004). Research ethics and animal welfare issues are taken into account at IPR, and according to their webpage, "Ethical and animal welfare concerns form a strong component of the Department's animal husbandry and research activities." Before any experimental procedure is carried out, review committees evaluate all research proposals for scientific merit and welfare concerns. Housing conditions at IPR meet European standards, and IPR is currently leading a multi-institutional effort to develop comprehensive guidelines for laboratory animals used in research and education in this part of the world. IPR has partnered with the National Council for Science and Technology, Kenya and the Consortium for National Health Research (a local funding agency for Health Research) for this task (Hau et al. 2014).

According to the Directive, NHPs used for scientific research should be captive-bred and reared on site to avoid transport stress, and where possible, their enclosures should include access to the outdoors. One of the advantages for the animals housed in NHP facilities in source countries, such as Kenya, is that it is possible to house the animals in outdoor, seminatural enclosures in their natural geographic range, which avoids subjecting the primates to long-distance transportation and having them make the adjustments necessary to acclimatize to a new climate and foreign biotic and abiotic environments (Schapiro et al. 2012).

Access to outdoor enclosures is important for the welfare of NHPs, and the newer guidelines emphasize this. The Directive states that (1) only purpose-bred animals should be used in research, (2) breeding systems should be designed to ensure good welfare, and (3) a combination of indoor-outdoor housing is recommended if it has no adverse welfare and health consequences for the animals, and is compatible with the goals of the study (EU 2002). The Directive contains an entire section on outdoor enclosures and also states that indoor enclosures, whenever possible, should be provided with windows, as they are a source of natural light and can provide environmental enrichment. This is in agreement with the Australian Policy on the Care and Use of Non-Human Primates for Scientific Purposes issued by the National Health and Medical Research Council (NHMRC 2003). The Australian policy states that the animals must be provided with daytime access to an outdoor enclosure that is freely available, if they are to be held for 6 weeks or longer, unless exempt under the terms of the policy (https://www.nhmrc.gov.au/_files_nhmrc/publications/attachments/ea14.pdf; also see the AAALAC International FAQ (http://www.aaalac.org/accreditation/faq_landing.cfm#E8)).

The Australian guidelines are fairly brief, but they are in very close agreement with the European and American guidelines. There are also Canadian guidelines for NHP care and use. The Canadian Council on Animal Care (CCAC) issued guidelines, *Guide to the Care and Use of Non-Human Primates* in 1984, and these are presently under revision. The 1984 guidelines are clearly in need of updating, and the upcoming version will demonstrate the positive advances that have taken place in NHP care during the past 30 years (http://www.ccac.ca/en_/standards/guidelines/underdevelopment).

Over the past 30 years, the ways in which we house and care for NHPs and the rules and guidelines regulating their housing and care have changed dramatically. In the past, extensive single housing in small, barren, squeeze-back cages, sometimes in multiple tiers, was the norm, while currently, social housing in large cages/enclosures equipped with furniture, fixtures, and toys is the default system. Additionally, attention is now paid to the foraging and locomotor needs of the animals, and management routines have been refined to provide the animals with opportunities to more frequently perform a greater variety of natural behaviors. Effective behavioral management and educational programs for animal care staff, for instance, developed and implemented at the National Primate Research Centers in the United States and European NHP centers, EUPrimNet (http://www.euprim-net.eu/), have resulted in widespread training and gentling programs for the animals, that now benefit from cognitive stimulation, reduced fearfulness, and the ability to voluntarily participate in a variety of husbandry, veterinary, and research procedures.

According to the Directive's Annex, "care covers all aspects of the relationship between animals and man. Its substance is the sum of material and non-material resources provided by man to obtain and maintain an animal in a physical and mental state where it suffers least, and promotes good science. It starts from the moment the animal is intended to be used in procedures, including breeding or keeping for that purpose."

Captive care of NHP and behavioral management strategies are constantly evolving, and new refinements are continually being developed, discovered, and implemented by the leading facilities in the field. Many of these refinements are quickly adopted as part of other NHP programs all over the world. Improving the ways in which we care for NHPs will never be finished, and the next generation of guidelines will no doubt reflect this continuing process.

In the interests of medical progress and our responsibilities to future generations, NHPs will continue to be vital as models for studies of debilitating diseases. It is thus of the utmost importance that the care of the animals needed for this activity is continuously improved and resources allocated to studies that strive to ensure optimum physical and mental well-being of the animals in our care.

CONCLUSION

There is considerable agreement among the various guidelines (European, American, Japanese, and Australian), which have been developed and implemented in recent years. Together, they admirably capture (1) the modern trends in housing and care of NHPs and (2) the importance of a stimulating environment and training of the animals to lower stress associated with husbandry, veterinary, and research procedures. To ensure high-quality behavioral management, both the EU guidelines and the *Guide* state that a person who understands the behavior of NHPs should be available for advice on issues related to social behavior, environmental enrichment strategies, and management.

Adherence to the excellent guidance in The Directive and the *Guide* will ensure that laboratory NHPs are housed and cared for in a manner that meets and exceeds today's best practice standards, standards based on analyses of the complex behavioral needs of NHPs in captive environments, while accounting for the requirements of the research projects.

REFERENCES

Abbott, A. 2014. Biomedicine: The changing face of primate research. A hard-won political victory for primate research is at risk of unravelling in pockets of Europe. *Nature* 506:24–26.

Baker, K.C. 2016. Survey of 2014 behavioral management programs for laboratory primates in the United States. *American Journal of Primatology* 78:780–796.

Baker, K.C., J.L. Weed, C.M. Crockett, and M.A. Bloomsmith. 2007. Survey of environmental enhancement programs for laboratory primates. *American Journal of Primatology* 69:377–394.

Bayne, K. 1998. Developing guidelines on the care and use of animals. In Fishman, J., D. Sachs, and R. Shaikh (Eds), *Xenotransplantation: Scientific Frontiers and Public Policy*, Vol. 862. Annals of the New York Academy of Sciences, New York, 105–110.

Bayne, K. 2014. A historical perspective on social housing of laboratory animals. *The Enrichment Record* Winter:8–11.

Bayne, K., J. Hurst, and S. Dexter. 1992. Evaluation of the preference to and behavioral effects of an enriched environment on male rhesus monkeys. *Laboratory Animal Science* 42(1):38–45.

Bayne, K., and C. McCully. 1989. The effect of cage size on the behavior of individually housed rhesus monkeys. *Laboratory Animals* 18(7):25–28.

Bayne, K., and T.H. Morris. 2012. Laws, regulations and policies relating to the care and use of nonhuman primates in biomedical research. In Abee, C., K. Mansfield, S. Tardif, and T. Morris (Eds), *Nonhuman Primates in Biomedical Research: Biology and Management*, Elsevier, New York.

Bayne K., G.S. Ramachandra, E.A. Rivera, and J. Wang. 2015. The evolution of animal welfare and the 3Rs in Brazil, China, and India. *Journal of the American Association for Laboratory Animal Science* 54(2):181–191.

Bayne, K. and P. Turner. 2013. Animal environments and their impact on laboratory animal welfare. In Bayne, K. and P. Turner (Eds), *Laboratory Animal Welfare*, Elsevier, New York, 77–93.

Boccia M.L., M.L. Laudenslager, and M.L. Reite. 1995. Individual differences in macaques' response to stressors based on social and physiological factors: Implications for primate welfare and research outcomes. *Laboratory Animals* 29:250–257.

Burm, S.M., J.B. Prins, J. Langermans, and J.J. Bajramovic. 2014. Alternative methods for the use of nonhuman primates in biomedical research. *ALTEX* 31(4):520–529.

Carlsson, H.-E., S.J. Schapiro, I. Farah, and J. Hau. 2004. Use of primates in research: A global overview. *American Journal of Primatology* 63:225–237.

CCAC—The Canadian Council on Animal Care. 1984. Guide to the care and use of non-human primates. http://www.ccac.ca/Documents/Standards/Guidelines/Vol2/non_human_primates.pdf, and these are presently under revision.

Dawkins, M.S. 1983. Battery hens name their price: Consumer demand theory and the measurement of ethological "needs." *Animal Behaviour* 31:1195–1205.

Dawkins, M.S. 1990. From an animal's point of view: Motivation, fitness and animal welfare (with commentaries). *Behavioral and Brain Sciences* 13:1–61.

EU Directive. 2010. Directive 2010/63/EU of the European Parliament and of the Council of 22 September 2010 on the protection of animals used for scientific purposes. *Official Journal of the European Union* L 276/33 (October 20, 2010).

European Convention. 1986. European Convention for the Protection of Vertebrate Animals Used for Experimental and Other Scientific Purposes. Appendix A of the European Convention for the Protection of Vertebrate Animals used for Experimental and Other Scientific Purposes (ETS NO. 123), Guidelines for Accommodation and Care of Animals (Article 5 of the Convention).

European Union. 2002. European Commission Health and Consumer Protection Directorate-General Directorate C—Scientific Opinions. C2—Management of scientific committees; scientific co-operation and networks. The welfare of non-human primates used in research. Report of the Scientific Committee on Animal Health and Animal Welfare 2002.

European Union. 2009. Scientific Committee on Health and Environmental Risks. SCHER: The need for non-human primates in biomedical research, production and testing of products and devices.

Ferdowsian, H., and A. Fuentes. 2014. Harms and deprivation of benefits for nonhuman primates in research. *Theoretical Medicine and Bioethics* 35:143–156.

Fraser, D., P.A. Phillips, and B.K. Thompson. 1993. Environmental preference testing to assess the well-being of animals—An evolving paradigm. *Journal of Agricultural and Environmental Ethics* 6:104–114.

Hagelin, J., H.-E. Carlsson, M.A. Suleman, and J. Hau. 2000. Swedish and Kenyan medical and veterinary students accept nonhuman primate use in medical research. *Journal of Medical Primatology* 29:431–432.

Hansen, A.K., and V. Baumans. 2007. Housing, care and environmental factors. In E. Kaliste (Ed.), *The Welfare of Laboratory Animals*, Springer, Dordrecht, the Netherlands, 37–50.

Hau, A.R., F.A. Guhad, M.E. Cooper, I.O. Farah, O. Souilem, and J. Hau. 2014. Animal experimentation in Africa: Legislation and guidelines. In J. Guillen (Ed.), *Laboratory Animals, Regulations and Recommendations for Global Collaborative Research*. Elsevier, New York, 205–216.

Hau, J., and S.J. Schapiro. 2004. The welfare of non-human primates. In E. Kaliste (Ed.), *The Welfare of Laboratory Animals*, Springer, Dordrecht, the Netherlands, 291–314.

Honess, P., and S. Wolfensohn. 2010. The extended welfare assessment grid: A matrix for the assessment of welfare and cumulative suffering in experimental animals. *ALTA—Alternatives to Laboratory Animals* 38:205–212.

Kagira, J.M., M. Ngotho, J.K. Thuita, N.W. Maina, and J. Hau. 2007. Hematological changes in vervet monkeys (*Chlorocebus aethiops*) during eight months' adaptation to captivity. *American Journal of Primatology* 69:1053–1063.

Lambeth, S.P., J. Hau, J.E. Perlman, M. Martino, and S.J. Schapiro. 2006. Positive reinforcement training affects hematologic and serum chemistry values in captive chimpanzees (*Pan troglodytes*). *American Journal of Primatology* 68:245–256.

McCann, C., H. Buchanan-Smith, L. Jones-Engel, K. Farmer, M. Prescott, et al. 2007. *IPS International Guidelines for the Acquisition Care and Breeding of Nonhuman Primates*. International Primatological Society. Available: http://www.internationalprimatologicalsociety.org/publications.cfm. Accessed April 28, 2016.

Mench, J.A., and G.J. Mason. 1997. Behavior. In M.C. Appleby and Hughes, B.O. (Eds), *Animal Welfare*, CAB International, Wallingford, UK, 127–141.

Moinde, N.N., M.A. Suleman, H. Higashi, and J. Hau. 2004. Habituation, capture and relocation of Sykes monkeys (*Cercopithecus mitis albotorquatus*) on the coast of Kenya. *Animal Welfare* 13:343–353.

National Institutes of Health. 1987. Nonhuman Primate Management Plan. http://netvet.wustl.edu/species/primates/nihprim.txt

National Institutes of Health. 2013. Announcement of agency decision: Recommendations on the use of chimpanzees in NIH-supported research. http://dpcpsi.nih.gov/council/pdf/NIH_response_to_Council_of_Councils_recommendations_62513.pdf.

National Research Council. 1996. *Guide for the Care and Use of Laboratory Animals*, 7th Edition. National Academies Press, Washington, DC.

National Research Council. 2011. *Guide for the Care and Use of Laboratory Animals*, 8th Edition, National Academies Press, Washington, DC.

NC3Rs. 2006. NC3Rs Guidelines: Primate accommodation, care and use. https://www.nc3rs.org.uk/sites/default/files/documents/Guidelines/NC3Rs%20guidelines%20-%20Primate%20accommodation%20care%20and%20use.pdf.

NHMRC. 2003. Australian Policy on the Care and Use of Non-Human Primates for Scientific Purposes to be read in conjunction with The Australian Code of Practice for the Care and Use of Animals for Scientific Purposes. Endorsed 6 June 2003. Prepared by the Animal Welfare Committee of the NHMRC.

Reinhardt, V. 1991. Training adult male rhesus monkeys to actively cooperate during in-homecage venipuncture. *Animal Technology* 42:11–17.

Ross, S.R., S. Calcutt, S.J. Schapiro, and J. Hau. 2011. Space use selectivity by chimpanzees and gorillas in an indoor–outdoor enclosure. *American Journal of Primatology* 73:197–208.

Ross, S.R., S.J. Schapiro, J. Hau, and K.E. Lukas. 2009. Space use as an indicator of enclosure appropriateness: A novel measure of captive animal welfare. *Applied Animal Behaviour Science* 121:42–50.

Schapiro, S.J., S.P. Lambeth, K.R. Jacobsen, L.E. Williams, B.N. Nehete, and P.N. Nehete. 2012. Physiological and welfare consequences of transport, relocation, and acclimatization of chimpanzees (*Pan troglodytes*). *Applied Animal Behaviour Science* 137:183–193.

Schapiro, S.J., J.E. Perlman, and B.A. Boudreau. 2001. Manipulating the affiliative interactions of group-housed rhesus macaques using positive reinforcement training techniques. *American Journal of Primatology* 55:137–149.

Suleman, M.A., E. Wango, I.O. Farah, and J. Hau. 2000. Adrenal cortex and stomach lesions associated with stress in wild male African green monkeys (*Cercopithecus aethiops*) in the post-capture period. *Journal of Medical Primatology* 29:338–342.

Suleman, M.A., E. Wango, R.M. Sapolsky, H. Odongo, and J. Hau. 2004. Physiologic manifestations of stress from capture and restraint of free-ranging male African green monkeys (*Cercopithecus aethiops*). *Journal of Zoo and Wildlife Medicine* 35:20–24.

Suran, M., and H. Wolinsky. 2009. The end of monkey research? New legislation and public pressure could jeopardize research with primates in both Europe and the USA. *EMBO Reports* 10:1080–1082.

Uno, H., R. Tarara, J.G. Else, M.A. Suleman, and R.M. Sapolsky. 1989. Hippocampal damage associated with prolonged and fatal stress in primates. *The Journal of Neuroscience* 9:1705–1711.

USDA Animal Care Policies. http://www.aphis.usda.gov/animal_welfare/downloads/animalcarepolicymanual. pdf.

US Department of Agriculture (USDA). (1991) Code of Federal Regulations, Title 9, Part 3, Animal Welfare; Standards; Final Rule. *Federal Register* 56(32), 1–109.

van Vlissingen, J.M.F. 1997. Welfare implications in biomedical research. *Primate Reports* 49:81–85.

Weatherall, D. 2006. *The Use of Non-Human Primates in Research*, Academy of Medical Sciences, Medical Research Council, Royal Society and Wellcome Trust, London.

Wolfensohn, S., S. Sharpe, I. Hall, S. Lawrence, S. Kitchen, and M. Dennis. 2015. Refinement of welfare through development of a quantitative system for assessment of lifetime experience. *Animal Welfare* 24:139–149.

World Organisation for Animal Health (OIE). 2012. Use of animals in research and education. Terrestrial Animal Health Code, Chapter 7.8. http://www.oie.int/index.php?id=169&L=0&htmfile=chapitre_aw_ research_education.htm.

CHAPTER 4

Behavioral Management
The Environment and Animal Welfare

Tammie L. Bettinger, Katherine A. Leighty,
Rachel B. Daneault, and Elizabeth A. Richards
Disney's Animal Kingdom

Joseph T. Bielitzki
Consultant

CONTENTS

The environment in which an animal is housed is the geography in which it lives its life. The ability to exhibit species-typical behaviors provides animals with the tools to manage features of life in a captive setting. Most primates are social; however, the ability to navigate complex group dynamics can be challenging and requires the development of appropriate social skills and a habitat that allows a complex array of behaviors to be exhibited. The social environment for singly housed animals is also challenging, in that it lacks conspecifics, and so behaviors that might normally be used to engage with other group members can become self-directed. Overlaying these issues is captivity itself. The captive environment contains human intrusion into the primate's space, management decisions to remove/add individuals to/from an established group, and decreased opportunities for an animal to make choices when it comes to the selection of food, mates, and spatial location.

The ability to interpret animal behavior is arguably one of the most important skills of a caregiver or manager. Behavior is the external manifestation of everything occurring inside of the animal, ranging from the physiological to the psychological to the social. For example, an animal with diabetes will often drink more water; an animal needing to emigrate from its group may start trying to escape from its enclosure; or a shift in dominance hierarchy may result in individuals isolating themselves or intently watching certain individuals. However, behavior is complex with many underlying factors contributing to its cause. For this reason, it is important to thoroughly assess a situation to ensure that any intervention or manipulation is addressing the real issue.

A habitat that allows for a wide range of behaviors improves the ability of the caregiver/manager to detect changes in behavior more quickly. An enclosure does not need to be huge or natural; the key is that the enclosure has structures and furnishings that allow the animal to exhibit a wide range of behaviors. In the 1980s, there was a shift to more "naturalistic" exhibits (Coe 1989), with the goal of providing animals a more complex environment. However, it soon became apparent that even naturalistic enclosures become static to the inhabitants who may live their entire life in the space. Nuttall (2004) proposed the "animal as client" model that links exhibit design to animal welfare and an animal's ability to express a range of natural behaviors. Appropriate complexity of the space, regardless of its size, is key to meeting the needs of the animal. We must keep in mind that the environment in nature is not static, nor is it nonresponsive to the animal's actions. For example, in the wild, a monkey will locomote through trees on branches that sway and bounce. This motion, provided by the environment, stimulates the vestibular, as well as visual, senses of the animal. More terrestrial primate species, such as baboons, may dig in soil in search of roots or grubs, creating holes and depressions in the substrate. The captive managers' ability to provide environments that can respond to the animal's manipulations can be challenging; however, this important aspect of feedback between animals and the environment should be considered a critical component of any behavioral management plan. Additionally, changes to the environment or addition of items meant to stimulate behavior must have functional relevance to the animal (Newberry 1995). The development of an environmental enrichment program should be focused on the specific species and individuals involved, as well as guided by the ultimate goal of improving animal welfare.

Barber and Mellen (2008) described the components of animal welfare as seven programs that together make up the "welfare infrastructure" (animal training, environmental enrichment, habitat, husbandry, nutrition, research, and veterinary care). These programs are interrelated pieces that together build a wall. As you begin removing or omitting pieces, the wall loses stability and starts to crumble. Pulling the aforementioned concepts together, it seems apparent that emphasis should be placed on ensuring that animals live in a functionally appropriate environment that facilitates a wide range of behaviors, provides feedback to the actions of the animals, and allows the animals some control over their surroundings. A "functionally appropriate" environment for a species depends on the natural and individual histories of the animals involved.

The physical environment is the primary aspect of the animal's life that we, as caregivers/managers, can control and manipulate. Further, most behavior challenges require modification (simple to complex) of the physical environment in order for the animals to exhibit appropriate behaviors for the given situation. We strive to evoke or facilitate the expression of a range of behaviors by focusing on the animal's sensory systems. By modifying sensory inputs through alterations to the existing environment, we have found that we can encourage most animals to exhibit species-typical behavioral responses in a wide range of situations.

SENSES AND THE ENVIRONMENT

Dominy et al. (2004) refer to senses as "the anatomical interface between the environment and the behaving organism." It is through their senses that animals receive a continuous stream of information about stimuli in their environment. This information is filtered and organized by the brain in a higher-order process called perception. For the purpose of this chapter, we use the term "perception" to refer to those sensory stimuli to which animals attend. Classically, the five senses are discussed as individual inputs to perception, but in the animal, often they are linked to provide greater situational awareness. The senses are an input system for life, whereas behaviors are the output system necessary for adaptation to the environment and social situations. Food acquisition, mating, rest, activity, and social interactions are all behaviors that depend on appropriate sensory inputs. While different authors refer to a variety of senses, we focus on the five most commonly referenced ones: smell, sight, hearing, taste, and touch. For primates, we also discuss motion and orientation. These inputs (motion and orientation) provide added information for coordinating motor activities, balance, and survival in the three-dimensional environment in which primates exist in the wild. As caregivers of nonhuman primates, it is important to keep in mind that the perceptual abilities of all primates, including humans, are not the same. That is to say, primates in your care may perceive environmental stimuli that you, their caregiver, do not, or vice versa. They may perceive lighting differently from you; you may be able to tune out a sound that they cannot; and they may smell a scent that you cannot detect. These differences in our sensory abilities and perceptual processes should always be considered (Poole 1998).

The conditions of captivity are limited in complexity and variation when compared with the environment in which primates live in the wild. Even the most complex captive environments are static to the primates that live in them every day of their lives. Compared to the wild, captivity provides benefits such as increased longevity by removing predators, provision of a well-balanced diet, and state-of-the art health care (Hediger 1964, 1968). However, captivity must also provide ongoing sensory input in order to provide the physical and psychological stimulation primates need, if we are to ensure their overall quality of life while in our care.

Environmental enrichment programs focus on ways to stimulate behavior, often by activating the various senses. However, programs may overlook assessments of the ways in which the senses are being stimulated inadvertently. It is important to keep in mind that, in captivity, animals are exposed to a variety of unintentional sensory stimulation, such as air handling systems, night lights from security checks, and even someone microwaving their lunch in a kitchen down the hall. Some of this sensory stimulation may be imperceptible to the human observer, or we may be habituated to its presence. Additionally, while the care staff leave at the end of the day, the primates do not. Some of the intrusions into the animal's sensory world may occur at night, when care staff and researchers are not present. Understanding what occurs at night, through sound recording, motion detectors, or video monitoring, can provide valuable information for assessing the impacts of the activities that occur after daytime working hours at your facility.

When assessing an animal's physical environment, it is important to identify these unintentional stimulations, as they may be a source of behavior problems or may cause stress to the animal. While information exists on the benefits of various sensory inputs, less information is available on the deleterious effects. For example, studies have shown that there are benefits from exposure to auditory stimuli, such as music (O'Neill 1989; Ogden et al. 1994; Brent and Weaver 1996; Howell et al. 2003). However, there is currently insufficient information available to determine the amount of quiet time an animal needs each day to achieve adequate rest. An additional challenge to keep in mind when evaluating the environment is determining the boundaries of an animal's overall sensory environment; is it the enclosure, or the building in which the enclosure is located, or are impacts coming from even farther away?

While humans have the ability to improve the life of primates in captivity, we can also inadvertently cause stress. Examples include medical staff who visit only for immobilizations, a caregiver passing through the area with a capture net, visitors who bang on the glass front of an exhibit, or humans who "smile and yawn" at the monkeys to elicit a response. Studies have shown that in some cases, the presence of caregivers can be enough to incite aggression in chimpanzees (Lambeth et al. 1997), and in rhesus macaques, heart rates remained elevated while caregivers were present, even after the stressful event had passed (Line et al. 1991). Although some stressful situations are unavoidable, as caregivers, we must be vigilant to ensure that we are not condoning activities that cause unnecessary agitation or stress.

Overview of Senses in Primates

Smell

Although primates may not have the olfactory acuity of taxa such as canids, they do rely on their sense of smell for interacting with the environment. Their sense of smell is well developed and provides information on the environment and about individuals. Smell is used for a variety of functions, from evaluating food to reproductive cueing to communication. While the exact role and importance of smell/olfaction in primates remain poorly defined (Fleagle 1988), there are many examples of its use. The arrival of a new infant in the group often triggers the behavior of monkeys/ apes sniffing the location where the new baby has come into contact with the substrate. Female genitalia are inspected by both visual and olfactory senses to assess reproductive status. Buchanan-Smith et al. (1993) saw increased anxiety-related responses in cotton-top tamarins when exposed to the fecal odor of potential predators, as compared to nonpredators. Capuchins and spider monkeys smell fruit to test for ripeness, and capuchin monkeys have been observed to increase their rate of sniffing when unpalatable compounds are added to desirable foods (see Nevo and Heymann 2015 for a review).

Western human cultures have, for the most part, neutralized many of the odors that were once likely used in communication. If we detect body odor, it is unlikely that we would draw inferences about the line of work or diet of the odorous person. It is more likely that we would conclude that the person needs to bathe or invest in deodorant. Similarly, our obsession with eliminating odors carries over to how we care for our animals as well. Further, we do not have an understanding of the impact of activities, such as changing cleaning agents or bringing in an unfamiliar worker to mow an outdoor yard, on the primates in our care. As caregivers, we should recognize when we alter an environment, and assess both the need for the change and the impact it might have on individuals and groups (Wells 2009).

In managed settings, we have the opportunity to utilize the sense of smell in a variety of positive ways. We can use smell to encourage exploration of the habitat by providing strongly scented extracts or perfumes. We can encourage foraging by concealing foods with strong odors, such as onions or cabbage, in the habitat. Predator or conspecific scents may increase vigilance behavior (Caine 2017). Scents of the opposite sex may encourage species-specific vocalizations/behaviors (but must be used carefully in bachelor groups to prevent problems with aggression) or stimulate scent-marking behavior. For some taxa, anointing behavior is common (Fragaszy et al. 2004); therefore, items that can be used for this activity should be included in the environment. We can also use scents to prepare groups and new individuals for introductions, by providing them with access to bedding of one another. Many facilities use scents ranging from herbal extracts, such as peppermint and lavender, to perfumes and colognes, to stimulate behavior, or to promote calmness.

We modify the environment, by removing the animal's natural scent, every time we disinfect its living quarters and change out substrates. Consideration should be given to how thoroughly and how frequently an area should be cleaned and scents removed. Not only do cleaning agents neutralize animal odors, but we also often use very odiferous products that are designed to smell appealing to humans. The scents of the cleaning agents should also be considered to determine whether the animals find them aversive. To our knowledge, no preference tests have been conducted with primates to determine whether they share our choices in scents used in cleaning agents. Overcleaning may cause exuberant scent marking, which could result in stereotypic, and even pathologic, behavior.

Sight

Primates rely on their well-developed sense of vision to gather information about the world around them. There is variation across taxa with respect to visual ability. Some taxa have vision specialized for nocturnal life, while others for diurnal activity. Nocturnal species may possess only rods, while apes have trichromatic vision similar to their human caregivers. Stereoscopic vision is of great importance to most primate taxa (Mittermeier et al. 1999), facilitating their arboreal lifestyle. The visual system of a primate is primed for the detection of motion. The ability to detect and interpret motion is key to avoiding predators, reading social cues from conspecifics, as well as navigating through the forest canopy.

Light cycles, both natural and artificial, can have physiological impacts on the life of an animal. Day length can affect reproductive cycles, even in equatorial species, where differences in day length between the summer and winter solstices are very small. Light intensity is another variable that requires consideration. The light levels on a savannah differ from those in the upper canopy of a forest, which differ significantly again from light levels on the forest floor. Nocturnal species may require totally different types of illumination compared with predominantly diurnal species.

Most primates spend considerable amounts of time scanning the environment. Scene complexity is important for variation in the visual field. Exposure to a changing scene adds novelty and stimulation to the environment. The ability to (1) see what is going on outside of the cage, (2) determine sources of other sensory input, and (3) be exposed to novel visual stimuli is important to keep the animal engaged with its surroundings.

So, where does the environment end? A primate's environment extends beyond the boundaries of containment and is defined by the limits of what it can perceive via its senses. As captive managers, we should strive to find ways to help provide primates with "explanations" for what happens in their world. We must provide mechanisms for the animals to fill in the gaps related to the "whys and wheres" of sensory perception. Examples of this might include the use of (1) closed circuit televisions, (2) mirrors to facilitate seeing down hallways or around corners, (3) windows into service areas or outdoor areas, or (4) elevated viewing perches. The ability to see what is happening outside of the enclosure, to be able to anticipate the arrival of food or care staff, or to see what is occurring in other groups provides the animals with opportunities to alter their behavior in meaningful ways.

Conversely, there are times when primates may want to be visually isolated, and so ensuring the environment contains areas with visual barriers is important. Adequate structures within the enclosure for resting (perching), avoiding aggressive cagemates (escape routes), and providing visual barriers for privacy/isolation are critical components of a functional environment, and address many relevant issues associated with group life. Bettinger et al. (1994) found that each female chimpanzee within a social group preferred locations, formed by concrete culverts, that provided areas of visual isolation from the rest of the group and that these areas were used consistently, each day, over the course of many years. Solitary monkeys that are in visual contact with dominant monkeys can be bullied into not consuming their food, even though the threat of physical harm does not exist (O'Neill 1989). In confined spaces, without proper visual barriers, subordinate animals within the group may not have safe access to food, which may prevent them from eating and can increase stress responses. While seeing neighboring groups can be stimulating, it can also be stressful; therefore, visual barriers should be available for the times when animals choose not to engage with neighboring groups.

For captive primates, the ability to manage their proximity to, and isolation from, humans is also important. Hebert and Bard (2000) found that orangutans preferred those areas of their enclosure where they were not visible to humans. The option to move to the back of the enclosure and top of the enclosure and/or to visually isolate themselves from caregivers or visitors is an aspect of choice that should be provided in all captive settings.

Returning to the concept that, in captivity, humans consistently insert themselves into the environment of the animals, we should be cognizant of the timing of visible human activity occurring around enclosures. What does the 24-h human activity pattern around the enclosure look like? When does cleaning of hallways or other service areas occur? When does HVAC, electrical, or plumbing maintenance take place? Do security guards come through the area and shine lights on the animals at night? At Disney's Animal Kingdom, we have found that when there is atypical activity at night, some animals appear more agitated the following day (T.L. Bettinger, personal observation). With this knowledge, we now proactively adjust routines to accommodate such potential behavioral changes.

Hearing

Hearing plays an important role in the lives of primates. More developed in nocturnal species, it should not be undervalued in any of the taxa. Hearing allows an individual primate to gain information about what is occurring outside of its visual range. This is particularly true when we think about the boundaries of the sensory environment. For example, an eyebrow flash from a monkey three enclosures over will have little effect on animals out of visual range; however, a threat vocalization from the same individual is likely to be heard and to evoke a response. Vocalizations convey a variety of information within social groups, as well as between social groups. Many primates utilize long-distance calls for communicating with group members not in the immediate vicinity (chimpanzees) or to signal that the territory is occupied (gibbons, siamangs, ruffed lemurs, howler monkeys). Shepherdson et al. (1989) found that Lar gibbons consistently responded to recordings of unfamiliar pairs by duet calling and brachiating around their enclosure. Berntson et al. (1989) reported that the sound of chimpanzee screams evoked cardiac responses in young chimpanzees, regardless of their early experience with other chimpanzees. Additionally, primates learn to recognize alarm calls of other taxa sharing their environment and develop appropriate behavioral responses, such as predator avoidance or moving toward a food source. Similarly, captive primates have learned the meaning of "management" sounds within their environment and have developed behavioral responses to these human-made stimuli. These may include the rattling of keys prior to a door opening, the noise from an approaching vehicle prior to delivery of the morning feeding, or the sound of a hose unwinding before cleaning an enclosure, among others. This pairing of audio cues with specific events is the basis of operant conditioning. These techniques can be systematically incorporated as part of behavioral management programs in which the caregiver provides specific cues to request the primate to engage in particular behaviors that facilitate their care. Such behaviors can include voluntarily shifting from one enclosure to another, opening the mouth for visual inspection, or presenting a shoulder to receive an injection.

In addition to hearing, the auditory system includes vestibular functions associated with balance and orientation, both skills necessary for arboreal life. In groups with young individuals, it is important to promote stimulation of the vestibular organ by providing the primates with opportunities for climbing, swaying, and suspension in three-dimensional space.

Sound presents some interesting challenges when managing an animal's environment. As discussed in the section on sight, it is once again difficult to define what constitutes the animal's environment. Noises produced at some distance can be heard by the primates in their enclosure. Unlike vision, however, sounds can be sudden in onset, causing a startle response; they may be constant across the 24-h period; or they may be of a frequency that humans cannot even detect. Additionally, sounds may be made in locations to which the primate has no visual access, thus preventing the animal from making decisions concerning the meaning of the noise. As captive managers, sounds may present one of the most nebulous challenges we face. As with noise pollution in the human environment, it is difficult to evaluate the true impacts on animal well-being, as there are examples of organisms adapting to noise, as well as examples of noise being detrimental.

Noise can produce a variety of responses in nonhuman primates, ranging from inducing tranquility to inducing stress. Some deleterious effects include increased stereotypies, elevated stress hormones, and decreased reproduction (Morgan and Tromborg 2007; Orban et al. submitted). Vibrations are associated with sounds and, even if not noticeable to humans, may disrupt the animals' sleep, potentially resulting in increases in anxiety. Unpredictability in the frequency and level of sounds can cause harmful reactions in captive primates, while predictability in noises, even aversive sounds, can help alleviate negative responses (Morgan and Tromborg 2007). For example, Westlund et al. (2012) studied stress hormones in cynomolgus macaques during construction activity. The macaques were divided into two groups: one group was given a cue prior to dynamite detonations, while the other group received no warning of the impending noise. Fecal cortisol levels in the group with no warning signal were significantly increased over the group that had the cue paired with the explosions. Thus, the predictability of the warning signal in alerting the macaques to the impending disturbance may help to mitigate stress in the animals.

Data pertaining to the impacts of sound are contradictory and counterintuitive. Sounds humans may perceive as calming or soothing may actually cause agitation. For example, Ogden et al. (1994) noted that adult gorillas exposed to rainforest noises, used to mask other perceived unnatural noises, became agitated. However, infants in this same group showed fewer stress behaviors and seemed calmer when the rainforest sounds were played, indicating that the effects of sounds vary with individuals, even within a species. Furthermore, Wells et al. (2007) found that rainforest noises and classical music were associated with increased relaxation behaviors and fewer stress behaviors in the gorillas they studied. These findings suggest that primates do not react in the same way to the same sounds, and care should be taken to examine how each individual is affected by various noises. Howell et al. (2003) indicated significant positive impacts on chimpanzee behavior with the use of a stereo system that played various types of music. Aggressive behaviors and agitation decreased, while social and relaxed behaviors increased. Brent and Weaver (1996) found a correlation between the use of radio music and decreased heart rate in four singly housed baboons. While there is potential for sound to be used as enrichment, there are also opportunities to promote desirable behavior by decreasing ambient noise (Wells 2009).

Sound is an area that we have spent a great deal of time investigating in recent months (Orban et al. submitted). In order to monitor sound resulting from some upcoming construction work, we began by obtaining baseline sound readings in some of our animal areas prior to the initiation of construction activity. Our first finding was that actions we had previously implemented in an attempt to make the animals more comfortable were, in fact, creating a great deal of noise. For example, fans installed to cool the areas were quite loud. The animals were using some of the items provided for enrichment in their social displays, which created a considerable amount of sudden-onset noise. These findings have prompted us to explore sound levels more closely in most of our animal areas. In addition to the level of the noise, we found that duration of the sound and an animal's ability to escape the sound strongly influence an animal's tolerance. In exhibits with more exposure to sound, we are working to create "sound sanctuaries" that provide an escape from noisy areas. Our next step is to try to determine the appropriate amount of quiet time that animals should have in order to obtain adequate rest. This question has proven to be very complicated, in large part, due to the myriad of factors that can influence an individual's response to various sound-related stimuli.

Taste

Taste is the sensory modality that directs organisms to identify and consume nutrients, while avoiding toxins and indigestible materials (Chaudhari and Roper 2010). More specifically, it is the ability to distinguish between sweet, sour, savory, and bitter characteristics of substances via the taste buds on the tongue. However, the perception of flavor, often confused with taste, requires the senses of smell and touch to be fully processed by the brain (Rolls and Baylis 1994). As humans, we most often associate taste with something we might consume. However, in nonhuman animals, taste is but one additional mechanism they use to interact with and interpret their environment. Most captive managers and caregivers have multiple examples of primates consuming many different items, from blankets to concrete to wood mulch. In addition to actually consuming problematic items, licking behavior can be equally detrimental when cleaning agents, disinfectants, or other compounds are introduced into the environment.

Wene et al. (1982) found that individual baboons demonstrated marked and consistent flavor preferences when offered chow treated with different flavorings, resulting in increased consumption and weight gain. Thus, by using flavoring, caregivers can increase the value and palatability of food items; however, caution should be taken in order to prevent overeating and obesity. Different flavors and textures will influence acceptability of commercially prepared diets, with many primates showing a preference for sweet items (National Research Council 2003). When planting (or placing) browse in a habitat, care must be taken to ensure that items are not toxic, because animals may not have learned to avoid them. Further, in captivity, where live vegetation may be limited, or when animals have access to items they would not have encountered in the wild, they may consume atypical items. Additionally, preferences shown by animals in the wild may not carry over to captivity. For example, colobus monkeys in the wild have been shown to select browse low in fiber. However, at the Denver Zoo, they were shown to prefer a higher fiber browse, presumably based on taste (Kirschner et al. 1999). At Disney's Animal Kingdom, we planted camphor trees in several of our primate exhibits, because the animals typically do not like the taste of camphor, which we thought would allow the trees to grow and provide shade and environmental complexity. Unfortunately, at times, we have observed particular individuals eating components of the camphor trees. Similarly, primates have rarely been observed to eat pine in the wild, but in an enclosure setting, they have been observed to eat pine sap and bark (T.L. Bettinger, personal observation).

When items are added to an enclosure, animals may interact with them differently from what we expect. Therefore, any time novel items are added, it is important to observe the animals to determine their reaction. Adding straw and forage materials are effective for increasing the amount of time a primate spends gathering food (Baker 1997); however, the animals may ingest the substrate in addition to the food items. When adding new plantings to an area, or providing substrates, perching, climbing structures, and even bedding, we have observed primates consuming such items (examples include straw, clothing, cardboard, concrete, and gunite, etc.; T.L. Bettinger, personal observation). Many factors contribute to the ways in which different taxa and individuals respond to novel items. Whether primates consume what humans would consider a nonfood item for the taste or for tactile stimulation of the oral cavity is not a critical distinction, yet it demonstrates overlap between the senses and leads us to our consideration of touch in the next section.

Touch

Tactile perception is complex and gives an animal important information about the environment it is occupying. While humans think of touch in the context of their skin coming into contact with an object, for most animals, this extends to oral exploration as well. Oral and tactile inspection of food appears to influence food acceptance. Many primates use touch as a means to assess food prior to consumption (Dominy 2004), and food texture can influence palatability or be used to facilitate medical treatments.

In addition to feeding, touch is an important mechanism for assessing aspects of the environment. Primates may prefer or dislike specific textures, and individuals will show different preferences. At Disney's Animal Kingdom, we have a male mandrill that will avoid blankets in his environments to the extent that he will sleep on the floor rather than his typical sleeping perch when it is covered in blankets (E.A. Richards, personal observation). One of our silverback gorillas selects silky fabrics, twirling them around in his hands and rubbing them repeatedly on his face (R.B. Daneault, personal observation).

Touch is a sense that is relatively simple to manipulate within an environment. It transcends most of the senses and can be used to encourage animals to use areas, avoid areas, explore, and rest. In addition to varying texture, the environment can be manipulated for touch by varying (1) the temperature of cage/substrates/perches, (2) pressure, and (3) angles. In primates, touch also has an important social function. Grooming oneself or one another and anointing the body with odors by rubbing on items or simply maintaining contact with a peer can all be manipulated by providing the appropriate items/individuals within the habitat. Whether it is an area to groom with a cagemate, an area to bask in the sun on a cool day, or an area with damp, cool dirt on a hot day, all of these stimulate the primates' sense of touch and promote meaningful choices based on the use of their environment for a variety of functions.

CONCEPT IMPLEMENTATION

Where to Get Started

Understanding an animal's natural and individual histories is one of the first steps in creating an appropriate environment for the animal. Such an environment will provide behavioral opportunities that are both species appropriate and allow an animal to understand and have some degree of control over its world. At Disney's Animal Kingdom, whenever we bring in a new species, new individual animal, or construct a new animal habitat, we utilize a formalized planning tool that provides a framework to discuss aspects of natural history, individual history, and environmental parameters (see Appendix 4.1 for questions used in this planning process). These planning meetings are attended by all relevant staff, from animal caregivers to researchers to veterinarians, to make sure that all opinions and ideas are included in the discussion. These meetings ensure that we are utilizing our collective knowledge and cooperatively devising methods for establishing an environment that affords appropriate sensory stimulation.

Managing Behavioral Problems

When housing primates in captive environments, either singly or in social groups, animal management challenges inevitably arise. Oftentimes, mitigating these challenges requires the assessment of complex and nuanced social, behavioral, medical, and animal management components. While some problems are solved simply using past experience or common sense, others may require an organized and objective examination of the issue. Using a problem-solving model can be an effective way of generating possible solutions. The model we utilize at Disney's Animal Kingdom employs the following categories to organize information for discussion: Goal, facts, hypotheses, learning issues, and action plans (see Appendix 4.2 for model outline and a supplemental case study to exemplify its application in practice). This model has been used successfully for a variety of animal management issues, including reducing conspecific aggression, encouraging animals to shift, and facilitating introductions.

CONCLUSIONS

The behavior of an animal is defined by its natural history, life experiences, and the environment in which it is housed. The stark environments seen in Harlow's early experiments (Harlow and Harlow 1962) clearly demonstrated the adverse effects of inadequate sensory and social stimulation on the development and maintenance of normative nonhuman primate behaviors. Our task as managers and care providers, regardless of institutional function, must be to meet the requirements of the captive animal through the application of sound behavioral management principles. To accomplish this, we must continually work to provide for the needs of the animals, recognizing that problems will arise and when they do, we must quickly identify and address the sources of the problems. Additionally, inherent in what we do are issues related to the effects of captivity and limited space. We must continually work to mitigate the effects captivity and confined space can have upon behavior, at both the group and individual levels. The importance of the physical and social environment is not a new concept (Hediger 1964, 1968); however, it is often taken for granted. While many aspects of the animal's life under managed care are outside the control of the caregiver, providing an appropriate and behaviorally supportive environment is one area where we can positively influence the welfare of the animal with creativity and resourcefulness.

Diligent attention to the sensory environment of the animal, focusing on smell, sight, hearing, taste, and touch, provides a springboard to enhancing the physical environment for the animal, while providing novelty in a static enclosure. Sensory modulation creates a variety of environmental situations that can be adapted by the behavioral management team. The animal's responses can be quantified to determine benefit to the animal versus management concerns of staff time, cost, and secondary impacts on other environmental factors. Focusing on the senses of the individual primate provides staff a starting point to actively engage in a program that can systematically provide variety for individuals as their needs and situations change.

The environment in which an animal lives its life should be dynamic, changing over time and in response to the life stage of the individual. The richness of the captive environment should not be constrained by a lack of creativity or complacency. Our responsibility as captive managers and caregivers should be to continue to expand our fundamental knowledge of the nonhuman primates in our care. We accept responsibility for the well-being of each animal held in captive circumstances from the moment of its birth or from the moment it comes under our care and consideration until it dies or leaves our institution. Captive management is a complex problem set requiring multidisciplinary solutions that will enhance the quality of life of each animal across the full spectrum of its life stages, infancy through reproductive adulthood and finally into old age. Management is equally dynamic, changing as experience and new knowledge contribute to the problem-solving exercise of providing optimal care to each animal under our stewardship.

APPENDIX 4.1   ANIMAL PLANNING MODEL

The following questions are a tool by which to gather information on an animal's natural history, individual history, and current housing conditions. This information is used to facilitate discussion among staff as to the species-typical behaviors that we want to encourage, as well as methods of setting up the environment to promote appropriate sensory stimulation.

Natural History

1. What is this species' wild habitat (e.g., desert, tropical rainforest, cover, moisture, concealment/camouflage options, temperature ranges, barriers from conspecifics)? If specific information on a particular species is unknown, provide information on closely related species/genus/family.
2. How does the animal in the wild behave in response to changes in temperature and weather? What temperature/humidity range does it experience in the wild?
3. What are some self-maintenance/comfort behaviors (e.g., grooming, proximity to conspecifics)?
4. When is it most active (diurnal, nocturnal, crepuscular)? Why (e.g., predator avoidance, food availability)? Does the activity pattern change seasonally?

5. Does the species in the wild inhabit primarily arboreal or terrestrial environments or does it switch between them at times?
6. What are the main threats to the animal in the wild? What is it likely to be afraid of (e.g., conspecifics, humans)? What different types of predators does it have to look out for in the wild? Are there antipredator behaviors? Where and how does the animal seek refuge in the wild from fearful situations (e.g., loud noises like thunder)? What does fearful behavior look like?
7. What are its primary sensory modalities (e.g., sight, smell, sound) for communicating, detecting predators, and for finding food, mates, or other social partners?
8. What is the social structure of this species (e.g., solitary, monogamous pairs, single male/multiple females, multiple males/multiple females)? What is the average/typical group size?
9. What is the average distance between social group members and from neighboring conspecifics?
10. Describe the primary social behaviors of this species (e.g., aggression, breeding, affiliation, play). What are the social roles of adult males, adult females, juveniles? Does the species' behavior change markedly based on estrus, pregnancy?
11. Does the social structure change (e.g., hordes, fission–fusion, bachelor groups)?
12. Does this species defend territories? Does it maintain a home range? What is the size of the home range/territory?
13. How does this animal advertise its home range or territory (e.g., scent marking, vocal displays)? How does the animal attract a mate (e.g., physical or vocal displays, scent marks)? Who displays?
14. Are both sexes involved in rearing young? Are the young precocial or altricial? How are the young fed?
15. How does the animal locomote through its habitat?
16. What is the animal's diet type (e.g., omnivore, herbivore, insectivore) in the wild? Does diet change seasonally?
17. What does the animal feed on in the wild? What variety of foods does it need to eat? What behaviors does it use to locate and procure the different types of food it needs (e.g., leaf-litter foraging, hunting, termite fishing)? Does it use tools to obtain food?
18. Where does the animal sleep or rest? Does that change seasonally? Does the species build night nests?
19. Any other considerations?

Individual History

20. Does this animal have any medical problems (e.g., arthritis, obesity, missing digits)?
21. Does this animal have any behavioral problems (e.g., fearful/aggressive to humans, stereotypy)?
22. Any other considerations (e.g., solitarily housed, hand-raised)?

Enclosure Design

23. What is the size of the animal's enclosure (exhibit and holding area)? What are the containment barriers (e.g., 2 × 2 inch mesh, bars)?
24. Can the animal use all components of its enclosure? Can it hide? For example, how many places could this animal be out of view from its cagemates?
25. How functional is the current enclosure? Does the exhibit facilitate/allow the animal to exhibit natural behaviors? Does the enclosure utilize available vertical space? How does the animal interact with enclosure elements?
26. Where and how is the animal's food (normal diet, enrichment, browse) provided? Does the animal have the opportunity to forage for food? Does the animal have a preference for one feeding site over another? Do all animals have adequate access to food based on social dynamics?
27. Does the physical environment contain elements of novelty (e.g., weather changes, can furniture be changed easily)?
28. What are the animal's opportunities to feed/forage, breed, and socialize in species-appropriate ways? Do/can/should the animal interact with other species in the exhibit?
29. Can the animal exhibit normal patterns of behavior? Are components of the physical environment available for this to occur?

30. Can the animal make choices about where and how it spends its time? Does the animal have control over acquisition of food? access to hiding places? protection from the elements?
31. Are there any hazards to this enclosure?
32. Any other considerations?
33. Given these considerations (natural history, individual history, and current enclosure), what behaviors should we attempt to encourage? discourage?

APPENDIX 4.2   PROBLEM SOLVING MODEL

The following framework is utilized to facilitate solution-oriented discussion when behavioral problems arise. Problem-solving meetings should incorporate all related care staff to allow for the expression of diverse opinions and collective knowledge to be taken into account when devising your action plan. Use of a neutral facilitator is recommended to lead the meeting in order to guide staff through discussion of the following items. All comments and opinions expressed should be documented, and the determined course of action should be shared with all parties.

- Goal: What do you want to achieve with your plan?
- Facts: What do you know about the problem that would inform decision making?
- Hypotheses: What are the suspected causes of the problem? What are possible solutions and potential animal reactions to these solutions?
- Learning Issues: What information needs to be gathered in order to make appropriate decisions?
- Action Items: Concrete steps and plans are decided and tasks are assigned to individuals with due dates.

APPLICATION OF THE PROBLEM-SOLVING MODEL—GORILLA CASE STUDY

Background: On February 19, 2010, a nulliparous western lowland gorilla (*Gorilla gorilla gorilla*) at Disney's Animal Kingdom gave birth to a single female offspring. Once the infant reached 2 weeks of age, animal care staff began to notice deficiencies in her development. With the help of human pediatric specialists, it was determined that she was significantly developmentally delayed. Although no diagnosis could be determined, it was discovered that the team could treat the infant's clinical signs using occupational therapy. The problem-solving model was utilized to facilitate discussion, set goals, and determine courses of action to accomplish the objectives. The end result was that after 8 months of therapy and environmental manipulation, the infant exhibited no signs of developmental delay. This was accomplished without removing the infant from the mother, or humans and gorillas sharing space.

Goals

- Provide occupational therapy to the developmentally delayed infant gorilla without separating mother and infant and without humans entering the gorilla environment.
- Allow infant gorilla to grow up in her social group in order to provide her with the necessary social skills to live in a group setting.
- Provide developmentally delayed infant gorilla with as normal a life as possible with conspecifics in a social group.

Facts

- No medical diagnosis was determined to explain the developmental delays.
- The infant gorilla showed the following clinical signs: shallow/labored breathing, "see-saw" breathing, spending a lot of time in extension posture, decreased head and neck control, decreased suckling

and oral exploration, poor coordination/motor planning, unable to roll or sit without assistance, minimal play/exploration, decreased muscle mass compared to conspecifics, inappropriate response to being put on the ground.
- The mother was mother reared herself, had been exposed to infants in her previous social group, and was experienced with learning through operant conditioning.
- The infant's mother was compensating for the infant's deficiencies but also appeared to challenge the infant physically when appropriate.
- The gorilla group was accepting the infant, and the mother gorilla was taking excellent care of her.

Hypotheses

- If we provide the infant with occupational therapy, her developmental deficiencies will diminish.
- If we leave the infant with her natal group, she will be more likely to develop normal gorilla social behavior.
- She will have a better quality of life if reared within her social group than if hand-reared by caregivers.

Learning Issues

- Would the mother reliably bring the infant to therapy sessions?
- How invasive was therapy and would the mother tolerate our interactions with the infant?
- Would therapy used for humans work on an infant gorilla?
- Could we alter the environment enough to be effective in working on the infant's delays while maintaining safety for animals and caregivers?

Action Items

- Develop a training strategy for the gorillas and acclimate them to the new routine in order to accomplish two therapy sessions per day.
  - Morning sessions and midday sessions
  - Separate mother and infant from rest of social group during training sessions
  - Increase length of training sessions
  - Ensure sessions involve a lot of interaction between the infant and the therapist
- Schedule weekly appointments for the therapist to attend training sessions.
- Write training plans to train mother and infant to participate in specific exercises.
- Obtain items suggested by the therapist to elicit limbic system stimulation, visual tracking and stimulation, auditory stimulation, jaw and mouth muscle strengthening, gustatory and oral texture stimulation, hand–eye coordination activities, limited physical manipulation.
- Encourage physical activity by augmenting the environment.
  - Add small toys with a variety of colors, textures, and sounds in areas where the mother frequented so that the infant could see them and attempt to grab and mouth them.
  - Add climbing features that require her to stretch her arms and legs as well as fully open her toes and fingers for her to get on top of them.
  - Add items that would encourage her to pull her body weight up and then maintain her balance to remain on the items providing strength training.

REFERENCES

Baker, K.C. 1997. Straw and forage material ameliorate abnormal behaviors in adult chimpanzees. *Zoo Biology* 16:225–236.

Barber, J.C.E. and J. Mellen. 2008. Assessing animal welfare in zoos and aquariums: Is it possible? In *The Well-Being of Animals in Zoo and Aquarium Sponsored Research: Putting Best Practices Forward*, eds Bettinger, T. and J. Bielitzki, 39–52. Greenbelt, MD: Scientists Center for Animal Welfare.

Berntson, G.G., S.T. Boysen, H.R. Bauer, and M.W. Torello. 1989. Conspecific screams and laughter: Cardiac and behavioral reactions of infant chimpanzees. *Developmental Pyschobiology* 22:771–787.

Bettinger, T., J. Wallis, and T. Carter. 1994. Spatial selection in captive adult chimpanzees. *Zoo Biology* 13:167–176.

Brent, L. and D. Weaver. 1996. The physiological and behavioral effects of radio music on singly housed baboons. *Journal of Medical Primatology* 35:370–374.

Buchanan-Smith, H.M., D.A. Anderson, and C.W. Ryan. 1993. Responses of cotton-top tamarins (*Saguinus oedipus*) to faecal scents of predators and non-predators. *Animal Welfare* 2:17–32.

Caine, N.G. 2017. Anti-predator behavior: Its expression and consequences in captive primates, Chapter 9. In *Handbook of Primate Behavioral Management* 127–138, ed. Schapiro, S.J. Boca Raton, FL: CRC Press.

Chaudhari, N. and S.D. Roper. 2010. The cell biology of taste. *The Journal of Cell Biology* 190:285–296.

Coe, J.C. 1989. Naturalizing habitats for captive primates. *Zoo Biology* 1:117–125.

Dominy, N. J. 2004. Fruits, fingers, and fermentation: The sensory cues available to foraging primates. *Integrative and Comparative Biology* 44:295–303.

Dominy, N.J., C.F. Ross, and T.D. Smith. 2004. Evolution of the special senses in primates: Past, present, and future. *The Anatomical Record* 281A(1):1078–1082.

Fleagle, J.G. 1988. *Primate Adaptation & Evolution*. San Diego, CA: Academic Press.

Fragaszy, D.M., E. Visalberghi, and L.M. Fedigan. 2004. *The Complete Capuchin: The Biology of the Genus Cebus*. Cambridge, UK: Cambridge University Press.

Harlow, H.S. and M.K. Harlow. 1962. Social deprivation in monkeys. *Scientific American* 207:137–146.

Hebert, P.L. and K. Bard. 2000. Orangutan use of vertical space in an innovative habitat. *Zoo Biology* 19:239–251.

Hediger, H. 1964. *Wild Animals in Captivity: An Outline of the Biology of Zoological Gardens*. New York: Dover Publications, Inc.

Hediger, H. 1968. *The Psychology and Behavior of Animals in Zoos and Circuses*. New York: Dover Publications, Inc.

Howell, S., M. Schwandt, J. Fritz, E. Roeder, and C. Nelson. 2003. A stereo music system as environmental enrichment for captive chimpanzees. *Lab Animal* 32:31–36.

Kirschner, A.C., L. A. Putnam, A.A. Calvin, and N.A. Irlbeck. 1999. Browse species preference and palatability of *Colobus guereza kikuyuensis* at the Denver Zoological Gardens. In *Proceedings of the Third Conference on Zoo and Wildlife Nutrition*, AZA Nutrition Advisory Group, Columbus, OH.

Lambeth, S.W., M.A. Bloomsmith, and P.L. Alford. 1997. Effects of human activity on chimpanzee wounding. *Zoo Biology* 16:327–333.

Line, S.W., H. Markowitz, K.N. Morgan, and S. Strong. 1991. Effects of cage size and environmental enrichment on behavioral and physiological responses of rhesus macaques to the stress of daily events. In *Through the Looking Glass: Issues of Psychological Well-being in Captive Non-human Primates*, eds Novak, M.A. and A.J. Petto, 160–179. Washington DC: American Psychological Association.

Mittermeier, R.A., A.B. Rylands, and W.R. Konstant. 1999. Primates of the world: An introduction. In *Walker's Primates of the World*, ed. R.M. Nowak, 1–52. Baltimore, MD: The Johns Hopkins University Press.

Morgan, K.N. and C.T. Tromborg. 2007. Sources of stress in captivity. *Applied Animal Behavior Science* 102:262–302.

National Research Council. 2003. Diet formulation, effects of processing, factors affecting intake, and dietary husbandry, Chapter 10. In *Nutrient Requirements of Nonhuman Primates*, 2nd Revised Edition, 182–190. Washington, DC: National Academies Press.

Nevo, O. and E.W. Heymann. 2015. Led by the nose: Olfaction in primate feeding ecology. *Evolutionary Anthropology* 24:137–148.

Newberry, R.C. 1995. Environmental enrichment: Increasing the biological relevance of captive environments. *Applied Animal Behaviour Science* 44:229–243.

Nuttall, D.B. 2004. An animal-as-client (AAC) theory for zoo exhibit design. *Landscape Research* 29:75–96.

Ogden, J.J., D.G. Lindburg, and T.L. Maple. 1994. A preliminary study of the effects of ecologically relevant sounds on the behaviour of captive lowland gorillas. *Applied Animal Behaviour Science* 39:163–176.

O'Neill, P. 1989. *Housing, Care and Psychological Well Being of Captive and Laboratory Primates*, 135–160. Amsterdam: Elsevier.

Orban, D.A., Soltis, J., Perkins, L., and Mellen, J.D. Submitted. Sound at the zoo: Using animal monitoring, sound measurement, and noise reduction in zoo animal management.

Poole, T.B. 1998. Meeting a mammal's psychological needs: Basic principles. In *Second Nature: Environmental Enrichment for Captive Animals*, eds Shepherdson, D.J., J.D. Mellen, and M. Hutchins, 83–94. Washington, DC: Smithsonian Institution Press.

Rolls, E.T. and L.L. Baylis. 1994. Gustatory, olfactory, and visual convergence within the primate orbitofrontal cortex. *The Journal of Neuroscience* 14:5437–5452.

Shepherdson, D.J., K. Carlstead, J. Mellen, and J.D. Reynolds. 1989. Auditory enrichment for Lar gibbons. *International Zoo Yearbook* 28:256–260.

Wells, D.L. 2009. Sensory stimulation as environmental enrichment for captive animals: A review. *Applied Animal Behavior Science* 118:1–11.

Wells, D.L., P.G. Hepper, D. Coleman, and M.G. Challis. 2007. A note of the effect of olfactory stimulation on the behavior and welfare of zoo-housed gorillas. *Applied Animal Behavior Science* 106:155–160.

Wene, J.D., G.M. Barnwell, and D.S. Mitchell. 1982. Flavor preferences, food intake, and weight gain in baboons (*Papio* sp.). *Physiology & Behavior* 28:569–573.

Westlund, K., A.-L. Fernström, E.-M. Wergård, H. Fredlund, J. Hau, and M. Spångberg. 2012. Physiological and behavioural stress responses in cynomolgus macaques (*Macaca fasciculais*) to noise associated with construction work. *Laboratory Animals* 46:51–58.

PART II

Content Areas with Behavioral Management Implications

CHAPTER 5

Variation in Biobehavioral Organization

John P. Capitanio
University of California, Davis

CONTENTS

INTRODUCTION

The focus of this chapter is on individual variation in biobehavioral organization, a term that reflects the organization of the individual that encompasses behavioral systems, physiological systems, and their integration. It is my thesis that understanding this variation, measuring it, and using that information can improve both the science that is done with nonhuman primates and the ability to care for our animals. In this chapter, I will discuss first some theoretical issues on behavior, physiology, and their interrelations. Then, I will describe the assessment program that we developed at the California National Primate Research Center and some of the studies we have done relating specifically to behavioral management. But first, I think it is important to provide a bit of background on the entire idea of "individual differences."

Biology has always had a love–hate relationship with individual variation, and how one feels about individual variation affects not only the way one looks at one's data but also how one analyzes one's data. On the one hand, biologists who conduct experiments want to minimize, as much as possible, differences between individuals. Consider a simple experiment in which a researcher wants to test the effect of a drug on the behavior of mice. Two groups of mice might be used, one of which is exposed to the drug and the other of which is exposed only to a control substance. In order to demonstrate the efficacy of the drug, the researcher needs to show that the variation *between* the groups is greater than the variation *within* the groups. A common statistical procedure that is used in such experiments is the analysis of variance, in which a test statistic, F, is calculated by dividing a measure of variation (referred to as a "mean square" or MS) between the groups by a measure of variation found within groups. Figure 5.1A depicts this relationship and suggests that F will be large (which could lead to the conclusion that the drug had an effect on our outcome measure) if the numerator (MS_b) is large and/or if the denominator (MS_w) is small. The effect of the drug is what determines MS_b. But the F ratio can also be large if individual variation within the two groups (MS_w) is small. (Interestingly, MS_w is often referred to as the "error term," a phrase that suggests that individual variation is not the experimenter's friend.) Researchers employ a variety of techniques to minimize MS_w, such as the use of genetically inbred strains of mice (reducing genetic variation should reduce MS_w), ensuring all animals are housed/treated/handled in exactly the same way (reducing environmental contributions to variation can also minimize MS_w), etc. In short, because individual differences can make it harder to determine the effects of a treatment, an experimental biologist might endorse the statement "individual variation is bad."

On the other hand, there are fields in biology in which individual variation is desired—genetics is certainly about individual differences, and so is personality psychology (in which causes and consequences of variation in personality are studied) and other areas in psychology, such as psychometrics, which is focused on abilities testing (e.g., measuring one's intelligence quotient, or IQ). There are statistical approaches for this view of individual variation, too, usually involving looking at covariances between two variables X and Y (Figure 5.1B). The test statistic one might calculate is r, a correlation coefficient, which is just a standardized version of a covariance. The word "covariance" clearly expresses this perspective: How does X covary with Y? How does height vary with weight? How does the amount of energy present in mother's milk vary with personality? From this perspective, it is precisely the differences between individuals that one wants to measure and understand. Scientists who specifically examine variation might be said to endorse the statement that "individual variation is good."

In fact, variation between individuals (individual differences) is fundamental to the fields of biology and medicine. The importance to biology was made clear by Charles Darwin in *The Origin of Species*:

(A) Analysis of variance

$$F = MS_b/MS_w$$

$$MS_w = \sigma^2_e$$

(B) Correlation and regression

$$cov(X, Y)$$

Figure 5.1 Statistical representations of two views of individual variation.

The many slight differences… being observed in the individuals of the same species inhabiting the same confined locality, may be called individual differences. No one supposes that all the individuals of the same species are cast in the same actual mould. These individual differences are of the highest importance for us, for they are often inherited… and they thus afford materials for natural selection to act on and accumulate…

(Darwin, 1859/1975, Chap. 2)

Darwin was arguing that variation is really what drives natural selection, the principal mechanism of evolution. Darwin was clearly an "individual variation is good" guy.

More recently, the importance of individual variation has become clear in medicine. We all know that people can respond differently to the same medication; for example, some people need a stronger dose than others, some people experience side effects, and for some, the drug may not work at all. In 2006, Elias Zerhouni, then Director of the National Institutes of Health, described his "3P's" approach to medicine:

…NIH is strategically investing in research to further our understanding of the fundamental causes of diseases at their earliest molecular stages so that we can reliably *predict* how and when a disease will develop and in whom. Because we now know that individuals respond differently to environmental changes according to their genetic endowment and their own behavioral responses, we can envision the ability to precisely target treatment on a *personalized* basis. Ultimately, this individualized approach, completely different than how we treat patients today, will allow us to *preempt* disease before it occurs.

(Zerhouni, 2006)

Zerhouni's testimony before Congress gave rise to the field of personalized medicine. Zerhouni may not have had the perspective that "individual variation is good," but he recognized that it is omnipresent and that better treatments will only be possible by understanding and measuring this variation so that treatments can be tailored to the particular characteristics of the individual. From the perspective of behavior management of captive nonhuman primates, Zerhouni's thinking is very applicable; in fact, if you substitute the phrase "behavior problem" for the word "disease" in his statement, you can clearly see the direction that I believe we need to go.

THEORY

The essence of behavior management is, of course, behavior. But what exactly is behavior, and what is it for? A biomedical scientist might have little interest in behavior, but have more interest in physiological processes. How does behavior (which can be easily observed) relate to those processes (which are usually much harder to observe)? If behavior is the output of psychological processes, but unrelated to physiological processes, then why should we care whether our animals are behaviorally or psychologically healthy? I will attempt to address some of these issues in this section.

Why Behave?

Why do animals behave at all? Certainly, there are living organisms on our planet that show no (or at best, very rudimentary) behavior, and they seem to do just fine: These are plants. In fact, some plants are hundreds, and even thousands, of years old, suggesting that behavior is not necessary for life; indeed, about the only animal species that show longevity that is comparable to long-lived plant species are very simple animals like sponges and corals that show minimal behavior. Clearly, plants are capable of achieving the basic hallmarks of life—reproduction, metabolism, growth, responsiveness to the environment, adaptation—without having to behave (see also Koshland 2002). What does "behavior" get us?

Organisms, whether animals or plants, exist in environments, and to accomplish the basic functions of life (described in the previous paragraph), organisms must engage with the environment in some way. For plants, this engagement takes place in a single spot. For animals, behavior opens up the world: One is not restricted only to the resources in the immediate vicinity—one can move to a different area and access the resources there. Some have tried to define behavior formally (e.g., Levitis et al. 2009), but for our purposes, we can simply state that "behavior refers to actions that mediate the needs and wants of the animal with the opportunities present in the environment." This is a functional definition, emphasizing what behavior does for the animal; it implies that animals have "needs and wants" and that the environment affords opportunities for those needs and wants to be realized. Obviously, an animal does not have to be consciously aware of these needs and wants, although because much behavior is goal directed, it can often look as if the animal does have that awareness. And what are these "needs and wants?" To a large extent, they revolve around the basic functions of life described earlier—a need for food (metabolism), for finding a mate (reproduction), for dealing with adverse aspects of the environment, such as bad weather conditions or the presence of predators (adaptation), and so on. This list may look mostly like "needs," but there are also "wants"—an animal may want to play, for example (the question of whether play is a "need" is not clear!). Or, in a social species, an animal's "need" for affiliation may be combined with a "want" to be near animal X instead of animal Y.

How Is Behavior Organized?

Apart from some very simple behaviors, such as reflexes that are mediated at the spinal cord level, behavior originates in the brain. But how? The relationship between behavior and particular brain regions is not especially close. Rather, psychologists have long used the concept of an "intervening variable" (sometimes referred to as a "latent variable") to describe the relationship between brain and behavior. Intervening variables are not directly observable. A good example of an intervening variable is personality (in the animal literature, the words "personality" and "temperament" are often used interchangeably, and I will adhere to that convention). Gordon Allport, a prominent personality psychologist, succinctly expressed the concept of an intervening variable with reference to personality:

> Personality *is* something and *does* something. It is not synonymous with behavior or activity… It is what lies *behind* specific acts and *within* the individual.
> **(Allport, 1937, p. 48, emphasis in the original)**

If we take the view that behavior (i.e., the acts of an individual that are usually observable) is a manifestation of a higher-order disposition/intervening variable, such as personality, then this suggests that the brain is not so much about specific acts as it is about dispositions; in fact, measures of brain function, such as neurochemistry, are typically mapped successfully onto dispositions, not specific behaviors (Pickering and Gray 1999). To be sure, the brain orchestrates the motor movements involved in behavioral acts, but the acts that one displays in response to something in the environment are likely to be a reflection of the dispositions one has. For example, a rhesus monkey with a timid disposition may respond to the presence of a technician who is doing a morning health check by withdrawing to the back of the cage and perhaps displaying a grimace; another animal with a confident disposition may act generally uninterested and not even change her position; while a third animal with an impulsive or aggressive disposition may charge the technician and shake the cage. Of course, dispositions are not usually thought of as binary: It is not the case that either you are confident or you are not. Rather, dispositions are thought of as continuous traits: You may be high, intermediate, or low in confidence (and impulsivity, and fearfulness, etc.). The particular mix of personality characteristics that individuals have is what can give rise to the different behaviors that one observes in response to the same stimulus.

Because all members of the same species generally have the same behavioral repertoire, and because all animals also have the same challenges to solve (find food, find a mate, etc.), differences between animals in the ways in which they solve those challenges presumably reflect differences in their dispositional organization (e.g., an animal high in confidence and low in fearfulness might show different behaviors in obtaining food than an animal that is low in confidence and high in fearfulness). So what leads to differences in dispositional organization? What contributes to some animals being high rather than low in aggressiveness, or fearfulness, or impulsivity, or confidence? We know that genetic factors play a role (e.g., anxious temperament is affected by CRHR1 polymorphism: Rogers et al. 2013), that the early environment plays a role (e.g., isolation rearing leads to a more aggressive and fearful temperament: Capitanio 1986), and that even the environments experienced by parents play a role (e.g., rhesus monkeys reared in a nursery show greater emotionality if they had a sire that was himself nursery reared, compared to nursery-reared animals whose sire was reared in a large social group: Kinnally and Capitanio 2015). These and other influences act via their effects on the brain.

What Is the Relationship between Behavior and Physiology?

There is more to an animal than just its behavior, of course. An animal comprises a variety of physiological systems that can influence the expression of behavior. Reproductive hormones, for example, are not continuously secreted at the same rate across days, months, and the year; sexual motivation typically tracks fluctuations in these hormones (Dixson 2012). Physiological systems are regulated; the classic example of this is the hypothalamic–pituitary–adrenal (HPA) system (or axis). The hypothalamus receives input from limbic structures and secretes corticotropin-releasing hormone (CRH), which travels to the anterior pituitary, stimulating the release of adrenocorticotrophic hormone (ACTH), which travels via the circulation to the adrenal cortex, stimulating the release of glucocorticoids (cortisol being the principal one in primates) into the circulation. As cortisol levels rise, glucocorticoid receptors on the hypothalamus and pituitary detect the increase and dampen the release of CRH and ACTH. This process is known as negative feedback and works much like the thermostat in your home. Cortisol is often considered a "stress hormone," but its relation to stress really is incidental; it is first and foremost a metabolic hormone, stimulating the production of glucose. Because stressful situations have, at least in our evolutionary past, typically been associated with the need to analyze the situation and take action, it made sense for glucose, which is fuel for the body's cells (and especially the brain), to increase in concentration in the blood in association with stress.

While a stressful situation may increase both the frequency of anxious behaviors and the levels of cortisol in the blood, the two are not tightly linked. This was evident in a study in which adult male rhesus monkeys were placed in a primate chair for 2 h/day for 7 consecutive days. It was clear that, over the first few days of experiencing chair restraint, vocalizations declined in frequency, as did behavioral indicators of agitation. Indeed, by the third or fourth day of restraint, the animals appeared calm, moved voluntarily from their cages to the restraint devices, sat still while the collars were attached, etc. Examination of cortisol concentrations, however, revealed that, although levels did decline, they did not return to baseline by the end of the 7-day period (Ruys et al. 2004). These data demonstrate a disconnect between behavioral adaptation and physiological adaptation. In other cases, however, behavior does track physiology more closely: Administration of pro-inflammatory cytokines, either via injection or through exposure to an inflammatory stimulus (e.g., a virus), leads to sickness behavior, a pattern of behavior associated with malaise, fatigue, reduced appetite, etc. (Dantzer, 2009). Numerous other examples exist: Infant rhesus monkeys with a nervous temperament show evidence of glucocorticoid desensitization—their glucocorticoid receptors do not appear to respond as strongly to elevated levels of cortisol as do the receptors of monkeys that are not nervous (Capitanio et al. 2011). Also, animals that are low in sociability, a personality dimension reflecting a tendency to affiliate, show an increased density of sympathetic nervous system nerve fibers in their lymph nodes (Sloan et al. 2008), a situation that can affect viral replication (Sloan et al. 2006).

We developed the concept of "biobehavioral organization" to highlight the interrelations of behavioral processes (especially those related to temperament) and physiological processes; an individual of any species is an organized entity (that is, a system) and comprises a variety of subsystems that more or less work together. Just as temperament is a higher-level construct that influences which behavioral endpoints are displayed in any given situation, so too are these physiological systems organized and regulated in ways that affect the concentrations of physiological endpoints, like cortisol or norepinephrine concentrations, in response to particular situations. Phenomena, such as glucocorticoid desensitization and innervation of lymphoid tissue by fibers of the sympathetic nervous system, represent different types of organization of these physiological systems, and evidence indicates that different patterns of physiological organization can be associated with temperament. That is, nervous temperament is not simply associated with changes in the concentrations of cortisol in the blood; it is associated with fundamentally altered regulation of the HPA-immune axis (Capitanio et al. 2011). Given that glucocorticoids generally have an anti-inflammatory effect, one could well imagine that a study of how a viral infection affects the expression of inflammatory genes could be influenced by whether one has some (or any) animals with a nervous temperament in the study.

Summary

Behavior is the principal means by which animals get their needs and wants met. The expression of specific behaviors in any given situation is a function of many things, but an important one is the animal's temperament, a construct that describes the set of dispositions that the animal possesses and that affects the expression of specific behaviors [but which is fundamentally different from behavior, as Allport (1937) indicates]. It is now becoming clear that temperament can be associated with different patterns of regulation of physiological systems that can be important influences on the animals' health and usefulness for scientific research, and different patterns of behavior that affect an animal's adaptation to the situations it encounters (including captivity). While the link between behavior and physiology is not always simple or direct, behavior can sometimes be an indicator that one or more of a particular animal's physiological systems might be organized in an unusual way, which could lead to poor health and/or poor research outcomes.

METHODS

In 2000, we devised a BioBehavioral Assessment (BBA) program at the California National Primate Research Center (CNPRC) that involved implementation of a highly standardized protocol that had as its aim the quantification of various measures of biobehavioral organization: activity, emotionality, HPA axis regulation, temperament, behavioral responses to social and nonsocial stimuli, etc. We also obtained a complete blood count for each animal to give us leukocyte subset numbers and other measures of hematologic function, and genotyped the subjects for two genes of neuropsychiatric interest, namely 5-HTTLPR (serotonin transporter promoter polymorphism) and MAOA-LPR (monoamine oxidase A promoter polymorphism). The overarching goal of the BBA program was to provide this quantified information (1) to staff at our facility for use in making behavioral management decisions and (2) to scientists for use in subject selection and assignment to groups, and for data mining. Each year, data that are collected and summarized are put onto the

CNPRC's internal server for employees to access. Our first year of data collection occurred in 2001, and to date, more than 4000 animals have been assessed. Since the start of the program, one focus has been on understanding the causes of the great variation that we see in our measures; a second focus, which we emphasize in the next section of this chapter, has been to use BBA data to identify characteristics of animals that are at risk for a variety of colony management-related outcomes (described in the final section of the chapter).

The subjects that participate in the BBA program are infant monkeys, 90–120 days of age, from each of the four "colonies" at CNPRC:

1. Half-acre outdoor field cages (FCR, field cage reared) that each house up to 200 animals of all age/sex classes
2. Outdoor corncrib (CCR, corncrib reared) structures that each house up to 30 animals
3. Our indoor colony, in which infants are housed in standard-sized cages with their mothers and, at most, one additional adult and infant pair (IMR, indoor mother reared)
4. Our indoor nursery, in which animals are relocated on the day of birth and individually housed in incubators until 3 weeks of age, at which point, they are given visual access to an infant of the same age with whom they are subsequently paired at 5 weeks of age (NR, nursery reared). Eventually, NR animals are housed in a corncrib with other monkeys that were similarly reared, and may eventually form a new group with other animals in a field cage.

Individuals possess their own characteristics that they bring to each "situation" they encounter. In a research facility, commonly encountered situations are numerous and varied, involving the following: relocations (moves from one cage/room to another); separations from cagemates; changes in the amount of available space (e.g., being relocated from a field cage to an indoor cage); encounters with unfamiliar animals (e.g., pairing for socialization purposes, group formations, being housed across a room from unfamiliar animals); husbandry routines (daily cage cleaning, feeding, cage changes); daily health checks by technicians; enrichment activities; treatments that may be received either for medical/health reasons or for experimental reasons (blood samples, injections, etc.); training for a variety of research and husbandry procedures (e.g., extending an arm for cooperative blood sampling); frequent disturbances in the rooms owing to treatments that the "other" animals are receiving; and so on. We were interested in identifying the individual characteristics that facilitated or hindered an animal's adaptation to some or all of these situations. Consequently, we made the decision to test our animals in the BBA program while they were individually housed, involving relocation to an unfamiliar room and a separation from mother (and other companions) for FCR, CCR, and IMR animals, and a separation from the pairmate for NR animals. These are challenging circumstances, and some of our assessments (described in the next section) provided an additional level of challenge, albeit more carefully controlled and of short duration. It is important to note that, broadly speaking, our procedures are not dissimilar from those that are sometimes used to assess temperament in human infants and children (Rothbart 1988; Gagne et al. 2011).

Specific Assessments in the BBA Program

Animals are tested in cohorts of five to eight animals. Upon arrival in our testing area, each animal is placed in an individual cage that contains a towel, a stuffed toy, and a novel object (see later section). Food and water are available ad libitum. For each cohort, a randomly generated testing sequence determines the order of testing for each animal across all the individual assessments; each animal in a cohort is tested in one assessment before any animal experiences the next assessment. Behavioral data collection utilizes ethograms that focus on indicators of activity, emotionality, anxiety, and environmental exploration.

Here we briefly describe each assessment, and the principal measures that have been found to be useful. Because of our large sample size ($N > 4000$), and the large number of individual behaviors that we record (e.g., up to 30 behaviors, depending on the specific assessment), we have employed factor analysis (Costello and Osborne 2005) to identify the latent traits that underlie the behaviors seen. Our strategy was to conduct an exploratory factor analysis for a given assessment using several hundred subjects to identify the most useful factor structure, and then conduct a confirmatory factor analysis on a separate set of individuals to determine if the structure is robust. Once the factors have been identified, each animal's score on these factors is computed as a z-score against all of the other animals tested in the same year. z-Scoring enables one to immediately see how a given animal "measures up" against the others tested that year. Details of these analyses are in the citations below.

A final note about the data—many of our assessments are relatively brief; often, however, they do not last for exactly the same amount of time for each animal. For example, each human intruder trial (described later) should last for 60 s; however, for some animals, it lasts for 57.5 s, and for other animals, it may last 63.4 s. Consequently, we transform our duration measures (e.g., locomotion) into the proportion of total time observed, and our frequency measures (e.g., scratch) into a rate per 60 s.

Focal Animal Observations

Approximately 15 min after placement in their holding cages, 5-min focal animal samples of behavior are recorded. A second round of focal observations is conducted ~22 h after the first; the two rounds of observation are referred to as day 1 observations and day 2 observations, respectively. Day 1 observations capture the monkeys' (relatively) immediate response to the separation and relocation, whereas day 2 observations capture the abilities of the animals to adapt to the situation.

Factor analyses of these data revealed two latent traits that underlie the behaviors seen during these observations (Golub et al. 2009). The first trait, *Activity*, reflects the proportion of observation time the animals spent locomoting; the proportion of time they were NOT hanging from the top or side of the cage; the rate (per 60 s) of environmental exploration; and dichotomous variables indicating whether the animals ate food, drank water, or were seen crouching in the cage. Scores for each animal were calculated for Day 1 Activity and Day 2 Activity. The behaviors that loaded onto the second trait, *Emotionality*, were the rate of cooing; rate of barking; and dichotomous codes of whether the animals scratched, displayed threats, or lipsmacked. Again, scores for each animal were calculated for Day 1 Emotionality and Day 2 Emotionality. Note that some of the behaviors used in the factor analyses were dichotomized. These were behaviors that occurred rarely; while an occasional animal might display these behaviors at high frequency, most often, they were recorded only once or twice for a small subset (e.g., fewer than 5%) of the animals. We decided that it was more important to know whether or not the animals ever displayed the behavior, rather than how frequently; consequently, we dichotomized these variables.

Visual Recognition Memory

The next test involves relocating the animals to a test cage in an adjacent room, and presenting them with previously recorded stimuli (still pictures of unfamiliar monkeys) that should address visual recognition memory. Each animal experiences seven problems. For each problem, the subject is presented with identical pictures side-by-side on a monitor for 20 s, after which the screen goes blank. Next are two 8-s test trials, in which the now-familiar picture and a novel picture are presented; the two trials differ only in the placement of the stimuli. The typical response of young monkeys is to spend more time looking at the novel picture in each of the two test trials; in fact, this test is sensitive to a variety of adverse experiences, including damage to limbic structures (Bachevalier et al. 1993), prenatal exposure to methyl mercury (Gunderson et al. 1988), and high-risk pregnancies and births (Gunderson et al. 1987).

Video Playback

In this assessment (also conducted in the test cage in an adjacent room), each animal is presented with a 10-min videotape of an unfamiliar adult male rhesus monkey alternately displaying nonsocial behavior (relaxed looking, environmental exploration), and viewer-directed aggressive behavior (threats, lunges). Data are recorded separately for the nonsocial and aggressive segments. Video playbacks have a long history in captive primate research (Plimpton et al. 1981; Capitanio et al. 1985), and present a standardized social stimulus to which the animals readily respond.

Human Intruder Test

Our abbreviated version of the human intruder test (see also Kalin and Shelton 1989, who describe a lengthier version) comprises four 1-min trials, in which a technician presents her (1) *profile* face from a far position (~1 m) and (2) a near position (~0.5 m), followed by (3) a full *frontal* face from the far and (4) near positions (Gottlieb and Capitanio 2013). This test has been used extensively to identify anxious behavior.

Preliminary analyses on this data set showed that animals displayed some consistency in their pattern of responses across the four trials; consequently, we took a mean for each behavior across the four conditions and used these data in our analyses. The factor analyses (Gottlieb and Capitanio 2013) revealed a four-factor structure: Activity (proportion of time spent active; rate of environment exploration; whether cage shake was recorded or not), Emotionality (rate of fear grimace; rate of coo vocalization; and dichotomized codes of whether convulsive jerk or self-clasp was recorded), Aggression (rate of threat; rate of bark; whether other vocalizations were recorded), and Displacement (rate of tooth grind; whether yawn was recorded).

Blood Sampling

Blood is collected on four occasions during the BBA program, and plasma cortisol concentrations are measured from each. The first sample is obtained after day 1 focal animal observations, ~2 h after the animals have been in our testing room. This sample is also used to assess numbers of red and white blood cells [and subsets] via a complete blood count and flow cytometry, concentrations of C-reactive protein, and genotype for 5-HTTLPR and MAOA-LPR. The second sample is obtained 5 h after the first sample, and is immediately followed by an injection of dexamethasone. The third sample is collected after day 2 focal animal observations and is immediately followed by an injection of ACTH; 30 min later, the fourth sample is collected. Dexamethasone is a synthetic glucocorticoid that should work, via negative feedback, to dampen the release of endogenous corticosteroid from the adrenal cortex. ACTH, which is normally the proximate stimulus causing release of corticosteroids, permits assessment of the rebound of the adrenal cortex from the earlier dexamethasone blockade. Both tests are frequently used clinical tests for humans to assess the proper functioning of the HPA axis.

Novel Objects

Each holding cage contains a novel object constructed of PVC that measures ~9 cm in length and 3.8 cm in diameter. Inside each novel object is an actimeter that records any force exerted on the object. This object remains in the cage until just after the second blood sample of day 1 (which is the last assessment on that day), at which point, the original object is swapped for a second, similarly sized object (also containing an actimeter), which remains in the cage until the end of the testing session on day 2. The actimeter data are downloaded into a reader and summarized to indicate the number of 15-s intervals, throughout the 25-h period, in which force was exerted on the object.

Temperament

At the end of the 25-h period, a technician rates each animal on a list of 16 trait adjectives. This is designed to provide an overall "thumbnail" portrait of the animal based on all of the experiences that the technician had with the animal: observing, testing, handling, feeding, changing towels and novel objects, etc.

Factor analyses (Golub et al. 2009) revealed a four-factor structure. Each scale was named for the adjective that had the highest positive loading. (Adjectives preceded by "NOT" indicate that they were reverse scored for the analysis.) The factors are Vigilant (vigilant, NOT depressed, NOT tense, NOT timid), Gentle (gentle, calm, flexible, curious), Confident (confident, bold, active, curious, playful), and Nervous (nervous, fearful, timid, NOT calm, NOT confident).

Are There Lasting Consequences of Participation in the BBA Program?

Clearly, the separation and relocation for a 25-h period, in addition to some of our assessments, like the human intruder test, are challenging to the animals. While challenging a system is perhaps the best way to understand how a system works, we were naturally concerned that our procedures might have a lasting impact on the animals. After all, previous work had shown that monkeys that underwent separations from their mother exhibited a variety of subtle deficits later in life, in domains ranging from personality to endocrine to immunological to social network structure (Mineka and Suomi 1978; Caine et al. 1983; Capitanio and Reite 1984; Laudenslager et al. 1985; Capitanio, et al. 1986). We did not expect that participation in the BBA program would have any adverse lasting consequences, however, for three reasons. First, our separation was of relatively short duration, 25 h. In contrast, the duration of separations in most of the studies referenced earlier typically ranged from 10 to 14 days, a much more extreme experience. Second, the separation that our animals experienced was accompanied by relocation to a novel environment. Studies (e.g., Hinde and Davies 1972) have demonstrated that the despair/depressive response that can result from separation can be delayed if the separation is accompanied by relocation to a novel environment. Finally, the separation studies referred to earlier were largely carried out when animals were about 6 months of age. This is typically when macaques are weaned, and their mothers resume cycling. There is a large theoretical and empirical literature showing that this time can be an especially difficult time for a young animal (e.g., Berman et al. 1994). We specifically selected an earlier age to avoid layering stress from our manipulation onto a backdrop of naturally occurring distress in animals of this age.

We devised a series of follow-up tests to determine whether there were lasting consequences of participation in the BBA program. These tests were modeled on the studies, referenced earlier, that demonstrated persisting consequences of the lengthier separations. While we cannot determine statistically whether the null hypothesis of "no difference" between BBA participants and nonparticipants is true, we could select a sample size that would enable us to find smaller effects than the earlier studies demonstrated. Results of this analysis are currently being prepared for publication; however, the results were very clear in showing no significant differences between those animals that participated in the BBA program and matched controls that were not participants (Capitanio et al. in preparation). As such, our results are consistent with findings from a thorough, recent review of the primate mother–infant separation literature on factors promoting resilience:

> Even though wild primates may not experience mother–infant separations for the duration of those used in some laboratory paradigms, a short mother–infant separation is a naturally occurring event that is likely to occur in the lifetime of every primate and for which individuals are likely to have pre-programmed adaptive coping responses.
>
> **(Parker and Maestripieri 2011, p. 1475)**

Examples of Variation in Some BBA Measures

Before describing how some of our measures relate to behavioral management outcomes, it may be useful to simply see the variation that we have uncovered in the BBA program, after having tested more than 4000 animals in this highly standardized program. Recall that all of the factors that we identified through factor analysis are z-scored; while this allows staff to immediately see how extreme an animal may be relative to other animals tested in the same year, from the perspective of looking at variation, the raw data are likely to be more informative. Because our measures are continuous, we have binned them for ease of viewing.

Coo vocalizations are one of the measures comprising days 1 and 2 Emotionality factors and are sometimes referred to as contact (or clear) calls, in that they are given when animals are separated from one another. As Figure 5.2A shows, the modal response of our animals on day 1 is 0 coos/min; in fact, 36.5% of the sample gave no coos during the entire 5-min observation period. The median response, however, was 2.78 coos/min, and the range was from 0 to 20.51 coos/min. About 1% of the animals cooed at the rate of 15 coos/min or more—that is 1 coo every 4 s on average.

Animals are given a rating for the trait, nervous (as well as 15 other traits), at the end of the BioBehavioral Assessment, and this single item loads most strongly on the Nervous temperament scale. Figure 5.2B shows that the modal response on this single item is 1, which refers to "total

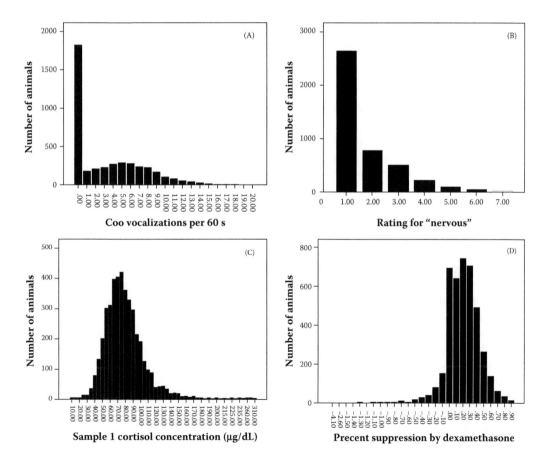

Figure 5.2 (A) Frequency of coo vocalizations during the focal observations. (B) Ratings for individual nervous item from the Nervous temperament scale. (C) Cortisol concentration at the first time point during the BBA. (D) Percent suppression in cortisol values by dexamethasone during the BBA.

absence" of this trait (defined as jittery, anxious, seems to be anxious about everything). About 62% of the sample was given a score of 1. The median score was 1.70; about 8% of the animals got a score of 4 or greater.

Cortisol Sample 1 is taken approximately 2 h following the separation and relocation to our testing room. Figure 5.2C shows a broad range of values, ranging from 12.7 to 312 µg/dL. The median value is 77.9 µg/dL. More interesting, however, is the animals' *Response to Dexamethasone*. Recall that immediately after the second blood sample, at approximately 4 PM, animals are injected with a standardized (based on weight) dose of dexamethasone, which should reduce endogenous cortisol output. A third blood sample is drawn the next morning. We subtracted the cortisol value for the third sample from the cortisol value for the second sample and divided this by the value for the second sample to get an index of the percentage suppression. As Figure 2D shows, about 75% of the animals showed suppression of at least 10% of their Sample 2 value (i.e., a value of 0.10 or greater). Approximately 16% of the sample, however, showed negligible suppression (values ranging from −0.10 to +0.10). Another 9% of the sample not only showed no suppression but also actually showed increased concentrations of cortisol (values below −0.10 in the figure), suggesting a dysregulated HPA axis.

These data demonstrate substantial variation in biobehavioral measures in infant rhesus monkeys. It is beyond the scope of this chapter to discuss factors that influence this variation, but some of our previous work has implicated rearing history (Capitanio et al. 2005), constituents of mother's milk (Hinde et al. 2015), genotype (Kinnally et al. 2010; Sorenson et al. 2013), prenatal exposure to ketamine (Capitanio et al. 2012), degree of Chinese ancestry (Jiang et al. 2013), timing of birth (Vandeleest et al. 2013), and even rearing history of the animals' sires (Kinnally and Capitanio 2015).

RELEVANCE TO BEHAVIORAL MANAGEMENT OF NONHUMAN PRIMATES

One of the original goals for the BBA program was to provide quantitative information on biobehavioral organization to help manage the colony better. As a first step in achieving this goal, we have conducted a series of analyses looking at the ways in which BBA measures relate to colony management outcomes. In this section, I will focus on our published work in this area; the complete references to the relevant papers appear at the end of the chapter.

Motor Stereotypy

Motor stereotypic behaviors (MSB), which involve repetitive motor movements such as pacing, twirling, swinging, rocking, etc., are relatively commonly performed behaviors by indoor- and/or singly-housed animals. While it appears that the presence of stereotypic behavior among animals housed in a particular environment seems to indicate that that environment is suboptimal, studies have strongly suggested that, *within* that environment, individuals displaying more MSB appear to have better well-being than those that do not display MSB (or that display low levels of MSB). This suggests that MSB may be serving as an active coping mechanism. Can infant measures of biobehavioral organization identify individuals that are more likely, at later ages, to develop MSB? We used a sample of 1145 animals with a mean age of 5.52 years (range of 1–10 years), comprising more than 13,000 observations, to answer this question (Gottlieb et al. 2013). Our analysis was comprehensive, and a variety of factors were found to be related to MSB, such as sex, age, rearing, frequency of room moves, and current housing situation. Beyond these, however, three BBA measures, associated with a predisposition toward an active temperament, were also associated with MSB: low scores on our Gentle temperament factor, higher scores for Activity on the human intruder test, and greater contact with the novel objects in the animals' cages. Our data confirmed suggestions by others that MSB seems to be associated with animals' pre-existing general activity levels, and suggest that particular attention should be paid to animals relocated into indoor/single housing that are not naturally very active prior to relocation—these animals, who often do not show MSB, may be experiencing reduced well-being and may require additional enrichment to forestall poor outcomes.

Depressive Behavior

In captive colonies of macaques, one sometimes sees monkeys sitting in a hunched posture: head lower than shoulders, arms drawn in toward the center of the body, but with eyes open. This hunched posture is widely considered to be an indicator of depressed mood, and considerable work has been done by Carol Shively's research group (Shively 2017) in examining the biological underpinnings of this behavior. In collaboration with Dr. Michael Hennessy (Hennessy et al. 2014), we were interested in developing a naturally occurring model of depressive behavior, and so we observed 26 adult male rhesus monkeys (5.6–7.8 years of age) that had been relocated from our half-acre outdoor corrals to individual housing indoors (Hennessy et al. 2014). All 26 macaques had participated in the BBA program at 3–4 months of age.

Behavioral observations were conducted using a camera, recording live from an adjacent room, as we have found that the presence of a human, no matter how unobtrusive, in indoor housing rooms subtly organizes the behavior of the animals. This often results in them "putting on their game face," even if they are ill or depressed. When observed via camera, we found 18 of the 26 animals displayed the hunched posture, indicative of depressed mood, during the first week following their relocations. Other animals were observed lying down on the cage bottom with eyes open, or daytime sleeping (eyes closed, head above shoulders). Together, we considered these three behaviors "depressive-like," and 21 of the 26 animals showed one or more of these behaviors. Importantly, the duration of depressive-like behavior during the first week was significantly associated with Day 1 Emotionality from the focal animal observations in the BBA program—animals that responded to the initial separation and relocation during BBA with more vocalization, scratching, lipsmacking, and threatening, were more likely to show depressive-like behavior several years later, during their first week following relocation from outdoor social caging to indoor individual caging (Figure 5.3). When we examined depressive-like behavior in the animals' second week of indoor housing (fewer animals displayed such behavior in Week 2), we found no relationship with BBA measures; this may not be surprising, as the BBA program is of short duration, and day 1 factors likely reflect an

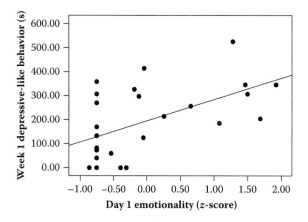

Figure 5.3 Correlation between time spent in depressive-like behavior (week 1) by z-score for Day 1 Emotionality.

immediate response to separation and relocation, which may parallel more closely the experience of the adult animals during their first, but not their second, week after relocation/separation. We believe these results (Hennessy et al. 2014) suggest that animals that have a tendency to respond to psychosocial challenges with emotional behavior when young may be most at risk for similar types of responses (though the specific behaviors may be different) as they age. From a behavioral management point of view, such animals might benefit from additional enrichment particularly, if they have to be individually housed.

Diarrhea

Diarrhea is one of the most ubiquitous sources of morbidity and mortality in captive nonhuman primate colonies, and often, no pathogen can be cultured (Ardeshir et al. 2013; Prongay et al. 2013). Evidence suggests that indoor-rearing (and especially nursery-rearing: Elmore et al. 1992) is a particular risk factor for diarrhea incidence. Still, considerable variation exists, even among indoor-reared monkeys. We examined a cohort of 353 monkeys that had been coded as either nursery-reared (NR) or indoor mother-reared (IMR) at the time of BBA testing, following them into their third year of life, or until they were shipped to another facility or assigned to a research protocol (Elfenbein et al. 2016). All occurrences of diarrhea were recorded for these animals. Our particular interest was in the prenatal environment of these animals; consequently, we coded a number of variables reflecting various experiences that *their dams* might have encountered while pregnant that could be indicators of "gestational stress," such as the number of relocations, number of partner changes, etc. We were especially interested in gestation location—whether at least 25% of gestation occurred outdoors (classified as "outdoor-gestated") versus pregnancies that occurred primarily indoors ["indoor-gestated," defined as 75% or more of gestation occurring indoors). Finally, we examined the role played by variation in measures of biobehavioral organization from the infants' BBA assessments.

Our sample was approximately evenly split between animals that were gestated indoors versus outdoors; postnatally, ~2/3 of our subjects were NR, with the remaining third IMR. Overall, animals experienced a mean of 1.35 diarrhea events, although almost half of the subjects (46%) never experienced a clinically relevant diarrhea event. Gestational environment was a strong predictor of diarrhea risk; animals gestated outdoors had a significantly lower risk. In fact, this variable was a stronger predictor of diarrhea than was postnatal environment (NR vs. IMR). The presence of a significant interaction between sex and gestation location indicated that outdoor gestation was especially protective for female fetuses. Several other variables were associated with diarrhea risk, including number of housing locations experienced prenatally and genotype. Importantly, Nervous temperament was also a significant risk factor; animals with higher scores were at greater risk. Again, an interaction was found that indicated that NR animals that were high in Nervous temperament were especially at risk. These data suggest that risk for diarrhea can begin prenatally, and that animals described as fearful and timid (i.e., Nervous temperament) are especially susceptible. [Recall that earlier, we indicated that animals that were high in Nervous temperament showed evidence of glucocorticoid desensitization (Capitanio et al. 2011); that is, once an inflammatory process begins, cells may be relatively unresponsive to the anti-inflammatory effects of glucocorticoids. Because much of chronic diarrhea involves inflammation of the colon, this result might provide a mechanism for the association of Nervous temperament and diarrhea.]

Social Pairing

In recent years, it has become clear that the various agencies that oversee animal research facilities have come to a consensus that nonhuman primates should be socially housed to facilitate their psychological well-being. While this is accomplished for the majority of animals by having them live in social groups that contain a species-typical mix of age/sex classes, it is often necessary to house animals, particularly those that are on biomedical research projects, in conditions that provide more experimental control and easier access to the animals. In the past, these research needs often resulted in indoor single housing of individuals in standard-sized (based on regulations) cages. The shift in emphasis to social housing as the default housing condition has resulted in laboratory facilities devoting substantial effort and resources to the task of pairing compatible animals. It is my belief that having quantitative data on enduring biobehavioral characteristics can be of great value to behavioral managers in improving the chances that two animals will be paired successfully. The value of pre-pairing data has been the subject of several studies (see reviews by DiVincenti and Wyatt 2011, and Truelove et al. 2017), many of which, unfortunately, have only been presented in the scientific literature as abstracts from professional meetings. Moreover, the results of these studies are often contradictory.

We were interested in determining whether pairing success could be predicted from measures obtained from the BBA program when the animals were infants (Capitanio et al. 2017). We identified 340 isosexual pairs (169 female pairs) in which both members of the pair had participated in the BBA program as infants. Animals ranged from 1.2 to 11.1 years of age; 121 pairings were unsuccessful, and 219 were successful. Because we suspected that different factors might be important for males and females, we conducted separate analyses by sex. We also did not incorporate all of the measures (such as serotonin transporter promoter genotype, or cortisol concentrations) that might have had an effect on pairing success, inasmuch as we were interested in utilizing measures that individuals at other facilities might be able to easily and inexpensively obtain. Finally, our focus was on measures that reflected the combined characteristics of the two individuals in each pairing attempt. Consequently, for each BBA measure (we utilized data from the focal animal observations, the human intruder test, and the temperament ratings), we calculated two measures: a mean and a difference score. For example, if one animal had a score of −1.34 on Nervous temperament and the second had a score of 0.78, the mean score for this pair would be −0.28, and the difference score would be 2.12.

Our expectation that different factors might be influential for males and females was confirmed. For females, the results were simple: Success was predicted by the difference scores for three measures of emotional functioning (Day 1 Emotionality, Emotionality from the human intruder test, and Nervous temperament), such that smaller differences between the two females significantly increased the chances of a successful pairing. In contrast, success in male pairings was greater for animals that were younger, that had been reared indoors, that showed a greater difference in weights between the two pairmates, and that had similar rearing histories. None of these factors were influential for female pairings. In terms of BBA measures, the two significant predictors for male pairs were the mean scores for Gentle temperament and for Nervous temperament—the lower the means, the greater was the likelihood of success. What this suggests, for example, is that if one attempts to pair a female with a high Nervous score, success will be more likely if her potential pairmate also has a high Nervous score (leading to a smaller difference score). In contrast, if one's animals are males, success in pairing a high-Nervous animal will be facilitated by pairing him with a low-Nervous animal (leading to a low mean score). This study, then, demonstrates that information on biobehavioral organization can be useful in increasing the chances that a particular pairing attempt will be successful.

Positive and Negative Reinforcement Training

Anyone who has trained multiple animals on a task has probably been struck by the variation that exists among animals in terms of speed of learning, attentiveness to the task, etc. As part of a completely separate study, we needed to train 30 adult male, individually housed rhesus monkeys to come up to the front of their cages to take a drug orally from a syringe. Positive reinforcement

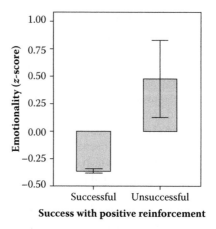

Figure 5.4 Emotionality score (z-score) for monkeys that were successfully or unsuccessfully trained using positive reinforcement training techniques.

training (PRT) was employed to shape the animals' behavior for accepting and drinking from the syringe, and 24 of the 30 animals were easily trained; six animals were refractory, however, even after many training sessions. To train those six animals, we used negative reinforcement—the animal was partially restrained using the intra-cage squeeze mechanism, and when he made the proper response, he was released. All six of these animals learned the task quickly once negative reinforcement training had been implemented. Because all 30 animals had participated in the BBA program as infants (5–8 years earlier), we examined the data from the 24 PRT-successful and the 6 PRT-unsuccessful animals, and discovered a highly significant difference between them; animals that were refractory using positive reinforcement had significantly higher Emotionality scores on the human intruder test (Figure 5.4). Recall that the behaviors included in this factor included rates of fear grimace and coo vocalization, and dichotomized codes of whether convulsive jerk or self-clasp were recorded. This result suggests that these animals were unusually reactive to the presence of the human, and their emotionality concerning the human's presence interfered with their ability to learn the task. It is possible that such emotionally reactive animals could eventually be habituated to the presence of humans, but the time frame for this process is unknown. Because we needed to get all of the animals trained in order to begin the study, we relied on negative reinforcement training. By making the animals just a bit more uncomfortable for a few seconds by use of the squeeze mechanism, and then releasing the mechanism once the animal made the desired response, training was able to proceed rapidly. We recognize that this is merely an anecdote, though some subsequent empirical work (manuscript in preparation) supports the conclusion that highly emotional animals may learn more quickly with negative, rather than with positive, reinforcement training.

CONCLUSION

Individual animals have enduring behavioral characteristics that they bring with them to each situation they encounter. These behavioral characteristics are often related in meaningful ways to physiological functioning, giving rise to the concept of biobehavioral organization. We believe that having quantitative measures of biobehavioral organization can be useful for three major reasons. As our review indicated earlier, knowledge of an individual's biobehavioral status can be useful to behavioral managers in predicting which animals might be especially at risk for the development of poor colony management-related outcomes, putting the animal's psychological well-being in jeopardy. A second use for this information, discussed much less in this chapter, is for animal selection for research studies. In an infectious disease study, for example, one might want to exclude animals with extreme scores for Nervous temperament, as such animals seem to have physiological characteristics that could impact their inflammatory responses. A third use for data on biobehavioral organization is that they enable us to understand the ways in which colony management procedures might be contributing to variation. For example, our data on diarrhea showed that animals that were gestated outdoors for at least 25% of their gestation had a lower risk for developing diarrhea later in life. This type of information can improve colony management practices in ways that could result in the production of animals that are less likely to demonstrate poor welfare-related outcomes.

Finally, while our BBA program might best be described as "comprehensive," in that we assess a variety of phenomena over a 25-h period, behavioral managers at other facilities may not need such an extensive program. For example, we noted earlier that Nervous temperament is associated with some health (i.e., diarrhea) and behavioral (i.e., social pairing) outcomes. It is possible that there are existing staff in one's facility (e.g., animal care staff who work daily with the same set of animals) that could identify animals that are timid and fearful, which are defining characteristics of a Nervous temperament. Moreover, behavioral managers can implement testing schemes that are relatively simple to accomplish, such as conducting a human intruder test with primates, to see which animals are especially emotional in their responses; such information can be used by staff responsible for pairing animals [see also related comments by Coleman (2017)].

Life in captivity involves many relatively routine challenges—relocations, separations from companions, pairing with new companions, husbandry procedures, etc., as described earlier. At an important level, behavioral management involves helping animals to adapt to these challenges, or correcting poor adaptation. Consequently, understanding and measuring the enduring characteristics of individuals in challenging situations (such as in our BBA program, or in smaller-scale programs by presenting animals with a human intruder or novel object) would be especially useful in helping to identify (even at an early age, as in the BBA program) animals that might be at risk for poor outcomes later in life, and then proactively working to prevent the development of poor outcomes. Ultimately, such an approach can lead to both improvement in the animals' lives, and better data by the scientists that utilize these animals for research purposes.

REFERENCES

Allport, G.W. 1937. *Personality: A Psychological Interpretation*. New York: Henry Holt.

Ardeshir, A., K.L. Oslund, F. Ventimiglia, J. Yee, N.W. Lerche, and D.M. Hyde. 2013. Idiopathic microscopic colitis of rhesus macaques: Quantitative assessment of colonic mucosa. *The Anatomical Record* 296:1169–1179.

Bachevalier, J., M. Brickson, and C. Haggar. 1993. Limbic-dependent recognition memory in monkeys develops early in infancy. *NeuroReport* 4:77–80.

Berman, C.M., K.L.R. Rasmussen, and S.J. Suomi. 1994. Responses of free-ranging rhesus monkeys to a natural form of social separation. I. Parallels with mother-infant separation in captivity. *Child Development* 65:1028–1041.

Caine, N.G., H. Earle, and M. Reite. 1983. Personality traits of adolescent pig-tailed monkeys (*Macaca nemestrina*): An analysis of social rank and early separation experience. *American Journal of Primatology* 4:253–260.

Capitanio, J.P. 1986. Behavioral pathology. In *Comparative Primate Biology, Volume 2A: Behavior, Conservation, and Ecology*, eds Mitchell, G. and J. Erwin, 411–454. New York: Alan R. Liss.

Capitanio, J.P., S.A. Blozis, J. Snarr, A. Steward, B.J. McCowan. 2017. Do "birds of a feather flock together" or do "opposites attract"? Behavioral responses and temperament predict success in pairings of rhesus monkeys in a laboratory setting. *American Journal of Primatology* 79(1):1–11.

Capitanio, J.P., M.A. Boccia, and D.J. Colaiannia. 1985. The influence of rank on affect perception by pigtail macaques. *American Journal of Primatology* 8:53–59.

Capitanio, J.P., L.A. Del Rosso, L.A. Calonder, S.A. Blozis, and M.C.T. Penedo. 2012. Behavioral effects of prenatal ketamine exposure in rhesus macaques are dependent on MAOA genotype. *Experimental and Clinical Psychopharmacology* 20:173–180.

Capitanio, J.P., S.P. Mendoza, and S.W. Cole. 2011. Nervous temperament in infant monkeys is associated with reduced sensitivity of leukocytes to cortisol's influence on trafficking. *Brain, Behavior, and Immunity* 25:151–159.

Capitanio, J.P., S.P. Mendoza, W.A. Mason, and N. Maninger. 2005. Rearing environment and hypothalamic-pituitary-adrenal regulation in young rhesus monkeys (*Macaca mulatta*). *Developmental Psychobiology* 46:318–330.

Capitanio, J.P., K.L.R. Rasmussen, D.S. Snyder, M. Laudenslager, and M. Reite. 1986. Long-term follow-up of previously separated pigtail macaques: Group and individual differences in response to novel situations. *Journal of Child Psychology and Psychiatry* 27:531–538.

Capitanio, J.P. and M. Reite. 1984. The roles of early separation experience and prior familiarity in the social relations of pigtail macaques: A descriptive multivariate study. *Primates* 25:475–484.

Coleman, K. 2017. Individual differences in temperment and behavioral management, Chapter 7. In *Handbook of Primate Behavioral Management*, ed. Schapiro, S.J., 95–113 Boca Raton, FL: CRC Press.

Costello, A.B. and J.W. Osborne. 2005. Best practices in exploratory factor analysis: Four recommendations for getting the most from your analysis. *Practical Assessment, Research & Evaluation* 10:1–9.

Dantzer, R. 2009. Cytokine, sickness behavior, and depression. *Immunology and Allergy Clinics of North America*. 29:247–264.

Darwin, C. 1859/1975. *The Origin of Species*. Franklin, PA: Franklin Library.

DiVincenti, L., Jr. and J.D. Wyatt. 2011. Pair housing of macaques in research facilities: A science-based review of benefits and risks. *Journal of the American Association for Laboratory Animal Science* 50:856–863.

Dixson, A.F. 2012. *Primate Sexuality: Comparative Studies of the Prosimians, Monkeys, Apes, and Humans*, 2nd Edition. Oxford, UK: Oxford University Press.

Elfenbein, H.A., L.D. Rosso, B. McCowan, and J.P. Capitanio. 2016. Effect of indoor compared with outdoor location during gestation on the incidence of diarrhea in indoor-reared rhesus macaques (*Macaca mulatta*). *Journal of the American Association for Laboratory Animal Science* 55(3):277–290.

Elmore, D.B., J.H. Anderson, D.W. Hird, K.D. Sanders, and N.W. Lerche. 1992. Diarrhea rates and risk factors for developing chronic diarrhea in infant and juvenile rhesus monkeys. *Laboratory Animal Science* 42:356–359.

Gagne, J.R., C.A. Van Hulle, N. Aksan, M.J. Essex, and H.H. Goldsmith. 2011. Deriving childhood temperament measures from emotion-eliciting behavioral episodes: Scale construction and initial validation. *Psychological Assessment* 23:337–353.

Golub, M.S., C.E. Hogrefe, K.F. Widaman, and J.P. Capitanio. 2009. Iron deficiency anemia and affective response in rhesus monkey infants. *Developmental Psychobiology* 51:47–59.

Gottlieb, D.H. and J.P. Capitanio. 2013. Latent variables affecting behavioral response to the human intruder test in infant rhesus monkeys (*Macaca mulatta*). *American Journal of Primatology* 75:314–323.

Gottlieb, D.H., J.P. Capitanio, and B. McCowan. 2013. Risk factors for stereotypic behavior and self-biting in rhesus macaques (*Macaca mulatta*); Animal's history, current environment, and personality. *American Journal of Primatology* 75:995–1008.

Gunderson, V., K. Grant-Webster, T. Burbacher, and N. Mottet. 1988. Visual recognition memory deficits in methylmercury-exposed *Macaca fascicularis* infants. *Neurotoxicology and Teratology* 10:373–379.

Gunderson, V., K. Grant-Webster, and J.F. Fagan. 1987. Visual recognition memory in high- and low-risk infant pigtailed macaques (*Macaca nemestrina*). *Developmental Psychology* 23:671–675.

Hennessy, M.B., B. McCowan, J. Jiang, and J.P. Capitanio. 2014. Depressive-like behavioral response of adult male rhesus monkeys during routine animal husbandry procedure. *Frontiers in Behavioral Neuroscience* 8:309.

Hinde, R.A. and L. Davies. 1972. Removing infant rhesus from mother for 13 days compared with removing mother from infant. *Journal of Child Psychology and Psychiatry* 13:227–237.

Hinde, K., A.L. Skibiel, A.B. Foster, L. Del Rosso, S.P. Mendoza, and J.P. Capitanio. 2015. Cortisol in mother's milk across lactation reflects maternal life history and predicts infant temperament. *Behavioral Ecology* 26:269–281.

Jiang, J., S. Kanthaswamy, and J.P. Capitanio. 2013. Degree of Chinese ancestry affects behavioral characteristics of infant rhesus macaques (*Macaca mulatta*). *Journal of Medical Primatology* 42:20–27.

Kalin, N.H. and S.E. Shelton. 1989. Defensive behaviors in infant rhesus monkeys: Environmental cues and neurochemical regulation. *Science* 243:1718–1721.

Kinnally, E.L. and J.P. Capitanio. 2015. Paternal early experiences influence infant development through nonsocial mechanisms in rhesus macaques. *Frontiers in Zoology* 12(Suppl 1):S14.

Kinnally, E.L., G.M. Karere, L.A. Lyons, S.P. Mendoza, W.A. Mason, and J.P. Capitanio. 2010. Serotonin pathway gene-gene and gene-environment interactions influence behavioral stress response in infant rhesus macaques. *Development and Psychopathology* 22:35–44.

Koshland, D.E. 2002. The seven pillars of life. *Science* 295(5563):2215–6.

Laudenslager, M., J.P. Capitanio, and M. Reite. 1985. Possible effects of early separation experiences on subsequent immune function in adult macaque monkeys. *American Journal of Psychiatry* 142:862–864.

Levitis, D.A., W.Z. Lidicker, and G. Freund. 2009. Behavioural biologists do not agree on what constitutes behaviour. *Animal Behaviour* 78:103–110.

Mineka, S. and S.J. Suomi. 1978. Social separation in monkeys. *Psychological Bulletin* 85:1376–1400.

Parker, K.J., and D. Maestripieri. 2011. Identifying key features of early stressful experiences that produce stress vulnerability and resilience in primates. *Neuroscience & Biobehavioral Reviews*. 35:1466–1483.

Pickering, A.D., and J.A. Gray. 1999. The neuroscience of personality. In *Handbook of Personality: Theory and Research*, 2nd Edition, eds Pervin, L.A. and O.P. John, 277–299. New York: The Guilford Press.

Plimpton, E.H., K.B. Swartz, and L.A. Rosenblum. 1981. Response of juvenile bonnet macaques to social stimuli presented through color videotapes. *Developmental Psychobiology* 14:109–115.

Prongay, K., B. Park, and S.J. Murphy. 2013. Risk factor analysis may provide clues to diarrhea prevention in outdoor-housed rhesus macaques (*Macaca mulatta*). *American Journal of Primatology* 75:872–82.

Rogers, J., M. Raveendran, G.L. Fawcett, A.S. Fox, S.E. Shelton, J.A. Oler, J. Cheverud, et al. CRHR1 genotypes, neural circuits and the diathesis for anxiety and depression. *Molecular Psychiatry* 18:700–707.

Rothbart, M.K. 1988. Temperament and the development of inhibited approach. *Child Development* 59:1241–1250.

Ruys, J.D., S.P. Mendoza, J.P. Capitanio, and W.A. Mason. 2004. Behavioral and physiological adaptation to repeated chair restraint in rhesus macaques. *Physiology and Behavior* 82:205–213.

Shively, C.A. 2017. Depression in captive nonhuman primates: Theoretical underpinnings, methods, and application to behavioral management, Chapter 8. In *Handbook of Primate Behavioral Management*, ed. Schapiro, S.J., 115–125. Boca Raton, FL: CRC Press.

Sloan, E.K., C.T. Nguyen, B.F. Cox, R.P. Tarara, J.P. Capitanio, and S.W. Cole. 2008. SIV infection decreases sympathetic innervation of primate lymph nodes: The role of neurotrophins. *Brain, Behavior, and Immunity* 22:185–194.

Sloan, E.K., R.P. Tarara, J.P. Capitanio, and S.W. Cole. 2006. Enhanced replication of simian immunodeficiency virus adjacent to catecholaminergic varicosities in primate lymph nodes. *Journal of Virology* 80:4326–4335.

Sorenson, A.N., E.C. Sullivan, S.P. Mendoza, J.P. Capitanio, and J.D. Higley. 2013. Serotonin transporter genotype modulates HPA axis output during stress: Effect of stress, dexamethasone test and ACTH challenge. *Translational Developmental Psychiatry* 1:21130.

Truelove, M.A., A.L. Martin, J.E. Perlman, J.S. Wood, and M.A. Bloomsmith. 2017. Pair housing of macaques: A review of partner selection, introduction techniques, monitoring for compatibility, and methods for long-term maintenance of pairs. *American Journal of Primatology* 79(1):1–15.

Vandeleest, J.J., S.P. Mendoza, and J.P. Capitanio. 2013. Birth timing and the mother–infant relationship predict variation in infant behavior and physiology. *Developmental Psychobiology* 55:829–837.

Zerhouni, E. 2006. Testimony before U.S. House of Representatives Subcommittee on Labor April 2006, http://www.nih.gov/about-nih/who-we-are/nih-director/fy-2007-directors-budget-request-statement (Accessed April 15, 2016).

CHAPTER 6

The Role of Stress in Abnormal Behavior and Other Abnormal Conditions Such as Hair Loss

Melinda A. Novak, Amanda F. Hamel, Amy M. Ryan,
Mark T. Menard, and Jerrold S. Meyer
University of Massachusetts, Amherst

CONTENTS

INTRODUCTION

It has been 25 years since federal legislation was enacted requiring facilities to promote the psychological well-being of nonhuman primates in research and other captive settings (Animal Welfare Act 1991). A significant challenge at the time concerned the use of engineering standards versus performance standards. In the former, psychological well-being was to be achieved by creating a set of rules regarding cages and features. In the latter, environmental changes were to be evaluated in terms of their impact on the animals and implemented after efficacy was established. Today, some aspects of psychological well-being are regulated by engineering standards; however, many aspects of well-being are achieved through performance standards. In this regard, the use of performance standards requires defining what psychological well-being means for nonhuman primates, both behaviorally and physiologically, and determining what changes might succeed in promoting well-being. It also requires a better understanding of environmental stressors and their possible adverse effects on well-being.

The complexity of these issues has led to the creation of behavioral management units at many primate facilities. These units are tasked with assessing and promoting psychological well-being in all the captive primates at their facilities. Additionally, the staff in these units are responsible for identifying problematic behaviors, determining possible environmental stressors, and evaluating the efficacy of various kinds of treatment. Over the last several decades, significant progress has been made in reducing stress and increasing psychological well-being, thereby meeting behavioral management objectives.

Standard definitions of psychological well-being typically include both behavioral and physiological assessments (Novak and Suomi 1988). In the physiological realm, the focus has been on standard daily health assessments, including coat condition and clinical signs, and more recently, on stress assessments, primarily using the hypothalamic–pituitary–adrenocortical (HPA) axis, which is one arm of the stress response system. In the behavioral realm, three features are emphasized: (1) species-typical range—the primate should show diversity of species-typical behavior, (2) species-typical levels—the primate should show the normal expression of species-typical behavior (e.g., threats and some aggressiveness is normal; hyperaggressiveness is not), and (3) abnormal behavior—the primate should show low levels of stereotypic and other forms of abnormal behavior. This chapter addresses several issues as they pertain to the behavioral management of well-being in captive macaques:

1. What is stress, and how do we measure it?
2. What constitutes an inherently abnormal condition that requires therapeutic intervention?
3. What is the relationship between stress and various abnormal conditions?

Underscored in this discussion is the view that stress and abnormality are not discrete entities, but rather continuous variables, where only one, or both, extremes represent possible threats to well-being. Much of the research presented in the following sections is based on studies of macaques, the primates that are most commonly used in research.

STRESS IN NONHUMAN PRIMATES

Stress exposure is an important component of reduced well-being. It is therefore imperative for behavioral management staff to understand what stress is and how it can be measured. Stress can be defined physiologically as any disruption of an organism's behavioral and physiological equilibrium or homeostasis. Stressors are variables that have the potential to disrupt homeostasis. Although brief perturbations can be associated with positive events, most efforts are directed to understanding perturbations that are associated with adverse events leading to distress (Selye 1975). The perturbation is considered the stress response, and one way that it can be measured is by examining changes in activity of the HPA axis. It is worth noting that stressors can be physical (e.g., excessive heat or cold), physiological (e.g., hypoglycemia or hemorrhage), or psychosocial (e.g., social deprivation or agonistic encounters). All types of stressors elicit HPA activation. Thus, from a behavioral management perspective, activation, in and of itself, does not tell us what event or circumstance has elicited the stress response.

The length of the stress exposure and the perception of the intensity of that exposure can lead to different outcomes. Brief, mild exposure is typically associated with a short-acting perturbation, after which homeostasis is restored. However, prolonged stress exposure over days and months leads to an increase in allostatic load (cumulative "wear and tear on the body"; McEwen 1998), which, in turn, may cause chronic disruption of the HPA axis. Such a disruption can be manifested in multiple ways including alteration of the basal set point of the HPA axis and/or a blunting of the system's reaction to a new challenge (i.e., failure to mount an adequate stress response).

Measuring HPA Activity and the Stress Response

The HPA axis consists of a circuit comprising the hypothalamus, pituitary gland, and adrenal cortex. Each of these structures produces a hormone that theoretically could be measured as an indicator of the stress response. Exposure to a stressor activates the release of corticotrophin-releasing hormone (CRH) from the hypothalamus; in turn, CRH travels down the pituitary stalk to target cells in the anterior pituitary gland, prompting the release of adrenocorticotropic hormone (ACTH) into the bloodstream; and finally, ACTH stimulates the secretion of cortisol from the adrenal cortex. Although both cortisol and ACTH can be obtained from peripheral blood samples, stress-induced changes in ACTH levels (both the rise and subsequent decline) are much more rapid than for cortisol. ACTH is also less stable in blood than cortisol. For these reasons, cortisol is typically the hormone of choice for measuring the HPA stress response (O'Connor et al. 2000).

Accurate assessment of HPA axis activity requires consideration of two additional features of the system. First, the secretion of CRH, ACTH, and cortisol is governed by circadian rhythm. In diurnal organisms, which include most nonhuman primates, secretory activity is highest in the early morning hours and lowest in the evening (Heintz et al. 2011; Urbanski and Sorwell 2012). Consequently, comparing cortisol or ACTH levels in stressed versus nonstressed animals usually requires sample collection at the same time of day to control for this rhythmicity. Importantly, the diurnal cortisol rhythm itself can be affected by chronic stress. For example, dampening of the rhythm has been noted in monkeys reared under adverse conditions compared to normally reared controls (Sanchez et al. 2005). The second key feature of the HPA axis is that blood-borne cortisol acts on glucocorticoid receptors in the pituitary gland and the brain (including the hypothalamus) to exert a negative feedback action on subsequent CRH and ACTH release (Keller-Wood and Dallman 1984). Negative-feedback-mediated termination of the HPA response to an acute stressor is adaptive because it prevents the organism from experiencing prolonged exposure to elevated cortisol levels when such exposure is not needed. Reduced efficiency of the negative feedback mechanism, manifested as a delayed termination of the stress response, is yet another way (besides altered set point or blunted stress response) that the HPA axis can become dysregulated.

Investigators and behavioral management staff currently have several options for collecting stress hormone data. However, these options are not equivalent and are dependent, in large part, on the questions being asked. Cortisol concentrations can be obtained from several different matrices (Novak et al. 2013). For many years, cortisol was routinely measured in blood samples (Bowman and De Luna 1968) and, to a lesser extent, urine samples (Setchell et al. 1977; Ziegler et al. 1996). Development of subsequent technology permitted the assay of cortisol from saliva (Boyce et al. 1995; Lutz et al. 2000; Tiefenbacher et al. 2003) and from feces (Millspaugh and Washburn

2004; Keay et al. 2006; Romano et al. 2010). Very recent advances in assaying cortisol from hair (Davenport et al. 2006, 2008) and fingernails (Izawa et al. 2015; Veronesi et al. 2015) now allow for an estimate of long-term cortisol secretion spanning weeks to months. Each of these sampling matrices has strengths and weaknesses and may be more suited to some kinds of clinical assessment and/or experimental research than others.

Plasma/Serum Samples

Total cortisol, consisting both of the free and the bound portions, is obtained from blood samples. Approximately 5%–20% of total cortisol is biologically active (free), with the remainder bound mainly to cortisol-binding globulin (CBG). One limitation is that variation in CBG can affect the levels of total cortisol (Le Roux et al. 2003). To avoid this problem, a free cortisol index can be obtained by measuring CBG concentrations and dividing total cortisol by CBG levels. Although this approach is seldom applied in nonhuman primate studies, it can provide important information when manipulations occur that alter CBG (e.g., see Davenport et al. 2008).

Blood samples provide cortisol levels at the moment of collection, and this can be useful for many purposes. For example, multiple samples collected at appropriate time points within or across days can provide information on basal HPA activity and the ways in which it may be altered over time due to environmental changes; the shape of the diurnal rhythm; and the magnitude and time course of the HPA response to, and recovery from, an acutely stressful event (e.g., brief veterinary procedures; brief separations of infants from mothers). Nevertheless, there are obvious limitations to the application of blood sampling methods to nonhuman primates. Behavioral management staff will need to consider the fact that obtaining even a single blood sample can, itself, be stressful, if it is associated with restraint (Crockett et al. 1993, 2000) or with seeing other animals provide blood samples (Flow and Jaques 1997). Training animals to present a limb for blood sampling (Coleman et al. 2008) may reduce the need for restraint and/or sedation and may be less stressful overall (Lambeth et al. 2005). When samples must be obtained across multiple time points, for example, when investigating possible changes in the diurnal cortisol rhythm, it is advisable either to use animals trained for blood collection or to employ remote monitoring via an indwelling subclavian venous catheter (see Downs et al. 2007 for an example of this last approach).

Acute stressors are often studied by first collecting a baseline sample prior to the stressor, followed by samples collected at varying time points after the imposition of the stressor. The post-stress cortisol levels are then compared to the baseline levels (see Short et al. 2014 for an example of this approach). In other cases, however, cortisol levels are evaluated in different groups of monkeys after exposure to a stressor and without any comparative baseline samples, either because of the difficulty of such collection or safety risks to the participants. Instead, the blood samples are always collected at the same time point after initiation of the stressor, and the assumption is that the first sample reflects a maximal "stress response." This approach was used in a study by Capitanio et al. (2005), in which the authors characterized reactions to the acute stress of separation of infants reared in different environments, one of which included mothers and infants in outdoor corrals. Blood samples have also been used to compare groups living in different environments. Using a between-subjects design, Schapiro et al. (1993) showed that cortisol concentrations were higher in indoor-housed as opposed to outdoor-housed monkeys.

Saliva Samples

Saliva samples provide an important alternative to blood samples. Like blood samples, saliva samples represent HPA axis activity for a very short period of time prior to the sample collection (minutes) and can be employed to evaluate the effects of acute stressors. Unlike blood samples, salivary cortisol contains only the biologically active or free component, which has been shown to correlate with free cortisol obtained from plasma samples (Lane 2006). If training is a significant component of a behavioral management program, saliva can be collected from awake, unrestrained monkeys by training them to chew on dental rope impregnated with sweetener (Lutz et al. 2000; Tiefenbacher et al. 2003). Although this technique is most often used with captive animals, Higham et al. (2010) successfully used a similar approach to collect saliva for cortisol assay from free-ranging rhesus macaques on the island of Cayo Santiago. Alternatively, saliva can be collected following sedation by placing dental rope in the cheek pouch of a monkey and removing it minutes later. In the former approach, some monkeys may be difficult to train, whereas in the latter approach, some monkeys may not provide sufficient saliva for assay. Low saliva production with associated dry mouth (xerostomia) has been shown to correlate with high plasma cortisol concentrations in monkeys (Davenport et al. 2003), consistent with findings of high salivary cortisol concentrations in humans reporting lack of salivary flow and dry mouth (Shigeyama et al. 2008). Thus, monkeys that produce insufficient saliva for assay may differ from other monkeys in their HPA axis activity. Special attention must be paid to possible contamination of saliva samples with minor bleeding in the mouth, possibly associated with tooth infections or inflammation of the gums. Inasmuch as free salivary cortisol concentrations are much lower than the total cortisol obtained from blood, blood contamination will cause a considerable inflation in the measured levels, invalidating the results.

Salivary cortisol has been measured in several different contexts and in a variety of nonhuman primate species, including rhesus macaques, squirrel monkeys, spider monkeys, baboons, chimpanzees, bonobos, gorillas, and orangutans. Generally, salivary cortisol has been used to determine the effects of an acute stressor, by comparing post-stress to prestress hormone levels. Research has shown that salivary cortisol levels in baboons were elevated during crowding (Pearson et al. 2015). Similarly, exposure to an unfamiliar room, with or without an unfamiliar conspecific, was associated with increased salivary cortisol concentrations in rhesus macaques (Lutz et al. 2003). More recently, salivary cortisol concentrations were used to evaluate the possible stressfulness of medical procedures that had previously been shaped through positive reinforcement training (PRT). Neither bonobos nor orangutans showed any significant variation in salivary cortisol concentrations across baseline, during PRT, or immediately thereafter (Behringer et al. 2014), suggesting that training may be effective in mitigating the stress associated with various medical procedures. As with blood samples, saliva samples have also been used to examine circadian rhythm activity (Boyce et al. 1995).

Urine Samples

Urine samples are sometimes used as an alternative to blood samples in nonhuman primate research. Unlike blood samples, urinary cortisol concentration reflects a longer period of time (from several hours to a day) and is not confounded by restraint stress that is often associated with blood sampling. However, because urinary output varies considerably across individuals, urinary cortisol concentrations have to be adjusted by correcting for creatinine content. Creatinine, which is a product of muscle metabolism, is excreted at a relatively steady rate independent of urine volume. Urine collection can be challenging when applied to captive nonhuman primates. In macaques, it usually involves separating pair housed monkeys overnight (a possible stressor) and ensuring that the urine remains relatively uncontaminated by other secretions, such as feces or blood. An alternative that has been used very successfully with socially housed marmosets and tamarins is to collect urine from individuals immediately after light onset (first morning void). Once habituated to the entry of an observer into a pen, or an observer's arm into a cage, the tamarins readily urinate into a cup (Ziegler et al. 1996). Using positive reinforcement training, chimpanzees will urinate into a cup on command (Anestis and Bribiescas 2004; Bloomsmith et al. 2015). Urine samples are not particularly suited to evaluating the effects of an acute stressor, but they may have some value in assessing the effects of different environmental or physiological conditions. For example, monkeys treated chronically with high-dose exogenous cortisol showed increases in urinary cortisol as expected; however, these artificially induced elevations in cortisol were not associated with hippocampal neuronal loss in the absence of an actual environmental stressor (Leverenz et al. 1999).

Fecal Samples

Fecal samples have been commonly used to assay hormone levels in free-ranging animals because of the ease of sample collection (see Fürtbauer et al. 2014; Mendonca-Furtado et al. 2014). An important consideration for HPA assessment using this sample matrix is that feces contain mostly metabolites of cortisol and other glucocorticoids (sometimes simply termed fecal glucocorticoid metabolites), which complicates the choice of an appropriate assay method (see, for example, Heistermann et al. 2006). As in the case of urinary cortisol concentrations, measuring either fecal cortisol itself or fecal glucocorticoid metabolite concentrations is not particularly useful for assessing the effects of acute stressors. However, because of the ease of sample collection, even under laboratory conditions, analyses of fecal cortisol or glucocorticoid metabolite concentrations have been used to assess the impact of alopecia and housing conditions on captive rhesus macaques (Steinmetz et al. 2006), evaluate the effects of environmental enrichment in brown capuchins (Boinski et al. 1999), examine the role of weekday and weekend activities on stress levels in common marmosets (Barbosa and Mota 2009), and understand the ways in which HPA axis activity is related to pair housing attempts in male rhesus monkeys (Doyle et al. 2008).

Hair Samples

A significant advance in the field of stress hormone research has been the development of novel measures of chronic HPA axis activation using either hair (Davenport et al. 2006) or fingernail samples (Veronesi et al. 2015). Like salivary cortisol, hair cortisol is thought to reflect the biologically active or free fraction of the cortisol molecule (Meyer and Novak 2012). Unlike salivary cortisol, hair cortisol measurements cannot reveal the effects of an acute stressor or altered circadian variation. However, hair cortisol concentrations can be used to examine the long-term effects of different kinds of environmental change without the need for extensive repetitive sampling, as would be required with any other sample matrix. Additionally, unlike blood, saliva, or urine, cortisol concentrations in hair are unaffected by the potential stress of sample collection and can be collected at any time of the day, either in trained animals or in animals restrained and sedated for sample collection.

An additional benefit of measuring cortisol in hair is that segments of hair differing in their distance from the skin can be used as a retrospective calendar of hormone deposition, thereby providing data on adrenocortical activity prior to, during, and after exposure to a significant stressor. The *calendar method* is based on two premises: (1) that cortisol in the hair shaft is fixed in place once it has been deposited, and (2) that hair growth rate for a given species is relatively consistent across individuals and conditions. Increasingly in human studies, hair is being used as a retrospective calendar to evaluate adrenocortical activity during pregnancy (Kirschbaum et al. 2009; Braig et al. 2016) and following major disasters, such as earthquakes (Gao et al. 2014) and war (Etwel et al. 2014). This calendar approach has also been used to track stress exposure and its effect on captive orangutans (Carlitz et al. 2014).

Since its inception, hair cortisol concentrations have been used by our group and others to evaluate the effects of environmental disruption, and to assess the relationship between chronic activation of the HPA axis and other biobehavioral variables, including temperament. For example, adult rhesus monkeys permanently relocated to a new environment showed elevated hair cortisol concentrations in response to the move (Davenport et al. 2008). Hair cortisol concentrations also increased concomitant with increased population density in monkeys housed in a semi-natural habitat (Dettmer et al. 2014). Other studies have shown that high levels of hair cortisol predict the development of anxiety responses to relocation and psychosocial stress in infant macaques (Dettmer et al. 2012), and we recently demonstrated that adult monkeys with high hair cortisol profiles showed significantly more anxious behavior in the human intruder test of anxiety than adult monkeys with low cortisol profiles (Hamel et al. 2017). Consistent with these findings, Laudenslager et al. (2011) found an association of high hair cortisol with low levels of novelty-seeking behavior in adult vervet monkeys.

Hair Cortisol Values and Reference Ranges

One of the challenges in measuring cortisol concentrations in monkeys is to determine what the levels actually represent. In human health monitoring, a reference range for a particular variable, like cortisol or white blood cell count, is the set of values that represent 95% of the population sampled (or statistically, the equivalent of two standard deviations from the mean). Values outside the range of two standard deviations represent a disease state or at least an elevated risk for disease. For example, *both very high and very low serum cortisol values in humans are pathological*, representing a state of hypercortisolemia (Cushing's syndrome) or a state of adrenal insufficiency (Addison's disease), respectively. Adrenal insufficiency has not been reported to occur in nonhuman primates; however, hyperadrenocorticism has been identified in a Japanese macaque (Kimura 2008) and a rhesus macaque (Wilkinson et al. 1999). In human clinical medicine, reference ranges for cortisol are available for serum or plasma concentrations of the hormone. A reference range for hair cortisol concentrations in nonclinical human populations has been proposed by Sauvé et al. (2007); although to date, this information has only been applied for research purposes, not for clinical diagnosis.

To assist behavioral management staff in interpreting hair cortisol concentrations in adult rhesus macaques, we provide hair cortisol data on 296 indoor-housed monkeys from five different facilities located across the United States (Figure 6.1). In the graphs, we show the 95% cutoffs represented by the two thick vertical lines. It should be noted that females have significantly higher hair cortisol concentrations than males, which is consistent with previous findings on plasma cortisol in male and female macaques tested under control conditions (Lado-Abeal et al. 2005). It may be prudent for researchers and colony managers to resample any animal outside the 95% population range and then, if the concentration remained unchanged, refer the animal to veterinarians for additional assessments.

In human health monitoring, values at the high or low end of "normal" (i.e., within the 95% range of a clinical parameter) are sometimes referred to as "high normal" or "low normal." These designations are warranted under at least two conditions. First, it may be the case that such relatively extreme values are indicative of suboptimal functioning, even if a diagnosable disease state has not

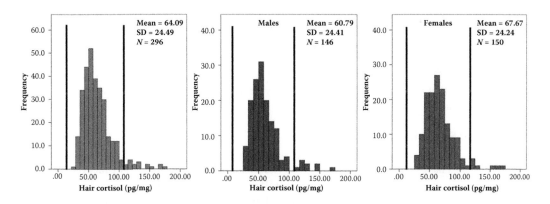

Figure 6.1   Hair cortisol distributions in the total population (left), males only (center), and females only (right).

been attained. Second, very high or very low clinical values for an individual may reveal a trajectory toward a disease state when compared to values obtained earlier. In such cases, the physician continues to monitor the patient to see whether the boundary between normal and abnormal has been breached, in which case treatment may be needed. The relevance for hair cortisol, whether in humans or nonhuman primates, is that very high values, even if not quite outside of the reference range, may be indicative of a state of chronic stress in the individual. Persistently elevated cortisol can be one of the manifestations of stress-induced increases in allostatic load, which over time confer greater disease vulnerability upon the organism (McEwen and Seeman 1999). Although the presence of high normal hair cortisol levels has not yet been shown to be a predictor of future health problems in either human or nonhuman primate populations, we believe that this represents a fertile area for future research.

BEHAVIORAL AND PHYSICAL DISORDERS IN NONHUMAN PRIMATES

A primary focus of behavioral managers at primate facilities is to deal with the prevention, development, and expression of abnormal behaviors in the primates under their care. Abnormal behavior includes not only highly ritualistic stereotyped movements, but also species-typical behavior that is expressed in an abnormal manner. Thus, behavioral management staff must have a working knowledge of all forms of species-typical behavior and all types of stereotypic behavior. For example, rhesus macaques housed in captivity can develop a variety of bizarre and unusual patterns of behavior (Novak et al. 2012). These range from repetitive, stereotypic activities, such as pacing, rocking, self-mouthing, and eye covering, to more serious behaviors, such as hair plucking and self-inflicted wounding. However, even in the case of a species-typical behavior, rhesus macaques may exhibit behavior in inappropriate contexts or at levels that are either excessively high (e.g., aggressive behavior) or excessively low (e.g., locomotion). Although all these forms of abnormality can be present in captive primates, most behavioral management efforts have been focused on self-injurious and stereotypic behaviors, both of which are also a focus of animal model research (Lutz 2014; Novak et al. 2014a).

Self-Injurious Behavior

Self-injurious behavior (SIB) is a serious problem that is generally considered a marker for reduced well-being, both in nonhuman primates and in humans. In macaques, SIB generally takes the form of bites directed primarily to arms and legs, which can yield wounds requiring veterinary care (Novak 2003). Although SIB is relatively uncommon, occurring in only 5%–10% of the captive population (Lutz et al. 2003; Rommeck et al. 2009), the impact of SIB can be substantial, adversely affecting the monkeys, the behavioral management staff who must cope with such cases, and the research scientists who may lose subjects from their studies. Both the development and exacerbation of SIB have been associated with major life stress events, such as early social separation and relocation to new environments (Lutz et al. 2003; Davenport et al. 2008). Thus, it is not unreasonable to consider that the stress response system may be dysregulated in this population, an issue we consider in the last section of this chapter.

Stereotypic Behavior

Stereotypies have been defined as repetitive, ritualistic behaviors that appear functionless. They represent a challenge to behavioral management staff due to both the large number of different kinds of stereotypies that have been observed and the relatively high prevalence of stereotypic behavior in captive primates. Stereotypies can range from whole body movements (e.g., rocking, pacing, somersaulting) to self-directed actions involving the hands or feet (e.g., eye covering, digit suck-ing, hair pulling, self-grasping) (Novak et al. 2012). Complicating the picture is that some monkeys perform only one kind of stereotypic behavior, whereas others can exhibit multiple forms (Bellanca and Crockett 2002; Lutz et al. 2003). Stereotypies are much more prevalent in monkey populations than SIB. Indeed, in certain studies, nearly all *singly housed* monkeys (90%) were found to exhibit some form of stereotypic behavior (Lutz et al. 2003; Camus et al. 2013).

In contrast to SIB, which is dangerous and can result in repeated tissue damage, stereotypies are not typically dangerous to the animal expressing them. Thus, the mere presence of stereotypic behavior may not be a useful sign that animals have reduced well-being. Instead, behavioral manag-ers should take into account the severity of stereotypic behavior. The importance of severity is illus-trated by studies of stereotypic behavior in humans. Humans often engage in one or more forms of mild ritualistic behavior. These activities include, but are not limited to, hair and skin manipulation (e.g., hair twirling, cheek pulling, beard tugging), eye covering, flexion–extension of legs, tapping of limbs against a surface or one's own body, repetitive object manipulation, pacing, and rocking. Self-report survey data from adult humans in nonclinical populations (typically college students) support the notion that mild forms of stereotypic behavior are relatively common and presumably harmless (Hansen et al. 1990; Woods and Miltenberger 1996; Rafaeli-Mor et al. 1999). Indeed, stereotypic behavior may have a calming effect on the individual. For example, Soussignan and Koch (1985) observed a reduction in heart rate during leg swinging in normal children. Thus, in most instances, these rituals are minor, having relatively little negative impact, or even a positive impact, on the indi-vidual. However, under some circumstances, such rituals can progress to the point where they become pathological in nature (i.e., obsessive and compulsive), reducing individual well-being and requiring treatment. It is reasonable to assume that this same continuum exists for nonhuman primates, such that low levels of stereotypic behavior may have little impact, whereas high levels of stereotypic behavior, which interfere with basic biological processes (e.g., exploration, attention to environmental change, and self-maintenance behaviors like grooming and eating), may constitute a marker for diminished well-being (see Lutz 2014 for a discussion of the causes and treatments for stereotypic behavior).

Hair Loss or Alopecia

Hair loss (alopecia), particularly in macaques, is an abnormal physical condition that has received increased scrutiny from United States Department of Agriculture-Animal and Plant Health Inspection Service (USDA-APHIS) inspectors. Although alopecia was originally thought to result from the self-directed stereotypy of hair pulling by singly housed monkeys, it is now clear that many cases of hair loss are due to factors unrelated to abnormal behavior. Nonetheless, behavioral management staff is usually called upon to deal with this problem. Lutz et al. (2013) report relatively high prevalences of hair loss at four different primate facilities across the United States, with alope-cia occurring in 35%–89% of the monkeys in the indoor colonies at these facilities. Although hair loss is common in macaque species, there is considerable variability across individuals in the sever-ity of hair loss. Severity, as reflected in the percentage of the body in which hair is missing, typically ranges along a continuum from 0% to 90% when measured digitally using ImageJ NIH software (Novak et al. 2014b). Very few indoor-housed macaques show no hair loss whatsoever, and we have typically defined a healthy control group as having <5% hair loss. In contrast, some macaques show nearly full body hair loss of the dark coat, as seen in the sedated monkey in Figure 6.2. Hair loss is associated with many different factors, including aging, hormonal dysregulation, dermatitis, nutri-tional deficiency, autoimmune diseases, etc. (see Novak and Meyer 2009 for a discussion of these factors). However, even after ruling out these conditions, otherwise healthy monkeys may still show substantial hair loss. In these populations, the development of hair loss in macaques has been noted to vary by sex (hair loss being more prevalent in females than in males; Lutz et al. 2013), species (lowest in *Macaca fascicularis*, intermediate in *M. mulatta*, and highest in *M. nemestrina*; Kroeker

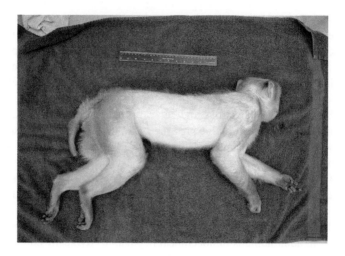

Figure 6.2 Sedated rhesus monkey with nearly full-body hair loss.

et al. 2014), and housing facility (Novak et al. 2014b). In the absence of any known precipitating factor, significant hair loss in otherwise healthy monkeys is generally thought to be a manifestation of stress exposure. We examine this premise in the following sections.

ABNORMALITY AND STRESS ASSESSMENT: CAUTIONARY TALES

Any discussion of the relationship between the stress response system and the abnormalities described earlier has to be considered in the context of three issues: (1) problems in comparing and interpreting cortisol values from different sampling matrices, (2) the difficulty of establishing causal relationships, and (3) the challenge in identifying the source of the stress. With respect to the first issue, several different sampling matrices may be employed, either by the same investigator or by several investigators to test specific hypotheses. However, the resulting cortisol concentrations, either in the same study or across studies, cannot simply be compared in a straightforward manner. For example, if an investigator examines the response of monkeys to relocation at three time points (prior to relocation, 2 months later, and 1 year later), measuring both hair and salivary cortisol concentrations, a comparison of the two matrices is complicated because they reflect different measuring units and different processes. Although both hair and saliva are thought to contain only the free or active form of the cortisol molecule, the measuring unit for hair is weight, whereas the measuring unit for saliva is volume. More importantly, salivary cortisol reflects HPA axis activation at a single point in time, whereas hair cortisol reflects an average value across the range of HPA activity over a period of several months. Although significant positive correlations between these two sampling matrices have occasionally been reported, there is no reason to expect *a priori* that they will be related. Thus, the finding of an effect in one sampling matrix, along with the finding of no effect in another matrix, is not necessarily an example of "contradictory findings." It is certainly possible that monkeys might show differences in chronic concentrations of hair cortisol that might not be apparent when salivary cortisol was measured, or vice versa.

Typically, cause and effect relationships cannot be discerned in the abnormalities described earlier, because the monkeys who have these conditions are identified only *after* the condition is well established. Any elevation in stress hormone levels in monkeys with a particular physiological or behavioral abnormality, as compared to monkeys without the abnormality in question, does not prove that stress caused the abnormality to develop. For example, the finding of higher hair cortisol concentrations in monkeys with hair loss could suggest that monkeys responded to some stress exposure by losing their hair. Alternatively, hair loss (caused by some other factor, such as prolonged antibiotic treatment) could subsequently result in an elevation in cortisol concentrations, perhaps because of alterations in gut microbial populations.

A significant challenge for behavioral management staff is to identify possible environmental stressors that may have led to the development of the various abnormalities described earlier. Such identification might guide therapeutic interventions. For example, removal of the stressor might reduce abnormalities in monkeys with existing conditions and/or might prevent the development of these conditions in other monkeys. A retrospective analysis of colony health/management records can be used to identify possible differences in stress exposure; however once again, these findings are only suggestive and cannot imply cause and effect. Alternatively, manipulations in which the environment is changed and the abnormal behavior is exacerbated provide much stronger evidence of an effect. For example, using a within-subjects design, both SIB and hair cortisol concentrations became elevated in monkeys moved to a new environment (Davenport et al. 2008).

SIB and Stress

Stress Exposure

Considerable evidence suggests that early life stress is a significant risk factor for the development of SIB in rhesus macaques. SIB has been associated with restrictive early rearing environments in which infant monkeys are reared with other infants (Lutz et al. 2007; Gottlieb et al. 2013), but SIB vulnerability is not limited to rhesus macaque infants reared in restrictive environments. Social separation in young, maternally reared monkeys can increase SIB vulnerability if it occurs during the juvenile period. Maternally reared monkeys that were individually housed at an early age (~14 months on average) were much more likely to develop SIB than were monkeys that remained with companions until 25 months of age on average (Lutz et al. 2003). SIB is also positively associated with the length of single-cage housing and the number of veterinary procedures (Lutz et al. 2003). These findings have been confirmed for pigtailed and rhesus macaques at several facilities, thus supporting the widespread generality of these results (Bellanca and Crockett 2002; Rommeck et al. 2009; Gottlieb et al. 2013).

HPA Axis Activity

Adverse early experiences have been linked to long-lasting changes in peripheral HPA axis and central (i.e., brain corticotropin-releasing factor) stress response systems in both monkeys (Coplan et al. 1996; Clarke et al. 1998) and humans (Davidson et al. 2004). Because some monkeys with early adverse experiences develop SIB, we examined HPA axis functioning in maternally reared, SIB-positive rhesus macaques that had been separated from their social group as juveniles. We predicted that monkeys with a history of self-inflicted wounding would show elevations in plasma cortisol compared to controls. However, contrary to predictions, SIB monkeys actually showed significantly lower cortisol concentrations compared to control monkeys to the mild stress of blood sampling (Tiefenbacher et al. 2000). Additionally, cortisol concentrations were negatively related to rates of self-directed biting.

To determine whether the stress response system was blunted, HPA axis activity was examined in response to an adrenocorticotrophic (ACTH) hormone challenge. If the system was downregulated, then ACTH should be less effective in eliciting a cortisol response. Indeed, SIB monkeys showed a reduced response to the ACTH challenge compared to controls, and this effect was strongest in the high frequency biters (Tiefenbacher et al. 2004). These findings suggest that SIB in some rhesus macaques is associated with complex and persistent changes in HPA activity that are related to both the outcome (i.e., wounding), and the expression (i.e., self-directed biting), of the behavioral pathology. These findings do not address the issue of cause and effect; namely, whether altered HPA axis activity increases vulnerability to SIB or whether monkeys develop SIB through some other mechanism, and altered HPA axis activity is the outcome.

Caveats

The aforementioned findings should not be taken to mean that all monkeys with SIB have a dysregulation of the HPA axis. In the population of monkeys that showed early life stress, both HPA axis activity and anxiety appeared to play some role. These monkeys showed a decrease in wounding in response to the anxiolytic drug, diazepam, and an increase in mild biting in response to the anxiogenic drug, FG7142 (Major et al. 2009).

However, not all monkeys with SIB appear to have significant early life stress. For these animals, treatment with diazepam actually made them *worse* (i.e., wounding increased). Furthermore, biting did not change in response to FG7142. These findings suggest that two subtypes of SIB may exist in macaques, one in which anxiety and HPA axis dysregulation are key features, and another which is presently uncharacterized. These findings are consistent with SIB in humans. Nonsuicidal self-injury (NSSI) in humans is a heterogeneous disorder, with at least two apparent subtypes; a reactive form, typically manifested as cutting or burning in response to increased anxiety; and a compulsive form of self-inflicted wounding, often taking the form of hair plucking, scratching, and nail biting (Favazza 1998; Black and Mildred 2013).

Environmental change has relatively little impact on SIB (Novak et al. 1998), with the exception of housing monkeys in outdoor environments (Fontenot et al. 2006). Consequently, the most successful treatments for SIB involve pharmacotherapy. Efficacy has been reported for the benzodiazepine, diazepam (Tiefenbacher et al. 2005); the selective serotonin reuptake inhibitor, fluoxetine (Fontenot et al. 2009); the α_2-adrenergic agonist, guanfacine (Freeman et al. 2015); and the opioid receptor antagonist, naltrexone (Kempf et al. 2012). However, as noted with diazepam earlier, efficacy varies across individuals and relapse potential has not been adequately evaluated. In most cases, monkeys either were not assessed when treatment was removed or were only followed for a brief period posttreatment (e.g., 4 weeks). Most significantly from the standpoints of behavioral management and scientific utility, pharmacotherapeutic treatment may interfere with research protocols.

Stereotypy and Stress

Stress Exposure

As in the case of SIB, stereotypic behavior is associated with adverse early experiences. Monkeys that are nursery-reared show higher levels of both motor and self-directed stereotypies than mother-reared macaques, an effect that persists into adolescence and adulthood (Cross and Harlow 1965; Feng et al. 2011). These findings are consistent across a wide range of settings. In a survey of zoo primates, risk for the development of stereotypies was increased in animals reared by hand, including those reared in a nursery (Marriner and Drickamer 1994). Even if monkeys are normally reared during the first year of life, the type of housing later in life is important, based on findings that singly housed adult monkeys have higher rates of stereotypic behavior than monkeys housed in pairs or in larger social groups (Bayne et al. 1992; Gottlieb et al. 2013).

The most comprehensive study of stereotypic behavior to date involved a retrospective analysis of the risk factors associated with these behaviors in 4000 macaques (Gottlieb et al. 2013). Stressful events, such as pair separation, frequency of room moves, and number of projects were reliable predictors of motor stereotypies. Additionally, Gottlieb et al. (2013) found that sex (being a male) and temperament (high reactivity) were associated with high rates of motor stereotypy.

HPA Axis Activity

Currently, there is only one study reporting a relationship between stereotypic behavior and cortisol concentrations. In this study, the focus was on a particular kind of stereotypy—pacing associated with head twirls. Pomerantz et al. (2012) showed that tufted capuchin monkeys that exhibited head twirling behavior had significantly higher fecal glucocorticoid metabolite concentrations than monkeys that did not perform this behavior.

Caveats

Unlike SIB, which is dichotomous (monkeys either do or do not have the condition), the presence of stereotypic behavior provides two additional challenges, both for the behavioral management team and for investigators who study abnormal behavior. First, there are many different kinds of stereotypic behavior, ranging from whole body motor movements (e.g., pacing and rocking) to self-directed behavior (e.g., digit sucking and eye covering). It is not clear whether all these patterns should be grouped together to create a single stereotypy score or whether they should be differentiated into separate categories. If the latter, what is the best approach to achieve this differentiation (e.g., motor versus self-directed)? Alternatively, are all the various kinds of stereotypies different, requiring their own evaluation? Second, whether one lumps or splits stereotypies into categories, there is a wide range of expression, from low frequencies in some monkeys to high frequencies in others. At what point does the rate of stereotypical behavior become pathological? In this regard, other measures may provide clarity. In the Pomerantz et al. (2012) study described earlier, head twirling behavior was significantly associated with negative cognitive bias, which suggests that the stereotypy may have been indicative of suboptimal welfare in this instance. To assess bias, monkeys were initially trained to associate two different reward sites with two stimuli that varied in size. The highly positive reward site was associated with a large rectangle, whereas the low reward site was associated with a small rectangle. During probe trials, monkeys were shown an ambiguous stimulus (a medium-sized rectangle) and their choice of site was recorded. Accordingly, monkeys selecting the high reward site showed positive bias, whereas monkeys selecting the low reward site showed negative bias. More generally, cognitive bias tasks have been recently used to assess welfare in nonhuman primates by measuring shifts in positive and negative affect (Bethel et al. 2012).

Hair Loss and Stress

Stress Exposure

The commonly held belief is that hair loss in healthy monkeys is a manifestation of stress exposure. In evaluating this premise, we examined the colony health management records of monkeys with and without hair loss at four national primate research centers, examining variables such as number of relocations, veterinary procedures, rearing history when known, and duration of time spent in individual cages. In contrast to our findings on SIB, no significant associations were discerned. Additionally, monkeys with hair loss behaved differently on the human intruder test than fully haired controls (Coleman et al. 2017). Alopecic monkeys showed *lower rates* of self-directed activity, freezing, and defensive reactions, consistent with less inhibited temperaments.

HPA Axis Activity

We subsequently compared hair cortisol concentrations in alopecic and fully haired monkeys at three different facilities. Although alopecic monkeys showed elevated hair cortisol concentrations compared to fully haired controls, this effect varied widely across facilities, yielding a major hair cortisol difference at one facility, a minor difference at a second facility, and a lack of any effect at the third facility (Novak et al. 2014b). The presence or absence of hair cortisol differences between alopecic and fully haired monkeys was unrelated to variation in chow diet, day/night cycles, or enrichment programs across the three facilities. However, it is interesting to note that the facility yielding the strongest effect also had a somewhat higher prevalence of the more severe kinds of alopecia, which can be seen in Figure 6.2. Because monkeys can regrow hair or lose hair over time, a second hair sample was collected 7 months after the original sample to test the prediction that monkeys losing hair would show increased hair cortisol concentrations, whereas monkeys that gained hair from the first to the second sample would show a reduction in hair cortisol concentrations. Over the 7-month sampling period, the majority of monkeys did not change their hair status (alopecic or fully haired). In particular, the first prediction could not be tested because hardly any of the previously fully haired animals became alopecic. On the other hand, we confirmed the second prediction, in that regrowth of hair was significantly associated with a reduction in hair cortisol concentrations (Novak et al. 2017). However, for reasons not yet known, this effect only occurred in the female monkeys.

SUMMARY AND CONCLUSIONS

Measurement of cortisol responses is well suited for the assessment of physiological stress in non-human primates. Cortisol or its metabolites can be assayed in a variety of different sample matrices; plasma or saliva are the most appropriate sample types for assessing cortisol responses to an acute stressor, whereas hair sampling works well for assessing long-term changes in cortisol secretion in response to a repeated or persistent stressor. Future studies should aim to establish reference ranges for hair cortisol concentrations in captive primates, both in zoo and research settings, and test the hypothesis that cortisol concentrations, either above or below the reference range, are predictive of increased vulnerability to the development of somatic or behavioral disorders. Stressful life experiences occurring either during early development or later in life have been linked to the onset and maintenance of behavioral abnormalities, such as SIB and stereotypies, in at least some animals. Dysregulation of the HPA axis has similarly been associated with SIB, but it is not yet clear whether this is also the case for stereotypies. Indeed, exploration of the causes and consequences of stereotyped behaviors in primates has been hampered by the diversity of such behaviors and the recognition that low levels of stereotypy are not necessarily indicative of reduced well-being. Alopecia is yet another condition that varies considerably in severity and across housing facilities, with increased hair cortisol being associated with more severe degrees of hair loss. Taken together, these findings confirm the long-held notion that major life stresses can contribute to both behavioral pathology and HPA axis dysregulation. However, the association between these outcomes is not a simple one.

Behavioral management units are charged with the responsibilities of assessing and promoting well-being in captive nonhuman primates. From the assessment perspective, behavioral management staff must make complicated decisions about what constitutes behavior that is sufficiently problematic to require intervention. We recommend that a severity index be used to guide these decisions. Thus, the focus should be on: (1) dangerous behavior (e.g., SIB) that can produce tissue damage, (2) stereotypic behavior that reaches a threshold such that it disrupts species-typical behavior patterns, training, and/or basic husbandry procedures, and (3) abnormal levels of aggressive behavior that can severely compromise social housing attempts. With respect to hair loss, the emphasis should be on those monkeys that show substantial (>50%) hair loss that, if left untreated, does not improve over a 6-month period of time.

Physiological assessment of the stress response system (e.g., measurements of cortisol) is yet another tool that can assist in determining the severity of a condition and the possible need for treatment. We recommend the use of hair as a sampling matrix for the following reasons: (1) hair cortisol reflects long-term changes in HPA axis function that may be associated with housing conditions, experimental treatments, and basic husbandry procedures, (2) cortisol concentrations in hair are unaffected by the sampling procedure (i.e., restraint and anesthetization) or by circadian rhythms, (3) hair samples can be obtained routinely during regular health exams, and (4) some information is available to guide the interpretation of what constitutes a high-risk range for cortisol (Figure 6.1). With respect to this last point, monkeys with extreme hair loss can show chronically elevated concentrations of hair cortisol. If these concentrations are more than 2 standard deviations away from the mean, then the animals should be resampled and examined for health-related conditions that could cause hypo- or hypercortisolemia.

The second mandate of behavioral management units, namely promoting well-being, is a two-stage effort. For monkeys with behavioral and physical disorders, the first objective is to identify the possible environmental factors that have contributed to the disorder in question. Then, based on that information, the second objective is to implement relevant treatments. Identifying possible environmental stressors is easier said than done. However, some well-established risk factors for rhesus macaques, identified previously, that contribute to the development and exacerbation of SIB and stereotypic behavior include: (1) nursery rearing from birth, (2) social separation from peers or group members and placement into single housing within the first 2 years of life, and (3) relocation of adults to new rooms or new facilities, which tends to increase biting in monkeys with a preexisting SIB condition and also increases stereotypic behavior. Whenever possible, these potential stressors should be avoided. Prevention is clearly a better strategy than trying to cope with pathological behavior. If these conditions cannot be avoided, then plans should be put in place to monitor the at-risk animals carefully. It is also the case that monkeys may show abnormal behavior and have none of the predisposing conditions described earlier. Increased behavioral monitoring may help identify possible stressors (e.g., a particular monkey nearby or inappropriate approach behavior by care staff). It is also useful to examine colony records for past exposure to stressors. Several scientific articles in which associations between colony health record variables and abnormal behavior were assessed are available to assist behavioral management units in this process (Bellanca and Crockett 2002; Lutz et al. 2003; Rommeck et al. 2009; Gottlieb et al. 2013).

The last challenge is to identify effective treatments that can fit within the realm of biomedical research. In some cases, removal of a stressor might suffice (e.g., changing care staff behavior or moving monkeys from single to pair housing). However, severely pathological behavior, such as SIB, often requires pharmacotherapy, which can compromise the research enterprise. A number of options are available to treat SIB, but the outcomes may depend on characteristics of the individual. Diazepam may be employed in emergency situations requiring rapid intervention to prevent serious wounding; however, we advise against chronic treatment with this compound because of inconsistent efficacy in different monkeys (Tiefenbacher et al. 2005), and because of its sedating side effects and the development of tolerance. Naltrexone, guanfacine, or fluoxetine may be more desirable, but these agents will all require gradual dose reduction and eventual treatment cessation after a period of time, and questions about subsequent relapse remain.

With the maturation of behavioral management units and an increased focus on behavioral indices of well-being, great strides have been made in improving the welfare of captive nonhuman primates. Additionally, as new tools have come on line (e.g., assessment of cortisol in hair as a biomarker of chronic HPA activity), the ability, both to identify problematic animals and to treat them, has become much more sophisticated.

ACKNOWLEDGMENTS

The writing of this chapter was supported by NIH Grant OD011180 to the first author.

REFERENCES

Anestis, S. F. and R. G. Bribiescas. 2004. Rapid changes in chimpanzee (*Pan troglodytes*) urinary cortisol excretion. *Horm. Behav.* 45:209–213.

Animal Welfare; Standards: Final Rule. 1991. United States Department of Agriculture, Animal and Plant Health Inspection Service. *Federal Register*. Vol. 55, 32, Part 3, pp. 6426–6505.

Barbosa, M. N. and M. T. Mota. 2009. Behavioral and hormonal response of common marmosets, *Callithrix jacchus*, to two environmental conditions. *Primates* 50:253–260.

Bayne, K., S. Dexter, and S. Suomi. 1992. A preliminary survey of the incidence of abnormal behavior in rhesus monkeys (*Macaca mulatta*) relative to housing condition. *Lab Anim.* 21:38–46.

Behringer, V., J. M. Stevens, G. Hohmann, et al. 2014. Testing the effect of medical positive reinforcement training on salivary cortisol levels in bonobos and orangutans. *PLoS One* 9:e108664. doi:10.1371/journal.pone.0108664.

Bellanca, R. and C. Crockett. 2002. Factors predicting increased incidence of abnormal behavior in male pigtailed macaques. *Am. J. Primatol.* 58:57–69.

Bethel, E. J., A. Holmes, A. MacLarnon, et al. 2012. Cognitive bias in a non-human primate: Husbandry procedures influence cognitive indicators of psychological well-being in captive rhesus macaques. *Anim. Welfare* 21:185–195.

Black, E. B. and H. Mildred. 2013. Predicting impulsive self-injurious behavior in a sample of adult women. *J. Nerv. Ment. Dis.* 201:72–75.

Bloomsmith, M., K. Neu, A. Franklin, et al. 2015. Positive reinforcement methods to train chimpanzees to cooperate with urine collection. *J. Am. Assoc. Lab Anim. Sci.* 54:66–69.

Boinski, S., S. P. Swing, T. S. Gross, et al. 1999. Environmental enrichment of brown capuchins (*Cebus apella*): Behavioral and plasma and fecal cortisol measure of effectiveness. *Am. J. Primatol.* 48:49–68.

Bowman, R. E. and R. F. De Luna. 1968. Assessment of a protein-binding method for cortisol determination. *Anal. Biochem.* 26:465–469.

Boyce, W. T., M. Champoux, S. J. Suomi, et al. 1995. Salivary cortisol in nursery-reared rhesus monkeys: Reactivity to peer interactions and altered circadian activity. *Dev. Psychobiol.* 28:257–267.

Braig, S., F. Grabher, C. Ntomchukwu, et al. 2016. The association of hair cortisol with self-reported chronic psychosocial stress and symptoms of anxiety and depression in women shortly after delivery. *Paediatr. Perinat. Epidemiol.* 30(2):97–104. doi:10.1111/ppe.12255.

Camus, S. M. J., C. Blois-Heulin, Q. Li, et al. 2013. Behavioural profiles in captive-bred cynomolgus macaques: Towards monkey models of mental disorders? *PLoS One* 8:1–14.

Capitanio, J. P., S. P. Mendoza, W. A. Mason, et al. 2005. Rearing environment and hypothalamic-pituitary-adrenal regulation in young rhesus monkeys (*Macaca mulatta*). *Dev. Psychobiol.* 46:318–330.

Carlitz, E. H., C. Kirschbaum, T. Stalder, et al. 2014. Hair as a long-term retrospective cortisol calendar in orang-utans (*Pongo* spp.): New perspectives for stress monitoring in captive management and conservation. *Gen. Comp. Endocrinol.* 195:151–156.

Clarke, A. S., G. W. Kraemer, and D. J. Kupfer. 1998. Effects of rearing condition on HPA axis response to fluoxetine and desipramine treatment over repeated social separations in young rhesus monkeys. *Psychiatry Res.* 79:91–104.

Coleman, K., C. K. Lutz, J. M. Worlein, et al. 2017. The correlation between alopecia and temperament in rhesus macaques (*Macaca mulatta*) at four primate facilities. *Am. J. Primatol.* 79(1):1–10. doi:10.1002/ajp.22504.

Coleman, K., L. Pranger, A. Maier, et al. 2008. Training rhesus macaques for venipuncture using positive reinforcement techniques: A comparison with chimpanzees. *J. Am. Assoc. Lab. Anim. Sci.* 47:37–41.

Coplan, J. D., M. W. Andrews, L. A. Rosenblum, et al. 1996. Persistent elevations of cerebrospinal fluid concentrations of corticotropin-releasing factor in adult nonhuman primates exposed to early life stressors: Implications for the pathophysiology of mood and anxiety disorders. *Proc. Natl. Acad. Sci. USA* 93:1619–1623.

Crockett, C. M., C. L. Bowers, G. P. Sackett, et al. 1993. Urinary cortisol responses of longtailed macaques to five cage sizes, tethering, sedation, and room change. *Am. J. Primatol.* 30:55–74.

Crockett, C. M., M. Shimoji, and D. M. Bowden. 2000. Behavior, appetite, and urinary cortisol responses by adult female pigtailed macaques to cage size, cage level, room change, and ketamine sedation. *Am. J. Primatol.* 52:63–80.

Cross, H. A. and H. F. Harlow. 1965. Prolonged and progressive effects of partial isolation on the behavior of macaque monkeys. *J. Exp. Res. Pers.* 1:39–49.

Davenport, M. D., C. K. Lutz, S. Tiefenbacher, et al. 2008. A rhesus monkey model of self injury: Effects of relocation stress on behavior and neuroendocrine function. *Biol. Psychiatry* 68:990–996.

Davenport, M. D., M. A. Novak, J. S. Meyer, et al. 2003. Continuity and change in emotional reactivity in rhesus monkeys throughout the prepubertal period. *Motiv. Emot.* 27:57–76.

Davenport, M. D., S. T. Tiefenbacher, C. K. Lutz, et al. 2006. Analysis of endogenous cortisol concentrations in the hair of rhesus macaques. *Gen. Comp. Endocrinol.* 147:255–261.

Davidson, J. R. T., D. J. Stein, A. Y. Shalev, et al. 2004. Posttraumatic stress disorder: Acquisition, recognition, course, and treatment. *J. Neuropsychiatry Clin. Neurosci.* 16:135–147.

Dettmer, A. M., M. A. Novak, J. S. Meyer, et al. 2014. Population density dependent hair cortisol concentrations in rhesus monkeys (*Macaca mulatta*). *Psychoneuroendocrinology* 42:59–67.

Dettmer, A. M., M. A. Novak, S. J. Suomi, et al. 2012. Physiological and behavioral adaptation to relocation stress in differentially reared rhesus monkeys: Hair cortisol as a biomarker for anxiety-related responses. *Psychoneuroendocrinology* 37:191–199.

Downs, J. L., J. A. Mattison, D. K. Ingram, et al. 2007. Effect of age and caloric restriction on circadian adrenal steroid rhythms in rhesus macaques. *Neurobiol. Aging* 29:1412–1422.

Doyle, L. A., K. C. Baker, and L. D. Cox. 2008. Physiological and behavioral effects of social introduction on adult male rhesus macaques. *Am. J. Primatol.* 70:542–550.

Etwel, F., E. Russell, M. J. Rieder, et al. 2014. Hair cortisol as a biomarker of stress in the 2011 Libyan war. *Clin. Invest. Med.* 37:403–408.

Favazza, A. R. 1998. The coming of age of self-mutilation. *J. Nerv. Ment. Dis.* 186:259–268.

Feng, X., L. Wang, S. Yang, et al. 2011. Maternal separation produces lasting changes in cortisol and behavior in rhesus monkeys. *Proc. Natl. Acad. Sci. USA* 108:14312–14317.

Flow, B. L. and J. T. Jaques. 1997. Effect of room arrangement and blood sample collection sequence on serum thyroid hormone and cortisol concentrations in cynomolgus macaques (*Macaca fascicularis*). *Cont. Topics Lab. Anim. Sci.* 36:65–68.

Fontenot, M. B., M. W. Musso, R. M. McFatter, et al. 2009. Dose-finding study of fluoxetine and venlafaxine for the treatment of self-injurious and stereotypic behavior in rhesus macaques (*Macaca mulatta*). *J. Am. Assoc. Lab Anim. Sci.* 48:176–184.

Fontenot, M. B., M. N. Wilkes, and C. S. Lynch. 2006. Effects of outdoor housing on self-injurious and stereotypic behavior in adult male rhesus macaques (*Macaca mulatta*). *J. Am. Assoc. Lab Anim. Sci.* 45:35–43.

Freeman, Z. T., K. A. Rice, P. L. Soto, et al. 2015. Neurocognitive dysfunction and pharmacological intervention using guanfacine in a rhesus macaque model of self-injurious behavior. *Transl. Psychiatry* 5:e567. doi:10.1038/tp.2015.61.

Fürtbauer, I., M. Heistermann, O. Schülke, et al. 2014. Low female stress hormone levels are predicted by same- or opposite-sex sociality depending on season in wild Assamese macaques. *Psychoneuroendocrinology* 48:19–28.

Gao, W., P. Zhong, Q. Xie, et al. 2014. Temporal features of elevated hair cortisol among earthquake survivors. *Psychophysiology* 51:319–26.

Gottlieb, D. H., J. P. Capitanio, and B. McCowan. 2013. Risk factors for stereotypic behavior and self-biting in rhesus macaques (*Macaca mulatta*): Animal's history, current environment, and personality. *Am. J. Primatol.* 75:995–1008.

Hamel, A. F., C. K. Lutz, K. Coleman, et al. 2017. Response to the Human Intruder Test is related to hair cortisol phenotype and sex in rhesus macaques (*Macaca mulatta*). *Am. J. Primatol.* 79(1):1–10.

Hansen, D. J., A. C. Tishelman, R. P. Hawkins, et al. 1990. Habits with potential as disorders. Prevalence, severity, and other characteristics among college students. *Behav. Modif.* 14:66–80.

Heintz, M. R., R. M. Santymire, L. A. Parr, et al. 2011. Validation of a cortisol enzyme immunoassay and characterization of salivary cortisol circadian rhythm in chimpanzees (*Pan troglodytes*). *Am. J. Primatol.* 73:903–908.

Heistermann, M., R. Palme, and A. Ganswindt. 2006. Comparison of different enzyme immunoassays for assessment of adrenocortical activity in primates based on fecal analysis. *Am. J. Primatol.* 68:257–273.

Higham, J. P., A. Vitale, A. M. Rivera, et al. 2010. Measuring salivary analytes from free-ranging monkeys. *Physiol. Behav.* 101:601–607.

Izawa, S., K. Miki, M. Tsuchiya, et al. 2015. Cortisol measurements in fingernails as a retrospective index of hormone production. *Psychoneuroendocrinology* 54:24–30.

Keay, J. M., J. Singh, M. C. Gaunt, et al. 2006. Fecal glucocorticoids and their metabolites as indicators of stress in various mammalian species: A literature review. *J. Zoo Wildl. Med.* 37:234–244.

Keller-Wood, M. E. and M. F. Dallman. 1984. Corticosteroid inhibition of ACTH secretion. *Endocr Rev.* 5:1–24.

Kempf, D. J., K. C. Baker, M. H. Gilbert, et al. 2012. Effects of extended-release injectable naltrexone on self-injurious behavior in rhesus macaques (*Macaca mulatta*). *Comp. Med.* 62:209–217.

Kimura, T. 2008. Systemic alopecia resulting from hyperadrenocorticism in a Japanese monkey. *Lab. Primate Newsl.* 47:5–9.

Kirschbaum, C., A. Tietze, N. Skoluda, et al. 2009. Hair as a retrospective calendar of cortisol production: Increased cortisol incorporation into hair in the third trimester of pregnancy. *Psychoneuroendocrinology* 34:32–37.

Kroeker, R., R. U. Bellanca, G. H. Lee, et al. 2014. Alopecia in three macaque species housed in a laboratory environment. *Am. J. Primatol.* 76:325–334.

Lado-Abeal, J., J. J. Robert-McComb, X. P. Qian, et al. 2005. Sex differences in the neuroendocrine response to short-term fasting in rhesus macaques. *J. Neuroendocrinol.* 17:435–44.

Lambeth, S. P., J. E. Perlman, E. Thiele, et al. 2005. Changes in hematology and blood chemistry parameters in captive chimpanzees (*Pan troglodytes*) as a function of blood sampling technique: Trained vs. anesthetized samples. *Am. J. Primatol.* 68:245–256.

Lane, J. 2006. Can non-invasive glucocorticoid measures be used as reliable indicators of stress in animals? *Anim. Welfare* 15:331–342.

Laudenslager, M. L., M. J. Jorgensen, R. Grzywa, et al. 2011. A novelty seeking phenotype is related to chronic hypothalamic-pituitary-adrenal activity reflected by hair cortisol. *Physiol. Behav.* 104:291–295.

Le Roux, C. W., G. A. Chapman, W. M. Kong, et al. 2003. Free cortisol index is better than serum total cortisol in determining hypothalamic-pituitary-adrenal status in patients undergoing surgery. *J. Clin. Endocrinol. Metab.* 88:2045–2048.

Leverenz, J. B., C. W. Wilkinson, M. Wamble, et al. 1999. Effect of chronic high dose exogenous cortisol on hippocampal neuronal number in aged nonhuman primates. *J. Neurosci.* 19:2356–2361.

Lutz, C. K. 2014. Stereotypic behavior in nonhuman primates as a model for the human condition. *ILAR J.* 55:284–296.

Lutz, C. K., K. Coleman, J. Worlein, et al. 2013. Hair loss and hair pulling in rhesus monkeys (*Macaca mulatta*). *J. Am. Assoc. Lab. Anim. Sci.* 52:454–457.

Lutz, C. K., E. B. Davis, A. M. Ruggiero, et al. 2007. Early predictors of self-biting in socially-housed rhesus macaques (*Macaca mulatta*). *Am. J. Primatol.* 69:584–590.

Lutz, C., S. Tiefenbacher, M. Jorgensen, et al. 2000. Techniques for collecting saliva from awake, unrestrained, adult macaque monkeys for cortisol assay. *Am. J. Primatol.* 52:93–99.

Lutz, C., A. Well, and M. Novak. 2003. Stereotypic and self-injurious behavior in rhesus macaques: A survey and retrospective analysis of environment and early experience. *Am. J. Primatol.* 60:1–15.

Major, C. A., B. J. Kelly, M. A. Novak, et al. 2009. The anxiogenic drug FG7142 increases self-injurious behavior in male rhesus monkeys (*Macaca mulatta*). *Life Sci.* 85:753–758.

Marriner, L. M. and L. C. Drickamer. 1994. Factors influencing stereotyped behavior of primates in a zoo. *Zoo Biol.* 13:267–275.

McEwen, B. S. 1998. Stress, adaptation, and disease: Allostasis and allostatic load. *Ann. N. Y. Acad. Sci.* 840:33–44.

McEwen, B. S. and T. Seeman. 1999. Protective and damaging effects of mediators of stress: Elaborating and testing the concepts of allostasis and allostatic load. *Ann. N. Y. Acad. Sci.* 896:30–47.

Mendonça-Furtado, O., M. Edaes, R. Palme, et al. 2014. Does hierarchy stability influence testosterone and cortisol levels of bearded capuchin monkeys (*Sapajus libidinosus*) adult males? A comparison between two wild groups. *Behav. Processes* 109(Pt A):79–88.

Meyer, J. S. and M. A. Novak. 2012. Mini-review: Hair cortisol: A novel biomarker of hypothalamic-pituitary-adrenocortical activity. *Endocrinology* 153:4120–4127.

Millspaugh, J. J. and B. E. Washburn. 2004. Use of fecal glucocorticoid metabolite measures in conservation biology research: Considerations for application and interpretation. *Gen. Comp. Endocrinol.* 138:189–199.

Novak, M. A. 2003. Self-injurious behavior in rhesus monkeys: New insights on etiology, physiology, and treatment. *Am. J. Primatol.* 59:3–19.

Novak, M. A., S. N. El-Mallah, and M. T. Menard. 2014a. Use of the cross-translational model to study self-injurious behavior in human and non-human primates. *ILAR J.* 55:274–283.

Novak, M. A., A. F. Hamel, K. Coleman, et al. 2014b. Hair loss and hypothalamic-pituitary-adrenal axis activity. *J. Am. Assoc. Lab. Anim. Sci.* 53:261–266.

Novak, M. A., A. F. Hamel, B. J. Kelly, et al. 2013. Stress, the HPA axis, and nonhuman primate well-being: A review. *Appl. Anim. Behav. Sci.*143:135–149.

Novak, M. A., B. J. Kelly, K. Bayne, et al. 2012. Behavioral pathologies. In *Nonhuman Primates in Biomedical Research* (Abee, C. R., K. Mansfield, S. D. Tardif, et al., Eds). Elsevier, New York, 177–196.

Novak, M. A., J. H. Kinsey, M. J. Jorgensen, et al. 1998. The effects of puzzle feeders on pathological behavior in individually housed rhesus monkeys. *Am. J. Primatol.* 46:213–227.

Novak, M. A., M. T. Menard, S. N. El-Mallah, et al. 2017. Assessing significant (>30%) alopecia as a possible biomarker for stress in captive rhesus monkeys (*Macaca mulatta*). *Am. J. Primatol.* 79(1):1–8.

Novak, M. A. and J. S. Meyer. 2009. Alopecia: Possible causes and treatments with an emphasis on captive nonhuman primates. *Comp. Med.* 59:18–26.

Novak, M. A. and S. J. Suomi. 1988. Psychological well-being of primates in captivity. *Am. Psychol.* 43:765–773.

O'Connor, T. M., D. J. O'Halloran, and F. Shanahan. 2000. The stress response and the hypothalamic-pituitary-adrenal axis: From molecule to melancholia. *Q. J. Med.* 93:323–333.

Pearson, B. L., D. M. Reeder, and P. G. Judge. 2015. Crowding increases salivary cortisol but not self-directed behavior in captive baboons. *Am. J. Primatol.* 77:462–467.

Pomerantz, O., J. Terkel, S. J. Suomi, et al. 2012. Stereotypic head twirls, but not pacing, are related to a "pessimistic"-like judgment bias among captive tufted capuchins (*Cebus apella*). *Anim. Cogn.* 15:689–698.

Rafaeli-Mor, N., F. Foster, and G. Berkson. 1999. Self-reported body-rocking and other habits in college students. *Am. J. Ment. Retard*: 104:1–10.

Romano, M. C., A. Z. Rodas, R. A. Valdez, et al. 2010. Stress in wildlife species: Noninvasive monitoring of glucocorticoids. *Neuroimmunomodulation* 17:209–212.

Rommeck, I., K. Anderson, A. Heagerty, et al. 2009. Risk factors and remediation of self-injurious and self-abuse behavior in rhesus macaques. *J. Appl. Anim. Welfare Sci.* 12:61–72.

Sanchez, M. M., P. M. Noble, C. K. Lyon, et al. 2005. Alterations in diurnal cortisol rhythm and acoustic startle response in nonhuman primates with adverse rearing. *Biol. Psychiatry* 57:373–381.

Sauvé, B., G. Koren, G. Walsh, et al. 2007. Measurement of cortisol in human hair as a biomarker of systemic exposure. *Clin. Invest. Med.* 30:E183–E191.

Schapiro, S. J., M. A. Bloomsmith, A. L. Kessel, et al. 1993. Effects of enrichment and housing on cortisol response in juvenile rhesus monkeys. *Appl. Anim. Behav. Sci.* 37:251–263.

Selye, H. 1975. *Stress without Distress*. New American Library, New York.

Setchell, K. D. R., K. S. Chua, and R. L. Himsworth. 1977. Urinary steroid excretion by the squirrel monkey (*Saimuri sciureus*). *J. Endocrinol.* 73:365–375.

Shigeyama, C., T. Ansai, S. Awano, et al. 2008. Salivary levels of cortisol and chromogranin A in patients with dry mouth compared with age-matched controls. *Oral Surg. Oral Med. Oral Pathol. Oral Radiol. Endod.* 6:833–839.

Short, S. J., G. R. Lubach, E. A. Shirtcliff, et al. 2014. Population variation in neuroendocrine activity is associated with behavioral inhibition and hemispheric brain structure in young rhesus monkeys. *Psychoneuroendocrinology* 47:56–67.

Soussignan, R. and P. Koch. 1985. Rhythmical stereotypies (leg-swinging) associated with reductions in heart-rate in normal school children. *Biol. Psychol.* 21:161–167.

Steinmetz, H. W., W. Kaumanns, I. Dix, et al. 2006. Coat condition, housing condition and measurement of faecal cortisol metabolites—A non-invasive study about alopecia in captive rhesus macaques (*Macaca mulatta*). *J. Med. Primatol.* 35:3–11.

Tiefenbacher, S., M. Fahey, J. K. Rowlett, et al. 2005. The efficacy of diazepam treatment for the management of acute wounding episodes in captive rhesus macaques. *Comp. Med.* 55:387–392.

Tiefenbacher, S., B. Lee, J. S. Meyer, et al. 2003. Noninvasive technique for the repeated sampling of salivary free cortisol in awake, unrestrained squirrel monkeys. *Am. J. Primatol.* 60:69–75.

Tiefenbacher, S., M. Novak, M. Jorgensen, et al. 2000. Physiological correlates of self-injurious behavior in captive, socially-reared rhesus monkeys. *Psychoneuroendocrinology* 25:799–817.

Tiefenbacher, S., M. A. Novak, L. M. Marinus, et al. 2004. Altered hypothalamic-pituitary-adrenocortical function in rhesus monkeys (*Macaca mulatta*) with self-injurious behavior. *Psychoneuroendocrinology* 29:500–514.

Urbanski, H. F. and K. G. Sorwell. 2012. Age-related changes in neuroendocrine rhythmic function in the rhesus macaque. *Age* 34:1111–1121.

Veronesi, M. C., A. Comin, T. Meloni, et al. 2015. Coat and claws as new matrices for non-invasive long-term cortisol assessment in dogs from birth up to 30 days of age. *Theriogenology* 15:791–796.

Wilkinson, A. C., L. D. Harris, G. A. Saviolakis, et al. 1999. Cushing's syndrome with concurrent diabetes mellitus in a rhesus monkey. *J Am. Assoc. Lab. Anim. Sci.* 38:62–66.

Woods, D. W. and R. G. Miltenberger. 1996. Are persons with nervous habit nervous? A preliminary examination of habit function in a nonreferred population. *J. Appl. Behav. Anal.* 29:259–61.

Ziegler, T. E., F. H. Wegner, and C. T. Snowdon. 1996. Hormonal responses to parental and nonparental conditions in male cotton-top tamarins, *Saguinus oedipus*, a New World primate. *Horm. Behav.* 30:287–297.

CHAPTER 7

Individual Differences in Temperament and Behavioral Management

Kristine Coleman
Oregon National Primate Research Center

CONTENTS

INTRODUCTION

As anyone working with captive nonhuman primates is keenly aware, individual primates can differ vastly with respect to their behavioral responses to stressful or novel stimuli. Walk into a room of unknown rhesus macaques, and you will undoubtedly be greeted with an array of responses, from threats to fear grimaces to seeming indifference. There are many reasons for these disparate behavioral responses, including past experience, current emotional state, and the stimulus itself. However, one of the major forces underlying these different reactions is biological predisposition or temperament. Once considered "noise" around an adaptive mean (Francis 1990), these individual differences in temperament are now generally accepted as interesting and important in their own right (Clark and Ehlinger 1987).

My own interest in temperament began as an undergraduate, when I participated in a study examining resource distribution in hermit crabs (*Pagurus longicarpus*). The crabs lived in individual plastic cups in a large seawater aquarium. My job on this project was to clean debris from the cups using a rigid suction catheter, similar to the kind one sees at the dentist. While perhaps not all that glamorous, this job allowed me to get to know the animals. I noticed that the crabs had different, but predictable, responses to the cleaning procedure. Some crabs retreated into their shells every time I put the catheter into their cup, while others consistently attacked it. Still others seemed undisturbed as I cleaned their enclosure, neither attacking nor retreating from the cleaning device. While I was fascinated by the consistency in their responses to the potential threat of my suction catheter, I failed to appreciate the nature of these individual differences.

Somewhat coincidentally, my doctoral research focused on these same sorts of behavioral differences. I studied individual differences in shyness and boldness in a population of pumpkinseed sunfish (*Lepomis gibbosus*). In a series of studies (Wilson et al. 1993; Coleman and Wilson 1998), we found that fish differed in their propensity to inspect novel objects and that these differences correlated with a host of other traits, including acclimation to laboratory conditions, choice of prey, and microhabitat usage. These studies were among the first to examine individual differences in temperament in a nonprimate species.

Since these early studies in animal temperament, similar differences have been found in a variety of diverse taxa, from octopus to fish to birds and reptiles. Strikingly, in every species in which differences in temperament have been investigated, they have been found, indicating the conserved nature of this trait. The impact of animal temperament can be seen in the wide range of academic disciplines in which it is now studied. While once studied predominantly by psychologists, temperament is now examined by researchers in disparate fields, including neuroscience (e.g., Roseboom et al. 2014; Fox et al. 2015), evolutionary ecology (e.g., Reale et al. 2007), and conservation biology (e.g., McDougall et al. 2006). Despite this increased interest in temperament, one field in which it has not gained broad acceptance is laboratory animal care and behavioral management. Indeed, the typical nonhuman primate housing room is filled with cages containing a more or less homogenous array of toys and enrichment devices, even though it is well known that an individual's behavioral needs can differ due to a variety of factors, including temperament. Therefore, it stands to reason that knowledge about individual differences in temperament should help guide decisions about how we manage the care of captive animals. In this chapter, I will discuss how temperament can be used to enhance the way we manage the behavior of captive primates.

WHAT IS TEMPERAMENT?

Generally speaking, temperament describes an individual's nature. It has been defined as the "stable behavioral and emotional reactions that appear early in life and influenced in part by genetic constitution" (Kagan 1994, p. 40). Today, the term "temperament" is often used interchangeably with "personality" (Capitanio 2011), although this has not always been the case. Historically, distinctions were made between the terms, with "temperament" being used to describe behavioral differences in animals and children and "personality" restricted to human adults (Watters and Powell 2012). Further, some researchers have argued that temperament reflects genetic behavioral differences, while personality reflects nongenetic differences (e.g., Buss and Plomin 1986), and others maintained that temperament reflects an individual's response to novelty in a nonsocial context, while personality refers to specific social characteristics (Clarke et al. 1995). Little distinction is made between the terms today. Both terms refer to an individual's basic position toward environmental change and challenge (Lyons et al. 1988), which emerges early in life and remains relatively consistent throughout development (McCall 1986). Further, there is general agreement that both have a genetic component (Bouchard and Loehlin 2001). For consistency, I will use the term temperament in this chapter (unless describing a study in which the term "personality" is used).

One reason for the current increased interest in temperament and personality is their role in various behavioral and/or health outcomes in humans and other species (Miller et al. 1999; Roberts et al.

2007; Deary et al. 2010; Capitanio 2011). As an example, behavioral inhibition, a temperamental construct defined as withdrawal from or timidity toward the unfamiliar (Kagan et al. 1988), has been linked to a vulnerability to stress-induced behavioral problems in human populations. Children who are behaviorally inhibited are more likely than others to suffer allergic disorders (Kagan et al. 1991) and respiratory illnesses (Boyce et al. 1995) and are at a greater risk for developing anxiety disorders and other psychopathologies later in life (Hirshfeld et al. 1992; Rosenbaum et al. 1993; Schwartz et al. 1999). Behaviorally inhibited rhesus macaque infants show greater hypothalamic–pituitary–adrenal (HPA) axis activation and behavioral responses to stresses, such as separation from peers, compared with others (Suomi 1991). They are also more likely than noninhibited individuals to develop airway hyperresponsiveness, a characteristic of asthma (Chun et al. 2013). Socially inhibited rhesus monkeys (i.e., those who scored low on the personality trait of "sociability") have lower antibody response to immunization and social relocation compared with highly sociable monkeys (Capitanio et al. 1999; Maninger et al. 2003). Given its strong influence on behavioral and health-related outcomes, it follows that temperament might also influence psychological well-being and welfare.

HOW IS TEMPERAMENT ASSESSED?

There are many methods by which temperament is assessed in both humans and nonhuman animals (See Freeman and Gosling 2010 for a comprehensive review). Most of these involve observing subjects either in a home environment or in a provoked situation (i.e., providing them with a stimulus designed to elicit a response). Despite the disparate methodologies used to assess temperament, the underlying dimensions are usually relatively similar (e.g., Konecna et al. 2008; Bergvall et al. 2011), and characterize how individuals deal with various challenges. I describe some commonly used methods in the following sections.

Home Environment Assessments

One way in which temperament can be evaluated (e.g., Capitanio 1999) is by observing subjects in their home environment to assess responses to everyday, naturalistic events (e.g., interactions with conspecifics, introduction to new food or caretakers). Individuals typically respond to these sorts of events with a range of behaviors rooted, in large part, in their temperament. For example, behaviorally inhibited individuals might respond to naturalistic challenges with heightened fear responses, increased vigilance and/or displacement behaviors. This kind of assessment is often done with children, either at home or in the school setting.

Assessing primate temperament or personality in the home environment often involves observer ratings (e.g., Capitanio 1999). Rating instruments typically involve two or more observers who score subjects based on a number of predefined traits or adjectives, such as "apprehensive," "active," "playful," and "curious" (e.g., Stevenson-Hinde and Zunz 1978). Scores are then analyzed via factor analysis in order to uncover various dimensions of behavior. Common factors that emerge from these studies include "sociability," "confidence," "fearfulness," "curiosity," and "excitability" (Freeman and Gosling 2010; Capitanio 2011). See Freeman et al. (2013) for a comparison of various approaches to these rating instruments.

Response to Challenge

Temperament in nonhuman primates is commonly assessed by evaluating the subject's response to some sort of purposeful environmental challenge or potentially threatening stimuli. These stimuli typically involve a degree of novelty, such as a new situation, object, or conspecific (e.g., Williamson et al. 2003).

One of the most widely used tests to measure temperament in macaques, specifically the construct of behavioral inhibition is the human intruder test (HIT; Kalin and Shelton 1989). It was designed to measure an individual's response to the potentially threatening social stimulus of an unfamiliar human intruder. The HIT was originally developed to assess behavior in infant macaques, although it has been used for animals of all age groups (e.g., Coleman et al. 2011; Corcoran et al. 2012). In the original studies, the subject was brought to a cage in a novel room where it remained alone for a period of time (e.g., 10 min). The infant was then exposed to a human intruder, with whom the subject had no prior experience. The intruder first stood by the subject's cage with his or her profile to the infant, taking care to avoid eye contact for 10 min. This stimulus was designed to represent a potential social threat (i.e., the "threat" was present, but had not yet noticed the subject). The intruder then turned his or her head and made direct eye contact, a threatening situation, with the subject for 10 min, after which the test ended and the infant was returned to its dam. While there have been various iterations of this test, they all have similar components (e.g., an unfamiliar human who makes direct eye contact with the subject). Subjects display a wide range of behavioral responses to this test. Generally, individuals who show excessive freezing behavior (e.g., a behavior in which the subject remains completely motionless, except for slight movements of the eyes; Kalin and Shelton 1989) when the intruder is not making direct eye contact and/or those showing excessive anxious behavior (e.g., scratching, distress behaviors) in the presence of the intruder are considered more "inhibited" than others (see Coleman and Pierre 2014 for review).

Another relatively common method for measuring temperament in primates and other species involves assessing response to novel objects, such as toys or food. These "novel object tests" may be performed in conjunction with the human intruder test (e.g., Williamson et al. 2003). Subjects exhibit a spectrum of responses to these novel stimuli. Bolder, more exploratory individuals quickly inspect the objects, while more inhibited individuals avoid the objects and may show distress behavior (see Figure 7.1).

Temperament Assessment at the Oregon National Primate Research Center (ONPRC)

Over the past 15 or so years, we have developed relatively simple cage-side versions of the human intruder and novel object tests that can be easily performed on large numbers of monkeys. Our tests can be modified depending on the scientific question being addressed.

In our temperament assessment, an intruder approaches the subject's cage, standing approximately 1 m from the front. Pair-housed monkeys are temporarily separated with a protected contact slide (which allows visual and some tactile access between partners) for the duration of the assessment. The

Figure 7.1 Example of a monkey inspecting (A) and avoiding (B) a brightly colored bird toy placed on the cage as part of a novel object test.

intruder maintains this position for 5 min, taking great care to avoid direct eye contact with the monkey. Using peripheral vision, the intruder takes instantaneous focal observations on the subject to record location in the cage (e.g., back, front) and behaviors of interest (e.g., freeze, stereotypic movement) every 15 s. At the end of the 5-min time period, the intruder then makes direct eye contact with the monkey for 2 min, again recording behavior every 15 s. The intruder then presents the subject with various novel objects, one at a time, for 3 min each. Objects include new foods, brightly colored objects, or objects with potentially threatening features (e.g., large eyes, such as Mr. Potato Head). These objects are either hung directly onto the cage or placed on a tray that is temporarily attached to the cage. The observer, who again avoids direct eye contact, records the latency to (1) inspect (within 10 cm), (2) intentionally touch, and (3) manipulate (displace from original location) each object. Overt aggression, urination, defecation, or signs of stress are also recorded. See Figure 7.2 for an example of a scoring sheet.

In an effort to be as consistent as possible, all of our intruders are females, and they all wear the same teal outer coat (a color that is not used by animal care staff) during the testing. We test no more than one animal in a room on any given day, and all testing is performed within 2 h of the morning meal, to ensure hunger is not a factor in the outcome.

Although we typically take observations in real time, the assessments can also be videotaped. Unlike real-time observations, videos can be watched again, thus ensuring that all behaviors are captured. However, scoring recorded behavior can take a great deal of time. The decision about whether or not to videotape should depend on the goals of the assessment (i.e., why you are taking the observations). If the goal is to compare subtle behavioral differences across groups, then recording the assessment will be important. If, however, the goal is to compare more conspicuous differences (e.g., comparing a monkey that quickly inspects objects and one that does not touch objects),

Monkey ID: **Male / Female**
Observer: **Date:**
Location: **Time:**

Accl		Beh	Loc	Yawn	Scratch	Fear	Threat	Lip-smack	Body Shake	vocal
	0:00									
	0:15									
	0:30									
	0:45									
	1:00									
	1:00									
	1:15									
	1:30									
	1:45									
	2:00									
	2:15									
	2:30									
	2:45									
	3:00									
	3:15									
	3:30									
	3:45									
	4:00									
	4:15									
	4:30									
	4:45									
	5:00									

DEC		Beh	Loc	Y	Sc	FG	Th	LS	BS	V
	0:00									
	0:15									
	0:30									
	0:45									
	1:00									
	1:00									
	1:15									
	1:30									
	1:45									
	2:00									

Comments:

(Continued)

Monkey ID: **Date:**   **Observer:**   **Comments**

1. **Initial response to novel object 1** _____ **(3 minutes):**

   Fear   Threat   Lipsmack Ignore   Avoid   Other:

   Latency to: Inspect _____   Touch _____   Manipulate _____

   Scratch? Y/N   Yawn? Y/N   Body shake? Y/N

2. **Initial response to novel object 2** _____ **(3 minutes):**

   Fear   Threat   Lipsmack Ignore   Avoid   Other:

   Latency to: Inspect _____   Touch _____   Manipulate _____

   Scratch? Y/N   Yawn? Y/N   Body shake? Y/N

3. **Initial response to novel object 3**_____ **(3 minutes):**

   Fear   Threat   Lipsmack Ignore   Avoid   Other:

   Latency to: Inspect _____   Touch _____   Manipulate _____

   Scratch? Y/N   Yawn? Y/N   Body shake? Y/N

   Urinate? Y/N   When?_____
   Defecate? Y/N   When?_____

Figure 7.2 Example of a scoring sheet for cage-side temperament assessment. The observer enters the room and stands by the monkey's cage, avoiding direct eye contact, for a 5-min acclimation period. During this time, the observer records behavior (stationary, location, stereotypy, freeze, vigilant) and location in cage (back, middle, front) every 15 s. Events (yawn, scratch, fear grimace, threat, lipsmack, body shake, vocalize) are recorded as 0/1 (i.e., if the behavior occurs during that 15 s interval, a score of 1 is recorded; otherwise, a score of 0 is recorded). After acclimation, the observer makes direct eye contact with the monkey for 2 min, and records the response of the animal in the same manner. After this period, the observer then puts novel objects on the animal's cage, recording the initial response (e.g., fear, threat, lipsmack, etc.). The observer then records the latency, in seconds, to inspect, touch and manipulate each item. The observer also notes whether or not the animal displayed any anxiety behaviors.

then live observations may serve that purpose. As with all behavioral tests, highly trained personnel are key to obtaining reliable data, regardless of whether the observations are live or recorded.

We have found a wide range of responses to the various stimuli of this temperament assessment. Most monkeys show some sort of reaction to the human intruder making direct eye contact, including threats, fear grimaces, or lipsmacking. These responses vary by species, as well as by temperament. For example, female cynomolgus macaques (*Macaca fascicularis*) tend to lipsmack more than female rhesus macaques (Coleman k., personal observation). Monkeys also vary in their response to novel objects. While most animals will touch most of the objects, a small percentage of animals (~10%–25%) avoid them, often remaining in the back of the cage (Figure 7.1B). Subjects occasionally become somewhat agitated by the objects; our Mr. Potato Head has been knocked off of the cage more than once.

It is important to note that the facility in which primates live can affect behavioral outcomes on this test. We recently undertook a study (Coleman et al. 2017) examining temperament in caged rhesus macaques at four US National Primate Research Centers. We utilized a cage-side version of the human intruder test, similar to the one detailed earlier. In this study, the tests were recorded and sent to a single laboratory for quantification. Even though the procedures were identical at each center, we found significant facility differences with respect to response to the human intruder. Animals at "Facility A" were somewhat more reactive than those at other centers, displaying higher amounts of threat and defensive behavior. Monkeys at "Facility B" were more likely than other monkeys to freeze. Each of these behavioral profiles has been associated with an anxious or inhibited temperament (Coleman and Pierre 2014). There are several potential reasons for these interfacility differences, including husbandry practices (e.g., stability of care staff, enrichment practices), early experience, and genetics. Regardless, these results highlight the importance of local conditions in data interpretation. As an example, while excessive freezing is often used as a measure of inhibition (Kalin et al. 1998), what constitutes "excessive" might not be the same at all facilities. In other words, behavioral outcomes on these tests may be relative to the reference population, rather than representing an absolute value (Coleman et al. 2017).

TEMPERAMENT AND BEHAVIORAL MANAGEMENT

As indicated elsewhere in this book, behavioral management is a comprehensive strategy for promoting psychological well-being, and includes factors such as socialization, nonsocial enrichment, and positive reinforcement training (Keeling et al. 1991; Whittaker et al. 2001; Weed and Raber 2005). The main goal of behavioral management is to produce animals that are in good physical condition, display a variety of species-typical behaviors, are resilient to stress, and that easily recover, both behaviorally and physiologically, from aversive stimuli (Novak and Suomi 1988). At most facilities, behavioral management plans are tailored to the unique behavioral patterns of each individual species (Lutz and Novak 2005; Jennings et al. 2009; National Research Council 2011). For instance, owl monkeys (*Aotus* spp.) and other species that utilize nests in the wild are typically provided with nest boxes in captivity; such nest boxes are not provided to macaque species, as they would be of little value. However, within a species, behavioral management is often provided with a "one size fits all" approach—what is good for one is assumed to be good for all. Socialization, environmental enrichment, and positive reinforcement training are generally assumed to be equally beneficial for all individuals, at least within a given age group. Young primates tend to be more exploratory than aged primates, and are often provided with additional enrichment and socialization opportunities. Still, factors such as personality or temperament are rarely specifically accounted for in behavioral management plans, despite the effect they can have on well-being. In this section, I describe studies that examine the relationship between individual differences in temperament and behavioral management. While the majority of studies described in the following will involve macaques, the principles can be applied to other species as well.

Socialization

Socialization, including pair or group housing, is a critical part of the behavioral management of captive primates. Socially housed primates display higher levels of species-appropriate behavior and fewer abnormal behaviors compared to singly housed primates (e.g., Schapiro et al. 1996). Further, socially housed monkeys show fewer signs of stress compared to their singly housed counterparts (Reinhardt et al. 1991; Eaton et al. 1994; Reinhardt 1999; Schapiro et al. 2000). However, socialization can result in aggression and trauma if the partners are not compatible. Finding compatible partners is not always easy or straightforward. While there is a paucity of published studies examining factors that can predict compatibility in pairs of animals, most published studies have focused on variables such as weight, age, and gender (Majolo et al. 2003; Truelove et al. 2017). Less attention is paid to behavioral characteristics such as temperament (although see McGrew 2017), despite knowledge that personality can play a role in compatibility in humans (e.g., Kelly and Conley 1987).

We examined the influence of temperament on compatibility and pairing success in female rhesus macaques (*Macaca mulatta*) in a series of studies. In our first study (McMillan et al. 2003), we examined 12 adult monkeys that had a successful and an unsuccessful pairing attempt within the previous year. Successful pairing attempts were defined as those in which the partners were co-housed for at least 3 months without any injuries or overt aggression. Attempts in which the monkeys fought immediately or within the first 2 weeks of introduction were considered unsuccessful. For the purposes of this study, we did not consider pairs that did not fall into one of these categories (>2 weeks, but <3 months of compatible housing). We assessed temperament in the 12 focal individuals, their successful partners and their unsuccessful partners by measuring response to novel objects presented to them in their home cages. Each monkey was given an "inhibition score" based on her reaction to the novelty. The scores went from 0 (low inhibition: the monkey inspected each object immediately) to 6 (high inhibition: the monkey did not inspect any of the items). We then calculated the difference in "inhibition scores" between the focal monkey and each of her partners (i.e., successful and unsuccessful). The inhibition score differences also varied from 0 (the two monkeys received the same inhibition score) to 6 (one animal was highly inhibited while the other was not at all inhibited). We found that the average difference in scores between the focal monkeys and their successful partners was significantly lower than the difference between the focal monkeys and their unsuccessful partners. In other words, the temperaments of successful partners were more similar than the temperaments of unsuccessful partners. Capitanio and colleagues found comparable results in a recent study (Capitanio et al. 2017); female, but not male, rhesus macaques were more likely to be successfully paired with partners that had similar temperaments. Temperament was assessed when the monkeys were infants in this study, using a standardized battery of tests (the Infant Biobehavioral Assessment; Capitanio 2017; Capitanio et al. 2005). Interestingly, despite the different assessment methodologies between the two studies, including differences in time between assessment and pairing attempt [e.g., several weeks in the case of McMillan et al. (2003) and several years in Capitanio et al. (2017)], the results were quite similar.

At approximately the same time that we were completing our study, the Oregon National Primate Research Center received a large group of adult, female rhesus macaques from another facility. These monkeys presented the perfect opportunity to test our findings in a prospective study (McMillan et al. 2003). We assessed temperament in these monkeys ($n = 74$) by measuring their response to novel objects presented to them in their home cage. As in our first study, we gave individuals an inhibition score based on their responses to the novelty. Animals given a score of 0–1 were categorized as "exploratory," those with a score of 5–6 were categorized as "inhibited," and the others were categorized as "moderate." We then attempted pair introductions according to these categorizations; 18 pairs consisted of individuals with similar temperaments (e.g., exploratory animals with exploratory animals) and the remaining 19 pairs contained animals with different temperaments (e.g., inhibited and exploratory). Somewhat surprisingly, the vast majority of these pairs (34 of the 37, 92%) were successful (i.e., there was no overt fighting between partners), regardless of the temperament of the partners. While this result was good for the welfare of the monkeys, it did not help us answer our scientific question. We therefore examined the pairs more closely, in an effort to determine the *quality* of the pair. We conducted focal observations on the pairs, recording behaviors such as grooming, huddling, and aggression. Pairs were then classified based on their level of social behavior; pairs in which the partners spent at least 50% of the observations in close social contact (e.g., grooming or huddling) were labeled "highly compatible pairs" ($n = 5$), while pairs in which partners showed aggression toward each other and spent <25% of time in close social contact (including the three unsuccessful pair attempts) were termed "incompatible pairs" ($n = 7$). The other 25 pairs showed varying levels of compatibility. For this study, we compared the difference in temperament between the highly compatible and incompatible pairs. We found that inhibition score differences between monkeys in the highly compatible pairs were significantly less than inhibition score differences between partners in the incompatible pairs. That is, pairs in which the partners had similar temperaments engaged in more affiliative and less aggressive behavior than pairs consisting of partners with disparate temperaments. These results are consistent with those of Weinstein and Capitanio (2008), who assessed temperament in infant monkeys and then measured social behavior of the subjects when they were 1 year of age. They found that the yearlings were more likely to affiliate with monkeys of a similar, rather than a dissimilar, temperamental style (Weinstein and Capitanio 2008). While these two studies examined different temperamental constructs, the results, along with the previously mentioned studies, suggest that various aspects of temperament can help guide choices about potential social partners for captive monkeys.

Environmental Enrichment

Nonsocial enrichment, including items such as toys and foraging devices, are among the most commonly used forms of enrichment (Baker et al. 2007). Because animals tend to lose interest in items with continuous exposure (e.g., Lutz and Novak 2005), these manipulanda are often rotated with the goal of promoting novelty. However, exposure to novelty can be potentially anxiogenic for highly inhibited animals. For example, a study of orange-winged Amazon parrots (*Amazona amazonica*) found that highly fearful birds showed increased anxiety in response to rotating enrichment, compared to nonfearful birds (Fox and Millam 2007). While there are few, if any, studies specifically examining the relationship between temperament and environmental enrichment use in a primate species, it stands to reason that fearful monkeys might respond to new objects in a manner similar to fearful parrots. Further, there is some evidence to suggest that temperament may affect the use of cognitive enrichment.

In an effort to try to provide animals with a more enriching environment, we have begun to utilize tablets as enrichment. Tablets, such as iPads and Kindles, offer an opportunity to provide nonhuman primates (NHPs) access to a wide range of games, known as apps, which can be easily changed and updated. We examined the use of a tablet (Kindle Fire, Amazon.com) as enrichment for captive rhesus macaques (O'Connor et al. 2015). As a first step, we examined preference for various types of apps. Specifically, we compared passive apps, in which the animal does not interact with the tablet, with interactive apps, in which the monkey does something to the tablet that results in some sort of a change (e.g., a balloon pops after being touched). As the interactive apps provide more choice and control than passive apps, we hypothesized that monkeys would prefer to use these games. We were also interested in determining whether shy, inhibited monkeys used the tablet in the same way as bold, exploratory monkeys.

The subjects in this study were 16 adult male rhesus macaques. They were all singly housed for scientific reasons. Prior to implementing the tablet enrichment, we assessed temperament using our cage-side test (see earlier section). The tablet was placed in a protective cover and hung on the outside of the monkey's cage (Figure 7.3). Monkeys were first acclimated to the tablet with a blank screen for 3 days, after which we provided them with the opportunity to use a passive (a colorful screen saver-like app) and two interactive apps (one in which brightly colored balloons "popped" when touched, and one that allowed monkeys to "paint" on the touchscreen). Each app was presented to monkeys for three 30-min sessions for 3 days in a row. We recorded the amount of time the monkeys spent touching (intentionally making contact with the screen) and closely inspecting (within 5 cm) the tablet.

Figure 7.3   Example of a monkey using a Kindle Fire tablet as enrichment.

As expected, animals tended to spend more time touching and inspecting the tablet when presented with interactive, compared to passive, apps ($p < .01$). Interestingly, inhibited monkeys ($n = 3$) did not touch the tablet at any point during the study, even after several days of acclimation. These shy monkeys did not show excessive distress; rather, they simply avoided the tablet. While the sample size is low, this result illustrates the fact that not all enrichment items are equally suitable for all individuals.

Other studies have also found that temperament can affect (or temperamental differences in) responses to cognitive enrichment. Yamanashi and Matsuzawa (2010) examined the effect of cognitive enrichment in six chimpanzees. Half of these animals were classified as "stress sensitive," because they displayed self-directed behaviors, such as scratching, while performing various cognitive tasks. The stress-sensitive chimpanzees were more likely to become agitated when they got a wrong response on the cognitive tasks compared to the stress-resistant chimpanzees (Yamanashi and Matsuzawa 2010). While cognitive tasks such as these can be utilized as enrichment (Washburn and Rumbaugh 1992), these two studies suggest that such tasks may not confer the same benefits to all individuals and, in fact, may even increase distress in some individuals.

Positive Reinforcement Training

The use of positive reinforcement training (PRT) in biomedical facilities has increased dramatically in the past decade (Baker 2016). PRT, a type of training in which subjects are rewarded for performing various behaviors (see Laule et al. 2003 for more detail), is known to reduce stress associated with various husbandry, veterinary, and research procedures and, as such, is a significant refinement to animal care practices (see Graham 2017; Magden 2017). Yet, it is clear to anyone who trains primates that there can be vast differences across subjects in terms of "trainability."

Some individuals learn tasks very quickly, while progress is much slower for others. Still others never seem to "get it." These kinds of differences can have substantial implications in a biomedical setting, where only a limited amount of time is typically allocated to train a primate to criterion. Further, some research protocols necessitate a homogenous experience for the animals. For example, if some individuals are not trained for a task (such as remaining stationary for a blood draw), then all subjects must be treated as if they are untrained (in the aforementioned example, all animals might have to be restrained for a blood draw). Therefore, it is important to identify factors that can predict trainability. One such factor is temperament, specifically the construct of behavioral inhibition. Given that inhibited individuals may be more wary of a human, it follows that training these monkeys may be more difficult than training less inhibited monkeys.

We have assessed the influence of temperament on positive reinforcement training in a series of studies. In our first study (Coleman et al. 2005), we examined the relationship between temperament and training in female rhesus macaques (*Macaca mulatta*). We used a simple novel object test to assess temperament in 60 monkeys. Subjects were classified as exploratory, inhibited, or moderate, based on their latency to touch the object (a piece of novel food). Exploratory monkeys touched the food immediately (i.e., within 10 s), inhibited monkeys did not touch the food within the allotted 3-min time period, and moderate monkeys touched the food, but with a latency of longer than 10 s. Interestingly, the vast majority of animals clearly fell into one of the two extremes (over 60% of the sample were exploratory and 25% were inhibited). Only 13% of animals were classified as "moderate."

We selected 20 animals as subjects, based on their temperament. We used PRT to train animals to (1) take treats from the trainer (i.e., "clicker train") and (2) touch a target (a piece of PVC tube) attached to the front of the home cage. Training animals to touch a target is a relatively straightforward task, and is often used as an early step in PRT (Laule et al. 2003). Subjects were considered trained for each of these tasks when they performed the behavior on command for three consecutive training sessions.

As expected, there were vast individual differences with regard to training success. While all of the exploratory and moderate monkeys immediately took treats from the trainer, fewer than half of the inhibited monkeys were successfully clicker trained, even after 12 training sessions. Further, we were only able to train two of the inhibited animals to touch the target. In contrast, 82% of the exploratory and moderate monkeys reliably performed this task within the allotted time. Thus, temperament correlated with training success.

This study was one of the first to use individual differences in temperament as a tool for behavioral management. Using a relatively simple screening test, we were able to predict whether or not monkeys would prove difficult to train. This knowledge can assist in animal assignments; for example, inhibited animals may not do well on studies in which training is required.

While informative, the outcome behavior for that study (Coleman et al. 2005) was target training, a relatively straightforward task. Few projects have target training as their ultimate training goal. Further, we only trained those animals for 12 training sessions. It is possible that had we trained longer, the shy monkeys would have become habituated to the trainer, which might have allowed for better success. We therefore repeated the study, using a more complicated task and a different species. In our second study, we assessed temperament using our cage-side test (see earlier section) in adult, female cynomolgus monkeys (*M. fascicularis*). Animals were given an "inhibition" score from 0 to 9, based on their response to the novel objects. We randomly chose 20 animals as subjects. Training objectives for this study included (1) touching a target and (2) turning and presenting the hindside at the front of the cage and allowing vaginal swabbing. As before, animals were trained with PRT for 5–10 min per session, 2–3 sessions per week, although in this study animals were trained for a total of 33 sessions.

Eleven out of 20 monkeys were labeled as "inhibited" on the temperament test, based on their "inhibition score." Unlike our previous study, we were able to train all of these monkeys to touch the target, although it took exploratory animals ($n = 9$) significantly fewer sessions to do so ($p = .005$). Not every animal reliably allowed vaginal swabbing, however. While 89% of exploratory animals reliably performed this behavior, only 45% of inhibited animals did so. Interestingly, when we compared the amount of time it took the 13 reliable (8 exploratory and 5 inhibited) animals to learn the task, we found they did not differ.

We re-examined our temperament data to determine whether other variables might be informative, and created a "reactive" score based on the number of times an animal was recorded as either showing fear or submissive behavior to the observer. We then compared this reactive score between inhibited animals that were ultimately trained for vaginal swabbing and those that were not. Inhibited monkeys that were trained showed less fear and/or submission than those that were not trained. These results suggest that there may be different populations of inhibited animals; some individuals were wary of the training initially, but habituated to the situation and, once they were less apprehensive about the training process, they learned at the same rate as exploratory monkeys. Other inhibited animals did not habituate to the training, and thus, never performed the task.

What to Do with the Really Shy Monkeys

Results of these and other studies (e.g., Houser L. A., and Coleman k., in preparation) suggest that a simple temperament test can identify individuals that may be difficult to train. However, it is not practical, or in many cases desirable, to train only exploratory animals. Further, trainers in our original study (Coleman et al. 2005) noted that the inhibited monkeys often stayed in the back corner of the cage during the training sessions (Coleman 2012). For these inhibited animals, training may not have provided the same psychological well-being benefits afforded to other, more exploratory individuals. Instead, training may actually *increase* stress for these animals. Thus, there is a need to develop alternate training techniques to help reduce associated stress for inhibited monkeys.

One potential method to increase trainability in inhibited monkeys may be to desensitize them to the presence of the trainer. Desensitization is an active process in which an aversive stimulus is paired with a positive reward, thereby reducing anxiety toward the stimulus (Laule et al. 2003). There is evidence suggesting that desensitization (Clay et al. 2009) and repeated provision of food treats (Baker and Springer 2006) may help reduce fear toward caretakers in macaques. We examined whether desensitization would improve training success for inhibited animals.

In this study (L.A. Houser, in preparation), we examined the length and frequency of desensitization in 12 female rhesus macaques identified as being overly fearful and inhibited toward care staff. Monkeys were randomly placed into one of two treatment groups, "5/day" or "1/day." In both groups, a trainer went into each subject's room and handed her food treats for 10 min per day, 3 times per week for a total of 2 months. Monkeys in the 5/day group received several short habituation sessions (2 min, 5 times per treatment day) while those in the 1/day group received fewer, longer sessions (10 min, 1 time per treatment day). After 2 months, we used PRT to train monkeys to touch a target on the outside of their cage for a total of 20 sessions. We recorded the latency (in days) for the monkeys to take a treat from the observer during the desensitization sessions. We also recorded latency to reliably touch the target, as well as behavior during training (fear, threats, etc.).

Although not statistically significant, monkeys in the 5/day group tended to take a treat in the presence of the observer sooner than those in the 1/day group (Figure 7.4A). A total of eight monkeys (5 from the 5/day group, 3 from the 1/day group) reliably touched the target within the 20 training sessions. There was no difference in the amount of time that it took these animals to perform this task. However, there was a trend ($p = .087$) for monkeys in the 1/day group to exhibit more fear during training than those in the 5/day group (Figure 7.4B). While the sample size is low, these results suggest that desensitization techniques, particularly when accomplished in short, frequent sessions, may be an effective strategy for reducing fear toward a trainer, ultimately improving training success.

Ideally, every monkey living in a cage would be desensitized to the presence of care staff, including trainers; veterinarians and veterinary technicians; and investigators. However, this kind

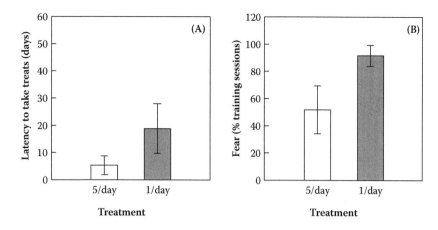

Figure 7.4 Latency to take treats from an observer during desensitization sessions (A) and number of training sessions in which subjects exhibited fear (B) for animals that underwent desensitization five times per day (2 min each time) or one time per day (for 10 min).

of desensitization may not be feasible at all facilities. In these cases, there may be value in conducting temperament assessments on all animals and proactively desensitizing inhibited individuals.

MANAGING BEHAVIORAL ISSUES

A major goal of behavioral management programs is to reduce the occurrence of abnormal or undesired behaviors, such as stereotypical or self-injurious behavior. Because these behavioral problems are typically difficult to attenuate once they have begun (Novak 2003), efforts are often focused on prevention. Understanding factors that may predispose individuals to develop behavioral problems can help us identify animals who might be at risk. Several studies have examined external risk factors, such as rearing history, management decisions, or research procedures. For example, inadequate socialization early in life is known to be a risk factor for self-injurious behavior (SIB) and stereotypical behavior (Bellanca and Crockett 2002; Novak 2003). Fewer studies have examined the role of intrinsic factors, such as temperament, in the development of these behavioral problems.

We examined whether temperament influenced behavioral response to artificial weaning in young rhesus macaques. At most primate facilities, infants reared in cages with their mothers are removed from their dams at some point (e.g., 6–12 months of age) for management purposes. This decrease in maternal attention can be stressful for young monkeys, and can lead to distress. Studies have shown that some, but not all, young macaques respond to maternal separation, such as weaning, with an increase in depressive behaviors, including decreases in play and eating, and increases in self-directed behavior (e.g., Rosenblum and Paully 1987; Boccia et al. 1994). We examined the role temperament plays in these behavioral responses to decreased maternal attention. We used a novel object test (Williamson et al. 2003) to assess temperament in 39 infant (3–4 months old) macaques living in cages with their mothers. At the time that this study was conducted, infants at our facility were weaned from their dams at 6 months of age and were then housed socially with other recently weaned monkeys. Clinical staff carefully assessed the infants for the first few days after weaning, and recorded signs of stress, such as inappetence or a depressed posture, in the animal's health records. While the majority of the infants that we assessed adjusted quickly to their new environment following weaning, five displayed distress behavior. These monkeys had significantly longer latencies to touch the novel fruit (i.e., were more inhibited) compared to the other infants (unpublished data). Results from this study suggest that inhibited infants may have more difficulty with stressful situations, such as weaning. Based on the results of this study, the ONPRC has increased the weaning age for indoor-reared infants to a minimum of 1 year. However, infant temperament is assessed prior to weaning, and those that appear highly inhibited may be left with their mother for longer than 1 year to minimize the adverse effects of weaning.

While we did not look for long-term behavioral outcomes in the aforementioned study, there is evidence that early temperament may correlate with the development of some abnormal behaviors. Two recent studies (Vandeleest et al. 2011; Gottlieb et al. 2013) examined this relationship in rhesus macaques. In these studies, temperament was assessed when the monkeys were infants, using a battery of tests (the Infant Biobehavioral Assessment; Capitanio 2017; Capitanio et al. 2005). Both active (Gottlieb et al. 2013) and nervous/gentle (Vandeleest et al. 2011) temperaments were found to positively correlate with the later development of stereotypic behavior. Interestingly, stereotypic behavior was also found to positively correlate with exploration of a novel stimulus, both when the monkeys were tested as infants (Gottlieb et al. 2013) and as adults (Gottlieb et al. 2015). In other words, animals that were inhibited with respect to this stimulus were *less* likely to exhibit stereotypic behavior than exploratory animals. However, in both studies, stereotypic behavior was assessed by an observer who stood in front of the monkey's cage. It is possible that the highly inhibited monkeys either froze or retreated to the back of the cage in the presence of the observer (who was not a familiar caregiver), behaviors that are mutually exclusive with stereotypic behavior. Different results may have been found if the animals had been assessed remotely, without a person present. Regardless, these results indicate that early temperament may provide information about the development of stereotypic behavior.

More recently, we examined the relationship between temperament and alopecia in rhesus macaques. Alopecia, or hair loss, is a significant problem at facilities caring for captive macaques. There are several potential etiologies for hair loss, including environmental factors, infection, and immune dysregulation (e.g., Novak and Meyer 2009; Kramer et al. 2011; Kroeker et al. 2014; Novak et al. 2017). However, few studies have examined the relationship between various temperamental traits and vulnerability to the development of alopecia. In our study (Coleman et al. 2017), we quantified hair loss in 101 monkeys living at four National Primate Research Centers. We then assessed temperament using a cage-side version of the human intruder test. Surprisingly, monkeys with an inhibited temperament had *less* alopecia than exploratory monkeys (Coleman et al. 2017). At first, these results seemed somewhat counterintuitive; we had expected inhibited monkeys to have higher or equivalent levels of alopecia compared to noninhibited monkeys. However, we did not examine patterns of alopecia in this study. There is evidence that some macaques pick their hair in a fashion similar to that seen in human patients with trichotillomania (Ellison et al. 2006), a condition associated with anxiety and obsessive–compulsive personality disorders in which patients pull out their own hair (Christenson and Crow 1996). This hair-pulling behavior often results in limited hair loss, such as small patches of alopecia, often on an arm or leg (personal observation). Thus, it is possible that the highly inhibited/anxious individuals in our study did have alopecia, but it might have presented as small localized patches, as opposed to large areas, of hair loss (Coleman et al. 2017).

Understanding factors that predispose some individuals to develop behavioral problems can help behavioral managers identify those individuals and provide early preventative measures or interventions.

SUMMARY

While the study of animal temperament has seen dramatic increases over the past several years, there is still a relative paucity of studies examining the ways in which temperament might influence the psychological wellbeing of captive primates or other animals. But, the studies that have been published to date support the idea that temperament can help inform choices about behavioral management practices. Knowing the temperament of nonhuman primates can assist in social introductions. Pairs in which the partners have similar temperaments may be more compatible than pairs in which partners are not alike, at least for females. An individual's propensity to use environmental enrichment can be influenced by its temperament. Shy, inhibited, or highly stress-sensitive individuals may not utilize or get the same benefits from some common environmental enrichment options compared to more exploratory individuals. Temperament can also affect an individual's trainability. Simple temperament assessments can help behavioral managers predict animals that will be relatively easy to train. Equally important, temperament assessments can also help identify inhibited animals that may need alternate training techniques, such as desensitization or a combination of negative and positive reinforcement training (Hannibal et al. 2013; Wergård et al. 2015). Lastly, knowing temperamental characteristics that correlate with the development of behavioral problems can greatly aid our ability to care for nonhuman primates, especially those that may be affected by such problems.

ACKNOWLEDGMENTS

I thank Dr. Daniel Gottlieb and Diana Gordon for useful comments on the manuscript, as well as Jennifer McMillan, Leigh Ann Tully, Jillann Rawlins-O'Connor, Adriane Maier, Lisa Houser, and Allison Heagerty for their assistance on the studies conducted at the ONPRC. Support is acknowledged from the Oregon National Primate Research Center, 8P51OD011092.

REFERENCES

Baker, K. C. 2016. Survey of 2014 behavioral management programs for laboratory primates in the United States. *Am J Primatol* 78:780–796.

Baker, K. C. and D. A. Springer. 2006. Frequency of feeding enrichment and response of laboratory nonhuman primates to unfamiliar people. *J Am Assoc Lab Anim Sci* 45(1):69–73.

Baker, K. C., J. L. Weed, C. M. Crockett, et al. 2007. Survey of environmental enhancement programs for laboratory primates. *Am J Primatol* 69(4):377–394.

Bellanca, R. U. and C. M. Crockett. 2002. Factors predicting increased incidence of abnormal behavior in male pigtailed macaques. *Am J Primatol* 58(2):57–69.

Bergvall, U. A., A. Schäpers, P. Kjellander, et al. 2011. Personality and foraging decisions in fallow deer, *Dama dama*. *Anim Behav* 81:101–112.

Boccia, M. L., M. L. Laudenslager, and M. L. Reite. 1994. Intrinsic and extrinsic factors affect infant responses to maternal separation. *Psychiatry* 57(1):43–50.

Bouchard, T. J., Jr. and J. C. Loehlin. 2001. Genes, evolution, and personality. *Behav Genet* 31(3):243–273.

Boyce, W. T., M. Chesney, A. Alkon, et al. 1995. Psychobiologic reactivity to stress and childhood respiratory illnesses: Results of two prospective studies. *Psychosom Med* 57(5):411–422.

Buss, A. H. and R. Plomin. 1986. The ESA approach to temperament. In *The Study of Temperament: Changes, Continuities and Challenges*, eds Plomin, R. and A. H. Buss, 67–79. Hillsdale, NJ: Lawrence Erlbaum Associates.

Capitanio, J. P. 1999. Personality dimensions in adult male rhesus macaques: Prediction of behaviors across time and situation. *Am J Primatol* 47(4):299–320.

Capitanio, J. P. 2011. Individual differences in emotionality: Social temperament and health. *Am J Primatol* 73(6):507–515.

Capitanio, J.P. 2017. Variation in biobehavioral organization, Chapter 5. In *Handbook of Primate Behavioral Management,* 55–73, ed Schapiro, S. J. Boca Raton, FL: CRC Press.

Capitanio, J. P., S. A. Blozis, J. Snarr, et al. 2017. Do "birds of a feather flock together" or do "opposites attract"? Behavioral responses and temperament predict success in pairings of rhesus monkeys in a laboratory setting. *Am J Primatol* 79:e22464.

Capitanio, J. P., S. P. Mendoza, and K. L. Bentson. 1999. The relationship of personality dimensions in adult male rhesus macaques to progression of simian immunodeficiency virus disease. *Brain Behav Immun* 13:138–154.

Capitanio, J. P., S. P. Mendoza, W. A. Mason, et al. 2005. Rearing environment and hypothalamic-pituitary-adrenal regulation in young rhesus monkeys (*Macaca mulatta*). *Dev Psychobiol* 46(4):318–330.

Christenson, G. A. and S. J. Crow. 1996. The characterization and treatment of trichotillomania. *J Clin Psychiatry* 57(Suppl. 8):42–49.

Chun, K., L. A. Miller, E. S. Schelegle, et al. 2013. Behavioral inhibition in rhesus monkeys (*Macaca mulatta*) is related to the airways response, but not immune measures, commonly associated with asthma. *PLoS One* 8(8):e71575.

Clark, A. B. and T. J. Ehlinger. 1987. Pattern and adaptation in individual behavioral differences. In *Perspectives in Ethology*, eds Bateson, P. P. G. and P. H. Klopfer, 1–47. New York: Springer.

Clarke, A. S. and S. Boinski. 1995. Temperament in nonhuman primates. *Am J Primatol* 37:103–125.

Clay, A. W., M. A. Bloomsmith, M. J. Marr, et al. 2009. Habituation and desensitization as methods for reducing fearful behavior in singly housed rhesus macaques. *Am J Primatol* 71(1):30–39.

Coleman, K. 2012. Individual differences in temperament and behavioral management practices for nonhuman primates. *Appl Anim Behav Sci* 137(3–4):106–113.

Coleman, K., C. K. Lutz, J. M. Worlein, et al. 2017. The correlation between alopecia and temperament in rhesus macaques (*Macaca mulatta*) at four primate facilities. *Am J Primatol* 79:e22504.

Coleman, K. and P. J. Pierre. 2014. Assessing anxiety in nonhuman primates. *ILAR J* 55(2):333–346.

Coleman, K., N. D. Robertson, and C. L. Bethea. 2011. Long-term ovariectomy alters social and anxious behaviors in semi-free ranging Japanese macaques. *Behav Brain Res* 225(1):317–327.

Coleman, K., L. A. Tully, and J. L. McMillan. 2005. Temperament correlates with training success in adult rhesus macaques. *Am J Primatol* 65(1):63–71.

Coleman, K. and D. S. Wilson. 1998. Shyness and boldness in pumpkinseed sunfish: Individual differences are context-specific. *Anim Behav* 56:927–936.

Corcoran, C. A., P. J. Pierre, T. Haddad, et al. 2012. Long-term effects of differential early rearing in rhesus macaques: Behavioral reactivity in adulthood. *Dev Psychobiol* 54(5):546–555.

Deary, I. J., A. Weiss, and G. D. Batty. 2010. Intelligence and personality as predictors of illness and death: How researchers in differential psychology and chronic disease epidemiology are collaborating to understand and address health inequalities. *Psychol Sci Public Interest* 11(2):53–79.

Eaton, G. G., S. T. Kelley, M. K. Axthelm, et al. 1994. Psychological well-being in paired adult female rhesus (*Macaca mulatta*). *Am J Primatol* 33:89–99.

Ellison, R., T. Hobbs, A. Maier, et al. 2006. An assessment of temperament and behavior in rhesus macaques with alopecia. *Am J Primatol* 68:107.

Fox, R. A. and J. R. Millam. 2007. Novelty and individual differences influence neophobia in orange-winged Amazon parrots (*Amazona amazonica*). *Appl Anim Behav Sci*.104(1–2):107–115.

Fox, A. S., J. A. Oler, A. J. Shackman, et al. 2015. Intergenerational neural mediators of early-life anxious temperament. *Proc Natl Acad Sci USA* 112(29):9118–9122.

Francis, R. C. 1990. Temperament in a fish: A longitudinal study of the development of individual differences in aggression and social rank in the Midas cichlid. *Ethology* 86:311–325.

Freeman, H. D., S. F. Brosnan, L. M. Hopper, et al. 2013. Developing a comprehensive and comparative questionnaire for measuring personality in chimpanzees using a simultaneous top-down/bottom-up design. *Am J Primatol* 75:1042–1053.

Freeman, H. D. and S. D. Gosling. 2010. Personality in nonhuman primates: A review and evaluation of past research. *Am J Primatol* 72(8):653–671.

Gottlieb, D. H., J. P. Capitanio, and B. McCowan. 2013. Risk factors for stereotypic behavior and self-biting in rhesus macaques (*Macaca mulatta*): Animal's history, current environment, and personality. *Am J Primatol* 75(10):995–1008.

Gottlieb, D. H., A. Maier, and K. Coleman. 2015. Evaluation of environmental and intrinsic factors that contribute to stereotypic behavior in captive rhesus macaques (*Macaca mulatta*). *Appl Anim Behav Sci* 171:184–191.

Graham, M. L. 2017. Positive reinforcement training and research, Chapter 12. In *Handbook of Primate Behavioral Management,* 187–200, ed. Schapiro, S. J. Boca Raton, FL: CRC Press.

Hannibal, D., D. Minier, J. Capitanio, et al. 2013. Effect of temperament on the behavioral conditioning of individual rhesus monkeys (*Macaca mulatta*). *Am J Primatol* 75:66.

Hirshfeld, D. R., J. F. Rosenbaum, J. Biederman, et al. 1992. Stable behavioral inhibition and its association with anxiety disorder. *J Am Acad Child Adolesc Psychiatry* 31(1):103–111.

Jennings, M., M. J. Prescott, H. M. Buchanan-Smith, et al. 2009. Refinements in husbandry, care and common procedures for non-human primates: Ninth report of the BVAAWF/FRAME/RSPCA/UFAW Joint Working Group on refinement. *Lab Anim* 43(Suppl. 1):1–47.

Kagan, J. 1994. *Galen's Prophecy: Temperament in Human Nature*. New York: Basic Books.

Kagan, J., J. S. Reznick, and N. Snidman. 1988. Biological bases of childhood shyness. *Science* 240(4849):167–171.

Kagan, J., N. Snidman, M. Julia-Sellers, et al. 1991. Temperament and allergic symptoms. *Psychosom Med* 53(3):332–340.

Kalin, N. H. and S. E. Shelton. 1989. Defensive behaviors in infant rhesus monkeys: Environmental cues and neurochemical regulation. *Science* 243(4899):1718–1721.

Kalin, N. H., S. E. Shelton, M. Rickman, et al. 1998. Individual differences in freezing and cortisol in infant and mother rhesus monkeys. *Behav Neurosci* 112(1):251–254.

Keeling, M. E., P. L. Alford, and M. A. Bloomsmith. 1991. Decision analysis for developing programs of psychological well-being: A bias-for-action approach. In *Through the Looking Glass*, eds Novak, M. A. and A. J. Petto, 57–65. Washington, DC: American Psychological Association.

Kelly, E. L. and J. J. Conley. 1987. Personality and compatibility: A prospective analysis of marital stability and marital satisfaction. *J Pers Soc Psychol* 52(1):27–40.

Konecna, M., S. Lhota, A. Weiss, et al. 2008. Personality in free-ranging Hanuman langur (*Semnopithecus entellus*) males: Subjective ratings and recorded behavior. *J Comp Psychol* 122(4):379–389.

Kramer, J. A., K. G. Mansfield, J. H. Simmons, et al. 2011. Psychogenic alopecia in rhesus macaques presenting as focally extensive alopecia of the distal limb. *Comp Med* 61:263–268.

Kroeker, R., R. U. Bellanca, G. H. Lee, et al. 2014. Alopecia in three macaque species housed in a laboratory environment. 2014. *Am J Primatol* 76:325–334.

Laule, G. E., M. A. Bloomsmith, and S. J. Schapiro. 2003. The use of positive reinforcement training techniques to enhance the care, management, and welfare of primates in the laboratory. *J Appl Anim Welfare Sci* 6(3):163–173.

Lutz, C. K. and M. A. Novak. 2005. Environmental enrichment for nonhuman primates: Theory and application. *ILAR J* 46(2):178–191.

Lyons, D. M., E. O. Price, and G. P. Moberg. 1988. Individual differences in temperament of domestic dairy goats: Constancy and change. *Anim Behav* 36:1323–1333.

Majolo, B., H. M. Buchanan-Smith, and K. Morris. 2003. Factors affecting the successful pairing of unfamiliar common marmoset (*Callithrix jacchus*) females: Preliminary results. *Anim Welfare* 12:327–337.

Maninger, N., J. P. Capitanio, S. P. Mendoza, et al. 2003. Personality influences tetanus-specific antibody response in adult male rhesus macaques after removal from natal group and housing relocation. *Am J Primatol* 61(2):73–83.

McCall, R. B. 1986. Issues of stability and continuity in temperament research. In *The Study of Temperament: Changes, Continuities and Challenges*, eds Plomin, R. and J. Dunn, 13–25. Hillsdale, NJ: Lawrence Erlbaum Associates.

McDougall, P. T., D. Réale, D. Sol, et al. 2006. Wildlife conservation and animal temperament: Causes and consequences of evolutionary change for captive, reintroduced, and wild populations. *Anim Conserv* 9:39–48.

McGrew, K. 2017. Pairing strategies for cynomolgus macaques, Chapter 17. In *Handbook of Primate Behavioral Management*, 255–264, ed. Schapiro, S. J. Boca Raton, FL: CRC Press.

McMillan, J., A. Maier, L. Tully, et al. 2003. The effects of temperament on pairing success in female rhesus macaques. *Am J Primatol* 60(Suppl. 1):95.

Magden, E. R. 2017. Positive reinforcement training and health care, Chapter 13. In *Handbook of Primate Behavioral Management*, 201–216, ed. Schapiro, S. J. Boca Raton, FL: CRC Press.

Miller, G. E., S. Cohen, B. S. Rabin, et al. 1999. Personality and tonic cardiovascular, neuroendocrine, and immune parameters. *Brain Behav Immun* 13(2):109–123.

National Research Council. 2011. *Guide for the Care and Use of Laboratory Animals*. Washington, DC: National Academies Press.

Novak, M. A. 2003. Self-injurious behavior in rhesus monkeys: New insights into its etiology, physiology, and treatment. *Am J Primatol* 59(1):3–19.

Novak, M. A., A. F. Hamel, A. M. Ryan, et al. 2017. The role of stress in abnormal behavior and other abnormal conditions such as hair loss, Chapter 6. In *Handbook of Primate Behavioral Management*, 75–94, ed. Schapiro, S. J. Boca Raton, FL: CRC Press.

Novak, M. A. and J. S. Meyer. 2009. Alopecia: Possible causes and treatments, particularly in captive nonhuman primates. *Comp Med* 59(1):18–26.

Novak, M. A. and S. J. Suomi. 1988. Psychological well-being of primates in captivity. *Am Psychol* 43(10):765–773.

O'Connor, J. R., A. Heagerty, M. Herrera, et al. 2015. Use of a tablet as enrichment for adult rhesus macaques. *Am J Primatol* 77:122.

Reale, D., S. M. Reader, D. Sol, et al. 2007. Integrating animal temperament within ecology and evolution. *Biol Rev Camb Philos Soc* 82(2):291–318.

Reinhardt, V. 1999. Pair-housing overcomes self-biting behavior in macaques. *Lab Primate Newsl* 38(1):4–5.

Reinhardt, V., D. Cowley, and S. Eisele. 1991. Serum cortisol concentrations of single-housed and isosexually pair-housed adult rhesus macaques. *J Exp Anim Sci* 34(2):73–76.

Roberts, B. W., N. R. Kuncel, R. Shiner, et al. 2007. The power of personality: The comparative validity of personality traits, socioeconomic status, and cognitive ability for predicting important life outcomes. *Perspect Psychol Sci* 2(4):313–345.

Roseboom, P. H., S. A. Nanda, A. S. Fox, et al. 2014. Neuropeptide Y receptor gene expression in the primate amygdala predicts anxious temperament and brain metabolism. *Biol Psychiatry* 76(11):850–857.

Rosenbaum, J. F., J. Biederman, E. A. Bolduc-Murphy, et al. 1993. Behavioral inhibition in childhood: A risk factor for anxiety disorders. *Harv Rev Psychiatry* 1(1):2–16.

Rosenblum, L. A. and G. S. Paully. 1987. Primate models of separation-induced depression. *Psychiatr Clin North Am* 10(3):437–447.

Schapiro, S. J., M. A. Bloomsmith, S. A. Suarez, et al. 1996. Effects of social and inanimate enrichment on the behavior of yearling rhesus monkeys. *Am J Primatol* 40:247–260.

Schapiro, S. J., P. N. Nehete, J. E. Perlman, et al. 2000. A comparison of cell-mediated immune responses in rhesus macaques housed singly, in pairs, or in groups. *Appl Anim Behav Sci* 68(1):67–84.

Schwartz, C. E., N. Snidman, and J. Kagan. 1999. Adolescent social anxiety as an outcome of inhibited temperament in childhood. *J Am Acad Child Adolesc Psychiatry* 38(8):1008–1015.

Stevenson-Hinde, J. and M. Zunz. 1978. Subjective assessment of individual rhesus monkeys. *Primates* 19:473–482.

Suomi, S. J. 1991. Uptight and laid-back monkeys: Individual differences in the response to social challenges. In *Plasticity of Development*, eds Brauth, S. E., W. S. Hall, and R. J. Dooling, 27–56. Cambridge, MA: MIT Press.

Truelove, M. A., A. L. Martin, J. E. Perlman, et al. 2017. Pair housing of macaques: A review of partner selection, introduction techniques, monitoring for compatibility, and methods for long-term maintenance of pairs. *Am J Primatol* 79:e22485.

Vandeleest, J. J., B. McCowan, and J. P. Capitanio. 2011. Early rearing interacts with temperament and housing to influence the risk for motor stereotypy in rhesus monkeys (*Macaca mulatta*). *Appl Anim Behav Sci* 132(1–2):81–89.

Washburn, D. A. and D. M. Rumbaugh. 1992. Testing primates with joystick-based automated apparatus: Lessons from the language research center's computerized test system. *Behav Res Methods Instrum Comput* 24(2):157–164.

Watters, J. V. and D. M. Powell. 2012. Measuring animal personality for use in population management in zoos: Suggested methods and rationale. *Zoo Biol* 31(1):1–12.

Weed, J. L. and J. M. Raber. 2005. Balancing animal research with animal well-being: Establishment of goals and harmonization of approaches. *ILAR J* 46(2):118–128.

Weinstein, T. A. R. and J. P. Capitanio. 2008. Individual differences in infant temperament predict social relationships of yearling rhesus monkeys, *Macaca mulatta*. *Anim Behav* 76(2):455–465.

Wergård, E. M., H. Temrin, B. Forkman, et al. 2015. Training pair-housed rhesus macaques (*Macaca mulatta*) using a combination of negative and positive reinforcement. *Behav Process* 113:51–59.

Whittaker, M., G. Laule, J. Perlman, et al. 2001. A behavioral management approach to caring for great apes. *The Apes: Challenges for the 21st Century Conference Proceedings*, Brookfield, IL.

Williamson, D. E., K. Coleman, S. A. Bacanu, et al. 2003. Heritability of fearful-anxious endophenotypes in infant rhesus macaques: A preliminary report. *Biol Psychiatry* 53:284–291.

Wilson, D. S., K. Coleman, A. B. Clark, et al. 1993. Shy-bold continuum in pumpkinseed sunfish (*Lepomis gibbosus*): An ecological study of a psychological trait. *J Comp Psychol* 107(3):250–260.

Yamanashi, Y. and T. Matsuzawa. 2010. Emotional consequences when chimpanzees (*Pan troglodytes*) face challenges: Individual differences in self-directed behaviours during cognitive tasks. *Animal Welfare* 19(1):25–30.

CHAPTER 8

Depression in Captive Nonhuman Primates
Theoretical Underpinnings, Methods, and Application to Behavioral Management

Carol A. Shively
Wake Forest School of Medicine

CONTENTS

INTRODUCTION

The purposes of this chapter are to describe depressive behavior in laboratory primates, describe how depressive behavior can be measured, and describe other behavioral and physiological characteristics of depressed primates. It is important to understand depressive behavior in nonhuman primates (NHPs) and the associated differences in physiology and neurobiology between NHPs that exhibit depressive behavior and those that do not, in order to manage these monkeys effectively in captive environments. Proximate and ultimate factors that may contribute to the likelihood of depression are discussed, as well as similarities to human depression, and potential therapeutic interventions. The available data derive only from macaques (*Macaca mulatta* and *Macaca fascicularis*); thus, the generalizability of these data to other species is an open question. Much of the data come from our laboratory, where we have studied depressive behavior in adult female cynomolgus monkeys for 28 years. It is important to keep in mind throughout this chapter that these monkeys consumed a semipurified diet, designed to mimic a typical Western diet, with significant amounts of mostly saturated fat (percent of calories = 42–44) and cholesterol (0.25–0.29 mg/kcal, similar to a human consumption of ~500 mg/day; Shively et al. 1997b, 2008, 2015). Reports of depression from other laboratories discussed in this chapter were from macaques fed standard monkey chow (10% of calories from fat, trace amounts of cholesterol).

WHAT DOES DEPRESSIVE BEHAVIOR LOOK LIKE?

In 1988, we began observing and recording the frequency and percent of time spent in a behavior termed "depressive," in which the monkeys sit in a slumped or collapsed body posture, which may include the arms across the body or self-clasping (Figure 8.1A), accompanied by a lack of responsivity to environmental events. We only record this behavior as "depressive" if the animal's eyes are open, in order to distinguish it from resting. This behavior is distinct from submissive behavior, in which the monkey's eyes are up; the monkey is attentive to social signals; the eyebrows are often raised; the monkey may have a closed mouth, silent bared-teeth display, or "grimace";

Figure 8.1 Definition of depressive behavior. (A) Depressive behavior defined as sitting in a slumped or collapsed body posture, which may include the arms across the body or self-clasping, accompanied by a lack of responsivity to environmental events. This behavior is only recorded as depression if the eyes are open to distinguish it from resting. (B) Depressed female cynomolgus monkeys also spend more time in body contact with another monkey (Shively et al. 2005). (C) Depressive behavior is distinct from submissive behavior in which the monkey's eyes are up; the monkey is attentive to social signals; the eyebrows are often raised; the monkey may have a closed mouth, silent bared-teeth display, or "grimace"; may lipsmack; and the body is made small in posture. (D) Typical dominance behavior, shown to give an example of the range of species-typical behavior which contrasts with "depressive" behavior. Dominance behavior usually includes eye contact, lowered eyebrows, piloerection, a forward and open stance, and the mouth may be open, in an open-mouth threat, as depicted.

may lipsmack; and the body is made small in posture (Figure 8.1C). Figure 8.1D shows typical dominance behavior to provide an example of the range of species-typical behaviors which contrast with "depressive" behavior. Dominance behavior usually includes eye contact, lowered eyebrows, piloerection, a forward and open stance, and the mouth may be open in an open-mouth threat, as in this example. The depressive behavior we were studying was reminiscent of that described in infant macaques removed from their mothers, and adults following separation from their family environment, or the loss of a cage mate (Seay et al. 1962; Suomi et al. 1975; Rasmussen and Reite 1982). We have observed this depressive behavior in three separate groups of adult female cynomolgus monkeys (a total of 120 animals) (Shively et al. 1997b, 2005, 2015). Interobserver agreement in the identification of depressive behavior was >92% in all experiments. A similar behavior has been recently described by other laboratories for adult male and female cynomolgus, and for male rhesus macaques (Camus et al. 2013a,b, 2014; Hennessy et al. 2014).

HOW IS DEPRESSIVE BEHAVIOR MEASURED?

Definitions of types of *observational sampling methods* may be found in Altmann (1974). Hennessy et al. (2014) studied adult male rhesus in single cages that had been brought indoors from outdoor social groups. They used *focal animal observations*, during the first week after moving the animals, to record a hunched posture, defined as sitting with head level lower than the shoulders, limbs at the center of the body, no movement (except yawning or scratching), with the eyes open or unable to determine whether they were open or not. Using this definition, they categorized 18 of 26 animals as depressed. When they also included daytime sleeping or lying down during the first 2 weeks, 23 of the 26 adult males were categorized as depressed. The high levels of depression were attributed to the use of a video camera with no person present since the presence of a human observer is arousing, and thus likely to reduce observations of depressive behavior.

Camus et al. (2013a,b, 2014) used *scan sampling* (240 30-s scans, every 2 min during a daily 2-h block for 6 days) to record a broad array of activity, postural, social, and stereotypic behaviors. Factor analysis revealed five factors, one of which (Factor D) included high levels of inactivity, slumped posture (head below shoulders), and low levels of cage investigation and self-directed behaviors. Using this approach, 4 of 37 single-caged male and 5 of 80 socially housed female cynomolgus monkeys, and 18 of 40 single-caged male and 6 of 35 socially housed female rhesus macaques met the definition of depressed.

We consider depressive behavior a state (Altmann 1974), rather than an instantaneous behavior, like aggression, and thus, usually report our results as percent of time spent in depressive behavior. However, frequency is also always recorded, and as one might expect, the frequency and duration of depressive behavior are highly positively correlated ($r = 0.82$, $n = 42$, $p < .001$, unpublished).

As discussed above, several behavioral sampling strategies can be successfully used to study depressive behavior. Scan sampling is especially effective when you are recording a limited number of behaviors performed by all animals living in small groups. If a sufficient number of scans are performed in a given time period, the data should approximate the time spent in the target behaviors. As described above, Camus et al. (2013a,b, 2014) recorded 90 scans every 2 min, in 30-min observation sessions once per day, for 6 days in 40 single-caged macaques. We have recorded 180 scans, every 20 s in 10-min observations, counterbalanced for time of day, over 1 week in 42 macaques housed in social groups of 4. This is an example of a high-density, short-term observation strategy. Alternatively, focal sampling may be used, which allows for the recording of both frequency and duration of a larger number of target behaviors. In our laboratory, we have performed 10-min focal animal samples three times every 2 weeks, counterbalanced for time of day, over a 3-year period. This is an example of a low-density, long-term observation strategy.

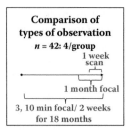

Figure 8.2 Experiment comparing types of observation appropriate to measure depressive behavior. We compared the percent of time spent in the depressed behavior as measured by: (1) three 10-min focals every 2 weeks for 18 months, (2) three 10-min focals every 2 weeks for the last month only of the 18-month time period, and (3) scans every 20s for 10min daily during the last week of the 18-month time period.

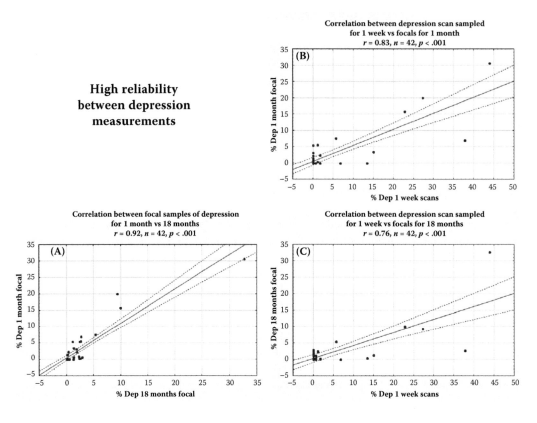

Figure 8.3 Correlations between types of observation used to measure depressive behavior. Different observation methods, density, and total time yielded similar results. (A) Focal samples for 1 and 18 months: $r = 0.92$, $p < .001$; (B) 1 week of scans and 1 month of focal: $r = 0.83$, $p < .001$; and (C) 1 week of scans and 18 months of focal: $r = 0.76$, $p < .001$ (all p-values are the result of two-sided tests, unpublished data).

We were curious about whether the observational strategy employed might affect the assessment of depression. To test this, we observed 42 monkeys (Figure 8.2) and compared the percentage of time spent in the depressed behavior as measured by: (1) three 10-min focal animal samples every 2 weeks for 18 months, (2) three 10-min focals every 2 weeks only for the last month of the 18-month time period, and (3) scans every 20s for 10min daily during the last week of the

18-month time period. Correlations among all methods were positive and high [focal samples for 1 and 18 months: $r = 0.92$, $p < .001$; 1 week of scans and 18 months of focals: $r = 0.76$, $p < .001$; and 1 week of scans and 1 month of focals: $r = 0.83$, $p < .001$ (all p-values are the result of two-sided tests, Shively et al., unpublished data; Figure 8.3)]. Thus, observational methods, observational densities, and total observation time did not significantly affect the recording of depressive behaviors.

We also examined monthly patterns of depressive behavior over a 1-year period. We found that there were three patterns of depressive behavior. Twenty-five percent of the monkeys displayed depressive behavior during 2 or fewer months throughout the year (categorized as "rarely or never"). Thirty-three percent of the monkeys displayed depressive behavior during 3–9 months throughout the year (categorized as "intermittently"), while the remaining 42% of the monkeys displayed depressive behavior during 10–12 months of the year (categorized as "most of the time"). The number of months in which a monkey was observed to display depressive behavior was highly positively correlated with the average time spent in depressive behavior over a 3-year period ($r = 0.84$, $p < .001$, two-sided test, Shively et al., unpublished data).

OTHER BEHAVIORS OF DEPRESSED MONKEYS

Depressed monkeys also have other behavioral characteristics. Depressed male and female macaques spend less time in locomotion, have higher levels of immobility, and lower rates of all species-typical behaviors than nondepressed macaques (Shively et al. 2005; Camus et al. 2013a; Hennessy et al. 2014). We also observed that depressed female cynomolgus monkeys were less aggressive, spent less time alone (out of arm's reach of another monkey), less time close to, and more time in body contact with another monkey than nondepressed female cynos as shown in Figure 8.1B (Shively et al. 2005). Since depressed monkeys are relatively inattentive to environmental events, it may be that sitting in body contact allows the depressed animal to respond to the other animals' startle and orienting behaviors to potentially threatening environmental events. Similarly, since depressed animals have such low activity levels, body contact may aid in thermoregulation.

PATHOPHYSIOLOGY OF DEPRESSED MONKEYS

There are sweeping differences in the physiology of depressed versus nondepressed monkeys, defined here as those falling above (depressed) or below (nondepressed) the mean of the distribution of time spent in depressive behavior. The hypothalamic–pituitary–adrenal (HPA) axis is perturbed in depressed compared to nondepressed monkeys. Depressed monkeys are insensitive to glucocorticoid-negative feedback in dexamethasone suppression tests and have higher cortisol responses to corticotropin-releasing hormone (CRH) challenge (Shively et al. 1997b, 2002, 2005). Depressed cynomolgus monkeys also exhibit ovarian dysfunction with preserved menses. Progesterone concentrations in the luteal phase are lower than those of nondepressed monkeys. Monkeys with low luteal phase progesterone concentrations also have comparatively low follicular phase estradiol concentrations, suggesting that depressed monkeys are estrogen-deficient (Adams et al. 1985; Shively et al. 2005).

Relative to their nondepressed counterparts, depressed monkeys have higher 24-h heart rates recorded via telemetry, higher total plasma cholesterol (TPC), lower high-density lipoprotein (HDL) cholesterol, and a higher ratio of circulating omega 6:omega 3 fatty acids (Shively et al. 2005; Chilton et al. 2011). In addition, depressed monkeys have lower body weights, lower body mass indices, and lower circulating insulin-like growth factor 1 (IGF-1) concentrations (Shively et al. 2005). The skeletal system is also perturbed. Depressed monkeys have higher bone alkaline phosphatase and lower whole body bone mineral content, as measured by dual X-ray absorptiometry (Shively et al. 2005).

High cortisol concentrations, poor ovarian function, lower activity levels, high TPC, low HDL cholesterol concentrations, and depression are all risk factors for atherosclerosis, which causes atherothrombotic coronary heart disease (Shively et al. 2009). We recorded behavior and measured the extent of coronary artery atherosclerosis in 36 adult female cynomolgus monkeys that consumed a Western-like diet for 4 years. Depressive behavior accounted for 53% of the variance in the extent of coronary artery atherosclerosis (plaque size and mean percent of time depressed: $r = 0.73$, $p < .001$, two-sided test) (Shively et al. 2008). Monkeys that fell above the mean time spent in depressed behavior had almost four times the atherosclerosis of those that fell below the mean (Shively et al. 2008). These monkeys were divided into quartiles based on the distribution of time spent depressed. Over the course of the experiment, 56% of the monkeys in the highest quartile of depression died (5/9), whereas only one (11%) or two (22%) died in each of the other three quartiles. Thus, behavioral depression was associated with increased mortality (correlation between days lived and percent of time spent in depressed behavior: $r = -0.41$, $p = .02$). Causes of death varied; for example, two of the animals died of trauma inflicted by cage mates, two did not recover from sedation, and one became ketotic and died. Although complete gross and histological pathology evaluations were done, in some cases, causes of death were not clear (Shively et al. 2005).

These observations indicate that the HPA, reproductive, cardiovascular, and autonomic nervous systems, and energy, lipid, and bone metabolism are perturbed, and mortality is increased in monkeys that exhibit depressive behavior. These differences may affect the outcomes of many studies, confounding research results, or producing results that are not replicable. These concerns underscore the importance of prevention, diagnosis, and treatment of depression in laboratory primates. Prevention strategies are presented in the Factors That Increase the Likelihood of Depression section, and treatment strategies are discussed in the Therapeutic Interventions section. Next, we describe observed perturbations in the central nervous system.

NEUROBIOLOGY OF DEPRESSED MONKEYS

Several studies comparing depressed and nondepressed monkeys reveal differences in structure and function of the brain. Using positron emission tomography (PET), we found reduced 5-HT1A receptor binding potential in behaviorally depressed compared with nondepressed animals. This reduction was observed bilaterally across 11 areas, including the raphe nucleus, amygdala, hippocampus (HC), and anterior cingulate cortex, and seemed to be linearly related to the severity of depression (Shively et al. 2006).

The glucocorticoid-negative feedback insensitivity in depressed NHPs mentioned in the previous section suggests the possibility of stress effects in the HC. The HC is rich in glucocorticoid receptors. These receptors may be downregulated in response to chronic stress, resulting in less sensitivity of the HPA axis to circulating glucocorticoids (Sapolsky 2000). The HC is functionally heterogeneous; the anterior HC plays a larger role in mood- and emotion-related functioning, whereas the posterior HC is engaged more in memory functions (Bannerman et al. 2004). We measured postmortem HC volume unilaterally and found reduced anterior HC volume in behaviorally depressed compared with nondepressed adult female cynomolgus monkeys ($n = 6$/group) matched on circulating estradiol, progesterone, and basal cortisol levels; social status; estimated age; and body weight (Willard et al. 2009). We then used magnetic resonance imaging to noninvasively assess HC volume bilaterally in ovariectomized cynomolgus females that were or were not depressed ($n = 6$/group), matched on social status, body weight (BW), and basal cortisol. There were global reductions in HC volume bilaterally in depressed versus nondepressed monkeys. The volumes of the posterior, as well as the anterior, portions were reduced, raising the possibility that posterior HC volume may be relatively protected by ovarian steroids in intact females (Willard et al. 2011).

The changes in serotonin neurotransmission and smaller HC volumes are consistent with human depression. Also, similar to human depression, there is no single biomarker for the diagnosis of depression. These differences in central nervous system structure and function in depressed versus nondepressed monkeys may also affect the outcome of many studies, underscoring again the importance of prevention, diagnosis, and treatment of depression in laboratory primates. Prevention strategies are discussed in the following section.

FACTORS THAT INCREASE THE LIKELIHOOD OF DEPRESSION

Depressive behavior is more common in subordinate than dominant female cynomolgus macaques. Weekly focal behavior observations of 42 females that lived for 26 months in social groups of about 4 females each showed that 61% of subordinates displayed depressive behavior, whereas only 10% of dominants were ever observed in this behavior (Shively et al. 1997b). The stress associated with low social status may increase the likelihood of depressive behavior (Shively et al. 1997a; Shively 1998). However, social subordination and depression are not homologous as 39% of subordinates did not display depressive behavior, and a few dominants did, suggesting that there are individual differences in stress sensitivity and resilience (Bethea et al. 2008). It is notable that rates of depression in the human population are inversely related to socioeconomic status (Lorant et al. 2003; Adler and Rehkopf 2008).

We characterized the behavior and physiology of 36 adult female cynomolgus macaques housed in single cages for 12 months, *prior* to moving them to social group housing for 12 months. Monkeys that fell above the mean of the distribution of percent of time spent in depressed behavior ("depressed") *when subsequently socially housed* were compared to those who fell below the mean ("nondepressed"). Compared to their nondepressed counterparts, depressed monkeys also were less physically active, had lower IGF-1, a lower cortisol response to CRH, higher TPC, a higher TPC:HDLC ratio, and higher average heart rates recorded over 24h while singly housed during the previous year. Single caging is stressful; thus, it may be that individuals that are more physiologically reactive to environmental stressors are at greater risk for depression. Consistent with this hypothesis, the trait-like characteristic of higher emotionality in infancy predicted depressive responses to single caging in adulthood in a study of rhesus males (Hennessy et al. 2014). Interestingly, depressive behavior while single-caged did not predict subsequent depressive behavior while socially housed ($r = 0.20$, NS, 1 year in each condition) (Shively et al. 2002, 2005).

Other factors that may increase the likelihood of depressive behavior include separation from long-term cage mates (Suomi et al. 1975; Rasmussen and Reite 1982), single caging (Suomi et al. 1975; Hennessy et al. 2014), and parturition (Chu et al. 2014). In addition, there may be species differences (rhesus versus cynomolgus; Camus et al. 2014), and differences between captive and wild-born animals (Camus et al. 2013b), although more data are needed concerning these two factors.

Thus, in terms of behavioral management approaches to prevention, the likelihood of depressive behavior may be minimized by avoiding adverse social situations that promote depressive behavior, including separation from long-term cage mates, and single caging.

ADAPTIVENESS OF DEPRESSIVE BEHAVIOR

Darwin (1887) said that "Pain or suffering of any kind, if long continued, causes depression and lessens the power of action; yet it is well adapted to make a creature guard itself against any great or sudden evil" (Darwin 1887, pp. 51–52). Studying depression from an evolutionary perspective may contribute to the understanding of the ways in which low mood might increase an organism's ability to cope under adverse conditions. Depression has been hypothesized to have several adaptive functions, including communicating the need for help to solicit resources, fostering disengagement from commitments to unreachable goals, regulating patterns of investment, conserving or reallocating energy and effort, signaling social defeat, avoiding risks, minimizing losses, promotion of analysis (rumination in humans), and promoting recovery from illness (Nesse 2000; Durisko et al. 2015).

PARALLELS BETWEEN HUMAN AND NONHUMAN PRIMATE DEPRESSION

Depression in humans and NHPs appears to share several characteristics. These include *risk factors*, such as low social status, early experience, and temperament (Lorant et al. 2003; Compas et al. 2004; Adler and Rehkopf 2008; Lopizzo et al. 2015), and *behaviors*, such as low social interaction, anhedonia, and immobility (American Psychiatric Association 2000). Shared characteristics also include aspects of physiology, including perturbations in lipid metabolism, and the HPA, autonomic, cardiovascular, skeletal, and central nervous systems (Lamers et al. 2013; Villanueva 2013; Gan et al. 2014; Bassett 2015; Rosenblat et al. 2016). These far-reaching perturbations in multiple systems likely account, at least in part, for the high morbidity and mortality (excluding suicide) associated with depression in humans, as well as NHPs (Kouzis et al. 1995; Brådvik and Berglund 2001; von Ammon Cavanaugh et al. 2001; Connerney et al. 2010).

THERAPEUTIC INTERVENTIONS

Therapeutic interventions for human depression have modest efficacy (Hansen et al. 2005; Turner et al. 2008; Pradhan et al. 2015). Little is known about successful interventions for depression in NHPs. We treated 21 of 42 adult female cynomolgus monkeys with sertraline HCl (Zoloft) for 18 months. Half of those treated were depressed, and half were not. Sertraline reduced anxious, but not depressive behavior (Shively et al. 2015). Since social disruption and the social isolation of single caging increase the likelihood of depression, manipulating the social environment may be a more efficacious approach for laboratory primates. We manipulated social status in 42 adult female cynomolgus monkeys and observed that subordinates who became dominant had lower rates of depressive behavior than subordinates who remained subordinates (Shively et al. 1997b). As mentioned above, subordinates are more likely than dominants to display depressive behavior. The fact that changing subordinates to dominants reduces depressive behavior suggests that this is a causal relationship. It appears that the stress associated with low social status causes depression in susceptible animals, and making them dominants reduces that stress (Shively 1997, 1998).

In human beings, higher rates of depression are associated with consumption of a Western diet, and lower rates of depression are associated with adherence to a Mediterranean diet pattern. The Mediterranean diet is rich in monounsaturated fats and has a low omega 6:omega 3 fatty acid ratio, both of which have been associated with low rates of depression. Monkey chow has amounts of monounsaturated fats and an omega 6:omega 3 fatty acid ratio similar to a Western diet and dissimilar to a Mediterranean diet. It is possible that a simple diet change (e.g., providing avocado as an enrichment food as it is high in omega 3 and monounsaturated fat) could reduce depression in laboratory primates. Thus, one behavioral management strategy that may reduce depressive behavior is diet manipulation.

Exercise is a relatively effective antidepressant in people, and mechanisms linking muscle function to neural serotonergic pathways have been identified (Agudelo et al. 2014). Laboratory primates often have very low activity levels. Finding ways to increase exercise levels may combat depression. Similarly, obesity is associated with depression in human beings and is also prevalent among laboratory primates (Mansur et al. 2015). Behavioral management strategies that increase exercise and combat obesity through diet change may be effective in reducing depressive behavior.

SUMMARY

In summary, there is no single biomarker for diagnosis of depression in humans or NHPs. However, depression can be relatively easy to assess using dense scan sampling observational techniques for as little as a week. Depressed animals have perturbed HPA, autonomic, and ovarian functions; alterations in lipid metabolism; altered neurobiological structure and function; more coronary artery atherosclerosis; and higher mortality. These multisystem perturbations underscore the importance of diagnosing and treating depression in laboratory primates as perturbed physiology may negatively impact research, as well as health and psychological well-being. Therapeutic interventions that may be helpful include behavioral management strategies that (1) reduce stress via social engineering to make subordinates dominant or by providing long-term stable companionship, (2) increase dietary omega 3 fatty acids and monounsaturated fats, (3) increase opportunities for exercise, and (4) reduce obesity.

ACKNOWLEDGMENTS

This work was supported by NIH Grants RO1 HL39789, RO1 HL87103, RO1 MH56881, R21 MH086731, RO1 HL087103, and the John D. and Catherine T. MacArthur Foundation.

REFERENCES

Adams, M. R., J. R. Kaplan, T. B. Clarkson, and D. R. Koritnik. 1985. Ovariectomy, social status, and athero-sclerosis in cynomolgus monkeys. *Arteriosclerosis* 5 (2):192–200.

Adler, N. E. and D. H. Rehkopf. 2008. U.S. disparities in health: Descriptions, causes, and mechanisms. *Annu Rev Public Health* 29:235–52. doi:10.1146/annurev.publhealth.29.020907.090852.

Agudelo, L. Z., T. Femenia, F. Orhan, M. Porsmyr-Palmertz, M. Goiny, V. Martinez-Redondo, J. C. Correia et al. 2014. Skeletal muscle PGC-1α1 modulates kynurenine metabolism and mediates resilience to stress-induced depression. *Cell* 159 (1):33–45. doi:10.1016/j.cell.2014.07.051.

Altmann, J. 1974. Observational study of behavior: Sampling methods. *Behaviour* 49 (3/4):227–67.

American Psychiatric Association. 2000. *Diagnostic and Statistical Manual of Mental Disorders*, 4th Edition. Washington, DC: American Psychiatric Publishing, Inc.

Bannerman, D. M., J. N. Rawlins, S. B. McHugh, R. M. Deacon, B. K. Yee, T. Bast, W. N. Zhang, H. H. Pothuizen, and J. Feldon. 2004. Regional dissociations within the hippocampus—Memory and anxiety. *Neurosci Biobehav Rev* 28 (3):273–83. doi:10.1016/j.neubiorev.2004.03.004.

Bassett, D. 2015. A literature review of heart rate variability in depressive and bipolar disorders. *Aust N Z J Psychiatry* 50 (6):511–9. doi:10.1177/0004867415622689.

Bethea, C. L., M. L. Centeno, and J. L. Cameron. 2008. Neurobiology of stress-induced reproductive dysfunction in female macaques. *Mol Neurobiol* 38 (3):199–230. doi:10.1007/s12035-008-8042-z.

Brådvik, L. and M. Berglund. 2001. Late mortality in severe depression. *Acta Psychiatr Scand* 103 (2):111–6.

Camus, S. M., C. Blois-Heulin, Q. Li, M. Hausberger, and E. Bezard. 2013a. Behavioural profiles in captive-bred cynomolgus macaques: Towards monkey models of mental disorders? *PLoS One* 8 (4):e62141. doi:10.1371/journal.pone.0062141.

Camus, S. M., C. Rochais, C. Blois-Heulin, Q. Li, M. Hausberger, and E. Bezard. 2013b. Birth origin differentially affects depressive-like behaviours: Are captive-born cynomolgus monkeys more vulnerable to depression than their wild-born counterparts? *PLoS One* 8 (7):e67711. doi:10.1371/journal.pone.0067711.

Camus, S. M., C. Rochais, C. Blois-Heulin, Q. Li, M. Hausberger, and E. Bezard. 2014. Depressive-like behavioral profiles in captive-bred single- and socially-housed rhesus and cynomolgus macaques: A species comparison. *Front Behav Neurosci* 8:47. doi:10.3389/fnbeh.2014.00047.

Chilton, F. H., T. C. Lee, S. L. Willard, P. Ivester, S. Sergeant, T. C. Register, and C. A. Shively. 2011. Depression and altered serum lipids in cynomolgus monkeys consuming a western diet. *Physiol Behav* 104 (2):222–7. doi:10.1016/j.physbeh.2011.01.013.

Chu, X. X., J. Dominic Rizak, S. C. Yang, J. H. Wang, Y. Y. Ma, and X. T. Hu. 2014. A natural model of behavioral depression in postpartum adult female cynomolgus monkeys (*Macaca fascicularis*). *Dongwuxue Yanjiu* 35 (3):174–81. doi:10.11813/j.issn.0254-5853.2014.3.174.

Compas, B. E., J. Connor-Smith, and S. S. Jaser. 2004. Temperament, stress reactivity, and coping: Implications for depression in childhood and adolescence. *J Clin Child Adolesc Psychol* 33 (1):21–31. doi:10.1207/s15374424jccp3301_3.

Connerney, I., R. P. Sloan, P. A. Shapiro, E. Bagiella, and C. Seckman. 2010. Depression is associated with increased mortality 10 years after coronary artery bypass surgery. *Psychosom Med* 72 (9):874–81. doi:10.1097/PSY.0b013e3181f65fc1.

Darwin, F., ed. 1887. *The Life and Letters of Charles Darwin, Including an Autobiographical Chapter.* London: John Murray.

Durisko, Z., B. H. Mulsant, and P. W. Andrews. 2015. An adaptationist perspective on the etiology of depression. *J Affect Disord* 172:315–323. doi:10.1016/j.jad.2014.09.032.

Gan, Y., Y. Gong, X. Tong, H. Sun, Y. Cong, X. Dong, Y. Wang et al. 2014. Depression and the risk of coronary heart disease: A meta-analysis of prospective cohort studies. *BMC Psychiatry* 14:371. doi:10.1186/s12888-014-0371-z.

Hansen, R. A., G. Gartlehner, K. N. Lohr, B. N. Gaynes, and T. S. Carey. 2005. Efficacy and safety of second-generation antidepressants in the treatment of major depressive disorder. *Ann Intern Med* 143 (6):415–26.

Hennessy, M. B., B. McCowan, J. Jiang, and J. P. Capitanio. 2014. Depressive-like behavioral response of adult male rhesus monkeys during routine animal husbandry procedure. *Front Behav Neurosci* 8:309. doi:10.3389/fnbeh.2014.00309.

Kouzis, A., W. W. Eaton, and P. J. Leaf. 1995. Psychopathology and mortality in the general population. *Soc Psychiatry Psychiatr Epidemiol* 30 (4):165–70.

Lamers, F., N. Vogelzangs, K. R. Merikangas, P. de Jonge, A. T. Beekman, and B. W. Penninx. 2013. Evidence for a differential role of HPA-axis function, inflammation and metabolic syndrome in melancholic versus atypical depression. *Mol Psychiatry* 18 (6):692–9. doi:10.1038/mp.2012.144.

Lopizzo, N., L. Bocchio Chiavetto, N. Cattane, G. Plazzotta, F. I. Tarazi, C. M. Pariante, M. A. Riva, and A. Cattaneo. 2015. Gene-environment interaction in major depression: Focus on experience-dependent biological systems. *Front Psychiatry* 6:68. doi:10.3389/fpsyt.2015.00068.

Lorant, V., D. Deliege, W. Eaton, A. Robert, P. Philippot, and M. Ansseau. 2003. Socioeconomic inequalities in depression: A meta-analysis. *Am J Epidemiol* 157 (2):98–112.

Mansur, R. B., E. Brietzke, and R. S. McIntyre. 2015. Is there a "metabolic-mood syndrome"? A review of the relationship between obesity and mood disorders. *Neurosci Biobehav Rev* 52:89–104. doi:10.1016/j.neubiorev.2014.12.017.

Nesse, R. M. 2000. Is depression an adaptation? *Arch Gen Psychiatry* 57 (1):14–20.

Pradhan, B., T. Parikh, R. Makani, and M. Sahoo. 2015. Ketamine, transcranial magnetic stimulation, and depression specific yoga and mindfulness based cognitive therapy in management of treatment resistant depression: Review and some data on efficacy. *Depress Res Treat.* Article ID 842817. doi:10.1155/2015/842817.

Rasmussen, K. L. and M. Reite. 1982. Loss-induced depression in an adult macaque monkey. *Am J Psychiatry* 139 (5):679–81.

Rosenblat, J. D., J. M. Gregory, A. F. Carvalho, and R. S. McIntyre. 2016. Depression and disturbed bone metabolism: A narrative review of the epidemiological findings and postulated mechanisms. *Curr Mol Med* 16 (2):165–78.

Sapolsky, R. M. 2000. Glucocorticoids and hippocampal atrophy in neuropsychiatric disorders. *Arch Gen Psychiatry* 57 (10):925–35.

Seay, B., E. Hansen, and H. F. Harlow. 1962. Mother-infant separation in monkeys. *J Child Psychol Psychiatry* 3:123–32.

Shively, C. A. 1997. Behavior and physiology of social stress and depression in female cynomolgus monkeys. *Brain Res* 41 (8):871–82.

Shively, C. A. 1998. Social subordination stress, behavior, and central monoaminergic function in female cynomolgus monkeys. *Biol Psychiatry* 44 (9):882–91.

Shively, C. A., D. P. Friedman, H. D. Gage, M. C. Bounds, C. Brown-Proctor, J. B. Blair, J. A. Henderson, M. A. Smith, and N. Buchheimer. 2006. Behavioral depression and positron emission tomography-determined serotonin 1A receptor binding potential in cynomolgus monkeys. *Arch Gen Psychiatry* 63 (4):396–403. doi:10.1001/archpsyc.63.4.396.

Shively, C. A., K. A. Grant, R. L. Ehrenkaufer, R. H. Mach, and M. A. Nader. 1997a. Social stress, depression, and brain dopamine in female cynomolgus monkeys. *Ann N Y Acad Sci* 807:574–7.

Shively, C. A., K. Laber-Laird, and R. F. Anton. 1997b. Behavior and physiology of social stress and depression in female cynomolgus monkeys. *Biol Psychiatry* 41 (8):871–82. doi:10.1016/s0006–3223(96)00185-0.

Shively, C. A., D. L. Musselman, and S. L. Willard. 2009. Stress, depression, and coronary artery disease: Modeling comorbidity in female primates. *Neurosci Biobehav Rev* 33 (2):133–44. doi:10.1016/j.neubiorev.2008.06.006.

Shively, C. A., T. C. Register, M. R. Adams, D. L. Golden, S. L. Willard, and T. B. Clarkson. 2008. Depressive behavior and coronary artery atherogenesis in adult female cynomolgus monkeys. *Psychosom Med* 70 (6):637–45. doi:10.1097/PSY.0b013e31817eaf0b.

Shively, C. A., T. C. Register, S. E. Appt, and T. B. Clarkson. 2015. Effects of long-term sertraline treatment and depression on coronary artery atherosclerosis in premenopausal female primates. *Psychosom Med* 77 (3):267–78. doi:10.1097/psy.0000000000000163.

Shively, C. A., T. C. Register, D. P. Friedman, T. M. Morgan, J. Thompson, and T. Lanier. 2005. Social stress-associated depression in adult female cynomolgus monkeys (*Macaca fascicularis*). *Biol Psychol* 69 (1):67–84. doi:10.1016/j.biopsycho.2004.11.006.

Shively, C. A., J. K. Williams, K. Laber-Laird, and R. F. Anton. 2002. Depression and coronary artery atherosclerosis and reactivity in female cynomolgus monkeys. *Psychosom Med* 64 (5):699–706.

Suomi, S. J., C. D. Eisele, S. A. Grady, and H. F. Harlow. 1975. Depressive behavior in adult monkeys following separation from family environment. *J Abnorm Psychol* 84 (5):576–8.

Turner, E. H., A. M. Matthews, E. Linardatos, R. A. Tell, and R. Rosenthal. 2008. Selective publication of antidepressant trials and its influence on apparent efficacy. *N Engl J Med* 358 (3):252–60. doi:10.1056/NEJMsa065779.

Villanueva, R. 2013. Neurobiology of major depressive disorder. *Neural Plast.* Article ID 873278. doi:10.1155/2013/873278.

von Ammon Cavanaugh, S., L. M. Furlanetto, S. D. Creech, and L. H. Powell. 2001. Medical illness, past depression, and present depression: A predictive triad for in-hospital mortality. *Am J Psychiatry* 158 (1):43–8. doi:10.1176/appi.ajp.158.1.43.

Willard, S. L., J. B. Daunais, J. M. Cline, and C. A. Shively. 2011. Hippocampal volume in postmenopausal cynomolgus macaques with behavioral depression. *Menopause* 18 (5):582–6. doi:10.1097/gme.0b013e3181fcb47e.

Willard, S. L., D. P. Friedman, C. K. Henkel, and C. A. Shively. 2009. Anterior hippocampal volume is reduced in behaviorally depressed female cynomolgus macaques. *Psychoneuroendocrinology* 34 (10):1469–75. doi:10.1016/j.psyneuen.2009.04.022.

Antipredator Behavior
Its Expression and Consequences in Captive Primates

Nancy G. Caine
California State University San Marcos

CONTENTS

Considering that primates have lived alongside formidable predators throughout their evolutionary history, it is surprising that the importance of predator-related selection pressures in shaping the behavior of primates has not always been acknowledged. Early field studies of primates tended to focus on dominance, aggression, and mating—behaviors that are both frequent and readily observable. In contrast, predation events are rarely observed, in part because they are not common (if they were common, the prey species would be extinct) and in part because the presence of human observers probably deters the predatory activity. Furthermore, nocturnal predation is rarely, if ever, witnessed by diurnal human investigators. Consequently, one of the most influential early textbooks on primate behavior (Jolly, 1972) devoted only 6 of 356 pages to its chapter on predation; indeed, there was relatively little to say about the topic, considering that there had been so little research aimed at understanding it.

The situation has changed. Once primatologists began to accept the fact that primates are vulnerable to predatory threats from the air, the ground, and even from the water, attention was turned to understanding the behavioral consequences of predation-related selection pressures. The literature is now filled with interesting and provocative research about the ways in which primates detect, avoid, and react to their predators (Fichtel, 2012). In this chapter, I hope to convince readers that primates maintained in captivity for research purposes have not lost their most basic antipredator instincts and that the overall management, and especially the behavioral management, of captive primates should take into account the ways in which monkeys and apes may be affected by research- and husbandry-related activity that taps into deeply ingrained antipredator reactions.

I begin by describing the types of animals that serve as predators of primates and the responses of monkeys and apes when they encounter one of these predators. Next, I identify three ways in which an understanding of predation-related adaptations by primates can enhance the well-being of laboratory primates and guide our research.

THE PREDATORS OF PRIMATES: RAPTORS, FELIDS, AND SNAKES

Raptors (hawks, eagles, owls, and related birds) across the world include primates in their diet. In Manu Park, Peru, harpy eagles (*Harpia harpyja*) are thought to be the most significant predators of capuchin monkeys (*Cebus* spp.) (Terborgh, 1983). Bones found in or beneath 16 nests of African crowned eagles (*Stephanoaetus coronatus*) over the course of 34 months represented at least 204 individual monkeys from at least 7 species of primates in the Ivory Coast (McGraw et al., 2006). Goodman et al.'s (1993) catalog of reported incidents of avian predation on lemurs presents a horrifying picture of the vulnerability of these primates when they are both awake and asleep. For instance, the authors estimate that *Microcebus murinus* may lose up to 25% of their population every year owing to predation from two owl species (*Tyto alba* and *Asio madagascariensis*).

Raptors pose a potent threat to primates because they attack quietly, at great speed and with great accuracy, under cover of the canopy. Tamarins (*Saguinus mystax* and *S. fuscicollis*) that see an approaching raptor at close range sometimes simply let go of their perch and drop far to the ground rather than facing an almost definite capture in the location at which the raptor is aiming. Alternatively, they may freeze in place and remain frozen for many minutes, not certain whether the raptor has perched nearby, readying for another attack (Heymann, 1990). Given the risk, it is not surprising that primates have learned to react to raptors even before they see them; many primate species recognize and respond defensively to the calls of aerial predators (e.g., Hauser & Wrangham, 1990; Fichtel & Kappeler, 2002).

Felids and other carnivores also prey on primates. In Africa, leopards are a real threat to, among other species, baboons (*Papio* spp.; Hamilton, 1982) and even bonobos (D'Amour et al., 2006). A variety of felids in South America, including nocturnal species such as ocelots (*Leopardus pardalis*) and margays (*L. wiedii*), eat arboreal primates (Miranda et al., 2005; Bianchi & Mendes, 2007); leopards and tigers in India and Southeast Asia attack large primates, including humans (Fichtel, 2012). Two non-felid carnivores, the tayra (*Eira barbara*) and the fossa (*Cryptoprocta ferox*), are potent predators of Neotropical (Bezerra et al., 2009) and Malagasy (Wright et al., 1997) primates, respectively. The nocturnal and stealthy hunting strategies of many carnivores make it particularly difficult to know when a predation event has occurred, but the defensive reactions of primates to these cats (alarm calls, fleeing, mobbing, etc.) provide evidence of the threat they pose to primates.

One of the most significant recent developments in the study of primates as prey is the Snake Detection Theory (SDT; Isbell, 2006, 2009). The SDT proposes that early primates, particularly those that evolved in areas with venomous snakes, were under powerful selection pressures to detect and avoid potentially fatal encounters with snakes. Isbell proposes that many of the characteristics that define the visual system of the primate order, including improved orbital convergence, retinal specializations, and central perceptual systems that specialize in detection of detail and movement, evolved in response to these snake-related selection pressures. What evidence is there to support these claims?

Humans and nonhuman primates are quick to detect snakes. A long list of studies using a variety of experimental paradigms (e.g., computer-based visual search and eye-tracking) has shown that laboratory-reared, snake-naïve Japanese macaques (*Macaca fuscata*; Shibasaki & Kawai, 2009) and humans, from infants (LoBue & DeLoache, 2010) to adults (LoBue & DeLoache, 2008; Yorzinski et al., 2014), detect images of snakes faster than nonthreatening objects. Under challenging attentional conditions, snakes are detected faster than other dangerous, animate objects (spiders)

(Soares et al., 2014). Because laboratory-reared macaques, human infants, and young children lack experience with snakes, it is likely that rapid snake detection is innate.

Noticing a snake is obviously important, but the next step is equally important too: reacting to the snake that has been detected. Fear is a mechanism by which animals avoid or retreat from threatening stimuli, and thus we would predict that primates would react to snakes with a behavior that suggests fear. Indeed, they do. Many anecdotes reported by field primatologists (e.g., Tello et al., 2002), along with a growing number of field experiments (e.g., Quattara et al., 2009), confirm strong defensive reactions displayed by monkeys and apes when they discover a nearby snake. The most common initial reaction is vocal: monkeys issue loud alarm calls, and in many cases, these elicit and/or are accompanied by the approach of group mates that mob the snake, sometimes for long periods of time. While mobbing, the monkeys tend to be piloerected and highly agitated; following the exposure, the monkeys may remain very vigilant and alter their typical movement patterns for hours (e.g., Crowfoot, 2012).

THE IMPORTANCE OF BEING AFRAID

Neuroscientists have described a pathway in the visual system that allows for detection of and appropriate reaction to predators (i.e., startle, fear, and immediate avoidance). The mammalian "fear module," believed to have been shaped by selection pressures specifically associated with snakes (Öhman & Mineka, 2001), consists of a largely unconscious pathway that includes the superior colliculus, pulvinar, and amygdala (Öhman, 2005). The superior colliculus is known for its role in subcortical processing of visual stimuli coming from the retina and for its contributions to behavioral reactions to threat (e.g., startle). The superior colliculus communicates with the pulvinar, an area of the thalamus that includes nuclei that are unique to primates (Grieve et al., 2000; Berman & Wurtz, 2010). The critical role of snake-related selection pressures in primate visual evolution is supported by the fact that macaque (*M. fuscata*) pulvinar neurons are selectively responsive to photos of snakes, more so than to images of monkey faces, monkey hands, or geometric shapes (Van Le et al., 2013).

The fear module is completed by the amygdala, which is essential to the emotional reaction to threat. The importance of the amygdala in this module is demonstrated by the fact that damage to the amygdala results in diminished, altered, or absent fear responses in rhesus monkeys (e.g., Prather et al., 2001; Antoniadis et al., 2007). When presented with threatening or fear-related stimuli (e.g., snakes, fearful facial expressions), the human amygdala shows activation prior to conscious awareness, even when the images are presented in the peripheral visual field (Bayle et al., 2009; for a review see Öhman et al., 2007). The automatic activation of the amygdala when viewing threatening stimuli is important for an emotional reaction that, in turn, leads to protective actions such as freezing or jumping away from the stimulus.

ANTIPREDATOR BEHAVIOR IN CAPTIVE PRIMATES

Many people are surprised to know that captive primates retain most of the same reactions to predators as do their wild counterparts. In the 1980s, a series of studies at the University of Wisconsin showed that, for rhesus monkeys, fear of snakes is not exactly innate, but learning to be afraid happens quickly and easily. Captive rhesus exhibited a strong fear response when confronted with a snake even after many years of captivity (Cook et al., 1985). The fear response was characterized by behaviors such as withdrawal, cage shaking, staring, vocalizing, and piloerection. These monkeys were also significantly more hesitant to reach for a piece of food in the presence of both animate and inanimate snake stimuli. Laboratory-reared, snake-naïve rhesus monkeys did not show the same fear reactions to snake stimuli when exposed to those stimuli for the first time, but they quickly learned to fear snakes after two brief observations of a wild-reared monkey's fearful response toward the snake stimuli. The laboratory monkeys also continued to exhibit a fear response toward snakes several months after the observational conditioning. Furthermore, they selectively learned to fear snakes as opposed to fabric flowers, when observation videos were edited to depict a monkey responding fearfully to either a snake or a bouquet of flowers (Cook & Mineka, 1989).

New World monkeys of the genus *Saguinus* do not need experience with felid predators to respond defensively to them. Caine and Weldon (1989) prepared samples of fecal odors that were applied to wooden dowels in the indoor cages of two groups of captive-born red-bellied tamarins (*S. labiatus*). Control stimuli included fecal odors from noncarnivorous sympatric species, such as pacas (*Cuniculus paca*). The tamarins spent more time sniffing and inspecting the dowels laced with predator odors, but they also spent less time at each visit to those dowels as compared to dowels with control odors. One tamarin began to alarm call after she sniffed the predator feces. The findings from this study were later replicated with captive cotton-top tamarins (*S. oedipus*) by Buchanan-Smith et al. (1993).

Primates living in outdoor cages provide us with an opportunity to study animals that probably see an occasional predator (e.g., a raptor, snake, or cat) but have never been attacked. Geoffroy's marmosets (*Callithrix geoffroyi*) that were forced by darkness to retire to their sleeping areas while a rattlesnake model was outside of their cage woke up the following morning and immediately inspected the precise location where the rattlesnake had last been seen (Hankerson & Caine, 2004). Semi-free-ranging lemurs (Macedonia & Yount, 1991) and outdoor-housed marmosets (Searcy & Caine, 2003) react strongly to the calls of raptors. For marmosets, even a perched raptor is perceived as threatening; marmosets respond to owl and hawk models in perched postures with alarm calls and mobbing (Petracca & Caine, 2013; Neal & Caine, 2016). Outdoor-housed rhesus monkeys (*M. mulatta*) react to snake models with cautious approach and inspection, especially if the snakes are in a threatening posture (Etting & Isbell, 2014).

APPLIED ETHOLOGY: ANTIPREDATOR BEHAVIOR AND CAPTIVE MANAGEMENT

The inclusion of a chapter on antipredator behavior in a book on behavioral management of captive primates is predicated on the notion that one or more aspects of captive management can benefit from managers' understanding of primates as prey. I propose three such benefits, each of which applies both to animal welfare and potential concerns about confounding variables in research designs. Specifically, I believe that managers should consider the importance of providing safe (from the animals' point of view) sleeping options within the captive environment. Second, I suggest that stimuli that seem benign from a human's point of view may generate defensive or fearful reactions from captive primates, which can affect their well-being and potentially interfere with the results of experiments. Finally, I encourage investigators to take advantage of antipredator adaptations to create ecologically valid measures of fear and anxiety and, potentially, to make use of certain antipredator behaviors to enrich the captive environment.

Sleeping

Most primate species are highly diurnal. Their foveas are dominated by cones, which can only operate under conditions of rather bright light. Consequently, cones become useless (more or less so, as moonlight and cloud cover varies) at night. Rods function under conditions of low light, but they do not allow for detailed vision. The dangers of being awake and active at night, when primates are nearly blind, are obvious, and this danger is exacerbated by the fact that some of the most potent predators of primates are nocturnal hunters. The solution, of course, is to find the safest place to spend the night; failing to do so can have dire consequences. In Poco das Antas reserve in Brazil, the population of golden lion tamarins (*Leontopithecus rosalia*) declined from 330 to 220 in just 5 years, probably because tayras were able to penetrate the tree holes in which the tamarins were roosting (Franklin et al., 2007).

Studies of wild primates confirm that monkeys and apes tend to prefer sleeping sites that are off the ground, minimizing their susceptibility to predators that are unable to climb. Some species look for concealed spots in high locations, such as tree holes or tangles of vines, whereas others prefer more open perches (Anderson, 1998). Baboons (*Papio* spp.) are known to choose sleeping sites that minimize predation by leopards, climbing steep cliffs or emergent trees at dusk and staying there until morning (Hamilton, 1982; Matsumoto-Oda, 2015). Pruetz et al. (2008) studied the nest-building behavior of wild chimpanzees (*Pan troglodytes verus*) and found that within Assirik National Park, where leopards and lions (and, to a lesser extent, hyenas and wild dogs) pose significant risk to chimpanzees, the nests are built higher in the trees than in another, similar location in Senegal (Fongoli), where large predators have been eradicated. However, even in the Fongoli site, the chimpanzees preferred to sleep above the ground on most of the nights.

Considering the strong preferences that most wild primates show for sleeping locations that are safely elevated, it would not be surprising if captive primates maintain such a preference. Caine et al. (1992) gave three groups of captive red-bellied tamarins (*S. labiatus*) the option of three sleeping boxes on each of 42 (experiment 1) and 17 (experiment 2) nights. The boxes differed in how concealed the monkeys would be once inside the box (experiment 1), and also in height above the cage floor (experiment 2). The results were quite striking: the monkeys strongly and significantly preferred the boxes that were most concealing overall (the only opening was a hole in the front of the box), and they also preferred boxes that were not open from above. The strongest preference of all was for height: when given the choice of two otherwise identical boxes, they always chose the higher option, without exception (on 17 of 17 nights). Considering that these monkeys had always lived in indoor laboratory environments, the strength of these preferences is rather remarkable.

An interesting study of human sleeping preferences is also relevant to this topic. Spörrle and Stitch (2010) tested the hypothesis that human vulnerability to nocturnal attack from predators, and later from human enemies, in our ancestral environment led to strong behavioral tendencies to prefer sleeping places that afford a view of the entrance to the sleeping area (e.g., a cave), while remaining somewhat concealed from the entrance. Giving participants the opportunity to arrange furniture on floor plans with respect to windows and doors, the authors showed that people prefer to sleep facing, but not close to, the entrances to a room. Thus, despite the fact that doors to bedrooms are not usually the primary barrier between people and potential aggressors, people are most comfortable sleeping in a location within a room that gives them the best opportunity to detect and react to a breach of the room's security.

To my knowledge, there are no studies other than the one described here (Caine et al., 1992) that systematically evaluate multiple variables associated with sleeping preferences in captive nonhuman primates. However, to the extent that we can provide maximally elevated sleeping platforms with characteristics that best conform to the sleeping preferences exhibited by their wild counterparts (e.g., matching substrate preferences; Samson & Hunt, 2014), this should enhance the well-being of the animals. Furthermore, one could argue that the Spörrle and Stitch's (2010) study reflects the adaptations of "captive" humans, and if it does, the orientation of sleeping areas for captive nonhuman primates should also warrant our attention. But in addition to the reasonable, if subjective, conclusion that captive primates will enjoy greater well-being if they can sleep where they feel most safe, we must also assess the potential consequences to research projects that involve captive primates if these preferences are ignored.

To the extent that animals do not feel safe in their sleeping site, the amount or quality of sleep they get might be affected. As described in the following section, Caine (1990) found that tamarins entered their sleeping boxes later when they were being watched, and it seems logical that they spent less time asleep on those nights. If, in fact, other species of captive primates show preferences for certain sleeping conditions, and if, in fact, sleeping in less-preferred locations reduces sleep time (an empirical question that should be investigated), then the animals are potentially subject to all of the many physical and psychological consequences that have been identified in sleep-deprived human and nonhuman animals, including increased risk for obesity (Greer et al., 2013), memory deficits (Prince et al., 2014), immune system disruptions (Besedovsky et al., 2012), reduced testosterone levels (Cote et al., 2013), and many others. It behooves us to consider the possibility that failure to provide preferred sleeping conditions might introduce unrecognized sources of systematic or random error in our studies of metabolic function, cognition, immunology, and other common areas of research. Samson and Shumaker (2015) have studied sleep quality in captive orangutans and baboons with noninvasive monitoring of awakenings, postural shifts, sleep duration, and other measures taken from infrared videography. Behavioral management programs are in a position to use these and other methods to determine sleep preferences, and to then develop and implement desirable sleep-related "strategies" for captive primates.

Humans as Predators: The Assumption of Habituation

Captive primates, at least those who live indoors, rarely, if ever, come into contact with their natural predators. However, captive primates are in contact with humans on a daily basis, and humans sometimes function as predators insofar as they pose perceived threats to the animals they control. Although great strides have been made in training laboratory primates to tolerate procedures that might otherwise be quite stressful (e.g., voluntarily presenting a limb for a blood draw; Coleman et al., 2008), not every uncomfortable procedure or stressful husbandry-related activity or procedure (e.g., noisy cleaning equipment) involving laboratory staff can be avoided in research settings. Sometimes, the animals in our care appear to be quite habituated to humans and the activities they perform, and therefore we might be tempted to assume that humans are benign elements in our animals' environment. But are primates truly habituated to us?

As described earlier, primates tend to be very careful about where they sleep. Callitrichid primates (tamarins and marmosets), both in the wild (Dawson, 1979) and in captivity (Caine, 1987), approach their sleeping sites silently and stealthily, presumably with the intention of disguising the location of the site from predators. What happens if a human is watching the tamarins as they prepare to retire for the night? If the monkeys are, in fact, very habituated to the presence of humans, they should not behave differently when a human is, or is not, present. I tested this assumption in a rather simple experiment (Caine, 1990). On some evenings, I sat quietly in the colony room during the hour that the monkeys typically began and completed the process of entering their sleeping boxes for the night. On other nights, I set up a video camera behind a blind and recorded the monkeys as they retired. Despite the fact that the monkeys were very familiar with me (I was in the colony rooms multiple times a day, almost every day), they took longer to settle into their sleeping boxes when I was in the room than when I was not. They entered and exited the boxes multiple times, obviously agitated, before finally choosing to stay. Apparently, from the monkeys' point of view, my familiarity did not fully override my potential as a "predator." It is well known that certain common and "routine" laboratory procedures, such as handling and blood draws, are associated with changes in various physiological parameters associated with stress (reviewed in Balcombe et al., 2004). Insofar as humans are associated with those procedures, the humans themselves may retain their status as potential predators in the eyes of laboratory primates. Indeed, the human intruder paradigm (Kalin & Shelton, 1989) is commonly used to evaluate emotional responses of primates (e.g., Gottlieb & Capitanio, 2013; Capitanio, 2017; Coleman, 2017) and has proved valuable in the assessment of the efficacy of anxiolytic drugs (reviewed in Barros & Tomaz, 2002). Sorge et al. (2014) recently shook the foundations of rodent research when they discovered that the mere presence of male (but not female) experimenters elicited a stress response in mice and rats that was strong enough to cause stress-induced analgesia and increased thigmotaxis, leading the authors to conclude that "… stress caused by male researchers may represent a confound of much existing animal research, extending even to nonbehavioral studies in which tissues were obtained from live rodents euthanized by either male or female personnel." (p. 631). The simple lesson is that laboratory personnel, from cleaning and security staff to senior scientists, should be aware of the possibility that the animals we care for and study are not necessarily habituated to our presence, even if they seem to be, and that we may affect their behavior (and, in turn, our data) in ways we have failed to recognize and control. Once again, behavioral management personnel are in an excellent position to address this concern by identifying and carefully assessing the effects of human presence on captive primates. Types and frequency of vocalizations, postural changes, direction of gaze, and self-directed behaviors are examples of behaviors that might change in response to human activity. Results of such assessments may lead to relatively simple modifications of laboratory routines that reduce experimental error and promote well-being.

Using Antipredator Reactions as a Measure of Fear, Anxiety, and Coping

Anxiety disorders, including panic disorder and phobias, are the most commonly reported class of psychological disorders, affecting about one in five adults in the United States (Kessler et al., 2005). If severe enough, these disorders are debilitating, reducing quality of life in many ways.

Fear is adaptive, insofar as it motivates us to steer clear of dangerous objects and situations. We tend to generalize from dangerous stimuli to similar stimuli, considering that, for instance, a berry that looks much like a known poisonous berry might very well be related to, and thus equally as poisonous as, the original. Unfortunately, human beings are also good at associating harmless stimuli with dangerous stimuli, leading to the potential for irrational and debilitating reactions to things that either cannot or are extremely unlikely to hurt us. In this way, a person may become afraid of all dogs after a negative experience with one aggressive dog, or may react to a hypodermic needle in the hands of a qualified nurse in a similar way that he/she reacts to a sharp object in the hand of a threatening stranger (Davey, 1992).

Fear of predators is ancient and rooted in our most basic biology. Consequently, invoking, and then studying, predator-related fear gives us a way of understanding fear in an ecologically relevant context. For example, as described earlier, Susan Mineka et al. learned that rhesus monkeys are prone to learning by social observation that snakes (but not other stimuli, such as flowers) are frightening (Cook & Mineka, 1989). This contributed to the very useful concept of "prepared classical conditioning," whereby the prevalence and origins of phobias are better understood (Seligman, 1971).

Barros and Tomaz (2002) illustrate the value of using nonhuman primate antipredator behavior in studying fear and anxiety with their "Marmoset Predator Confrontation Test" (MPCT) paradigm. The authors point out that "ethological" methods are most likely to present us with the range of natural behaviors that are associated with fear and anxiety in animals. Consequently, the opportunities for assessing some of the varied, and sometimes subtle, ways in which fear and anxiety are expressed can enhance the value of an animal model for understanding anxiety disorders in humans. The MPCT makes use of taxidermied felids (other predator models have been tested as well) placed in one section of a multichambered apparatus; marmosets (*Callithrix penicillata*) come upon the cat as they are exploring the chambers. Proximity to the stimulus, scratching, scent marking, and exploration are scored from videotapes of the trials. Barros et al. (2000, 2001) validated the MPCT by showing that administration of anxiolytic drugs (diazepam and buspirone) changed the marmosets' behavior in ways consistent with the anxiety-reducing effects of the drugs (e.g., more time spent in proximity to the cat). The authors point to various advantages associated with this testing paradigm, including the fact that the monkeys could be habituated to the apparatus itself with four or five sham exposures preceding the actual test, such that fear and anxiety reactions associated with exposure to novel environments are less likely to compound the reactions to the predator itself.

Finally, a largely unexplored question is whether or not we could use engineered predator encounters to promote species-typical behavior that might be enriching [as proposed for tamarins by Moodie and Chamove (1990)], promote cooperative social behavior, and, counterintuitively, reduce stress. For instance, as described earlier, a common response to terrestrial predators is mobbing, in which groupmates are recruited to approach and harass the predator with loud calls and agitated movements. It certainly appears that marmosets are in a highly aroused state when they are mobbing a predator (or predator model), but Cross and Rogers (2006) reported an unexpected and provocative phenomenon in their study of mobbing behavior by common marmosets. The authors discovered that cortisol levels, taken from saliva and thus collected noninvasively, decreased below control levels (control = marmosets not exposed to the snake model) at the first cortisol reading taken 5 min after withdrawal of the stimulus, and remained significantly below baseline for 55 min. Further analysis showed that higher numbers of *tsik* calls (mobbing calls) emitted by focal subjects in response to the stimulus were associated with greater reductions from baseline cortisol levels measured at the 5-min postpresentation mark. This finding led to a second experiment, in which *tsik* calls recorded from group mates were played to marmosets that were temporarily isolated from their groups. Marmosets are highly social and exhibit behavioral signs of stress (agitation, calling, etc.) and increased levels of salivary cortisol when isolated from their groupmates (Cross et al., 2004). However, the marmosets in the condition that included *tsik* calls displayed reductions in salivary cortisol (compared to preisolation baseline measures) over the 30 min following isolation, whereas the control group (isolated with no *tsik* call playbacks) showed increases over baseline. Unfortunately, there was no control condition in which other types of vocalizations were played to isolated individuals, and so we do not know whether the reduction in cortisol was specific to mobbing calls. However, taken together, the results of these two experiments suggest that marmosets that participate in species-typical antipredator behavior may actually benefit from the experience, to the extent that cortisol reductions indeed reflect a benefit, and calls associated with group-mediated antipredator behavior might serve to calm marmosets that are exposed to experimentally mandated stressors in research protocols (Cross & Rogers, 2006; but see Snowdon & Teie, 2010). Whether or not similar phenomena might apply to other commonly used laboratory primates, which have very different social systems than marmosets, is a question that warrants future study.

CONCLUSION

Primates are prey to many types of predators and have been throughout their evolutionary history. Living in captivity does not eradicate the antipredator reactions that have arisen in response to the significant selection pressures posed by predators over millions of years. Recognizing and understanding those deeply ingrained reactions can: (1) contribute to the management and welfare of captive primates; (2) point to ways in which antipredator behavior may unknowingly influence our data; and (3) enhance our appreciation of the range of adaptive behaviors of the animals under our care. As discussed in this chapter, antipredator adaptations associated with the need to sleep safely can be targeted for behavioral management assessment and interventions that identify optimal sleeping locations. The extent to which primates perceive humans to be predatory threats can be evaluated by looking for changes (sometimes subtle) in the behavior of animals in the presence of humans and human activity, and procedures to minimize those reactions can be put into place with behavioral management plans. Finally, behavioral managers may be able to exploit the fact that fearful reactions of primates to predators often involve cooperative social behaviors. Use of predator exposures to build, maintain, or restore social bonds is a counterintuitive, but intriguing, possibility for behavioral managers to consider.

REFERENCES

Anderson, J.R. (1998). Sleep, sleeping sites, and sleep-related activities: Awakening to their significance. *American Journal of Primatology, 46*, 63–75.

Antoniadis, E.A., Winslow, J.T., Davis, M., & Amaral, D.G. (2007). Role of the primate amygdala in fear-potentiated startle: Effects of chronic lesions in the rhesus monkey. *The Journal of Neuroscience, 27*(28), 7386–7396.

Balcombe, J.P., Barnard, N.D., & Sandusky, C. (2004). Laboratory routines cause animal stress. *Journal of the American Association for Laboratory Animal Sciences, 43*(6), 42–51.

Barros, M., Boere, V., Huston, J.P., & Tomaz, C. (2000). Measuring fear and anxiety in the marmoset (*Callithrix penicillata*) with a novel predator confrontation model: Effects of diazepam. *Behavioral and Brain Research, 108*, 205–211.

Barros, M., Mello Jr., E.L., Huston, J.P., & Tomaz, C. (2001). Behavioral effects of buspirone in the marmoset employing a predator confrontation test of fear and anxiety. *Pharmacology, Biochemistry, and Behavior, 68*, 255–262.

Barros, M. & Tomaz, C. (2002). Nonhuman primate models for investigating fear and anxiety. *Neuroscience & Biobehavioral Reviews, 26*, 187–201.

Bayle, D.J., Henaff, M.A., & Krolak-Salmon, P. (2009). Unconsciously perceived fear in peripheral vision alerts the limbic system: A MEG study. *PLoS One, 4*(12), e8207.

Berman, R.A. & Wurtz, R.H. (2010). Functional identification of a pulvinar path from superior colliculus to cortical area MT. *The Journal of Neuroscience, 30*(18), 6342–6354.

Besedovsky, L., Lange, T., & Born, J. (2012). Sleep and immune function. *European Journal of Physiology, 463*, 121–137.

Bezerra, B.M., Barnett, A.A., Souto, A., & Jones, G. (2009). Predation by the tayra on the common marmoset and the pale-throated three-toed sloth. *Journal of Ethology, 27*(1), 91–96.

Bianchi, R.D.C. & Mendes, S.L. (2007). Ocelot (*Leopardus pardalis*) predation on primates in Caratinga Biological Station, southeast Brazil. *American Journal of Primatology, 69*(10), 1173–1178.

Buchanan-Smith, H.M., Anderson, D.A., & Ryan, C.W. (1993). Responses of cotton-top tamarins (*Saguinus oedipus*) to faecal scents of predators and non-predators. *Animal Welfare, 2*(1), 17–32.

Caine, N.G. (1987). Vigilance, vocalizations, and cryptic behavior at retirement in captive groups of red-bellied tamarins (*Saguinus labiatus*). *American Journal of Primatology, 12*, 241–250.

Caine, N.G. (1990). Unrecognized anti-predator behavior can bias observational data. *Animal Behaviour, 39*(1), 195–197.

Caine, N.G., Potter, M.P., & Mayer, K.E. (1992). Sleeping site selection by captive tamarins (*Saguinus labiatus*). *Ethology, 90*, 63–71.

Caine, N.G. & Weldon, P.J. (1989). Responses by red-bellied tamarins (*Saguinus labiatus*) to fecal scents of predatory and non-predatory neotropical mammals. *Biotropica, 21*(2), 186–189.

Capitanio, J.P. (2017). Variation in biobehavioral organization. In: Schapiro, S.J. (Ed.) *Handbook of Primate Behavioral Management,* 55–73. Boca Raton, FL: CRC Press.

Coleman, K. (2017). Individual differences in temperament and behavioral management. In: Schapiro, S.J. (Ed.) *Handbook of Primate Behavioral Management,* 95–113. Boca Raton, FL: CRC Press.

Coleman, K., Pranger, L., Maier, A., Lambeth, S.P, Perlman, J.E., Thiele, E., & Schapiro, S.J. (2008). Training rhesus macaques for venipuncture using positive reinforcement techniques: A comparison with chimpanzees. *Journal of the American Association for Laboratory Animal Science, 47*(1), 37–41.

Cook, M. & Mineka, S. (1989). Observational conditioning of fear to fear-relevant versus fear-irrelevant stimuli in rhesus monkeys. *Journal of Abnormal Psychology, 98*(4), 448–459.

Cook, M., Mineka, S., Wolkenstein, B., & Laitsch, K. (1985). Observational conditioning of snake fear in unrelated rhesus monkeys. *Journal of Abnormal Psychology*, *94*(4), 591–610.

Cote, K.A., McCormick, C.M., Geniole, S.N., Renn, R.P., & MacAulay, S.D. (2013). Sleep deprivation lowers reactive aggression and testosterone in men. *Biological Psychology, 92*, 249–256.

Cross, N., Pines, M.K., & Rogers, L.J. (2004). Saliva sampling to assess cortisol levels in unrestrained common marmosets and the effect of behavioral stress. *American Journal of Primatology*, *62*, 107–114.

Cross, N. & Rogers, L.J. (2006). Mobbing vocalizations as a coping response in the common marmoset. *Hormones and Behavior, 49*, 237–245.

Crowfoot, M.C. (2012). Why mob? Reassessing the costs and benefits of primate predator harassment. *Folia Primatologica, 83*, 252–273.

D'Amour, D.E., Hohmann, G., & Fruth, B. (2006). Evidence of leopard predation on bonobos (*Pan paniscus*). *Folia Primatologica, 77*(3), 212–217.

Davey, G.C.L. (1992). Classical conditioning and the acquisition of human fears and phobias: A review and synthesis of the literature. *Advances in Behaviour Research and Therapy, 14*, 29–66.

Dawson, G.A. (1979). The use of time and space by the Panamanian tamarin, *Saguinus oedipus*. *Folia Primatologica, 31*, 253–284.

Etting, S.F. & Isbell, L.A. (2014). Rhesus macaques (*Macaca mulatta*) use posture to assess level of threat from snakes. *Ethology, 120*, 1177–1184.

Fichtel, C. (2012). Predation. In Mitani, J.C., Call, J., Kappeler, P.M., Palombit, R.A., & Silk, J.B. (Eds), *The Evolution of Primate Societies* (pp. 169–194). Chicago, IL: University of Chicago Press.

Fichtel, C. & Kappeler, P.M. (2002). Anti-predator behavior of group-living Malagasy primates: Mixed evidence for a referential alarm call system. *Behavioral Ecology and Sociobiology, 51*, 262–275.

Franklin, S.P., Hankerson, S.J., Baker, A.J., & Dietz, J.M. (2007). Golden lion tamarin sleeping site use and preretirement behavior during intense predation. *American Journal of Primatology, 69*(3), 325–335.

Goodman, S.M., O'Connor, S., & Langrand, O. (1993). A review of predation on lemurs: Implications for the evolution of social behavior in small, nocturnal primates. In Ganzhorn, J. & Kappeler, P.M. (Eds), *Lemur Social Systems and Their Ecological Basis* (pp. 51–66). New York: Springer.

Gottlieb, D.H. & Capitanio, J.P. (2013). Latent variables affecting behavioral response to the human intruder test in infant rhesus macaques (*Macaca mulatta*). *American Journal of Primatology, 75*(4), 314–323.

Greer, S.M., Goldstein, A.N., & Walker, M.P. (2013). The impact of sleep deprivation on food desire in the human brain. *Nature Communications*, 4:2259.

Grieve, K.L., Acuña, C., & Cudeiro, J. (2000). The primate pulvinar nuclei: Vision and action. *Trends in Neurosciences, 23*(1), 35–39.

Hamilton, W.J. (1982). Baboon sleeping site preferences and relationships to primate grouping patterns. *American Journal of Primatology, 3*, 41–53.

Hankerson, S.J. & Caine, N.G. (2004). Pre-retirement predator encounters alter the morning behavior of captive marmosets (*Callithrix geoffroyi*). *American Journal of Primatology, 63*, 75–85.

Hauser, M.D. & Wrangham, R.D. (1990). Recognition of predator and competitor calls in nonhuman primates and birds: A preliminary report. *Ethology, 86*(2), 116–130.

Heymann, E. (1990). Reactions of wild tamarins, *Saguinus mystax* and *Saguinus fuscicollis* to avian predators. *International Journal of Primatology, 11*, 327–337.

Isbell, L.A. (2006). Snakes as agents of evolutionary change in primate brains. *Journal of Human Evolution*, *51*, 1–35.

Isbell, L.A. (2009). *The Fruit, the Tree, and the Serpent: Why We See So Well*. Cambridge, MA: Harvard University Press.

Jolly, A. (1972). *The Evolution of Primate Behavior*. New York: Macmillan.

Kalin, N.H. & Shelton, S.E. (1989). Defensive behaviors in infant rhesus monkeys: Environmental cues and neurochemical regulation. *Science, 243*(4899), 1718–1721.

Kessler, R.C., Chiu, W.T., Demler, O., Merikangas, K.R., & Walters, E.E. (2005). Prevalence, severity, and comorbidity of 12-month DSM-IV disorders in the National Comorbidity Survey Replication. *Archives of General Psychiatry, 62*, 617–627.

LoBue, V. & DeLoache, J.S. (2008). Detecting the snake in the grass. *Psychological Science, 19*(3), 284–289.

LoBue, V. & DeLoache, J.S. (2010). Superior detection of threat-relevant stimuli in infancy. *Developmental Science, 13*(1), 221–228.

Macedonia, J.M. & Yount, P.L. (1991). Auditory assessment of avian predator threat in semi-captive ringtailed lemurs (*Lemur catta*). *Primates*, *32*(2), 169–182.

Matsumoto-Oda, A. (2015). How surviving baboons behaved after leopard predation: A case report. *Anthropological Science*, *123*(1), 13–17.

McGraw, W.S., Cook, C., & Shultz, S. (2006). Primate remains from African crowned eagle (*Stephanoaetus coronatus*) nests in Ivory Coast's Tai Forest: Implications for primate predation and early hominid taphonomy in South Africa. *American Journal of Physical Anthropology*, *131*, 151–165.

Miranda, J.M.D., Bernardi, I.P., Abreu, K.C., & Passos, F.C. (2005). Predation on *Alouatta guariba clamitans* (Primates, Atelidae) by *Leopardus pardalis* (Linnaeus) (Carnivora, Felidae). *Revista Brasileira de Zoologia*, *22*(3), 793–795.

Moodie, E.M. & Chamove, A.S. (1990). Brief threatening events beneficial for captive tamarins? *Zoo Biology*, *9*, 275–286.

Neal, S.J. & Caine, N.G. (2016). Scratching under positive and negative arousal in common marmosets (*Callithrix jacchus*). *American Journal of Primatology*, *78*, 216–226.

Öhman, A. (2005). The role of the amygdala in human fear: Automatic detection of threat. *Psychoneuroendocrinology*, *30*(10), 953–958.

Öhman, A., Carlsson, K., Lundquist, D., & Ingvar, M. (2007). On the unconscious subcortical origin of human fear. *Physiology & Behavior*, *92*, 180–185.

Öhman, A. & Mineka, S. (2001). Fear, phobias, and preparedness: Toward an evolved module of fear and fear learning. *Psychological Review*, *108*, 483–552.

Petracca, M.M. & Caine, N.G. (2013). Alarm calls of marmosets (*Callithrix geoffroyi*) to snakes and perched raptors. *International Journal of Primatology*, *34*, 337–348.

Prather, M.D., Lavenex, P., Mauldin-Jourdain, M.L., Mason, W.A., Capitanio, J.P., Mendoza, S.P., & Amaral, D.G. (2001). Increased social fear and decreased fear of objects in monkeys with neonatal amygdala lesions. *Neuroscience*, *106*(4), 653–658.

Prince, T.M., Winner, M., Choi, J., Havekes, R., Aton, S., & Abel, T. (2014). Sleep deprivation during a specific 3-h time window post-training impairs hippocampal synaptic plasticity and memory. *Neurobiology of Learning and Memory*, *109*, 122–130.

Pruetz, J.D., Fulton, S.J., Marchant, L.F., McGrew, W.C., Schiel, M., & Waller, M. (2008). Arboreal nesting as anti-predator adaptation by savanna chimpanzees (*Pan troglodytes verus*) in southeastern Senegal. *American Journal of Primatology*, *70*, 393–401.

Quattara, K., Lemasson, A., & Zuberbühler, K. (2009). Anti-predator strategies of free-ranging Campbell's monkeys. *Behaviour*, *146*(12), 1687–1708.

Samson, D.R. & Hunt, K.D. (2014). Chimpanzees preferentially select sleeping platform construction tree species with biomechanical properties that yield stable, firm, but compliant nests. *PLoS One*, *9*(4), e95361.

Samson, D.R. & Shumaker, R.W. (2015). Orangutans (*Pongo* spp.) have deeper, more efficient sleep than baboons (*Papio papio*) in captivity. *American Journal of Physical Anthropology*, *157*(3), 421–427.

Searcy, Y. & Caine, N.G. (2003). Hawk calls elicit alarm and defensive reactions in captive Geoffroy's marmosets (*Callithrix geoffroyi*). *Folia Primatologica*, *74*, 115–125.

Seligman, M.E.P. (1971). Phobias and preparedness. *Behavior Therapy*, *2*(3), 307–320.

Shibasaki, M. & Kawai, N. (2009). Rapid detection of snakes by Japanese monkeys (*Macaca fuscata*): An evolutionarily predisposed visual system. *Journal of Comparative Psychology*, *123*, 131–135.

Snowdon, C.T. & Teie, D. (2010). Affective responses in tamarins elicited by species-specific music. *Biology Letters*, *6*(1), 30–32.

Soares, S.C., Lindstrom, B., Esteves, F., & Öhman, A. (2014). The hidden snake in the grass: Superior detection of snakes in challenging attentional conditions. *PLoS One*, *9*(12), e114724.

Sorge, R.E., Martin, L.J., Isbester, K.A., Sotocinal, S.G., Rosen, S., Tuttle, A.H., Wieskopf, J.S. et al. (2014). Olfactory exposure to males, including men, causes stress and related analgesia in rodents. *Nature Methods*, *11*(6), 629–632.

Spörrle, M. & Stitch, J. (2010). Sleeping in safe places: An experimental investigation of human sleeping place preferences from an evolutionary perspective. *Evolutionary Psychology*, *8*(3), 405–419.

Tello, N.S., Huck, M., & Heymann, E.W. (2002). Boa constrictor attack and successful group defence in moustached tamarins, *Saguinus mystax*. *Folia Primatologica*, *73*(2–3), 146–148.

Terborgh, J. (1983). *Five New World Primates*. Princeton, NJ: Princeton University Press.

Van Le, Q., Isbell, L.A., Matsumoto, J., Nguyen, M., Hori, E., Maior, R.S., Tomaz, C., Tran, A.H., Ono, T., & Nishijo, H. (2013). Pulvinar neurons reveal neurobiological evidence of past selection for rapid detection of snakes. *Proceedings of the National Academy of Sciences*, *110*(47), 19000–19005.

Wright, P.C., Heckscher, S.K., & Dunham, A.E (1997). Predation on Milne-Edward's sifaka (*Propithecus diadema edwardsi*) by the fossa (*Cryptoprocta ferox*) in the rain forest of southeastern Madagascar. *Folia Primatologica*, *68*, 34–34.

Yorzinski, J.L., Penkunas, M.J., Platt, M.L., & Coss, R.G. (2014). Dangerous animals capture and maintain attention in humans. *Evolutionary Psychology*, *12*(3), 534–548.

CHAPTER 10

Future Research with Captive Chimpanzees in the United States
Integrating Scientific Programs with Behavioral Management

William D. Hopkins and Robert D. Latzman
Georgia State University

CONTENTS

Historical and recent behavioral, cognitive, neurological, and molecular biological studies with primates generally, and with chimpanzees specifically, have contributed and advanced scientific theory in two critical ways. First, because of their genetic and biological similarities to humans, studies with nonhuman primates, including the great apes, have been vital for testing evolutionary theories on the origins of human-specific skills and specializations. For example, comparative studies of primates, notably chimpanzees, have been vital for testing theories on the origins of language; tool use and tool making; culture; theory of mind and deception; complex emotional systems, such as empathy; and cortical organization and lateralization, to name a few (Savage-Rumbaugh and Lewin 1994; Byrne 1995; Van Schaik et al. 1999; Rumbaugh and Washburn 2003; Byrne and Corp 2004; Whiten and Mesoudi 2008; Hopkins 2013; Hopkins et al. 2014b). Second, studies in nonhuman primates have been critically important for modeling human disease and psychological problems in an effort to improve our understanding of mechanisms underlying dysfunction, and potential interventions and treatments (Phillips et al. 2014).

In the past 5 years, however, the value and use of nonhuman primates in biomedical and behavioral research has come under increased scrutiny. Indeed, the scientific community is at a critical juncture in shaping the current and future use of nonhuman primates, specifically chimpanzees, in behavioral, genomic, and neuroscience research in the United States. Because the need for chimpanzees in biomedical research programs has fallen over the past 40 years, both scientific and public interest in how chimpanzees behave and think, as well as how their brains and genomes are constructed, have grown exponentially (see Figure 10.1). The dramatic rise in research on chimpanzee behavior, cognition, and neurogenomics has rightfully led to ethical and moral questions regarding the ways in which they are treated and used for the purposes of benefitting human health and, specifically, their use in biomedical and psychological research. Nowhere has this discussion been more visibly elevated than within the National Institutes of Health (NIH) where, over the past 5 years, the scientific community has witnessed a number of policy and administrative decisions regarding the care and management of captive chimpanzees, as well as decisions concerning guidelines and constraints on the types of research that will be allowable. In our view, some of these decisions, if not changed, will have a significant negative impact on future translational research with chimpanzees that has the potential to provide information on human health in meaningful ways (Latzman and Hopkins 2016). In addition, many of the policies currently being advocated have the potential to negatively affect chimpanzee well-being and conservation efforts. Here, we argue that rather than shutting down or severely minimizing research with captive chimpanzees, the NIH should support the creation of a National Chimpanzee Behavioral, Cognition, and Neurogenomic Research Program (NCBCN). This program would be designed to maximally utilize this invaluable, yet diminishing, resource for the purposes of promoting scientific inquiry that would benefit our understanding of human behavior within an evolutionary and biological framework in addition to advancing public interest and awareness of the need for chimpanzee conservation in captivity and in the wild.

Importantly, we believe the successful development of this program can be accomplished by combining the efforts and objectives of the scientific community with existing and developing behavioral management programs within the organizations that currently house and care for chimpanzees. Specifically, the goal of many behavioral management programs is to train nonhuman primates, and in this case chimpanzees, to engage in a variety of husbandry activities and to comply with procedures that enhance our ability to assess behavioral and health-related variables. Such procedures include voluntary presentation for venipuncture, saliva samples, or cheek swabs, or other behaviors that promote the safe collection of biomaterials. Biomaterials, such as blood, can be used to clinically assess

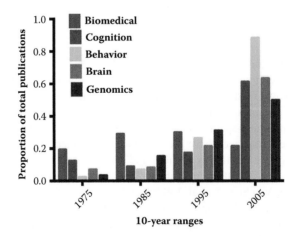

Figure 10.1 Using PubMed, we calculated the total number of publications in 10-year intervals beginning in 1975 using the keywords chimpanzees and either (1) Cognition, (2) Behavior, (3) Brain, (4) Genomics, or (5) Biomedical (this included the specific words malaria, hepatitis, and HIV). The number of publications on each topic within each 10-year interval was divided by the total number of publications to derive the proportion presented on the y-axis. As can be seen, research publications on behavior, cognition, brain, and genomics have been on an exponential rise since 1975, whereas research classified as biomedical has remained constant or declined over this same 40-year period.

the health of the chimpanzees by performing assays of chemistry- and metabolism-related parameters; however, these same biomaterials can be used for additional research-related purposes, such as analyses of physiological or hormonal measures, as well as for extraction of DNA for genomic studies. In this chapter, we describe efforts by our research group over the past 10–15 years that illustrate how integrating and working directly with behavioral management programs have enhanced our scientific goals. We also discuss how findings from our research can enhance and improve behavioral management and veterinary science programs. Our research group has been at the forefront of this effort, and our goal in this chapter is to illustrate our approach and to describe ways in which it could be applied in the future.

BEHAVIORAL STUDIES ON MANUAL MOTOR SKILL AND HANDEDNESS

Handedness

Historically, population-level handedness has been considered a hallmark of human evolution (Warren 1980; Ettlinger 1988; Corballis 1992). Indeed, it is only during the past 25 years that views regarding the uniqueness of human handedness have begun to change, based, in part, on findings in nonhuman primates, specifically chimpanzees. Our laboratory has assessed handedness for a variety of different behaviors and tasks in captive chimpanzees, including (1) simple reaching, (2) bimanual feeding, (3) coordinated bimanual actions, (4) throwing, (5) manual gestures, (6) grooming, (7) coconut opening, and (8) tool use (see Hopkins 2013 for review). Measures of handedness, such as simple reaching, bimanual feeding, and tool use are typically obtained during routine enrichment or feeding programs. For example, many institutions housing chimpanzees provide foraging and tool-use devices as a form of environmental enrichment, and we have simply taken advantage of these circumstances to obtain measures of hand use and skill (Hopkins et al. 2009). In addition to the scientific question of whether nonhuman primates show population-level handedness, collection of these data, from relatively large samples of individuals with known pedigrees, has allowed us to test for the influence of genetic and nongenetic factors on individual handedness. For instance, recent studies by our group have shown that hand preference and motor skill for tool use are significantly heritable (Hopkins et al. 2013a, 2015).

In contrast, as a component of our scientific aims, we have designed and implemented specific types of handedness tasks that not only elicit specific motor actions but also serve as enrichment for the animals. For example, data from wild populations of chimpanzees have shown that they crack nuts using "hammers" and "anvils," and have hand preferences for these behaviors (Boesch 1991; McGrew et al. 1999; Biro et al. 2006). To simulate the motor actions of nut-cracking or anvil use (ballistic movements), we provided captive chimpanzees with coconuts (see Figure 10.2) and quantified (1) the number of times they hammered the coconut, (2) their hand use (right or left), and (3) their proficiency in opening the coconuts. To assess proficiency, we simply recorded whether or not the chimpanzees successfully opened the coconut. Although we found no differences in proficiency between wild-caught and captive-born individuals, we did find a sex difference [X^2 (1, $N = 181$) = 5.53, $p = .018$], with a higher proportion of males (66/85) successfully opening the coconuts than females (59/96). Interestingly, males and females did not differ in the number of bouts in which they hammered the coconuts; thus, the increased proficiency by males is not attributable to them performing more hammering bouts. There was no evidence for a species-level bias toward one hand or the other; there were 67 left-handed chimpanzees and 68 right-handed chimpanzees, while the remaining 37 were ambiguously-handed for coconut cracking (i.e., showed no preference). Furthermore, the distribution of hand preference was not significantly related to either the sex or rearing history of the animals.

Another example of how the design of our testing materials has contributed to the behavioral management of chimpanzees comes from our studies on handedness for coordinated bimanual actions, specifically for what is referred to as the TUBE task (Hopkins 1995). The apparatus for the TUBE task includes a PVC pipe that is ~20–30 cm in length and 2.5 cm in diameter. Adhesive foods

Figure 10.2   Examples of chimpanzees engaged in different handedness tasks.

(e.g., peanut butter) are placed on the inside edges of the pipe, preventing the subjects from licking the food off, and instead, requiring them to hold the pipe with one hand and extract the food with a digit (usually the index finger) on their other hand. Testing the primates simply involves baiting the TUBE and handing it to them. Observers record the frequency and the number of bouts of left and right hand use to extract the food. To date, the TUBE task has been administered to more than 1100 nonhuman primates, including more than 500 chimpanzees (Hopkins et al. 2011; Meguerditchian et al. 2013). Thus, the TUBE task is a simple and efficient way of assessing handedness in nonhuman primates *and*, at the same time, provides a foraging enrichment opportunity for the animals. Further, for apes or other tool-using species, a simple change in the location of the adhesive food can alter the motor and cognitive demands of the task. Specifically, when the length of the PVC pipe is extended to >30 cm and the adhesive food is placed in the inside center, rather than on the inside edges, the device now requires the use of a tool to extract the food. In this circumstance, the subjects must hold the pipe with one hand and with their other hand, use a tool (rather than a finger) that can be inserted into the pipe to extract the food (Hopkins and Rabinowitz 1997). In this case, the task still requires a coordinated bimanual action, but also requires the use of a tool. In summary, our group has clearly demonstrated the mutually beneficial impact that handedness research can have, both for the advancement of science and for the enhancement of the well-being of captive chimpanzees.

SOCIAL AND PHYSICAL COGNITION

As part of a larger research program aimed at assessing the role of genetic and nongenetic factors on social and physical cognition in chimpanzees, our group has administered a number of tests designed to measure cognitive abilities in the social and nonsocial domains. This has included studies aimed at quantifying abilities such as chimpanzees' (1) responses to joint attention cues, such as gaze and pointing (Hopkins et al. 2013a, 2014a); (2) initiation of joint attention (Leavens et al. 2015); (3) imitation recognition (Pope et al. 2015); and (4) delay of gratification (Beran et al. 2014; Latzman et al. 2015c). We have also administered the Primate Cognition Test Battery [PCTB; a battery of different tasks that broadly aim to quantify social and physical cognition abilities in apes and monkeys

(Herrmann et al. 2007, 2010; Schmitt et al. 2011)] to more than 150 chimpanzees. The social cognitive tasks measure abilities such as social learning, pointing, comprehension of communication cues, gaze following, and understanding of causality. Physical cognition tasks include measures of object permanence, spatial memory, transposition, and quantity discrimination, among others. To date, we have found that the overall performance on the PCTB task is significantly heritable in chimpanzees, with ~50% of the variance in performance attributable to genetic factors (Hopkins et al. 2014d). Furthermore, we find that nongenetic factors, such as sex or the rearing history of the chimpanzees (mother- or human-reared), only account for a small and nonsignificant portion of the variance. The one caveat to this conclusion is that the so-called language-trained or encultured apes perform significantly better on all measures, particularly on the social cognition tasks, when compared to the average mother- and human-reared chimpanzees. Indeed, performance on the PCTB tasks by these encultured apes rivals those of 3-year-old children (Lyn et al. 2010; Russell et al. 2011).

From the standpoints of chimpanzee welfare and behavioral management, to administer the PCTB, chimpanzees are often temporarily separated from their groups. We mention this fact because there seems to be a common misconception in both the animal welfare and animal rights communities that separating chimpanzees or other apes for the purposes of cognitive testing induces unreasonable duress and stress in the animals. There could be nothing further from the truth with respect to this claim, and, indeed, we would argue that providing chimpanzees with the option to leave their social group for the purposes of cognitive testing is inherently enriching and important for a number of reasons. First, wild chimpanzees live in fission–fusion societies and do not remain in constant, fixed group settings 24 h/day, 365 days a year. Thus, offering the apes the opportunity to leave their group for cognitive testing, for which they receive food rewards, is entirely consistent with the natural conditions in which they have evolved. Second, we would argue that by virtue of the fact that most chimpanzees will voluntarily leave their group to participate in our cognitive studies, the chimpanzees are demonstrating that these opportunities are inherently enriching to them. In simple operant conditioning terms, behaviors that are followed by positive consequences (rewarded) are more likely to be repeated in the future, whereas behaviors followed by either no reward or punishment are less likely to occur in the future. Thus, the fact that chimpanzees will leave their group to participate in a cognitive task for a second time suggests that the first experience was rewarding, and therefore, likely to be enriching. Third, it is of no scientific value whatsoever for distressed chimpanzees to participate in cognitive or behavioral tasks. Thus, even if testing chimpanzees alone was a scientific priority, the data obtained from an isolated ape that was stressed or under duress would be severely compromised and of little value. In short, in our view, it is in the best interests of both the animals' well-being and the scientific enterprise to provide chimpanzees with cognitive tasks that allow them to choose to participate by leaving their social group as part of a comprehensive behavioral management program.

COMPUTER-BASED COGNITIVE TESTING

In the late 1980s, Rumbaugh and colleagues were the first to develop a computer-based automated system for testing learning and other cognitive processes in chimpanzees and rhesus monkeys (Rumbaugh et al. 1989; Washburn et al. 1989; Hopkins et al. 1992; Washburn and Rumbaugh 1992; Hopkins and Washburn 1994). Since then, many other laboratories and zoos have developed similar systems for use with other nonhuman primate species. Scientifically, the automated test systems offer many advantages over methods using manual apparatus, such as the Wisconsin General Test Apparatus (WGTA), including better control over stimulus presentation, intertrial intervals, and data collection, as well as enhanced efficiency. One of the most interesting observations from these early studies, and from more contemporary studies as well, was the high levels of motivation to work on the computer-generated tasks exhibited by the primates. Indeed, many animals will perform hundreds, or thousands, of trials per day, an extraordinary improvement in the volume of data compared to manual testing. This suggests, consistent with the reasoning described earlier, that working on these computer tasks is inherently interesting and rewarding to the animals, and therefore, enriching. More recently, the use of automated, computerized test systems has been extended beyond singly housed or pair-housed individuals to include subjects living in larger social groups (Andrews and Rosenblum 2001; Fagot and Bonté 2010; Gazes et al. 2012). For example, Fagot and colleagues have 10 computer-based work stations for a social group consisting of more than 35 baboons, and these primates voluntarily choose to leave the social group to work on these computer systems. Collectively, these animals, on average, perform more than 1000 trials per day, and the presence of these systems has no effect on social behavior, social dynamics, or abnormal behavior (Fagot et al. 2014). Further, testing the baboons in social groups has no apparent influence on motor or cognitive performance (Gazes et al. 2012). Thus, like manual testing procedures, computerized testing procedures provide both enrichment for the animals (potentially even more than that provided by manual tests) and the opportunity to collect scientifically meaningful data (Whitehouse et al. 2013).

PERSONALITY

Stable social groups with individualized relationships, such as those within which chimpanzees live, lead to a range of social personality traits. As anyone who has worked with chimpanzees can attest, each ape shows a relatively consistent pattern of individual differences in behavior across time and situation; such individual variation can be labeled as "personality." Indeed, animals vary in their responses to the experiences they may encounter in captive environments, and, as a result of this variability, the degree of well-being experienced by individuals is not uniform (King and Figueredo 1997; Weiss et al. 2002, 2007; King et al. 2005; Freeman and Gosling 2010). Hence, the assessment of individual personality can be useful in providing critical information concerning subjective experiences, individual tendencies, and dispositions, and can help to guide important management decisions relevant to the welfare of captive animals. Thus, the importance of considering individual variation in personality in apes' responses across an array of animal welfare contexts should be clear.

In support of this assertion, a growing literature has documented the importance of animal personality, in general, and chimpanzee personality more specifically, and its implications for the breeding-, management-, and welfare-related behaviors of captive animals. In chimpanzees, individual variation in personality has been found to be a predictor of a range of social behaviors and of subjective well-being (Weiss et al. 2002). In addition, personality has been found to predict, for example, the chimpanzee's motivation to perform on certain cognitive tasks (Herrelko et al. 2012), success in problem-solving tasks (Hopper et al. 2014), responses to inequity and receiving less than expected in social situations (Brosnan et al. 2015), and training success of blood glucose testing (Reamer et al. 2014), among many other management- and welfare-related outcomes.

In addition to empirical demonstrations of the predictive validity of personality in chimpanzees, approaches to the assessment of traits have been found to be reliable, valid, and important psychometric considerations when evaluating the ultimate utility of an assessment approach. When considering the human caretaker rating approach, currently the most common method of personality assessment in the literature, converging data have demonstrated strong interrater reliability coefficients [most typically assessed using ICC (3,k)] across raters (Weiss et al. 2007; Latzman et al. 2015a). Further, with regard to the basic organization of personality in chimpanzees, although the ultimate number of factors found across studies has varied to some degree, the existence of a reliable and stable set of dispositional traits in chimpanzees is clear, with a number of studies confirming the existence of a replicable set of dispositional traits (Weiss et al. 2007) organized in a coherent, hierarchical manner (Latzman et al. 2014). Taken together, we now have valid and reliable means to assess variation in personality among chimpanzees—a powerful set of tools for professionals interested in animal management and welfare issues.

In addition to strong psychometrics, studies have shown that at least some of the variability in chimpanzee personality has a neurobiological foundation. For example, chimpanzee personality traits have been shown to be heritable and potentially linked to specific genetic polymorphisms (Weiss et al. 2000; Hong et al. 2011; Hopkins et al. 2012a; Staes et al. 2014; Latzman et al. 2015a). Further, neuroscientific studies have demonstrated significant neuroanatomical contributions to chimpanzee personality traits (Latzman et al. 2015b). Taking all aforementioned details into consideration, individual variability in chimpanzee personality appears to have a strong neurobiological foundation covarying within families, underscoring the importance of considering this variation when making management and welfare decisions.

Although there is clear support in the empirical literature for a core set of neurobiologically based personality traits with important external predictive correlates, there is also burgeoning literature suggesting that these traits can be organized into meaningful dimensions that reflect various levels of psychological health and well-being. For example, Latzman et al. (2016a) have recently developed a chimpanzee operationalization of psychopathic personality (psychopathy) tendencies. Specifically, through the development of the CHMP-Tri scales, scales developed using items from existing personality instruments through a mixed expert-consensus and psychometric approach, core biobehavioral dispositions can be assessed, including: boldness (dispositional fear/fearless), meanness (predatory disaffiliation), and disinhibition (weak inhibitory control). Consistent with some conceptions of psychopathy (Latzman et al. 2016b), these tendencies focus partially on adaptive features (associated with the boldness dispositional facets), as well as more typically maladaptive features (associated with the meanness and disinhibition facets). In addition to strong psychometric properties and clear parallels to tendencies studied in humans, these dispositions also evidenced significant associations with overt behavioral indicators, suggesting strong predictive ability for these scales. Further, consistent with findings for broad dispositional personality trait dimensions, individual variation across CHMP-Tri scales show significant heritability (Latzman et al. in press). Given the clear relevance of these dispositions, and the utility of the CHMP-Tri scales in assessing individual variability, applicability to management and welfare seems apparent.

Further underscoring the importance of chimpanzee personality to animal management and welfare is the large human literature confirming the importance of various personality traits to both psychological and physical health. Until recently, however, the transportability of this literature to chimpanzees was largely assumed, rather than empirically demonstrated. Latzman et al. (2016c), however, have empirically confirmed the transportability of the human literature to chimpanzees. Specifically, in a sample of humans, Latzman et al. (2016c) demonstrated that chimpanzee-derived personality scales, such as the ones used in the research reviewed earlier, translate well to humans and operate quite similarly to the established human-derived personality scales used in the human personality literature. These findings provide compelling evidence not only of the translational value of chimpanzee personality research to studies of humans, but also the value of human personality research to studies of chimpanzees.

There are a number of important implications that can be taken from the chimpanzee personality literature with regard to animal management and welfare. One of the potentially most meaningful relates to the social housing and relocation of apes. It is clear that the personality of individuals within a social group can affect social compatibility, as well as the stability and success of that group. Thus, for example, personality assessments can be used to inform decisions concerning which individuals could be placed together when planning the introduction of chimpanzees into social groups. Such considerations may be particularly important in the current situation, as the NIH attempts to "retire" their chimpanzees, that is, moving them to federal sanctuaries, accommodations unfamiliar to these apes. Further, as alluded to earlier, understanding the personality traits of individual apes may help to predict varying responses, the resulting effects on well-being of relocation, and the host of experiences that the animals will likely encounter as a result of a move. The assessment of personality may help to identify those apes that are most likely to struggle with relocation and allow captive managers to devise "personalized" prevention and/or intervention plans.

NEUROIMAGING

Studies of comparative neuroscience, particularly within primates, have a long history in the sciences. Although early studies primarily utilized postmortem brains for comparative morphological and cytoarchitectonic analyses, more recent studies in nonhuman primates have taken advantage of the development of noninvasive *in vivo* imaging methods, such as magnetic resonance imaging (MRI), functional MRI (fMRI), and positron emission tomography (PET) to examine the neural bases of behavior and cognition (Hopkins and Phillips 2016). Our laboratory has primarily used MRI and PET imaging techniques to assess the mechanisms underlying individual and phylogenetic variation in cortical morphology and connectivity. In addition, we, and others, have used PET imaging as a method for assessing the neural basis of different perceptual, motor, and cognitive processes in chimpanzees (Taglialatela et al. 2008, 2009, 2011; Parr et al. 2009; Hopkins et al. 2010; Hecht et al. 2013).

MRI is a noninvasive, *in vivo* imaging method that allows for collection of a three-dimensional (3D) picture of the brain from which measures of different structures can be obtained. For example, from MRI scans, we have quantified the volume of a variety of chimpanzee brain regions, including the inferior frontal gyrus, planum temporale, superior temporal gyrus, and the motor-hand area of the precentral gyrus. From these data, we have (1) assessed the role of genetic and nongenetic factors in explaining individual and phylogenetic variation in chimpanzees and humans (Gomez-Robles et al. 2013; Autrey et al. 2014; Bogart et al. 2014), and (2) correlated anatomical or morphological measures with different behavioral and cognitive functions, such as handedness, tool-use proficiency, receptive joint attention abilities, personality, and delay of gratification (Hopkins and Taglialatela 2012a,b; Hopkins et al. 2012b, 2014c; Phillips et al. 2013; Lacreuse et al. 2014). The MRI scans are an invaluable scientific resource, but they have also proven to be useful from a clinical perspective. Specifically, brain tumors, cerebral infarcts, or anomalies can be detected from MRI scans, and we have discovered that ~10% of our sample of 224 chimpanzees have some type of neural anomaly. For instance, we have observed small lesions in both white and gray matter in a number of chimpanzees, particularly elderly individuals (see Figure 10.3). We have also identified tumors in two individuals and have found at least four chimpanzees who are missing their entire fornix (see Figures 10.4 and 10.5).

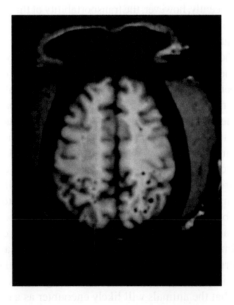

Figure 10.3 Axial view of a chimpanzee MRI scan. Dark circles within the image reflect small lesions or holes that are filled with CSF.

These neurological problems likely have behavioral or cognitive consequences, and knowing about them allows colony managers to implement specific social or housing accommodations to help these individual apes. For example, as noted earlier, we previously quantified handedness and skill in grasping small food items by chimpanzees (Hopkins et al. 2002; Hopkins and Russell 2004). In this behavioral paradigm, we station the animals and then present food items to the left or right of the subjects to encourage the use of the ipsilateral hand to reach and grasp the food. We record the number

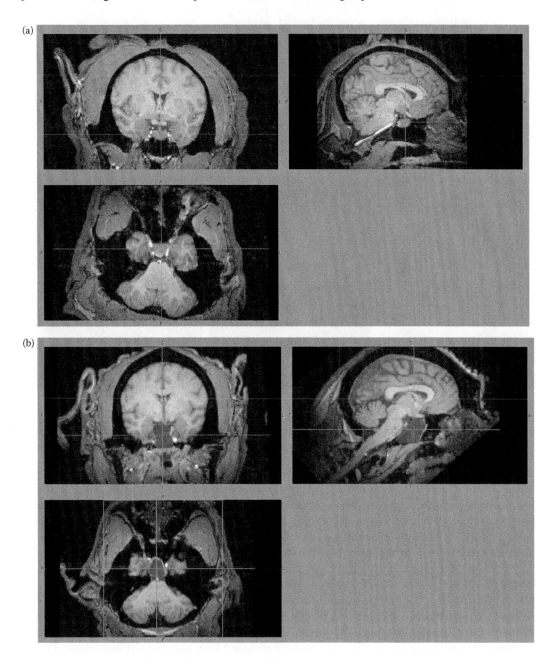

Figure 10.4 Panel of coronal, sagittal, and axial views of (a) a normal chimpanzee and (b) a chimpanzee with a tumor on the pituitary gland. Cross-hairs on each region mark the typical (a) and affected (b) regions.

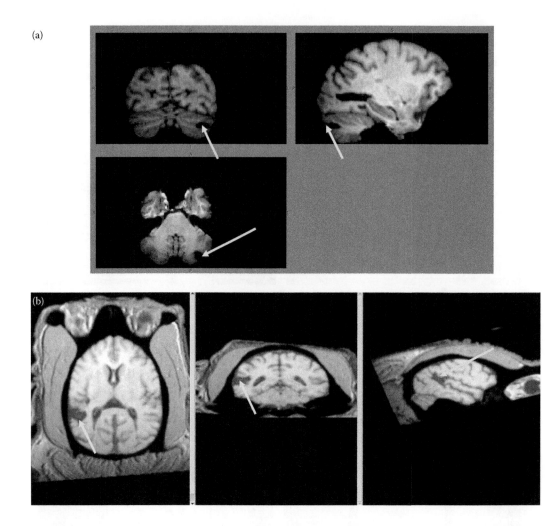

Figure 10.5 Panel of coronal, sagittal, and axial views of relatively large lesions located in (a) the left cerebellum and (b) the posterior temporal lobe in the right hemisphere.

of errors they make when grasping for each food item (quartered peanuts). Presenting the food items to each side of the subject allows us to obtain an equal number of responses for each hand. We have tested >400 chimpanzees and have always been able to eventually obtain responses from the animals using both hands. However, we had one subject that would never use his left hand to grasp the food items and, if he attempted, was largely unsuccessful. Aside from his unusual behavior in this paradigm, he was otherwise normal in terms of his social and affective behavior. This animal was one of our neuroimaging subjects, and as can be seen in Figure 10.6 [a series of images in the axial plane (Connor-Stroud et al. 2014)], this subject had experienced significant vascular mineralization at some point in his life, resulting in significant lesions throughout the cortex, including the region of the right primary motor cortex (Figure 10.7), likely explaining his grasping difficulties with his left hand.

   MRI scans are typically obtained from awake humans, while for nonhuman primates, including chimpanzees, the scans are almost exclusively obtained from anesthetized individuals. In our laboratory, with rare exception, we have obtained MRI scans from the chimpanzees when they were scheduled for their annual physical examination, so as to minimize the number of anesthetic events that they experience. Although we frequently initially relied on the veterinarians to anesthetize the

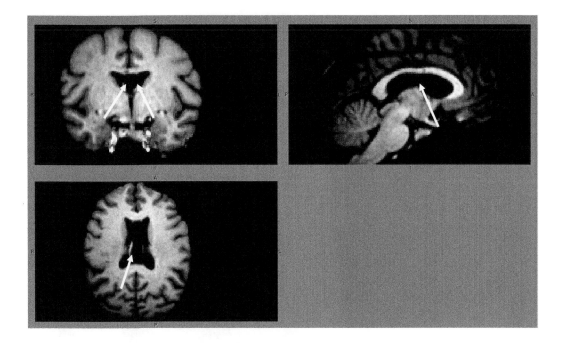

Figure 10.6 Panel of coronal, sagittal, and axial views of a chimpanzee scan missing the fornix.

chimpanzees, we subsequently refined our anesthetic protocol to match our behavioral management program, in which we have trained the chimpanzees to voluntarily accept an anesthetic injection. Thus, rather than using more stressful techniques to immobilize the apes, such as darting, we have tried to reduce stress by training the animals to voluntarily present for an anesthetic injection. The chimpanzees were not trained to voluntarily accept an injection for scientific reasons, but rather for pragmatic reasons. Chimpanzees are quite trainable and will participate in these events if they are worked with, and their behavior shaped, in a systematic fashion. Training chimpanzees to voluntarily present for sedation makes all of the events associated with annual physical exams less stressful for the apes and safer for both the staff and the animals.

CONCLUDING THOUGHTS

Although it is clear that research with chimpanzees has greatly benefitted our understanding of human health and behavior, as well as the evolution of human-specific specializations, the recent decision by NIH to retire the chimpanzees that are theirs means that scientists are now faced with the possibility that continued captive research with this amazing species will come to a halt or, at a minimum, become significantly reduced. Indeed, in May of 2016, NIH quietly outlined the provisions for acceptable research with chimpanzees and effectively eliminated the potential for neuroimaging, and included stipulations that severely compromise the collection of sophisticated cognitive data from chimpanzees. In our opinion, for many of the reasons described earlier, eliminating neuroimaging and controlled, sophisticated, experimental cognitive research with chimpanzees would be a tragedy from scientific, animal welfare, and fiscal perspectives.

Rather than simply retiring and warehousing the existing captive chimpanzee population, we are of the view that, for several reasons, NIH- and other Public Health Service (PHS)- supported programs need to create the equivalent of a NCBCN. First, in 2005, NIH funded and published the chimpanzee

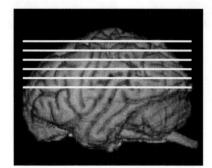
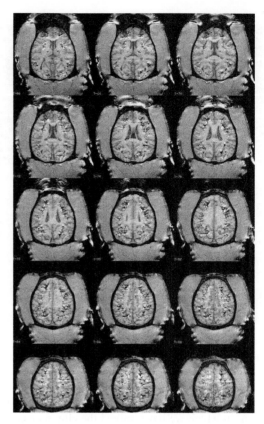

Figure 10.7   Axial view of a chimpanzee's MRI scan. This chimpanzee participated in a number of studies on hand preference and motor skill. In one study, we tested for performance difference in grasping small food items by the left and right hand. After more than 400 trials, and repeated attempts to "encourage" him to use his left hand, we were never able to get him to use it for grasping. A subsequent MRI scan revealed widespread lesions throughout the cortex due to extensive vascular mineralization.

genome (The Chimpanzee Sequencing and Analysis Consortium 2005), which is a valuable genomic resource used by the scientific community. More recently, in 2015, the National Institute of Neurological Disorders and Stroke (NINDS) committed funds to create an initial National Chimpanzee Brain Resource (NCBR), which will serve as a resource for scientists to request postmortem materials, as well as MRI brain scans of chimpanzees that have been acquired as data from previous NIH-funded research projects. At present, facilities that hold NIH-owned chimpanzees are participating in the NCBR, by contributing the brains of any individuals that die to the NCBR. The scientific value of both the chimpanzee genome and the NCBR could be strengthened considerably by the inclusion of behavioral, cognitive, and neuroimaging data, all of which can be obtained noninvasively or by the use of the minimally invasive methods described earlier.

The fact that NIH owns ~350 chimpanzees that will continue to be financially supported by NIH until they die is a second justification for creating an NCBCN. Further, there are no future plans for the breeding of captive chimpanzees supported by NIH, making NIH-owned chimpanzees a dwindling resource that the American taxpayer has supported, and will continue to support. In our view, this is an incredible and unjustifiable financial waste of a clearly valuable scientific resource. As we have outlined, with advances in minimally invasive molecular biological, neurogenomic, neuroimaging, and cognitive research methods, many types of research can be performed ethically, without compromising the well-being and care of the chimpanzees. Indeed, both the Institute of Medicine

(IOM; Altevogt et al. 2011) and the NIH Working Group (http://grants.nih.gov/grants/guide/notice-files/NOT-OD-12-025.html) explicitly recognized that engaging chimpanzees in cognitive research or training them to participate in husbandry behaviors, such as voluntarily presenting for venipuncture, was inherently enriching to the animals. Moreover, all of these kinds of research projects can be accomplished while the animals are living in conditions that satisfy the social grouping, housing, and welfare standards that NIH has adopted.

A third justification for the creation of an NCBCN is that, in addition to the NIH-owned captive chimpanzee resource dwindling in numbers, the population of wild chimpanzees is dwindling also. Chimpanzees are an endangered species, and, by many estimates, their numbers in the wild are diminishing rapidly, primarily due to habitat loss and poaching. Neurogenomic, molecular biology, and neuroimaging are difficult, if not impossible, to carry out with wild chimpanzees. Thus, obtaining these types of data in captive populations would complement the behavioral, ecological, and, arguably, the conservation efforts of many scientists working with wild populations, potentially benefitting research efforts with wild chimpanzees (Lonsdorf 2007). Although conservation of endangered species may not be part of the mission of NIH, there is certainly no harm associated with data collected from NIH-supported chimpanzee research being used to help conserve this endangered species. Indeed, in light of the fact that NIH bred these chimpanzees for research purposes, it could be argued that NIH has a moral obligation to continue to use them in research, which may enhance their preservation in the wild.

In summary, given the circumstances and dwindling numbers of chimpanzees in captivity and the wild, now is not the time for us to decrease research with them, but, instead, we should be increasing our efforts to study chimpanzees. As noted earlier, NIH will have to continue to support the chimpanzees that were bred under their previous funding initiatives, and simply warehousing them until they die 30 years from now is both irresponsible and unjustifiable. In short, attrition until extinction is not, and should not be, a retirement and management plan for the captive chimpanzee population. We have an opportunity, and indeed, we would argue an obligation, to continue to study them for the purposes of benefitting both human and chimpanzee physical and psychological health. This goal can be accomplished without the use of invasive methods and by adhering to the highest ethical standards of research with primates, particularly if scientific initiatives are integrated with existing behavioral management programs. None of the procedures and tests that we have described in this chapter are considered invasive, and are readily used with human participants, including children and infants. Further, none of the ideas or procedures described in this chapter should be considered objectionable by any reasonable individual concerned with the welfare of captive chimpanzees. In sum, scientists and animal welfare advocates should embrace this opportunity to do something scientifically, ethically, and financially beneficial for the long-term survival and care of captive and wild chimpanzees.

ACKNOWLEDGMENTS

This research was supported in part by NIH grants NS-42867, NS-73134, HD-60563, and HD-56232. American Psychological Association guidelines for the ethical treatment of animals were adhered to during all aspects of the studies described.

REFERENCES

Altevogt, B.M., D.E. Pankevich, M.K. Shelton-Davenport, and J.P. Kahn. 2011. *Chimpanzees in Biomedical and Behavioral Research: Assessing the Necessity.* Washington, DC: The National Academies Press.

Andrews, M.W. and L.A. Rosenblum. 2001. New methodology applied to bonnet macaques (*Macaca radiata*) to address some contradictory evidence of manual asymmetries in old world monkeys. *Journal of Comparative Psychology* 115:1123–1130.

Autrey, M.M., L.A. Reamer, M.C. Mareno, C.C. Sherwood, J.G. Herndon, T.M. Preuss, S.J. Schapiro, and W.D. Hopkins. 2014. Age-related effects in the neocortical organization of chimpanzees: Gray and white matter volume, cortical thickness, and gyrification. *NeuroImage* 101:59–67.

Beran, M.J., T.A. Evans, F. Paglieri, J.M. McIntyre, E. Addessi, and W.D. Hopkins. 2014. Chimpanzees (*Pan troglodytes*) can wait, when they choose to: A study with the hybrid delay task. *Animal Cognition* 17 (2):197–205.

Biro, D., C. Sousa, and T. Matsuzawa. 2006. Ontogeny and cultural propagation of tool use by wild chimpanzees at Bossou, Guinea: Case studies in nut cracking and leaf folding. In *Cognitive Development of Chimpanzees*, eds Matsuzawa, T., T. Tomonaga, and M. Tanaka, 476–507. New York: Springer.

Boesch, C. 1991. Handedness in wild chimpanzees. *International Journal of Primatology* 12 (6):541–558.

Bogart, S.L., A.J. Bennett, S.J. Schapiro, L.A. Reamer, and W.D. Hopkins. 2014. Different early rearing experiences have long term effects on cortical organization in captive chimpanzees (*Pan troglodytes*). *Developmental Science* 17 (2):161–174.

Brosnan, S.F., L.M. Hopper, S. Richey, H.D. Freeman, C.F. Talbot, S.D. Gosling, S.P. Lambeth, and S.J. Schapiro. 2015. Personality influences responses to inequity and contrast in chimpanzees. *Animal Behaviour* 101:75–87.

Byrne, R.W. 1995. *The Thinking Ape: Evolutionary Origins of Intelligence*. Oxford, UK: Oxford University Press.

Byrne, R.W. and N. Corp. 2004. Neocortex size predicts deception rate in primates. *Proceedings of the Royal Society B: Biological Sciences* 271 (1549):1693–1699.

Connor-Stroud, F.R., W.D. Hopkins, T.M. Preuss, Z. Johnson, E. Breding, Z. Zhang, and P. Sharma. 2014. Extensive vascular mineralization in the brain of a chimpanzee (*Pan troglodytes*). *Journal of Comparative Medicine* 64 (3):224–229.

Corballis, M.C. 1992. *The Lopsided Brain: Evolution of the Generative Mind*. New York: Oxford University Press.

Ettlinger, G.F. 1988. Hand preference, ability and hemispheric specialization: In how far are these factors related in the monkey? *Cortex* 24:389–398.

Fagot, J. and E. Bonté. 2010. Automated testing of cognitive performance in monkeys: Use of a battery of computerized test systems by a troop of semi-free-ranging baboons (*Papio papio*). *Behavior Research Methods* 42 (2):507–516.

Fagot, J., J. Gullstrand, C. Kemp, C. Defilles, and M. Mekaouche. 2014. Effects of freelay accessible computerized test systems on the spontaneous behaviors and stress level of Guinea baboons (*Papio papio*). *American Journal of Primatology* 76 (1):56–64.

Freeman, H.D. and S.D. Gosling. 2010. Personality in nonhuman primates: A review and evaluation of past research. *American Journal of Primatology* 72 (8):653–671.

Gazes, R.P., E.K. Brown, B.M. Basile, and R.R. Hampton. 2012. Automated cognitive testing of monkeys in social groups yields results comparable to individual laboratory-based testing. *Animal Cognition* 16:445–458.

Gomez-Robles, A., W.D. Hopkins, and C.C. Sherwood. 2013. Increased morphological asymmetry, evolvability, and plasticity in human brain evolution. *Proceedings of the Royal Society B: Biological Sciences* 280 (1761):20130575.

Hecht, E.E., L.E. Murphy, D.A. Gutman, J.R. Votaw, D.M. Schuster, T.M. Preuss, G.A. Orban, D. Stout, and L.A. Parr. 2013. Differences in neural activation for object-directed grasping in chimpanzees and humans. *The Journal of Neuroscience* 33 (35):14117–14134.

Herrelko, E.S., S-J. Vick, and H. Buchanan-Smith. 2012. Cognitive research in zoo-housed chimpanzees: Influence of personality and impact on welfare. *American Journal of Primatology* 74:828–840.

Herrmann, E., J. Call, M.V. Hernandez-Lloreda, B. Hare, and M. Tomasello. 2007. Humans have evolved specialized skills of social cognition: The cultural intelligence hypothesis. *Science* 317:13601366.

Herrmann, E., B. Hare, J. Call, and M. Tomasello. 2010. Differences in the cognitive skills of bonobos and chimpanzees. *PLos One* 5 (8):e12438.

Hong, K.W., A. Weiss, N. Morimura, T. Udono, I. Hayasaka, T. Humle, Y. Murayama, S. Ito, and M. Inoue-Murayama. 2011. Polymorphism of the tryptophan hydroxylase 2 (TPH2) gene is associated with chimpanzee neuroticism. *PLoS One* 6 (7):e22144.

Hopkins, W.D. 1995. Hand preferences for a coordinated bimanual task in 110 chimpanzees: Cross-sectional analysis. *Journal of Comparative Psychology* 109:291–297.

Hopkins, W.D. 2013. Behavioral and brain asymmetries in chimpanzees: A case for continuity. *Annals of the New York Academy of Sciences* 1288:27–35.

Hopkins, W.D., M. Adams, and A. Weiss. 2013a. Genetic and environmental contributions to the expression of handedness in chimpanzees (*Pan troglodytes*). *Genes, Brain and Behavior* 12:446–452.

Hopkins, W.D., C. Cantalupo, M.J. Wesley, A. Hostetter, and D. Pilcher. 2002. Grip morphology and hand use in chimpanzees (*Pan troglodytes*): Evidence of a left hemisphere specialization in motor skill. *Journal of Experimental Psychology: General* 131:412–423.

Hopkins, W.D., Z.R. Donaldson, and L.Y. Young. 2012a. A polymorphic indel containing the RS3 microsatellitein the 5′ flanking region of the vasopressin V1a receptor gene is associated with chimpanzee (*Pan troglodytes*) personality. *Genes, Brain, and Behavior* 11:552–558.

Hopkins, W.D., A.C. Keebaugh, L.A. Reamer, J. Schaeffer, S.J. Schapiro, and L.J. Young. 2014a. Genetic influences on receptive joint attention in chimpanzees (*Pan troglodytes*). *Scientific Reports* 4 (3774):1–7.

Hopkins, W.D., A. Meguerditchian, O. Coulon, S.L. Bogart, J.F. Mangin, C.C. Sherwood, M.W. Grabowski, et al. 2014b. Evolution of the central sulcus morphology in primates. *Brain, Behavior and Evolution* 84:19–30.

Hopkins, W.D., M. Misiura, L.A. Reamer, J.A. Schaeffer, M.C. Mareno, and S.J. Schapiro. 2014c. Poor receptive joint attention skills are associated with atypical grey matter asymmetry in the posterior superior temporal gyrus of chimpanzees (*Pan troglodytes*). *Frontiers in Cognition* 5 (7):1–8.

Hopkins, W.D. and K.A. Phillips. 2016. Noninvasive imaging technologies in primates. In *Lateralzed Brain Functions*, eds Rogers, L.J. and G. Vallortigara. New York: Springer.

Hopkins, W.D., K.A. Phillips, A. Bania, S.E. Calcutt, M. Gardner, J. Russell, J. Schaeffer, E.V. Lonsdorf, S.R. Ross, and S.J. Schapiro. 2011. Hand preferences for coordinated bimanual actions in 777 great apes: Implications for the evolution of handedness in hominins. *Journal of Human Evolution* 60 (5):605–611.

Hopkins, W.D. and D.M. Rabinowitz. 1997. Manual specialization and tool-use in captive chimpanzees (*Pan troglodytes*): The effect of unimanual and bimanual strategies on hand preference. *Laterality* 2:267–277.

Hopkins, W.D., L. Reamer, M.C. Mareno, and S.J. Schapiro. 2015. Genetic basis for motor skill and hand preference for tool use in chimpanzees (*Pan troglodytes*). *Proceedings of the Royal Society: Biological Sciences B* 282 (1800):20141223.

Hopkins, W.D. and J.L. Russell. 2004. Further evidence of a right hand advantage in motor skill by chimpanzees (*Pan troglodytes*). *Neuropsychologia* 42:990–996.

Hopkins, W.D., J.L. Russell, and J.A. Schaeffer. 2012b. The neural and cognitive correlates of aimed throwing in chimpanzees: A magnetic resonance image and behavioural study on a unique form of social tool use. *Philosophical Transactions of the Royal Society B: Biological Sciences* 367 (1585):37–47.

Hopkins, W.D., J.L. Russell, and J. Schaeffer. 2014d. Chimpanzee intelligence is heritable. *Current Biology* 24 (14):1–4.

Hopkins, W.D., J.L. Russell, J.A. Schaeffer, M. Gardner, and S.J. Schapiro. 2009. Handedness for tool use in captive chimpanzees (*Pan troglodytes*): Sex differences, performance, heritability and comparison to the wild. *Behaviour* 146 (11):1463.

Hopkins, W.D. and J.P. Taglialatela. 2012a. A preliminary report on the neural correlates of joint attention in chimpanzees. In *Joint Attention* eds Seeman, A. Cambridge, MA: MIT Press.

Hopkins, W.D. and J.P. Taglialatela. 2012b. Initiation of joint attention is associated with morphometric variation in the anterior cingulate cortex of chimpanzees (*Pan troglodytes*). *American Journal of Primatology* 75 (5):441–449.

Hopkins, W.D., J.P. Taglialatela, J.L. Russell, T.M. Nir, and J. Schaeffer. 2010. Cortical representation of lateralized grasping in chimpanzees (*Pan troglodytes*): A combined MRI and PET study. *PLoS One* 5 (10):e13383.

Hopkins, W.D. and D.A. Washburn. 1994. Do right- and left-handed monkeys differ on cognitive measures? *Behavioral Neuroscience* 108:1207–1212.

Hopkins, W.D., D.A. Washburn, L. Berke, and M. Williams. 1992. Behavioral asymmetries of psychomotor performance in rhesus monkeys (*Macaca mulatta*): A dissociation in hand preference and performance. *Journal of Comparative Psychology* 106:392–397.

Hopper, L.M., S.A. Price, H.D. Freeman, S.P. Lambeth, S.J. Schapiro, and R.L. Kendall. 2014. Influence of personality, age, sex, and estrous state on chimpanzee problem-solving success. *Animal Cognition* 17:835–847.

King, J.E. and A.J. Figueredo. 1997. The five-factor model plus dominance in chimpanzee personality. *Journal of Research on Personality* 31:257–271.

King, J.E., A. Weiss, and K.H. Farmer. 2005. A chimpanzee (*Pan troglodytes*) analogue of cross-national generalization of personality structure: Zoological parks and African sanctuary. *Journal of Personality* 73:389–410.

Lacreuse, A., J.L. Russell, W.D. Hopkins, and J.G. Herndon. 2014. Cognitive and motor aging in female chimpanzees. *Neurobiology of Aging* 35 (3):623–632.

Latzman, R.D., L.E. Drislane, L.K. Hecht, S.J. Brislin, C.J. Patrick, S.O. Lilienfeld, H.J. Freeman, S.J. Schapiro, and W.D. Hopkins. 2016a. A chimpanzee model of triarchic psychopathy constructs: Development and initial validation. *Clinical Psychological Science* 4 (1):50–66.

Latzman, R.D., H.D. Freeman, S.J. Schapiro, and W.D. Hopkins. 2015a. The contributions of genetics and early rearing experiences to hierarchical personality dimensions in chimpanzees (*Pan troglodytes*). *Journal of Personality and Social Psychology* 109 (5):889–900.

Latzman, R.D., L.K. Hecht, H.D. Freeman, S.J. Schapiro, and W.D. Hopkins. 2015b. Investigating the neuroanatomical correlates of personality in chimpanzees (*Pan troglodytes*): Associations between personality and frontal cortex. *NeuroImage* 123:63–71.

Latzman, R.D. and W.D. Hopkins. 2016. Letter to the Editor: Avoiding a lost opportunity for psychological medicine: Importance of chimpanzee research to the National Institutes of Health portfolio. *Psychological Medicine* 46 (11):2445–2447.

Latzman, R.D., W.D. Hopkins, A.C. Keebaugh, and L.J. Young. 2014. Personality in chimpanzees (*Pan troglodytes*): Exploring the hierarchical structure and associations with the vasopressin V1A receptor gene. *PLoS One* 9 (4):e95741.

Latzman, R.D., C.J. Patrick, H.D. Freeman, S.J. Schapiro, and W.D. Hopkins. in press. Etiology of triarchic psychopathy dimensions in chimpanzees (*Pan troglodytes*). Clinical Psychological Science.

Latzman, R.D., Sauvigné, K. C., and Hopkins, W.D. 2016c. Translating chimpanzee personality to humans: Investigating the transportability of chimpanzee-derived personality scales to humans. *American Journal of Primatology* 78:601–609.

Latzman, R.D., J.P. Taglialatela, and W.D. Hopkins. 2015c. Delay of gratification is associated with white matter connectivity in the dorsal prefrontal cortex: A diffusion tensor imaging (DTI) study in chimpanzees (*Pan troglodytes*). *Proceedings of the Royal Society of London: Biological Sciences B* 22 (282):1809–1814.

Latzman, R.D., L.J. Young, and W.D. Hopkins. 2016b. Displacement behaviors in chimpanzees (*Pan troglodytes*): A neurogenomics investigation of the RDoC negative valence systems domain. *Psychophysiology* 53 (3):355–363.

Leavens, D.A., L.A. Reamer, M.C. Mareno, J.L. Russell, D.C. Wilson, S.J. Schapiro, and W.D. Hopkins. 2015. Distal communication by chimpanzees (*Pan troglodytes*): Evidence for common ground? *Child Development* 86 (5):1623–1638.

Lonsdorf, E.V. 2007. The role of behavioral research in the conservation of chimpanzees and gorillas. *Journal of Applied Animal Welfare and Science* 10:71–78.

Lyn, H., J.L. Russell, and W.D. Hopkins. 2010. The impact of environment on the comprehension of declarative communication in apes. *Psychological Science* 21 (3):360–365.

McGrew, W.C., L.F. Marchant, R.W. Wrangham, and H. Klein. 1999. Manual laterality in anvil use: Wild chimpanzees cracking *Strychnos* fruits. *Laterality* 4:79–87.

Meguerditchian, A., J. Vauclair, and W.D. Hopkins. 2013. On the origins of human handedness and language: A comparative review of hand preferences for bimanual coordinated actions and gestural communication in nonhuman primates. *Developmental Psychobiology* 55 (6):637–650.

Parr, L.A., E. Hecht, S.K. Barks, T.M. Preuss, and J.R. Votaw. 2009. Face processing in the chimpanzee brain. *Current Biology* 19:50–53.

Phillips, K.A., K.L. Bales, J.P. Caitanio, A. Conley, P.W. Czoty, B.A. 't Hart, W.D. Hopkins, et al. 2014. Why primate models matter. *American Journal of Primatology* 76 (9):801–827.

Phillips, K.A., J. Schaeffer, E. Barrett, and W.D. Hopkins. 2013. Performance asymmetries in tool use are associated with corpus callosum integrity in chimpanzees (*Pan troglodytes*): A diffusion tensor imaging study. *Behavioral Neuroscience* 127 (1):106–113.

Pope, S.M., J.L. Russell, and W.D. Hopkins. 2015. The association between imitation recognition and socio-communicative competencies in chimpanzees (*Pan troglodytes*). *Frontiers in Psychology* 6:188.

Reamer, L.A., R.L. Haller, E.J. Thiele, H.D. Freeman, S.P. Lambeth, and S.J. Schapiro. 2014. Factors affecting initial training success of blood glucose testing in captive chimpanzees (*Pan troglodytes*). *Zoo Biology* 33:212–220.

Rumbaugh, D.M., W.K. Richardson, D.A. Washburn, E.S. Savage-Rumbaugh, and W.D. Hopkins. 1989. Rhesus monkeys (*Macaca mulatta*), video tasks, and implications for stimulus-response contiguity. *Journal of Comparative Psychology* 103:32–38.

Rumbaugh, D.M. and D.A. Washburn. 2003. *The Intelligence of Apes and Other Rational Beings.* New Haven, CT: Yale University Press.

Russell, J.L., H. Lyn, J.A. Schaeffer, and W.D. Hopkins. 2011. The role of socio-communicative rearing environments in the development of social and physical cognition in apes. *Developmental Science* 14 (6):1459–1470.

Savage-Rumbaugh, E.S. and R. Lewin. 1994. *Kanzi: The Ape at the Brink of the Human Mind.* New York: John Wiley & Sons.

Schmitt, V., B. Pankau, and J. Fischer. 2011. Old World monkeys compare to apes in the primate cognition test battery. *PLoS One* 7 (4):e32024.

Staes, N., J.M.G. Stevens, P. Helsen, M. Hillyer, M. Korody, and M. Eens. 2014. Oxytocin and vasopressin receptor gene variation as a proximate base for inter- and intraspecific behavioral differences in bonobos and chimpanzees. *PLoS One* 9 (11):e113364.

Taglialatela, J.P., J.L. Russell, J.A. Schaeffer, and W.D. Hopkins. 2008. Communicative signaling activates "Broca's" homolog in chimpanzees. *Current Biology* 18:343–348.

Taglialatela, J.P., J.L. Russell, J.A. Schaeffer, and W.D. Hopkins. 2009. Visualizing vocal perception in the chimpanzee brain. *Cerebral Cortex* 19 (5):1151–1157.

Taglialatela, J.P., J.L. Russell, J.A. Schaeffer, and W.D. Hopkins. 2011. Chimpanzee vocal signaling points to a multimodal origin of human language. *PLoS One* 6 (4):e18852.

The Chimpanzee Sequencing and Analysis Consortium. 2005. Initial sequence of the chimpanzee genome and comparison with the human genome. *Nature* 437:69–87.

Van Schaik, C.P., R.O. Deaner, and M.Y. Merrill. 1999. The conditions for tool use in primates: Implications for the evolution of material culture. *Journal of Human Evolution* 36:719–741.

Warren, J.M. 1980. Handedness and laterality in humans and other animals. *Physiological Psychology* 8:351–359.

Washburn, D.A., W.D. Hopkins, and D.M. Rumbaugh. 1989. Automation of learning-set testing: The video-task paradigm. *Behavior Research Methods, Instruments, & Computers* 21:281–284.

Washburn, D.A. and D.M. Rumbaugh. 1992. A comparative assessment of psychomotor performance: Target prediction by humans and macaques. *Journal of Experimental Psychology: General* 121:305–312.

Weiss, A., J.E. King, and M.R. Enns. 2002. Subjective wellbeing is heritable and genetically correlated with dominance in chimpanzees. *Journal of Personality and Social Psychology* 83 (5):1141–1149.

Weiss, A., J.E. King, and A.J. Figuerdo. 2000. The heritability of personality factors in chimpanzees (*Pan troglodytes*). *Behavior Genetics* 30 (3):213–2221.

Weiss, A., J.E. King, and W.D. Hopkins. 2007. A cross-setting study of chimpanzee (*Pan troglodytes*) personality structure and development: Zoological parks and Yerkes National Primate Research Center. *American Journal of Primatology* 69:1264–1277.

Whitehouse, J., J. Micheletta, L.E. Powell, C. Bordier, and B.M. Waller. 2013. The impact of cognitive testing on the welfare of group housed primates. *PLos One* 8 (11):e78308.

Whiten, A. and A. Mesoudi. 2008. Establishing an experimental science of culture: Animal social diffusion experiments. *Philosophical Transactions of the Royal Society B: Biological Sciences* 363 (1509):3477–3488. doi:10.1098/rstb.2008.0134.

CHAPTER 11

Utility of Systems Network Analysis for Understanding Complexity in Primate Behavioral Management

Brenda McCowan and Brianne Beisner
University of California, Davis

CONTENTS

I think the next century will be the century of complexity.

—Stephen Hawking
January 2000

INTRODUCTION

Primate Social Networks and Their Management

Many nonhuman primates are highly social species and live in groups known as social networks. These networks can be relatively simple, comprising one male and one female and their offspring, or highly complex, such as the multimale, multifemale groups characteristic of many primates living in captivity (Sussman 2003). The challenge we face as managers of these larger captive primate populations is properly understanding the complex, multiplex structure of these groups' behavioral networks (e.g., aggression, coalitions, policing, subordination, grooming, huddling) and how they relate to social group stability, and the rate and severity of aggression. Network analysis has significant potential to improve the management of these captive nonhuman primate populations (really animal populations, in general), including those in laboratories, zoos, and sanctuaries, because it provides tools that allow us to reveal "hidden" dynamical patterns in the relationships among individuals in these complex social groups (see section "Why Is Network Analysis Useful for Primate Social Management, Specifically, and Primate Behavioral Management, in General?").

Network Analysis and Its Uses

Network analysis is a major approach for examining and quantifying complexity in systems. Its focus is on the patterns of relationships that arise among interacting social and other physical entities. An important working assumption of this approach is that indirect relationships (i.e., "friends of friends") in networks matter. This focus on indirect relationships, therefore, necessarily relies on data that are deeply connected within multiple layers in systems and thus are not easily observed directly. The multilayered, interconnected nature of these relationships is ubiquitous across all human and nonhuman systems. Therefore, a particular strength of network analysis is that it provides mathematical methods for calculating measures of direct and indirect relationships (social or otherwise) across multiple levels of organization, from the ecosystem level to the population and group levels, and to the level of the individual (Freeman 1984; McCowan et al. 2011; Sih et al. 2009).

The most common understanding for the use of network analysis is for *social* network analysis. In this context, these quantitative measures can be used to characterize and compare components of social network structures. They allow us to statistically test hypotheses about the ways in which individual attributes or group factors influence social structures of groups, and how the dynamics of group social structure affect individual or collective group behavior. For example, at the group level, these metrics can be used to evaluate the effect of the physical and social environment on social network structure; track changes in social network structures over time owing to changes in group composition or environmental circumstances; or statically or dynamically compare social structures across groups, populations, or species. At the individual level, these measures allow for the quantitative characterization of the variation in social and other environmental experiences by individuals within groups. In addition, they can be used to test hypotheses concerning the ways in which an individual's social role affects social stability or flow of information throughout the group or network (Beisner et al. 2011a,b; Flack et al. 2006; Lusseau 2007; McCowan et al. 2011).

Therefore, as the term seems to imply, *social* network analysis has a clear role in deciphering our understanding of social group structure and dynamics. However, network analysis has many other applications relevant to a diversity of disciplines, because it is designed to reveal emergent structure in *any* type of system. In fact, the use of network analysis to understand emergent structure has a long history (Freeman 1984). Network analysis stems from mathematical graph theory and has been employed in a number of disciplines, including studies of transportation systems (Sen et al. 2003), business and economics (Levine 1972), and the spread of both information (Albert et al. 1999; Liben-Nowell and Kleinberg 2008) and computer viruses (Newman et al. 2002) through the Internet. In the biological sciences, network approaches have been used to explore diverse phenomena, ranging from cellular organization (reviewed by Barabási and Oltvai 2004; Bullmore and Sporns 2009) to food webs (reviewed by Ings et al. 2009) to disease spread (reviewed by Martínez-López et al. 2009).

Unsurprisingly, the use of network analysis for studying sociality also has a long history in anthropology, social psychology, and sociology (reviewed by Scott 2000; Wasserman and Faust 1994). However, its value for studying social interactions of animals, although recognized more than three decades ago (Wilson 1975), was not realized until the mid-1990s, when the methodology increased in popularity (Sade and Dow 1994). Over the past 25 years, network analysis has become an increasingly common tool for studying animal behavior, not only in the field of primatology (Beisner and McCowan 2013, 2014a; Beisner et al. 2011a,b, 2015, 2016; Flack and de Waal 2004; Flack and Krakauer 2006; Flack et al. 2006; Fushing et al. 2014; McCowan et al. 2008, 2011) but also in other areas, including behavioral ecology (Croft et al. 2004; Lusseau 2003a) and epidemiology (Corner et al. 2003; Otterstatter and Thomson 2007).

This recent upsurge in interest in network analysis for understanding patterns in animal behavior and behavioral ecology is partly owing to the widespread availability of comprehensive network analysis software and the publication of a series of review articles highlighting the utility of network analysis approaches in animal behavior research (Krause et al. 2007, 2009, 2015; Sih et al. 2009; Wey and Blumenstein 2010). These reviews have especially emphasized the value of network analysis for assessing the effects of social network structure on disease spread, information flow, group dynamics, and future reproductive fitness. The utility of network analysis for assessing the heterogeneity of social environments experienced by individual group members, and for identifying individuals who may play especially important social roles, for example, in maintaining group stability (Beisner et al. 2011a,b; Flack et al. 2006; McCowan et al. 2008, 2011), has also been highlighted in the literature. Notably, only a few studies have been conducted addressing the role of network analysis in animal management and welfare (but see Beisner et al. 2011a,b; Böhm et al. 2009; Durell et al. 2004; Jones et al. 2010; McCowan et al. 2008, 2011; Webb 2005), despite suggestions that the network approach could be used to provide new insights into these areas (Beisner and McCowan 2014a; Krause et al. 2007; Makagon et al. 2012).

Why Is Network Analysis Useful for Primate Social Management, Specifically, and Primate Behavioral Management, in General?

Network analysis has proven quite useful in applications related to primate social group management (Beisner and McCowan 2014a; Beisner et al. 2011a,b, 2016; Flack and de Waal 2004; Flack and Krakauer 2006; Fushing et al. 2014; McCowan et al. 2008, 2011). Network analysis, in this social context, allows for the detection and description of the patterns and quality of interactions among animals, facilitating the identification of the following:

1. Group compositions that increase the risk of deleterious aggression, wounding, and social collapse.
2. Individuals who either contribute to the maintenance of social stability or facilitate the continuance of problematic aggression or social instability.
3. The social dynamics underlying group stability.
4. Avenues for management intervention to increase social stability or to prevent the development of social structures associated with social instability. Examples of intervention include targeted removal of individuals that instigate or perpetuate deleterious aggression (e.g., high-ranking natal males) while simultaneously avoiding removal of key individuals (e.g., conflict policers).

Network analysis has the advantage of being well suited for quantifying indirect relationships or associations, which allows for the detection of emergent network patterns and structures, and the characterization of the different social roles that exist in the group. Standard analytical approaches, by contrast, are typically limited to rates of behaviors and direct interactions.

While the potential for its use in managing large social groups may be obvious, network analysis can be used to examine a number of other important aspects of primate behavioral management. Potential applications include the use of massive databases ("Big Data"), containing social pairing history information, to examine these historical networks against a variety of attributes (e.g., personality, sex, age, rearing history, etc.). This network approach could allow us to use patterns in large quantities of data (rather than often-biased recollection of immediate history) to determine which animals to use in a pairing attempt, based upon clusters or suites of attributes or experiences that led to successful pairings in the past. Another potential application using these large databases would be the use of a series of bipartite networks (see definitions in the Bipartite Network section) to examine the suite of social and environmental risk factors that lead to clustering of abnormal behaviors. Patterns in the suite of past and current experiences of animals, along with genotype data, could be used to predict, and thus prevent, through interventions, the development of deleterious abnormal behaviors.

Goals for This Chapter

In this chapter, we provide an overview of the potential importance of network analysis for primate welfare and management science. The sections that follow present an introduction to social network analysis, followed by a discussion of the utility of these tools for captive primate behavioral management, and in particular, social group management.

PRIMER ON SOCIAL NETWORK ANALYSIS

Network Visualization

Social networks are typically represented as sociograms. They can be represented both in matrix (or edge list) form (see Figure 11.1A,B) and as a graph (see Figure 11.2A through D). Graphical representation facilitates recognition of patterns: Interacting entities are drawn as "nodes" or "actors" and are linked by "edges" or "ties." In behavioral studies, nodes typically represent individuals. However, importantly, nodes can represent any interacting entities, such as molecules, cells, individuals, groups of individuals, or populations—anything that is connected to other entities in some fashion.

Relationships between the nodes are represented by edges. The most basic sociogram, therefore, simply shows the presence or absence of a relationship between each pair of nodes in the network,

(A)

	A	B	C	D	E	F	G	H	I	J	K	L	M	N	O	P	Q
A					10				5		5						5
B							5										
C					5				1		1						
D											1				1		1
E	10		5						10		10						5
F		1	1							1							
G		5							5								
H													10		5		5
I	5		1		5						1						
J							5										
K	5		1		5				1								
L																	
M								10							5		5
N		1	1														
O							5						5				
P		1	1														
Q	5				5				5				1		5		

(B)

ID1	ID2	Total
A	E	10
A	I	5
A	K	5
A	Q	5
B	F	1
B	G	5
B	N	1
B	P	1
C	E	5
C	F	1
C	I	1
C	K	1
C	N	1
C	P	1
E	A	10
E	C	5
E	I	5
E	K	5
E	Q	5
G	B	5
G	J	5
H	M	10

Figure 11.1 (A) A sample sociogram depicted as a matrix. Nodes (letters) represent interacting individuals. Edges (numbers) represent relationships between the nodes. (B) Same sociogram (partial) represented as an "edge list," where each dyad is represented in a row in columns A and B, with the frequency of their interaction in column C.

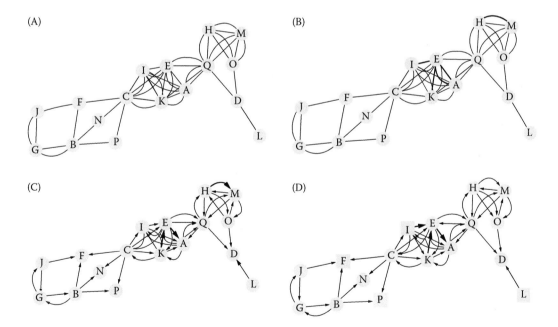

Figure 11.2 (A) A sample binary network. Nodes (circles) represent interacting individuals. Edges (lines) represent relationships between the nodes. (B) A weighted network in which interaction strengths are represented by the thickness of the edge. (C) A weighted and directed network where direction is indicated using arrows, which point away from the actor and toward the receiver. (D) A weighted and directed network highlighting node attributes. Characteristics of the nodes are drawn on by manipulating node size, shape, or color. In this example, shape has been used to denote gender: females (circles) versus males (squares), and color indicates age: adults (blue) versus juveniles (yellow).

known as a binary or dichotomous network (Figure 11.2A). Complexity can be added to the graph by mapping on relationship properties and node attributes. For example, the thickness of each edge in a weighted graph (Figure 11.2B) can represent relative relationship strength, such as interaction frequency, rate, or duration. Arrows can be used to indicate the directionality of the relationships (Figure 11.2C). Such directed graphs are especially useful for exploring interaction data, such as grooming events or instances of aggression. When the individuals' attributes (e.g., sex, age, relatedness, dominance status, etc.) are known and are of interest, these can be represented by node size, shape, or color (Figure 11.2D).

Direct and Indirect Pathways

As repeatedly emphasized, perhaps one of the most important aspects of social network analysis is its emphasis on both direct *and* indirect pathways. Links or edges between nodes create pathways that can be used to measure the flow of a behavior (or other phenomenon) among group members. Pairs of individuals in the network graph that do not have a direct link (i.e., were not observed to interact) may still have an indirect pathway connecting them, if they have interacted with the same third party (Figure 11.3; e.g., A → B → C is an indirect pathway linking A and C). By providing a quantitative way of representing both direct and indirect social connections, network analysis can be used for studying higher-order, emergent properties of a system (Lusseau 2003b; Pasquaretta et al. 2014). Such emergent properties may then provide important insights into the relative cohesion and stability of the social group, thereby indicating potential avenues for management intervention if necessary.

A directed pathway of interactions, such as social signals, from A > B > C, is a mathematical representation of a potential path through which information may be transmitted in the network. Pairs of nodes with no direct connection (e.g., individuals that do not directly interact) may gather information by observing group members' interactions. Many social vertebrates, including primates, rats, birds, and fish, can use known relationships to deduce unknown relationships (known as transitive inference) such as deducing A > C from A > B and B > C (Bond et al. 2003; Durell et al. 2004; Grosenick et al. 2007; McGonigle and Chalmers 1977). Members of a social group may learn, or reaffirm, their dominance position relative to an animal with whom they have had no recent interactions, if they both have a dominance interaction with the same third party. For example, in captive groups of rhesus macaques, we have found that dyads connected by only indirect pathways of subordination signals have the same level of dominance certainty as dyads that signal directly, and both of these types of dyads have more definite dominance relationships than pairs with no connection in the subordination signaling network (Beisner et al. 2016). Therefore, networks can also be used to represent cognitive processes that animals may use to gather information from their social environment.

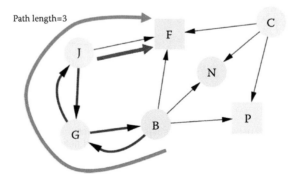

Figure 11.3 Examples of direct (red line) and indirect paths (green line). Indirect path shown has a path length of 3.

Table 11.1   Summary of Network Measures

Terminology	Description	Level of Analysis
Node	Social entity (i.e., individual, group)	
Edge	Relationship between two nodes	
Measures of prominence		
Degree centrality	Number of direct connections a focal node has to other nodes in the network	Individual
Degree centralization	Variation in degree centralities of nodes within the network	Network
Eigenvalue centrality	A weighted version of degree centrality	Individual
Closeness centrality	Sum of shortest paths that connect a focal node to all other nodes in the network	Individual
Closeness centralization	Variation in closeness centrality measures of nodes within the network	Network
Betweenness centrality	Proportion of shortest paths between any two nodes in the network that pass through the focal node	Individual
Betweenness centralization	Variation in betweenness centrality measures of nodes within the network	Network
Measures of range		
Reach	Number of edges separating the focal node from other nodes of interest	Individual
Diameter	Longest distance between any two nodes in the network	Network
Measures of cohesion		
Density	Proportion of all possible connections that are present in the network	Network
Network reciprocity	Extent to which pairs of nodes make reciprocal connections to each other	Network
Network clustering coefficient	Tendency of groups of nodes to be interconnected with each other	Network
Network fragmentation	An inverse measure of the amount of connection redundancy in a network	Network
Assortativity	Tendency of nodes with certain attributes (sex, age, node degree, and centrality) to interact	Network
Bipartite networks	A type of network that explores relationships between entities, attributes, and events	

Standard Network Metrics

Social network analysis is composed of five major principles and a battery of quantitative techniques that measure variables derived from each of these principles at multiple levels of analysis [(Wasserman and Faust 1994); summarized in Table 11.1]. These principles include

1. *Prominence* or key players, which is an indicator of who is "in charge"
2. *Range*, which is an indicator of the extent of the node's network or a network's overall reach
3. *Cohesion*, which is the grouping of nodes according to strong common relationships with each other
4. *Structural equivalence*, which is the grouping of nodes according to similarity in their overall social environments
5. *Brokerage*, which indicates the bridging of otherwise unconnected components within or between parts of a network

Using these principles, network analysis can explore relational properties of social (and other) networks, such as how cohesive a group is or what subgroups of interconnected nodes exist, and positional properties, such as who occupies different roles and positions in a network. As such, roles and positions in networks are identified by examining similarities in connections among the nodes in a network.

For primatologists, the most useful types of measures are found under the principles of prominence, range, and cohesion. In addition, structural equivalence has potential value in assessing dominance or other hierarchical relationships in some animal groups (e.g., how similarities in social connections of different individual primates in a group relate to their individual positions in rank). Brokerage is useful for looking at the bridges of connections between animal subgroups with different attributes (e.g., the role specific matrilines play in connecting other matrilines in a large macaque group). These measures can be quantified at multiple levels of analysis but are most commonly used at the individual and group (or subgroup) levels. They can be calculated for both binary and weighted networks, as well as undirected and directed networks. In the rest of the sections under this heading, we define and describe the utility of some of the more common measures under each principle; brief definitions are also provided in Table 11.1 (also in Makagon et al. 2012). The mathematical derivations and detailed descriptions of these measures can be found in many excellent previous publications (Carrington et al. 2005; Croft et al. 2008; Wasserman and Faust 1994; Whitehead 2008).

Prominence: Centrality and Centralization

A primary use of graph theory in social network analysis is to identify "important" or "key" actors. Centrality concepts quantify theoretical ideas about an individual actor's or node's prominence within a network. Group-level measures of centrality, known as centralization, assess the inequality among the importance of all individuals across a network. These measures quantify the extent to which a set of actors are organized around a central node. Especially useful metrics of prominence include degree, closeness, betweenness, and eigenvalue centrality.

Degree

Degree refers to the number of direct connections a node has with other nodes in the network. For example, in Figure 11.2B, individual B is connected to four other individuals, and therefore has a node degree of 4. In directed networks, a node's indegree, or the number of connections incoming from others to the node, can be distinguished from its outdegree, or the number of connections that are outgoing from that node to others. For example, in Figure 11.2C, individual B has an indegree of 1 and an outdegree of 4, while individual L has an indegree of 0 and an outdegree of 1. At the individual level, degree centrality can be used to look at the differential roles that individuals play in behavioral networks (e.g., if certain individual primates are overall more involved in aggressive or affiliative behavior than others). Analyses of interaction asymmetries (indegree versus outdegree) may uncover important differences in an individual's roles within the network (e.g., how incoming and outgoing grooming behavior in nonhuman primate groups relates to dominance interactions). At the group level, degree centralization can be used to compare the extent to which groups are centered around one or two highly connected individuals, representing a high level of centralization, or in which dispersion of degree centrality is more uniformly distributed across individuals or nodes, representing a low level of centralization. The extent of centralization in a network can have a profound effect on its underlying cohesiveness and thus could be used to compare stability across different networks or groups.

Closeness

A node that is close to many others can quickly interact without going through many intermediaries. Interactions of individuals with high closeness are more efficient and accurate, resulting in less information loss due to disruption of transmission. Closeness is a measure of the minimum cumulative distance at which a node is connected to others in the network. It differs from degree because, while degree takes into account only direct relationships, closeness takes into account both direct and indirect connections. Closeness centrality measures might be used to model disease transmission or social transmission of abnormal behaviors in captive primate social groups, or to identify high-risk individuals in these systems (e.g., those with relatively high closeness centrality). Closeness centralization measures might be used to evaluate how different groups are similar in behaviors that affect health outcomes in captive groups. For example, transmission of *Shigella* or other common pathogens may be expected to occur faster in primate groups with low closeness centralization of grooming behavior (closeness is more widespread across group members, thereby encouraging quicker transmission) than in those with higher closeness centralization measures (closeness is concentrated on one or two individuals in the group, thereby limiting transmission).

Betweenness

A node with high betweenness occupies a "between" position on the paths connecting many pairs of other nodes in the network. This measure differs from closeness, because it is the number of additional paths through which others in the network can travel that is the important metric, not the directness or distance of edges. If the number of alternate paths (i.e., paths that do not include the focal node) is low, the individual has a high betweenness. Cutpoint individuals, known as cutpoint potential in some contexts (VanderWaal et al. 2014a), that connect subgroups within the network (e.g., individuals C and Q in Figure 11.4A, and then removed in Figure 11.4B) represent a case of

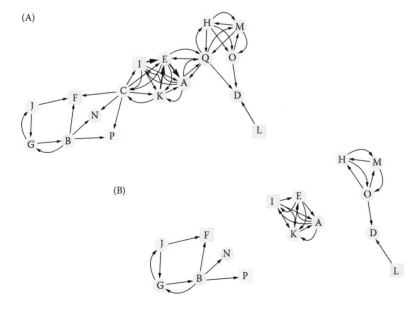

Figure 11.4 Example of high betweenness individuals that have cutpoint potential. If individuals C and Q are removed from (A), a previously connected graph with only one component becomes unconnected with three components (B).

extremely high betweenness centrality. They are necessarily on all paths between members of the different subgroups, thereby creating multiple components when removed from the graph. Clearly, cutpoints and other high betweenness individuals are important, because they may control the flow of information or the exchange of resources between group members. Betweenness centrality may, therefore, be especially useful for looking at the roles individuals play in maintaining cohesiveness of a network (Lusseau 2007; but also see Flack et al. 2006). At the group level, betweenness centralization measures the degree to which a few individuals are fundamental to maintaining group cohesiveness relative to other groups (see Figure 11.4). Betweenness centralization may, therefore, be useful in testing predictions about the effects of age or group membership on group cohesiveness.

Notably, betweenness (and other centrality measures) are often highly correlated with degree, because those who are well connected are also likely to play "between" roles in networks. Indeed, it is when these measures are not highly correlated that network structure tends to become more interesting (as shown in Figure 11.4) (also see VanderWaal et al. 2014a).

Eigenvalue Centrality

Eigenvalue centrality (as well as Boncich Power) takes into account not only how well an individual or node is connected (degree centrality) but also how well connected each of its neighbors are. That is, a node's centrality is dependent on the centrality of its neighbors. This measure of centrality highlights that, everything else being equal, an individual is more likely to play a key role in disease or information flow when his/her immediate neighbors are themselves well connected. Other scenarios may dictate that the degree to which one's friends are *not* connected is the important metric (which Boncich Power can measure). One can imagine multiple scenarios in captive primate groups where such indirect degree information would be critical to the understanding of patterns of relationships. For example, individuals may be expected to preferentially direct their affiliative interactions at group members that have influence over many others in the network.

Range

At the individual level, range or the extent of the network can be evaluated using a variety of measures. One of these is reach, which calculates the number of network nodes that a focal individual can reach within a specified distance (i.e., within a specified number of edges). For example, in Figure 11.4A, individual C, at distance = 2, has a reach of 4. Special cases of these measures are used to calculate closeness, degree, and other measures as described earlier. The most common metric of range at the network level is diameter. The diameter of a network is an indicator of the network's size, that is, how many steps along edges are necessary to get from one side of a network to the other. Diameter is defined as the shortest path between the most distant nodes in the network. (In Figure 11.4A, the diameter of the network is 8; the shortest path between L and G.) Reach can be used to test hypotheses about the influence of indirect relationships on behavior or dominance rank. For example, it can be used to evaluate whether only direct or secondary connections matter, or whether more distantly connected nodes also have an impact on the focal individual's behavior or rank. Reach and diameter are also useful quantities in that they can be used to set an upper bound on the lengths of connections under study. Many researchers limit their explorations of the connections among actors to involve those that are no longer than the diameter of their networks.

Cohesion

As discussed earlier, many measures can be used to examine the role(s) that specific individuals play in maintaining cohesion in a network, such as betweenness and cutpoints. At the group level, centralization (degree to which a set of nodes are organized around a central node; see Table 11.1

Summary of Network Measures) can be used as a measure of network cohesion. Other measures that can provide insight into animal behavior cohesion include density, reciprocity, transitivity (as represented by the clustering coefficient), fragmentation, assortativity, and community structure.

Density and Reciprocity

Density measures the degree to which all members of a population interact with all other members. Dense networks contain redundant connections and may therefore be less likely to be affected by the removal of a randomly selected node. Reciprocity represents the degree to which individuals in a group reciprocate the connections among one another. It is useful for measuring cohesion in groups, because when the connections are mutual, there are more paths and directions through which information and other phenomena can flow in the network.

Clustering Coefficient

The clustering coefficient is a measure of the degree to which nodes in a graph tend to cluster together. As a measure of transitivity (A connected to B who is connected to C who is connected to A), it can be used to measure an individual's neighborhood "cliquishness" using the local clustering coefficient. The network's overall cliquishness can be determined using an averaged measure of the local clustering coefficient across individuals. The global clustering coefficient measures the degree to which the entire network is composed of transitive relationships and thus measures the degree to which the network is connected. These measures of clustering could be used to evaluate robustness or cohesiveness in grooming or other affiliative networks and consequently, both individuals' positions in such networks, as well as overall group stability.

Fragmentation

Using the concept of cutpoint, fragmentation is defined as the proportion of mutually reachable nodes as each node is removed from the network (Borgatti 2003). Fragmentation is an inverse measure of the amount of connectedness or connection redundancy in a network. If multiple paths (redundant) exist between any two nodes in a network, then fragmentation will be low and cohesion high. Fragmentation can be used to evaluate cohesiveness in any number of social relationships, including grooming, dominance, and subordination networks.

Assortativity

The assortativity coefficient is a measure of the amount of mixing between and across subgroups of animals with certain attributes (sex, age, and node degree) as compared to that expected by chance. This is a measure of how homogenous or heterogenous relationships are in a population and can be used to determine whether certain relationships are more homophilic, and others more heterophilic, under various social circumstances.

Communities

Communities are structural relationships in groups that are above the individual level, but below the population level. Girvan and Newman (Newman et al. 2002) developed an algorithm to quantify the number of communities in a network, using the notion of edge betweenness, which is the number of shortest paths between nodes that make use of an edge or connection. They determined the number of communities by successively removing the edge with the highest betweenness until each individual in a network was its own community. To determine whether the communities detected using this method were meaningful, they calculated a quality parameter, known as modularity or Q. Q measures the extent to which edges between individuals are intracommunity, rather than intercommunity, and is defined as the number of edges within the community minus the expected number of edges within the community if connections were random (Newman and Girvan 2004). These types of community measures have been used to evaluate a number of networks, such as collaboration and airport networks, and have particular utility for examining grooming and other affiliative relationships in the context of nonhuman primate management.

Bipartite Networks

A special type of social network is the bipartite network. Unlike previously described "one-mode" social networks that focus on relationships between interacting entities (e.g., individuals and groups), bipartite or "two-mode" networks explore relationships that arise between one or more *networks*. To that end, bipartite network analysis simultaneously concentrates on the interacting entities (individuals, groups, etc.) and the attributes they share, or the events in which those entities are mutually involved. In bipartite networks, both the actors and the attributes or events are represented as nodes; the edges connect the actors to the events or to attributes they share (see Figure 11.6).

Bipartite networks can, therefore, be viewed from the perspective of the nodes (because coparticipation in events, or sharing an attribute, links nodes together) or the perspective of the attributes of events (because sharing of attributes or participation of the same nodes in multiple events links the attributes or events together; Faust 2005). Such approaches have previously been used to examine a variety of social affiliations, including patterns of connections between boards of major banks and major industries (Levine 1972), membership in voluntary associations (McPherson 1982), and coauthorship of scientific publications (Newman 2001). These types of networks can also be used to look at how animals' attributes (e.g., personality, sex, age, matriline, etc.) interact with components of both their social and physical environments to produce negative or positive health and well-being outcomes (see Figure 11.6).

Computational Network Metrics

Percolation and Conductance

Percolation and Conductance is a method for characterizing directional flow through a directed, weighted network, and it is perhaps most useful for determining the hierarchical structure of a society, because dominance interactions are typically unidirectional within a dyad (Fujii et al. 2014). The analysis begins by applying a percolation algorithm to the network, which allows us to efficiently use multistep pathways of relatively higher order (network pathways that are up to seven steps long). This algorithm performs a set of random walks through the network by randomly selecting a starting node (A) and calculating the probability of interacting with a candidate neighbor node (B, C, or D). The conductance principle is then applied to explore all potential flow pathways, weighting the contribution of each path to the imputed matrix by its likelihood of being successfully traversed during the random walk. The output matrix contains pairwise dominance probabilities (i.e., the probability that the row individual outranks the column individual), given the consistency (or inconsistency) of dominance pathways between each pair.

There are several aspects of hierarchical structure that can be quantified using this method that are not possible using standard methods. First, by applying the principle of transitive inference, we can glean dominance information from indirect pathways in the network to estimate pairwise dominance relationships for pairs that do not interact with one another (i.e., missing data in the win/loss matrix). Indeed, the indirect relationships add a substantial amount of information that can be used to infer dominance relationships. For example, one way of illustrating the importance of indirect pathways in a dominance network is to evaluate the number of cells of missing data in a standard win/loss matrix that can be filled with imputed data. We performed percolation and conductance on dominance networks from three of our captive rhesus macaque groups (ranging in size from 54 to 101 animals of 3 years of age and older) using pathways of up to four steps (A → B → C → D → E). In each of these groups, 29%–43% of the cells with missing data were filled in with imputed wins, such that their dyadic dominance probabilities were greater than 70% (i.e., given the imputed wins and losses estimated from dominance pathways, there is a 70% chance the row animal outranks the column animal; Figure 11.5).

This is incredibly useful for managing social groups of primates because hierarchical structure is the backbone of many, if not most, primate societies (Beisner and McCowan 2014b; Beisner et al. 2016; Fushing et al. 2014). In addition, the presence of missing data in the win/loss matrix can impair one's ability to determine whether there are truly unsettled dominance relationships (which may be harmful to social stability), as opposed to dominance relationships that are very clear but rarely expressed through dominance interactions. Furthermore, by virtue of being able to gather dominance information from transitive dominance pathways, smaller quantities of data on

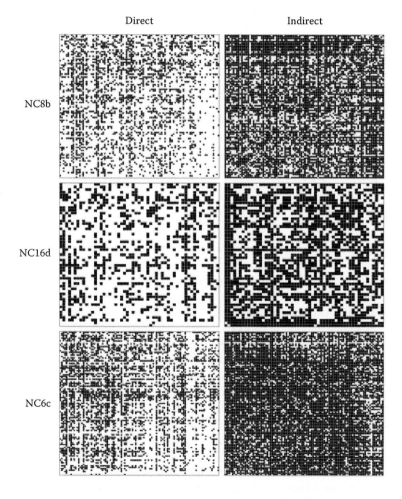

Figure 11.5 Comparison of aggression network matrices (i.e., win/loss matrices) to show the difference between raw win/loss matrices that have only direct relationships (left panel) and imputed win/loss matrices that use both direct and indirect relationships (right panel) for three different field cage enclosures. Darker color indicates higher dominance probability, which can be used to infer certainty. Note how much darker the right panels are than the left panels.

dominance interactions (which can be time consuming to collect; see *Practical Usage of Network Analysis: How Much Data Are Enough?*) are required to accurately estimate the dominance hierarchy. Proper application of percolation and conductance could allow colony and behavioral managers to (1) gain a better understanding of their social groups' dominance hierarchies from the limited data currently available; (2) potentially decrease the quantity and duration of observations currently done and still gain estimates of dominance relationships that are similar to those previously collected; or (3) determine that adding a program of observation is now feasible, because a smaller number of hours of observation is required to successfully determine dominance relationships.

Data Cloud Geometry

A large and complex collection of data, usually called a data cloud, naturally exhibits multiscale (e.g., individual, family, community) characteristics and features, generically termed geometry. Characterizing this geometry allows us to extract valuable information from the data. A new procedure called Data Cloud Geometry (Chen and Fushing 2012; Fushing et al. 2013) identifies community structure at multiple levels by performing a random walk through a network, where the probability of each step in the random walk is guided by the data. Cumulatively, these random walks produce a similarity matrix describing the extent to which each pair of animals has similar social connections. Individuals with greater similarity in social connections are considered to be "closer," and thus cluster together at a lower level of the hierarchical tree than individuals with fewer similarities in their connections. The unique advantage of using this method over other hierarchical clustering methods is that it assigns cluster membership more accurately for outliers. By transforming the hierarchy into an "ultrametric space" (i.e., a special kind of metric space, with triangle inequality, where points can never fall "between" other points) and then representing it via an ultrametric tree or a Parisi matrix, outliers will not, by default, be assigned to the same cluster simply because each of them significantly differ from other clusters. Similar to percolation and conductance, this new method for identifying community structure will prove to be highly useful for detecting and monitoring social relationships in nonhuman primates.

Joint Modeling

All social systems have interconnected behavioral, biological, and/or physical networks, and the synergistic interactions among these networks yield emergent properties of complexity, such as stability and the characteristic dynamics of a social group. However, most networks are constructed from single behaviors (Beisner et al. 2011a,b; Flack et al. 2006; VanderWaal et al. 2014b; Wey and Blumenstein 2010), because methodologies and computational algorithms that can examine interdependencies among multiple networks have been developed only recently (Barrett et al. 2012; Chan et al. 2013). Joint network modeling (JNM) is a data-driven, iterative modeling approach in which multiple networks are empirically constructed to quantify the interdependence among them. First, the raw data are used to calculate expected probabilities of jointly observing two behaviors (from two behavioral networks, such as groom and aggression) for a given dyad under the null hypothesis that these behaviors are independent. Then, the model is built by sequentially applying constraint functions (each of which is based upon a hypothesis derived from existing theory, data, or expertise), and these constraint functions adjust the expected probabilities of jointly observing two behaviors (which began at independence). Constraint functions are applied until the expected probabilities match the observed network data (Chan et al. 2013).

All social systems are composed of multiple interconnected networks, and joint modeling can be used to monitor the complex social dynamics of any captive or wild social system. JNM may be of greatest utility in quantifying the impact of environmental, social, or ecological change on the underlying structure of a social group. For example, zoos, sanctuaries, and conservation organizations might use JNM to monitor the social health of reintroduced or translocated social groups that are adjusting to unfamiliar environments to determine whether the group has established a normal social dynamic (Pinter-Wollman et al. 2009; Ruiz-Miranda et al. 2006). Further, the joint relationship between multiple networks can be a powerful tool for detecting changes in social dynamics in research facilities owing to management or research actions, such as temporary or permanent removal of animals for veterinary care or enrollment in a research project.

Data Mechanics

Data Mechanics constructs bipartite networks (e.g., subject ID × personality profile; Figure 11.6) and then shuffles the ordering of rows and columns, using principles of thermodynamics, to find the "lowest energy" state of the network matrix. This shuffling endeavors to cluster together blocks of nodes that are similar to one another. For example, given a data set of N subjects that have been evaluated across M behavioral measures, the $N \times M$ matrix is shuffled (first rows, then columns, then rows again, and so on) to visualize sets of subjects that have similar behavioral or personality profiles. Notably, this method generates a hierarchical tree of profiles that allows for the identification of higher-order behavioral or even health profiles (if using health measures as column traits). Data mechanics promises to revolutionize the method by which we glean "big data" for patterns, and can be used for a number of applications in primate behavioral management. One significant example of its utility is understanding the high-dimensional interplay of the multiple risk factors (e.g., individual attributes such as personality, genetic predispositions, early and current experience, social and environment contexts) that lead to negative and positive health/well-being in nonhuman primates (McCowan et al. 2016).

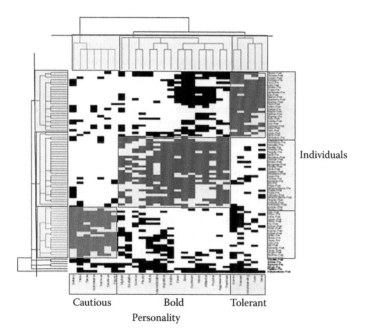

Figure 11.6   Data Mechanics tree and heat map representing bipartite network analysis of clusters of individuals (macaques) set on the Y-axis with clusters of personality type on the X-axis. Note the blocks of individuals sharing personality profiles on the Y-axis as well as the blocks of personality traits on the X- axis. For illustration for the level in the tree where the red line is located, we can block personality according to individuals and individuals according to personality (highlighted yellow, blue, and green boxes). Note how individuals who are similar in personality profiles group together and personality traits corresponding to personality types group together (Partial text and figure from this section on computational approaches were reprinted from McCowan et al., *Frontiers in Psychology*, 7, 433, 2016).

APPLICATION TO PRIMATE SOCIAL GROUP MANAGEMENT

As mentioned previously, there are multiple applications for network analysis in primate (behavioral) management, but we will focus this section on the body of work that addresses management-related issues for large social groups of nonhuman primates living in laboratory settings. For more than 10 years, our research group has been applying network analysis toward developing proactive, adaptive management strategies for large breeding groups of rhesus macaques, using both the standard and computational network approaches described earlier. The first goal of this work was to determine the risk factors leading to deleterious aggression (e.g., wounding, social relocations) in these captive groups, using both observational and experimental knockout approaches. The second goal of this research was to use the identified risk factors to develop an adaptive management plan to proactively reduce such aggression and trauma by identifying specific management practices (e.g., removal of high-ranking natal males). We also seek to identify the dynamic processes underlying the network tipping (critical) points that would predict and thus serve as an early warning system to stop deleterious aggression before it occurs. We review these findings in all of the sections under this heading.

The Problem of Aggression

Management of aggression in large groups of captive primates has been a major challenge for laboratories, zoos, and sanctuaries. In most social primates, low levels of aggression serve to maintain social order. Yet, in some especially despotic species, such as rhesus macaques, which heavily rely on aggression to mediate their dominance relationships, there is enormous potential for aggression to escalate out of control when social groups become unstable (Bernstein and Ehardt 1985a). This instability can lead to severe wounding, especially in the confined spaces of a captive environment. The degree to which intense aggression plays a role in dominance interactions is dependent upon a number of interconnected, within-group, and group-level factors, including: the personality and temperament of key or dominant individuals (Capitanio 1999), the diversity and rate of interactions within the social network, and the underlying composition of the social network (Beisner and McCowan 2013, 2014a,b; Beisner et al. 2011a,b, 2015, 2016; Flack and de Waal 2004; Flack and Krakauer 2006; Fushing et al. 2013, 2014; McCowan et al. 2008, 2011; Oates-O'Brien et al. 2010). Adult females are known to be the primary instigators of intense aggression (Bernstein and Ehardt 1985b), and when key individuals, such as high-ranking adult males (Beisner and McCowan 2013; Flack et al. 2006; McCowan et al. 2011), fail to successfully intervene, large outbreaks of aggression can occur, leading to severe wounding and death of multiple animals. In captive colonies of group-housed rhesus macaques, these severe outbreaks of aggression are known as "social overthrows" or "cage wars" (Beisner et al. 2015; Fushing et al. 2014). Interestingly, these "social overthrows" or "cage wars" have historically been deemed unpredictable by captive primate managers. Little or no predictive power results from monitoring simple rates of aggression or other behaviors in these groups. Instead, a deeper understanding of the patterns of relationships, which network analysis captures, is needed to uncover the pathways that lead to instability in social groups, such that accurate predictions can be made at least to intervene, and preferably, to prevent such instability from occurring.

Linking Network Structure to Stability via Three Major Pathways

Through a combination of standard and computational network analyses, we found three major pathways by which measures of social stability, such as rates of wounding and social relocations, are linked directly or indirectly to network structure in rhesus macaques. We label these pathways according to the key variables (e.g., demographic or genetic factors) found to influence network and behavioral measures of stability: (1) Presence of natal males, (2) matrilineal genetic fragmentation, and (3) the power structure and conflict policing behavior supported by this structure. We discuss these three major pathways in the following sections.

Natal Male Presence

The presence of natal males in the group is unique to captive social groups, because colony management protocols of captive groups often restrict the male dispersal pattern typical of wild, free-ranging groups (but see Kaufman 1976; Koford 1963; Tilford 1982, for wild, free-ranging examples of males remaining in their natal groups). This presence of natal males affects male individual rank, aggressive behavior, as well as position within the male displacement (i.e., approach–avoid) network via their alliances with others, particularly maternal kin (Beisner et al. 2011b). Natal males vary in their frequency of kin alliances, forming kin alliances more often when their female relatives are high-ranking. Natal males that frequently cooperate with their maternal kin attain higher individual rank, use intense aggression more frequently, and also have higher Bonacich power (a network measure of power or influence like eigenvalue centrality) in the male displacement network. Topological knockout of these high-ranking, natal juvenile males from the data revealed that group-level rates of *intense aggression decreased in all groups upon removal of these males* from the data (range: 4%–14% decrease).

To experimentally examine the influence of natal males on network structure, aggression, and trauma in their social groups, we permanently removed 1–2 high-ranking natal males (2–6 years of age) from the alpha matrilines of five social groups. We first examined the structure of subordination signaling networks (i.e., silent-bared-teeth signals, pSBT, in peaceful contexts), because pSBTs are a good measure of group stability; the frequency and diversity of pSBTs received predicts policing ability, and the loss of hierarchical structure in the pSBT network has been associated with social collapse in one of our study groups (Fushing et al. 2014). Experimental removal of natal males, who (behaviorally) represented a threat to the alpha male's status during baseline periods (i.e., receiving many subordination signals, having an ambiguous dominance relationship with the alpha male), resulted in increased pSBT network complexity (i.e., more first- and second-order pathways) and improved overall policing success. Notably, natal males that threatened the alpha male's status showed other behavioral evidence of being problematic: They participated disproportionately in fights and interventions (problem natal males = 5%–7% participation; benign natal males = 2%–3% participation) (Beisner et al. in preparation).

These results suggest that the strategic, proactive removal of high-ranking natal males, generally, and specifically those that pose a threat to the alpha male's status, will lead to greater stability in social groups, at least in rhesus, and likely in other macaque species as well.

Matriline Defragmentation

The importance of matrilines and their structure is a hallmark feature of macaque societies (Bernstein and Ehardt 1985b). Adult females and their female offspring provide the backbone of macaque society; so the necessity of cohesion within these social entities would seem paramount to the stability of social groups. Our research has indicated that this is indeed the case (Beisner et al. 2011a). Matrilines within groups that had lower genetic fragmentation, as measured by a greater average matrilineal degree of relatedness, had more cohesive grooming relationships (fewer grooming communities), less intense aggression among kin, and received less wounding than matrilines that had higher genetic fragmentation (Beisner et al. 2011a). In addition, matrilines whose matriarch (the oldest female that serves as the head of the matriline) was present received less wounding. Topological knockouts in the groom networks of each matriline (removal of an individual's data from the data set) showed that removal of matriline fragments decreased the number of groom communities. Experimental knockouts (physical permanent removals) of individuals to reduce matriline genetic fragmentation resulted in a reduction of wounding approximately 12 weeks post removal (McCowan, unpublished data). These results indicate that selective proactive removal of individuals from matrilines to increase intramatriline genetic relatedness can reduce aggression via cohesion in grooming relationships.

Power Structure and Conflict Policing

Finally, we looked at what is perhaps the most compelling feature of social stability: social power and conflict policing. In a series of analyses that build upon one another, our research has shown that social power is critical to social stability for several reasons such as the following: (1) pSBT signals given in peaceful contexts are formal subordination signals that communicate acceptance of one's subordinate role, and giving subordination signals is associated with lower rates of aggression (presumably because there is less need to aggressively reinforce the relationship; Beisner and McCowan 2014a); (2) the network of subordination signals (i.e., pSBT signals given in peaceful contexts) generates a skewed distribution of social power where subordination signals (and pathways of signals) converge on a small number of individuals (Beisner et al. 2016); and (3) individuals that receive frequent subordination signals from many different subordinates (including transitive network pathways of signals) are better able to police others' conflicts, because there is greater group consensus that they are powerful (Beisner et al. 2016; McCowan et al. 2011). Adult males, particularly males unrelated to the group, have the highest social power and are the most frequent and successful policers (mean policing frequency: unrelated males = 13.9; natal males from alpha matriline = 7.52; other natal males = 1.72). Because of the policing role that adult males play in the group, low adult female-to-unrelated adult male sex ratio is associated with high intervention success and less trauma (Beisner et al. 2012). Furthermore, social groups in which policing is more frequent and/or average social power is higher show signs of greater social stability, including (1) lower rates of contact aggression, trauma, and social relocations (Beisner and McCowan 2013; McCowan et al. 2011); (2) greater cohesion in their status/displacement networks (McCowan et al. 2011); and (3) less frequent escalation of conflict (i.e., shorter average conflict length: (McCowan et al. 2011)).

Social Instability: Finding the Critical Tipping Point

The final piece to the puzzle of social stability lies in being able to accurately assess a target social group for its current level of stability. Because stability is an emergent property of a complex social system, pinpointing when a social group is unstable and at risk of collapse is challenging. For example, outbreaks of severe aggression lead to social overthrow or social collapse (McCowan et al. 2008; Oates-O'Brien et al. 2010). Yet, as stated earlier, rates of aggression show no predictable relationship with group stability. Until now, one could only be sure of the lack of stability when a group had fissioned or experienced a social overthrow or collapse. JNM has proven to be incredibly useful in detecting social instability *prior to collapse*, as it readily reveals aberrant patterns of social dynamics. Using JNM, we found that the interdependence between aggression and status signaling networks was consistent among stable groups but different among unstable groups during the period leading up to their collapse (Beisner et al. 2015). The most prominent feature of aggression–status network interdependence in stable social groups was the presence of more pairs than expected that displayed opposite direction status–aggression (i.e., A threatens B, and B signals acceptance of subordinate status). On the other hand, unstable groups showed a relatively smaller magnitude of these opposite direction aggression–status pairs (but still higher than expected under the null hypothesis), as well as a greater-than-expected number of pairs with bidirectional aggression. JNM also confirms the importance of the power structure to group stability, because deterioration of the joint relationship between aggression and status networks is linked to, and is likely driven by, significant loss of subordination signaling interactions (Beisner et al. 2015; Chan et al. 2013; Fushing et al. 2014).

The selection of which networks to include in JNM is an important step. The networks most relevant to social stability will vary from one system to the next. It is, therefore, important to distinguish keystone networks from other more subsidiary networks. Keystone networks represent the most fundamental relationships, because they govern or influence the manner in which other networks in the system interact (Fushing et al. 2014). As such, the presence and direction of links in these networks are less likely to be influenced by situational variables than other networks, which tends to generate a more rigid structure.

Practical Usage of Network Analysis: How Much Data Are Enough?

One important question in applying network analysis to various aspects of primate behavior management is how much data are enough for practical and effective usage of these approaches in addressing management issues. Although several recent publications have addressed questions related to network data reliability (Croft et al. 2011; Feczko et al. 2015; Lusseau et al. 2008; Silk et al. 2015; Wey et al. 2008), we wish to address one specific area that is highly relevant to primate social group management, namely, how much data are needed to represent network structures of large social groups so that social management issues can be adequately addressed?

As discussed earlier, network structure in large social groups comprises the relationships between individuals. Frequently, these relationships are parsed in terms of either aggression or affiliation. For aggression relationships, we often wish to examine dominance interactions, and sparse or missing interactions are a common problem in these types of behavioral observational data, particularly in primatology (Lusseau et al. 2008). Although it is impossible to tell the difference after the fact, unknown relationships may reflect either a true lack of interaction or sampling limitations in the experimental design (Klass and Cords 2011). Yet, transitive inference suggests animals may not need direct interaction to glean information about the relationships among their surrounding conspecifics (McGonigle and Chalmers 1977; Whitaker et al., Effects of data density on the percolation and conductance method, 2016).

Percolation and Conductance, as mentioned previously, is a new network-based method for quantifying dominance relationships that addresses these potential methodological and theoretical causes for ambiguous and missing interactions (Fujii et al. 2014). Using this method, a dyadic relationship with little direct interaction can be modified, based upon its agreement with the direction of other indirect pathways. If there is a dyadic relationship with no interactions, the relationship can be imputed from the relative consistency in direction of the dominance pathways. Utilizing indirect dominance pathways to address ambiguous and missing interactions makes this method more robust to the effects of sparse data sets.

We, therefore, examined, using percolation and conductance, the effects of varying amounts of data on dominance certainty, and specifically, on our standard and computation network measures as described earlier. Our approach was to evaluate how changes in data density affect (1) measures of dominance certainty and (2) global and individual network measures, using a dense data set of aggressive interactions from a social group of rhesus macaques. We simulated varying degrees of data density by sampling interactions randomly without replacement from a full data set (6 h/day on 4 days/week for 6 weeks for a total of 144 h over 6 weeks or 24/week) and replicating each sample type (e.g., 10% of the full data set sampled, 20% sampled … 90% sampled) 100 times.

The results of these tests varied depending on the type of network measure. For direct relationships, such as degree at the individual level (Figure 11.7A,B), and density at the group level (Figure 11.8A), systematic subsampling from the data set resulted in a monotonic decrease in

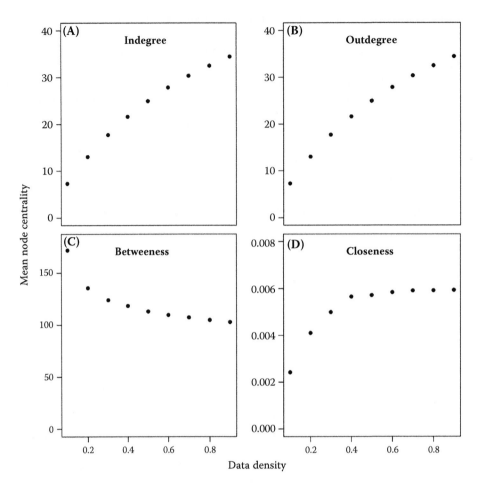

Figure 11.7 Individual-level standard network measures against data density for (A) indegree, (B) outdegree, (C) betweenness, and (D) closeness.

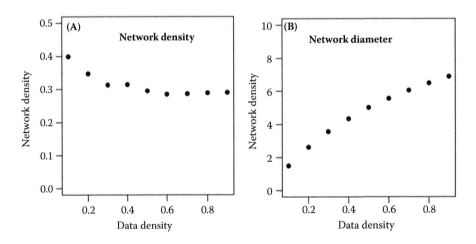

Figure 11.8 Group-level standard network measures against data density for (A) network density and (B) network diameter.

standard network measures. In contrast, for indirect relationships, such as betweenness and closeness at the individual level (Figure 11.7C,D) and diameter at the group level (Figure 11.8B), a threshold effect was observed, such that a certain percentage of the data (~40%–50%) was needed to produce reliable network measures (Whitaker et al., Effects of data density on the percolation and conductance method, 2016). We found a similar threshold relationship for data density for our computational measure of dominance certainty, which relies in part on indirect pathways (Figure 11.9A through D).

These observed differences make sense; because indirect measures rely on the connectedness of graphs through their indirect pathways, one would expect that a certain threshold of data is needed to generate reliable connectivity. Overall, these findings suggest that approximately 50% of the data originally collected (or 12 h/week per cage, or an average of 43 interactions per animal) in these large social groups are needed to sufficiently represent dominance interaction networks in these social groups (Whitaker et al., Effects of data density on the percolation and conductance method, 2016). For generalizing to other group sizes, we would expect necessary sampling effort to decrease according to group size or network density, but this relationship is likely not linear, as shown in Figure 11.8.

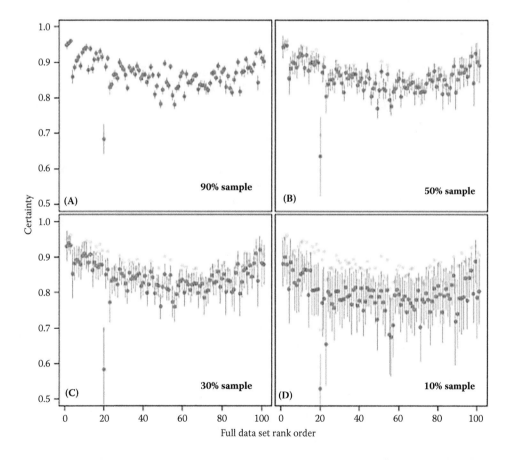

Figure 11.9 Computational dominance certainty against rank order for (A) 90% of full data set, (B) 50% of full data set, (C) 30% of full data set, and (D) 10% of full data set. Note that error (blue lines) around 50% is closer to 90% than 30% or 10%, suggesting that a threshold effect of data density is in operation for this measure. Gray data points represent full data set for comparison.

SUMMARY AND CONCLUSION

Network Analysis Utility

Network analysis focuses on the patterns of relationships that arise among interacting social and other physical entities. Its ability to mathematically represent both direct and indirect connections has made social network analysis a key approach for examining and quantifying complexity in systems, including the complexity of social relationships and social dynamics found in primate groups. To date, both standard and computational network approaches have been used to successfully identify (1) risk factors leading to deleterious aggression in captive groups of macaques, such as matrilineal fragmentation, presence of natal males, and the absence of a skewed power structure that supports conflict policing; and (2) the social dynamics, including interdependence across aggression and status networks, which underlie a critical tipping point in network structure that is predictive of deleterious aggression and social collapse.

These findings suggest a number of improvements that can be made to current behavioral/colony management practices. First, removal of natal males, particularly those from high-ranking families (who cooperate with their maternal kin to attain high rank) and/or those who occupy a similar network position as the alpha male, can prevent these males from becoming a source of social agitation and instability. In our experience, high-ranking natal males that become problematic do so around puberty (~3–5 years old), further suggesting that such males be monitored more closely at this age. Second, improvement and maintenance of matrilineal cohesion can be achieved by establishing a long-term plan for gradual removal of females to prune the growth of genetic lineages that would otherwise contribute to a social disconnect within the family. For example, if matriarch A1 has four daughters (A2, A3, A4, and A5), a long-term plan for maintenance of genetic cohesion might include consistent removal of A2's and A3's female offspring to allow only a few of the daughters' lineages to grow. Finally, behavioral managers should regularly monitor the subordination signaling network (i.e., the power structure), sex ratio, and conflict policing behavior present in their social groups to determine whether each group has the appropriate group composition to promote their own natural conflict management mechanisms. Sex ratios that differ dramatically from those of wild, free-ranging groups are less able to manage their own internal social conflicts and are therefore at greater risk of becoming unstable (Beisner et al. 2012).

The importance of sex ratio to group stability brings into question the stability of unimale, multifemale breeding groups in rhesus macaques. Estimates from the California National Primate Research Center indicate that aggression requiring hospitalization in these smaller breeding groups can be 50%–100% higher on a per-animal basis compared to the larger, multimale, matrilineal groups (McCowan, personal observation). One solution (which has not yet been attempted or systematically studied) might be to house these smaller groups of females with more than one adult male, so that policing of aggression among females is maximized. An important caveat, however, is that these groups must have a sufficient number of females to reduce competition among the males during the breeding season, although social mechanisms are in place such as male-to-male SBT signaling that serve to manage this type of competition (Beisner et al. 2016). Indeed, field-based research indicates that a minimum sex ratio of 1 male per 5 females is optimal for wild macaque populations (Beisner et al. 2012). In addition, recent field-based studies with commensal rhesus macaques in Shimla, India, by our research group suggest that smaller breeding groups with two to three males are not only observed but are also quite common in these populations (McCowan and Beisner, personal observation). We can use this information to guide the management of captive rhesus macaques at biomedical and other captive facilities. However, systematic studies assessing potentially optimal configurations of these smaller breeding groups in captivity are important to determine whether, and when, such management strategies would be beneficial to primate breeding facilities.

Cost/Benefit Review

While the approaches described in this chapter clearly demonstrate the utility of network analysis for improving management of captive animal populations, gathering network data (particularly from observational behavior data collection) is time consuming and requires well-trained observational staff, experienced data analysts, and a scientific approach. However, the costs of hiring and training appropriate staff are likely to be offset by both the improvements in animal welfare (by maintaining individuals in naturalistic social groups for longer periods of time than previously possible) and the cost savings in veterinary care related to reduced occurrence and treatment of social and physical trauma. Further, longer-term maintenance of animals in social groups also benefits research centers in particular, by fostering the conduct of high-quality science with reduced concern about the confounding influences of less-natural social environments. In addition, computational network approaches can be used to take advantage of existing databases. Data Cloud Geometry and Data Mechanics can be applied to databases of social pairing history to examine historical networks (i.e., connections among animals that have been paired at some point in the past) to evaluate attributes of interest related to pairing success, such as personality, sex, age, and rearing history. Percolation and Conductance can be applied to sparse behavioral databases of agonistic interactions in social groups to improve understanding of dominance structure. In sum, the potential benefits of implementing a network-based behavioral management program are sufficiently high to offset the costs of implementation.

Future Applications for Primate Behavioral Management

The future of network approaches in primate (behavioral) management will likely stem from two major themes, as long as managers accept and embrace that captive primate systems are complex systems requiring examination of the multilevel, interacting variables that influence primate health and well-being. The first of these themes is continued development of methods relevant to the management context. For example, we might wish to expand joint network approaches to examine 3+ behavioral networks and develop statistical physics-inspired spillover techniques to look at networks more longitudinally. While JNM allows us to detect unstable social dynamics before collapse, this approach might be improved to detect such collapse even sooner, and to watch for potential signs of collapse and whether they resolve or accelerate. Another methodological improvement will be to adopt phase-transition approaches from statistical physics to improve our understanding of not only stability but also captive groups' responses to *any* type of management or research activity.

A second theme is continued encouragement to implement these methods at captive facilities, including other primate centers and sanctuaries. This chapter may serve to help convince NPRC directors that their behavior management programs should be expanded or transitioned to include hiring or training individuals on network data collection and analysis. We also hope to convince others involved in behavioral management that they can learn to use network techniques. All captive facilities have high turnover of individuals in social groups; zoos, in particular, are known for forming and reforming new groups, including the transfer of single individuals to another zoo for genetic purposes. For this reason, network analysis can be helpful in monitoring social dynamics throughout this process and maintaining the highest welfare standards for these animals.

Finally, with the amount of data that primate centers, sanctuaries, and zoos are currently archiving digitally, there is ample opportunity to refine methods such as Data Mechanics to look at management activities, choices, and predictions from a mile-high, data-driven perspective, permitting a very complex analysis of the ways in which animal welfare can be maximized, as well as improving the efficiency and effectiveness of managing nonhuman primates in captivity. Primate (behavioral) management is a complex system requiring the appropriate and effective methods that networks and other complexity science provide. We need to embrace these new tools of the 21st century to reap the benefits they have to offer.

ACKNOWLEDGMENTS

We would like to acknowledge all past and present observers and research staff in the McCowan Laboratory. These include Amy Nathman, Allison Barnard, Alyssa Maness, Esmeralda Cano, Jenny Greco, Tamar Boussina, Alison Vitale, Shannon Seil, Megan Jackson, Darcy Hannibal, Jessica Vandeleest, Jian Jin, and Krishna Balasubramaniam. We would also like to thank the CNPRC behavior management, animal care, and research services staff for their support in this research. A portion of the conceptual framework for this chapter, namely the discussion of standard network measures, came from a previous collaboration with Drs. Maja Makagon and Joy Mench. We thank Dr. Fushing Hsieh for his creativity and talent in developing computational metrics for evaluating robustness mechanisms. We thank Alex Whittaker for his work on evaluating data density on network measures. This research was funded by two NIH grants awarded to B. McCowan (R24-OD011136; R01-HD068335) as well as the CNPRC base grant (P51-OD01107-53).

REFERENCES

Albert, R., H. Jeong, and A.-L. Barabási. 1999. Internet: Diameter of the world-wide web. *Nature* 401:130–131.
Barabási, A.-L., and Z. N. Oltvai. 2004. Network biology: Understanding the cell's functional organization. *Nature Reviews Genetics* 2:101–113.
Barrett, L. F., S. P. Henzi, and D. Lusseau. 2012. Taking sociality seriously: The structure of multi-dimensional social networks as a source of information for individuals. *Philosophical Transactions of the Royal Society of London B Biological Sciences* 367:2108–2118.
Beisner, B. A., D. L. Hannibal, K. R. Finn, H. Fushing, and B. McCowan. 2016. Social power, conflict policing, and the role of subordination signals in rhesus macaque society. *American Journal of Physical Anthropology* 160:102–112.
Beisner, B. A., M. E. Jackson, A. Cameron, and B. McCowan. 2011a. Detecting instability in animal social networks: Genetic fragmentation is associated with social instability in rhesus macaques. *PLoS One* 6 (1):e16365.
Beisner, B. A., M. E. Jackson, A. Cameron, and B. McCowan. 2011b. Effects of natal male alliances on aggression and power dynamics in rhesus macaques. *American Journal of Primatology* 73:790–801.
Beisner, B. A., M. E. Jackson, A. Cameron, and B. McCowan. 2012. Sex ratio, conflict dynamics and wounding in rhesus macaques (*Macaca mulatta*). *Applied Animal Behaviour Science* 137:137–147.
Beisner, B. A., J. Jin, H. Fushing, and B. McCowan. 2015. Detection of social group instability among captive rhesus macaques using joint network modeling. *Current Zoology* 61:70–84.
Beisner, B. A., and B. McCowan. 2013. Policing in nonhuman primates: Partial interventions serve a prosocial conflict management function in rhesus macaques. *PLoS One* 8 (10):e77369.
Beisner, B. A., and B. McCowan. 2014a. Social networks and animal welfare. In *Animal Social Network*, edited by R. J. J. Krause, D. Franks, and D. Croft. Oxford, UK: Oxford University Press.
Beisner, B. A., and B. McCowan. 2014b. Signaling context modulates social function of silent bared teeth displays in rhesus macaques (*Macaca mulatta*). *American Journal of Primatology* 76:111–121.
Bernstein, I., and C. L. Ehardt. 1985a. Age-sex differences in the expression of agonistic behavior in rhesus monkey (*Macaca mulatta*) groups. *Journal of Comparative Psychology* 99:115–132.
Bernstein, I., and C. L. Ehardt. 1985b. Agonistic aiding: Kinship, rank, age, and sex influences. *American Journal of Primatology* 8:37–52.
Böhm, M., M. R. Hutchings, and P. C. L. White. 2009. Contact networks in a wildlife-livestock host community: Identifying high-risk individuals in the transmission of bovine TB among badgers and cattle. *PLoS One* 4:e5016.

Bond, A. B., A. C. Kamil, and R. P. Balda. 2003. Social complexity and transitive inference in corvids. *Animal Behaviour* 65 (3):479–487. doi: 10.1006/anbe.2003.2101.

Borgatti, S. 2003. The key player problem. In *Dynamic Social Network Modeling and Analysis: Workshop Summary and Papers*, edited by R. Breiger, K. Carley, P. Pattison, 241–252. Washington, DC: The National Academies Press.

Bullmore, E., and O. Sporns. 2009. Complex brain networks: Graph theoretical analysis of structural and functional systems. *Nature Reviews Neuroscience* 10:186–198.

Capitanio, J. P. 1999. Personality dimensions in adult male rhesus macaques: Prediction of behaviors across time and situation. *American Journal of Primatology* 47 (4):299–320. doi: 10.1002/(SICI)1098-2345(1999)47:4<299::AID-AJP3>3.0.CO;2-P.

Carrington, P., S. Wasserman, and J. Scott. 2005. *Models and Methods in Social Network Analysis.* Cambridge, MA: Cambridge University Press.

Chan, S., H. Fushing, B. A. Beisner, and B. McCowan. 2013. Joint modeling of multiple social networks to elucidate primate social dynamics: I. Maximum entropy principle and network-based interactions. *PLoS One* 8 (2):e51903.

Chen, C., and H. Fushing. 2012. Multiscale community geometry in a network and its application. *Physical Review E* 86:041120.

Corner, L. A., D. Pfeiffer, and R. S. Morris. 2003. Social network analysis of *Mycobacterium bovis* transmission among captive brushtail possums (*Trichosurus vulpecula*). *Preventive Veterinary Medicine* 59:147–167.

Croft, D. P., R. James, and J. Krause. 2008. *Exploring Animal Social Networks.* Princeton, NJ: Princeton University Press.

Croft, D. P., J. Krause, and R. James. 2004. Social networks in the guppy (*Peocilia reticulata*). *Proceedings of the Royal Society London B* 271:516–519.

Croft, D. P., J. R. Madden, D. W. Franks, and R. James. 2011. Hypothesis testing in animal social networks. *Trends in Ecology and Evolution* 26 (10):502–507. doi: 10.1016/j.tree.2011.05.012.

Durell, J. L., I. A. Sneddon, N. E. O'Connell, and H. Whitehead. 2004. Do pigs form preferential associations? *Applied Animal Behaviour Science* 89:41–52.

Faust, K. 2005. Using correspondence analysis for joint displays of affiliation networks. In *Models and Methods in Social Network Analysis*, edited by P. J. Carrington, J. Scott, and S. Wasserman, 117–147. Cambridge, MA: Cambridge University Press.

Feczko, E., T. A. J. Mitchell, H. Walum, J. M. Brooks, T. R. Heitz, L. J. Young, and L. A. Parr. 2015. Establishing the reliability of rhesus macaque social network assessment from video observations. *Animal Behaviour* 107:115–123 doi: 10.1016/j.anbehav.2015.05.014.

Flack, J. C., and F. B. M. de Waal. 2004. Dominance style, social power, and conflict. In *Macaque Societies: A Model for the Study of Social Organization*, edited by B. Thierry, M. Singh, and W. Kaumanns, 157–182. Cambridge, UK: Cambridge University Press.

Flack, J. C., M. Girvan, F. B. M. de Waal, and D. C. Krakauer. 2006. Policing stabilizes construction of social niches in primates. *Nature* 439:426–429.

Flack, J. C., and D. C. Krakauer. 2006. Encoding power in communication networks. *The American Naturalist* 168:E87–E102.

Freeman, L. C. 1984. Turning a profit from mathematics: The case of social networks. *Journal of Mathematical Sociology* 10:343–360.

Fujii, K., H. Fushing, B. A. Beisner, and B. McCowan. 2014. Computing power structures in directed biosocial networks: Flow percolation and imputed conductance. Technical report. Department of Statistics, UC Davis.

Fushing, H., Ò. Jordà, B. Beisner, and B. McCowan. 2014. Computing systemic risk using multiple behavioral and keystone networks: The emergence of a crisis in primate societies and banks. *International Journal of Forecasting* 30 (3):797–806. doi: 10.1016/j.ijforecast.2013.11.001.

Fushing, H., H. Wang, K. VanderWaal, B. McCowan, and P. Koehl. 2013. Multi-scale clustering by building a robust and self-correcting ultrametric topology on data points. *PLoS One* 8:e56259.

Grosenick, L., T. S. Clement, and R. D. Fernald. 2007. Fish can infer social rank by observation alone. *Nature* 445 (7126):429–432. doi: 10.1038/nature05511.

Ings, T. C., J. M. Monotya, J. Bascompte, N. Blüthgen, L. Brown, C. R. Dormann, F. Edwards, et al. 2009. Ecological networks—Beyond food webs. *Journal of Animal Ecology* 78:253–269.

Jones, H. A. C., L. A. Hansen, C. Noble, B. Damsgård, D. M. Broom, and G. P. Pearce. 2010. Social network analysis of behavioural interactions influencing fin damage development in Atlantic salmon (*Salmo salar*) during feed-restriction. *Applied Animal Behaviour Science* 127:139–151.

Kaufman, J. H. 1976. Social relations of adult males in a free-ranging band of rhesus monkeys. In *Social Communication among Primates*, edited by S. A. Altmann, 73–98. Chicago, University of Chicago Press.

Klass, K., and M. Cords. 2011. Effect of unknown relationships on linearity, steepness and rank ordering of dominance hierarchies: Simulation studies based on data from wild monkeys. *Behavioural Processes* 88:168–176. doi: 10.1016/j.beproc.2011.09.003.

Koford, C. B. 1963. Rank of mothers and sons in bands of rhesus monkeys. *Science* 141 (3578):356–357.

Krause, J., D. P. Croft, and R. James. 2007. Social network theory in the behavioural sciences: Potential applications. *Behavioral Ecology and Sociobiology* 62:15–27.

Krause, J., R. James, D. W. Franks, and D. P. Croft. 2015. *Animal Social Networks*. Oxford and New York: Oxford University Press.

Krause, J., D. Lusseau, and R. James. 2009. Animal social networks: An introduction. *Behavioral Ecology and Sociobiology* 63:967–973.

Levine, J. H. 1972. The sphere of influence. *American Sociological Review* 37:14–27.

Liben-Nowell, D., and J. Kleinberg. 2008. Tracing information flow on a global scale using internet chain-letter data. *PNAS* 105:4633–4638.

Lusseau, D. 2003a. The emergent properties of a dolphin social network. *Proceedings of the Royal Society of London B* 270:186–188.

Lusseau, D. 2007. Evidence for social role in a dolphin social network. *Evolutionary Ecology* 21:357–366.

Lusseau, D., H. Whitehead, and S. Gero. 2008. Incorporating uncertainty into the study of animal social networks. *Animal Behaviour* 75:1809–1815. doi: 10.1016/j.anbehav.2007.10.029.

Makagon, M., B. McCowan, and J. Mench. 2012. How can social network analysis contribute to social behavior research in applied ethology? *Applied Animal Behaviour Science* 138:152–161.

Martínez-López, B., A. M. Perez, and J. Sánczez-Vizcaíno. 2009. Social network analysis: Review of general concepts and use in preventative veterinary medicine. *Transboundary and Emerging Diseases* 56:109–120.

McCowan, B., A. Anderson, A. Heagarty, and A. Cameron. 2008. Utility of social network analysis for primate behavioral management and well-being. *Applied Animal Behaviour Science* 109:396–405.

McCowan, B., B. A. Beisner, J. P. Capitanio, M. E. Jackson, A. Cameron, S. Seil, E. R. Atwill, and H. Fushing. 2011. Network stability is a balancing act of personality, power, and conflict dynamics in rhesus macaque societies. *PLoS One* 6 (8):e22350.

McCowan, B., B. Beisner, E. Bliss-Moreau, J. Vandeleest, J. Jin, D. Hannibal, F. Hsieh. 2016. Connections matter: Social networks and lifespan health in primate translational models. *Frontiers in Psychology* 7:433. doi: 10.3389/fpsyg.2016.00433.

McGonigle, B. O., and M. Chalmers. 1977. Are monkeys logical? *Nature* 267 (5613):694–696.

McPherson, J. M. 1982. Hypernetwork sampling: Duality and differentiation among voluntary organizations. *Social Networks* 3:225–249.

Newman, M. E. J. 2001. Scientific collaboration networks. 1. Network construction and fundamental results. *Physical Review E* 64:016131.

Newman, M. E. J., S. Forrest, and J. Balthrop. 2002. Email networks and the spread of computer viruses. *Physical Review E* 66:035101(R).

Newman, M. E. J., and M. Girvan. 2004. Finding and evaluating community structure in networks. *Physical Review E* 69:026113.

Oates-O'Brien, R. S., T. B. Farver, K. C. Anderson-Vicino, B. McCowan, and N. W. Lerche. 2010. Predictors of matrilineal overthrows in large captive breeding groups of rhesus macaques (*Macaca mulatta*). *Journal of the American Association for Laboratory Animal Science* 49:196–201.

Otterstatter, M. C., and J. D. Thomson. 2007. Contact networks and transmission of an intestinal pathogen in bumble bee (*Bombus impatiens*) colonies. *Oecologia* 154:411–421.

Pasquaretta, C., M. Leve, N. Claidiere, E. van de Waal, A. Whiten, A. J. J. MacIntosh, M. Pele, et al. 2014. Social networks in primates: Smart and tolerant species have more efficient networks. *Scientific Reports* 4. doi: 10.1038/srep07600. http://www.nature.com/srep/2014/141223/srep07600/abs/srep07600.html#supplementary-information.

Pinter-Wollman, N., L. A. Isbell, and L. A. Hart. 2009. Assessing translocation outcome: Comparing behavioral and physiological aspects of translocated and resident African elephants (*Loxodonta africana*). *Biological Conservation* 142 (5):1116–1124. doi: 10.1016/j.biocon.2009.01.027.

Ruiz-Miranda, C. R., A. G. Affonso, M. M. D. Morais, C. E. Verona, A. Martins, and B. B. Beck. 2006. Behavioral and ecological interactions between reintroduced golden lion tamarins (*Leontopithecus rosalia* Linnaeus, 1766) and introduced marmosets (*Callithrix spp*, Linnaeus, 1758) in Brazil's Atlantic Coast forest fragments. *Brazilian Archives of Biology and Technology* 49:99–109.

Sade, D. S., and M. Dow. 1994. Primate social networks. In *Advances in Social Network Analysis: Research in the Social and Behavioral Sciences*, edited by S. Wasserman, and J. Galaskiewicz, 152–166. Thousand Oaks, CA: Sage.

Scott, J. 2000. *Social Network Analysis*. London: Sage.

Sen, P., S. Dasgupta, A. Chatterjee, P. A. Sreeram, G. Mukherjee, and S. S. Manna. 2003. Small-world properties of the Indian railway network. *Physical Review E* 67:036106.

Sih, A., S. F. Hanser, and K. A. McHugh. 2009. Social network theory: New insights and issues for behavioral ecologists. *Behavioral Ecology Sociobiology* 83:975–988.

Silk, M. J., A. L. Jackson, D. P. Croft, K. Colhoun, and S. Bearhop. 2015. The consequences of unidentifiable individuals for the analysis of an animal social network. *Animal Behaviour* 104:1–11.

Sussman, R.W. 2003. *Primate Ecology and Social Structure*, Chapter 1: Ecology: General principles. Needham Heights, MA: Pearson Custom Publishing.

Tilford, B. L. 1982. Seasonal rank changes for adolescent and subadult natal males in a free-ranging group of rhesus monkeys. *International Journal of Primatology* 3:483–490.

VanderWaal, K., E. R. Atwill, L. A. Isbell, and B. McCowan. 2014a. Quantifying microbe transmission networks in wild and domestic ungulates in Kenya. *Biological Conservation* 169:136–146.

VanderWaal, K. L., E. R. Atwill, L. A. Isbell, and B. McCowan. 2014b. Linking social and pathogen transmission networks using microbial genetics in giraffe (*Giraffa camelopardalis*). *Journal of Animal Ecology* 83 (2):406–414. doi: 10.1111/1365-2656.12137.

Wasserman, S., and K. Faust. 1994. *Social Network Analysis: Methods and Applications*. Cambridge, UK: Cambridge University Press

Webb, C. R. 2005. Farm animal networks: Unraveling the contact structure of the British sheep population. *Preventive Veterinary Medicine* 68:3–17.

Wey, T., D. T. Blumstein, W. Shen, and F. Jordan. 2008. Social network analysis of animal behaviour: A promising tool for the study of sociality. *Animal Behaviour* 75:333–344.

Wey, T. W., and D. T. Blumenstein. 2010. Social cohesion in yellow-bellied marmots is established through age and kin structuring. *Animal Behaviour* 79:1343–1352.

Whitehead, H. 2008. *Analyzing Animal Societies: Quantitative Methods for Vertebrate Social Analysis*. Chicago, IL: University of Chicago Press.

Wilson, E. O. 1975. *Sociobiology: The New Synthesis*. Cambridge, MA: Harvard University Press.

Application and Implementation in Behavioral Management

Part III

Positive Reinforcement Training and Research

Melanie L. Graham
University of Minnesota

CONTENTS

Biomedical research is necessary to advance therapeutics and interventions that reduce morbidity and mortality as well as enhance quality of life (QOL). Novel technologies that provide alternatives to certain conventional animal models, have evolved and more are forthcoming, but successful advancement of basic research to the clinical patient is still heavily reliant on carefully selected, valid animal models to evaluate specific aspects of the target disease or therapeutic (Graham and Prescott 2015). Validity in animal models is multifactorial, and the scientific community has been charged with improving the design, conduct, and analysis of *in vivo* research, including giving greater emphasis to the specific interplay between animal welfare and scientific outcomes as it relates to construct validity of animal models (Bailoo et al. 2014; Collins and Tabak 2014). Though nonhuman primates (NHPs) represent the smallest fraction of animals used in biomedical research (<1%), they play a predominant role in translational investigations (Carlsson et al. 2004). NHPs have been used to accurately model certain age-associated diseases, behavior, immune-mediated diseases or processes, infectious diseases, neurological diseases, reproductive biology, and pharmacologic safety (Price et al. 1991; Kennedy et al. 1997; Lane 2000; Brok et al. 2001; Betarbet et al. 2002; Preston and de Waal 2002; Barr et al. 2003; Brosnan and de Waal 2003; Buse et al. 2003; Coffman and Hessel 2005; Buckley et al. 2011; Messaoudi et al. 2011; Nout et al. 2012; Verdier et al. 2015). The key similarities in biological complexity, immune response, physiology, and behavioral repertoire that exist between humans and NHPs, because of their close phylogenetic relationship, enable experiments that are otherwise impossible with other species or nonanimal alternatives (Phillips et al. 2014). These similarities introduce a much higher degree of complexity in husbandry and clinical care. NHPs develop rich social relationships, and their environments must be designed with sufficient ingenuity to elicit a diverse behavioral repertoire (Novak and Suomi 1991; Schapiro and Bloomsmith 1994, 1995; Schapiro 2002; Lutz and Novak 2005; Olsson and Westlund 2007).

The clinical management of a highly representative NHP model can actually be as sophisticated as its human counterparts and often involves substantial handling, especially in overt disease models (Graham and Schuurman 2015).

The diversity of research using NHPs exposes animals to a variety of medical experiences ranging from universal practices, such as blood collection and simple injections to more complicated disease management or intensive therapeutic interventions. The majority of preclinical studies rely on frequent diagnostic and treatment procedures. Some studies require that the animals be positioned in a certain way and not interfere with the procedure; the procedure may involve some level of discomfort; and may even be briefly painful. Apart from the physical pain that the animals may experience, the procedure itself may cause distress to the extent that levels of fear and anxiety are increased in subsequent interactions. In clinical patients, it has been documented that these types of treatments and procedures often elicit anxiety, fear, and avoidance (Kennedy et al. 2008; Noel et al. 2010; Majidi et al. 2015). Managing QOL in patients is double-edged: a careful balance of positive and negative experiences. Often patients must comply with necessary test and treatment strategies that are intrinsically aversive, in order to realize the positive effects of limiting complications of disease. However, the burden of frequent medical intervention can also negatively affect QOL, causing some patients to avoid treatments, thus exacerbating disease (Jacobsen et al. 1990; Phipps and DeCuir-Whalley 1990; Rudman et al. 1999). In case of animal models, the term well-being is typically substituted for QOL. The promotion of psychological well-being of NHPs, especially related to experimentally induced stressors, is a provision of United States and European Union standard regulations for NHPs that participate in biomedical research. Prior to initiation of experimentation, the study protocol should be fully evaluated for opportunities for reduction and refinement in areas where no replacement can be foreseen, in accordance with the principles of Replacement, Reduction, and Refinement that are collectively known as the "3Rs" (Russell and Burch 1959). Researchers should appreciate refinement at its most basic level as a tool to reduce the negative experiences of the animal, but also consider the more progressive interpretation that seeks to increase positive experiences, thereby allowing the animal to flourish and enhance scientific aims by addressing factors related to validity.

Behavioral management techniques can be used to reduce or eliminate (1) restraint, which is typically used for routine physical examinations or procedures, and also (2) more complicated, protocol-specific research interventions, presenting a major opportunity for refinement.

Positive reinforcement training (PRT) can be used to help animals effectively cope with research interventions, and in case of disease modeling, participate in their own medical care through cooperation (Bloomsmith et al. 1998; Laule 2003, 2010; Laule et al. 2003; Prescott and Buchanan-Smith 2003; Schapiro et al. 2003; Coleman et al. 2008; Fernström et al. 2009; Hill et al. 2011; Graham et al. 2012). The design of training program should address the spectrum of animal welfare by reducing the negative experiences (e.g., restraint and invasive procedures) or affective states (e.g., anxiety and fear) and enhancing positive experiences (e.g., control, consistency, trust, and reward; Spruijt et al. 2001). Training complex behaviors to NHPs is an important opportunity for the animals to respond to meaningful challenges, to apply cognitive skills to determine actions, and to be active participants in their environments, all contributing to reduction in stress. Because of frequent contact for interventions and medical care, human caregivers develop familiarity and relationships with NHPs, providing a perfect opportunity to customize training strategies to decrease the animals' pain and psychological distress (Waitt et al. 2002).

Along with the welfare-related incentives to implement PRT, there are also compelling rationales for implementing PRT from the scientific perspective. Considerable concern has been raised by the growing number of examples of failed translation between successful preclinical modeling and the clinic. Improperly characterized models or those that fail to faithfully represent the clinical situation can seriously compromise the value of preclinical experiments (Hackam and Redelmeier 2006; Hackam 2007; van der Worp et al. 2010; Henderson et al. 2013; Denayer et al. 2014; Hay et al.

2014; McGonigle and Ruggeri 2014). In this context, experimental handling of NHPs deserves special emphasis because conventional approaches, such as chemical or physical restraint, influence the accuracy of hematological, biochemical, clinicopathological, and metabolic parameters (Bush et al. 1977; Goosen et al. 1984; Adams et al. 1988; Landi 1990; Reinhardt et al. 1990; Bennett et al. 1992; Lemieux et al. 1996; Hassimoto and Harada 2003; Lee et al. 2003; Buynitsky and Mostofsky 2009). A representative example can be found for Type 1 diabetes (T1D) modeling in NHPs, where intensive handling is required due to the need for frequent glucose monitoring and insulin administration, similar to human patients. Most experimental interventions introduce complex drug regimens, routine blood collection, and prolonged metabolic tests. Because it has been shown that restraint stress can impair glucose tolerance and significantly increase glucose levels, bias in the resulting data may underestimate therapeutic value (Yasuda et al. 1988; Shirasaki et al. 2013). Stress can also prompt changes in appetite, lead to gastrointestinal distress (e.g., nausea, vomiting, and diarrhea), and depress typical immune function (Morrow-Tesch et al. 1993; Mayer 2000). Consequently, adverse events as well as the potential for incorrect attribution to the investigational product may increase, even if they are unrelated. This may lower the predictive value in these models and potentially overestimate risk, which could prematurely end the development of a relevant therapeutic. Suboptimal handling techniques used in studies with NHPs might confound those data to the extent that a potentially useful therapeutic is actually blocked from reaching clinical patients, which should be deeply concerning to scientists pursuing research questions in the NHP model. However, refinement using PRT can introduce features into the model to make it more "clinical trial-like," avoid model-induced stress confounding, and improve the quality of management in disease models. Then, NHP studies can essentially be conducted with the same rigor as a human clinical trial, thereby accelerating the decision to advance to the clinic with additional predictive confidence.

BASIS OF TRAINING TECHNIQUES

Caregivers should be encouraged to review established behavioral principles to design a behavioral management plan based on anticipated medical management and research procedures. The principles of operant conditioning, applied behavioral analysis, and behavioral medicine can be used together to prevent or minimize nonproductive behavior or increase adaptive coping behavior. Reviewing these principles specifically with the perspective of training animals to cooperate with aversive medical intervention is worthwhile. Nonproductive behaviors typically arise from anxiety, for example, fleeing or fear-induced grabbing of caregivers or medical equipment, and can jeopardize the safety of the animal and caregivers. Structured PRT promotes adaptive coping behavior by preparing animals for planned procedures so that they can anticipate what the procedure will be like and by converting the experience into an overall pleasurable activity. Desensitization by counterconditioning is arguably the most effective tool caregivers can employ to prepare animals for aversive research procedures. Desensitization aims to convert anxiety or fear to relaxation, and even pleasure (Clay et al. 2009). Trainers use PRT to help animals tolerate progressive exposure to an aversive stimulus, only increasing the exposure or complexity when animals demonstrate a noticeable decrease in anxiety and successfully complete the task. Desensitization can be informal or formal, and more detailed examples are provided.

All NHPs experience "informal" learning as part of their interactions with the environment, the surrounding circumstances, caregivers, and/or conspecifics. A prime example of informal learning occurs in most captive NHP colonies, where the animals learn to tolerate humans during routine husbandry procedures. Inadvertent habituation occurs when the daily cleaning and maintaining of enclosures briefly brings animals into close proximity with caregivers. This contact, or exposure, is relatively unobtrusive and typically includes the provision of food (a primary reinforcer) and enrichment, stimulating the development of positive associations. However, the natural instinct of NHPs to respond with vigilance or even avoidance (flight; Caine 2017) generally persists when actions or events deviate from the expected routine. The way in which individual animals respond to caregivers can range from relatively confident to anxious and is dependent on the situation. This response probably reflects the animal's historical experiences with caregivers but also reflects the personality and temperament, also termed as "behavioral reactivity" (see Chapters by Capitanio 2017 and by Coleman 2017). It is essential for caregivers to evaluate the animal's temperament prior to, and during, planned research procedures to "personalize" the behavioral management approach and timing to the individual animal's needs.

Pre-assessments, Caregiver Responsibility, and Skill Acquisition Planning

Studies on NHPs have shown a relationship among temperament and training efficiency and success (Coleman et al. 2005; Reamer et al. 2014; Bliss-Moreau and Moadab 2016). It is useful to perform a baseline assessment of the animal's temperament or affective state prior to beginning a study, and also to reassess as training progresses.

Elegant tools to assess personality and temperament have been designed and validated by Capitanio and Freeman to enable successful identification of animals as more "open" versus "inhibited" or "reactive," giving trainers the opportunity to customize the training such that reinforcement can be deliberately applied in a way that allows the animal to perform most productively (Capitanio 1999; Freeman et al. 2013). Preferred items or activities that NHPs will seek, such as treats, toys, and specially enriched environments, have the best potential as (positive) reinforcers to strengthen desirable behaviors. Individual animal preferences should be determined by noting which activities or items (rewards) elicit the most enthusiasm. Lower value rewards can be used for simple tasks, whereas high-value rewards can be reserved for more complicated tasks. All animals benefit when caregivers create opportunities for pleasant daily interactions (i.e., warm communication, treat, or simple toy) prior to the initiation of specific skills training. Animals that display traits consistent with inhibition (Table 12.1) are often reluctant to interact with humans or are suspicious of novel items/situations. For these animals, caregivers should consider more frequent interactions, which can be initially brief but can be extended as the animals demonstrate increasing comfort with the situations. Therefore, building and maintaining rapport with the animals in our care is the basis for the cooperation and effective coping that are necessary for the NHPs, people, and research projects to benefit both animal welfare and scientific outcomes (Figure 12.1).

It is clear that the behavior and attitude of the caregiver have a pivotal role in the success or failure of a training program (Waitt et al. 2002). Skilled caregivers are "self-aware" and adjust

Table 12.1   Traits Used in Rating Non-human Primate Temperament and Personality

Open	Inhibited
• *Affiliative* (sociable)	• *Active* (engages in overt, energetic behavior)
• *Bold* (not restrained, tentative, shy, or coy)	• *Aggressive* (high frequency of displays, threats)
• *Calm* (reacts in an even calm way, not easily agitated)	• *Alarmed/anxious* (hesitant, indecisive, tentative, fearful, alarm calls, maximizes space to caregivers or novel items)
• *Confident* (behaves in a positive assured manner)	• *Depressed* (isolated, withdrawn, sullen, brooding)
• *Curious/exploratory* (readily explores and investigates novel items)	• *Excitable/vigilant* (easily aroused to an emotional state, attentive, watchful)
• *Playful* (engages in self-play)	

Source: Freeman, H.D., et al. *Am. J. Primatol.,* 75(10), 1042–1053, 2013.

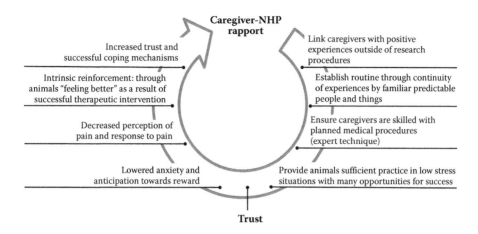

Figure 12.1 The relationship cycle between caregivers and nonhuman primates (NHPs).

their own behaviors (e.g., body language, proximity, tone, or reactions) based on the association with the animal's behavior. Likewise, they create a relaxed, familiar learning environment for the animal and their social cohort (i.e., direct partner or group, as well as other animals in visual contact). NHPs are quite capable of vicarious experience, and so caregivers should also consider taking advantage of informal observational learning by deliberately working with experienced NHPs to demonstrate target behaviors to animals in the early formal training phase (Chang et al. 2011; Falcone et al. 2012).

Once the caregiver has established formal training readiness, a step-wise training plan should be constructed to provide consistency among caregivers. This is formally defined as task analysis in which a behavior is broken down into steps and presented to the learner sequentially (Table 12.2). When training begins, the caregiver should document the behavioral response to evaluate the (1) efficacy of the approach and (2) progress of the animal. Once a task analysis plan is determined, the caregiver team should consider which behavioral approaches are most appropriate for application at each step, with the remark that *positive reinforcement* (PR) should form the foundation for the overall interaction. PR is the presentation of a preferred item or activity immediately following a target behavior. Conversely, *negative reinforcement* (NR) is the withdrawal of a nonpreferred item or activity immediately following a target behavior. Both PR and NR increase the likelihood of the performance of the target behavior, and using them together can strengthen the reliability of the behavior. Caregivers should be reminded that positive punishment (PP), the application of a stimulus to reduce a behavior, should essentially never be used with laboratory animals (with perhaps the extremely rare exception of preventing imminent danger to an animal or a person). Apart from the ethical considerations, PP breaks trust and is therefore, not relevant to discuss in the context of these training recommendations. Negative punishment (NP) uses withdrawal of a reinforcing activity or item to decrease an undesirable behavior. This is often referred to as omission training or "time-out from PR" to avoid confusion with the colloquial use of the term punishment. NP can occasionally be useful in certain situations, such as when an animal is presenting behavior that is inconsistent with the behavior that the trainer is attempting to elicit (e.g., inattentiveness or grabbing medical equipment). The trainer may decide to temporarily remove access to rewards, putting the animal in "time out." In such a situation, the trainer may decide to ignore the animal until the undesired behavior ceases and is replaced by the desired behavior (e.g., focused and calm behavior), for which the animal then earns a reward (PR). Together, these concepts should be systematically applied; generally rewarding behaviors consistent with compliance and cooperation with care and not rewarding those that caregivers want to decrease. This is an example of a *differential reinforcement* schema.

Table 12.2   Task Analysis

Activity	Tools	Purpose
Takes treats reliably from caregivers directly facing the cage in close proximity	Positive reinforcement training (PRT)	Demonstrates effective desensitization and basic trust has been established between NHPs and caregivers
Introduce medical equipment in rooms [e.g., treatment carts, intravenous (IV) pumps, analyzers, scales, monitors]	PRT	This process uses desensitization to progressively introduce and then bring in close proximity medical equipment that will be utilized in the study
Retrain experience with dividing panels or squeeze-back panel (SBP) mechanism engaged	PRT/negative reinforcement training (NRT)	Most NHPs have a negative experience with cages designed to limit space or, in the case of SBP, forcefully restrain animals. Most animals have experience with dividing panels or SBP (usually to apply a painful stimulus like a needlestick) and become extremely anxious when they are used. It is important to recognize how valuable it is to remove anxiety or fear of these aspects constantly present in the animal's homecage environment. Desensitization is used to create a different association and give the animal control over its environment with its actions. Dividers or SBP is applied and the animal is rewarded for calm behavior with PR for 30–60 s, then also negative reinforcement (NR) at the conclusion of each trial, when the divider or SBP is removed because the animal has completed the entire behavior. Dividers or SBP can then be used to block escape tendency, rather than provoking fear associated with restraint.
Sits calmly in front of cage opening with dividing panels or SBP engaged 10%–50%	PRT/NRT	
Sits calmly in front of cage opening with dividing panels or SBP engaged 50%–90%	PRT/NRT	
Sits calmly in front of cage opening while touching or resting against the dividing panels or SBP	PRT/NRT	
Allows trainer to touch limbs	PRT/NRT	As the animal is being rewarded (also in earlier phases depending on comfort level) the trainer should begin gently touching hands and feet at the cage opening. At completion of the session, the dividers or SBP are removed.
Allows trainer to manipulate limbs	Prompting/PRT/NRT	Prompting is used to demonstrate the target behavior. The trainer grasps the target limb, guides it through the opening, and holds for a brief interval. Substantial trust between the trainer and the monkey is required and the animal receives a jackpot reward for completing this step.
Animal presents limb to caregiver on cue	PRT	Once animals understand the full behavior (present at the front of the cage and extend the limb on cue), trainers can fade out prompts and fully rely on PR.
Introduce changes in position, manipulation of the limb for planned tasks	PRT	Animals present limb on cue, the trainer prepares the animal for the movements and manipulation that will occur during more challenging tasks. This includes switching hands, very light hold on the toes/fingers, skin tent, site palpation, heel tap/squeeze, and limb movement to position. Also the "startle" reaction, due to, for example, a door slamming while working with the caregiver, is assessed and calm reactions rewarded.
Introduce contact medical equipment	PRT/distraction	Trainers introduce skin disinfectants, meters with sound, mock needles (blunt lancets, blunt syringes), stethoscope, blood pressure cuff, O_2 sensor, and temperature probe. Contact times increase across sessions.
Performs medical task	PRT/distraction	The trainer interacting with animals on a designated task should demonstrate the highest skill level. This ensures the task is performed with confidence and as quickly as possible, giving confidence to the animal. Steady soft praise can be given to indicate to animals they are holding the behavior as desired.
Long duration or increased complexity homecage medical tasks	PRT/distraction	Once an animal demonstrates a high level of skill, trainers can increase the length of time that an animal works with caregivers and also the degree of handling. This facilitates long duration (~30 min) fluid infusion and clinically relevant metabolic testing. During metabolic testing, food and drink rewards cannot be used, so it is important to prepare manipulanda that will prove interesting over the course of interaction.

Most procedures in research and medical care related to modeling a disease condition involve aversive stimuli (e.g., discomfort, loss of control, or pain) that are likely to elicit avoidance and anxiety. Anxiety can actually increase the perception of pain during medical procedures and subsequently, this procedure-associated pain strengthens anxiety in similar situations in the future, a counterproductive cycle that should be deliberately prevented (Cornwall and Donderi 1988; Bowers et al. 1998; Kain et al. 2006). Using planned desensitization, animals should be carefully and gradually exposed to these stimuli while occupied with distracting and pleasurable activity. As animals demonstrate relaxation (or even enthusiastic anticipation) during procedures, the duration and intensity can be increased. When aversive stimuli are introduced, they often provoke an *escape response*, which is a powerful negative reinforcer. It is critical that trainers provide an environment that blocks escape, maintaining the animal in the situation, which will weaken this behavior. The trainer can then redirect the animal to a more productive behavior, providing an opportunity for the animals to be positively reinforced for performing an approximation of the target behavior.

Oral drug administration, blood collection, parenteral drug administration by injection, and hands-on physical examination are the most common aversive procedures encountered by NHPs participating in research. Accordingly, each situation is discussed in the following paragraphs to empower caregivers to better understand the practical application of the principles to decrease anxiety and prompt compliance that are discussed above. These approaches can serve as examples for other model-specific behaviors that facilitate care and are not specifically outlined in this chapter.

HAND FEEDING AND ORAL DRUG ADMINISTRATION

Conventional gavage, typically in restraint, can be avoided if animals are trained to voluntarily accept drugs or fluids cooperatively *and* reliably. Gavage can be stressful and can introduce safety issues to both the animal (e.g., incorrect tube placement) and the caregiver (e.g., bite or scratch). It is a natural behavior for animals to accept palatable food and drink, but prior to training animals may exhibit variable degrees of ease toward caregivers. Based on personality, some animals will maximize flight distance from trainers, while others will immediately take items directly. The trainer should determine the point at which the animal is comfortable on this continuum by starting in the most anxiety-provoking position (facing the cage in direct proximity to the animal) and then backing off until the animal is confident enough to accept the item. Using this point as the baseline for training, trainers can subsequently progress animals by using frequent brief interactions, each time moving closer to the animal and eliciting more contact in order for the subject to earn the reward. This is also excellent practice for caregivers and helps in increasing their awareness of individual NHP preferences and their levels of motivation for obtaining a preferred reward. It is important that caregivers recognize that preferences and motivation may differ at any given point in time, thus to be effective, they should customize the rewards to the individual and the situation. One way to determine preference is by offering animals a choice condition of at least 3–5 different food or drink items in this phase. Because highly palatable food items are powerful primary motivators, caregivers should pair the food reinforcer with praise (vocal or click), attention, and physical affection (grooming and touching) as the animal feels more comfortable. Praise, attention, and physical affection will then become useful conditioned reinforcers. Clickers have been successfully used with NHPs, but alternatively, a vocal marker, such as "yes" or "good," in a warm praising tone, can be used as a means of letting the animal know that the desired behavior has been performed (Schapiro et al. 2001; Fernström et al. 2009; Baker et al. 2010). Praising correct behavior comes naturally to most caregivers, so for many groups, a vocal mark may be simpler than teaching the entire care team to use the clicker correctly. The utility of the clicker is diminished (1) when caregivers use it inconsistently, either missing or delaying marks, and especially, (2) when caregivers are working with animals on medical tasks that require both hands.

After animals readily accept food and drink from caregivers, the introduction of tart flavors can be attempted to simulate the taste of drugs and to identify individual preferences for masking flavors. Many drugs can be crushed (with the exception of control-release formulations or enteric coating) and blended with a masking food or flavoring agent for administration. Anecdotally, our group has had considerable success with "chasing" extremely unpalatable drugs with a highly preferred flavor. Some caregivers make the mistake of not spending additional training time with those animals that readily accept food and drink on the first trial. These animals may subsequently refuse medicated food items when they are given for the first time "when it counts" in the study. In addition, these animals may rapidly become suspicious of food offered by caregivers and form a negative association with the procedure. In situations in which study timing or medical need dictates that the drug must be administered "on time" (with limited temporal flexibility), non-compliant animals may be subjected to more invasive approaches or simply classified as non-informative (i.e., completely confounded) if levels are unstable. Noninformative animals affect both study interpretation (i.e., lost data, insufficient number of subjects at endpoint) and also welfare (i.e., re-enrollment).

Though the hand feeding and oral administration task can generally be classified as one of the most straightforward to train, it is critical in establishing the important link between behavior and reinforcement. Subsequently, this training momentum can be used to introduce more complicated or higher effort tasks to achieve a higher probability of success.

COOPERATIVELY PRESENTING FOR HANDS-ON MEDICAL PROCEDURES: BLOOD COLLECTION, INJECTION, AND PHYSICAL EXAMINATION

Sedation or physical restraint is often used in NHPs for procedures such as blood collection, injections, and examination. Sedation has unwanted side effects, including removal from social partners, changes in appetite, vomiting, and the potential to further compromise sick animals. Mechanical restraint can be very stressful in animals that are not properly accustomed to restraint. Even animals acclimated to chair restraint have demonstrated physiologic disturbances attributed to removal from their group.

Most procedures can effectively be performed cooperatively with animals in the familiar homecage, including abbreviated physical examination (i.e., abdominal palpation, auscultation, skin turgor assessment, blood pressure, oxygen saturation, and cursory examination of oral cavity, nasal passages, and eyes), capillary blood collection, intramuscular injection, subcutaneous injection, and intravascular access for blood collection or drug administration. Training for this procedure is divided into a number of steps (Table 12.2) at a pace that challenges the NHP, but does not overwhelm its ability to cope (gradual exposure). These procedures require that caregivers have direct, hands-on (grasping) contact with the animals. Grasping the animal is a behavior that, in the absence of training, is almost always perceived as threatening by the animal. In a situation perceived as threatening, where there is an initial tendency to choose escape even over highly palatable food reinforcers, it is important that the trainer provides a secure learning environment that blocks avoidance or escape behavior. This allows the animal to face the situation and recognize that it is not actually threatening and also to experience positive reinforcement. If this is done in a homecage system where space can be manipulated (e.g., adding a panel to bring the animal in close proximity to the trainer), contingent on the presentation of the target behavior, the trainer can also (negatively) reinforce by removing the panel. The animal earns a food reward for their success as well as control over their space by combining contingencies, which may be advantageous by increasing the individual reinforcer value.

Desensitization is used consistently to pair PR with anxiety-provoking stimuli throughout skills training. The trainer should be sure that the animal experiences some success in every session and that the sessions conclude on a positive note. Because many of these behaviors are complicated to shape, prompting the behavior by progressing through multiple steps of the movement or activity with the NHP ensures that the animal has an opportunity to be successful and obtain rewards. As the animals' confidence increases, so does cooperation; they will reliably work with caregivers, even if the procedure is briefly painful or uncomfortable (Graham et al. 2012). Caregivers should be reminded to consistently use distraction to decrease distress during these procedures, which can also reduce the perception of discomfort or pain. Distraction can be an activity that the animal enjoys like playing with or chewing on a toy, shredding paper, grooming or being groomed, or even listening to music.

NHP RESEARCH PROGRAMS BENEFIT FROM PRT

PRT is an essential component of a holistic approach to improving animal well-being and can be used to enhance construct validity of the animal model.

The most common deterrent for adopting a training program is the misperception that the time investment is prohibitive. Time invested in training cooperative behavior is more than recovered by the time saved when performing experimental procedures with trained subjects, especially in long-term studies. NHPs learn over time, and behavior change is rarely instantaneous, but the behaviors described in this chapter can generally be trained in 3 months or less (Graham et al. 2012). Slow progress and regression in training are typically related to the training method or caregiver inexperience, which can be usually overcome by systematic evaluations of the training to determine whether the reinforcers being utilized are appropriate/desirable for each individual subject.

In our research, we use streptozotocin-induced diabetic NHPs to study efficacy and safety of islet transplantation using innovative immunosuppressive strategies to prolong graft survival. We implemented a behavioral management program to train NHPs to cooperate with medical management associated with their chronic disease state. This includes training in hand feeding and drinking, shifting, and limb presentation utilizing predominately positive reinforcement. We formally evaluated our experience in advancing this program of study behavior acquisition in 52 macaques, focusing on a subset of 14 animals that were demographically similar. Over 90% of the NHPs were successful in behavior acquisition following an average total time investment of 12 h, distributed over 3 months. In this study, the presence of acute or chronic involution in thymic histology was significantly reduced in diabetic and immunosuppressed NHPs successfully trained to cooperate. Glucocorticosteroids, also called stress hormones, are well known for their direct and chronic effect on thymus histology, in addition to glycemic control. We have detected important differences in responses to glucose when comparing a cohort of cynomolgus macaques trained for in-homecage cooperation ($n = 10$) with a historical control group ($n = 10$) anesthetized with ketamine (Figure 12.2).

The glucose peak was higher in cooperating animals, and the glucose disappearance rate (between 10 and 30 min, $K_G = 4.1 \pm 0.6$) was significantly higher than animals sedated with ketamine ($K_G = 2.4 \pm 0.4$, $p = .029$). This suggested that handling techniques play a role in glucose metabolism, most prominently first phase response. Because our animals cooperated with metabolic testing, we were also able to perform mixed meal and oral glucose tolerance tests for the first time, yielding a relevant physiological stimulus to evaluate graft function. Implementing a behavioral management program was a pivotal refinement in our diabetic macaque model and was successful both in improving well-being of the animals and in eliminating certain confounding variables (e.g., stress hormones with immunosuppressive and metabolic activity) that interfere with proper safety and efficacy evaluation of cell therapy products and immunosuppressive regimens.

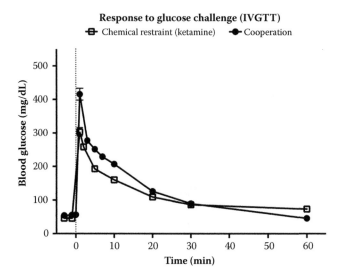

Figure 12.2   Metabolic testing. Blood glucose response to intravenous (IV) glucose challenge in cooperating
animals (closed circle) versus animals sedated with ketamine (open square).

Trained animals have broad experiences with medical procedures, making them better equipped
to cope with novel situations that they may encounter while on study. The training relationship that
has been established between animals and caregivers is routinely rewarding, so even in novel situations, the animals know that they will be predictably rewarded for their efforts. There are practical
considerations from a caregiver perspective as well; a positive relationship with NHPs improves caregiver attitude and reduces occupational health exposures (e.g., scratches, bites, or needlestick injuries).

Trained behavior contributes to excellent medical care, acting as a sensitive early indicator of
clinical symptoms or impending adverse events, allowing for prompt assessment and the initiation
of treatment. For example, if an animal that normally cooperates suddenly refuses to do so, the
animal should be carefully evaluated for an underlying health issue. However, it is possible that
the NHP is simply experiencing an "off day," in which case, the caregivers simply need to be more
patient and give the animal sufficient time to act on the cue. Providing sufficient time and flexibility,
such that the animal can accomplish the target behavior in a way that it prefers, is an important way
to give the animal a perceived sense of control and choice in their cooperative interactions with
caregivers (Bowers 1968). Examples include allowing the animals to (1) choose which limb they
prefer to present, (2) determine the treatment order in the room (submissive animals may prefer to
go after dominant animals), or (3) select their reward.

The broad impact on the scientific program should be equally appreciated. Cooperation enables
full experimental data collection through more productive interactions with animals mimicking
the complexity of the intended clinical situation. Likewise, cooperation avoids exacerbating issues
in animals whose health is compromised in disease models that accurately represent the clinical
manifestations of the disorder. A potential experimentally confounding behavior associated with
restraint stress can be eliminated, which should yield reliable clinical observations in safety studies and supporting laboratory evaluations. By decreasing the number of noninformative animals,
smaller number of animals can be used in experiments in accordance with the principle of reduction.

The improvements in welfare associated with the aforementioned cooperative techniques
contribute to the validity of scientific results, improve the translational value of the model, and offer
benefits to both animals and caregivers.

REFERENCES

Adams, M. R., J. R. Kaplan, S. B. Manuck, B. Uberseder, and K. T. Larkin. Persistent sympathetic nervous system arousal associated with tethering in cynomolgus macaques. *Laboratory Animal Science* 38, no. 3 (1988): 279–281.

Bailoo, J. D., T. S. Reichlin, and H. Wurbel. Refinement of experimental design and conduct in laboratory animal research. *ILAR Journal* 55, no. 3 (2014): 383–391. doi:10.1093/ilar/ilu037.

Baker, K. C., M. A. Bloomsmith, K. Neu, C. Griffis, and M. Maloney. Positive reinforcement training as enrichment for singly housed rhesus macaques (*Macaca mulatta*). *Animal Welfare (South Mimms, England)* 19, no. 3 (2010): 307.

Barr, C. S., T. K. Newman, M. L. Becker, C. C. Parker, M. Champoux, K. P. Lesch, D. Goldman, S. J. Suomi, and J. D. Higley. The utility of the non-human primate model for studying gene by environment interactions in behavioral research. *Genes, Brain and Behavior* 2, no. 6 (2003): 336–340.

Bennett, J. S., K. A. Gossett, M. P. McCarthy, and E. D. Simpson. Effects of ketamine hydrochloride on serum biochemical and hematologic variables in rhesus monkeys (*Macaca mulatta*). *Veterinary Clinical Pathology/American Society for Veterinary Clinical Pathology* 21, no. 1 (1992): 15–18.

Betarbet, R., T. B. Sherer, and J. T. Greenamyre. Animal models of Parkinson's disease. *Bioessays* 24, no. 4 (2002): 308–318.

Bliss-Moreau, E. and G. Moadab. Variation in behavioral reactivity is associated with cooperative restraint training efficiency. *Journal of the American Association for Laboratory Animal Science* 55, no. 1 (2016): 41–49.

Bloomsmith, M. A., A. M. Stone, and G. E. Laule. Positive reinforcement training to enhance the voluntary movement of group-housed chimpanzees within their enclosures. *Zoo Biology* 17, no. 4 (1998): 333–341. doi:10.1002/(SICI)1098–2361(1998)17:4<333::AID-ZOO6>3.0.CO;2-A.

Bowers, K. S. Pain, anxiety, and perceived control. *Journal of Consulting and Clinical Psychology* 32, no. 5, Pt. 1 (1968): 596–602.

Bowers, C. L., C. M. Crockett, and D. M. Bowden. Differences in stress reactivity of laboratory macaques measured by heart period and respiratory sinus arrhythmia. *American Journal of Primatology* 45, no. 3 (1998): 245–261.

Brok, H. P. M., J. Bauer, M. Jonker, E. Blezer, S. Amor, R. E. Bontrop, J. D. Laman, and B. A. 't Hart. Non-human primate models of multiple sclerosis. *Immunological Reviews* 183, no. 1 (2001): 173–185.

Brosnan, S. F. and F. B. M. de Waal. Monkeys reject unequal pay. *Nature* 425, no. 6955 (2003): 297–299.

Buckley, L. A., K. Chapman, L. A. Burns-Naas, M. D. Todd, P. L. Martin, and J. A. Lansita. Considerations regarding nonhuman primate use in safety assessment of biopharmaceuticals. *International Journal of Toxicology* 30, no. 5 (2011): 583–590. doi:10.1177/1091581811415875.

Buse, E., G. Habermann, I. Osterburg, R. Korte, and G. F. Weinbauer. Reproductive/developmental toxicity and immunotoxicity assessment in the nonhuman primate model. *Toxicology* 185, no. 3 (2003): 221–227.

Bush, M., R. Custer, J. Smeller, and L. M. Bush. Physiological measures of nonhuman primates during physical restraint and chemical immobilization. *Journal of the American Veterinary Medical Association* 171, no. 9 (1977): 866–869.

Buynitsky, T. and D. I. Mostofsky. Restraint stress in biobehavioral research: Recent developments. *Neuroscience & Biobehavioral Reviews* 33, no. 7 (2009): 1089–1098.

Caine, N. G. Anti-predator behavior: Its expression and consequences in captive primates, Chapter 9. In Schapiro, S. J. (ed.) *Handbook of Primate Behavioral Management* (pp. 127–138). CRC Press, Boca Raton, FL, 2017.

Capitanio, J. P. Personality dimensions in adult male rhesus macaques: Prediction of behaviors across time and situation. *American Journal of Primatology* 47, no. 4 (1999): 299–320.

Capitanio, J. P. Variation in biobehavioral organization, Chapter 5. In Schapiro, S. J. (ed.) *Handbook of Primate Behavioral Management* (pp. 55–73). CRC Press, Boca Raton, FL, 2017.

Carlsson, H.-E., S. J. Schapiro, I. Farah, and J. Hau. Use of primates in research: A global overview. *American Journal of Primatology* 63, no. 4 (2004): 225–237.

Chang, S. W., A. A. Winecoff, and M. L. Platt. Vicarious reinforcement in rhesus macaques (*Macaca mulatta*). *Frontiers in Neuroscience* 5, no. 27 (2011): 1–10.

Clay, A. W., M. A. Bloomsmith, M. J. Marr, and T. L. Maple. Habituation and desensitization as methods for reducing fearful behavior in singly housed rhesus macaques. *American Journal of Primatology* 71, no. 1 (2009): 30–39.

Coffman, R. L. and E. M. Hessel. Nonhuman primate models of asthma. *The Journal of Experimental Medicine* 201, no. 12 (2005): 1875–1879. doi:10.1084/jem.20050901.

Coleman, K. Individual Differences in temperment and behavioral management, Chapter 7. In: Schapiro, S.J. (ed.) *Handbook of Primate Behavioral Management* (pp. 95–113). CRC Press, Boca Raton, FL, 2017.

Coleman, K., L. Pranger, A. Maier, S. P. Lambeth, J. E. Perlman, E. Thiele, and S. J. Schapiro. Training rhesus macaques for venipuncture using positive reinforcement techniques: A comparison with chimpanzees. *Journal of the American Association for Laboratory Animal Science* 47, no. 1 (2008): 37–41.

Coleman, K., L. A. Tully, and J. L. McMillan. Temperament correlates with training success in adult rhesus macaques. *American Journal of Primatology* 65, no. 1 (2005): 63–71. doi:10.1002/ajp.20097.

Collins, F. S. and L. A. Tabak. Policy: NIH plans to enhance reproducibility. *Nature* 505, no. 7485 (2014): 612–613.

Cornwall, A. and D. C. Donderi. The effect of experimentally induced anxiety on the experience of pressure pain. *Pain* 35, no. 1 (1988): 105–113.

Denayer, T., T. Stöhr, and M. Van Roy. Animal models in translational medicine: Validation and prediction. *New Horizons in Translational Medicine* 2, no. 1 (2014): 5–11.

Falcone, R., E. Brunamonti, and A. Genovesio. Vicarious learning from human models in monkeys. *PLoS One* 7, no. 7 (2012): e40283.

Fernström, A-L., H. Fredlund, M. Spångberg, and K. Westlund. Positive reinforcement training in rhesus macaques: Training progress as a result of training frequency. *American Journal of Primatology* 71, no. 5 (2009): 373–379. doi:10.1002/ajp.20659.

Freeman, H. D., S. F. Brosnan, L. M. Hopper, S. P. Lambeth, S. J. Schapiro, and S. D. Gosling. Developing a comprehensive and comparative questionnaire for measuring personality in chimpanzees using a simultaneous top-down/bottom-up design. *American Journal of Primatology* 75, no. 10 (2013): 1042–1053.

Goosen, D. J., J. H. Davies, M. Maree, and I. C. Dormehl. The influence of physical and chemical restraint on the physiology of the chacma baboon (*Papio ursinus*). *Journal of Medical Primatology* 13, no. 6 (1984): 339–351.

Graham, M. L. and M. J. Prescott. The multifactorial role of the 3 Rs in shifting the harm-benefit analysis in animal models of disease. *European Journal of Pharmacology* 759 (2015): 19–29.

Graham, M. L., E. F. Rieke, L. A. Mutch, E. K. Zolondek, A. W. Faig, T. A. DuFour, J. W. Munson, J. A. Kittredge, and H. J. Schuurman. Successful implementation of cooperative handling eliminates the need for restraint in a complex non-human primate disease model. *Journal of Medical Primatology* 41 (2012): 89–106.

Graham, M. L. and H-J. Schuurman. Validity of animal models of type 1 diabetes, and strategies to enhance their utility in translational research. *European Journal of Pharmacology* 759 (2015): 221–230.

Hackam, D. G. Translating animal research into clinical benefit. *British Medical Journal* 334, no. 7586 (2007): 163–164.

Hackam, D. G. and D. A. Redelmeier. Translation of research evidence from animals to humans. *JAMA* 296, no. 14 (2006): 1727–1732.

Hassimoto, M. and T. Harada. Use of a telemetry system to examine recovery of the cardiovascular system after excitement induced by handling stress in a conscious cynomolgus monkey (*Macaca fascicularis*). *Journal of Medical Primatology* 32 (2003): 346–352.

Hay, M., D. W. Thomas, J. L. Craighead, C. Economides, and J. Rosenthal. Clinical development success rates for investigational drugs. *Nature Biotechnology* 32, no. 1 (2014): 40–51.

Henderson, V. C., J. Kimmelman, D. Fergusson, J. M. Grimshaw, and D. G. Hackam. Threats to validity in the design and conduct of preclinical efficacy studies: A systematic review of guidelines for in vivo animal experiments. *PLoS Medicine* 10, no. 7 (2013): e1001489.

Hill, M., P. Thebault, M. Segovia, C. Louvet, G. Beriou, G. Tilly, E. Merieau, I. Anegon, E. Chiffoleau, and M. C. Cuturi. Cell therapy with autologous tolerogenic dendritic cells induces allograft tolerance through interferon-γ and Epstein-Barr virus-induced gene 3. *American Journal of Transplantation* 11 (2011): 2036–2045.

Jacobsen, P. B., S. L. Manne, K. Gorfinkle, O. Schorr, B. Rapkin, and W. H. Redd. Analysis of child and parent behavior during painful medical procedures. *Health Psychology* 9, no. 5 (1990): 559.

Kain, Z. N., L. C. Mayes, A. A. Caldwell-Andrews, D. E. Karas, and B. C. McClain. Preoperative anxiety, postoperative pain, and behavioral recovery in young children undergoing surgery. *Pediatrics* 118, no. 2 (2006): 651–658. doi:10.1542/peds.2005-2920.

Kennedy, R. M., J. Luhmann, and W. T. Zempsky. Clinical implications of unmanaged needle-insertion pain and distress in children. *Pediatrics* 122, no. Suppl 3 (2008): S130–S133. doi:10.1542/peds.2008-1055e.

Kennedy, R. C., M. H. Shearer, and W. Hildebrand. Nonhuman primate models to evaluate vaccine safety and immunogenicity. *Vaccine* 15, no. 8 (1997): 903–908.

Landi, M. S. The effects of four types of restraint on serum alanine aminotransferase and aspartate aminotransferase in the *Macaca fascicularis*. *International Journal of Toxicology* 9, no. 5 (1990): 517–523.

Lane, M. A. Nonhuman primate models in biogerontology. *Experimental Gerontology* 35, no. 5 (2000): 533–541.

Laule, G. E. Positive reinforcement training and environmental enrichment: Enhancing animal well-being. *Journal of the American Veterinary Medical Association* 223, no. 7 (2003): 969–973.

Laule, G. Positive reinforcement training for laboratory animals. In *The UFAW Handbook on the Care and Management of Laboratory and Other Research Animals*, 206–218. John Wiley & Sons, Oxford, UK, 2010. doi:10.1002/9781444318777.ch16.

Laule, G. E., M. A. Bloomsmith, and S. J. Schapiro. The use of positive reinforcement training techniques to enhance the care, management, and welfare of primates in the laboratory. *Journal of Applied Animal Welfare Science* 6, no. 3 (2003): 163–173.

Lee, J. I., S. H. Hong, S. J. Lee, Y. S. Kim, and M. C. Kim. Immobilization with ketamine HCl and tiletamine-zolazepam in cynomolgus monkeys. *Journal of Veterinary Science (Suwon-Si, Korea)* 4, no. 2 (2003): 187–191.

Lemieux, A. M., C. L. Coe, W. B. Ershler, and J. W. Karaszewski. Surgical and psychological stressors differentially affect cytolytic responses in the rhesus monkey. *Brain, Behavior, and Immunity* 10, no. 1 (1996): 27–43.

Lutz, C. K. and M. A. Novak. Environmental enrichment for nonhuman primates: Theory and application. *ILAR Journal* 46, no. 2 (2005): 178–191.

Majidi, S., K. A. Driscoll, and J. K. Raymond. Anxiety in children and adolescents with type 1 diabetes. *Current Diabetes Reports* 15, no. 8 (2015): 1–6.

Mayer, E. A. The neurobiology of stress and gastrointestinal disease. *Gut* 47, no. 6 (2000): 861–869.

McGonigle, P. and B. Ruggeri. Animal models of human disease: Challenges in enabling translation. *Biochemical Pharmacology* 87, no. 1 (2014): 162–171. doi:10.1016/j.bcp.2013.08.006.

Messaoudi, I., R. Estep, B. Robinson, and S. W. Wong. Nonhuman primate models of human immunology. *Antioxidants & Redox Signaling* 14, no. 2 (2011): 261–273.

Morrow-Tesch, J. L., J. J. McGlone, and R. L. Norman. Consequences of restraint stress on natural killer cell activity, behavior, and hormone levels in rhesus macaques (*Macaca mulatta*). *Psychoneuroendocrinology* 18, no. 5–6 (1993): 383–395.

Noel, M., C. M. McMurtry, C. T. Chambers, and P. J. McGrath. Children's memory for painful procedures: The relationship of pain intensity, anxiety, and adult behaviors to subsequent recall. *Journal of Pediatric Psychology* 35, no. 6 (2010): 626–636. doi:10.1093/jpepsy/jsp096.

Nout, Y. S., E. S. Rosenzweig, J. H. Brock, S. C. Strand, R. Moseanko, S. Hawbecker, S. Zdunowski, J. L. Nielson, R. R. Roy, and G. Courtine. Animal models of neurologic disorders: A nonhuman primate model of spinal cord injury. *Neurotherapeutics* 9, no. 2 (2012): 380–392.

Novak, M. A. and S. J. Suomi. Social interaction in nonhuman primates: An underlying theme for primate research. *Laboratory Animal Science* 41, no. 4 (1991): 308–314.

Olsson, I. A. S. and K. Westlund. More than numbers matter: The effect of social factors on behaviour and welfare of laboratory rodents and non-human primates. *Applied Animal Behaviour Science* 103, no. 3 (2007): 229–254.

Phillips, K. A., K. L. Bales, J. P. Capitanio, A. Conley, P. W. Czoty, B. A. 't Hart, W. D. Hopkins, S-L. Hu, L. A. Miller, and M. A. Nader. Why primate models matter. *American Journal of Primatology* 76, no. 9 (2014): 801–827.

Phipps, S. and S. DeCuir-Whalley. Adherence issues in pediatric bone marrow transplantation. *Journal of Pediatric Psychology* 15, no. 4 (1990): 459–475.

Prescott, M. J. and H. M. Buchanan-Smith. Training nonhuman primates using positive reinforcement techniques. *Journal of Applied Animal Welfare Science* 6, no. 3 (2003): 157–161.

Preston, S. D. and F. B. M de Waal. Empathy: Its ultimate and proximate bases. *Behavioral and Brain Sciences* 25, no. 1 (2002): 1–20.

Price, D. L., L. J. Martin, S. S. Sisodia, M. V. Wagster, E. H. Koo, L. C. Walker, V. E. Koliatsos, and L. C. Cork. Aged non-human primates: An animal model of age-associated neurodegenerative disease. *Brain Pathology* 1, no. 4 (1991): 287–296.

Reamer, L. A., R. L. Haller, E. J. Thiele, H. D. Freeman, S. P. Lambeth, and S. J. Schapiro. Factors affecting initial training success of blood glucose testing in captive chimpanzees (*Pan troglodytes*). *Zoo Biology* 33, no. 3 (2014): 212–220.

Reinhardt, V., D. Cowley, J. Scheffler, R. Vertein, and F. Wegner. Cortisol response of female rhesus monkeys to venipuncture in homecage versus venipuncture in restraint apparatus. *Journal of Medical Primatology* 19, no. 6 (1990): 601–606.

Rudman, L. A., M. H. Gonzales, and E. Borgida. Mishandling the gift of life: Noncompliance in renal transplant patients. *Journal of Applied Social Psychology* 29, no. 4 (1999): 834–851.

Russell, W. M. S and R. L. Burch. *The Principles of Humane Experimental Technique.* Universities Federation for Animal Welfare. Potters Bar, UK, 1959.

Schapiro, S. J. Effects of social manipulations and environmental enrichment on behavior and cell-mediated immune responses in rhesus macaques. *Pharmacology, Biochemistry, and Behavior* 73, no. 1 (2002): 271–278.

Schapiro, S. J. and M. A. Bloomsmith. Behavioral effects of enrichment on pair-housed juvenile rhesus monkeys. *American Journal of Primatology* 32, no. 3 (1994): 159–170.

Schapiro, S. J. and M. A. Bloomsmith. Behavioral effects of enrichment on singly-housed, yearling rhesus monkeys: An analysis including three enrichment conditions and a control group. *American Journal of Primatology* 35, no. 2 (1995): 89–101.

Schapiro, S. J., M. A. Bloomsmith, and G. E. Laule. Positive reinforcement training as a technique to alter non-human primate behavior: Quantitative assessments of effectiveness. *Journal of Applied Animal Welfare Science* 6, no. 3 (2003): 175–187.

Schapiro, S. J., J. E. Perlman, and B. A. Boudreau. Manipulating the affiliative interactions of group-housed rhesus macaques using positive reinforcement training techniques. *American Journal of Primatology* 55, no. 3 (2001): 137–149.

Shirasaki, Y., N. Yoshioka, K. Kanazawa, T. Maekawa, T. Horikawa, and T. Hayashi. Effect of physical restraint on glucose tolerance in cynomolgus monkeys. *Journal of Medical Primatology* 42, no. 3 (2013): 165–168.

Spruijt, B. M., R. van den Bos, and F. T. A. Pijlman. A concept of welfare based on reward evaluating mechanisms in the brain: Anticipatory behaviour as an indicator for the state of reward systems. *Applied Animal Behaviour Science* 72, no. 2 (2001): 145–171.

van der Worp, H. B., D. W. Howells, E. S. Sena, M. J. Porritt, S. Rewell, V. O'Collins, and M. R. Macleod. Can animal models of disease reliably inform human studies? *PLoS Medicine* 7, no. 3 (2010): e1000245.

Verdier, J-M., I. Acquatella, C. Lautier, G. Devau, S. Trouche, C. Lasbleiz, and N. Mestre-Francés. Lessons from the analysis of nonhuman primates for understanding human aging and neurodegenerative diseases. *Frontiers in Neuroscience* 9 (2015): 64.

Waitt, C., H. M. Buchanan-Smith, and K. Morris. The effects of caretaker-primate relationships on primates in the laboratory. *Journal of Applied Animal Welfare Science* 5, no. 4 (2002): 309–319.

Yasuda, M., J. Wolff, and C. F. Howard. Effects of physical and chemical restraint on intravenous glucose tolerance test in crested black macaques (*Macaca nigra*). *American Journal of Primatology* 15, no. 2 (1988): 171–180.

CHAPTER 13

Positive Reinforcement Training and Health Care

Elizabeth R. Magden
The University of Texas MD Anderson Cancer Center

CONTENTS

INTRODUCTION

We know how to provide appropriate health care for nonhuman primates (NHPs), but our goal is always to make it better. We are constantly looking for new ways to improve their health care, whether that involves the use of new analgesics, the integration of new therapeutics, or improved diagnostic testing. One tool we have to enhance the health care of NHPs is the implementation of positive reinforcement training (PRT). PRT is a powerful tool that can be and is being used to enhance the health care and well-being of NHPs.

PRT is a form of operant conditioning in which an animal is rewarded for performing a desired behavior (Pryor 1999). It has become an established and integral component of success-ful behavioral management programs (Schapiro et al. 2003, 2005, 2017; Bloomsmith and Else 2005; Coleman et al. 2008; Rogge et al. 2013). The use of this training technique enhances NHP welfare and facilitates health care by encouraging the animals to voluntarily participate in their own care (Schapiro and Lambeth 2007). Should the animals choose to participate, PRT tech-niques are initially used to train the animals to present various body parts. This is accomplished by choosing a conditioned reinforcer (bridge), such as a clicker, that is used to precisely indicate the exact moment the animal performs the desired behavior. The animals are then trained to perform a specific behavior through the reinforcement of successive approximations of that behavior [such as touching a target (e.g., a dowel or stick)], which facilitates the shaping of the desired behavior.

Our goals when using PRT to facilitate NHP health care are ultimately to improve animal welfare. Specifically, we want to train the animals to cooperate with health-care procedures. This cooperation with their own health care has several benefits, including:

1. *Improved veterinary care.* Animals are trained to present body parts with injuries, or areas that need to be monitored or rechecked. We can also train animals to cooperate with veterinary procedures such as auscultation, oral/ophthalmic/otologic examinations, and biological sample collection for various diagnostic tests or health monitoring.
2. *Treatments in home enclosures.* Once animals are trained to present various body parts, we can administer treatments in a manner that reduces stress while the animals remain in their home enclosures. Topical wound care and the initiation of complementary therapeutics (i.e., acupuncture) are just some of the many possible applications.
3. *Reduced anesthetic events.* While sedations are often required for physical exams and health-care procedures/treatments, they can be stressful for the animal being sedated (Lambeth et al. 2006). An additional drawback is that anesthetic drugs pose a risk (albeit small) for the animals being sedated, particularly geriatric or otherwise health-compromised animals. Efforts to reduce anesthetic events through cooperation with health-care procedures decrease overall stress and minimize potential adverse drug effects, ultimately benefiting the animals.
4. *Enrichment opportunities.* The process of PRT promotes mental stimulation for the animals being trained. A unique aspect of this training is that it provides the animal with a sense of control—participation is voluntary and the animals can choose if they want to participate. PRT has also been shown to help resolve social issues and facilitate animal introductions, and to reduce abnormal behaviors (Laule et al. 2003).

When training animals to cooperate with health-care procedures, there is a significant amount of desensitization that must occur. Desensitization is the process of pairing positive rewards with potentially fear-invoking stimuli to make these stimuli less aversive (Laule et al. 2003). While a stethoscope or syringe may be viewed negatively by an animal the first time it is viewed, through the process of desensitization, we can eliminate the fear associated with these objects. There are several resources available that describe methods on how to appropriately implement PRT techniques (Pryor 1999; Laule et al. 2003; Prescott and Buchanan-Smith 2003; Schapiro et al. 2003; Laule and Whittaker 2007).

VETERINARY CARE

Examinations

Veterinarians can closely examine NHPs in their home enclosures with the use of PRT techniques. PRT enhances the level of trust among the animals, trainers, and veterinarians. This high level of trust enables us to closely visualize the animals. With close inspection, we can confirm continued good health or detect potential health issues early.

Examinations often start by having the animals present various body parts. Using PRT techniques, animals are trained to present their chest for auscultation (Figure 13.1), which allows veterinarians to assess cardiorespiratory function. This is particularly beneficial when animals have a known disease process (e.g., cardiomyopathy, pneumonia, etc.). The trained behavior of auscultation enables us to monitor the disease state without the risks or stress associated with anesthesia. Minimizing or avoiding anesthetic events for animals with compromised health status provides benefits in terms of overall health and the reduction of anesthesia-related stress.

Figure 13.1 Adult male chimpanzee (*Pan troglodytes*) trained to voluntarily present his chest for auscultation.

Training animals to open their mouths, or present ears and eyes, can also provide numerous health benefits. Veterinarians can assess dental disease, tooth fractures, or oral trauma when an animal is trained to open its mouth (Figure 13.2). We can also evaluate ears for possible infection or foreign bodies (generally sticks, rocks, or orange seeds) by doing an otoscopic exam (Figure 13.3). We can even assess eyes for possible trauma or ulceration with a thorough ophthalmic examination (Figure 13.4). Infections can be monitored by routine measurement of body temperature, taken rectally when animals voluntarily present their backsides, or via ear presentation for otic temperature assessment.

NHPs have also been trained to present for ultrasound examination. Using PRT, they can be desensitized to the ultrasound machine and the various probes. They are then trained to present either their chest for an echocardiogram to evaluate cardiac function, or their ventral abdomen for assessment of internal organs (Drews et al. 2011; Magden et al. 2015). In the past, all of these assessments have required sedation.

The procedures described here clearly indicate that NHPs can be trained to cooperatively participate in these types of health care-related behaviors. Advancements in behavioral management strategies have facilitated health-care practices and enabled us to raise the standard of veterinary care. We now practice a higher level of care due to the true synergism between health-care management and behavioral management, another step toward our goal of taking captive NHP care to the "next level."

Sample Collection

NHPs can be trained using PRT techniques to voluntarily allow the collection of a variety of biological samples, which are crucial veterinary diagnostic tools. Through the collection of various biological samples, we can assess health status and begin appropriate treatments. We can obtain cultures from sites of infection (such as an infected wound), from nasal discharge, or from the back of the throat (Figure 13.5). These cultures help us to determine whether there is an active infection, to identify the infectious organism (if present), and allow us to start appropriate (antibiotic) treatment.

Collection of voluntary urine and semen samples is also possible in NHPs. Urine collection using positive reinforcement training has proven successful in common marmosets (*Callithrix jacchus*) and chimpanzees (*Pan troglodytes*) (Bassett et al. 2003; McKinley et al. 2003; Bloomsmith et al. 2015). Urine samples provide a noninvasive diagnostic test for a variety of possible health

Figure 13.2 "Open mouth" behavior trained for assessment of the oral cavity.

Figure 13.3 Otoscopic examination trained using PRT techniques.

conditions (Figure 13.6). The veterinarian can assess the urine samples for possible urinary tract infections, monitor protein loss/renal disease, measure glucose in diabetic cases, or perform pregnancy tests (Laule et al. 1996; Schapiro et al. 2005; Lammey et al. 2011a). Semen collection aids in the assessment of fertility or provides specimens for storage for future breeding purposes. While semen can be collected using a rectal probe to stimulate electroejaculation in sedated animals (VandeVoort 2004), we can avoid the stress and risks of anesthesia by training animals to voluntarily provide semen samples (Figure 13.7). This behavior is initially shaped by asking the animal to present its abdomen. As male chimpanzees often have erections during training sessions, the trainer must work to desensitize the animal to the artificial vagina (a PVC tube with penrose tubing)

Figure 13.4 Ophthalmic examination of an adult male chimpanzee (*Pan troglodytes*).

Figure 13.5 Nasal culture (A) and throat culture (B) obtained from chimpanzees (*Pan troglodytes*).

and coordinate thrusting direction to capture the ejaculate (Schapiro et al. 2005). In other NHP species, the animals can be trained to cooperate with direct penile electrostimulation, which does not require anesthesia (VandeVoort 2004).

Blood sample collection is a particularly valuable trained behavior. Anesthetic events are associated with stress, and stress affects hematologic parameters. Therefore, the most accurate blood values can be obtained in samples that are collected without sedation (Lambeth et al. 2006). Both chimpanzees and rhesus macaques can be trained for voluntary venipuncture and blood collection (Lambeth et al. 2006; Coleman et al. 2008) (Figure 13.8). The blood collected can be analyzed for general health status, to monitor disease progression, or to evaluate the success of treatments (i.e., infections treated with antibiotics). Capillary blood samples can also be obtained, and these smaller blood samples can be used to test blood glucose levels via a glucometer (Figure 13.9; Reamer et al. 2014). This is very valuable when providing the best care for diabetic animals.

Figure 13.6 Adult female chimpanzee (*Pan troglodytes*) providing a voluntary urine sample.

Treatments

While many therapeutics are successfully administered orally or via injectable routes, there are some that have been more difficult to administer to nonsedated NHPs. Treatment of wounds is one of the most common veterinary interventions required by socially housed NHPs. Nonhuman primates are social animals with dynamic social relationships. Their interactions with one another occasionally result in wounding; however, the relatively low risk of injury is outweighed by the numerous benefits of social housing. If wounds do occur, we can optimize our treatment and minimize anesthesia, if animals have been trained to accept wound care in their home enclosures. Animals that are trained to present various body parts can also be trained to present wounds for topical treatments. When positive reinforcement is used, wounds can be cleaned with various cleansing solutions, and topical antibiotics or salves can be applied (Figure 13.10). PRT also facilitates the treatment of wounds with low-level laser therapy (LLLT) that enhances tissue healing (see section "Complementary Therapeutics").

In addition to treatment for wound care, other medications can be administered via behaviors shaped with PRT. In cases of eye trauma, ulceration, or infection, ophthalmic ointment or drops can be administered as the animal presents for an ophthalmic exam. For ear infections, antibiotic drops can be administered into the ear canal as an animal presents its ear(s) (Figure 13.11). These types of treatments are often administered multiple times each day. It would not be feasible to sedate an animal that frequently for the necessary treatment. Administration of ophthalmic and otic medications is a prime example of how PRT can be used to provide required therapies that would not otherwise be possible if the animals were not trained.

Figure 13.7 Semen collection in an adult male chimpanzee (*Pan troglodytes*).

Figure 13.8 Venipuncture in a rhesus macaque (*Macaca mulatta*).

Nebulization is another example of a treatment modality that becomes feasible as a result of PRT. Nebulization involves inhaling a mist containing medications (antibiotics or other therapeutics) targeted to the lungs for treatment of respiratory disease. NHPs have been trained to accept nebulizer therapy (Figure 13.12). This has been accomplished by making use of a previously trained

Figure 13.9   Capillary blood sample using a lancet (A) and glucometer for glucose testing (B) in a cynomolgus macaque (*Macaca fascicularis*).

Figure 13.10   Topical wound care on the sex skin of an adult female chimpanzee (*Pan troglodytes*) using dilute (0.05%) chlorhexidine.

behavior—mouth presentation. Once the animal has been desensitized to the nebulizer, it only needs to present its mouth to the nebulizer for the prescribed treatment duration. In one case, the nebulizer was used to treat a case of chronic air sacculitis in a chimpanzee (Gresswell and Goodman 2011). In a second case, a chimpanzee was treated for pneumonia with nebulized antibiotics (SJ Buchl, personal communication). In a research context, oxytocin has been administered to rhesus macaques trained to accept the nebulized drug (Parr et al. 2013).

Complementary Therapeutics

There are several complementary therapeutic techniques that can benefit the health of NHPs. Acupuncture, LLLT, and physical therapy can all enhance conventional pharmacologic treatments when used in conjunction with such treatments. Chimpanzees have been trained using PRT

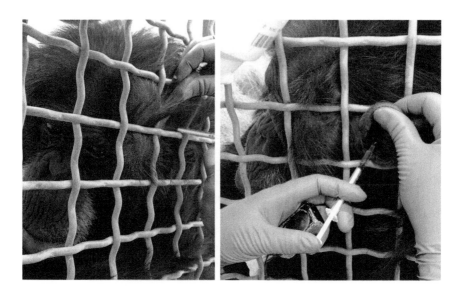

Figure 13.11   Adult male chimpanzee (*Pan troglodytes*) receiving antibiotic ear drops.

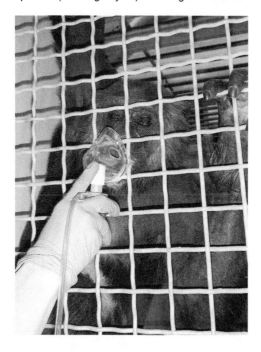

Figure 13.12   Chimpanzee (*Pan troglodytes*) participating in nebulization therapy.

to voluntarily participate in acupuncture therapy, LLLT, and physical therapy, providing proven benefits to the animals that choose to participate (Magden et al. 2013, 2016).

Acupuncture

Acupuncture is a therapy with ancient roots, having been practiced for over 2000 years in China (Ma 1992; White and Ernst 2004). Acupuncture involves the insertion of thin, sterile needles into defined anatomic sites, which then stimulate physiologic processes through neural signaling (Robinson 2008).

The majority of acupuncture sites correspond to neurovascular bundle locations, blood plexuses, and motor endplate zones (Dung 1984; Ogay et al. 2009; Robinson 2009). Once the needles are inserted into these highly innervated and vascular acupuncture points, the needles are gently rotated, delivering a mechanical signal to the nervous tissue, ultimately leading to proven health benefits (Langevin and Yandow 2002). Both the World Health Organization (WHO) and the National Institutes of Health (NIH) have issued consensus statements in support of the efficacious use of acupuncture to treat a variety of conditions in humans (NIH 1997; WHO 1996). Some of the conditions that have been shown to improve with acupuncture treatment include migraines, osteoarthritis, orofacial pain, gastrointestinal disorders, cardiac arrhythmias, hypertension, and stress (Chiu et al. 1997; Middlekauff et al. 2001; Grober and Thethi 2003; Ouyang and Chen 2004; Goddard 2005; Linde et al. 2005; Romoli et al. 2005; Yin and Chen 2010; Lombardi et al. 2012; Mavrommatis et al. 2012; Manyanga et al. 2014).

In chimpanzees, acupuncture is being used to successfully alleviate pain associated with osteoarthritis (Magden et al. 2013). The treatments are performed in the animals' home enclosure, and the chimpanzees are trained using PRT techniques to present the body part at which the acupuncture needle is inserted. The chimpanzees are then trained to remain in position for ~10 min, the average duration of an acupuncture treatment session (Figure 13.13). Dental floss is tied around the acupuncture

Figure 13.13 Acupuncture and laser therapy (A, far view; B, near view) to alleviate pain associated with stifle osteoarthritis using acupuncture sites ST-34, ST-35, and ST-36 (ST = stomach).

needles to ensure the needles remain with the veterinarian, even if the animal gets up and moves from the front of the enclosure. All medications are continued unchanged (Nonsteroidal anti-inflammatory drugs, etc.), and mobility is assessed using a defined mobility scoring system. In other species of NHPs, mobility can be assessed using a pedometer worn by the animal that measures the number of steps the animal takes. As pain control improves with the acupuncture therapy, there should be a corresponding increase in activity and the average number of steps taken each day by the animals.

Acupuncture can also be used to improve wound healing, by increasing cell proliferation and angiogenesis, and decreasing pro-inflammatory cytokines (Park et al. 2012). By stimulating blood flow and decreasing inflammation, acupuncture can decrease the time required for an animal to fully heal from an injury or wound. The shorter recovery time will minimize any discomfort experienced by the animal.

We have also used acupuncture to decrease the incidence of a diagnosed cardiac arrhythmia (VPC—ventricular premature complexes) in a chimpanzee (Magden et al. 2016). The use of acupuncture to treat cardiac rhythm disturbances is based on several studies demonstrating success in human cases. One study evaluated 98 human cases with VPC and found 96.9% of the cases improved with an acupuncture treatment regimen that involved four defined acupuncture sites (Baojie and Feng 2008). Several other studies have also demonstrated decreases in various types of cardiac arrhythmias with the use of acupuncture therapy (VanWormer et al. 2008; Lomuscio et al. 2011).

Acupuncture is most effective when it occurs frequently (1–3 sessions per week) and when it is done with awake animals (i.e., not under anesthesia). PRT is critical for successful acupuncture treatment, as frequent treatments would not be practical if anesthesia was required. Acupuncture represents an excellent example of the delivery of veterinary treatments that would not be possible or feasible without the voluntary cooperation of trained animals.

Laser Therapy

LLLT refers to the use of red visible or near-infrared light (600–1000 nm) to reduce pain and inflammation, and to stimulate healing. The mechanism of action of LLLT involves the absorption of photons by cytochrome c oxidase in cell mitochondria. The photons displace nitric oxide from the mitochondria, which leads to increased ATP production and RNA/protein synthesis (Huang et al. 2011). LLLT has been shown to decrease time associated with wound healing, to decrease inflammation, and to promote myofibroblast proliferation in several studies of wound healing (Medrado et al. 2003; Hopkins et al. 2004; Desmet et al. 2006; Hawkins et al. 2006; Robinson 2014).

Most NHPs are socially housed in groups organized into dominance hierarchies. As group dynamics shift, fighting can occur, potentially resulting in wounding. The treatment of wounds is one of the most common veterinary procedures among NHPs. Given the relatively frequent occurrence of wounding, and the proven benefits of LLLT to enhance wound healing, training animals to voluntarily participate in LLLT improves both their health and well-being.

Once again, PRT techniques that are used for body part presentation can be used to train the animals to present for LLLT. After the animals are desensitized to the laser unit, they can be asked to present the body part of interest (e.g., a wound or arthritic joint). Depending on the type of laser used, the surface area of the treatment zone, and the condition being treated, the time required for LLLT ranges from a few seconds to a few minutes. Some lasers are equipped with attachment probes that can be used to target specific acupuncture sites, or as a physical extension of the laser to facilitate treatments through the enclosure mesh.

Physical Therapy

An additional complementary therapeutic that can enhance the welfare of captive NHPs is physical therapy. Physical therapy is especially important for animals that are geriatric, or have compromised mobility, due to osteoarthritis, obesity, stroke, etc. By using PRT, we can encourage a variety of movements, including standing, climbing, grasping, and walking around the enclosures. Animals that have previously been trained to target and cooperatively shift between enclosures (Schapiro et al. 2017) will already have the training background to successfully participate in physical therapy. Training an animal to voluntarily participate in physical therapy can enhance its mobility, provide an enriching activity, improve its welfare, and ultimately, serve to enhance the quality, and increase the length, of its life.

Anesthesia

There are two goals of PRT in relation to anesthesia. The first is to reduce the overall number of anesthetic events, by training animals to cooperate with procedures that can be performed in their home enclosures. By reducing anesthetic events, we minimize the stress associated with anesthesia. Some of the stressors associated with anesthetic events include (1) administration of an injectable anesthetic agent, (2) withdrawal of food for 12 h prior to the event to prevent aspiration of food particles, (3) physical separation from social partners immediately prior to and following the anesthetic event, and (4) recovery from anesthesia and the associated temporary disorientation. Additionally, anesthesia drugs carry a small risk of adverse side effects with their use in animal sedations. Reducing the number of sedations will reduce the risks associated with anesthesia.

Despite our best efforts to minimize the number of anesthetic events, there will continue to be a need to anesthetize animals for physical examinations and to complete various procedures (including surgery). This leads to our second anesthesia-related goal of PRT, to train animals to voluntarily present for an anesthetic drug injection. Lambeth et al. (2006) showed a significant increase in stress-associated hematologic values when animals were sedated using a non-voluntary anesthetic injection in comparison to a voluntary injection. Chimpanzees, macaques, and squirrel monkeys have been trained to present for a voluntary intramuscular anesthetic injection using PRT techniques (Schapiro et al. 2005; Videan et al. 2005; Russell et al. 2006; Coleman et al. 2008; Gillis et al. 2012). The use of PRT for voluntary injection is beneficial not only because it decreases animal stress but also because it provides more accurate measurements when analyzing blood parameters for clinical diagnoses and research studies.

An alternative to anesthetizing rhesus macaques is the use of the pole-and-collar to manually position animals in a restraint chair (McMillan et al. 2014). In the past, animals have been trained to acquiesce to this desired behavior using negative reinforcement. However, in a study by McMillan et al. (2014), rhesus monkeys were trained to voluntarily cooperate with this transfer and restraint method using PRT. With the use of PRT, the pole-and-collar restraint method can serve as an alternative to anesthesia (chemical restraint) or physical restraint in properly trained animals (McMillan et al. 2014). The use of PRT for a behavior that was previously accomplished using negative reinforcement demonstrates the evolution of animal care and the ways in which we can continually work to improve animal welfare.

BEHAVIORAL MANAGEMENT

A well-established behavioral management program and strategy will promote an effective positive reinforcement training program. Strong behavioral management programs thus facilitate animal health care and well-being by promoting positive reinforcement training. As mentioned previously, PRT permits the animals to make a *choice*. NHPs that participate in a PRT program are provided with a level of control over certain aspects of their captive environment. By allowing the animals to choose if they participate in their health care, we are providing an enriching experience for the animal. In addition, the provision of choice eliminates the stress that may otherwise be associated with the desired procedures, which should lead to more reliable research data.

A particularly appealing side effect of being able to perform health-care procedures using PRT is an increase in efficiency. The collection of samples is much faster when animals are willing and eager to cooperate with providing the samples. By increasing the ease of sample collection, we improve our ability to diagnose disease, resulting in better health care for the animals. The time gained from working with trained subjects allows staff to spend more time on other activities, such as creating new environmental enrichment devices, because the time needed to collect samples or conduct various other procedures (i.e., wound care, etc.) has gone down.

While the examples in this chapter focus on chimpanzees and macaques, many other species have been successfully trained for a variety of behaviors using PRT. Bonobos and orangutans have been trained to cooperate with medical procedures using PRT (Behringer et al. 2014). Neotropical primates, such as marmosets, squirrel monkeys, and owl monkeys, have also been trained to cooperate with various procedures using PRT techniques (Bassett et al. 2003; Rogge et al. 2013). Squirrel monkeys have been trained to voluntarily participate in husbandry and intramuscular injection procedures (Gillis et al. 2012), red-bellied tamarins can be trained to voluntarily enter transportation boxes (Owen and Amory 2011), and marmosets have been trained to provide voluntary urine samples using PRT techniques (Bassett et al. 2003; McKinley et al. 2003). In every instance, PRT has been an effective tool for improving the health care of NHPs, providing animals with control and choice, facilitating our movement to the next level of captive care.

REFERENCES

Baojie, H. and W. Feng. 2008. Acupuncture treatment for 98 cases of ventricular premature beats. *J Tradit Chin Med* 28(2):86–89.

Bassett, L., Buchanan-Smith, H. M., McKinley, J., and T. E. Smith. 2003. Effects of training on stress-related behavior of the common marmoset (*Callithrix jacchus*) in relation to coping with routine husbandry procedures. *J Appl Anim Welf Sci* 6:221–233.

Behringer, V., Stevens, J. M. G., Hohmann, G., Möstl, E., Selzer, D., and T. Deschner. 2014. Testing the effect of medical positive reinforcement training on salivary cortisol levels in bonobos and orangutans. *PLoS One* 9(9):1–5.

Bloomsmith, M. A. and J. G. Else. 2005. Behavioral management of chimpanzees in biomedical research facilities: The state of the science. *ILAR J* 46:192–201.

Bloomsmith, M., Neu, K., Franklin, A., Griffis, C., and J. McMillan. 2015. Positive reinforcement methods to train chimpanzees to cooperate with urine collection. *J Am Assoc Lab Anim Sci* 54(1):66–69.

Chiu, Y. J., Chi, A., and I. A. Reid. 1997. Cardiovascular and endocrine effects of acupuncture in hypertensive patients. *Clin Exp Hypertens* 19(7):1047–1063.

Coleman, K., Pranger, L., Maier, A., Lambeth, S. P., Perlman, J. E., Thiele, E., and S. J. Schapiro. 2008. Training rhesus macaques for venipuncture using positive reinforcement techniques: A comparison with chimpanzees. *J Am Assoc Lab Anim Sci* 47(1):37–41.

Desmet, K. D., Paz, D. A., Corry, J. J., Eells, J. T., Wong-Riley, M. T., Henry, M. M., Buchmann, E. V., et al. 2006. Clinical and experimental applications of NIR-LED photobiomodulation. *Photomed Laser Surg* 24(2):121–128.

Drews, B., Harmann, L. M., Beehler, L. L., Bell, B., Drews, R. F., and T. B. Hildebrandt. 2011. Ultrasonographic monitoring of fetal development in unrestrained bonobos (*Pan paniscus*) at the Milwaukee County Zoo. *Zoo Biol* 30(3):241–253.

Dung, H. C. 1984. Anatomical features contributing to the formation of acupuncture points. *Am J Acupunct* 12:139–143.

Gillis, T. E., Janes, A. C., and M. J. Kaufman. 2012. Positive reinforcement training in squirrel monkeys using clicker training. *Am J Primatol* 74:712–720.

Goddard, G. 2005. Short term pain reduction with acupuncture treatment for chronic orofacial pain patients. *Med Sci Monit* 11(2):CR71–CR74.

Gresswell, C. and G. Goodman. 2011. Case study: Training a chimpanzee (*Pan troglodytes*) to use a nebulizer to aid the treatment of airsacculitis. *Zoo Biol* 30:570–578.

Grober, J. S. and A. K. Thethi. 2003. Osteoarthritis: When are alternative therapies a good alternative? *Consultant* 43(2):197–202.

Hawkins, D., Houreld, N., and H. Abrahamse. 2006. Low level laser therapy (LLLT) as an effective therapeutic modality for delayed wound healing. *Ann N Y Acad Sci* 1056:486–493.

Hopkins, J. T., McLoda, T. A., Seegmiller, J. G., and G. D. Baxter. 2004. Low-level laser therapy facilitates superficial wound healing in humans: A triple-blind, sham-controlled study. *J Athl Train* 39:223–229.

Huang, Y.-Y., Sharma, S. K., Carroll, J., and M. R. Hamblin. 2011. Biphasic dose response in low level light therapy—An update. *Dose Response* 9:602–618.

Lambeth, S. P., Hau, J., Perlman, J. E., Martino, M., and S. J. Schapiro. 2006. Positive reinforcement training affects hematologic and serum chemistry values in captive chimpanzees. *Am J Primatol* 68:245–256.

Lammey, M. L., Ely, J. J., Zavaskis, T., Videan, E., and M. M. Sleeper. 2011a. Effects of aging and blood contamination on the urinary protein-creatinine ratio in captive chimpanzees (*Pan troglodytes*). *J Am Assoc Lab Anim Sci* 50:374–377.

Langevin, H. M. and J. A. Yandow. 2002. Relationship of acupuncture points and meridians to connective tissue planes. *Anat Rec* 269:257–265.

Laule, G. E., Bloomsmith, M. A., and S. J. Schapiro. 2003. The use of positive reinforcement training techniques to enhance the care, management, and welfare of primates in the laboratory. *J Appl Anim Welf Sci* 6(3):163–173.

Laule, G. E., Thurston, R. H., Alford, P. L., and M. A. Bloomsmith. 1996. Training to reliably obtain blood and urine samples from a diabetic chimpanzee (*Pan troglodytes*). *Zoo Biol* 15:587–591.

Laule, G. E. and M. Whittaker. 2007. Enhancing nonhuman primate care and welfare through the use of positive reinforcement training. *J Appl Anim Welf Sci* 10:31–38.

Linde, K., Streng, A., Jurgens, S., Hoppe, A., Brinkhaus, B., Witt, C., Wagenpfeil, S., et al. 2005. Acupuncture for patients with migraine—A randomized controlled trial. *JAMA* 293(17):2118–2125.

Lombardi, F., Belletti, S., Battezzati, P. M., and A. Lomuscio. 2012. Acupuncture for paroxysmal and persistent atrial fibrillation: An effective non-pharmacological tool? *World J Cardiol* 4(3):60–65.

Lomuscio, A., Belletti, S., Battezzati, P. M., Lombardi F. 2011. Efficacy of acupuncture in preventing atrial fibrillation recurrences after electrical cardioversion. *J Cardiovasc Electrophysiol* 22(3):241–247.

Ma, K.-W. 1992. The roots and development of Chinese acupuncture: From prehistory to early 20th century. *Acupunct Med* 10:92–99.

Magden, E. R., Haller, R. L., Thiele, E. J., Buchl, S. J., Lambeth, S. P., and S. J. Schapiro. 2013. Acupuncture as an adjunct therapy for osteoarthritis in chimpanzees (*Pan troglodytes*). *J Am Assoc Lab Anim Sci* 52(4):475–480.

Magden, E. R., Mansfield, K. M., Simmons, J. H., and C. R. Abee. 2015. Nonhuman Primates. In *Laboratory Animal Medicine*, 3rd edition. Anderson, L. C., Otto, G., Pritchett-Corning, K. R., Whary, M. T., and J. Fox (eds). San Diego, CA: Elsevier Academic Press.

Magden, E. R., Sleeper, M. M., Buchl, S. J., Jones, R. A., Thiele, E., and G. K. Wilkerson. 2016. Use of an implantable loop recorder in a chimpanzee (*Pan troglodytes*) to monitor cardiac arrhythmias and assess effects of acupuncture and laser therapy. *Comp Med* 66(1):52–58.

Manyanga, T., Froese, M., Zarychanski, R., Abou-Setta, A., Friesen, C., Tennenhouse, M., and B. L. Shay. 2014. Pain management with acupuncture in osteoarthritis: A systemic review and meta-analysis. *BMC Complement Altern Med* 14:312–320.

Mavrommatis, C. I., Argyra, E., Vadalouka, A., and D. G. Vasilakos. 2012. Acupuncture as an adjunctive therapy to pharmacological treatment in patients with chronic pain due to osteoarthritis of the knee: A 3-armed, randomized, placebo-controlled trial. *Pain* 153(8):1720–1726.

McKinley, J., Buchanan-Smith, H. M., Bassett, L., and K. Morris. 2003. Training common marmosets (*Callithrix jacchus*) to cooperate during routine laboratory procedures: Ease of training and time investment. *J Appl Anim Welf Sci* 6:209–220.

McMillan, J. L., Perlman, J. E., Galvan, A., Wichmann, T., and M. A. Bloomsmith. 2014. Refining the pole-and-collar method of restraint: Emphasizing the use of positive training techniques with rhesus macaques (*Macaca mulatta*). *J Am Assoc Lab Anim Sci* 53:61–68.

Medrado, A., Pugliese, L. S., Reis, S. R. A., and Z. A. Andrade. 2003. Influence of low level laser therapy on wound healing and its biological action upon myofibroblasts. *Lasers Surg Med* 32:239–244.

Middlekauff, H. R., Yu, J. L., and K. Hui. 2001. Acupuncture effects on reflex responses to mental stress in humans. *Am J Physiol Regul Integr Comp Physiol* 280(5):R1462–R1468.

NIH (National Institutes of Health). 1997. Acupuncture. NIH Consensus Statement. November 3–5, 15(5):1–34. Accessed December 30, 2015. https://consensus.nih.gov/1997/1997acupuncture107html.htm.

Ogay, V., Min, F., Kim, K. H., Kim, J. S., Bae, K. H., Han, S. C., and K. S. Soh. 2009. Observation of coiled blood plexus in rat skin with diffusive light illumination. *J Acupunct Meridian Stud* 2(1):56–65.

Ouyang, H. and J. D. Z. Chen. 2004. Review article: Therapeutic roles of acupuncture in functional gastro-intestinal disorders. *Aliment Pharmacol Ther* 20:831–841.

Owen, Y. and J. R. Amory. 2011. A case study employing operant conditioning to reduce stress of capture for red-bellied tamarins (*Saguinus labiatus*). *J Appl Anim Welf Sci* 14(2):124–137.

Park, S. I., Sunwoo, Y. Y., Jung, Y. J., Chang, W. C., Park, M. S., Chung, Y. A., Maeng, L. S., et al. 2012. Therapeutic effects of acupuncture through enhancement of functional angiogenesis and granulogenesis in rat wound healing. *J Evid Based Complement Altern Med* 2012(464586):1–10.

Parr, L. A., Modi, M., Siebert, E., and L. J. Young. 2013. Intranasal oxytocin selectively attenuates rhesus monkeys' attention to negative facial expressions. *Psychoneuroendocrinology* 38:1748–1756.

Prescott, M. J. and H. M. Buchanan-Smith. 2003. Training nonhuman primates using positive reinforcement techniques. *J Appl Anim Welf Sci* 6:157–161.

Pryor, K. 1999. *Don't Shoot the Dog!: The New Art of Teaching and Training.* New York: Simon and Schuster.

Reamer, L. A., Haller, R. L., Thiele, E. J., Freeman, H. D., Lambeth, S. P., and S. J. Schapiro. 2014. Factors affecting initial training success of blood glucose testing in captive chimpanzees (*Pan troglodytes*). *Zoo Biol* 33:212–220.

Robinson, N. G. 2008. The scientific basis of acupuncture. Colorado Veterinary Medical Association, Medical Acupuncture for Veterinarians, Course Handout. Fort Collins, CO, Spring 2011.

Robinson, N. G. 2009. Making sense of the metaphor: How acupuncture works neurophysiologically. *J Equine Vet Sci* 29(8):642–644.

Robinson, N. G. 2014. Laser acupuncture: Keep it scientific. *Photomed Laser Surg* 32(12):1–2.

Rogge, J., Sherenco, K., Malling, R., Thiele, E., Lambeth, S., Schapiro, S., and L. Williams. 2013. A comparison of positive reinforcement training techniques in owl and squirrel monkeys: Time required to train to reliability. *J Appl Anim Welf Sci* 16:211–220.

Romoli, M., Allais, G., Airola, G., and C. Benedetto. 2005. Ear acupuncture in the control of migraine pain: Selecting the right acupoints by the "needle contact test." *Neurol Sci* 26:S158–S161.

Russell, J. L., Taglialatela, J. P., and W. D. Hopkins. 2006. The use of positive reinforcement training in chimpanzees (*Pan troglodytes*) for voluntary presentation for intramuscular injections. *Am J Primatol* 68:122.

Schapiro, S. J., Bloomsmith, M. A., and G. E. Laule. 2003. Positive reinforcement training as a technique to alter nonhuman primate behavior: Quantitative assessments of effectiveness. *J Appl Anim Welf Sci* 6(3):175–187.

Schapiro, S. J. and S. P. Lambeth. 2007. Control, choice, and assessments of the value of behavioral management to nonhuman primates in captivity. *J Appl Anim Welf Sci* 10(1):39–47.

Schapiro, S. J., Magden, E. R., Reamer, L., Mareno, M. C., and S. P. Lambeth. 2017. Behavioral training as part of the health care program. In *Management of Animal Care and Use Programs in Research, Teaching, and Testing*, 2nd edition. Weichbrod, R. H., Thompson, G. A., and J. N. Norton (eds). Boca Raton, FL: CRC Press.

Schapiro, S. J., Perlman, J. E., Thiele, E., and S. Lambeth. 2005. Training nonhuman primates to perform behaviors useful in biomedical research. *Lab Anim (NY)* 34:37–42.

VandeVoort, C. A. 2004. High quality sperm for nonhuman primate ART: Production and assessment. *Reprod Biol Endocrinol* 2(33):1–5.

VanWormer, A. M., Lindquist, R., and S. E. Sendelbach. 2008. The effects of acupuncture on cardiac arrhythmias: A literature review. *Heart Lung* 37(6):425–431.

Videan, E. N., Fritz, J., Murphy, J., Borman, R., Smith, H. F., and S. Howell. 2005. Training captive chimpanzees to cooperate for an anesthetic injection. *Lab Anim (NY)* 34(5):43–48.

White, A. and E. Ernst. 2004. A brief history of acupuncture. *Rheumatology* 43:662–663.

WHO (World Health Organization). 1997. Acupuncture: Review and analysis of reports on controlled clinical trials. http://apps.who.int/medicinedocs/pdf/s4926e/s4926e.pdf. Accessed February 7, 2017.

Yin, J. and D. Z. Chen. 2010. Gastrointestinal motility disorders and acupuncture. *Auton Neurosci* 157:31–37.

CHAPTER 14

The Veterinarian–Behavioral Management Interface

Eric Hutchinson
National Institutes of Health

CONTENTS

The role of veterinarians in the behavioral management of nonhuman primates (NHPs) used in research has both legal and historical precedent. In 1985, the Animal Welfare Act (AWA) was amended to ensure that facilities housing primates develop a plan for promoting their psychological well-being, and the act further specified that this plan should be directed by the attending veterinarian. Meanwhile, veterinarians made many direct and indirect contributions that would shape primate behavioral management in the years to come. Viktor Reinhardt and Kathryn Bayne, for example, were involved directly in behavioral management, developing techniques and setting industry standards that remain relevant to this day (Reinhardt et al. 1989; Bayne et al. 1992). Others, such as Michale Keeling, recognized the need for formal research in behavioral management techniques (Weed 2006), assembling teams of behavioral scientists who could build the same evidence-based standards that veterinary medicine aspires to. The involvement of veterinarians in primate behavioral management programs continues even today. A survey conducted in 2014 (Baker 2016) reported that 37% of primate environmental enhancement programs are directly overseen by a veterinarian. Yet there are many anecdotal reports of veterinary and behavioral management staff either failing to communicate well or being actively at odds with one another—concerns that are often voiced in informal gatherings of one group, where the other group is absent. Indeed, there are some characteristics of laboratory veterinarians, behavioral specialists, and their respective professional training that likely serve as natural barriers to effective communication and cooperation. By acknowledging these barriers and taking steps to mitigate them, especially in areas where the two groups most commonly interact, veterinarians and behavioral specialists can maximize their cooperation in service of enhancing the lives of the NHPs in their care.

The professional training received by the typical laboratory animal veterinarian has several built-in features that may serve to create knowledge gaps when it comes to primate behavior or its importance. For instance, while many veterinary students are required to take a single class on animal behavior, these classes are a very small part of a larger curriculum and tend to focus on issues specific to companion animals, such as leash aggression and inappropriate elimination. Likewise, once a veterinarian moves on to a residency, some of the training programs approved by the American College of Laboratory Animal Medicine (ACLAM) have small NHP collections, have limited direct clinical contact between trainees and primates, or rely on rotations to other institutions for primate clinical experience. Thus, through no fault of their own, laboratory animal veterinarian residents may not gain experience or fully engage with the sort of well-developed behavioral management program that has become common at the National Primate Research Centers and other large primate vivaria. Further, the ACLAM board examination reference list does not include any behavior-specific journals or texts, and one of the most commonly utilized literature-search databases for veterinary medicine, MEDLINE, does not index many of the key journals in which primate behavioral scientists publish, including *Animal Behaviour*, the *American Journal of Primatology,* and *Ethology*. This means that lab animal veterinarians, either when studying for their board exam or while trying to keep up with current knowledge and best practices following ACLAM certification, must actively search beyond their usual resources for information on primate behavior.

Beyond this potential lack of primate behavior-specific training for laboratory animal veterinarians, there are also structural challenges to cooperation between veterinarians and behavioral specialists. As discussed earlier, the AWA states that the attending veterinarian should oversee the environmental enhancement program for NHPs. Given this, it is unsurprising that Baker et al. (2007) found that all facilities surveyed had a veterinarian serving as the organizational head of the behavioral program. Nevertheless, 59% of these facilities also reported having a behavioral scientist provide primary oversight for the program. In its best permutations, such an arrangement can maximize harmony between veterinary medicine and behavior; the attending veterinarian gives input or serves as final mediator when behavioral program goals come into conflict with research or medical plans, but in large part, a behavioral specialist serves as the functional director of the behavioral program. In a worst-case scenario, the organizational authority and natural hierarchy established by the AWA mandate manifests as an attending veterinarian and perhaps, other clinical veterinarians, exerting influence over a behavioral management program without seeking out, or properly considering, the expertise of behavioral scientists or practitioners. A similar dynamic long existed in human operating rooms, with organizational authority given to surgeons over nurses and other practitioners in the room, which led to a culture lacking in two-way information exchange—a dynamic that has received much-deserved scrutiny in recent years as a significant contributor to medical mistakes (Lingard et al. 2004). Such an approach to primate management may result in the use of outdated techniques or treatments, or behavior and veterinary programs working at cross purposes. It is important for veterinarians to note that behavioral management is a complicated field by itself, not necessarily a subfield of veterinary medicine, and that even in areas where veterinary medicine and behavioral management overlap, behavioral scientists bring unique experience and expertise to bear. A good example of this can be found in the literature covering pharmaceutical treatment of self-injurious behavior (SIB) in primates (Tiefenbacher et al. 2005; Novak et al. 2014). Though veterinarians are the ones most commonly prescribing these medications, much of the most enduring work examining causes and treatments of the behavior have been driven primarily by behavioral scientists (Weld et al. 1998; Fontenot et al. 2009).

For their part, veterinarians, aware of these potential pitfalls and interested in engaging effectively with behavioral management, can take concrete steps to fill in the behavioral knowledge gaps sometimes left by their professional training. For areas of overlap between veterinary medicine and behavioral management, it is useful for veterinarians to familiarize themselves, and update themselves, with the behavior literature, beyond the typical laboratory animal medicine resources. This will require utilizing literature databases with good coverage of behavioral sciences (i.e., other than MEDLINE) and could include utilizing a professional information service, such as the Animal Welfare Information Center, or setting up a citation alert to send notifications of new publications on relevant topics. It should also include keeping abreast of behavior-relevant trends in laboratory animal medicine by discussing these topics with veterinary, behavior, and regulatory colleagues, since many "best practices" may not be discussed in the literature. For example, exemptions to socially housing NHPs that have long been common place and considered acceptable by most Institutional Animal Care and Use Committees, such as for infection with simian immunodeficiency virus, external surgical implants, or indwelling catheters, are now being challenged in many facilities. This has led to socially housing primates in a wide range of circumstances—changes in practice that have been slow to be reflected in the literature and regulatory guidelines. Perhaps the most obvious and best way for veterinarians to become educated in the theory and current best practices of primate behavioral management is by interaction with behavioral scientists. This can be achieved by attending conferences or workshops specific to the field of primate behavior or, more efficiently, leaning on the experience and knowledge of the behavioral specialists at one's own institution. Efforts to open communication between behavior and veterinary staff, such as by attending one another's rounds and/or setting up mutually attended special topic lectures, will necessarily encourage understanding of one another's fields and expertise, in addition to providing a convenient venue for coordination of effort.

Behavioral specialists obviously must also play a role in ensuring cooperation with the veterinarians in their facilities. In addition to taking part in veterinary rounds and clinical meetings as suggested, behavioral specialists can make plain the value of their expertise by sharing it in settings beyond day-to-day case management. This could include providing seminars for veterinarians and care staff, offering to participate in the training of any laboratory animal medicine residents or visiting veterinary students, or serving as a member or consultant to the Institutional Animal Care and Use Committee while demonstrating thorough knowledge of the behavior-relevant provisions of the AWA and the *Guide for the Care and Use of Laboratory Animals* (the *Guide*, NRC 2011).

The approach that behavioral specialists use when discussing cases or more general behavioral management topics with veterinarians can also influence how their message is perceived and, thus, how seriously it is considered. Like behavioral scientists, veterinarians are trained to give priority to evidence-based solutions, but as discussed, veterinarians may not necessarily be aware of the evidence supporting certain behavioral management solutions. Behavioral specialists should thus be prepared, and plan, to discuss supporting facts from the literature for ideas, solutions, or policies they propose. Despite being trained to consider scientific evidence, practitioners vary individually in the ways in which they use and prioritize that evidence to make decisions. While some prefer hard data while making decisions, and find it hard to be satisfied with any amount of such data, others rely more on intuition and are likely to make a decision based on "gut feel" after considering the evidence. This decision-making dichotomy is captured by the commonly used Myers Briggs personality inventory (Carlyn 1977; Johnson et al. 2009) as a preference for "sensation" or "intuition" and could lead to miscommunication or frustration when behavioral specialists and veterinarians with opposite preferences interact. For example, a "sensing" behavioral specialist, who prefers gathering as much information as possible and who has compiled literature, case examples, and hard data in order to convince a veterinarian to accept a particular approach, may be confused or frustrated when, at the end of the conversation, the "intuitive" veterinarian still asks, "do you think it will work?" Being aware of these personal preferences and being prepared to at least acknowledge the other person's approach are key to ensuring mutual understanding.

Another important characteristic of most veterinarians, and one that could potentially lead to conflict, is their sensitivity toward public questioning of clinical choices. During veterinary school, second guessing, or even directly questioning, clinical choices or diagnoses in the presence of a client or other nonveterinarian is considered taboo, and this cultural norm is reinforced both formally through instruction and informally by peer interactions. A behavioral specialist who is unaware of this may inadvertently do damage to a relationship with a veterinarian if they are not sufficiently diplomatic when questioning clinical choices, such as which drugs to use for abnormal behavior cases or even whether a case is primarily behavioral or medical in nature.

Another way that veterinarians and behavioral specialists can encourage teamwork and cooperation is by ensuring they use a "whole animal" approach to medical and behavioral management. Such an approach has been encapsulated within the realm of animal ethics as the concept of *telos*. *Telos*, first introduced by Aristotle, is the idea that all things have a base purpose or nature. As applied to animal well-being—as proposed in Rollin's (1981) "Animal Rights and Human Morality"—*telos* is the idea that ethical animal use must seek to do more than simply minimize pain and suffering, instead it should aim to maintain an animal's basic nature, for example, the "monkeyness" of the monkey. Whether one prefers the term *telos* or the more familiar concept of the "whole animal," support for this general framework can be found in increasing measure through the evolution of regulations and guidelines governing laboratory animal use. When the 1985 amendment to the AWA added psychological well-being to pain and suffering on the list of things that must be considered for ethical primate research, it set the stage for considering more than just physical suffering or pain when discussing animal use. Though they did not explicitly refer to the concept of *telos*, the rules issued by USDA in 1991 requiring extra space for brachiating species could only be explained by a deference to these species' basic natures; they were made to brachiate, a fact that is obvious when noting their various physical adaptations for the activity; therefore, they should be provided with opportunities to brachiate. The 1996 edition of the *Guide for the Care and Use of Laboratory Animals* (NRC 1996) outlined the need to house animals "with a goal of maximizing species-specific behaviors and minimizing stress induced behaviors," again appealing to the animal's nature in guiding ethical care. Prohibitions against prolonged restraint and removal of canine teeth for anything other than dental disease are other examples of the ways in which the concept of *telos* is being applied, purposefully or not, to the ethical guidelines governing laboratory animal use.

Certainly, the concept of considering the "whole animal" when making clinical decisions is hardly foreign to laboratory animal veterinarians, as it is a concept taught in texts and lectures throughout veterinary school and beyond. Indeed, being able to predict the many ways an experimental manipulation will affect an animal is one of the primary duties of a laboratory veterinarian when reviewing research protocols or consulting with scientists. However, much of the actual practice of veterinary medicine is learned through the heuristic of animal-as-interconnected-organ-systems. This approach is further reinforced by specialized training in laboratory animal medicine; if one cannot be sure of the species of animal they need to care for on a given day, it makes sense to be prepared to study each animal for their component parts that are shared and similar among most species. It is easy to observe how, over time, veterinarians might lose sight of the importance of considering the monkeyness of a monkey in their decision-making processes, especially since teleological and "systems" approaches will lead to the same conclusion in many cases. A primate with a serious bacterial infection should receive antibiotics, regardless of whether one is focused on the effects of the infection on its organs or on its monkeyhood. The teleological approach, however, has many implications around the margins of the more obvious decisions. A monkey in a social group that is noted with a single day of mild diarrhea, for instance, may be removed from the social group for treatment, if the veterinarian only considers the impact of diarrhea on the animal's many organ systems. A teleological approach, on the other hand, would suggest that mild diarrhea is less a threat to the animal's overall well-being than removing it from its social group. Preserving *telos* does not imply that behavioral factors should be considered above all others; in the example given earlier, diarrhea may obviously progress in any number of ways to the point where the health of the animal is more of a concern than its continued social interactions. It does, however, provide a framework in which the duty of a veterinarian is elevated above the simple treatment of disease and management of pain and in which psychological well-being is considered on equal footing with physical health, freedom from pain, and experimental aims. Notably, the most recent edition of the *Guide* gives credence to this equal footing, referring nearly as many times to "stress" (132) as it does to "pain" (150) (NRC 2011).

Having discussed many of the ways in which harmony between veterinary medicine and behavioral management can be achieved, we can discuss how these concepts might apply to some of the areas where veterinarians and behavioral specialists interact most frequently. Environmental enrichment is one such area, where cooperation can maximize benefit to the animals, while minimizing conflict among the people caring for them. The AWA requires an environmental enhancement plan to cover all nonhuman primates but leaves facilities to determine the individual components that constitute the plan. By working together, veterinarians and behavioral specialists can strike a balance between enriching primates' enclosures and maintaining a safe environment, where necessary research and medical interventions can be completed. Many facilities achieve this by forming an enrichment committee that includes veterinarians and behavioral specialists and often includes care personnel and facility managers as well. It is important that all the members involved are committed to the goal of achieving whatever is reasonable to enhance the lives of primates. It is all too easy for such a committee to serve as a place for enrichment ideas to be nit-picked to death, but by establishing a few working principles for the group, this trap can be avoided. One such principle could be that new ideas of one kind or another will be implemented as a default. This "bias-for-action" (Keeling et al. 1991) frees a committee from adjudicating endlessly whether change should occur and instead allows it to focus on shaping and selecting which changes to implement. Another principle could be that convenience for the people working with primates will be considered when choosing new enrichments, but it should not serve as a break point for ideas of high enrichment value. A final principle could be that real safety concerns will serve as a potential veto point for any enrichment idea, but that the danger will be clearly articulated and an honest risk assessment will be made by the group. There are very few enrichments for which a potential health or safety hazard is inconceivable, but weighing the potential and seriousness of those hazards against their benefit is a crucial task in making progress toward more enriched environments and better preservation of *telos*. When an enrichment item in practice does cause a safety or health concern, a full accounting of the failure should take place, rather than a knee-jerk reaction to discontinue the enrichment item. Things to consider in this analysis should include the frequency of the hazardous consequence relative to the amount of exposure animals have had to the enrichment; for example, if a plastic toy has been in use for many years, enjoyed by thousands of primates, and was chewed to a dangerously sharp point exactly once, discontinuing the use of that toy without finding a functional replacement is less reasonable than for an experimental enrichment that immediately broke and left sharp edges. In any case, as the field of laboratory primate behavioral management evolves, so should environmental enhancement plans, and the manner in which that evolution can be achieved should involve cooperation between veterinarians and behavioral specialists.

Primate socialization is another obvious area where behavioral management and veterinary medicine must necessarily interact, similar to the enrichment example above. With the *Guide* (NRC 2011) putting forward social housing as the default condition for primates and other social animals, the ability to cooperate during the socialization process has become more important than ever, and there are several easy steps that can be taken to maximize this prospect. Discussing socialization prior to beginning is one of the simplest ways to avoid misunderstandings between veterinarians and behavioral specialists. All involved should discuss the relative risks, such as whether it is a pairing of juveniles, with high expectations of success, or of adult males, where serious fighting is more of a concern. There should be understanding of the short- and long-term goals of socialization; there may be more flexibility in options for socializing a group of animals going through quarantine prior to being used on experiment than there is for a planned breeding pair of known genetic makeup. A mutually beneficial time for socialization should be discussed as well, ensuring adequate veterinary coverage for socialization steps with the risk of wounding. Perhaps most importantly, thresholds for intervention should be discussed. Regardless of their commitment to a teleological approach to medicine, veterinarians will vary widely in their tolerance for fights, minor wounds, and other clinical problems, such as diarrhea, during the socialization process. Once a socialization is underway, the emotional weight of seeing a laborious process end due to clinical intervention can lead to frustration on the part of behavioral specialists and unproductive conflict with veterinarians. By discussing clinical "endpoints" beforehand, when the socialization is still an abstract plan, a more reasonable discussion can take place, ensuring that endpoints are properly weighed against the goals and potential teleologic benefits of a successful socialization. In facilities where multiple veterinarians have the potential to intervene, discussions with the attending veterinarian in advance of a socialization event that set firm (ideally in writing) endpoints are of increased importance and can prevent arbitrary and inconsistent clinical interventions later. When setting these parameters, it is important for all to note that clinical interventions often beget further clinical interventions, especially in the case of group socializations, where removing one animal may upset a delicate or nascent dominance hierarchy and lead to group conflict (see McCowan and Beisner 2017). As important as communication prior to a socialization is, further discussion following a socialization can be just as impactful. For instance, allowing all involved to discuss the reasons that they think a socialization failed can reveal which factors veterinarians and behavioral specialists see as relevant to the process and can lead to better understanding and practices for future socializations. Was there disruptive human activity going on in the room at the time a pair broke down? Perhaps that can be avoided the next time by enhanced communication. Was a fight allowed to progress too far, requiring surgical repair? The behavioral specialist may intervene sooner the next time in the face of observed antagonism. These informal, post hoc analyses can be very effective tools for improving the socialization process and enhancing cooperation among those involved.

The treatment of abnormal or undesired behaviors is an area of interaction where veterinary medicine and behavioral management each have their own distinct approaches. Each of these approaches in turn has distinct strengths, weaknesses, and areas of greatest influence, such that they may either be complementary or in conflict, depending upon the ability of veterinarians and behavioral specialists to coordinate their efforts. This becomes apparent when we examine the sequence of events that leads to the expression of an abnormal behavior, as represented in a very simplified manner in Figure 14.1. First, an external stimulus is perceived by the animal. That stimulus is then processed by the animal, producing emotional and/or physical changes, such as anxiety. This "internalization" step determines whether and how the animal will respond behaviorally. This behavioral response then leads to feedback, which goes back to be perceived and internalized by the animal, increasing or decreasing the likelihood of the behavioral sequence recurring in the future.

On its own, veterinary medicine's approach to treating abnormal behaviors is limited mostly to the internalization step in this sequence. This is certainly a "target-rich" step in the process that produces abnormal behavior. There are a number of drug therapies, each with their own literature support, strengths, and weaknesses, available to decrease the likelihood that the internalization of a stimulus will produce a behavior. Benzodiazepines, such as diazepam, may decrease the salience of an external, anxiety-producing stimulus (Tiefenbacher et al. 2005; Major et al. 2009). Treatments like guanfacine or those targeting serotonin signaling, like tryptophan and fluoxetine, may influence baseline temperament or anxiety thresholds, so anxiety is less likely to produce abnormal behavior (Weld et al. 1998; Freeman et al. 2015b). Veterinary medicine may also exert influence on the "feedback" step; naltrexone blocks internal opioid signaling that is thought to serve as an internal reinforcer in cases of self-injury, making self-injury less likely in the future (Kempf et al. 2012). These therapies, however, cannot influence the presence of external stimuli. Additionally, veterinary treatments cannot easily

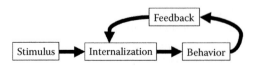

Figure 14.1 A simplified representation of the sequence of events that can lead to abnormal behavior.

influence the results of a behavior within ethical boundaries. For example, Elizabethan collars; dental interventions, such as canine blunting or removal; and physical or chemical restraint may all prevent the actual damage caused by self-injurious behavior, but all violate the principle of *telos* without addressing the underlying cause or distress associated with an abnormal behavior.

In contrast, behavioral management can influence all steps in the sequence. One of the first steps in the behavioral management of an abnormal behavior is to identify possible inciting stimuli, such as the presence of a person holding a syringe. Once identified, management steps can be taken to limit contact with that stimulus, or influence how it is perceived and processed by the animal, through techniques such as desensitization. Behavioral management can also influence the internalization stage, by the prevention of pathological states, for example, by ensuring social housing of juveniles and reducing the risk of serotonergic dysfunction (Kinnally et al. 2008). Behavioral management can also directly influence the expression of a behavior by conditioning an animal to respond in a way that is incompatible with the abnormal or undesired behavior; for example, in cooperative feeding training, the undesirable behavior of fighting over food is replaced with the behavior of physically moving to a spot separate from social partners (Laule et al. 2003). Finally, behavioral management can influence some of the feedback mechanisms by which abnormal behaviors are reinforced. An animal that sham bites its wrist in response to a direct stare, for instance, often learns to do so because it causes the stare to be broken off. In this case, the dual approach of limiting the stimuli by training people not to stare, and limiting positive feedback by teaching people not to look away immediately in response to the sham bite, would affect both the stimulus and feedback steps in the sequence. When used in concert to address abnormal behaviors, veterinary and behavioral management techniques can be synergistic, accomplishing more than either alone. For instance, the identification of a discrete stimulus triggering a behavior can guide veterinary treatment; primates afraid primarily of a once-every-2-weeks cage transfer process may be successfully managed with only a short acting anxiolytic, like midazolam, given just prior to the event. An animal identified by the behavioral staff as at risk for self-injurious behavior (SIB) due to abnormal rearing may be more likely to respond to treatments targeting serotonin signaling. Further, pharmacologic management of anxiety may actually facilitate desensitization, allowing the animal to tolerate more aversive stimuli as part of the positive reinforcement training process. In addition, both veterinary and behavioral inputs are useful in determining whether a treatment method is effective. For an abnormal behavior that is only expressed in certain contexts and that results in no physical symptoms, it may be necessary to enlist a behavioral specialist to determine whether a particular drug therapy is working. Likewise, in the case of infrequent SIB, wounds may be tracked as the only measurable way to determine whether a behavioral intervention is effective (Freeman et al. 2015a).

Although there may be many natural barriers to cooperation between veterinarians and behavioral specialists, the fields of laboratory animal medicine and behavioral management need not be at odds. By instituting practices that encourage open communication and mutual respect, these two groups can work together in many of the areas where both have influence on the lives of the primates in their care, whether it be socialization, enrichment, or the treatment of abnormal behaviors. Ultimately, a positive relationship between the veterinarians and behavioral specialists at an institution has the power to improve the lives of laboratory primates.

REFERENCES

Animal Welfare Act as Amended. 2007. 7 USC §2131–2159.

Baker KC. Survey of 2014 behavioral management programs for laboratory primates in the United States. *Am. J. Primatol.* 2016. doi:10.1002/ajp.22543.

Baker KC, Weed JL, Crockett CM, and Bloomsmith MA. Survey of environmental enhancement programs for laboratory primates. *Am. J. Primatol.* 2007;69:377–394.

Bayne KA, Hurst JK, and Dexter SL. Evaluation of the preference to and behavioral effects of an enriched environment on male rhesus monkeys. *Lab Anim Sci.* 1992;42(1):38–45. Erratum in: *Lab Anim Care* 1992;42(5):528.

Carlyn M. An assessment of the Myers-Briggs type indicator. *J Pers Assess.* 1977;41(5):461–73.

Fontenot MB, Musso MW, McFatter RM, and Anderson GM. Dose-finding study of fluoxetine and venlafaxine for the treatment of self-injurious and stereotypic behavior in rhesus macaques (*Macaca mulatta*). *J Am Assoc Lab Anim Sci.* 2009;48(2):176–84.

Freeman ZT, Krall C, Rice KA, Adams RJ, Metcalf-Pate KA, and Hutchinson EK. Severity and distribution of wounds in rhesus macaques (*Macaca mulatta*) correlate with observed self-injurious behavior. *J Am Assoc Lab Anim Sci.* 2015a;54(5):516–20.

Freeman ZT, Rice KA, Soto PL, Pate KA, Weed MR, Ator NA, DeLeon IG, et al. Neurocognitive dysfunction and pharmacological intervention using guanfacine in a rhesus macaque model of self-injurious behavior. *Transl Psychiatry.* 2015b;5:e567. doi:10.1038/tp.2015.61.

Johnson SW, Gill MS, Grenier C, and Taboada J. A descriptive analysis of personality and gender at the Louisiana State University School of Veterinary Medicine. *J Vet Med Educ.* 2009 Fall;36(3):284–90. doi:10.3138/jvme.36.3.284.

Keeling ME, Alford PL, and Bloomsmith MA. Decision analysis for developing programs of psychological well-being: A bias-for-action approach, pp. 57–65. In: *Through the Looking Glass. Issues of Psychological Well-Being in Captive Nonhuman Primates.* Novak MA and Petto AJ, (eds). Washington, DC: American Psychological Association. 1991.

Kempf DJ, Baker KC, Gilbert MH, Blanchard JL, Dean RL, Deaver DR, and Bohm RP Jr. Effects of extended-release injectable naltrexone on self-injurious behavior in rhesus macaques (*Macaca mulatta*). *Comp Med.* 2012;62(3):209–17.

Kinnally EL, Lyons LA, Abel K, Mendoza S, and Capitanio JP. Effects of early experience and genotype on serotonin transporter regulation in infant rhesus macaques. *Genes Brain Behav.* 2008;7(4):481–6.

Laule GE, Bloomsmith MA, and Schapiro SJ. The use of positive reinforcement training techniques to enhance the care, management, and welfare of primates in the laboratory. *J Appl Anim Welf Sci.* 2003;6(3):163–73.

Lingard L, Espin S, Whyte S, Regehr G, Baker GR, Reznick R, Bohnen J, Orser B, Doran D, and Grober E. Communication failures in the operating room: An observational classification of recurrent types and effects. *Qual Saf Health Care.* 2004;13(5):330–4.

Major CA, Kelly BJ, Novak MA, Davenport MD, Stonemetz KM, and Meyer JS. The anxiogenic drug FG7142 increases self-injurious behavior in male rhesus monkeys (*Macaca mulatta*). *Life Sci.* 2009;85(21–22):753–8. doi:10.1016/j.lfs.2009.10.003.

McCowan B and Beisner B. Utility of systems network analysis for understanding complexity in primate behavioral management, Chapter 11. In: *Handbook of Primate Behavioral Management.* Schapiro SJ (ed.). Boca Raton, FL: CRC Press. 2017.

National Research Council. *Guide for the Care and Use of Laboratory Animals*, 7th edition. Washington, DC: National Academy Press. 1996.

National Research Council. *Guide for the Care and Use of Laboratory Animals*, 8th edition. Washington, DC: National Academy Press. 2011.

Novak MA, El-Mallah SN, and Menard MT. Use of the cross-translational model to study self-injurious behavior in human and nonhuman primates. *ILAR J.* 2014;55(2):274–83. doi:10.1093/ilar/ilu001.

Reinhardt V, Houser D, and Eisele S. Pairing previously singly caged rhesus monkeys does not interfere with common research protocols. *Lab Anim Sci.* 1989;39(1):73–4.

Rollin BE. *Animal Rights and Human Morality.* Buffalo, NY: Prometheus Books. 1981.

Tiefenbacher S, Fahey MA, Rowlett JK, Meyer JS, Pouliot AL, Jones BM, and Novak MA. The efficacy of diazepam treatment for the management of acute wounding episodes in captive rhesus macaques. *Comp Med.* 2005;55(4):387–92.

Weed, JL and O'Neill-Wagner PL. Animal behavior research findings facilitate comprehensive captive animal care: The birth of behavioral management. In: *Animal Welfare Information Center, Environmental Enrichment for Nonhuman Primates Resource Guide.* 2006. Available at https://awic.nal.usda.gov/environmental-enrichment-nonhuman-primates-resource-guide-animal-behavior-research-findings. Accessed February 21, 2017.

Weld KP, Mench JA, Woodward RA, Bolesta MS, Suomi SJ, and Higley JD. Effect of tryptophan treatment on self-biting and central nervous system serotonin metabolism in rhesus monkeys (*Macaca mulatta*). *Neuropsychopharmacology.* 1998;19(4):314–21.

CHAPTER 15

Social Learning and Decision Making

Lydia M. Hopper
Lincoln Park Zoo

CONTENTS

INTRODUCTION

Experimental research of primate social learning is of academic interest as it allows for comparisons to be made between our own cultural world, which relies on social learning, and that of other species. Beyond this academic interest, can we apply primates' abilities to learn from one another to assist with their management in a captive setting? I believe so, and in this chapter, I outline the important facets of social learning, as well as the ways in which colony managers, trainers, and caregivers can implement what we know about primate social learning into daily practice. First, I describe social learning; second, I explain how we test it; and finally, I describe the conditions that promote it. After providing a theoretical overview, I then discuss how primates' abilities to socially learn from one another might aid their captive management. Specifically, I focus on how social learning by primates can be applied to training and to cultivate positive cultures. Although I discuss primates' abilities to learn from others in general terms, I also highlight that there appears to be some variance in different individual's and species' ability to learn socially, and that this might be further impacted by the social dynamics of the group in which the primates are housed, all of which should be considered by managers hoping to promote social learning in the daily lives of their animals.

WHAT IS (SOCIAL) LEARNING?

Shettleworth (1998) defines learning as a "change in state resulting from experience [which results in] a change in cognitive state, not just behavioral capacity" (p. 100). Furthermore, she notes that the outcome of learning can be inferred from changes in an animal's behavior but cautions that changes in the behavior do not always reflect learning. The flexibility of primates' learning specifically has been studied since the first half of the 20th century. Primatologists including Robert Yerkes, Wolfgang Köhler, and Harry Harlow, who conducted systematic and objective testing of primate cognition and behavior, observed that primates were flexible in their learning and did not simply follow conditioned responses (Köhler, 1925; Suomi & Leroy, 1982). Beyond learning individually via a trial-and-error method, primates are also capable of learning by observing others. This so-called *social learning* enables individuals to avoid the potential costs of individual learning (Bandura, 1977) and possibly gain the most relevant information that is currently in use by their group mates (Hoppitt & Laland, 2013). Some researchers have proposed that primates are only capable of copying behaviors that they could also invent by themselves (Köhler, 1925; Tennie et al., 2009), whereas others note that, by definition, social learning allows primates to gain new skills that they could not discover themselves, except in limited circumstances (e.g., if the individual is a "rare innovator," c.f. Hopper et al., 2007).

Social learning incorporates learning that arises from direct observation of another's actions (observational learning) and learning from the environmental changes caused by another's actions (emulative learning, Tennie et al., 2010; Caldwell et al., 2012). Consider a situation where, for example, if you offered two monkeys that are housed together in the same enclosure a choice of two novel foods: one of which tasted pleasant and a second that did not. The first monkey will have to try the foods by relying on its own trial-and-error learning to discover which of the two foods was palatable. The second monkey, however, could wait and learn from the experiences of the first to potentially avoid eating the unpleasant food (i.e., via social learning). The second monkey could either learn by observing the first monkey's preferential eating of one of the foods and then selecting that food itself (i.e., observational learning) or by seeing which food was depleted or by smelling the food in the breath and on the hair of the first monkey (i.e., emulative learning; Galef, 2001 provides detailed methods of such an experiment run with rodents). For a number of primate species, simply the presence of group mates can encourage them to eat novel foods (*Cebus apella*, Visalberghi & Addessi, 2000; Addessi & Visalberghi, 2001) or even less-preferred foods (*Pan troglodytes*, Hopper et al., 2011; Finestone et al., 2014). In these examples, the primates are not learning *how* to eat, but rather, they are socially learning *what* to eat (Hopper et al., 2015). Specifically, a familiar behavior is elicited by the actions of others (in this case, eating); a response termed as "social facilitation" (Zajonc, 1965) or "response facilitation" (Hoppitt et al., 2007).

Beyond encouraging the sampling of novel foods, social learning enables primates to learn a range of novel behavioral skills by observing others, including how to access food in foraging puzzles (e.g., *C. apella*, Dindo et al., 2008; *Saimiri boliviensis*, Hopper et al., 2013); how to order icons presented in a serial learning task (e.g., *Macaca mulatta*, Subiaul et al., 2004); and how to make and use tools (e.g., *P. troglodytes*, Whiten et al., 2005; Price et al., 2009). Given this, social learning has been dubbed a "second inheritance system" (Whiten, 2005) as it facilitates the transmission of behavioral traits among individuals within a group, independent of genetic inheritance.

Experimental studies of social learning by captive primates have used four key techniques to record the transmission of information between individuals and down the "generations": dyadic interactions, diffusion chains, replacement method, and open diffusion paradigms (each one is depicted in Figure 15.1). In all, researchers observe and record the transmission of a behavior performed by a specific individual to one or more naïve individuals that have the opportunity to observe the actions of the first.

In *dyadic interactions*, pairs of animals are tested together. One animal (the model) is trained out of sight of its partner to perform a novel behavior. The model animal is then reintroduced to its naïve partner to determine if the naïve individual will replicate the actions performed by the trained model (e.g., *M. mulatta*, Subiaul et al., 2004; *P. troglodytes*, Hopper et al., 2008). Expanding on this simple technique, in a *diffusion chain* paradigm (akin to the telephone game), an initial model animal is trained (A) and paired with a second naïve individual (B). If the second animal (B) learns from the trained model (A), the observer becomes the model and is then paired with another naïve individual (C) (e.g., *P. troglodytes*, Horner et al., 2006; *C. apella*, Dindo et al., 2008). This pair-wise process continues until all animals have been exposed to the introduced behavior. An extension of the diffusion chain is the *replacement method* in which small groups, instead of individuals, are successively exposed to trained individuals along a series of experimental "generations" (e.g., *P. troglodytes*, Menzel et al., 1972). Thus, individual A is trained and is introduced to individuals B and C. Individual A is then removed from the group and individual D is added to the pair, so the group now comprises individuals B, C, and D. This continues until the make-up of the group does not contain any "founder" individuals. In contrast to these three highly controlled methods, the *open diffusion* paradigm represents a more naturalistic method. Out of sight of its group, one animal is trained to perform the novel behavior and then is reintroduced to its group to determine if the members of their group will learn by observing the trained individual (*P. troglodytes*, Whiten et al., 2005; *S. boliviensis*, Hopper et al., 2013). Thus, all group members are simultaneously exposed to the new "invention."

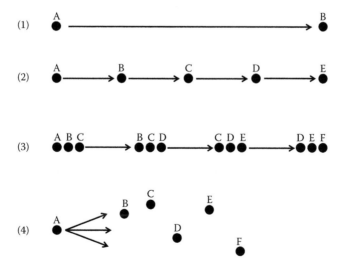

Figure 15.1 Experimental techniques used to test for primate social learning: (1) dyadic interactions, (2) diffusion chains, (3) the replacement method, and (4) open diffusion paradigms. In this figure, each dot represents an individual primate and the arrows show possible information transmission between individuals. For example, in the dyadic interaction (1), individual A is proficient in a specific behavior and individual B learns this behavior by observing A, thus, the behavior is transmitted from A to B.

Teaching

Although primates are capable of learning new skills via social learning, it should be noted that they do not typically engage in active teaching. Outside of a few reports that mothers encourage their offspring when learning new skills (e.g., *P. troglodytes*, Boesch, 1991; *Macaca nemestrina*, Maestripieri, 1996; see Maestripieri, 1995 for a review), primates do not appear to slow down or repeat their actions so that others can learn by observing them, even if their actions do promote learning by observers [although Humle & Snowdon (2008) provide a convincing case of teaching by tamarins, *Saguinus oedipus*]. This is critically important because according to the definition of teaching by Caro and Hauser (1992),

> an individual actor A can be said to teach if it modifies its behavior only in the presence of a naive observer, B, at some cost or at least without obtaining an immediate benefit for itself. A's behavior thereby encourages or punishes B's behavior, or provides B with experience, or sets an example for B. As a result, B acquires knowledge or learns a skill earlier in life or more rapidly or efficiently than it might otherwise do, or that it would not learn at all (p. 153).

Although active teaching appears to be rare among primates, perhaps because the need for targeted instruction is rare (Byrne & Rapaport, 2011), certain species are very tolerant of allowing group members to observe them from close proximity, which encourages social learning. This has been termed as "master-apprenticeship" (Matsuzawa et al., 2001). Although not the focus of this chapter, differentiating active teaching from social learning is important. They are clearly not mutually exclusive, but it is important to note that social learning does not require the expert or model animal to intentionally train the naïve observer, nor does it require the animals to understand the mental states of one another—the transmission of information can be passive on the part of the expert. (For more detailed discussions of teaching by nonhuman animals, see Thornton & Raihani, 2008 and Byrne & Rapaport, 2011).

FACTORS THAT INFLUENCE SOCIAL LEARNING AMONG PRIMATES

Considering the Species

Whether, and what, primates can socially learn has been studied extensively with both captive and wild populations. It was originally proposed that apes are more adept at social learning than monkeys (Fragaszy & Visalberghi, 2004); however, it is now known that both apes and monkeys appear equally good at social learning, although there is variation both within and across species. This variation in social learning abilities reported for species is likely due to the differences in experimental protocols, differences in environment and husbandry practices, differences in personalities, and varying difficulty in tasks. To date, social learning has been demonstrated by all four great ape species (*Pan paniscus*, Tomasello et al., 1993; *Gorilla gorilla*, Stoinski et al., 2001; *P. troglodytes*, Whiten et al., 2005; *Pongo pygmaeus*, Dindo et al., 2011), Old World monkeys (e.g., *Chlorocebus pygerythrus*, van de Waal et al., 2013; *Papio ursinus*, Carter et al., 2014), New World monkeys (e.g., *Callithrix jacchus*, Caldwell & Whiten, 2003; *Saimiri sciureus*, Claidière et al., 2013), and prosimians (e.g., *Lemur catta*, Kendal et al., 2010; *Varecia variegata*, Stoinski et al., 2011).

Although the social learning abilities of primates have been well documented, much of the research has focused only on a few species, including chimpanzees (*P. troglodytes*, e.g., Bonnie et al., 2007; Hopper et al., 2015), capuchins (*C. apella* e.g., Brosnan & de Waal, 2004), marmosets (*C. jacchus*, e.g., Caldwell & Whiten, 2003; Burkart et al., 2012), and squirrel monkeys (*S. sciureus*, Claidiere et al., 2013; *S. boliviensis*, Hopper et al., 2013). Given the large number of macaques housed in captivity, especially *M. fascicularis* and *M. mulatta*, it is surprising how few studies have been conducted on the investigation of social learning capabilities of macaques. One of the exceptions to this is a study performed by Subiaul et al. (2004) on rhesus macaques. They showed that the monkeys could learn by observing another solve a serial order task presented on a touchscreen. Outside of a captive setting, however, the social learning abilities of *Macaca* species have been inferred from the observations of wild Japanese macaques (*M. fuscata*), which are thought to learn socially grooming techniques, stone handling play behavior, and famously, food processing techniques. More research is required to untangle the similarities and differences across species' abilities to socially learn in a captive setting.

Considering the Social Landscape

Beyond interspecies differences in social learning abilities, intraspecies differences have also been reported. Therefore, we must also consider the social landscape in which animals live when assessing social learning. For the highly despotic rhesus macaque (*M. mulatta*), an individual's rank within its group can determine if it can observe social information and also whether it is able to use that information. This interplay was demonstrated elegantly in a study by Drea and Wallen (1999), who tested macaques with a discrimination task in which different colored boxes signaled the presence of food rewards (peanuts). When tested in a social setting with more dominant group members, subordinate monkeys failed to learn the discrimination. However, when they were tested only with other similarly low-ranked individuals, the monkeys were able to learn the discrimination task as proficiently as dominant monkeys. Instead of concluding that subordinates were unable to learn the task in the presence of dominant monkeys, the authors concluded that the low-ranking monkeys suppressed their performance in the presence of dominant group mates. Similarly, low-ranking baboons (*Papio papio*) tested in a group setting and presented with a test of inhibitory control on touchscreen computers performed less well when dominant group mates were nearby (Huguet et al., 2014).

These studies highlight the ways in which a primate's *individual* learning performance is influenced by its social environment, and research has shown that the social environment also influences primates' *social* learning. For example, if the inventor of a new behavior is low-ranking or peripheral, others in its group may not notice the invention, making it unlikely that the new behavior will spread within the group. This is most likely because primates do not monitor the behavior of low-ranking individuals (*P. troglodytes*, Biro et al., 2003). Conversely, if the inventor is high-ranking, low-ranking individuals may not be able to get close enough to learn about inventions; the dominant individual may monopolize access to an area of food source, or the low rank of a subordinate may inhibit its ability to exploit the invention (*P. troglodytes*, Cronin et al., 2014a).

Of course, primates' social learning can be simultaneously influenced by multiple factors and predicting who will learn from whom is difficult. However, it is likely that those individuals that have stronger affiliative relationships will be more likely to learn from one another (Bonnie & de Waal, 2006), if only because their increased time in proximity to one another will also increase opportunities for social learning. To document the transmission of novel behaviors within a group of primates, researchers have applied social network analyses, which can also reveal whether transmission occurs between more closely or less closely connected individuals. Social network analyses allow researchers to define and describe the social dynamics within a group using a variety of different metrics. Such networks can either describe the affiliative relationships among primates within a social group, by using data such as time spent in contact or time spent grooming, or the agonistic relationships within a group, by including measures of aggression and fighting (for a more detailed discussion of social network analysis, see McCowan & Beisner, 2017). In a study with captive squirrel monkeys (*S. sciureus*), Claidière et al. (2013) reported that monkeys that were more centrally integrated within their group, as identified by social network analysis, adopted the novel foraging technique of a trained group mate more rapidly than peripheral members of the group. Furthermore, the social network analysis revealed that the monkeys were likely to observe and learn from those individuals with whom they had strong affiliative bonds. Using comparable methods, researchers studying a wild group of chimpanzees in Uganda were able to track the transmission of a new "moss sponging" behavior within the group (Hobaiter et al., 2014). As has been reported for squirrel monkeys (Claidière et al., 2013), Hobaiter et al. (2014) demonstrated that closely affiliated chimpanzees learned from one another.

Outside of an experimental context, social network analyses have also allowed researchers to differentially define the strength of bonds among members within groups of captive primates. Specifically, for researchers studying four groups of sanctuary-housed chimpanzees in Zambia, social network analysis revealed that the groups were distinct in how interconnected individuals within each community were, such that two of the groups were highly interconnected and two were not (Cronin et al., 2014b). Interestingly, these group differences in social "culture," possibly influenced by the personality and leadership style of their alpha males, also positively correlated with responses on a test of social tolerance in which the groups were presented with a highly desired and easily monopolized food resource. Chimpanzees from the highly interconnected groups gathered in closer proximity around the food than those from the other two groups (Cronin et al., 2014b). This increased social tolerance might, in turn, increase the probability of social learning in other contexts.

Considering the Individual

By definition, social learning requires a social interaction, but whether learning occurs is also highly dependent on the individual characteristics of the naïve animal. Social learning is as much about an individual learning the new behavior as it is about the social group from whom it is learning. An individual's personality, rank, or relationship to its group mates all may influence its likelihood of using social information. This was demonstrated through a series of experiments conducted with wild Chacma baboons (*P. ursinus*) in Namibia. Carter et al. (2012) first tested the baboons' personalities by presenting them with novel foods (to assess "boldness") and with a taxidermic venomous snake (to assess "anxiety"). In a second study, these personality ratings were used to determine the ways in which the baboons' personality influenced their use of social information. To do this, researchers tested the baboons in two studies of social learning: one in which the monkeys could observe others eating a novel food and a second in which they could observe a group mate extract food hidden in a small cardboard box (Carter et al., 2014). Juvenile baboons spent longer times watching demonstrations than adults and appeared to be more likely to evidence social learning than adults. Specifically, juveniles manipulated the novel food for longer and showed greater persistence in the hidden-food task after observing a model than adults did. Considering the baboons' personality (when controlling for age), less anxious baboons spent longer times watching group mates interacting with the novel foods while conversely, bolder and more anxious animals were more likely to apply the social information, interacting more with the novel food after a demonstration. Thus, the individuals' age and personality influenced the likelihood of looking at others for information, as well as whether the animals then used that information (for a more detailed discussion of primate personality and temperament, see Capitanio, 2017 and Coleman, 2017).

The study by Carter et al. (2014) highlights that simply providing primates with an opportunity for social learning is not sufficient to ensure that information transmission occurs—certain individuals are more likely than others to observe and use the social information available. Indeed, such variability in copying is not just restricted to baboons but has been reported for a number of primate species. Furthermore, it is not just age and personality that predict whether an animal will copy others, rather, a suite of traits underlie these "transmission biases" (Rendell et al., 2011). Low-ranking, captive chimpanzees, for example, appear more likely than dominant ones to use social information (Kendal et al., 2015). Kendal et al. (2015) also showed that chimpanzees that were presented with a novel foraging device were more likely to rely on social information (i.e., watch and copy others) if they were uncertain how to solve the task. Thus, it might be imagined that in circumstances where animals are unsure what to do, they will look at others; whereas if they already have a solution, they will be less likely to use social information. This transmission bias has been termed *copy when uncertain* although discussed theoretically (e.g., Laland, 2004) and tested with nonprimate species (e.g., *Rattus norvegicus*, Galef et al., 2008), little empirical data exist concerning how widespread this is among primate species.

In addition to individuals' motivations to use social information, their relationship with the expert and the expert's identity also influence the likelihood of social learning. For example, chimpanzees appear more likely to copy dominant and older group mates (Biro et al., 2003; Horner et al., 2010), even if it garners them a less-preferred reward (Hopper et al., 2011), whereas vervet monkeys (*C. pygerythrus*) have been reported to replicate the behavioral techniques performed by their matrilineal kin (van de Waal et al., 2014). Such social learning patterns represent "model-based" transmission biases in contrast to the "state-based" *copy when uncertain* transmission bias (Rendell et al., 2011). Interestingly, although a kin-based transmission bias may promote learning among kin, it may inhibit the social transmission of information between nonkin or less-closely related kin (Brosnan & Hopper, 2014 provide a review).

APPLICATIONS FOR PRIMATE MANAGEMENT: TRAINING

As primates are able to learn new skills by observing others, might it be possible for managers to exploit captive primates' social learning abilities to aid training or introduce desired behaviors to a colony? In this section, I will discuss how primates' ability of social learning might be applied to the context of training.

Each of the experimental methods used to study social learning that have been described (Figure 15.1) has its own pros and cons in an applied context. Dyadic learning is obviously best suited for pair-housed primates or animals that are singly caged in adjacent cages. There is some evidence that primates learn from one another when there is a barrier between them (e.g., Brosnan & de Waal, 2004), so even singly housed animals may be able to learn from animals in adjacent cages with a mesh or clear barrier between them. However, more success for social learning has been reported when primates are housed in the same cage. For example, Caldwell and Whiten (2003) found that marmosets (*C. jacchus*) learned more effectively from others when they were tested in the same cage and could scrounge food rewards from trained individuals who already knew how to successfully obtain rewards.

Dyadic learning and diffusion chain methods are less suitable for animals housed in large social groups, because pairs of animals have to be separated from their group for each interaction, increasing management effort and risk to the animals upon reintroduction to their group. The benefits of the diffusion chain method, however, are that (1) it increases the control a researcher or trainer has over the number of demonstrations each animal sees and (2) it ensures that all animals will be exposed to the required behavior. When pairing animals to facilitate social learning, whether for dyadic learning or the diffusion chain method, trainers must take care to pair animals that are socially compatible and unlikely to fight when paired (as described in Dindo et al., 2008). If the trained model animal is too dominant over its partner, the partner might not approach to observe the model (e.g., Hopper et al., 2013). Conversely, if the trained model is too low ranking, it may not perform the behavior in front of its cage mate (Drea & Wallen, 1999).

The open diffusion method is only feasible with group-housed primates. It is the most naturalistic social learning protocol, but it provides the least control over which animals will observe and learn the introduced behavior.

Regardless of which method is used, whether social learning occurs is dependent on a number of factors, including the naïve learner, the expert model, the social environment, and the species in question. Social learning is never guaranteed! However, as I discuss below, even the presence of group mates may facilitate learning during training sessions. Therefore, consideration of the ways in which social learning may positively influence training is important.

Social Learning and Training with Group-Housed Primates

It has been shown that a number of primate species are less neophobic and more exploratory when they have social support, which, in turn, encourages (social) learning (e.g., Dindo et al., 2009). Therefore, simply training primates in a social setting may decrease training time. Indeed, when considering training primates in a group setting, Prescott and Buchanan-Smith (2003) noted that "... individual primates often are more relaxed when in groups than when isolated and can learn socially through observation of their conspecifics" (p. 159), and a recent study with laboratory-housed squirrel monkeys (*S. boliviensis*) demonstrated just this. In their study, Hopper et al. (2013) presented captive squirrel monkeys with a novel foraging device that required monkeys to push a door to either the left or right to retrieve a meal worm hidden behind the door. Monkeys were significantly more likely to solve this puzzle when they were tested with social support; some pair-housed and most group-housed monkeys solved the puzzle, whereas none of the monkeys tested alone did.

Beyond providing social support and encouraging exploration, training primates in a social setting may also facilitate the training process, as primates can learn by observing the actions of others. Supporting this, Prescott and Buchanan-Smith (2003) proposed that "allowing animals to be observers during training sessions may enable them to be trained more rapidly because of their familiarity with the training situation" (p. 159). Unfortunately, to my knowledge, no published study has yet investigated the training time required for singly versus socially housed primates, so whether social support facilitates training is only a theoretical idea at this stage. Indeed, as discussed above, when trained in a social setting, low-ranking individuals may be too inhibited to approach the trainer or engage in the session, suggesting that training in a social setting does not always facilitate learning. Because not all individuals are equally likely to learn by observing others when being trained in a group setting, trainers may be required to implement a modified training protocol. For example, Schapiro et al. (2003) noted that "when working with groups of animals... it is often easier to train certain group members than others" (p. 185) and they recommended that trainers be mindful of the dominance relationships among group members. Specifically, Schapiro et al. (2003) proposed that trainers reinforce dominant individuals both for performing their target behaviors and also when they allowed subordinate animals to receive their reinforcers following a cooperative feeding session, as described in Laule et al. (2003). For a more detailed discussion of applied positive reinforcement training with (socially housed) captive primates, see the chapters by Graham (2017) and Magden (2017).

Although much experimental work with captive primates has shown that individuals can learn from one another when tested in a group setting (e.g., Whiten et al., 2005; Bonnie et al., 2007; Hopper et al., 2013), it should be noted that typically, not all group members do so. In a clinical setting, when trainers are required to train all their animals on identical protocols and ensure that all animals perform the behavior at the same level, the introduction of novel behaviors via an open diffusion method may lack sufficient control. Not only may some group members not adopt the behavior (either because they did not learn it, or they did not choose to perform it), but also they may not copy the exact actions demonstrated but rather may emulate a solution using their own strategy (Hopper et al., 2007; Tennie et al., 2009). Such "corruptions" of the introduced behavior highlight the flexibility of primates' learning (i.e., their ability to innovate, Brosnan & Hopper, 2014), but it also reduces the homogeneity of behaviors performed by group members. Whether this is an issue for animal managers and trainers may depend on the aim of the training. For example, if you wish to train primates to use a new feeding device but are not interested in how they obtain the food, then training one group member and allowing the new behavior to spread via social learning would likely be sufficient. However, if you are training animals on a more precise behavior, such as presenting an arm to receive an injection, and the specific limb that is presented is important, each animal might need to be trained individually to ensure that they all learn the behavior precisely and present the correct limb on command. Training can still take place in a social setting, even if animals are trained individually.

Social Learning and Training with Singly Housed Primates

It may seem counterintuitive to discuss applying concepts of social learning to singly housed monkeys, however, if singly housed animals still have visual access to other monkeys living in the same room, social learning could occur. Coleman et al. (2013) experimentally tested the benefits of social learning on training time with captive monkeys housed in a laboratory setting. The researchers trained eight singly housed female rhesus macaques (*M. mulatta*) to touch a target. Once these monkeys were trained, the researchers then trained eight additional monkeys that were housed in the same room, directly across from the original eight trained on the target task. This meant that the second set of eight monkeys all had the opportunity to directly observe the training sessions received by the first set of eight monkeys. The researchers recorded the number of training sessions that each of the 16 naïve monkeys required to learn the target behavior and found that the second set of eight monkeys, who had watched their room mates' training sessions, touched the target significantly sooner than the first set of eight monkeys. Coleman et al. (2013) concluded that being able to observe other monkeys participate in training (i.e., social learning) reduced training time. It must be noted that this is a single study, and therefore, it is unclear how generalizable these results would be to other contexts, species, or training protocols. More detailed investigations regarding the influence of social learning on training performance of individually housed monkeys should be conducted in the future.

Copying on Command: Do-As-I-Do

The "do-as-I-do" technique derives from experimental work that investigated whether animals that were trained to copy a suite of actions performed by a human (e.g., raise left arm) when given the command "copy this," would also copy novel actions by the human when given the same command paired with an action that they had not witnessed before. However, there are only a few studies that have investigated primates' ability to copy novel actions on command, and the results are mixed. For example, Custance et al. (1995) showed that two chimpanzees were able to replicate novel actions when instructed to copy, whereas a more recent study revealed that three capuchin monkeys (*C. apella*) showed some aptitude for copying the experimenter's actions but were more successful when they were required to copy an action that incorporated an object (Fragaszy et al., 2011). The findings of Fragaszy et al. (2011) reflect research with chimpanzees that has shown they appear better able to match actions performed by conspecifics that involve an object (e.g., a tool, Whiten et al., 2005) than bodily gestures (Tennie et al., 2012). Indeed, like capuchins, and perhaps chimpanzees, the first experimental test of dogs' ability to copy on command showed that they were able to copy actions that involved objects (e.g., placing a bottle in a specific location, Topál et al., 2006).

Expanding upon do-as-I-do research with dogs, Fugazza and Miklósi (2014) tested the efficacy of the do-as-I-do method as a training technique for dogs compared to traditional clicker training. Their study revealed that for simple actions (e.g., knock over a bottle) both techniques were equally efficient, but the dogs trained using the do-as-I-do protocol outperformed those trained using clicker training methods on more complex actions (e.g., open a drawer) or sequences of actions (e.g., get on a chair and then ring a bell). Based on these findings, the do-as-I-do method is now being promoted as a useful dog training technique (Fugazza, 2014). Although the do-as-I-do method has been applied successfully for training dogs, a species renowned for being human oriented and adept at attending to humans providing cues (Kaminski & Marshall-Pescini, 2014), it has not been widely adopted with other species.

Although a number of primate species, including capuchins, chimpanzees, and orangutans, have been shown to learn from humans by observing, in most cases, the human in question was a primary trainer or foster parent with whom the animal had a strong, positive relationship (Russon & Galdikas, 1995; Huber et al., 2009). Primates' abilities to learn by observing less familiar humans appear limited (Ross et al., 2010; Hopper et al., 2015). Another restriction of the do-as-I-do method for training primates is that it is limited to actions that the trainer can physically demonstrate. In addition, this method requires a heavy initial investment to first train the primates to replicate actions on command before novel actions can be introduced, which is a very time-consuming process (Custance et al., 1995).

Observing Videos to Aid Training

Social learning of novel behaviors can also be encouraged by the provision of videos of conspecifics performing a behavior (Perlman et al., 2010). Experimental research has shown that both monkeys (e.g., *Colobus guereza kikuyuensis*, Price & Caldwell, 2007; *C. jacchus*, Gunhold et al., 2014) and apes (e.g., *P. troglodytes*, Price et al., 2009; Hopper et al., 2015) can learn by observing videos of conspecifics. A key benefit of video demonstrations is increased control over the content that each animal sees. Videos can be edited so that trainers can show an exact number of identical demonstrations of specific actions, and thus they can control which animals see the demonstrations and when (Hopper et al., 2012). Furthermore, because a single video demonstration involving the same animal(s) performing the same behavior(s) can be shown to multiple observers, fewer animals need to be trained as models (Price & Caldwell, 2007). Despite these benefits, it should be noted that primates do not appear to learn as well from videos of conspecifics compared to when they have the opportunity to observe live models (Ross et al., 2010; Hopper et al., 2012, 2015). Even though primates may not learn as well by seeing videos of conspecifics, playing video footage of a conspecific performing the desired behavior has been shown to speed up learning during training sessions (Perlman et al., 2010). In their study, Perlman et al. (2010) trained chimpanzees to urinate on command and animals that were also shown a video of a previously trained chimpanzee during training sessions learned to perform the behavior more quickly than those trained without seeing the video in conjunction with their training sessions.

APPLICATIONS FOR PRIMATE MANAGEMENT: CULTIVATING CULTURES

In the 1980s, half of the males in a troop of olive baboons (*P. anubis*) living in the Masai Mara Reserve of Kenya died from disease. It was the more aggressive males who died, leaving a cohort of unaggressive survivors, and therefore, a natural experiment was created. The troop, which had previously been characterized by the despotic dominant males, became more egalitarian and increased rates of affiliative behaviors were recorded (Sapolsky & Share, 2004). Specifically, females received and gave significantly more grooming after the death of the aggressive males. Interestingly, in the 1990s, these behavioral patterns persisted even though the original, more affiliative males that had become dominant in the 1980s had left the group. Sapolsky and Share (2004) observed that the troop's unique and "pacific" culture was adopted by new males joining the troop. Thus, the social dynamics of a group of primates can not only encourage and facilitate the transmission of behaviors and skills but also cultural styles. Beyond primates' skill of observing and replicating others' actions, their proclivity for social learning can also seed "positive" cultures. Could this be exploited by managers of captive primates?

An experimental test of socially learned positive behaviors was conducted by Menzel et al. (1972), who were interested to see whether they could create a new social tradition of play among groups of young chimpanzees. In their study, trios of 3-year-old chimpanzees were presented with two novel toys (a "satellite" and a "swing"). Using the replacement method (Figure 15.1), over a series of 17 test sessions, 19 chimpanzees were exposed to the new toys. When the first trio saw the novel toys, they were nervous and rarely touched them. Over time, however, the "generations" of chimpanzees became less neophobic and more likely to play with the toys. Reflecting upon the chimpanzees' individual responses, the authors remarked that

> one of the most striking phenomena here was that many 'timid' animals hid in a corner and whim-pered while the 'bold' animal in that trio played vigorously with the object…and yet, in the next grouping, the timid animal became the bold one and someone else cowered in the corner (p. 165).

Even though later trios contained none of the founder chimpanzees and were being shown the novel toys for the first time, they were less fearful than the first trio and were much more likely to play with the toys. By the end of the study, all members of the trios played with the toys and the authors reported that "an enduring tradition of boldness and play then ensued" (p. 161). The social support afforded to the chimpanzees by being tested with two familiar conspecifics allowed the chimpanzees to be less neophobic and more exploratory, revealing the interplay between such social support and the creation of socially learned customs for play.

Beyond socially learned customs of play, socially learned customs of sharing have also been induced among primates. Claidière et al. (2015) presented dyads of captive chimpanzees with an apparatus that one chimpanzee could use to deliver food to both themself and their test partner. Specifically, the chimpanzee could select whether they gave their partner a high-value food (a slice of pineapple) or a low-value food (a slice of carrot). In addition, and depending on the trial, the "donor" chimpanzee would either receive a high- or low-value food, regardless of which food they gave their partner. This was a test of prosocial behavior, and the question asked was would the chimpanzees give their partner a high-value food, and would they do so even if they themselves got a low-value food? Initially, the chimpanzees did not appear to be willing to give their partner the high-value food, opting to give them the low-value option. However, the authors were interested to see if the chimpanzees' behavior would change after they experienced another chimpanzee sharing high-value foods with them. This was a test of upstream, or generalized, reciprocity; chimpanzee A shares with B and, as a result of this experience, B is more likely to share with C in the future. Therefore, the researchers rigged the apparatus such that it always delivered the better food to the test partner. Strikingly, Claidière et al. (2015) reported that the social experience of receiving high-value rewards from one chimpanzee resulted in the partner chimpanzees being more likely to give high-value foods to a different chimpanzee when later paired with them. The authors concluded that "social experience of prosocial behavior increased the prosocial responses of chimpanzees…who had not hitherto shown significant prosociality" (p. 5). Thus, the chimpanzees learned to share via social learning.

Using a less intensive methodology, and perhaps one that would be more applicable for managers caring for captive primates, a recent study induced a "temporary cultural style of high affiliation" in captive marmosets (C. jacchus) using playbacks of affiliative (chirp) calls. Watson et al. (2014) played recordings of chirp calls at an above-average rate (mean 18 calls/5 min) to pairs of marmosets. The calls were recorded from animals within the same facility but not from individuals within the 19 study groups. After the playback period, which lasted 33 h over 11 days, the authors reported that affiliation among the monkeys who had heard the affiliative calls increased. Furthermore, intragroup and intergroup agonism and anxiety-related behaviors decreased. This change in cultural style was true both during and after the playback of affiliative calls but was not long-lasting, representing only a temporary shift. Although it did not have long-lasting effects, this technique is appealing because such a method would be relatively easy for managers to implement in a captive setting; once affiliative calls have been recorded, they could be played to monkeys repeatedly with little effort. It would be interesting to know whether such a method would continue to be influential, or whether the monkeys would become desensitized to the playbacks over time. Other applications might include reserving playbacks for times of transition (e.g., during introductions or if animals are moved to new rooms or cages) or used to target specific groups that evidence higher levels of aggression. Future work is required to determine how effective this method might be under different circumstances and for different species.

A NOTE OF CAUTION

Throughout this chapter, I have highlighted the many ways in which primates' proclivity to learn from one another can be exploited to introduce novel behaviors and cultural styles to captive populations of primates in ways that could aid management and improve welfare. However, I wish to also highlight that just as primates can learn "good" behaviors from one another, they can also adopt "bad" behaviors. Indeed, early work with monkeys demonstrated socially acquired object avoidance in several species (e.g., *Erythrocebus*, *Macaca*, and *Saimiri*) (Griffin, 2004), which is a socially learned response called "observational conditioning" (Mineka et al., 1984). From their research investigating observational conditioning, Cook and Mineka (1990) concluded that "naïve, laboratory-reared monkeys can acquire a strong and persistent fear of snakes and snake-like stimuli by watching videotapes of conspecifics reacting fearfully to these stimuli" (p. 377). Obviously, learning to avoid dangerous stimuli by observing the fear responses of group mates is extremely adaptive for primates. However, in a captive setting, this could result in a socially learned fear of stimuli that are perceived as negative but are a necessary part of care (e.g., needles). Therefore, managers and trainers should be careful when presenting such stimuli to primates to avoid a conditioned fear response spreading contagiously through their colony.

Although the data for this are less conclusive, it has also been proposed that certain abnormal behaviors, such as coprophagy and feces throwing, might be transmitted among captive primates via social learning (e.g., *P. troglodytes*, *M. mulatta*, Hook et al., 2002; Freeman & Ross, 2014; Hopper et al., 2016). The transmission of an abnormal behavior performed by a single female chimpanzee to members of her group was documented by researchers observing sanctuary-housed chimpanzees in Zambia (van Leeuwen et al., 2014). In 2010, an adult female chimpanzee was observed sticking blades of grass in one of her ears, a behavior not performed by any of her group mates. By 2012, 8 of the 12 chimpanzees in the group were observed to perform the "grass-in-ear" behavior but with varying frequency, whereas the "inventor" was observed to perform the "grass-in-ear" behavior 168 times; the number of times "observers" were recorded performing the behavior ranged from 36 times to just once.

CONCLUSION

Although primate social learning has been extensively studied for a number of years, the ways in which this ability can be applied to the behavioral management of primates and their care have received far less attention. In this chapter, I have outlined what we know about primate social learning and indicate the ways in which it has been tested and applied in a captive setting. However, I propose that future research should focus on the different ways in which theoretical applications are accurate and can be realized in a real-world setting. A number of key questions remain unanswered. For example, what are the limitations to primate social learning? What metrics can be developed to predict which animals will learn and when? What novel techniques can be developed to exploit primates' social learning skills to facilitate training and management? In the future, primatologists, researchers, behaviorists, and colony managers should work together to determine how our theoretical knowledge of primate behavior can be applied to enhance the care and management of nonhuman primates in captive settings.

ACKNOWLEDGMENTS

I wish to thank Steve Schapiro for inviting me to speak at the inaugural Primate Behavioral Management Conference in March 2015, where I first presented my ideas considered in this chapter, and for the opportunity to contribute to this volume. I would also like to thank Susan (Lambeth) Pavonetti, Jaine Perlman, Mollie Bloomsmith, and Kris Coleman for insightful discussions on the interplay between training and social learning. Thanks too to Steve Ross and Sarah Jacobson for comments on earlier drafts of this chapter.

REFERENCES

Addessi, E. & Visalberghi, E. (2001). Social facilitation of eating novel food in tufted capuchin monkeys (*Cebus apella*): Input provided by group members and responses affected in the observer. *Animal Cognition, 4*, 297–303.

Bandura, A. (1977). *Social Learning Theory*. Englewood Cliffs, NJ: Prentice-Hall.

Biro, D., Inoue-Nakamura, N., Tonooka, R., Yamakoshi, G., Sousa, C., & Matsuzawa, T. (2003). Cultural innovation and transmission of tool use in wild chimpanzees: Evidence from field experiments. *Animal Cognition, 6*(4), 213–223.

Boesch, C. (1991). Teaching among wild chimpanzees. *Animal Behaviour, 41*, 530–532.

Bonnie, K. E. & de Waal, F. B. M. (2006). Affiliation promotes the transmission of a social custom: Handclasp grooming among captive chimpanzees. *Primates, 47*, 27–34.

Bonnie, K. E., Horner, V., Whiten, A., & de Waal, F. B. M. (2007). Spread of arbitrary conventions among chimpanzees: A controlled experiment. *Proceedings of the Royal Society B, 274*(1608), 367–372.

Brosnan, S. F. & de Waal, F. B. M. (2004). Socially learned preferences for differentially rewarded tokens in the brown capuchin monkey (*Cebus apella*). *Journal of Comparative Psychology, 118*(2), 133–139.

Brosnan, S. F. & Hopper, L. M. (2014). Psychological limits on animal innovation. *Animal Behaviour, 92*, 325–332.

Burkart, J., Kupferberg, A., Glasauer, S., & van Schaik, C. (2012). Even simple forms of social learning rely on intention attribution in marmoset monkeys (*Callithrix jacchus*). *Journal of Comparative Psychology, 126*(2), 129–138.

Byrne, R. W. & Rapaport, L. G. (2011). What are we learning from teaching? *Animal Behaviour, 82*(5), 1207–1211.

Caldwell, C. A., Schillinger, K., Evans, C. L., & Hopper, L. M. (2012). End state copying by humans (*Homo sapiens*): Implications for a comparative perspective on cumulative culture. *Journal of Comparative Psychology, 126*(2), 161–169.

Capitanio, J. P. (2017). Variation in biobehavioral organization, Chapter 5. In Schapiro, S. J. (Ed.), *Handbook of Primate Behavioral Management* (pp. 55–73). CRC Press, Boca Raton, FL.

Caro, T. M. & Hauser, M. D. (1992). Is there teaching in nonhuman animals? *The Quarterly Review of Biology, 67*(2), 151–174.

Carter, A. J., Marshall, H. H., Heinsohn, R., & Cowlishaw, G. (2012). How not to measure boldness: Novel object and antipredator responses are not the same in wild baboons. *Animal Behaviour, 84*(3), 603–609.

Carter, A. J., Marshall, H. H., Heinsohn, R., & Cowlishaw, G. (2014). Personality predicts the propensity for social learning in a wild primate. *PeerJ, 2*, e283.

Claidière, N., Messer, E. J. E., Hoppitt, W., & Whiten, A. (2013). Diffusion dynamics of socially learned foraging techniques in squirrel monkeys. *Current Biology, 23*, 1251–1255.

Claidière, N., Whiten, A., Mareno, M. C., Messer, E., Brosnan, S. F., Hopper, L. M., Lambeth, S. P., Schapiro, S. J., & McGuigan, N. (2015). Selective and contagious prosocial resource donation in capuchin monkeys, chimpanzees and humans. *Scientific Reports 5*, 7631.

Coleman, K. (2017). Individual differences in temperment and behavioral management, Chapter 7. In Schapiro, S. J. (Ed.), *Handbook of Primate Behavioral Management* (pp. 95–113). CRC Press, Boca Raton, FL.

Coleman, K., Houser, L. A., & Maier, A. (2013). Improving the efficiency of positive reinforcement training for non-human primates. *American Journal of Primatology, 75*(S1), 115.

Cook, M. & Mineka, S. (1990). Selective associations in the observational conditioning of fear in rhesus monkeys. *Journal of Experimental Psychology: Animal Behavior Processes, 16*(4), 372–389.

Cronin, K. A., Pieper, B. A., van Leeuwen, E. J. C., Mundry, R., & Haun, D. B. M. (2014a). Problem solving in the presence of others: How rank and relationship quality impact resource acquisition in chimpanzees (*Pan troglodytes*). *PLoS One, 9*(4), e93204.

Cronin, K. A., van Leeuwen, E. J. C., Vreeman, V., & Haun, D. B. M. (2014b). Population-level variability in the social climates of four chimpanzee societies. *Evolution and Human Behavior, 35*(5), 389–396.

Custance, D. M., Whiten, A., & Bard, K. A. (1995). Can young chimpanzees (*Pan troglodytes*) imitate arbitrary actions? Hayes & Hayes (1952) revisited. *Behaviour, 132*, 837–859.

Dindo, M., Stoinski, T., & Whiten, A. (2011). Observational learning in orangutan cultural transmission chains. *Biology Letters, 7*(2), 181–183.

Dindo, M., Thierry, B., & Whiten, A. (2008). Social diffusion of novel foraging methods in brown capuchin monkeys (*Cebus apella*). *Proceedings of the Royal Society of London. Series B, 275*(1631), 187–193.

Dindo, M., Whiten, A., & de Waal, F. B. M. (2009). Social facilitation of exploratory foraging behavior in capuchin monkeys (*Cebus apella*). *American Journal of Primatology, 71*(5), 419–426.

Drea, C. M. & Wallen, K. (1999). Low-status monkeys "play dumb" when learning in mixed social groups. *Proceedings of the National Academy of Sciences of the United States of America, 96*(22), 12965–12969.

Finestone, E., Bonnie, K. E., Hopper, L. M., Vreeman, V. M., Lonsdorf, E. V., & Ross, S. R. (2014). The interplay between individual, social, and environmental influences on chimpanzee food choices. *Behavioural Processes, 105*, 71–78.

Fragaszy, D., Deputte, B., Cooper, E., Colbert-White, E., & Hemery, C. (2011). When and how well can human-socialized capuchins match actions demonstrated by a familiar human? *American Journal of Primatology, 73*(7), 643–654.

Fragaszy, D. & Visalberghi, E. (2004). Socially biased learning in monkeys. *Learning and Behavior, 32*(1), 24–35.

Freeman, H. D. & Ross, S. R. (2014). The impact of atypical early histories on pet or performer chimpanzees. *PeerJ, 2*, e579.

Fugazza, C. (2014). *Do As I Do: Using Social Learning to Train Dogs Book*. Dogwise Publishing, Wenatchee, WA.

Fugazza, C. & Miklósi, Á. (2014). Should old dog trainers learn new tricks? The efficiency of the Do as I do method and shaping/clicker training method to train dogs. *Applied Animal Behaviour Science, 153*, 53–61.

Galef, B. G. (2001). Social learning of food preferences in rodents: Rapid appetitive learning. *Current Protocols in Neuroscience*, doi:10.1002/0471142301.ns0805ds21.

Galef, B. G., Dudley, K. E., & Whiskin, E .E. (2008). Social learning of food preference in "dissatisfied" and "uncertain" Norway rats. *Animal Behaviour, 75*, 631–637.

Graham, M. L. (2017). Positive reinforcement training and research, Chapter 12. In Schapiro, S. J. (Ed.), *Handbook of Primate Behavioral Management* (pp. 187–200). CRC Press, Boca Raton, FL.

Griffin, A. S. (2004). Social learning about predators: A review and prospectus. *Learning and Behavior, 32*(1), 131–140.

Gunhold, T., Whiten, A., & Bugnyar, T. (2014). Video demonstrations seed alternative problem-solving techniques in wild common marmosets. *Biology Letters, 10*(9), 20140439.

Hobaiter, C., Poisot, T., Zuberbühler, K., Hoppitt, W., & Gruber, T. (2014). Social network analysis shows direct evidence for social transmission of tool use in wild chimpanzees. *PLoS Biology, 12*(9), e1001960.

Hook, M. A., Lambeth, S. P., Perlman, J. E., Stavisky, R., Bloomsmith, M. A., & Schapiro, S. J. (2002). Intergroup variation in abnormal behavior in chimpanzees (*Pan troglodytes*) and rhesus macaques (*Macaca mulatta*). *Applied Animal Behaviour Science, 76*(2), 165–176.

Hopper, L. M., Freeman, H. D., & Ross, S. R. (2016). Reconsidering coprophagy as an indicator of negative welfare for captive chimpanzees using a bottom-up approach. *Applied Animal Behaviour Science, 176*, 112–119.

Hopper, L. M., Holmes, A. N., Williams, L. E., & Brosnan, S. F. (2013). Dissecting the mechanisms of squirrel monkeys (*Saimiri boliviensis*) social learning. *PeerJ, 1*, e13.

Hopper, L. M., Lambeth, S. P., & Schapiro, S. J. (2012). An evaluation of the efficacy of video displays for use with chimpanzees (*Pan troglodytes*). *American Journal of Primatology, 74*(5), 442–449.

Hopper, L. M., Lambeth, S. P., Schapiro, S. J., & Whiten, A. (2008). Observational learning in chimpanzees and children studied through "ghost" conditions. *Proceedings of the Royal Society of London B: Biological Sciences, 275*(1636), 835–840.

Hopper, L. M., Lambeth, S. P., Schapiro, S. J., & Whiten, A. (2015). The importance of witnessed agency in chimpanzee social learning of tool use. *Behavioural Processes* 112, 120–129.

Hopper, L. M., Schapiro, S. J., Lambeth, S. P., & Brosnan, S. F. (2011). Chimpanzees' socially maintained food preferences indicate both conservatism and conformity. *Animal Behaviour, 81*, 1195–1202.

Hopper, L. M., Spiteri, A., Lambeth, S. P., Schapiro, S. J., Horner, V., & Whiten, A. (2007). Experimental studies of traditions and underlying transmission processes in chimpanzees. *Animal Behaviour, 73*, 1021–1032.

Hoppitt, W., Blackburn, L., & Laland, K. N. (2007). Response facilitation in the domestic fowl. *Animal Behaviour, 73*(2), 229–238.

Hoppitt, W. & Laland, K. N. (2013). *Social Learning: An Introduction to Mechanisms, Methods and Models*. Princeton University Press, Princeton, NJ.

Horner, V., Proctor, D., Bonnie, K. E., Whiten, A., & de Waal, F. B. M. (2010). Prestige affects cultural learning in chimpanzees. *PloS One, 5*(5), e10625.

Horner, V., Whiten, A., Flynn, E., & de Waal, F. B. M. (2006). Faithful replication of foraging techniques along cultural transmission chains by chimpanzees and children. *Proceedings of the National Academy of Science of the United States of America, 103*(37), 13878–13883.

Huber, L., Range, F., Voelkl, B., Szucsich, A., Virányi, Z., & Miklósi, Á. (2009). The evolution of imitation: What do the capacities of non-human animals tell us about the mechanisms of imitation. *Philosophical Transactions of the Royal Society of London B: Biological Sciences, 364*, 2299–2309.

Huguet, P., Barbet, I., Belletier, C., Monteil, J-M., & Fagot, J. (2014). Cognitive control under social influence in baboons. *Journal of Experimental Psychology: General, 143*(6), 2067–2073.

Humle, T. & Snowdon, C. T. (2008). Socially biased learning in the acquisition of a complex foraging task in juvenile cottontop tamarins, *Saguinus oedipus*. *Animal Behaviour* 75, 267–277.

Kaminski, J. & Marshall-Pescini, S. (2014). *The Social Dog: Behavior and Cognition*. Academic Press Inc., San Diego, CA.

Kendal, R. L., Custance, D. M., Kendal, J. R., Vale, G., Stoinski, T. S., Rakotomalala, N. L., & Rasamimanana, H. (2010). Evidence for social learning in wild lemurs (*Lemur catta*). *Learning & Behavior, 38*(3), 220–234.

Kendal, R., Hopper, L. M., Whiten, A., Brosnan, S. F., Lambeth, S. P., Schapiro, S. J., & Hoppitt, W. (2015). Chimpanzees copy dominant and knowledgeable individuals: Implications for cultural diversity. *Evolution and Human Behavior, 36*(1), 65–72.

Köhler, W. (1925). *The Mentality of Apes*. Translated from the second revised edition by Ella Winter. Liveright, New York.

Laland, K. N. (2004). Social learning strategies. *Learning and Behavior, 32*(1), 4–14.

Laule, G. E., Bloomsmith, M. A., & Schapiro, S. J. (2003). The use of positive reinforcement training techniques to enhance the care, management and welfare of primates in the laboratory. *Journal of Applied Animal Welfare Sciences, 6*(3), 163–173.

Maestripieri, D. (1995). Maternal encouragement in nonhuman primates and the question of animal teaching. *Human Nature, 6*(4), 361–378.

Maestripieri, D. (1996). Maternal encouragement of infant locomotion in pigtail macaques, *Macaca nemestrina*. *Animal Behaviour*, 51, 603–610.

Magden, E. R. (2017). Positive reinforcement training and health care, Chapter 13. In Schapiro, S. J. (Ed.), *Handbook of Primate Behavioral Management* (pp. 201–216). CRC Press, Boca Raton, FL.

Matsuzawa, T., Biro, D., Humle, T., Inoue-Nakamura, N., Tonooka, R., & Yamakoshi, G. (2001). Emergence of culture in wild chimpanzees: Education by master-apprenticeship. In Matsuzawa, T. (Ed.), *Primate Origins of Human Cognition and Behavior* (pp. 557–574). Springer, Tokyo.

McCowan, B. & Beisner, B. (2017). Utility of systems network analysis for understanding complexity in primate behavioral management, Chapter 11. In Schapiro, S. J. (Ed.), *Handbook of Primate Behavioral Management* (pp. 157–183). CRC Press, Boca Raton, FL.

Menzel, E. W., Davenport, R. K., & Rogers, C. M. (1972). Protocultural aspects of chimpanzees' responsiveness to novel objects. *Folia Primatologica, 17*, 161–170.

Mineka, S., Davidson, M., Cook, M., & Keir, R. (1984). Observational conditioning of snake fear in rhesus monkeys. *Journal of Abnormal and Psychology, 93*(4), 355–372.

Perlman, J. E., Horner, V., Bloomsmith, M. A., Lambeth, S. P., & Schapiro, S. J. (2010). Positive reinforcement training, social learning, and chimpanzee welfare. In Lonsdorf, E. V., Ross, S. R., Matsuzawa, T., & Goodall, J. (Eds.), *The Mind of The Chimpanzee: Ecological and Experimental Perspectives* (pp. 320–331). University of Chicago Press, Chicago, IL.

Prescott, M. J. & Buchanan-Smith, H. M. (2003). Training nonhuman primates using positive reinforcement techniques. *Journal of Applied Animal Welfare Science, 6*(3), 157–161.

Price, E. & Caldwell, C. A. (2007). Artificially generated cultural variation between two groups of captive monkeys, *Colobus guereza kikuyuensis*. *Behavioural Processes, 74*, 13–20.

Price, E. E., Lambeth, S. P., Schapiro, S. J., & Whiten, A. (2009). A potent effect of observational learning on chimpanzee tool construction. *Proceedings of the Royal Society of London B: Biological Sciences, 276*(1671), 3377–3383.

Rendell, L., Fogarty, L., Hoppitt, W. J. E., Morgan, T. J. H., Webster, M. M., & Laland, K. N. (2011). Cognitive culture: Theoretical and empirical insights into social learning strategies. *Trends in Cognitive Sciences, 15*(2), 68–76.

Ross, S. R., Milstein, M. S., Calcutt, S. E., & Lonsdorf, E. V. (2010). Preliminary assessment of methods used to demonstrate nut-cracking behavior to five captive chimpanzees (*Pan trogolodytes*). *Folia Primatologica, 81*, 224–232.

Russon, A. E. & Galdikas, B. M. F. (1995). Constraints on great apes' imitation: Model and action selectivity in rehabilitant orangutan (*Pongo pygmaeus*) imitation. *Journal of Comparative Psychology, 109*(1), 5–17.

Sapolsky, R. M. & Share, L. J. (2004). A pacific culture among wild baboons: Its emergence and transmission. *PLoS Biology, 2*(4), e106.

Schapiro, S. J., Bloomsmith, M. A., & Laule, G. E. (2003). Positive reinforcement training as a technique to alter nonhuman primate behavior: Quantitative assessments of effectiveness. *Journal of Applied Animal Welfare Science, 6*(3), 175–187.

Shettleworth, S. J. (1998). *Cognition, Evolution, and Behavior*. Oxford University Press, New York.

Stoinski, T. S., Drayton, L. A., & Price, E. E. (2011). Evidence of social learning in black-and-white ruffed lemurs (*Varecia variegata*). *Biology Letters, 7*(3), 376–379.

Stoinski, T. S., Wrate, J. L., Ure, N., & Whiten, A. (2001). Imitative learning by captive western lowland gorillas (*Gorilla gorilla gorilla*) in a simulated food-processing task. *Journal of Comparative Psychology, 115*(3), 272–281.

Subiaul, F., Cantlon, J. F., Holloway, R. L., & Terrace, H. S. (2004). Cognitive imitation in rhesus macaques. *Science, 305*(5682), 407–410.

Suomi, S. J. & Leroy, H. A. (1982). In memoriam: Harry F. Harlow (1905–1981). *American Journal of Primatology, 2*(4), 319–342.

Tennie, C., Call, J., & Tomasello, M. (2009). Ratcheting up the ratchet: On the evolution of cumulative culture. *Philosophical transactions of the Royal Society of London B: Biological Sciences, 364*, 2405–2415.

Tennie, C., Call, J., & Tomasello, M. (2010). Evidence for emulation in chimpanzees in social settings using the floating peanut task. *PLoS One, 5*(5), e10544.

Tennie, C., Call, J., & Tomasello, M. (2012). Untrained chimpanzees (*Pan troglodytes schweinfurthii*) fail to imitate novel actions. *PLoS One, 7*(8), e41548.

Thornton, A. & Raihani, N. J. (2008). The evolution of teaching. *Animal Behaviour, 75*, 1823–1836.

Tomasello, M., Savage-Rumbaugh, S., & Kruger, A. C. (1993). Imitative learning of actions on objects by children, chimpanzees, and enculturated chimpanzees. *Child Development, 64*(6), 1688–1705.

Topál, J., Byrne, R. W., Miklósi, Á., & Csányi, V. (2006). Reproducing human actions and action sequences: "Do as I Do" in a dog. *Animal Cognition, 9*(4), 355–367.

van de Waal, E., Borgeaud, C., & Whiten, A. (2013). Potent social learning and conformity shape a wild primate's foraging decisions. *Science, 340*(6131), 483–485.

van de Waal, E., Bshary, R., & Whiten, A. (2014). Wild vervet monkey infants acquire the food-processing variants of their mothers. *Animal Behaviour, 90*(0), 41–45.

van Leeuwen, E. C., Cronin, K., & Haun, D. M. (2014). A group-specific arbitrary tradition in chimpanzees (*Pan troglodytes*). *Animal Cognition, 17*(6), 1421–1425.

Visalberghi, E. & Addessi, E. (2000). Seeing group members eating a familiar food enhances the acceptance of novel foods in capuchin monkeys. *Animal Behaviour, 60*(1), 69–76.

Watson, C. F. I., Buchanan-Smith, H. M., & Caldwell, C. A. (2014). Call playback artificially generates a temporary cultural style of high affiliation in marmosets. *Animal Behaviour, 93*, 163–171.

Whiten, A. (2005). The second inheritance system of chimpanzees and humans. *Nature, 437*(7055), 52–55.

Whiten, A., Horner, V., & de Waal, F. B. M. (2005). Conformity to cultural norms of tool use in chimpanzees. *Nature, 437*, 737–740.

Zajonc, R. B. (1965). Social facilitation. *Science, 149*, 269–274.

CHAPTER **16**

Collaborative Research and Behavioral Management

Steven J. Schapiro
The University of Texas MD Anderson Cancer Center and University of Copenhagen

Sarah F. Brosnan and William D. Hopkins
The University of Texas MD Anderson Cancer Center and Georgia State University

Andrew Whiten
University of St. Andrews

Rachel Kendal
Durham University

Chet C. Sherwood
George Washington University

Susan P. Lambeth
The University of Texas MD Anderson Cancer Center

CONTENTS

INTRODUCTION

The behavioral management of captive nonhuman primates (NHPs) can be significantly enhanced through synergistic relationships with noninvasive research projects. Many behavioral and cognitive research procedures are challenging and enriching (physically, cognitively, and/or socially) for the animals (Hopper et al. 2016; Hopkins and Latzman 2017) without involving any invasive (surgical, biopsy, etc.) procedures. Noninvasive behavioral research programs present the primates with opportunities to choose to voluntarily participate (or not), providing them with greater control over their circumstances than they would have in the absence of such procedures. Providing NHPs with control is of importance when attempting to establish "functionally appropriate captive environments." However, it must be emphasized that research designs in which access to food and/ or fluid is restricted would not really satisfy the criteria associated with *voluntary* participation.

This chapter focuses on the behavioral management implications of the collaborative research projects that are ongoing at the National Center for Chimpanzee Care (NCCC) of The University of Texas MD Anderson Cancer Center in Bastrop, Texas. Formerly the Chimpanzee Biomedical Research Resource (CBRR), this colony of 132 chimpanzees (currently) has been involved in several biomedical, and many behavioral, research studies for almost 30 years. The National Institutes of Health (NIH) funded the CBRR, as it does all of its research resources, to facilitate the participation of appropriate animal models in critical research. The same animals that comprised the CBRR also comprise the NIH-funded NCCC, with the modified goal of enhancing captive care of this endangered species. The relationships between collaborative research and behavioral management to be discussed in this chapter would be relevant at any NHP research facility (NIH-funded or otherwise). While this chapter primarily emphasizes behavioral research projects focused on chimpanzees, the same benefits of collaborative research would accrue for other genera of NHPs and, to a certain extent, for other types of research projects.

COLLABORATIVE RESEARCH AND BEHAVIORAL MANAGEMENT

We have been collaborating with a core group of investigators for almost 15 years; scientific relationships that are attributable, at least in part, to the fact that we are "user-friendly" for such academic interactions. The NCCC provides access to the animals, logistical support, and perhaps most importantly, intimate knowledge concerning (1) the typical behaviors of each individual chimpanzee and (2) the relevant social dynamics in all groups.

It is our responsibility, as managers of National Resources (in addition to the chimpanzees of the NCCC, our facility also maintains the Rhesus Monkey, Squirrel Monkey, and Owl Monkey Breeding and Research Resources), to make them available for critical health- and behavior-focused investigations. Therefore, we do everything in our power to make this happen, and many of our collaborative projects have resulted in publications in peer-reviewed journals. The collaborative research programs involving the chimpanzees of the NCCC have now matured to the point where many of the "independent" data sets associated with specific projects can be integrated with one another. Such integration provides important cross-disciplinary insights into chimpanzee behavior and cognition (Hopper et al. 2014b; Brosnan et al. 2015; Latzman et al. 2015b), and most importantly for this chapter, guidance for the continuous refinement of behavioral management strategies for this species.

MONITORING HEALTH AND ACTIVITY

Working with NHPs on research projects provides personnel with many additional opportunities to *monitor the health and activity* of the animals. More frequent observations should not only promote the early identification of health or behavioral problems (NHPs participating in behavioral research projects may be observed by care staff, behavior staff, veterinary technicians, veterinarians, and/or researchers as frequently as 10 times each day), but the research procedures themselves may also allow for in-depth observations that facilitate the identification of subtle changes in condition (e.g., the animal performed the task many times with no hesitation in the past, but, for some reason, is hesitating today). Researchers working with study subjects develop intimate knowledge concerning the subjects' temperaments, response tendencies, and reinforcement preferences. Changes are quickly apparent, and in addition to triggering clinical intervention, such familiarity may be useful in developing individualized treatment programs (see Bloomsmith 2017), which may be especially important for geriatric animals. However, there is one caveat to mention in relation to additional research observations enhancing health monitoring, especially for animals that are on a quality-of-life watch (Lambeth et al. 2013). Judgments of health condition by researchers may occasionally be clouded by the researchers' attachment to the subject and the animal's performance on its study tasks. Animals (chimpanzees in our experience) may still perform their tasks, even as other indicators suggest that their health condition has significantly deteriorated.

COLLABORATIVE RESEARCH PROJECTS

During the past 15 years, we have published collaborative empirical work that has addressed behavioral cerebral laterality, hemispheric asymmetries, and the evolution of language (Gómez-Robles et al. 2015, 2016; Hopkins et al. 2017); behavioral economics (House et al. 2014; Brosnan et al. 2015; Claidière et al. 2015); social learning and cultural transmission (Silk et al. 2005; Hopper et al. 2007; Dean et al. 2012; Kendal et al. 2015; Davis et al. 2016; Vale et al. in press); and chimpanzee personality (Freeman et al. 2013a; Latzman et al. 2016, 2017). We will discuss each of these areas, briefly describing some of the techniques and findings and emphasizing the aspects of the work that are most applicable to behavioral management.

Behavioral Laterality, Cerebral Hemispheric Asymmetries, and the Evolution of Language

One of the first collaborative relationships established at the NCCC was with Bill Hopkins' group at Yerkes National Primate Research Center, to add the NCCC chimpanzees as subjects in his studies *of laterality, hemispheric asymmetries, and the evolution of language*. These studies primarily involved NHP subjects and were designed to provide insights into the evolution of human language abilities, utilizing a multitude of behavioral measures, in addition to anatomical and genetic measures.

Investigations of handedness were the initial emphasis of our collaboration, and we have published data derived from a number of unimanual, bimanual, and precision tasks, as well as studies of postural effects on handedness (Hopkins et al. 2003, 2011, 2013, 2015, 2017; Sherwood et al. 2010; Meguerditchian et al. 2012). All of these tasks have enhanced our behavioral management program by providing the chimpanzees with numerous foraging opportunities and related challenges, including opportunities to display complex species-typical behaviors, such as making and using tools. In addition, by understanding the lateralized nature of the animals' abilities, we have been able to provide individualized enrichment opportunities, based on the animals' hand preferences and proficiency on the tasks.

Our more recent collaborations with the Hopkins group have included Chet Sherwood's group as well (Teffer et al. 2013; Gómez-Robles 2016) and have focused on the relationships among the behavioral (i.e., gestures, vocalizations, responses on joint attention tasks, responses on a cognitive test battery), neuroanatomical (MRI scans and histological sections), and genetic (e.g., polymorphic vasopressin receptor genes) variables that appear to be associated with the communicative abilities of chimpanzees. As before, we are interested in those factors that might provide insight into the evolution of language, and we have conducted and published studies examining neocortical asymmetries and specializations (Sherwood et al. 2010; Wallez et al. 2012; Bianchi et al. 2013; Teffer et al. 2013; Autrey et al. 2014; Hopkins et al. 2014b, 2017) and the influence of specific genes on behavior, including the heritability of target responses (Hopkins et al. 2014a, 2015; Latzman et al. 2015; Gómez-Robles 2016; Evans et al. 2015). The data collected for these studies included blood samples and brain MRIs from almost all of the animals in the NCCC, resulting in the largest collection of chimpanzee brain MRIs in the world. As Hopkins and Latzman discuss in their chapter (2017), owing to restrictions imposed by NIH, it is now virtually impossible to collect these kinds of data. Studies of chimpanzee language-related cognitive abilities, brain structures, and genetic traits have contributed to our behavioral management program by identifying characteristics of individuals that are useful for managing their physical and social health. This work should provide us with information that is useful for the formation and maintenance of compatible social groups. In addition, the brain MRI data have been used to identify individuals at the NCCC that may be at risk for cognitive problems, in a manner similar to that described by Hopkins and Latzman (2017).

Behavioral Economics

Our collaboration with Sarah Brosnan's group at Georgia State University, addressing multiple questions associated with the general area of "behavioral economics," is also long-standing (Silk et al. 2005; Brosnan et al. 2015) and quite productive. These studies typically involve both NHP and human subjects, with the understanding of the evolution of human (1) economic behavior and (2) cooperation as the primary foci. The chimpanzees of the NCCC have participated in studies examining: inequity (Brosnan et al. 2010, 2015; Hopper et al. 2014a), performance on a variety of economic games (Brosnan et al. 2011), the concept of ownership (Brosnan et al. 2007, 2008, 2012), and issues related to conformity (Hopper et al. 2011). The prosocial tendencies of chimpanzees have also been studied as part of this collaborative program (Silk et al. 2005, 2013; Vonk et al. 2008; Brosnan et al. 2009; House et al. 2014; Claidière et al. 2015). Inequity studies typically involve sessions with a pair of chimpanzees, in which both chimpanzees perform a task and one chimpanzee receives a "high-value" reward, while the other receives a "medium-value" reward (Brosnan et al. 2010, 2015; Figure 16.1). A fairly

Figure 16.1 Chimpanzees exchanging tokens in an inequity study.

large number of variables can be manipulated in these studies to provide various control conditions to facilitate more fine-tuned analyses of the factors that contribute to the animals' responses to inequity (Hopper et al. 2014a). Studies of game theory, ownership, conformity, and prosociality typically use token exchange or related paradigms to assess levels of coordination and/or cooperation between animals (Silk et al. 2005; Brosnan et al. 2011, 2012; Hopper et al. 2011). Again, subtle manipulations of relevant experimental variables can be used to provide additional insight into the cognitive factors affecting the chimpanzees' responses (Hopper et al. 2014a).

Participation in these types of studies by individuals, pairs, and/or groups of chimpanzees is completely voluntary, providing the animals with opportunities to choose whether to participate or not. Some of the chimpanzees at the NCCC participate in five or more concurrent or consecutive studies. As Hopkins and Latzman (2017) attest, such opportunities are likely to be beneficial for the animals. Some animals are not comfortable as subjects in these experiments, and therefore choose to abstain from the studies. Again, as Hopkins and Latzman (2017) state, it would be counterproductive, from both welfare and scientific perspectives, to attempt to collect data from unwilling subjects. Many of the methods employed in the behavioral economic studies require the animals to make important social decisions (e.g., should they respond such that they alone receive a reward, or should they choose the response that rewards a social partner in addition to themselves). These types of decisions occur routinely for chimpanzees living in natural conditions, but may also occur considerably less frequently in captive situations. Hopper (2017) discusses some of the ways in which providing additional, research-related opportunities for social decision making in captivity may facilitate the provision of a functionally appropriate captive environment for the animals. In addition to the collaborative work performed with the chimpanzees of the NCCC, Brosnan and her group have conducted similar studies with the rhesus monkeys and squirrel monkeys (Brosnan et al. 2012; Freeman et al. 2013; Hopper et al. 2013) at the Keeling Center, as well as the capuchin monkeys (Brosnan and de Waal 2003; Brosnan and Bshary 2016) at the Language Research Center of Georgia State University.

Social Learning and Cultural Transmission

We have been investigating questions related to *social learning and cultural transmission* in collaboration with the research groups of Andrew Whiten and Rachel Kendal in the United Kingdom. Both social learning abilities and the nongenetic transfer of information are hallmarks of humans, and these studies have focused on assessing the abilities of chimpanzees to succeed at tasks that should indicate whether chimpanzees, our closest living relatives, possess these abilities as well. Many of these studies involve (1) teaching a "model" chimpanzee one of several solutions to a cognitive puzzle (often referred to as an artificial fruit; Figure 16.2), (2) allowing a group mate(s) to observe the model (the "observer") solving the puzzle, and then (3) studying the puzzle-solving technique(s) utilized by the observer(s) and additional members of the population. Other studies addressing these research questions involve token exchange procedures (see Figure 16.3) that are similar to those utilized for some of the behavioral economics studies mentioned earlier, whereas the remainder involve various food acquisition (tool making and tool use) or food selection tasks.

Use of these techniques and analyses of the data help elucidate, in general, whether chimpanzees learn from one another, and more specifically, help identify the social factors that influence the success of learning (Hopper et al. 2007; Whiten et al. 2007; Price et al. 2009; Dean et al. 2012; Vale et al. 2014, 2016, in press; Kendal et al. 2015; Davis et al. 2016). If conducted under ideal conditions, these studies will not only provide insights into intragroup social learning phenomena, but will also provide insights into intergroup social learning processes (Whiten et al. 2007). Hopper et al. (2008, 2015) have demonstrated that the *social* component of social learning is critical, as studies of "ghost" conditions did not result in learning, but observations of video recordings of trained animals did (Hopper et al. 2012). Virtually all of the techniques used to assess social learning and cumulative culture are enriching for the chimpanzees, providing them with multiple situations that

Figure 16.2   Chimpanzees using and observing an "artificial fruit" device in a cultural transmission study.

Figure 16.3   Chimpanzees exchanging tokens for rewards in a social learning study.

are cognitively and socially challenging. These techniques require the animals to solve problems that functionally simulate those they might encounter in the wild (e.g., "opening" artificial fruits to obtain treats, observing others perform complex and novel behaviors). They also provide the chimpanzees with multiple opportunities to make meaningful behavioral and social decisions. As Hopper (2017) points out, it may be valuable for behavioral management programs to establish conditions that promote social learning.

Personality

Our collaborative studies of chimpanzee *personality* or temperament (see Capitanio 2017 and Coleman 2017 for more complete discussions of NHP personality/temperament) have included both behavioral data and trait rating approaches to the assessment of personality (Freeman et al. 2013a). Our behavioral approach emphasized responses to novel items and novel situations, especially exposure to a human-like karate dummy in the animals' home enclosures. The trait rating approach was fairly typical for studies of primate personality, with caregivers, managers, and researchers who were familiar with the animals scoring each chimpanzee on a number of different adjectives. This research project has resulted in a well-cited methodology paper (Freeman et al. 2013a), a number of papers focusing on triarchic psychopathy constructs (Latzman et al. 2016, 2017), and a paper describing neuroanatomical correlates of personality (Latzman et al. 2015b). From a behavioral management perspective, these studies of chimpanzee personality have provided us with data that are valuable for managing and advancing our chimpanzee positive reinforcement training (PRT) program (Reamer et al. 2014); understanding individual personality characteristics that contribute to the compatibility of social groups; and refining subject selection practices for several types of behavioral and cognitive research projects (Hopper et al. 2014b). In general, understanding the personalities of the NCCC chimpanzees enhances our ability to provide them with environments and tasks that optimize their captive welfare.

Research Synergisms

As mentioned earlier in this chapter, we are now at the point where we have collected enough data from each of these collaborative endeavors to begin to integrate the findings to address issues of synergistic interest. This is the easiest to describe through the integration of our personality data with the other studies. We have discovered important relationships between personality and: hemispheric asymmetries and socio-communicative abilities (Latzman et al. 2015a); performance on behavioral economic tasks (Brosnan et al. 2015); responses in social learning and cumulative culture paradigms (Hopper et al. 2014); and factors related to training, and medical and behavioral management (Reamer et al. 2014b). In addition, the behavioral economics and the social learning and culture data are naturally quite closely related, and we have been able to address multiple questions of interest to both areas by combining the data (Claidière et al. 2015; Kendal et al. 2015).

ACCLIMATION AND TRAINING

Many behavioral and cognitive research procedures require the animals to (1) attentively interact with humans (caregivers and/or researchers) and (2) be trained to perform desired tasks. Honess (2017) and Fernandez et al. (2017) both emphasize the value of establishing conditions that provide macaques with opportunities to appropriately acclimate to interactions with humans. Although it is important for macaques, this may be even more important for those working with chimpanzees. We have been investing in the establishment of positive working relationships with our chimpanzees for many years, beginning our PRT program more than 25 years ago. We continue, and will continue, to refine these techniques for as long as we care for the animals. The evidence suggests that training NHPs using positive reinforcement techniques provides them with opportunities to "learn to learn" (Reamer et al. 2014, 2017); training them to perform research procedures can enhance training of applied management-related behaviors, and similarly, training them to perform applied behaviors can enhance training for research procedures. The chimpanzees of the NCCC have participated in a comprehensive "health-related" training program for more than two decades (Bloomsmith et al. 1994; Laule et al. 1996; Schapiro et al. 2005; Magden 2017), including training to present for acupuncture (Magden et al. 2013) and to present for low-level laser therapy (Magden et al. 2016). They have now also participated in a wide variety of cognitively challenging behavioral studies for 15 years. The cumulative effect of all of the attention that the NCCC chimpanzees have received for management and research purposes has been to yield a colony of animals that responds well to most management and research procedures.

Figure 16.4 Voluntary capillary blood sample from the toe of a chimpanzee for analysis of blood glucose levels.

There are many benefits associated with a combined applied and research training program. Such an approach enhances our abilities to work with the animals, both in health maintenance and research situations. Acupuncture and laser therapy would not be possible with animals that are unwilling to station (sit and stay), nor would data collection for some of the studies involving economic games (Brosnan et al. 2011). One of the most important benefits associated with having animals that are willing to work with their caregivers is to be able to provide individualized care to the animals (see Figure 16.4). We are in the process of testing aged chimpanzees' preferences for specific medications to treat their arthritis symptoms. Using a choice procedure (Schapiro et al. 2014), we are allowing the chimpanzees to choose which of two medications they prefer, opening a critical channel of communication between the animals and ourselves that provides them with the opportunity to make meaningful choices, while voluntarily participating in their own care. In our opinion, this is an important step in taking the behavioral management, and overall care, of captive NHPs to the "next level."

CONCLUSIONS

In conclusion, collaborative research projects can effectively contribute to successful captive management strategies, serving as effective components of a comprehensive behavioral management program. In addition to generating data for critical investigations (an appropriate and mandated function of NHPs supported as national research resources), collaborative projects provide the animals with important cognitive and social stimulation that (1) functionally simulates natural conditions, (2) is enriching, and (3) may otherwise be missing from even an enhanced captive environment. When chimpanzees participate in noninvasive research projects, the health care of the animals is likely to be augmented as they are observed more frequently by a greater number of people. Any change in response on the research tasks will alert researchers and caregivers alike that something may be wrong with the subject, allowing for rapid diagnosis and individualized treatment of the problem. Collaborative research projects, especially noninvasive projects that are cognitively challenging, are important tools for enhancing the welfare of captive NHPs.

ACKNOWLEDGMENTS

Much of the work described in this chapter has been supported by grants and subcontracts from: NIH to WDH (R01 MH092923, R01 NS073134, R01 HD05623204) or to Christian R. Abee (U42 RR015090, U42 OD-011197); NSF to SFB or CCS (SES 0729244, SES 0847351, INSPIRE

1542848), the John Templeton Foundation to AW (40128, Ref. 20721), and a Royal Society Dorothy Hodgkin Fellowship to RLK. Many postdoctoral fellows, graduate students, undergraduate students, and research assistants, especially Lydia Hopper, have been instrumental in the collection and analysis of the data included in the work mentioned in this chapter. None of this work would have been possible without the exceptional commitment, dedication, and expertise of the NCCC animal care staff and behavioral management team.

REFERENCES

Autrey MM, L Reamer, MC Mareno, et al. 2014. Age-related effects in the neocortical organization of chimpanzees: Gray and white matter volume, cortical thickness, and gyrification. *NeuroImage* 101:59–67.

Bianchi S, CD Stimpson, AL Bauernfeind, et al. 2013. Dendritic morphology of pyramidal neurons in the chimpanzee neocortex: Regional specializations and comparison to humans. *Cerebral Cortex* 23:2429–2436.

Bloomsmith MA. 2017. Behavioral management of laboratory primates: Principles and projections, Chapter 29. In *Handbook of Primate Behavioral Management*. Schapiro SJ (ed.). Boca Raton, FL: CRC Press.

Bloomsmith MA, GE Laule, PL Alford, et al. 1994. Using training to moderate chimpanzee aggression during feeding. *Zoo Biology* 13:557–566.

Brosnan SF and R Bshary. 2016. On potential links between inequity aversion and the structure of interactions for the evolution of cooperation. *Behaviour* 153:1267–1292.

Brosnan SF and FBM de Waal. 2003. Monkeys reject unequal pay. *Nature* 425:297–299.

Brosnan SF, M Grady, SP Lambeth, et al. 2008. Chimpanzee autarky. *PLoS One* 1:e1518.

Brosnan SF, J Henrich, MC Mareno, et al. 2009. Chimpanzees (*Pan troglodytes*) do not develop contingent reciprocity in an experimental task. *Animal Cognition* 12:587–597.

Brosnan SF, LM Hopper, S Richey, et al. 2015. Personality influences responses to inequity and contrast in chimpanzees. *Animal Behaviour* 101:75–87

Brosnan SF, OD Jones, M Gardner, et al. 2012. Evolution and the expression of biases: Situational value changes the endowment effect in chimpanzees. *Evolution and Human Behavior* 33:376–386.

Brosnan SF, OD Jones, SP Lambeth, et al. 2007. Endowment effects in chimpanzees. *Current Biology* 17:1704–1707.

Brosnan SF, A Parrish, MJ Beran, et al. 2011. Responses to the Assurance game in monkeys, apes, and humans using equivalent procedures. *Proceedings of the National Academy of Sciences of the United States of America* 108:3442–3447, 2011.

Brosnan SF, C Talbot, M Ahlgren, et al. 2010. Mechanisms underlying responses to inequitable outcomes in chimpanzees, *Pan troglodytes*. *Animal Behaviour* 79:1229–1237.

Capitanio JP. 2017. Variation in biobehavioral organization, Chapter 5. In *Handbook of Primate Behavioral Management*. Schapiro SJ (ed.). Boca Raton, FL: CRC Press.

Claidière N, A Whiten, MC Mareno, et al. 2015. Selective and contagious prosocial resource donation in capuchin monkeys, chimpanzees and humans. *Scientific Reports* 5:7631, 2015.

Coleman K. 2017. Individual differences in temperament and behavioral management, Chapter 7. In *Handbook of Primate Behavioral Management*. Schapiro SJ (ed.). Boca Raton, FL: CRC Press.

Davis SJ, GL Vale, SJ Schapiro, et al. 2016. Foundations of cumulative culture in apes: Improved foraging efficiency through relinquishing and combining witnessed behaviours in chimpanzees (*Pan troglodytes*). *Scientific Reports* 6:35953.

Dean LG, RL Kendal, SJ Schapiro, et al. 2012. Identification of the social and cognitive processes underlying human cumulative culture. *Science* 335:1114–1118.

Evans RE, JL Russell, S Bogart, et al. 2015. Polymorphic microsatellites in the 5′ flanking region of vasopressin receptor gene 1a (AVPR1a) impact sociability in captive chimpanzees (*Pan troglodytes*). *American Journal of Primatology* 77:65–66.

Fernandez L, M-A Griffiths, P Honess. 2017. Providing Behaviorally Manageable Primates for Research, Chapter 28. In *Handbook of Primate Behavioral Management*. Schapiro SJ (ed.). Boca Raton, FL: CRC Press.

Freeman HD, SF Brosnan, LM Hopper, et al. 2013a. Developing a comprehensive and comparative questionnaire for measuring personality in chimpanzees using a simultaneous top-down/bottom-up design. *American Journal of Primatology* 75:1042–1053.

Freeman HD, J Sullivan, LM Hopper, et al. 2013b. Different responses to reward comparisons by three primate species. *PLoS One* 8:e76297.

Gómez-Robles A, WD Hopkins, SJ Schapiro, et al. 2015. Relaxed genetic control of cortical organization in human brains compared with chimpanzees. *Proceedings of the National Academies of Science of the United States of America* 112:14799–14804.

Gómez-Robles A, WD Hopkins, SJ Schapiro, et al. 2016. The heritability of chimpanzee and human brain asymmetry. *Proceedings of the Royal Society B: Biological Sciences*. doi:10.1098/rspb.2016.1319.

Honess P. 2017. Behavioral management of long-tailed macaques (*Macaca fascicularis*), Chapter 20. In *Handbook of Primate Behavioral Management*. Schapiro SJ (ed.). Boca Raton, FL: CRC Press.

Hopkins WD, M Gardner, M Mingle, et al. 2013. Within- and between-task consistency in hand use as a means of characterizing hand preferences in captive chimpanzees (*Pan troglodytes*). *Journal of Comparative Psychology* 127:380–391.

Hopkins WD, MA Hook, S Braccini, et al. 2003. Hand preferences on a bimanual task in a large group of captive chimpanzees. *International Journal of Primatology* 24:677–689.

Hopkins WD, AC Keebaugh, L Reamer, et al. 2014a. Genetic influences on receptive joint attention in chimpanzees (*Pan troglodytes*). *Scientific Reports* 4:3774, doi:10.1038/srep03774.

Hopkins WD and RD Latzman. 2017. Future research with captive chimpanzees in the USA: Integrating scientific programs with behavioral management, Chapter 10. In *Handbook of Primate Behavioral Management*. Schapiro SJ (ed.). Boca Raton, FL: CRC Press.

Hopkins WD, A Meguerditchian, O Coulon, et al. 2017. Motor skill for tool-use is associated with asymmetries *in* Broca's area and the motor hand area of the precentral gyrus in chimpanzees (*Pan troglodytes*). *Behavioural Brain Research* 318:71–81.

Hopkins WD, M Misiura, LA Reamer, et al. 2014b. Poor receptive joint attention skills are associated with atypical gray matter asymmetry in the posterior superior temporal gyrus of chimpanzees. (*Pan troglodytes*). *Frontiers in Cognition*. doi:10.3389/fpsyg.2014.00007.

Hopkins WD, KA Phillips, A Bani. 2011. Hand preferences for coordinated bimanual actions in 777 great apes: Implications for the evolution of handedness in hominids. *Journal of Human Evolution* 60:605–611.

Hopkins WD, L Reamer, MC Mareno, et al. 2015. Genetic basis in motor skill and hand preference for tool use in chimpanzees (*Pan troglodytes*). *Proceedings of the Royal Society of London B: Biological Sciences* 282: 1223.

Hopper LM. 2017. Social learning and decision making, Chapter 15. In *Handbook of Primate Behavioral Management*. Schapiro SJ (ed.). Boca Raton, FL: CRC Press.

Hopper LM, SP Lambeth, and SJ Schapiro. 2012. An evaluation of the efficacy of video displays for use with chimpanzees (*Pan troglodytes*). *American Journal of Primatology* 74:452–459.

Hopper LM, SP Lambeth, SJ Schapiro, et al. 2008. Observational learning in chimpanzees and children studied through "ghost" conditions. *Proceedings of the Royal Society of London B: Biological Sciences* 275:835–840.

Hopper LM, SP Lambeth, SJ Schapiro, et al. 2013. The ontogeny of social comparisons in rhesus macaques (*Macaca mulatta*). *Journal of Primatology* 2:109. doi:10.4172/2167-6801.1000109.

Hopper LM, SP Lambeth, SJ Schapiro, et al. 2014a. Social comparison mediates chimpanzees' responses to loss, not frustration. *Animal Cognition* 17:1303–1311.

Hopper LM, SP Lambeth, SJ Schapiro, et al. 2015. The importance of witnessed agency in chimpanzee social learning of tool use. *Behavioural Processes* 112:120–129.

Hopper LM, SA Price, HD Freeman, et al. 2014b. Influence of personality, age, sex, and estrous state on chimpanzee problem-solving. *Animal Cognition* 17:835–847.

Hopper LM, SJ Schapiro, SP Lambeth, et al. 2011. Chimpanzees' socially maintained food preferences indicate both conservatism and conformity. *Animal Behaviour* 81:1195–1202.

Hopper LM, MA Shender, and SR Ross. 2016. Behavioral research as physical enrichment for captive chimpanzees. *Zoo Biology* 34:293–297.

Hopper LM, A Spiteri, SP Lambeth, et al. 2007. Experimental studies of traditions and underlying transmission processes in chimpanzees. *Animal Behaviour* 73:1021–1032.

House BR, JB Silk, SP Lambeth, et al. 2014. Task design influences prosociality in captive chimpanzees (*Pan troglodytes*). *PLoS One* 9(9):e103422.

Kendal RL, LM Hopper, A Whiten, et al. 2015. Chimpanzees copy dominant and knowledgeable individuals: Implications for cultural diversity. *Evolution and Human Behavior* 36:65–72.

Lambeth SP, SJ Schapiro, BJ Bernacky, et al. 2013. Establishing "quality of life" parameters using behavioral guidelines for the humane euthanasia of nonhuman primates. *Animal Welfare* 22:429–435.

Latzman RD, LE Drislane, LK Hecht, et al. 2016. A chimpanzee (*Pan troglodytes*) model of triarchic psychopathy constructs: Development and initial validation. *Clinical Psychological Science* 4:50–66. doi:10.177/2167702615568989.

Latzman RD, HD Freeman, SJ Schapiro. 2015a. The contribution of genetics and early-rearing experiences to hierarchical personality dimensions in chimpanzees *(Pan troglodytes). Journal of Personality and Social Psychology* 109(5):889–900. doi:10.1037/pspp0000040.

Latzman RD, LK Hecht, HD Freeman, et al. 2015b. Neuroanatomical correlates of personality in chimpanzees (*Pan troglodytes*): Associations between personality and frontal cortex. *NeuroImage* 123:63–71.

Latzman RD, CJ Patrick, HD Freeman, et al. 2017. Etiology of triarchic psychopathy dimensions in chimpanzees (*Pan troglodytes*). *Clinical Psychological Science*. DOI: 10.1177/2167702616676852

Laule GE, RH Thurston, PL Alford, et al. 1996. Training to reliably obtain blood and urine samples from a diabetic chimpanzee (*Pan troglodytes*). *Zoo Biology* 15:587–591.

Magden ER. 2017. Positive reinforcement training and health care, Chapter 13. In *Handbook of Primate Behavioral Management*. Schapiro SJ (ed.). Boca Raton, FL: CRC Press.

Magden ER, R Haller, E Thiele, et al. 2013. Acupuncture as an effective adjunct therapy for osteoarthritis in chimpanzees (*Pan troglodytes*). *Journal of the American Association for Laboratory Animal Science*, 52:475–480.

Magden, ER, MM Sleeper, SJ Buchl, et al. 2016. Use of an implantable loop recorder in a chimpanzee (*Pan troglodytes*) to monitor cardiac arrhythmias and assess the effects of acupuncture and laser therapy. *Comparative Medicine* 66:52–58.

Meguerditchian A, MJ Gardner, SJ Schapiro, et al. 2012. The sound of one hand clapping: Handedness and perisylvian neural correlates of a communicative gesture in chimpanzees. *Proceedings of the Royal Society of London B: Biological sciences* 279:1959–1966.

Price EE, SP Lambeth, SJ Schapiro, et al. 2009. A potent effect of observational learning on chimpanzee tool construction. *Proceedings of the Royal Society of London B: Biological Sciences* 276:3377–3383.

Reamer LA, R Haller, SP Lambeth, et al. 2017. Behavioral management of *Pan spp.*, Chapter 23. In *Handbook of Primate Behavioral Management*. Schapiro SJ (ed.). Boca Raton, FL: CRC Press.

Reamer LA, RL Haller, EJ Thiele, et al. 2014. Factors affecting initial training success of blood glucose testing in captive chimpanzees (*Pan troglodytes*). *Zoo Biology* 33:212–220.

Schapiro SJ, JE Perlman, E Thiele, et al. 2005. Training nonhuman primates to perform behaviors useful in biomedical research. *Laboratory Animal* 34(5):37–42, 2005.

Schapiro SJ, LA Reamer, MC Mareno, et al. 2014. Providing chimpanzees with opportunities to voluntarily participate in their own care: Choice of medications. *American Journal of Primatology* 76(Suppl 1):73.

Sherwood CC, T Duka, CD Stimpson, et al. 2010. Neocortical synaptophysin asymmetry and behavioral lateralization in chimpanzees (*Pan troglodytes*). *The European Journal of Neuroscience* 31(8):1456–1464.

Silk JB, SF Brosnan, J Henrich, et al. 2013. Chimpanzees share food for many reasons: The role of kinship, reciprocity, social bonds, and harassment on food transfers. *Animal Behaviour* 85:941–947.

Silk JB, SF Brosnan, J Vonk, et al. 2005. Chimpanzees are indifferent to the welfare of unrelated group members. *Nature* 437:1357–1359.

Teffer K, DP Buxhoeveden, CD Stimpson, et al. 2013. Developmental changes in the spatial organization of neurons in the neocortex of humans and common chimpanzees. *Journal of Comparative Neurology* 521:4249–4259.

Vale GL, SJ Davis, E van de Waal, et al. In press. Unique test for conformity to new local dietary preferences in migrating chimpanzees. *Proceedings of the Royal Society of London B: Biological Sciences.*

Vale GL, EG Flynn, SP Lambeth, et al. 2014. Public information use in chimpanzees (*Pan troglodytes*) and children (*Homo sapiens*). *Journal of Comparative Psychology* 128:215–223.

Vale GL, EG Flynn, L Pender, et al. 2016. Robust retention and transfer of tool construction techniques in chimpanzees. *Journal of Comparative Psychology* 130:24–35.

Vonk J, SF Brosnan, JB Silk, et al. 2008. Chimpanzees do not take advantage of very low cost opportunities to deliver food to unrelated group members. *Animal Behaviour* 75:1757–1770.

Wallez C, J Schaeffer, A Meguerditchian, et al. 2012. Contrast of hemispheric lateralization for oro-facial movements between learned attention-getting sounds and species-typical vocalizations in chimpanzees: Extension in a second colony. *Brain and Language* 123:75–79.

Whiten A, A Spiteri, V Horner, et al. 2007. High-fidelity transmission of multiple traditions within and between groups of chimpanzees. *Current Biology* 17:1038–1043.

CHAPTER 17

Pairing Strategies for Cynomolgus Macaques

Keely McGrew
Charles River Laboratories

CONTENTS

INTRODUCTION

The importance of social housing of captive nonhuman primates (NHPs) to their psychological well-being has been well established (Baker et al. 2012). In spite of the benefits, the ability to provide social contact can be difficult in the laboratory setting. As recently as 2007, more than 50% of indoor-housed primates in the United States were singly housed (Baker et al. 2007). Some of the impediments to social housing in research facilities include overestimation of the risks involved, underestimation of the benefits of social housing, equipment and space limitations, and real or perceived study constraints (DiVincenti and Wyatt 2011).

For social species, like cynomolgus monkeys (*Macaca fascicularis*), contact with a conspecific is highly valued (Bernstein 1998). Cynomolgus macaques naturally live in multimale, multifemale troops that are organized around matrilineal hierarchies. Males tend to stay in their natal group until fully mature and then quickly join a new group. They are not solitary creatures and seek out social contact (Aureli et al. 1989; Honess 2017).

Despite the perceived impediments, social housing of primates in research facilities is expected, and enforced, by regulatory agencies (USDA 1985). Pair housing is a common method of providing social contact within the constraints of the laboratory environment. Providing NHPs with social contact through pairing allows them to express natural behaviors such as grooming, playing, mounting, or huddling. Pair housing reduces the chance of abnormal behavior development (Reinhardt 1999) and allows for regulation of stress (Sachser et al. 1998; Weatherall et al. 2006). Providing social housing enhances the validity of the animal as a research subject by reducing the negative impact of stressful conditions (Mendoza 1991; Mendoza et al. 2000; Schapiro et al. 2000; Gilbert and Baker 2011).

By socially housing NHPs at import quarantine facilities, it may be possible to establish pairs earlier in the research "lives" of the animals and maintain them through subsequent transfers to research institutions. The aim of this chapter is to share pairing strategies (1) documented in past literature and (2) implemented in the evolution of a large social housing program at an import quarantine facility, including the methodologies implemented in choosing, introducing, and monitoring new sets of pairs of different age classes.

METHODOLOGY

Selection of Social Partners

Choosing the animals suitable for pair housing may involve evaluating age class, sex, weight, study design/intention, rearing and behavioral history, temperament, and viral status. Facility procedures and design may drive some of these decisions. Many facilities select isosexual pairs for social housing to prevent reproduction or the implementation of birth control methods.

It is important to keep in mind when designing a pair housing strategy that not all macaque species are identical in behaviors. Rhesus and cynomolgus macaques respond differently during the human intruder test (Clarke and Mason 1988). Slight differences in manifestations of aggression and affiliation between cynos, rhesus, and pigtails have been documented in the literature (Carlson 2008). Levels of CSF 5-HIAA (a metabolite of serotonin) in male monkeys may determine their response to appeasement behaviors in others. For example, low levels of CSF 5-HIAA may result in higher aggression in individuals (Westergaard et al. 1999). Likewise, individuals may differ in their activation of the hypothalamic-pituitary-adrenal (HPA) axis—some mild, some prolonged—leading to differences in reactivity and impulsivity. Therefore, a "one size fits all" strategy is unlikely to work for all macaques or even for all members of a single species housed at a specific institution. The processes described in the following text are processes that have worked at the author's site with cynomolgus macaques only, as rhesus monkeys are not typically housed at this site.

The author's site has created more than 3000 new, isosexual pairs of juvenile cynos of both genders with extremely high degrees of success (close to 100% for both genders). Success rates were 96%–97% for approximately 1000 isosexual pairs of slightly older, subadult animals of both sexes. Although both the juveniles and subadults come from different suppliers, the rearing histories and viral status of the monkeys are similar.

Initially, our partner selection method did not take into account the animals' study destiny as a pairing criterion. Unfortunately, this resulted in many pairs being disrupted so that one or both animals could go on study, with new pairs needing to be formed. To reduce the time burden associated with forming and re-forming pairs, and to increase the chances that pair mates would go on study together, we now use assigned order numbers (study destiny) as a selection criterion when establishing pairs.

Obtaining information from suppliers on previously compatible social partners and maintaining or reestablishing those pairs can facilitate earlier pairing of adult males, lowering the risk of fighting. Behavioral history and temperament were taken into account when choosing partners for adult males at our facility. A temperament assessment can also be a useful tool to choose partners that are the most likely to be compatible, thereby reducing the risks associated with pairing.

A technique for analyzing pre-pairing behavior using a modified version of the human intruder test (PAIR-T method) to select particular partners was developed for the adult males at the author's site. The PAIR-T method increased the percentage of successful pairs by 8%, to >90% (Harding 2012). To date, this method has been used in >250 pairings of adult males with continued high

Table 17.1 The Behaviors That Contribute to the Five Temperament Types

Temperament Type	Dominant	Neutral	Affiliative	Anxious	Fearful
Behavioral characteristics 1	Open-mouth threats	Scanning room	Lip smacking, coo calls	Combination of threats and lip smacking	Fast lip smacking (ears pulled back)
Behavioral characteristics 2	Lunging at cage front	Sitting calmly	Approaches cage front	Pacing or circling	Presses into back corner, freezes
Behavioral characteristics 3	Yawning, cage rattling, threat displays	Engages in normal behavior: playing, exploring	Peering, presenting, presses body to cage front	Increases conflicting facial gestures and pacing	May dart from side to side, fear grimace, continued vocalization

success rates. Our rationale was that matching more aggressive males with less aggressive males would result in more successful pairs, owing to the natural dominant–subordinate relationship that commonly exists with this type of pairing (Reinhardt 1989). The PAIR-T test uses elements of the human intruder test, a temperament assessment system that involves recording behaviors in the presence of a human intruder offering two phases of observation (profile and stare phase), widely used by Kalin and colleagues (Kalin and Shelton 1989; Kalin et al. 1991, 1998; see also chapters by Capitanio 2017 and Coleman 2017), to study fear responses in infant rhesus macaques. The PAIR-T method utilizes a modified cage-side version of this test.

The *profile* phase of the human intruder test consists of the observer positioning himself/herself ~3 ft in front of the cage and presenting his/her left profile to the animal (facing away from the animal). In the *stare* phase, the observer turns and looks directly at the animal. A score sheet is used that includes an ethogram of common behavioral responses, 17 of which were identified as significant behavioral markers for assessments of compatibility. The behaviors are recorded using a 0–1 scale, and the information is then used to categorize each animal into one of five different "temperament types": fearful, anxious, affiliative, neutral, and dominant (Table 17.1). The key behaviors that are recorded are outlined and defined in Table 17.2. The temperament types are then used to assess the risks associated with pairing. Typically, the information is used to reduce the chance of pairing two animals that display high rates of aggressive behaviors with one another, and instead increase the chance of choosing two monkeys that are likely to form a dominant–subordinate relationship.

Although pairing animals of a similar temperament has been successful for female cynomolgus (Coleman 2012), it has been postulated that this may not be the case for males. Focusing on the mean temperament value of the two individuals may work best for adult male pairings (Capitanio et al. 2017). During the development of the PAIR-T assessment technique, pairing two males that were similar in temperament and that scored high in aggression, low in fear, and low in affiliative behaviors was avoided; instead, these types of animals were paired with animals of dissimilar temperament (for example, choosing a partner low on aggression and high on affiliative behavior). To date, we have used this method with 272 pairs of adult males. The success rates of the different combinations of "temperaments" are presented in Table 17.3.

Equipment

There are several cages on the market that are designed to facilitate social housing. It is less labor intensive for technicians responsible for pairing if the cage is designed to facilitate the pairing process. Some aspects of caging that make pairing easier include (1) built-in tactile dividers, in addition to a solid panel; (2) removable floor panels that provide additional vertical space; and (3) flight zones that allow subordinate animals sufficient room to flee from dominants.

Table 17.2   Operational Definitions of the Behaviors That Contribute to the Five Temperament Types

Associated Temperament	Behavior	Definition
Dominant	Open-mouth threat	Aggressive gesture where jaw is open and teeth bared
	Stare threat	Assertive expression where head and neck are forward and eyes are open and wide
	Cage display	Assertive behavior involving grabbing a part of the cage and rocking quickly so that cage moves and rattles
	Aggressive charge/lunge	Rushing forward in an assertive manner, which may end with an abrupt stop
Fearful	Fear grimace	Submissive facial expression with the corners of lips drawn back, exposing the lower and upper teeth
	Freeze	Fixed or motionless stiff body posture, usually accompanied by averted gaze
	Crouch	A crouched position with legs and arms held beneath the body, head lowered; often in back corner of cage
	Alarm call	Loud, high-pitched fearful vocalization
	Submissive present	A rear-facing position where the hindquarters of the animal is displayed with tail raised or to the side
Anxious	Yawn	Inhalation of air through an open mouth with teeth exposed, usually eyes are averted
	Scratch	Vigorous movement of arms or legs with partly flexed fingernails being drawn deliberately across the skin
	Stereotypy	Rhythmic or repeated nonfunctional locomotion
Neutral	Scanning room	Maintaining an unfocused or static position with a normal/relaxed posture with no other simultaneous behavior. If not unfocused, may be observing environment in a calm manner
	Self-groom	Normal self-directed picking or spreading of fur with hands or mouth; cleaning of hands/nails with fingers or teeth
Affiliative	Lip smack	Mouth slightly opened and closed rhythmically. As the mouth is opened, there is a smacking sound when the tongue is drawn across the palate
	Present	Put forward belly, neck, or other body part for grooming, not a submissive rump present
	Coo call	Vocalization of medium pitch and intensity, mouth opened in circle or diamond shape

Introductions

There are various methods for initially introducing new social partners to one another discussed in the literature. Truelove et al. (2017) have identified the five main introduction techniques as follows: gradual steps, cage-run-cage, rapid steps, transport, and anesthetization (see Table 17.4). *Gradual step* introductions employ the use of visual contact and/or protected contact panels to introduce the animals slowly over time, monitoring each step along the way, whereas *rapid step* involves moving from visual contact to full contact more quickly (within 1 day). The *cage-run-cage* system allows the animals to be initially introduced in cages, followed by exposure in larger caging or in a run, before moving to full contact in caging. *Transport* introductions are conducted in novel cages

Table 17.3 Number of Successful and Unsuccessful Pairs by Temperament Combinations

Temperament Combination	Number of Successful Pairs	Number of Unsuccessful Pairs	Total Pairings	Success Rate (%)
Dominant-dominant	1	1	2	50
Neutral/dominant	14	1	15	93
Dominant/affiliative	18	2	20	90
Dominant/anxious	30	7	37	81
Dominant/fearful	8	1	9	89
Neutral/neutral	25	3	28	89
Neutral/affiliative	19	5	24	79
Neutral/anxious	20	4	24	83
Neutral/fearful	4	1	5	80
Affiliative/affiliative	19	1	20	95
Affiliative/anxious	23	3	26	88
Affiliative/fearful	8	2	10	80
Anxious/anxious	23	3	26	88
Anxious/fearful	17	1	18	94
Fearful/fearful	8	0	8	100

Table 17.4 Brief Descriptions of Different Introduction Methods

Introduction Method	Steps	Advantages	Disadvantages
Gradual steps	Visual, protected, full contact	Minimization of injury risk to individuals	Increased staff time for monitoring steps
Rapid steps	Visual, full contact	Quick, less staff time required	Increased short-term distress for individuals
Cage-run-cage	Visual, protected, full contact	Additional flight space	Increased staff time for monitoring steps
Transport	Transport box to full contact	Quick, potentially fewer wounding events	Increased cumulative stress at transport
Anesthetization	Full contact at anesthesia recovery	Quick, potentially fewer wounding events	Differing rates of anesthetic recovery may pose danger

and involve introductions at the time of transport procedures; finally, *anesthetization* introductions are full-contact introductions that occur during recovery from anesthesia in an animal's home cage.

With higher-risk pairs (involving adult males older than 5 years of age), the gradual steps or cage-run-cage methods are used by some facilities in an attempt to allow the animals to establish a dominant–subordinate relationship prior to full contact (Crockett et al. 1994). The gradual step method involves the use of dividers, allowing visual or tactile contact over a period of days or weeks, until it is determined that the animals are showing behaviors that indicate they will be socially compatible. Evaluating animals during this introduction period may eliminate incompatible animals from progressing to the full-contact phase, and, in some cases, has been demonstrated to reduce injury risk (West et al. 2009). This method of gradual introduction is a common pairing method utilized across laboratory animal facilities. This method can be more labor intensive than the rapid step or transport method, but may be more appropriate for adult males, given their higher risk of injury. The anesthetization method has been used with a degree of success in some populations of macaques (Nelsen et al. 2014). However, it is not widely used owing to concerns about differing recovery rates of animals under anesthesia.

We have used four of these methods to pair house cynomolgus macaques in the past several years; the only method we did not employ was cage-run-cage. The transport method, as defined by Truelove et al. (2017), most closely fits the pairing method that we use when monkeys initially arrive at our facility. The "novelty factor" of the transport method seems to help facilitate pair bonds without seeming to negatively impact the animals during their first days after arrival. This allows for the formation of pairs and trios by husbandry staff during the initial stages of the housing process, expediting operations and providing important, immediate social support for the monkeys to combat the stress associated with transport.

Social housing success rates were assigned based on the percentage of pairs that lived together for *at least 2 weeks* without being separated for incompatibility. This criterion was chosen to be consistent with other published literature (DiVincenti and Wyatt 2011) and with the demands of the particular study site. Our juvenile pairing attempts were successful nearly 100% of the time, in both sexes, when using both the rapid steps and the transport methods. Adult (more than 5 years of age) female pairs ($n = 73$) also had nearly 100% success with the rapid steps method. There were no differences in success rates between the rapid steps and the gradual step introduction methods for juvenile males, juvenile females, and adult females. Therefore, the rapid step method became the most commonly utilized technique, because it is the least labor intensive, placing the smallest time burden on staff. Although the gradual steps, transport methods, and anesthetization methods have been used to pair house adult males at our site, the success rate of the rapid step method was so high (83% success in more than 500 sets of pairs using this method) that it became the default pairing technique. Over time, this method was coupled with the PAIR-T method, increasing the success rate to 91% for 274 additional pairs of adult males in the past few years.

Regardless of the strategy used for adult male introductions, these types of pairings require more frequent monitoring during the initial phases of the introduction than do pairs of juveniles or subadults (see next paragraph). Relatively large numbers of pairs of juveniles or subadults can be formed and monitored simultaneously, because these types of pairings have been determined to be unlikely to result in fighting or injury requiring intervention. For juveniles and subadults, the initial continuous observation may be brief. Animals have access to each other day and night from the onset of the pairing.

Initial full-contact introductions of adult males should involve continuous observations. The question in offering full-contact access from a monitoring perspective usually involves determining when to move from continuous observation (observer in the room with the animals in close enough proximity to be able to intervene if a risk develops) to intermittent checks. In general, when pairing adult males, it is recommended that at least three positive interactions and no negative interactions be observed prior to switching from continuous cage-side monitoring to intermittent checks throughout the day. Typically, once the third positive interaction has occurred and there have been no negative interactions, the observer will move out of direct proximity to the cages but remain within earshot for another few minutes, taking the opportunity to log the introductions on the room record. If there are no alarm calls or indications of scuffling during this period, the animals are assigned a series of intermittent checks at a frequency of twice the regular rate of health observations for the next 3 days (example: check mid-morning and mid-afternoon, in addition to early morning and late afternoon). In our experience, if fighting is to occur, it generally occurs within the first 3 days of introduction, justifying the increase in monitoring frequency during the first 3 days of adult male introductions.

In some cases, the animals do not engage in interactions that are either obviously positive or negative before the continuous monitoring phase ends, or they engage in a mixture of negative and positive interactions. In these cases, intermittent monitoring checks are assigned across shorter time points (example: every half hour on the first day, and every 2 h throughout the day over the next 2 working days).

Because of the high-risk interval of 3 days, adult males are typically paired early in the week, allowing enough time for frequent monitoring while staffing levels are at their highest. With adult males, however, it is our experience that conducting no more than five simultaneous pairing attempts is the safest option, allowing the observer to focus full attention on that small number of pairs. In addition, it is prudent to coordinate with the veterinary department prior to pairing high-risk animals (or any animals, for that matter) to ensure medical attention, if needed, can be provided quickly.

Behaviors indicative of compatibility exhibited by new pairs of any age class include: monkeys moving in unison, hugging, allogrooming, playing, sharing resources (such as space, food, and enrichment; Watson 2002), and co-enlisting against perceived threats (threat signals to a common "enemy"). Behaviors indicative of incompatibility include: unequal food sharing, frequent bickering (e.g., chasing/grabbing), attempts to injure one another (biting, scratching, or hitting), avoidance of eye contact between pair mates, avoidance of space sharing, canine grinding, charging, threat yawning, or other aggressive displays. Additional signs of incompatibility, specific to adult males, include frequent mounting and chest-to-chest hugs. During introductions of adult males, an observer might notice the animals engaging in this chest-to-chest hug, which usually progresses to genital sniffing and grabbing, and then, possibly, biting at one another's face and shoulders. This behavior can result in fighting if one of the animals does not submit to the other during the first few minutes, making this a behavior pattern that must be closely monitored.

Pairing attempts do not need to be immediately halted if some of the aforementioned behaviors that may signal incompatibility are noted. Monkeys can be given three negative interaction attempts before the decision is made to halt the pairing and separate the animals. However, if the animals progress from a negative social interaction to actual fighting, the pairing is immediately halted and the divider is put back in place.

When two adult males begin fighting during a full-contact introduction, it can be difficult to insert the divider between them. It is best to prepare in advance for this outcome, so that one has the ability to separate the animals more quickly. Usually, a loud interruptive noise (hand clapping, verbal cue) is enough to break the animals' focus on one another and to allow the divider to be used to separate the animals into different cages. Presenting a negative reinforcer (placing a hand on the squeeze back mechanism) may also provide incentive for the animals to stop their behavior and move to different cages.

A pair-housing plan should have a process to it but should also remain fluid and flexible, depending on the state of the animals involved. If there is a high arousal level of other animals in the room (e.g., if animals are cage shaking, vocalizing loudly or frequently, or have high activity levels), it might make sense to discontinue the introduction(s) and try again another day. It can be helpful to use large areas (more than the minimum required for two animals) for introductions, particularly for adult males. Once paired, changes in cage location, changes in the composition of animals within the room, and periods of pair separation should be minimized to increase the probability of long-term success.

The observers who are assessing pair compatibility must have sufficient expertise at assessing the nuances of NHP behavior, particularly during introductions and pair assessments. To determine whether compatibility data are being collected consistently, interobserver reliability analyses should be performed with a minimum of 80% agreement between observers indicative of reliability. Interobserver reliability analyses may require a combination of videotaped and live observations to adequately train staff, and may be time consuming, but are absolutely imperative for collecting meaningful data and making appropriate decisions.

LONG-TERM MONITORING AND MAINTENANCE

Husbandry and behavior staff can work together to monitor new pairs for compatible/incompatible behaviors. At our facility, pair monitoring is conducted informally by husbandry staff during routine health checks and husbandry duties. The main observations made by the husbandry staff include health-related reports (injuries, etc.), but the staff is also trained to recognize behaviors indicative of compatibility. If injury appears imminent, the husbandry staff may separate the pair, but otherwise, the pair is kept together until further observations can be performed by one of the other groups (behavior, veterinary) involved in the pairing process. When a pair is noted to behave in a manner consistent with incompatibility, it is reported on the health observation form and communicated back to the behavior staff. A common reason that pairs are reported for incompatibility by husbandry staff include bickering over food or enrichment, which is not an automatic exclusion from social housing, unless the behavior and the veterinary staff agree that a significant risk of injury is present. Following a husbandry-generated incompatible behavior report, the behavior staff conducts a formal assessment of the pair.

The behavior group works with the veterinary staff to ensure input from all relevant parties before reaching a decision about a pair's long-term future. If a pair is separated for incompatible behaviors, a note is added to the animals' medical record to keep track of social housing outcomes for individuals. A few pairs have broken up after a long-term relationship (1–2 years), but the vast majority of pairs have been maintained intact for the duration of the animals' stay at our facility, which typically ranges between 80 and 200 days.

SPECIAL CONSIDERATIONS FOR RESEARCH

Numerous reports demonstrate research primates can be successfully pair housed under various conditions previously believed to preclude social housing, such as instrumentation with cranial implants (Reinhardt 1989; Roberts and Platt 2005), biotelemetry devices (Doyle et al. 2008), and postvascular access port surgery (Murray et al. 2002). Clinical observations for paired animals may need to be conducted slightly differently from that for individually housed animals, but they can be done. For example, when making cage or pan observations (excreta), determining which animal is affected may be difficult. Potential solutions include (1) attributing the finding to the pair rather than to an individual animal or (2) separating the animals for a brief period to see whether the finding is repeated, and then attributing it to the appropriate individual(s). For food intake and metabolic studies, the pair may need to be temporarily separated using a mesh divider (to allow for visual and tactile access to their partner) during food distribution and collection of urine and feces (Reinhardt and Reinhardt 2001).

Maintaining stable pairs is important for successful social housing, as well as for the psychological well-being of the NHPs. Randomizing pairs of primates as if they were an individual animal will allow for preservation of existing partnerships throughout an entire study. Although there are unique challenges associated with conducting studies when animals are socially housed, maintaining monkeys in social settings addresses appropriate standards in both animal welfare and scientific practices related to the design and conduct of reliable and valid research.

SUMMARY

Despite the challenges, successful pair housing of macaques has been demonstrated to be possible in the laboratory environment. It has been learned over the past decades of social housing in laboratories that each species and age group of macaques has different levels of pairing risk. A single approach is unlikely to work for all members of a species, or for all animals at a site. Juvenile and subadult cynomolgus macaques of both sexes, and adult females typically, can be paired with minimal risk. Pairing male cynomolgus macaques requires a balanced approach, but it is possible. Strategies that can be employed to decrease the risks associated with adult male pairings are gradual introductions, use of an ethogram to determine temperament and probable compatibility in advance (like the PAIR-T or a similar method), forming pairs under certain conditions (low arousal level in the room, increased floor or vertical space, etc.), and the use of increased, systematic monitoring schedules. The benefits of pair housing generally outweigh the potential risks. In the case of our facility, using all of these strategies has allowed us to maintain nearly 100% social housing, with partners staying together essentially from arrival until departure, despite a population of monkeys that arrives, and then departs, relatively quickly (80–200 days).

REFERENCES

Aureli, F., C. P. van Schaik, and J. A. R. A. M. van Hooff. Functional aspects of reconciliation among captive long-tailed macaques (*Macaca fascicularis*). *American Journal of Primatology* 19, no. 1 (1989): 39–51.

Baker, K. C., M. A. Bloomsmith, B. Oettinger, K. Neu, C. Griffis, V. Schoof, and M. Maloney. Benefits of pair housing are consistent across a diverse population of rhesus macaques. *Applied Animal Behaviour Science* 137, no. 3 (2012): 148–156.

Baker, K. C., J. L. Weed, C. M. Crockett, and M. A. Bloomsmith. Survey of environmental enhancement programs for laboratory primates. *American Journal of Primatology* 69, no. 4 (2007): 377–394.

Bernstein, I. S. *The Psychological Well-Being of Nonhuman Primates*. National Academy Press, Washington, DC, 1998.

Capitanio, J. P. Variation in biobehavioral organization, Chapter 5. In Schapiro, S. J. (ed.) *Handbook of Primate Behavioral Management*, 55–73. CRC Press, Boca Raton, FL, 2017.

Capitanio, J. P., S. A. Blozis, J. Snarr, A. Steward, and B. J. McCowan. Do "birds of a feather flock together" or do "opposites attract"? Behavioral responses and temperament predict success in pairings of rhesus monkeys in a laboratory setting. *American Journal of Primatology* 79, no. 1 (2017): 1–11.

Carlson, J. *Safe Pair Housing of Macaques*. Animal Welfare Institute, Washington, DC, 2008.

Clarke, A. S. and W. A. Mason. Differences among three macaque species in responsiveness to an observer. *International Journal of Primatology* 9, no. 4 (1988): 347–364.

Coleman, K. Individual differences in temperament and behavioral management practices for nonhuman primates. *Applied Animal Behavioral Science* 137, no. 3 (2012): 106–113.

Coleman, K. Individual differences in temperament and behavioral management, Chapter 7. In Schapiro, S. J. (ed.) *Handbook of Primate Behavioral Management*, 95–113. CRC Press, Boca Raton, FL, 2017.

Crockett, C. M., C. L. Bowers, D. M. Bowden, and G. P. Sackett. Sex differences in compatibility of pair-housed adult longtailed macaques. *American Journal of Primatology* 32, no. 2 (1994): 73–94.

DiVincenti, L., Jr. and J. D. Wyatt. Pair housing of macaques in research facilities: A science-based review of benefits and risks. *Journal of the American Association for Laboratory Animal Science* 50, no. 6 (2011): 856–863.

Doyle, L. A., K. C. Baker, and L. D. Cox. Physiological and behavioral effects of social introduction on adult male rhesus macaques. *American Journal of Primatology* 70, no. 6 (2008): 542–550.

Gilbert, M. H. and K. C. Baker. Social buffering in adult male rhesus macaques (*Macaca mulatta*): Effects of stressful events in single vs. pair housing. *Journal of Medical Primatology* 40, no. 2 (2011): 71–78.

Harding, K. Assessment of a temperament test for use in pairing adult male *Macaca fascicularis*. *Journal of the American Association for Laboratory Animal Science* 51, no. 5 (2012): 636.

Honess, P. Behavioral management of long-tailed macaques (*Macaca fascicularis*), Chapter 20. In Schapiro, S. J. (ed.) *Handbook of Primate Behavioral Management*, 305–338. CRC Press, Boca Raton, FL, 2017.

Kalin, N. H., C. Larson, S. E. Shelton, and R. J. Davidson. Asymmetric frontal brain activity, cortisol, and behavior associated with fearful temperament in rhesus monkeys. *Behavioral Neuroscience* 112, no. 2 (1998): 286.

Kalin, N. H. and S. E. Shelton. Defensive behaviors in infant rhesus monkeys: Environmental cues and neurochemical regulation. *Science* 243, no. 4899 (1989): 1718–1721.

Kalin, N. H., S. E. Shelton, and L. K. Takahashi. Defensive behaviors in infant rhesus monkeys: Ontogeny and context-dependent selective expression. *Child Development* 62, no. 5 (1991): 1175–1183.

Mendoza, S. P. Sociophysiology of well-being in nonhuman primates. *Laboratory Animal Science* 41, no. 4 (1991): 344–349.

Mendoza, S. P., J. P. Capitanio, and W. A. Mason. Chronic social stress: Studies in non-human primates. In Moberg, G., Mench, J. (eds.), The *Biology of Animal Stress: Basic Principles and Implications for Animal Welfare,* CABI Publishing, Wallingford, UK (2000), pp. 227–248.

Murray, L., M. Hartner, and L. P. Clark. Enhancing postsurgical recovery of pair-housed nonhuman primates (*M. fascicularis*). *Contemporary Topics in Laboratory Animal Science* 41 (2002): 112–113.

Nelsen, S. L., D. Bradford, and P. Houghton. A comparison of two social housing techniques for sexually mature male cynomolgus macaques (*Macaca fascicularis*). *American Journal of Primatology* 76 (2014): 104.

Reinhardt, V. Behavioral responses of unrelated adult male rhesus monkeys familiarized and paired for the purpose of environmental enrichment. *American Journal of Primatology* 17, no. 3 (1989): 243–248.

Reinhardt, V. Pair-housing overcomes self-biting behavior in macaques. *Laboratory Primate Newsletter* 38 (1999): 4–6.

Reinhardt, V. and Reinhardt, A. *Environmental Enrichment for Caged Rhesus Macaques (Macaca mulatta)—Photographic Documentation and Literature Review,* 2nd Edition. Animal Welfare Institute, Washington, DC, 2001.

Roberts, S. J. and M. L. Platt. Effects of isosexual pair-housing on biomedical implants and study participation in male macaques. *Journal of the American Association for Laboratory Animal Science* 44, no. 5 (2005): 13–18.

Sachser, N., M. Dürschlag, and D. Hirzel. Social relationships and the management of stress. *Psychoneuroendocrinology* 23, no. 8 (1998): 891–904.

Schapiro, S. J., P. N. Nehete, J. E. Perlman, and K. J. Sastry. A comparison of cell-mediated immune responses in rhesus macaques housed singly, in pairs, or in groups. *Applied Animal Behaviour Science* 68, no. 1 (2000): 67–84.

Truelove, M. A., A. L. Martin, J. E. Perlman, J. S. Wood, and M. A. Bloomsmith. Pair housing of macaques: A review of partner selection, introduction techniques, monitoring for compatibility, and methods for long-term maintenance of pairs. *American Journal of Primatology* 79, no. 1 (2017): 1–15.

U.S. Department of Agriculture. Animal Welfare Act, Food Security Act, title 17, subtitle F, section 1752. *Federal Register* 52 (1985): 10303.

Watson, L. M. A successful program for same-and cross-age pair-housing adult and subadult male *Macaca fascicularis*. *Laboratory Primate Newsletter* 41, no. 2 (2002): 6–9.

Weatherall, D., P. Goodfellow, and J. Harris. The use of non-human primates in research. A working group report Chaired by Sir David Weatherall FRS FMedSci, 2006. https://acmedsci.ac.uk/viewFile/publicationDownloads/1165861003.pdf

West, A. M., S. P. Leland, M. W. Collins, T. M. Welty, W. L. Wagner, and J. M. Erwin. Pair-formation in laboratory rhesus macaques (*Macaca mulatta*): A retrospective assessment of a compatibility testing procedure. *American Journal of Primatology* 71 (2009): 41.

Westergaard, G. C., P. T. Mehlman, S. J. Suomi, and J. D. Higley. CSF 5-HIAA and aggression in female macaque monkeys: Species and interindividual differences. *Psychopharmacology* 146, no. 4 (1999): 440–446.

CHAPTER 18

Managing a Behavioral Management Program

Susan P. Lambeth
The University of Texas MD Anderson Cancer Center

Steven J. Schapiro
The University of Texas MD Anderson Cancer Center and University of Copenhagen

CONTENTS

INTRODUCTION

In this chapter, we describe the tools necessary to build, maintain, and manage a behavioral management program for nonhuman primates, including (1) the types of approaches (project-oriented, section-wide, and facility-wide) that will help build a solid foundation and (2) the strategies (reactive and proactive) that can increase your success in managing. We provide examples of the use of both reactive and proactive strategies, underscore the need to understand how animal regulations can support your program, and discuss ways to overcome obstacles in managing.

Primate behavioral management is an integrative approach used to improve captive animal welfare, using environmental enrichment, operant conditioning, and socialization strategies intertwined with colony management and facility design. The goal of an effective behavioral management program is to optimize all aspects of care and provide for the primates' species-typical needs through functional simulations of their natural environment (Desmond, 1994; Schapiro, 2002;

Schapiro and Lambeth, 2007; Schapiro et al., 2001). Managing a successful behavioral management program includes understanding primate behavior, as well as understanding the drives and needs of the people who care for the animals. To establish and maintain an optimal program, it is imperative to stimulate, educate, and inspire the people looking after the primates to provide the best possible care.

WHAT IS MANAGEMENT?

Management is the ability to efficiently attend to, and metaphorically juggle, multiple "things" at one time. Juggling requires focus on the action necessary to toss the one "thing" in your hand, while keeping the other "things" up in the air. Overseeing a behavioral management program includes providing for the animals' physical and psychological needs; modifying and developing the program; educating and motivating personnel; and communicating with a multitude of people. *Effective* managing, like juggling, requires oversight of multiple aspects at the same time, along with the ability to manage one aspect (action of the thing in hand), while still keeping all of the other aspects in mind (the things in the air). This ensures program balance and programmatic growth. For example, you may prefer those aspects of your job that involve managing the care of your animals more than you like the people aspect of managing, but if you do not attend to both (among other things), your program is likely to deteriorate.

BEHAVIORAL MANAGEMENT PROGRAM APPROACHES

When developing or maintaining a behavioral management program, it is important to choose an approach and determine whether this approach facilitates your *most* effective and productive program. There are at least three distinct approaches to developing and maintaining behavioral management programs: (1) a project-based approach; (2) a section-wide approach; and (3) a facility-wide approach (Perlman et al., 2012; Whittaker et al., 2008). A project-based approach might be exemplified by a single caregiver/researcher who proposes a small-scale enrichment or training project, such as testing a new foraging puzzle device with two groups of "their" animals. This approach has the potential to increase primate well-being and may inspire other personnel to initiate additional projects. Success in such small-scale projects can provide the confidence to tackle larger endeavors. Potential disadvantages associated with this approach include a lack of institutional support and limited benefits to animal well-being (e.g., only a small number of animals benefit).

A section-wide approach is usually initiated by section managers. This approach has a bigger scope, can affect a greater number of animals, involves more staff, and if successful, may influence overall operations in other sections. A section-wide approach might be exemplified by one section of your facility, the chimpanzee section for example, proposing a section-wide enrichment or training project, such as cooperative feeding for all of the chimpanzees. For this illustration, cooperative feeding would not be implemented for the rhesus macaques in the rhesus macaque section. The potential disadvantages of implementing a section-wide approach may include limited support or collaboration from other sections, and/or a breakdown of established processes, if management changes.

Facility-wide approaches are initiated and supported at all levels within an organization. Clearly, this is the most beneficial behavioral management approach but can also be the most complicated of the three approaches because it requires extensive planning, resources, and time to develop. However, the distinguishable benefits of this approach are that all animals are included, and program sustainability is greatly enhanced as a result of extensive support from facility leadership. All three approaches can be successful, but the ultimate goal of any behavioral management program is to eventually gain facility support and implement strategies using a facility-wide approach.

At the Michale E. Keeling Center for Comparative Medicine and Research (KCCMR), the behavioral management program was developed using a facility-wide approach with support from the KCCMR director. The program was designed to have "dedicated" enrichment technicians (now behavioral specialists), research assistants, and animal trainers specifically to provide enrichment, assist with primate socialization procedures, collect behavioral information to determine that the implementation of behavioral management techniques are successful, and employ positive reinforcement training techniques to ensure voluntary (1) cooperation of the primates in their own care (husbandry and clinical) and (2) participation in noninvasive behavioral research procedures.

There are pros and cons associated with the dedication of full-time employees (FTEs) to the accomplishment of behavioral management goals, rather than assimilating behavioral management tasks into the job descriptions of all staff members. One advantage is the ability to acquire specially skilled individuals who are trained in the theory and implementation of enrichment, behavioral research, training, and socialization, and assign them specifically to complete these tasks. Disadvantages of having certain employees focused on behavioral management activities include the possibility that other members of the animal care staff may be less likely to participate in those behavioral management-related activities because they view it as "someone else's job," and the specialization may delay the achievement of the ultimate goal of integrating behavioral management responsibilities into all employees' job descriptions.

STRATEGIES TO "DRIVE" A BEHAVIORAL MANAGEMENT PROGRAM

In order to successfully manage a behavioral management program, it is important to identify the strategy that is driving your response to challenges. There are two main strategies to drive a program: (1) a **reactive strategy** that initiates solutions *after* issues or challenges arise and, (2) a **proactive strategy** that anticipates challenges and designs interventions to address them *in advance*. Both strategies may be used in behavioral management programs; however, using a proactive strategy, or turning a reactive strategy into a proactive strategy, builds a more dynamic program, allowing for the development of new and innovative ways to improve the care of captive primates.

A typical reactive strategy that is used in a behavioral management program is to present a solution (e.g., increase enrichment) after an animal has been identified as having a behavioral issue (e.g., exhibiting an abnormal behavior). To address this at the KCCMR, we have developed an electronic behavioral assessment process that can be initiated by *any staff member* who has observed an animal exhibiting a behavior of concern. This person enters the following data into the electronic database: the animal's identification number, a description of the behavior, when the issue occurred, potential triggers (i.e., what was happening in the environment when the behavior was exhibited), and the frequency of the behavior. The behavioral management coordinator then receives an e-mail notification, assigns a priority level, and the type and frequency of observations for the behavioral management specialist to complete. The behavioral management specialist enters historical animal information (rearing, housing, health issues, and research history), interviews the requester, and completes the assigned observations. The last step is to combine all of the information and utilize a problem-solving approach to formulate a solution(s).

Problem-Solving Approach

The problem-solving approach is a step-by-step technique that involves defining the challenge, establishing a goal, gathering all necessary information to form a hypothesis, devising a behavioral management solution(s), and evaluating the implemented solution(s). To maximize the problem-solving approach, it is essential to form a group (containing people who are familiar with the animal and the issue) to discuss and define the challenge. *Defining the problem* or challenge is often left out of the process, but staff may have differing perceptions of the challenge, making it imperative to clearly identify and agree on its definition in order to effectively solve it.

The next step is to summarize any and all relevant information, including when the behavior is occurring; who is involved (both human and animals); and the frequency, duration, and intensity of the behavior. It is important to identify the factors that may be influencing the behavior, such as environmental (i.e., weather, new enclosure, space limitations, new personnel, and animal sedation); behavioral (i.e., intimidating situations and boredom); facility (i.e., insufficient space, rigid schedule, and insufficient resources); social forces at work (i.e., new social member and changes in dominance); or animal health (i.e., illness, physical injury, and geriatric condition). After the relevant information has been gathered, a theory or hypothesis concerning the reason(s) for the problem can be proposed. This step is frequently introduced too early in the process, prior to the assessment of the relevant information, potentially neglecting a credible theory or hypothesis. The next step in the process is to devise and implement behavioral management solutions that are specifically tailored to address the characteristics of the problem. The final, and perhaps the most important, step in successfully addressing a behavioral management issue is to evaluate the "results" of the solution to determine whether the challenge has been satisfactorily addressed, or whether additional strategies need to be implemented (Schapiro and Lambeth, 2007).

COMPREHENSIVE BEHAVIORAL MANAGEMENT PROGRAMS

A proactive strategy to address the development of behavioral issues in captivity is to cultivate a *comprehensive behavioral management program* designed to *prevent* the development of behavioral issues before they arise. A comprehensive behavioral management program promotes species-typical behaviors and increases psychological well-being by going above and beyond the required minimum (see Bloomsmith, 2017). At a minimum, this is accomplished by establishing (1) an environmental enrichment program that addresses the social, physical, feeding, occupational, and sensory needs of the primates by simulating their natural activities; and (2) a positive reinforcement training (PRT) program to achieve the primates' voluntary cooperation with husbandry-, clinical- and, research-related procedures.

The most critical component of an environmental enrichment program is to address the social needs of the primate, including early social rearing experiences, which are essential for normal behavioral development (Anderson and Chamove, 1985; Davenport and Menzel, 1963; Harlow, 1958; Harlow and Harlow, 1962; Harlow et al., 1965; Mason et al., 1968; Turner et al., 1969). Introducing unfamiliar primates to one another in captivity is also extremely critical, and it involves more than simply understanding the methods necessary to familiarize animals with one another. Personnel performing these procedures must also be able to discern the intricacies of primate behavior (dominance, affiliation, submission, fear, aggression, avoidance, posturing, etc.). The introduction process begins with identifying the goals/benefits of the socialization effort; defining the plan; securing the necessary resources, including the space to safely implement the introduction procedures (protected contact across wire mesh and the ability to separate easily); and gaining support from others (veterinarians, colony managers, and husbandry and clinical staff) to plan timing, location, and for any necessary husbandry modifications. To successfully prepare for animal introductions, it is important to gather any available information on the animal's rearing history, previous social experience, personality, and dominance tendencies or current dominance level of the primate within its social group.

Utilizing a customized approach [as discussed by Bloomsmith (2017)] is a crucial aspect of constructing a comprehensive behavioral management program (Bloomsmith et al., 1991). Moreover, enrichment, training, and socialization should be routinely incorporated into every aspect of the animals' care, enabling all staff members to participate and to view their contributions as fundamental parts of optimizing the animals' captive experience.

Clearly, establishing a comprehensive behavioral management program can be a daunting, time-consuming process, but the amount of personnel time and effort that is expended to address behavioral issues *after they develop* is likely to far exceed the initial output of resources, time, and energy to establish the program. The investment in a proactive behavioral management program greatly reduces the need for later resource output and establishes an improved environment that increases animal well-being, establishes a precedent for continuous reevaluation and problem prevention, and sets the stage for future success.

EXAMPLES OF TURNING REACTIVE STRATEGIES INTO PROACTIVE STRATEGIES

Often, unforeseen circumstances require a reactive, rather than a proactive, strategy to *immediately* address an "acute" animal issue. If that issue is likely to reoccur, the reactive solutions employed should serve as "pilot data" for determining strategies to proactively reduce the likelihood of, or prevent, reoccurrence. For example, two overweight adult female chimpanzees at the National Center for Chimpanzee Care (NCCC) at the KCCMR developed type II diabetes. An immediate response, using a reactive strategy, was clearly warranted to address this health issue. We quickly modified and monitored their diet and prescribed medication in an effort to lower their blood glucose (BG) levels. To effectively manage a chimpanzee with type II diabetes and to determine whether the modifications implemented are working, it is imperative to be able to regularly monitor the animals' BG level. Therefore, we used PRT techniques to train these chimpanzees to voluntarily provide capillary blood samples for BG testing.

To provide some context, all of the chimpanzees at the NCCC willingly participate in our extensive PRT program, which is designed to allow the animals to voluntarily cooperate with a variety of husbandry, research, clinical, and colony management tasks (Laule et al., 2003; Schapiro et al., 2005). All of the animals present various body parts on cue to allow clinical care and wound treatment. They are trained to touch a target; desensitized to a variety of veterinary implements (stethoscope, otoscope, Q-tip, ophthalmoscope, tongue depressor, etc.), and trained to feed cooperatively (Bloomsmith et al., 1994) and shift on cue to locations within their enclosure. PRT has also been utilized to successfully train the chimpanzees to voluntarily accept subcutaneous injections (Perlman et al., 2004), station for acupuncture treatment (Magden, 2017; Magden et al., 2013), accept low-level light laser treatment (Magden, 2017; Magden et al., 2016), and allow reading of an implantable cardiac loop recorder in the animal's back (Magden et al., 2016). A majority of the chimpanzees consistently present voluntarily for an anesthetic injection (Lambeth et al., 2006), although neither of the diabetic chimpanzees had been previously trained to allow BG testing.

The animal trainers invested a significant amount of time (daily training sessions over a year) to train the two female type II diabetic chimpanzees to allow BG testing. The extensive time investment associated with this training was acceptable, because we knew that maintenance of this behavior would be critical to provide optimal care to the animals for the rest of their lives. In order to transition our reactive approach to a proactive strategy that addressed the probability that additional older chimpanzees would develop type II diabetes and require similar management practices, we prioritized three objectives as follows: (1) the identification of overweight chimpanzees that were at risk for developing type II diabetes, (2) the design of a multifaceted weight management program for these animals, and (3) the incorporation of a new complex behavior, BG testing, into the general training program for all animals.

Prior to actively training all of the chimpanzees for the BG testing procedure, we assessed each chimpanzee's level of cooperation when first asked to perform the testing procedures (again, prior to attempting to train the animals for the procedures). Nearly 30% of the chimpanzees allowed the entire BG testing procedure (presenting a finger/toe, disinfecting the digit, holding for the lancet device, and allowing blood to be collected on a glucometer test strip for analysis) without any prior training for this specific target behavior (Reamer et al., 2014). This finding prompted us to investigate factors that might affect initial cooperation with BG testing, and our data revealed that sex, personality, and past training performance all affected training success (Reamer et al., 2014). Possessing a repertoire of trained behaviors from which they could "generalize" helped chimpanzees succeed at BG testing, demonstrating that chimpanzees "learn to learn" and underlining the value of having a functional PRT program as part of a captive management system. Proactively including, within the overall training program, the steps in the complex behavior chain required for BG testing ensures that BG testing on those chimpanzees subsequently identified as at-risk for developing type II diabetes will not require any "special" training. This expansion of the repertoire of trained behaviors is likely to enhance our abilities to introduce even more clinical training procedures when needed.

To identify overweight and at-risk chimpanzees, we developed a chimpanzee body condition rating scale to visually monitor the chimpanzees and easily identify weight changes before potential health issues became a threat (Bridges et al., 2013; Lambeth et al., 2011). Captive chimpanzees receive a stable diet and are more restricted in their activity than are their wild counterparts. Consequently, serious weight-related health risks, including cardiovascular anomalies, respiratory issues when anesthetized, the development of type II diabetes, and obesity, can develop.

We identified seven chimpanzees at the NCCC as morbidly obese and designed a proactive weight management program that combined husbandry and behavioral management techniques to achieve acceptable body weights while maintaining the subjects' welfare (Lambeth et al., 2011). The suite of weight management changes included reducing the availability of calories from primate chow, adding fibrous produce to increase satiation, implementing activity-inducing enrichment procedures, using PRT techniques to increase cooperation during feeding, and regular monitoring of body weights. The subjects demonstrated significant clinical improvements, including lowered BG levels and the elimination of anesthesia-associated respiratory issues. Overall, this proactive weight management strategy, utilizing a combination of both husbandry and behavioral management techniques, improved chimpanzee health and welfare.

INITIATING A PROACTIVE STRATEGY

The establishment of a quality of life (QOL) program at the KCCMR that utilizes behavioral guidelines (Lambeth et al., 2013) to ensure that the best judgments are made for a primate with a chronic or debilitating condition is a good example of the initiation of a proactive strategy. This program does not change the professional responsibility of veterinarians to make the ultimate decision concerning the euthanasia of a nonhuman primate. However, it does augment the well-established euthanasia guidelines set forth by the American Veterinary Medical Association (AVMA Guidelines on Euthanasia, 2013) by utilizing observations of behavioral changes from a QOL *team* (animal care staff, behaviorists, trainers, and enrichment personnel, in addition to the veterinarian and other clinical staff) to monitor the primate's QOL. Sometimes, pain and distress are difficult to interpret using solely clinical parameters. Thus, we compiled a list of behaviors and questions designed to solicit information regarding each individual's behavioral characteristics (loves to train, not picky about any foods, etc.) to guide a QOL team to qualitatively and quantitatively assess deviations from the "normal behavioral repertoire" of that *individual* primate. The QOL team then determines the number of behavioral deviations necessary to trigger an immediate discussion of the animal's shifting (typically diminishing) QOL. The development and inclusion of behavioral guidelines in the QOL assessment process is a proactive step forward in improving animal welfare and in defining difficult concepts, such as quality of life, for all primates.

OVERCOMING OBSTACLES—UTILIZING ANIMAL WELFARE REGULATIONS

The USDA Animal Welfare Act (USDA, 1991), the only federal law that regulates the treatment of animals in research, exhibition, transport, and by dealers, defines minimum acceptable standards for primate environmental enhancement to promote psychological well-being (see Hau and Bayne, 2017 for additional discussion). Even though behavioral managers are aware of the governing principles regarding animal welfare, they are often underutilized when attempting to overcome some of the common obstacles associated with the development of behavioral management programs.

Capitalizing on the broadly written regulations can provide the necessary incentives to engage both staff and higher-ups in optimizing primate behavioral management and overall care practices. For example, the regulations require that each facility's enhancement plan "… must include specific provisions to address the social needs of nonhuman primates of species known to exist in social groups in nature" (USDA, 1991; §3.81, p. 100), although the regulations do allow for exceptions and exemptions to the environmental enhancement plan. An exemption from social housing would be applicable if an animal displays overly aggressive behavior; however, somewhat problematically, the definition of overly aggressive behavior is subject to interpretation. Considerations can include: Is the animal overly aggressive if it is "vicious" to another animal once or twice? How often does the primate exhibit this behavior? Is the primate overly aggressive to numerous social partners, or might the aggressive behavior be a result of housing, location, neighboring conspecifics, difficulty assimilating to a new environment, social anxiety, or other reason that could possibly be resolved using behavioral management techniques to alter or improve the behavior? If our commitment is to responsibly manage and maximize primate well-being, as emphasized in the guidelines, then we should prioritize a problem-solving approach (see the Problem-Solving Approach section above) and find a way to provide social options for the aggressive animal, even if it requires housing or location modifications.

The regulations also state that "certain nonhuman primates must be provided special attention regarding enhancement of their environment, based on the needs of the individual species and in accordance with the instructions of the attending veterinarian" (USDA, 1991; §3.81, p. 101). Primates requiring special attention include infants and young juveniles, animals displaying signs of psychological distress through behavior or appearance, animals with restricted activity on biomedical research projects, and animals that are unable to see and hear conspecifics, as well as great apes weighing more than 110 lbs. (50 kg). The regulations do not specifically mention the type of, or the amount of, specific attention. This is another area that is left to the interpretation of the reader and can be a way to maximize enrichment options according to the natural behavior of the animal (Bloomsmith, 2017).

Any nonhuman primate may be exempted from participation in the enrichment plan owing to research protocols, health issues, or considerations of well-being; however, the exemption must be reviewed every 30 days by the veterinarian, unless the exemption is permanent (USDA, 1991; §3.81). It is beneficial, in these cases, to create a systematic process to maintain this information and include a behaviorist reviewer, in addition to the required veterinarian. At the KCCMR, we created an enrichment exemption database, whereby exemption of any animal from the enrichment plan (most exemptions are social) must be documented in the database system. The database entry cues the program to send an e-mail to *both* the veterinarian and the behavioral manager for review and electronic signatures. Clearly, there are benefits to this increased communication between veterinarians and behavioral managers; it allows both entities to closely track the duration of the exemption and to work together to solve the problem.

Understandably, regulations have to be written in ways that do not impose undue burdens on those regulated. It is up to behavioral managers to treat regulations as *minimum* requirements that may require individualized solutions for individual animals. There may be many solutions to what appears to be a single problem, but there are rarely "one-size-fits-all" solutions that will work for all animals or at every facility [see many of the chapters in this volume, including especially Coleman (2017) and Jorgensen (2017)]. The broadly written regulations simultaneously allow situational interpretations by behavioral managers, while adding regulatory "weight" to initiatives to improve welfare.

OVERCOMING OBSTACLES

When managing behavioral management programs, it is not uncommon to be hindered by various obstacles, including garnering support from animal care staff, coworkers, and management; lack of personnel; lack of time; and negative attitudes. While we do not have a magic wand or a panacea for behavioral management obstacles, we provide some hints and examples that we hope are helpful.

Gaining Support

A common impediment to achieving the goals associated with a behavioral management program is the inability to take that first step, even though you have great ideas, have the best interests of the primates at heart, and "know" exactly what needs to be done. Similarly, you may have experienced other members of the primate management team presenting obstacles when you initiate new procedures, or attempt to change or improve existing programs. How can you turn these obstacles into opportunities and bring the naysayers on board? One way to accomplish this is to set yourself up for success in a manner similar to breaking down a primate PRT training plan (reinforcing successive approximations) for a new behavior. Proceed by setting and achieving reasonable, stepwise goals for your behavioral management program, rather than attempting to immediately achieve all of the goals. It can be difficult to convince others to get on board when your plans represent a substantial overhaul of the status quo. Change is difficult for those involved with the care and management of nonhuman primates, even if the change is "good." Therefore, be flexible, go slowly, and find a supportive person to help with, and champion, your idea. Most people working with primates in scientific settings respond to empirical data; therefore, proposing, conducting, and disseminating a small, but meaningful study, involving quantitative data and analysis, can help advance your idea.

As an example, at the NCCC, the husbandry staff currently provides fresh bedding material (e.g., excelsior or fleece) to the chimpanzees each evening to enable them to perform their species-typical nest-building behaviors (Boesch and Boesch-Achermann, 2000; Goodall, 1962, 1986; Maughan and Stanford 2001; Pruetz et al., 2008). Nesting materials were not always a part of the husbandry program for the chimpanzees; we had made a few attempts over the years to supply hay to encourage nesting, but it was not a routine component of the care program. Therefore, we conducted a small study and were able to demonstrate that hay significantly reduced the amount of feces smearing (Neu et al., 2001) performed by the chimpanzees. However, the hay blocked the drains, creating issues with our facility maintenance group, and more importantly, it was reported to be too much additional work for the animal care staff (picking up and properly disposing of soiled hay).

At this potential impasse in our attempts to satisfy the animals' behavioral needs (nesting) without overburdening dedicated employees, we decided to empirically assess the amount of time it took to clean up the hay, and to determine whether there were ways to prevent the hay from blocking the drains. We identified a supportive animal care staff member who volunteered to record information regarding hay disposal (pick up and cleaning times, as well as problems with it going down the drains) and to report the results. The provision of hay added approximately 3–5 min to baseline cleaning times per enclosure, due to the additional raking and build-up of trash it promoted. Raking and hand disposal of the hay greatly reduced the amount of hay that was washed into the drain; however, remnants still went down, causing an unacceptable number of blockages. In addition to the drain-related problems, some of the care personnel were allergic to the hay.

In our continuing quest for nesting materials for chimpanzees that addressed both the needs of human caregivers and of the animals, we conducted a pilot study of excelsior (i.e., wood wool), a material that does not cause human allergies nor break into small pieces likely to block drains, instead of hay as nesting material. Excelsior comes in bales of "flakes," like hay, and was immediately used by the chimpanzees to construct nests that functionally simulated those they build in the wild. Analyses of the costs and additional effort required to distribute and clean up the excelsior revealed that this material could be used in an appropriately cost-effective manner. For the past several years, chimpanzees have received fresh excelsior every evening to build new nests. This is an example of one way in which potential obstacles to behavioral management can be overcome: A problem was identified; potential solutions were proposed; points of resistance and problems with those solutions were identified; modifications were made; and a successful resolution was eventually achieved.

Interestingly, once excelsior for nesting became part of the routine care program, everyone associated with it became proud of this particular refinement. We have disseminated this refinement in many of our interactions in the laboratory animal community (workshops, consulting visits, veterinary residency programs, etc.). In fact, the NIH's working group (National Research Council, 2011) has recommended the provision of nesting materials for chimpanzees as an important component of what they call "ethologically appropriate environments," which we refer to in this volume as Functionally Appropriate Captive Environments. To us, this is a prime example of how systematically and vigorously pursuing animal welfare objectives can result in important enhancements for nonhuman primates living in captive settings.

Assuming the Negative in Others

It is not unusual to become frustrated with others who present obstacles and assume that these people do not care about the animals (or you), are not interested in improving the well-being of the animals, are simply resistant to change, or somehow revel in presenting obstacles. However, it does not yield productive results to assume negative intentions in others; it only builds a wall between you and them, instigating defensive actions and responses. It is important to consider a better alternative: that everyone working with primates cares about their health and well-being. It is then your responsibility to determine the driving force behind their opposition as a means of eventually resolving it.

Everyone working with nonhuman primates has a different motivation. Veterinarians are ultimately responsible for the health and well-being of the animals; so they may view an attempt to try something new as the introduction of new risks to the animals. Animal care staff are rarely well compensated and are infrequently recognized for their hard work and commitment; so asking them for extra help or to willingly accept a change in workload is commonly met with frustration. We try to be inclusive and to involve others in the problem-solving process, while truly listening to their concerns and ideas. We recommend a dialogue, rather than a debate, when you are discussing care changes with individuals who are reluctant to agree with your suggestions. A dialogue respectfully seeks to identify the underlying meaning and to recognize differences, whereas a debate creates barriers between yourself and others. Feeling "judged" and feeling "respected" do not usually occur simultaneously. Arriving at a workable solution does not have to involve agreeing on every single point. The motivations that bring different parties to *workable solutions* can result from what may initially appear to be conflicting goals.

Sometimes, people are not presenting obstacles or purposely weakening an implemented procedure, but rather misinterpreting what has been asked. An example of this at the NCCC involved the animal care staff's ability to maintain reliable shifting behavior (moving from one section of their enclosure to another, when requested to do so) from the chimpanzees. Through the use of problem-solving techniques and identifying reasonable goals, the two dedicated animal trainers were able to train our entire population (>150 chimpanzees) to shift between parts of their enclosure on cue and then transferred the trained shifting behavior to the animal care staff. However, the behavior became unreliable soon after it was transferred to the husbandry staff. We hypothesized that the husbandry staff was not consistently adhering to the shifting protocol and proposed the solution of simply strengthening the shifting protocol and retraining the staff to ensure a consistent shifting process.

Unfortunately, this solution did not result in reliable shifting behavior. The trainers worked again with the animal care staff to ensure that they were providing a consistent cue and were properly reinforcing the animals when they successfully shifted. Even so, the chimpanzees were still not reliably performing the desired behavior. The trainers then shadowed the care staff to determine whether they were inadvertently deviating from the shifting protocol. Despite the fact that the staff had been properly trained to shift animals outside and inside each day, and they had reviewed the improved protocol, there were still slight (undetectable to the care staff), but extremely important, variations in consistency related to the behavior–reward contingency. The procedure for shifting the chimpanzees from the inside to the outdoor portion of their enclosure was to speak the cue, "outside everyone," and then allow the animals a reasonable amount of time to move outside, before shutting the door between the inside and the outside. Once the *door was closed*, the animals would be immediately reinforced on the outside. Shadowing of the animal care staff by the trainers revealed that if a staff member walked into an animal area and all of the chimpanzees were already outside, care staff would shut the doors and start cleaning the inside, denying the animals the opportunity to earn a reinforcement as soon as the door closed. Because the chimpanzees had been trained to receive a cue, perform a behavior, and then receive reinforcement, their trained shifting behavior became erratic when they did not receive reinforcement after the *door closed*.

To proactively address this, we developed a "point person" program, because we only have two dedicated trainers, and they were spending too much time transferring and monitoring trained behaviors. Point people are those members of the animal care staff who display motivation and interest in training, who then receive supplemental coaching in the basic terms and concepts of PRT. They are appointed to monitor the maintenance of trained behaviors that have been transferred to the animal care staff. They are also trained in problem-solving techniques, and they can apply them when animals deviate and do not reliably perform the target behaviors.

Time and Personnel Limitations

You have developed an idea, or you are trying to initiate a program, but no one has the time to implement it. To succeed, you should start small and do everything you can to set yourself, and others, up for success. For instance, instead of trying for sweeping changes to the "way things have been done," simply choose one or two animal enclosures or groups in which to test your idea(s). Similarly, when attempting to convince people, start with those one or two individuals who are likely to be excited about the new idea.

Using PRT techniques can save significant time and resources in the long run, even when initial investments appear to be considerable. A fairly easy example to consider is training monkeys to test their water sources (lixits) on cue. This task must be completed for every enclosure, every day; so if the primates can do the work for you, care staff can complete all of their work more efficiently, and may even be able to undertake additional (behavioral management-related) tasks.

CONCLUSIONS

One of the best ways to overcome obstacles and ensure a successful and evolving behavioral management program that prioritizes primate well-being is to receive and give reinforcement for the program's incremental successes. It is productive to occasionally stop to review the program and identify and evaluate accomplishments. It is important to understand that laboratory animal facilities, especially those caring for nonhuman primates, may be resistant to change. These types of facilities operate most efficiently when care practices are consistent and empirically vetted. Exceptional behavioral management programs evolve and develop over time. It is important to think outside of the box, to learn from the experiences of others, and to keep improving captive care. This chapter has provided you with some practical guidance that you can apply to enhance the behavioral management of the primates under your care.

REFERENCES

Anderson, J.R. and A.S. Chamove. 1985. Early social experience and the development of self-aggression in monkeys. *Biology of Behaviour* 10:147–157.

AVMA Guidelines on Euthanasia. 2013. Available at: https://www.avma.org/KB/Policies/Documents/euthanasia.pdf. Accessed 12/2/16.

Bloomsmith, M.A. 2017. Behavioral management of laboratory primates: Principles and projections, Chapter 29. In Schapiro, S.J. (ed.). *Handbook of Primate Behavioral Management*, 497–513. Boca Raton, FL: CRC Press.

Bloomsmith, M.A., L. Brent, and S.J. Schapiro. 1991. Guidelines for developing and managing an environmental enrichment program for nonhuman primates. *Laboratory Animal Science* 41:372–377.

Bloomsmith, M.A., G.E. Laule, P.L. Alford, et al. 1994. Using training to moderate chimpanzee aggression during feeding. *Zoo Biology* 13(6):557–566.

Boesch, C. and H. Boesch-Achermann. 2000. *The Chimpanzees of Tai Forest: Behavioural Ecology and Evolution*, 316p. Oxford, UK: Oxford University Press.

Bridges, J.P., E.C. Mocarski, L.A. Reamer, S.P. Lambeth, and S.J. Schapiro. 2013. Weight management in captive chimpanzees (Pan troglodytes) using a modified feeding device. *American Journal of Primatology* 75:55.

Coleman, K. 2017. Individual differences in temperament and behavioral management, Chapter 7. In Schapiro, S.J. (ed.). *Handbook of Primate Behavioral Management*, 95–113. Boca Raton, FL: CRC Press.

Davenport, R.K. and E.W. Menzel. 1963. Stereotyped behavior of the infant chimpanzee. *Archives of General Psychiatry* 8(1):99–104.

Desmond, T. 1994. Behavioral management: An integrated approach to animal care. *Annual Proceedings of the American Zoological and Aquarium Association* 19–22.

Goodall, J. 1962. Nest-building behavior in the free ranging chimpanzee. *Annals of the New York Academy of Sciences* 102:455–568.

Goodall, J. 1986. *The Chimpanzees of Gombe: Patterns of Behavior*. Cambridge, MA: The Belknap Press of Harvard University Press.

Harlow, H.F. 1958. The nature of love. *American Psychologist* 13(12):673.

Harlow, H.F., R.O. Dodsworth, and M.K. Harlow. 1965. Total social isolation in monkeys. *Proceedings of the National Academy of Sciences USA* 54(1):90–97.

Harlow, H.F. and M. Harlow. 1962. Social deprivation in monkeys. *Scientific American* 207:136–146.

Hau, J. and Bayne K. 2017. Rules, regulations, guidelines, and directives, Chapter 3. In Schapiro, S.J. (ed.). *Handbook of Primate Behavioral Management*, 25–36. Boca Raton, FL: CRC Press.

Jorgensen, M. 2017. Behavioral management of *Chlorocebus* spp., Chapter 21. In Schapiro, S.J. (ed.). *Handbook of Primate Behavioral Management*, 339–365. Boca Raton, FL: CRC Press.

Lambeth, S.P., B.J. Bernacky, P. Hanley, et al. 2011. Weight management in a captive colony of chimpanzees (*Pan troglodytes*). *American Journal of Primatology* 73:40.

Lambeth, S.P., J. Hau, J.E. Perlman, et al. 2006. Positive reinforcement training affects hematologic and serum chemistry values in captive chimpanzees (*Pan troglodytes*). *American Journal of Primatology* 68:245–256.

Lambeth, S.P., S.J. Schapiro, B.J. Bernacky, et al. 2013. Establishing "quality of life" parameters using behavioural guidelines for humane euthanasia of captive non-human primates. *Animal Welfare* 22(4):429–435.

Laule, G., M.A. Bloomsmith, and S.J. Schapiro. 2003. The use of positive reinforcement training techniques to enhance the care, management, and welfare of primates in the laboratory. *Journal of Applied Animal Welfare Science* 6:163–173.

Magden, E.R. 2017. Positive reinforcement training and health care, Chapter 13. In Schapiro, S.J. (ed.). *Handbook of Primate Behavioral Management*, 201–216. Boca Raton, FL: CRC Press.

Magden, E.R., R. Haller, E. Thiele, et al. 2013. Acupuncture as an effective adjunct therapy for osteoarthritis in chimpanzees (*Pan troglodytes*). *Journal of the American Association for Laboratory Animal Science* 52:475–480.

Magden, E.R., M.M. Sleeper, S.J. Buchl, et al. 2016. Use of an implantable loop recorder in a chimpanzee (*Pan troglodytes*) to monitor cardiac arrhythmias and assess the effects of acupuncture and laser therapy. *Comparative Medicine* 66:52–58.

Mason, W.A., R.K. Davenport, and E.W. Menzel. 1968. Early experience and the social development of rhesus monkeys and chimpanzees. In Newton, G. and S. Levine (eds), 440–480. *Early Experience and Behavior*. Springfield III: Charles Thomas.

Maughan, J.E. and C.B. Stanford. 2001. Terrestrial nesting by chimpanzees in Bwindi Impenetrable National Park, Uganda. *American Journal of Physical Anthropology, Supplement* 32:104.

National Research Council. 2011. *Chimpanzees in Biomedical and Behavioral Research: Assessing the Necessity*. Washington, DC: The National Academies Press.

Neu, K., S. Lambeth, E. Toback, et al. 2001. Hay can be used to decrease feces smearing in groups of captive chimpanzees. *American Journal of Primatology* 54(1):78.

Perlman, J.E., M.A. Bloomsmith, M.A. Whittaker, et al. 2012. Implementing positive reinforcement animal training programs at primate laboratories. *Applied Animal Behaviour Science* 137:114–126.

Perlman, J.E., E. Thiele, M.A. Whittaker, et al. 2004. Training chimpanzees to accept subcutaneous injections using positive reinforcement training techniques. *American Journal of Primatology* 62(Suppl 1):96.

Pruetz, J.D., S.J. Fluton, L.F. Marchant, et al. 2008. Arboreal nesting as anti-predator adaptation by savanna chimpanzees (*Pan troglodytes verus*) in southeastern Senegal. *American Journal of Primatology* 70(4):393.

Reamer, L.A., R.L. Haller, E.J. Thiele, et al. 2014. Factors affecting initial training success of blood glucose testing in captive chimpanzees (*Pan troglodytes*). *Zoo Biology* 33:212–220.

Schapiro, S.J. 2002. Effects of social manipulations and environmental enrichment on behavior and cell-mediated immune responses in rhesus macaques. *Pharmacology, Biochemistry, and Behavior* 73:271–278.

Schapiro, S.J. and S.P. Lambeth. 2007. Control, choice, and assessments of the value of behavioral management to captive primates. *Journal of Applied Animal Welfare Science* 10:39–47.

Schapiro, S.J., J.E. Perlman, and B.A. Boudreau. 2001. Manipulating the affiliative interactions of group-housed rhesus macaques using positive reinforcement training techniques. *American Journal of Primatology* 55:137–149.

Schapiro, S.J., J.E. Perlman, E. Thiele, et al. 2005. Training nonhuman primates to perform behaviors useful in biomedical research. *Lab Animal* 34(5):37–42.

Turner, C.H., R.K. Davenport, Jr., and C.M. Rogers. 1969. The effect of early deprivation on the social behavior of adolescent chimpanzees. *American Journal of Psychiatry* 125(11):1531–1536.

United States Department of Agriculture. 1991. Subpart D, §3.81. Available at: https://www.aphis.usda.gov/animal welfare/downloads/Animal%20Care%20Blue%20Book%20-%202013%20-%20FINAL.pdf. Accessed 12/11/2016.

Whittaker, M., J. Perlman, and G. Laule. 2008. Facing real world challenges: Keeping behavioral management programs alive and well. In Hare, H.J. and J.E. Kroshko (eds). *Proceedings of the Eighth International Conference on Environmental Enrichment*, Vienna, Austria, pp. 87–89. San Diego, CA: The Shape of Enrichment.

Genera-Specific Behavioral Management

CHAPTER 19

Behavioral Management of *Macaca* Species (except *Macaca fascicularis*)

Daniel Gottlieb, Kristine Coleman, and Kamm Prongay
Oregon National Primate Research Center

CONTENTS

INTRODUCTION

In the past two decades, there has been a dramatic evolution in behavioral management programs designed to provide captive nonhuman primates (NHPs) with species-appropriate housing, environmental enrichment, and socialization. Such programs are designed to promote normal behavior, reduce stress, and improve NHPs' ability to cope with stress, with the ultimate goal of ensuring the animals' mental and physical health. Once considered "extra," these programs are now a fundamental part of animal care. This progress is, in large part, a result of increased research efforts examining behavioral needs of NHPs and how these needs might be met in captivity. With *Macaca* being the most commonly used genus of NHP in biomedical research (Carlsson et al. 2004), a great deal of research has focused on finding optimal behavioral management strategies for macaque species. Such research has led to innovations in socialization and rearing strategies, identification of efficacious enrichment tools and techniques, and an increased understanding of the role of positive reinforcement training (PRT) in promoting welfare. Macaque behavioral managers now have a large toolbox of techniques they can utilize to improve captive welfare; however, the optimal strategies for any given program vary based on facility size, resources, and research goals. In this chapter, we discuss many of these behavioral management tools, including socialization, environmental enrichment, and PRT. We further provide our recommendations for a successful macaque behavioral management program, recognizing that limitations common to most primate facilities (e.g., funding, staff, time, etc.) often impact management decisions.

Rhesus macaques (*Macaca mulatta*) are the most commonly used macaque species in biomedical NHP research, followed by cynomolgus macaques (*M. fascicularis*), Japanese macaques (*M. fuscata*), and pigtailed macaques (*M. nemestrina*) (Carlsson et al. 2004). Given their prevalence in biomedical research, a large portion of the literature cited in this chapter comes directly from rhesus macaques. However, due to the social, biological, and ethological similarities between the macaque species, most concepts and examples outlined in this chapter are directly applicable to all macaque species (for recommendations specific to cynomolgus macaque behavioral management, please see Chapter 20 by Honess, 2017).

NATURAL HISTORY

Macaca is the most geographically widespread NHP genus, and members of this taxon can be found in their natural habitat throughout Asia (Thierry 2007). Barbary macaques (*M. sylvanus*) are an exception and can be found in their natural habitat in North Africa (Thierry 2007). All species of macaques are highly social and live in groups generally characterized as being multimale and multifemale, with female philopatry and male emigration (Thierry 2007). Average group size varies considerably, with troop sizes ranging from a few to hundreds of individuals (Ménard 2004). Group size is thought to vary as a function of food availability, with the largest groups of macaques often living near rural and urban areas with abundant easily accessible food (Singh and Vinathe 1990; Chapais 2004; Hasan et al. 2013).

Because of female philopatry and male emigration, macaque troops generally contain subgroups of related females (known as matrilines) and unrelated males. Matrilines form the foundation for many macaque social interactions, as well as the basis of the dominance hierarchy; females are most likely to socialize and form coalitions with members of their own matriline, and females generally "inherit" a rank from their mother (Chapais 2004; Thierry 2007). As highly social species, macaques spend a significant amount of time in proximity to conspecifics, and, in the wild, are reported to spend 4%–43% of the day engaging in direct social behaviors, such as grooming and playing (Singh and Vinathe 1990; O'Brien and Kinnaird 1997; Hanya 2004). Macaques are generally active for the majority of the day, spending 13%–79% of the day feeding and foraging and 2%–61% of the day traveling (Chapais 2004). Activity budgets can differ greatly between troops of the same species and are thought to vary as a function of food density, quality, and availability (Hill 2004).

Macaque societies are known for having distinct dominance hierarchies within each sex; however, the level of aggression, dominance asymmetries, and formalized relationships vary considerably among species (Flack and de Waal 2004; Thierry 2007). The three most common macaques in biomedical research—rhesus, cynomolgus, and Japanese—are characterized as having a despotic hierarchy (i.e., formalized relationships with large dominance asymmetries reinforced through severe aggression). Other species, including pigtailed macaques, are thought to have a more tolerant hierarchy, in which not all relationships are formalized and large dominance asymmetries are reinforced through moderate-to-mild aggression (Flack and de Waal 2004). Differences in dominance styles among the macaque species are pronounced in both wild and captive populations.

GENERAL BEHAVIORAL MANAGEMENT STRATEGIES AND GOALS

Major goals for behavioral management programs include reducing stress, increasing the animals' ability to cope with stress, improving health, increasing species-normal behaviors, and decreasing abnormal behaviors. Common strategies utilized to achieve these goals include socialization, environmental enrichment, and PRT. In the following section, we will outline these strategies and review existing empirical research.

Socialization

Socialization is generally considered the single most important factor for the welfare of captive primates, particularly early in life (i.e., rearing) (Lutz and Novak 2005; Buchanan-Smith et al. 2009; Coleman et al. 2012). Social housing of NHPs can prevent and remediate development of abnormal behaviors (Bellanca and Crockett 2002; Rommeck et al. 2009a; Gottlieb et al. 2013a) (for a review, see Novak et al. 2006), alter immune function (Schapiro et al. 2000), and improve resiliency (Gilbert and Baker 2011). Further, both the Animal Welfare Act regulations (USDA 1991) and *The Guide for the Care and Use of Laboratory Animals* (Council 2011) specify that single housing for laboratory NHPs is only justified by veterinary-related welfare concerns, or when required by scientific protocols (and approved by the local Institutional Animal Care and Use Committee). The vast majority of macaques maintained in US laboratories are socially housed. A 2014 survey of 39 primate facilities, including data from over 50,000 macaques, found that over 80% of captive macaques were socially housed (Baker 2016).

Types of Socialization

There are many different ways in which socialization can be accomplished; macaques can be housed in large, outdoor corrals with hundreds of conspecifics (Figures 19.1 and 19.2), in indoor cages with a single social partner, or in a variety of social configurations in-between. For this review, we will discuss two broad categories of socialization: social groups and caged pairing.

Social Groups

As mentioned earlier, because macaque species are highly social and have adapted themselves to live in large, socially complex societies, group housing is the preferred method of maintaining most macaques. Depending on the facilities and equipment available, captive macaque social groups can range from small groups of three or four conspecifics to large breeding groups with hundreds of individuals. Compared with those housed in cages, macaques in social groups have increased opportunities for expressing species-normal behaviors, for developing cognitive and social skills (de Waal 1991), and for doing physical exercise. Further, they have altered immune function (Schapiro et al. 2000), are at lower risk for developing stereotypic and self-abusive behaviors (Bayne et al. 1992b; Gottlieb et al. 2013a), and are at lower risk for chronic diarrhea (Hird et al. 1984).

Figure 19.1   Rhesus macaques in a corral enclosure enriched with structural enrichment.

Figure 19.2   Japanese macaques in a corral enclosure enriched with structural enrichment.

Although there are many benefits to social housing, there are multiple management and welfare concerns associated with housing animals in groups, particularly for despotic species (e.g., rhesus and Japanese macaques). Frequent fighting, as well as the aggressive establishment and reinforcement of social hierarchies, can lead to injury and illness (McCowan et al. 2008; Beisner et al. 2012). Removal of animals from established groups for clinical, behavioral, or colony management purposes can cause social unrest and high levels of stress in the remaining members of the group and can lead to further aggression due to reshuffling of the social hierarchy (Belzung and Anderson 1986; Flack et al. 2005; Oates-O'Brien et al. 2010; Beisner et al. 2015). Even without overt aggression, living in a group can be stressful for some individuals, particularly subordinate members (Coe 1991). Despite these potential costs, group living is considered the best way to house most captive macaques, although some individuals might fare better in smaller groups or even paired housing, due to other factors, such as temperament or health.

Particularly concerning for socially housed macaques are matrilineal or social overthrows in which the dominant female or matriline is mobbed by multiple attackers (Chance et al. 1977;

Samuels and Henrickson 1983; Ehardt and Bernstein 1986; Hambright and Gust 2003; Oates-O'Brien et al. 2010). These overthrows, also known as "cage wars" (e.g., McCowan et al. 2008), can result in severe or fatal trauma to multiple animals, and often lead to destabilization of the social group (Gygax et al. 1997; Oates-O'Brien et al. 2010; Beisner et al. 2015). Such extreme cases of social aggression have been most commonly recorded in rhesus (McCowan et al. 2008 Samuels and Henrickson 1983; Ehardt and Bernstein 1986), Japanese (Gygax et al. 1997), and cynomolgus (Chance et al. 1977) macaques. In addition to the immediate negative welfare impacts of social overthrows (e.g., animal stress, injury, and death), these events have large financial costs (e.g., increased veterinary costs, increased staff time, loss of valuable research animals), and long-lasting impact on colony management and resources (e.g., reduced production of offspring, increased need for indoor caged housing, reduced genetic variability). All efforts should be made to prevent, rather than simply respond to, social overthrows. Although social overthrows can be extremely difficult to predict, social network analyses have demonstrated multiple predictors of social instability, including the sex ratio, degree of matriline fragmentation, changes in rate of signals of subordination, and frequency of bidirectional aggression (Beisner et al. 2011, 2012; McCowan et al. 2011; Chan et al. 2013; Beisner et al. 2015; McCowan and Beisner 2017). Monitoring groups for these precursors of instability, as well as utilizing a proactive and collaborative approach to social group management (see section "360° Communication"), can limit the occurrence of these overthrows.

Caged Pairing

While generally preferred, social group housing is not always possible, due to either research or facility constraints. For caged macaques, social housing is often accomplished by full contact pairing (i.e., providing two monkeys with two or more adjoining cages to which they both have full access). Compared to single-housed macaques, paired monkeys have increased opportunities to express species-normal social behaviors, show fewer stereotypic and self-abusive behaviors (Lutz et al. 2003; Weed et al. 2003; Baker et al. 2012a, 2014; Gottlieb et al. 2013a), and are better able to cope with stressful situations (Gilbert and Baker 2011) (for a review on the benefits of social housing, see DiVincenti and Wyatt 2011). Protected contact housing, in which animals have limited physical contact through a partially open divider, provides an alternate form of socialization when full contact pairing is not an option. Forms of protected contact housing include mesh dividers, solid metal dividers with small holes for grooming, or grooming contact dividers made of widely spaced bars (Crockett et al. 1997). Although protected contact provides opportunities for social behaviors and tactile contact otherwise unavailable in single housing (Crockett et al. 1997), it does not offer the same degree of socialization as full contact pairing. Rhesus macaques housed in protected contact display more abnormal behaviors than those housed in full contact, and similar levels of abnormal behaviors compared to those that are singly housed (Baker et al. 2012b, 2014; Gottlieb et al. 2013a, 2015). In contrast, multiple studies in cynomolgus macaques have found no difference in abnormal or tension-related behaviors between animals in grooming contact and full contact housing (Baker et al. 2012b; Lee et al. 2012). Although studies have not been performed comparing grooming contact to full socialization in other macaque species, as a general rule, protected contact should not be assumed to be better or even equivalent to full contact housing. That being said, in situations in which full contact pairing is not appropriate (e.g., for research-related or clinical reasons), protected contact is preferred over single housing.

Rearing

Appropriate socialization is particularly important during early development. Infant macaques can be raised in a variety of conditions, including mother rearing in groups (generally considered the optimal rearing environment for macaques); indoor–mother rearing, in which the infant is raised in a cage with its biological or foster mother and at the most one additional adult female and infant macaque pair; and nursery rearing, in which the infant is separated from the dam and reared in a nursery, usually with access to at least one other infant. Over the past few decades, our understanding of the impact of rearing environment on macaque development and well-being has greatly increased. Early research by Harlow et al. demonstrated the importance of proper socialization to behavioral development; infants reared without social contact demonstrated deficiencies in normal social behavioral development, and had high prevalences of abnormal and pathological behaviors (Cross and Harlow 1965; Harlow et al. 1965; Harlow and Suomi 1971). While isolation rearing is no longer considered ethically appropriate, nursery rearing is still commonly utilized in certain situations that include avoidance of maternally transmitted pathogens, maternal abandonment/death, and specific scientific project requirements. Nursery rearing affects multiple developmental outcomes in rhesus macaques, including physiology (for a review, see Novak et al. 2006), immunological responses (Coe et al. 1989; Lubach et al. 1995; Capitanio et al. 2006), and temperament (Capitanio et al. 2006; Rommeck et al. 2011; Gottlieb and Capitanio 2013). Nursery rearing also increases the risk of developing (1) colitis and chronic diarrhea (Hird et al. 1984; Elmore et al. 1992) and (2) abnormal behaviors, such as stereotypies, self-directed behaviors, floating limb, and self-abusive behaviors (Bellanca and Crockett 2002; Novak and Sackett 2006; Rommeck et al. 2009a; Gottlieb et al. 2013a, 2015). Similarly, indoor–mother reared infants also show altered physiology (Capitanio et al. 2006) and temperament (Capitanio et al. 2006; Gottlieb and Capitanio 2013), and are at increased risk for chronic diarrhea (Hird et al. 1984) and abnormal behaviors (Gottlieb et al. 2013a, 2015) compared to monkeys reared with their mothers in groups.

Not all nursery rearing is performed in the same manner. In many facilities, nursery infants are continuously paired in dyads together, which can lead to abnormal social behaviors, such as partner clinging, and excessive fear and withdrawal (Chamove et al. 1973; Ruppenthal et al. 1991; Rommeck et al. 2009b). Other facilities utilize intermittent pairing, in which infants are kept together during the day, but are separated at night; or playgroup socialization, in which infants are single housed for a portion of the day and intermittently placed in playgroups with multiple conspecifics (Ruppenthal et al. 1991; Rommeck et al. 2008, 2009b). While rearing methods involving intermittent socialization can decrease aggression and abnormal social behaviors, they can also increase highly concerning abnormal behaviors, including self-biting, floating limb syndrome, self-clasping, rocking, and stereotypic behaviors (Ruppenthal et al. 1991; Rommeck et al. 2008, 2009b). Providing continuously paired infants with lifelike surrogates that provide nutrition, as well as kinesthetic, vocal, and tactile stimulation, can help reduce some of the abnormal behaviors (Brunelli et al. 2014).

In some situations, providing abandoned or orphaned infants with a surrogate dam can reduce the need for nursery rearing. Females who recently lost their own infants, and are thus lactating, are often willing to "adopt" a new infant. If lactating females are unavailable, non-lactating females can be trained to allow infants to bottle feed, and can thus also serve as "surrogate" dams (Welch et al. 2010).

It is important to note that the relative impact of nursery rearing is dependent on the species in question. For example, many of the observed abnormal behaviors associated with nursery playgroup socialization appear to decrease, and potentially fully extinguish, with age in pigtailed macaques (Ruppenthal et al. 1991; Worlein and Sackett 1997). Further, Sackett et al. (1981) found differences in levels of abnormal, social, and exploratory behavior between nursery-reared rhesus, cynomolgus, and pigtailed macaques. Isolation-reared rhesus macaques displayed the most abnormal behaviors, cynomolgus macaques showed intermittent levels, and pigtailed macaques exhibited relatively few abnormal behaviors. Rhesus macaques also showed deficits in exploratory behavior, and both rhesus and pigtailed macaques showed deficits in social behavior compared to socially reared controls. In contrast, social behavior in isolation-reared cynomolgus raised in a nursery was similar to socially reared controls (Sackett et al. 1981). Thus, it is likely that not all species of macaque develop equivalent negative outcomes in response to nursery rearing.

Environmental Enrichment

Environmental enrichment can be broadly defined as "an animal husbandry principle that seeks to enhance the quality of captive animal care by identifying and providing the environmental stimuli necessary for optimal psychological and physiological well-being" (Shepherdson, 1998, p. 1). Enrichment is typically designed with the goal of providing opportunities to perform species-normal behaviors, which is important, as the inability to perform such behaviors can cause stress and frustration for captive animals (Petherick and Rushen 1997). To be ethologically appropriate, enrichment should be designed with species-specific behavioral goals in mind. Otherwise, enrichment can run the risk of being visually pleasing to the caretakers, but ineffective at improving animal welfare. Nonsocial environmental enrichment can be broadly categorized into four, non-independent categories: food, sensory, physical, and occupational (cognitive) (Bloomsmith et al. 1991; Keeling et al. 1991). In the following section, we will summarize various forms of enrichment used with captive macaques and present empirical evidence as to their efficacy.

Foraging (Food-Based) Enrichment

In the wild, macaques typically spend, on average, almost half of their time feeding and foraging for food (Chapais 2004). In contrast, in captivity, food is provided to animals once or twice daily, and consumption requires minimal searching and manipulation, greatly limiting foraging opportunities. To promote species-typical foraging behaviors, most, if not all, facilities provide captive macaques with foraging enrichment (Baker 2016). Effective foraging enrichment increases the amount of time animals spend finding, processing, and eating food, and is not simply the provision of extra produce or food items. This kind of enrichment is often provided in specialized devices, which can be placed in, or hung outside of, a cage. Examples of foraging devices include puzzle balls (hollow containers filled with food items), foraging boards (artificial turf, fleece, or plastic boards covered with small forage materials), and smeared objects (trays covered in a sticky substance, such as peanut butter or apple sauce). In the absence of a device, hiding food or placing it on the outside of an animal's cage in such a way that it promotes foraging (e.g., on top of the cage) can be an easy, yet effective form of foraging enrichment (Reinhardt 1993b).

In almost all studies evaluated to date, foraging enrichment has been shown to successfully increase species-typical foraging behavior in captive macaques (e.g., Bayne et al. 1991; Reinhardt 1993a; Schapiro and Bloomsmith 1995; Schapiro et al. 1996b; Reinhardt and Roberts 1997; Gottlieb et al. 2011 but see Schapiro and Bloomsmith 1994). Additionally, the provision of foraging enrichment has been shown to increase play and activity, decrease self-grooming, and modify social behaviors (Lam et al. 1991; Bayne et al. 1992a; Lutz and Novak 1995; Schapiro and Bloomsmith 1995; Schapiro et al. 1996a). However, the biological significance and long-term welfare implications of such behavioral changes are not always immediately apparent (Schapiro and Bloomsmith 1994).

In addition to promoting species-normal behaviors, foraging enrichment has been shown to significantly decrease stereotypic behaviors in rhesus macaques (Bayne et al. 1991, 1992a; Novak et al. 1998; Gottlieb et al. 2011, 2015), even with repeated exposure over several months (Bayne et al. 1991, 1992a). In these studies, the beneficial effects tended to be most pronounced when enrichment was first given, with monkeys largely ignoring devices once they were emptied of food. However, a recent retrospective study found that caged rhesus macaques were less likely to display stereotypic behaviors when a foraging device was present, as opposed to absent, even after the device had been depleted of food (Gottlieb et al. 2015), suggesting that the use of foraging devices can have benefits that extend beyond the period when food is present.

Although not directly a form of foraging enrichment, the substrate used to cover the floor of the cage or enclosure, including various types of bedding or grass, has also been found to increase opportunities for foraging. Unlike concrete or dirt flooring, bedding material, such as wood shavings, can increase the time it takes animals to forage for scattered food enrichment, increase species-typical behaviors, decrease aggression, and decrease over-grooming in group-housed macaques (Chamove et al. 1982; Byrne and Suomi 1991; Doane et al. 2013). In addition, a natural substrate, such as grass, can function as a form of natural forage material.

Physical Enrichment

Physical enrichment is a broad category of environmental enhancements designed to provide animals with opportunities to explore and/or manipulate. It includes durable enrichment, such as chew toys, mirrors, and other manipulatable objects; destructible enrichment, such as cardboard boxes filled with paper, old phone books, and magazines; and structural items, such as perches, swings, resting platforms, visual barriers, and pools. Physical enrichment affords individuals the opportunities to express species-normative behaviors such as play, locomotion, and exploration. Physical enrichment is one of the most commonly provided NHP enrichments; in a recent study of 39 primate facilities, 100% of respondents reported using both durable and structural enrichments, with the vast majority providing both forms of enrichment to all monkeys (Baker 2016).

Durable enrichment is typically located inside, or hanging outside, the primary enclosure. Items such as dog toys and wood blocks provide opportunities for chewing and biting behavior. Many animals have been reported to bite these items in a fashion similar to self-biting, and hence these toys may provide an opportunity for an alternate, less dangerous behavior (Crockett and Gough 2002). Mirrors, when hung within arms' reach of the primary enclosure, can increase an individual's field of vision within a room (Lutz and Novak 2005).

Destructible enrichment items are typically placed directly in the primary enclosure. These enrichment items are commonly filled with food, such as grains and seeds, making them both destructible and a form of foraging enrichment. Although relatively inexpensive, destructible enrichment is infrequently used at some facilities due to plumbing, drainage, and sanitation concerns (Baker 2016).

Structural enrichment, also known as "cage furniture," is typically built into the primary enclosure. Perches, one of the most common forms of structural enrichment, allow animals to sit above ground level and off the floor of the cage. While not arboreal per se, macaques often prefer to be above ground, particularly when facing a threat, and therefore perching is essential to the well-being of most NHPs (Reinhardt 1992). Further, perches have been found to reduce aggression in group-housed Japanese macaques (Nakamichi and Asanuma 1998). Porches, small cage extensions hung on the outside of a single cage, can further provide caged animals with opportunities to perch (for more detail, see section "Facilities and Equipment"). Other common forms of structural enrichment include visual barriers to reduce aggression, furnishings to allow climbing (e.g., wooden structures, plastic play sets, recycled fire hose, car wash strips), and pools to encourage swimming and play behavior (Figures 19.1 through 19.3). In addition to promoting species-typical behaviors, pools can allow outdoor-housed macaques opportunities to cool off on hot days (Robins and Waitt 2011).

Sensory Enrichment

Sensory enrichment includes environmental modifications that stimulate an animal's visual, auditory, tactile, and/or olfactory senses. In the following section, we will discuss three of the major forms of sensory enrichment provided to macaques: visual, auditory, and olfactory.

Figure 19.3   Rhesus macaques using a pool as enrichment.

Visual Enrichment

Many facilities use televisions as visual enrichment for macaques. Monkeys are typically exposed to videos of conspecifics, nature, humans, and/or cartoons (Lutz and Novak 2005; Reinhardt 2010). Anecdotally, television enrichment is thought to provide cognitive stimulation, present engaging species-appropriate stimuli, and help distract from stressful stimuli in the environment (Reinhardt 2010). It has been questioned, however, whether NHPs perceive videos in the same manner as humans (Reinhardt 2010). Further, unlike active enrichment, which encourages natural behaviors (e.g., foraging, climbing), visual enrichment may only encourage passive behaviors. Research has demonstrated that rhesus macaques will watch television enrichment when it is presented (Platt and Novak 1997), and that rhesus (Harris et al. 1999), bonnet (Brannon et al. 2004) and Japanese (Ogura 2012; Ogura and Matsuzawa 2012) macaques will perform operant tasks to access video clips. However, not all studies have demonstrated this interest in videos; rhesus macaques have shown a preference for a blank screen over video (Washburn et al. 1997), and bonnet macaques have shown either no preference between video clips and food rewards (Andrews and Rosenblum 2001) or a clear preference for food reward over video (Brannon et al. 2004). The manner in which visual stimulation is presented to the monkeys is likely a large contributing factor in the enrichment's over-all efficacy. Ogura and Matsuzawa found a decrease in abnormal behaviors in Japanese macaques given operant control over television enrichment (Ogura 2012; Ogura and Matsuzawa 2012), while similar findings were not observed for rhesus macaques passively presented television enrichment (Schapiro and Bloomsmith 1995). With many inconsistent findings regarding the benefits of television enrichment, it is unclear whether this enrichment is ethologically appropriate, or provides a tangible benefit to the animals.

Auditory Enrichment

Radios are frequently utilized as both a form of auditory stimulation and as a tool to mask stressful noises in the captive environment, such as husbandry movements and activity outside of the animals' room. Multiple studies have demonstrated that, when given operant choice tasks, rhesus macaques will actively choose to listen to music/radio enrichment (Drewsen 1989; Markowitz and Line 1989; Novak and Drewsen 1989; Line et al. 1990). Further, audio enrichment has been shown to decrease abnormal behaviors, decrease aggressive behaviors, and increase affiliative behaviors in rhesus macaques (Drewsen 1989; Novak and Drewsen 1989; O'Neill 1989; Graves 2011). That being said, inappropriate music (e.g., music that is too loud or abrasive) could unintentionally be a source of uncontrollable stress for some individuals. Radio stations and audio levels should be chosen with care, and the animals' responses should be monitored.

Olfactory

While olfactory enrichment, such as oils or scented candles, is occasionally provided for New World primates and for apes, it is less often provided for macaques or other Old World monkeys (Coleman et al. 2012). Unlike many New World primate species, which communicate through odor-producing skin glands and marking behaviors (Prescott 2006; Buchanan-Smith et al. 2009), macaques are not as reliant on communication through scent, thereby making aromatic enrichment less biologically relevant. Still, little research exists to either support or dismiss the use of olfactory enrichment for macaques. Potential forms of olfactory enrichment that may be biologically relevant to macaque species include food-based scents, "calming" oils such as lavender and bergamot, and aromas specifically designed to cover unpleasant environmental scents. Further research is needed to assess the potential benefits of olfactory enrichment for macaques.

Occupational (Cognitive) Enrichment

Cognitive enrichment is designed to mentally stimulate animals through tasks that require utilization of cognitive faculties, such as problem solving and memory. Examples of cognitive enrichment include puzzle feeders (devices that require monkeys to manipulate food through a simple maze before extraction; Bayne et al. 1991, 1992a; Lam et al. 1991; Schapiro and Bloomsmith 1995) and computerized tasks (e.g., Platt and Novak 1997). Tablets, such as iPads, have increasingly been used as a form of cognitive enrichment (O'Connor et al. 2015; Coleman, 2017). Unlike computer games, animals do not need to be trained to use simple games and apps on tablets. A recent study (O'Connor et al., 2015) examined the use of a Kindle Fire (Amazon) tablet as enrichment for male rhesus macaques. The monkeys in that study preferred interactive apps (those that responded to the monkeys; i.e., bubbles popped after being touched) to passive apps (i.e., one in which the monkeys were shown a colorful screen). See Coleman (2017) for more details on this study. Unfortunately, many current cognitive enrichment options can be both expensive and time-consuming to prepare, and further research is needed to identify cognitive enrichment devices/tasks that are both effective and practical.

Positive Reinforcement Training

Another component of behavioral management is PRT, a type of operant conditioning in which the animal gets rewarded (i.e., reinforced) for performing desired behaviors (such as touching a target or presenting for an injection; e.g., Schapiro et al. 2001; Laule et al. 2003; Graham 2017; Magden 2017). Positive reinforcement techniques have been used to successfully train macaques to perform various veterinary and/or research procedures (Reichard et al. 1992; Schapiro et al. 2003), such as moving a body part (e.g., thigh) close to the front of the cage for examinations or injections (e.g., Mueller et al. 2008), taking oral medications (e.g., Klaiber-Schuh and Welker 1997), and remaining still for blood samples (e.g., Coleman et al. 2008). For more detail on applications of PRT, see Chapters 12 (Graham 2017) and 13 (Magden 2017).

There are many welfare benefits associated with PRT. By desensitizing animals to potentially stressful stimuli (such as injections), PRT can reduce the fear and stress associated with common husbandry/experimental procedures (Moseley and Davis 1989). Studies on caged monkeys have shown that cortisol levels during venipuncture are lower for monkeys trained to cooperate with this procedure than for untrained individuals (Reinhardt et al. 1990). In addition, PRT provides individuals with the chance to cooperate with procedures, and thus may give animals a sense of control over their environment (Laule et al. 2003), which is known to reduce stress (Mineka et al. 1986) and be important for well-being.

While training does not necessarily directly increase species-normal behavior, it has been shown to provide mental stimulation for subjects (Laule et al. 2003), and therefore can be an effective form of psychological enrichment. PRT can help enhance the relationship between the subject and the trainer (Bloomsmith et al. 1997). Training can also facilitate group housing by allowing animals needing some types of medical attention to remain in groups. For example, rhesus macaques living in large social groups have been trained to approach a particular part of their enclosure and step on a scale or take medication (e.g., Padilla 2011). PRT has also been used to encourage prosocial behavior in group-housed rhesus macaques (Schapiro et al. 2001). Finally, PRT has been shown to reduce undesired behaviors, such as stereotypical behavior (Coleman and Maier 2010), although this finding is not universal (Baker et al. 2009).

Use of PRT is, perhaps, one of the fastest growing parts of behavioral management. In a 2003 survey on NHP behavioral management by Baker et al. (2007), 55% of facilities reported using PRT with their animals, and only 9% reported having a dedicated trainer. In a follow-up survey in 2014 (Baker 2016), 100% of facilities reported using PRT for their animals, and ~25% had a dedicated trainer. Further, while in the past, most dedicated trainers worked primarily with chimpanzees, many facilities now employ trainers focused on macaques.

FACILITIES AND EQUIPMENT

Innovations in caging equipment have increased opportunities for socialization (i.e., pairing), physical activity, and exploration that were previously unavailable with traditional indoor cages. At the very least, most current cages are built such that two cages can be connected horizontally, allowing two adult macaques to be housed together. In some cases, multiple cages can be connected horizontally, allowing multiple animals to be group housed together indoors. Numerous cages allow vertical connection, increasing opportunities for climbing and exploration. Such cages are not only ethologically appropriate, but can also aid in physical rehabilitation that requires vertical movement. However, it is important to note that cages in which top and bottom levels are connected by complete or partial removal of the middle floorboard often do not meet the minimum area requirements for two average-sized adult macaques. In contrast, cages that are vertically connected through a small (e.g., 1 ft^2) opening or by an external tunnel can legally house two average-sized adult macaques while still promoting climbing.

Exterior cage extensions can similarly promote perching and climbing behaviors. Tunnels, cage extensions that attach to top and bottom level cage doors, promote climbing, perching, and can provide physical rehabilitation following injury. Porches, small cage extensions hung on the outside of a single cage, are less bulky than tunnels, but still provide increased space, opportunities to perch, as well as a widened field of view (Figure 19.4). Porches have been shown to effectively decrease stereotypy (personal observation) and feces painting (Gottlieb et al. 2014), an abnormal behavior in which the animal smears and/or rubs feces on a surface, typically the side of the cage.

Exercise enclosures (i.e., large indoor cages or pens) can be used to provide opportunities for physical activity or rehabilitation for indoor-housed macaques (Baker 2016). Animals typically given exercise enclosures for medical reasons include obese animals that need to lose weight, aged animals with arthritis, and animals recovering from muscle surgery or trauma. Exercise enclosures can also be used to help transition animals from caged housing to large group housing. Minimal exercise before the release into a large outdoor enclosure has been used with the goals of building

Figure 19.4   Rhesus macaque sitting in porch enrichment. (Reprinted from Gottlieb, D. H. et al., *J. Am. Assoc. Lab. Anim. Sci.*, 53(6), 653–656, 2014; Figure 1. With permission.)

muscle mass, increasing bone density, and lowering overall risk of bone fractures in animals that have been cage housed for an extended period of time (Prongay personal observation). Finally, animals can be intermittently rotated into exercise enclosures as a form of enrichment (Griffis et al. 2013).

RESEARCH-IMPOSED RESTRICTIONS/EXEMPTIONS

There are times when research protocols may preclude the provision of social or environmental enrichment. While such research-imposed restrictions or exemptions present challenges by limiting traditional behavioral management techniques, it is necessary to balance the psychological well-being needs of the animals with the needs of the research studies in which they participate. For example, studies involving precise measurement of food or caloric intake may restrict certain types of feeding enrichment. However, it is often possible to use noncaloric items (e.g., ice cubes, commercially available noncaloric treats) and/or to provide the subjects' daily food ration in foraging devices. Perhaps the most common kind of research-imposed restriction is an exemption from social housing. For studies in which full contact socialization is contraindicated, protected contact or intermittent pairing may present an acceptable option. If not, then the animals may benefit from additional interaction with care staff. Finally, PRT can often help reduce the need for restrictions. For example, training macaques to come to the front of their enclosure and voluntarily present for an injection or other manipulation can allow subjects to remain in a social group while on study. Having behavioral management scientists on, or associated with, Institutional Animal Care and Use Committees (IACUCs) can facilitate the process of balancing the needs of the research and the needs of the animals.

ABNORMAL BEHAVIORS

Captive macaques can develop abnormal behaviors (Capitanio 1986), defined as behaviors that are statistically rare in wild populations, cause harm to the animal, or are the result of past damage or illness (Mench and Mason, 1997). Stereotypic behaviors, repetitive behaviors caused by central nervous system dysfunction, frustration, or repeated attempts to cope (Mason 2006), are some of the most commonly seen abnormal behaviors in macaques (Bellanca and Crockett 2002; Lutz et al. 2003, 2011). The most prevalent types of stereotypic behaviors exhibited by macaques are motor stereotypic behaviors (also known as "repetitive-motion stereotypies" or "cage stereotypies") (Bellanca and Crockett 2002; Lutz et al. 2003, 2011), which include full body repetitive behaviors, such as pacing, bouncing, twirling, swinging, rocking, and somersaulting (Capitanio 1986; Lutz et al. 2011; Gottlieb et al. 2013a). Other abnormal behaviors which macaques may engage in include self-abusive behaviors (self-bite, self-hit), self-directed behaviors (eye poke, digit suck, self-clasp, hair pluck), postural behaviors (leg lift, floating limb), ingestive behaviors (coprophagia, urophagia), and feces painting (smearing of feces on the cage wall) (e.g., Lutz et al. 2003; Novak 2003; Rommeck et al. 2009a; Gottlieb et al. 2014; for a review see Capitanio 1986). For a description of abnormal behaviors common in captive macaques, see Table 19.1.

Although their exact etiology and causation are not always well understood, numerous risk factors have been established for the development and expression of abnormal behaviors, including rearing condition, access to socialization, and exposure to chronic stressors. Macaques are at increased risk for stereotypic behaviors, self-abuse, self-directed behaviors, and postural behaviors when raised without a mother, or even without a social group (Bellanca and Crockett 2002; Novak and Sackett 2006; Novak et al. 2006; Lutz et al. 2007; Rommeck et al. 2009a; Gottlieb et al. 2013a).

Table 19.1 Abnormal Behaviors Commonly Seen in Captive Macaques

Category	Behavior	Behavior Description
Stereotypic behaviors	Pacing	Walking back and forth or in a circle repeatedly, in the exact same pattern.
	Bouncing	Jumping up and down repeatedly using a rigid posture. This behavior should not be confused with bouncing with a less rigid posture, which serves to make noise or shake the cage.
	Twirling	Repeatedly turning the body horizontally.
	Swinging	Grasping a part of the cage with hands or feet while repeatedly swinging in the exact same pattern.
	Rocking	Rhythmically moving either side-to-side or forward and backward repeatedly, in the exact same pattern.
	Somersaulting	Flipping forwards or backwards repeatedly.
Self-directed behaviors	Self-clasp	Monkey grasps self with hands and/or feet.
	Self-suck	Monkey sucks their own body, including their digits, tail, or genitals.
	Eye poke/cover	A "saluting" gesture of the hand over the eye which is often accompanied by pressing of the knuckle or finger into the orbital space above the eye socket.
	Hair pluck	Monkey forcefully removes their own hair with the hands or teeth. Behavior can result in the removal of individual hairs, or large amounts of hair at once. Behavior is often followed by ingestion of hair.
Postural behaviors	Floating limb	A limb (either arm or leg) drifts over or around the body. Limb is often reported to appear to be moving of its own accord.
	Leg lift	One or more legs is wrapped around the back of the body or propped on the neck for a prolonged period of time.
Self-abusive behaviors	Self-bite	Monkey bites his or her body. Often directed at arms and legs.
	Self-hit	Monkey hits or slaps any part of their own body.
Other abnormal behaviors	Coprophagia	Ingestion of feces
	Urophagia	Ingestion of urine
	Feces painting	Smearing and/or rubbing of feces on a surface, typically the side of the cage.

Indeed, there is an inverse relationship between large, complex rearing environments and the risk of developing stereotypic behaviors (Gottlieb et al. 2015). Socialization remains an important predictor of abnormal behaviors even after an animal is weaned. Continued social housing (group or pair housing) significantly reduces both current and future risk of stereotypic and self-abuse behaviors (Schapiro et al. 1996a; Lutz et al. 2003; Novak 2003; Baker et al. 2012a; Gottlieb et al. 2013a, 2015). Finally, stressful events such as relocations (i.e., moving to a new room), veterinary procedures, blood draws, and assignment to research protocols have been shown to positively predict abnormal behaviors (Lutz et al. 2003; Novak 2003; Rommeck et al. 2009a; Gottlieb et al. 2013a).

Abnormal behaviors, particularly stereotypic behaviors, are frequently associated with compromised welfare conditions (Mason 1991; Mench and Mason 1997; Mason and Latham 2004). Specifically, stereotypic behaviors have been shown to develop at a higher rate in environments associated with poor welfare (Mason and Latham 2004), and are more frequently elicited in animals experiencing frustration, lack of stimulation, lack of environmental control, and/or unavoidable stress (Mason 1991). Complicating the picture, however, is the idea that stereotypic behaviors may function as successful coping mechanisms in suboptimal environments. When comparing animals raised and housed in identical suboptimal environments, higher levels of stereotypic behavior often correlate with positive measures of welfare, such as lowered corticosteroids and heart rate (Mason and Latham 2004). In other words, when faced with suboptimal environments, the individuals that develop stereotypic behaviors may actually fare better than their non-stereotypic counterparts. *Therefore, expression of stereotypic behavior should never be the sole measure of an individual's welfare and overall enrichment needs.*

Despite the difficulty in utilizing stereotypic behavior to evaluate current individual welfare state, these behaviors can still be useful tools to compare the relative benefits of various environmental conditions. When comparing multiple environments, those with the lowest overall development and expression of stereotypic behavior are often considered the most effective in promoting animal well-being (Mason 1991; Mason and Latham 2004). Therefore, if an environment, housing condition, or experimental/management protocol consistently leads to a high level of stereotypic behavior development, it is appropriate to assume that this condition is less conducive to positive welfare than an environment with little-to-no stereotypic behavior development. For example, both indoor–mother reared and nursery-reared animals display significantly higher rates of stereotypic behaviors than social group-reared infants (Gottlieb et al. 2013a, 2015). These studies can be informative of the relative welfare conditions of various rearing and housing environments. Maintaining records of stereotypic behaviors in various housing conditions can help establish the best practices at any individual facility. Further, on an individual level, a change or the sudden emergence of stereotypic behavior can indicate a change in individual welfare. Abnormal behaviors are often reported to increase when macaques are relocated to a new location, have a change in husbandry staff, or are introduced to new, stressful environmental stimuli. For example, a normally calm individual may begin to pace a great deal after being moved across from an aggressive animal. Anecdotally, these stereotypic behaviors often decrease when the stressful stimulus is removed or modified (e.g., the aggressive animal is relocated). Knowing the baseline behavior of individuals can help identify sudden or marked changes in behavioral expression.

EXPERT RECOMMENDATIONS

While the literature discussed in the previous sections provides specific examples of macaque enrichment and behavioral management strategies, it is important to recognize that the appropriate tactics of any behavioral management program will depend on the facility's overall size, staffing, budget, available resources, and intended research use of the monkeys. With this in mind, the remaining sections will focus on five principles that we believe should be followed in any macaque behavioral management program, regardless of size or limiting factors.

Prioritize Socialization

All research to date has pointed to socialization as being the single most important factor in promoting positive macaque well-being and appropriate behavioral development. Thus, the priority for any behavioral management team should be to socialize macaques whenever possible. Group housing is most similar to the macaque natural environment, and large social groups are often considered the "gold standard" for macaque socialization. Outdoor field cages and corrals provide optimal opportunities for large social groups; however, innovations in caging also allow small groups to be housed indoors in pens or multiple connected cages. When not group housed, all captive macaques should be provided some form of socialization. Single housing should be avoided unless pair housing is explicitly prohibited due to IACUC-approved scientifically justified exemptions, or contraindicated due to clinical or behavioral needs of the individual (e.g., monkey is sick, or is highly aggressive or fearful when pair housed). Protected contact housing is better than single housing, but efforts should be made to provide full contact socialization to all caged individuals.

Socialization is of particular importance for young animals. Macaque infants should be allowed to stay with their mothers for at least 10–14 months, or longer if the animals are living in a group (McCann et al. 2007; Prescott et al. 2012). Early separation of infants from their mothers can lead to behavioral problems and alterations in both physiology and immunology (Capitanio et al. 2006; Novak and Sackett 2006; Novak et al. 2006; Prescott et al. 2012; Gottlieb et al. 2013a). Allowing infants at least 1 year with their mothers will afford them the opportunity to properly develop and learn appropriate monkey behavior. Once weaned, grouping young animals with an adult female or male can help reduce aggression and teach them proper adult behavior (Champoux et al. 1989; Prescott et al. 2012). While nursery rearing should be avoided whenever possible, factors such as loss of dam, failure of dam to nurse, or rejection of infant by dam can prevent traditional mother rearing. In these circumstances, one option is to find a lactating female, such as a dam that recently lost her infant to illness, to serve as "foster mother." While lactating females that recently lost an infant may be willing, and are often eager, to accept a new infant, the number of such females is often quite low. At the Oregon National Primate Research Center (ONPRC), we have used operant conditioning to train non-lactating females that are highly motivated to hold infants to allow the infants to come to the front of the cage to nurse from a bottle (Welch et al. 2010; K. Coleman and N. D. Robertson, in preparation). To date, we have raised 23 infants in this manner. We have found that this fostering works best when the infants are younger than 1 month of age. Further, we have created "doggie doors" for infants old enough to eat on their own (e.g., 3–4 months), but still young enough to benefit from living with an adult female. These "doggie doors" are transparent slides that contain an opening large enough for the infant to get through, but are too small for the foster mother to enter (Johnson 2013). The infant therefore gets access to both cages, while the foster mother only has access to one cage. This design allows us to provide food in the cage to which the adult does not have access. We have used this door with at least 10 infants. By providing these young macaques with a mother figure, we hope to decrease the common behavioral and physiological impacts of traditional nursery rearing.

There may be times when pairing efforts need to be prioritized, such as in large facilities where animals are frequently relocated to new caged housing. In these situations, efforts should be focused on pairing those that would benefit most from socialization. Beyond infants and juveniles, other monkeys for which pairing should be a priority include those on long-term protocols, those that have been singly housed for an extended period, and those that appear to have difficulty coping with caged housing. When pairing animals, it is important to know their intended future use (e.g., if they are likely to go on a specific research protocol) so as to find appropriate partners. Socializing animals that are separated shortly afterward for research protocols can cause unnecessary stress for the monkeys, and is not an optimal use of staff time.

Exposure to conspecifics is not the only way to provide socialization for macaques. Interaction with human caretakers is an alternate form of socialization that may be particularly beneficial for singly housed animals. Forms of human/macaque socialization include grooming, handing out treats, or presenting visual stimuli to the animals (Baker 2016). At the ONPRC, we have instituted a formalized "Human Interaction Program," in which husbandry staff are encouraged to provide a selection of sensory-stimulating objects (e.g., toys, nontoxic bubbles, etc.) to monkeys. We focus on high-priority rooms (e.g., stress-sensitive rooms, rooms containing single-housed animals, rooms on enrichment-restrictive projects). All individuals who engage in human interaction programs must be properly trained on safe and behaviorally appropriate macaque interactions.

Prevention before Remediation

Once an animal has developed abnormal behaviors, the behaviors can be very difficult to stop. Abnormal behavior expression is a function of both past experiences (e.g., rearing history, time single housed), and current environment (e.g., pairing status, current housing/enrichment). A high rate of individual abnormal behavior expression may be purely a consequence of past environmental conditions, and can be both unrelated and unaffected by current environmental conditions. Focusing efforts on decreasing established abnormal behaviors may be unsuccessful, as abnormal behaviors can persist even in ideal environments. Thus, prevention of abnormal behaviors through early socialization, avoidance of single housing (particularly during early developmental years), frequent environmental stimulation, and limitation of environmental stressors (e.g., frequent room moves, partner separations) is key to reducing their overall occurrence in captive macaque colonies.

Regardless of the program, macaques may still occasionally develop abnormal behaviors in captive environments. Self-abusive behaviors, such as self-biting, that cause physical damage to the animal, are of most concern, and behavioral remediation of these specific individuals should always be a top priority. Stereotypic behaviors, in contrast, are usually not inherently harmful or problematic to the individual. Rather, stereotypic behaviors are concerning only because they are correlated with compromised welfare conditions. All efforts to remediate stereotypic behaviors should be focused on the underlying environmental cause of the behavior (e.g., stress, frustration, fear, boredom), not on simply blocking or stopping behavioral expression. For example, placing obstacles in a cage that obstructs an animal from jumping may decrease the occurrence of stereotypic cage flipping; however, it does not address the root cause of the abnormal behavior.

Many animals express abnormal behaviors in direct response to aversive stimuli in their environment. Problem animals should be monitored, and husbandry and behavioral staff should be in frequent communication in order to identify and remove environmental triggers. Environmental stressors that can cause abnormal behaviors include aggressive or threatening conspecifics, frequent activity in the room (particularly if it is not predictable), losing a social partner, or changes in care staff. Individuals with abnormal behaviors often benefit from removal of the aversive stimulus, or relocation to a new, calmer environment. If the distressed animal responds poorly to animals or husbandry technicians in general, providing a visual barrier as a place to "hide" can be equally beneficial. Upper cages and cages further away from the door have been shown to be less stressful for macaques and may decrease self-biting behavior (Gottlieb et al. 2013a), and thus are optimal cages for highly sensitive individuals. Relocating sensitive animals, however, must be done with care; it is important to monitor rooms and communicate with husbandry technicians to ensure moves are chosen appropriately and are without negative consequence.

Occasionally, entire rooms of animals can become hypersensitive, or in need of environmental remediation. This can occur when specific projects are particularly restrictive or stressful, or when rooms contain a relatively high number of nursery- or indoor-reared animals. In these cases, modifying the environment of the entire room can be an efficient and time-effective strategy to prevent or remediate abnormal behaviors. For example, simply performing routine husbandry activities (e.g., feeding, cleaning, provision of enrichment) reliably at the same time daily can decrease stress and anxiety in captive macaques (Gottlieb et al. 2013b). While it is often not feasible to perform husbandry activities at the exact same time daily in all animal rooms in a facility, a reliable schedule can be prioritized in hypersensitive rooms.

Provide as Much Enrichment as Possible to All Individuals, with Novelty and Variety

A basic, yet important, goal of any behavioral management program is to provide all individuals with as much enrichment as possible. That being said, the number of animals in a given facility, as well as time, financial, and staffing limitations, will greatly impact the amount and types of enrichment options for that facility. While complex and cognitive enrichment, such as puzzle boxes and tablets, may be ideal, simple and easy to prepare foraging enrichment is often more feasible when facing tight budgets and limited staff time. Efforts should be made to design an enrichment program that balances complexity with feasibility, ensuring all animals receive the benefits of enrichment. Frequently rotating and providing novel enrichment can increase its utility, benefits, and functional lifespan.

Evaluate Enrichment and Do No Harm

The provision of enrichment or other behavioral management strategies should never come at the cost of animal or human health and safety. Behavioral management strategies and enrichment techniques implemented with the best intentions can have unforeseen negative consequences. For example, excessive foraging enrichment, particularly high caloric food, can lead to obesity and negative health outcomes; inappropriate audio enrichment, such as loud or aversive music, can be a source of uncontrollable stress for monkeys; frequent transfer of animals for pairing or management purposes can be stressful for some individuals; foraging enrichment for socially housed monkeys can lead to competition and fighting; unmonitored or poorly managed groups are at increased risk of high rates of injury and even death; and enrichment devices with chains or small pieces can be a choking hazard. Even with these potential risks, enrichment should not be avoided or banned. Rather, all forms of enrichment should be monitored, carefully evaluated, and strict regulations should be in place to ensure animal safety. For example, to prevent hanging devices from becoming a potential hazard, chains should be limited to a species-specific safe length, and all staff should be trained in proper device safety.

Any new enrichment or behavioral management technique should be tested and evaluated on a limited number of individuals before colony-wide implementation. Not only does this ensure animal safety, but it can also help determine whether the new enrichment technique is efficacious, beneficial, and a worthwhile use of resources. While a formalized, controlled experimental study will provide valuable non-biased information, simple observation of enrichment use on a few individuals can help behavioral managers evaluate enrichment safety, identify unforeseen negative consequences (e.g., enrichment clogs the drain or falls apart easily), and confirm basic levels of use by the animals.

360° Communication

It is important to recognize that behavioral management does not exist in a vacuum; all management decisions can directly impact clinical medicine, colony management, and research protocols. When developing behavioral management procedures and strategies, it is essential to communicate and receive feedback and support from other stakeholders. A transparent animal management process, where the goals, concerns, and needs of stakeholders can be discussed and collaborative solutions developed, ensures colony goals are met and animal welfare is maintained. As an example of an effective use of this process, we describe a team-approach that we developed at the ONPRC in which all decisions about socially housed animals are discussed and determined collaboratively and openly.

This team meets weekly to review current issues and long-term plans for group-housed animals. Social group dynamics and morbidity and mortality are objectively evaluated for positive and negative trends. Clinical cases of animals requiring veterinary care are discussed from clinical, behavioral, and animal resource management perspectives before determining whether an animal should return to a breeding group or be reassigned. This approach ensures social groups are actively managed to meet breeding and research goals, and that animals not suited for breeding and/or research groups are identified and reassigned to appropriate protocols, all while promoting optimal animal welfare. Each case is reviewed using several evaluation criteria: veterinary staff address cases from a basic health and functioning perspective, focusing on freedom from disease and injury; behavioral and husbandry staff provide both objective and subjective assessments of natural living and positive and negative affective states, as well as the individual's role in the social group; and colony management evaluates anticipated research utility. In discussions where stakeholders have trouble reaching agreement on whether or not an animal should return to its group, brainstorming is encouraged until an acceptable solution is reached. When an animal or group is on a research protocol, the discussion also includes information on scientific needs and anticipated challenges. Together, stakeholders develop a viable plan that ensures animal welfare without sacrificing research integrity.

In addition to the benefits outlined above, this collaborative process has reduced the time needed to make decisions about animals and groups, reduced the number of emails and telephone calls made to discuss problems, and decreased response time to major and unforeseen incidents. Building such a team is not without challenges; a purpose statement is maintained and regularly updated, communication ground rules are defined, and all stakeholders are held accountable. Complex issues may necessitate a separate meeting, focused on the particular issue or group. Issues not requiring immediate answers may be placed in the "parking lot" for later review. To build trust and ensure new individuals are comfortable with the process, new members receive a formal orientation and participate in follow-up discussions with the team leader after their initial meeting.

Teams like this are an effective way to make fully informed decisions that best benefit the animals, and ensure key players are both supportive and fully invested in decisions. While we specifically describe a collaborative approach to social housing, similar methods of inclusion and communication are equally important in all areas of captive animal care, including indoor animal care, husbandry practices, and project implementation.

CONCLUSIONS

In this chapter, we have outlined many tools for the behavioral management of captive macaques, including group socialization, pair housing, rearing strategies, foraging enrichment, sensory enrichment, physical enrichment, cognitive enrichment, PRT, special caging and equipment, and strategies for abnormal behavior prevention and remediation. While the first step in a successful macaque behavioral management program is understanding this vast toolbox of behavioral management techniques, it only takes a few days (if not minutes) in the real world to recognize that these tools are not always available; research exemptions may exclude animals from full contact pairing and provision of some forms of enrichment; facility size, caging, and available equipment can restrict group and socialization options; financial limitations often prevent the purchase of desired enrichment;

and short staffing limits enrichment preparation and distribution, the ability to perform PRT with all desired animals, and the general scope of the enrichment program. The five previously outlined behavioral management principles should be used to help guide your strategies and decisions when faced with restrictions and obstacles that limit your toolbox of behavioral management techniques.

As an example, suppose an enrichment team develops a large, novel foraging tray out of artificial turf for socially housed animals. While they wish to perform a controlled study to determine the efficacy of the new enrichment, they do not have the time or resources to perform such an experiment. In light of this limitation, should the new enrichment still be given to the animals without testing? If so, how should it be implemented? Here we can turn to the outlined principles: *Provide as much enrichment as possible to all individuals, with novelty and variety:* New and varied enrichment is key to any enrichment program, and the inability to perform a controlled experiment should not prevent the utilization of novel enrichment. *Evaluate enrichment and do no harm:* Even without a controlled experiment, it is still important to make certain that the enrichment is safe for the animals. The safety of the new enrichment should be discussed with veterinary staff before implementation, and initial use of the enrichment should be observed on one or two highly monitored individuals to evaluate basic use and safety. *360° communication:* Beyond communicating with veterinarians, the new enrichment should be presented to husbandry staff, relevant researchers, and any other staff who may be impacted by the new enrichment. These individuals can point out logistical issues that may affect implementation, can help monitor and communicate use and efficacy, and are more likely to support the new enrichment if they feel their needs and opinions are being heard and respected. *Prevention before remediation:* When incorporating the artificial turf into the enrichment program, the turf should not be reserved exclusively for "problem" groups. For example, if the turf is expected to decrease aggression, it can be given to groups at high risk for social instability, such as newly formed groups, or those that recently lost key individuals, as well as those with recent aggression or instability.

While we provide a long list of behavioral management techniques in this chapter, there is no single strategy that is appropriate for every captive macaque or every facility housing them. Not Facilities vary in terms of size, resources, and overall goals. Further, as socially complex, highly intelligent animals, the needs of macaques vary on an individual basis. Rather than outline the specific details of the "optimal" macaque behavioral management program, we hope this chapter has provided a large toolbox of management and enrichment options and strategies, so that the appropriate tool can be selected when a new problem or concern arises.

ACKNOWLEDGMENTS

We thank the ONPRC Behavioral Services Unit staff, Colony Epidemiology Group, Nonhuman Primate Resources Unit, Eileen Korey, and Allison Heagerty for help and support of this chapter. This work was supported by NIH P51OD011092.

REFERENCES

Andrews, M. W. and L. A. Rosenblum. 2001. Effects of change in social content of video rewards on response patterns of bonnet macaques. *Learning and Motivation* 32 (4):401–408.

Baker, K. C. 2016. Survey of 2014 behavioral management programs for laboratory primates in the United States. *American Journal of Primatology* 78:780–796.

Baker, K. C., M. Bloomsmith, K. Neu, et al. 2009. Positive reinforcement training moderates only high levels of abnormal behavior in singly housed rhesus macaques. *Journal of Applied Animal Welfare Science* 12 (3):236–252.

Baker, K. C., M. A. Bloomsmith, B. Oettinger, et al. 2012a. Benefits of pair housing are consistent across a diverse population of rhesus macaques. *Applied Animal Behaviour Science* 137 (3–4):148–156.

Baker, K. C., M. A. Bloomsmith, B. Oettinger, et al. 2014. Comparing options for pair housing rhesus macaques using behavioral welfare measures. *American Journal of Primatology* 76:30–42.

Baker, K. C., C. M. Crockett, G. H. Lee, et al. 2012b. Pair housing for female longtailed and rhesus macaques in the laboratory: Behavior in protected contact versus full contact. *Journal of Applied Animal Welfare Science* 15 (2):126–143.

Baker, K. C., J. L. Weed, C. M. Crockett, et al. 2007. Survey of environmental enhancement programs for laboratory primates. *American Journal of Primatology* 69 (4):377–394.

Bayne, K., S. Dexter, H. Mainzer, et al. 1992a. The use of artificial turf as a foraging substrate for individually housed rhesus monkeys (*Macaca mulatta*). *Animal Welfare* 1:39–53.

Bayne, K., S. Dexter, and S. Suomi. 1992b. A preliminary survey of the incidence of abnormal behavior in rhesus monkeys (*Macaca mulatta*) relative to housing condition. *Lab Animal* 21 (5):38, 40, 42–46.

Bayne, K., H. Mainzer, S. Dexter, et al. 1991. The reduction of abnormal behaviors in individually housed rhesus monkeys (*Macaca mulatta*) with a foraging grooming board. *American Journal of Primatology* 23 (1):23–35.

Beisner, B. A., M. E. Jackson, A. N. Cameron, et al. 2011. Detecting instability in animal social networks: Genetic fragmentation is associated with social instability in rhesus macaques. *PloS One* 6 (1):e16365.

Beisner, B. A., M. E. Jackson, A. N. Cameron, et al. 2012. Sex ratio, conflict dynamics, and wounding in rhesus macaques (*Macaca mulatta*). *Applied Animal Behaviour Science* 137 (3–4):137–147.

Beisner, B. A., J. Jin, H. Fushing, et al. 2015. Detection of social group instability among captive rhesus macaques using joint network modeling. *Current Zoology* 61 (1):70–84.

Bellanca, R. U. and C. M. Crockett. 2002. Factors predicting increased incidence of abnormal behavior in male pigtailed macaques. *American Journal of Primatology* 58 (2):57–69.

Belzung, C. and J. R. Anderson. 1986. Social rank and responses to feeding competition in rhesus monkeys. *Behavioural Processes* 12 (4):307–316.

Bloomsmith, M. A., L. Y. Brent, and S. J. Schapiro. 1991. Guidelines for developing and managing an environmental enrichment program for nonhuman primates. *Laboratory Animal Science* 41 (4):372–377.

Bloomsmith, M. A., S. P. Lambeth, A. M. Stone, et al. 1997. Comparing two types of human interaction as enrichment for chimpanzees. *American Journal of Primatology* 42:96.

Brannon, E. M., M. W. Andrews, and L. A. Rosenblum. 2004. Effectiveness of video of conspecifics as a reward for socially housed bonnet macaques (*Macaca radiata*). *Perceptual and Motor Skills* 98 (3):849–858.

Brunelli, R. L., J. Blake, N. Willits, et al. 2014. Effects of a mechanical response-contingent surrogate on the development of behaviors in nursery-reared rhesus macaques (*Macaca mulatta*). *Journal of the American Association for Laboratory Animal Science* 53 (5):464.

Buchanan-Smith, H. M., M. R. Gamble, M. Gore, et al. 2009. Refinements in husbandry, care and common procedures for non-human primates. *Laboratory Animals* 43:S1–S47.

Byrne, G. D. and S. J. Suomi. 1991. Effects of woodchips and buried food on behavior patterns and psychological well-being of captive rhesus-monkeys. *American Journal of Primatology* 23 (3):141–151.

Capitanio, J. P. 1986. Behavioral pathology. In *Comparative Primate Biology, Volume 2A: Behavior, Conservation, and Ecology*. Mitchell, G. and J. Erwin (eds). New York: Alan R. Liss.

Capitanio, J. P., W. A. Mason, S. P. Mendoza, et al. 2006. Nursery rearing and biobehavioral organzation. In *Nursery Rearing of Nonhuman Primates in the 21st Century*. Sackett, G. P., G. C. Ruppenthal, and K. Elias (eds), pp. 191–214. New York: Springer.

Carlsson, H. E., S. J. Schapiro, I. Farah, et al. 2004. Use of primates in research: A global overview. *American Journal of Primatology* 63 (4):225–237.

Chamove, A. S., J. R. Anderson, S. C. Morgan-Jones, et al. 1982. Deep woodchip litter: Hygiene, feeding, and behavioral enhancement in eight primate species. *International Journal for the Study of Animal Problems* 3 (4):308–318.

Chamove, A., L. Rosenblum, and H. Harlow. 1973. Monkeys (*Macaca mulatta*) raised only with peers. A pilot study. *Animal Behaviour* 21 (2):316–325.

Champoux, M., B. Metz, and S. J. Suomi. 1989. Rehousing nonreproductive rhesus macaques with weanlings: I. Behavior of adults toward weanlings. *Laboratory Primate Newsletter* 28 (4):4.

Chan, S., H. Fushing, B. A. Beisner, et al. 2013. Joint modeling of multiple social networks to elucidate primate social dynamics: I. Maximum entropy principle and network-based interactions. *PloS One* 8 (2):e51903.

Chance, M., G. Emory, and R. Payne. 1977. Status referents in long-tailed macaques (*Macaca fascicularis*): Precursors and effects of a female rebellion. *Primates* 18 (3):611–632.

Chapais, B. 2004. How kinship generates dominance structures: A comparative perspective. In *Macaque Societies*. Thierry, B., M. Singh, and W. Kaumanns (eds), pp. 186–209. Cambridge, UK: Cambridge University Press.

Coe, C. L. 1991. Is social housing of primates always the optimal choice? In *Through the Looking Glass: Issues of Psychological Well-Being in Captive Nonhuman Primates*. Novak, M. A. and A. J. Petto (eds), pp. 78–92. Washington, DC: American Psychological Association.

Coe, C. L., G. R. Lubach, W. B. Ershler, et al. 1989. Influence of early rearing on lymphocyte proliferation responses in juvenile rhesus monkeys. *Brain, Behavior, and Immunity* 3 (1):47–60.

Coleman, K. 2017. Individual differences in temperament and behavioral management, Chapter 7. In *Handbook of Primate Behavioral Management*, 95–113. Schapiro, S. J. (ed.). Boca Raton, FL: CRC Press.

Coleman, K., M. A. Bloomsmith, C. M. Crockett, et al. 2012. Behavioral management, enrichment, and psychological well-being of laboratory nonhuman primates. In *Nonhuman Primates in Biomedical Research: Biology and Management*. Abee, C. R., K. Mansfield, S. D. Tardif, et al. (eds), pp. 149–176. London, UK: Academic Press.

Coleman, K. and A. Maier. 2010. The use of positive reinforcement training to reduce stereotypic behavior in rhesus macaques. *Applied Animal Behavioral Science* 124 (3–4):142–148.

Coleman, K., L. Pranger, A. Maier, et al. 2008. Training rhesus macaques for venipuncture using positive reinforcement techniques: A comparison with chimpanzees. *Journal of the American Association of Laboratory Animal Science* 47 (1):37–41.

Crockett, C. M., R. U. Bellanca, C. L. Bowers, et al. 1997. Grooming-contact bars provide social contact for individually caged laboratory macaques. *Contemporary Topics in Laboratory Animal Science* 36 (6):53–60.

Crockett, C. and G. Gough. 2002. Onset of aggressive toy biting by a laboratory baboon coincides with cessation of self-injurious behavior. *American Journal of Primatology Supplement* 1 (57):39.

Cross, H. A. and H. F. Harlow. 1965. Prolonged and progressive effects of partial isolation on the behavior of macaque monkeys. *Journal of Experimental Research in Personality* 1 (1):39–49.

de Waal, F. B. 1991. The social nature of primates. In *Through the Looking Glass: Issues of Psychological Wellbeing in Captive Nonhuman Primates*. Novak, M. and A. J. Petto (eds), pp. 69–77. Washington, DC: American Psychological Association.

DiVincenti, L. and J. D. Wyatt. 2011. Pair housing of macaques in research facilities: A science-based review of benefits and risks. *Journal of the American Association for Laboratory Animal Science* 50 (6):856–863.

Doane, C. J., K. Andrews, L. J. Schaefer, et al. 2013. Dry bedding provides cost-effective enrichment for group-housed rhesus macaques (*Macaca mulatta*). *Journal of the American Association for Laboratory Animal Science* 52 (3):247.

Drewsen, K. H. 1989. The importance of auditory variaton in the homecage environment of socially housed rhesus monkeys (*Macaca mulatta*). Doctoral Dissertation. University of Massachusetts Amherst.

Ehardt, C. L. and I. S. Bernstein. 1986. Matrilineal overthrows in rhesus monkey groups. *International Journal of Primatology* 7 (2):157–181.

Elmore, D. B., J. H. Anderson, D. W. Hird, et al. 1992. Diarrhea rates and risk factors for developing chronic diarrhea in infant and juvenile rhesus monkeys. *Laboratory Animal Science* 42 (4):356–359.

Flack, J. and F. B. M. de Waal. 2004. Dominance style, social power, and conflict management: A conceptual framework. In *Macaque Societies*. Thierry, B., M. Singh, and W. Kaumanns (eds), pp. 157–186. Cambridge, UK: Cambridge University Press.

Flack, J. C., D. C. Krakauer, and F. B. M. de Waal. 2005. Robustness mechanisms in primate societies: A perturbation study. *Proceedings of the Royal Society B: Biological Sciences* 272 (1568):1091–1099.

Gilbert, M. H. and K. C. Baker. 2011. Social buffering in adult male rhesus macaques (*Macaca mulatta*): Effects of stressful events in single vs. pair housing. *Journal of Medical Primatology* 40 (2):71–78.

Gottlieb, D. H. and J. P. Capitanio. 2013. Latent variables affecting behavioral response to the human intruder test in infant rhesus macaques (*Macaca mulatta*). *American Journal of Primatology* 75 (4):314–323.

Gottlieb, D. H., J. P. Capitanio, and B. McCowan. 2013a. Risk factors for stereotypic behavior and self-biting in rhesus macaques (*Macaca mulatta*); animal's history, current environment, and personality. *American Journal of Primatology* 75 (10):995–1008.

Gottlieb, D. H., K. Coleman, and B. McCowan. 2013b. The effects of predictability in daily husbandry routines on captive rhesus macaques (*Macaca mulatta*). *Applied Animal Behaviour Science* 143 (2–4):117–127.

Gottlieb, D. H., S. Ghirardo, D. E. Minier, et al. 2011. Assessment of efficacy of three types of foraging enrichment in rhesus macaques (*Macaca mulatta*). *The Journal of the American Association for Laboratory Animal Science* 50 (6):1–7.

Gottlieb, D. H., A. Maier, and K. Coleman. 2015. Evaluation of environmental and intrinsic factors that contribute to stereotypic behavior in captive rhesus macaques (*Macaca mulatta*). *Applied Animal Behaviour Science* 171:184–191.

Gottlieb, D. H., J. R. O'Connor, and K. Coleman. 2014. Using porches to decrease feces painting in rhesus macaques (*Macaca mulatta*). *Journal of the American Association for Laboratory Animal Science* 53 (6):653–656.

Graham, M. L. 2017. Positive reinforcement training and research, Chapter 12. In *Handbook of Primate Behavioral Management*, 187–200. Schapiro, S. J. (ed.). Boca Raton, FL: CRC Press.

Graves, L. M. 2011. The effect of auditory enrichment on abnormal, affiliative, and aggressive behavior in laboratory-housed rhesus macaques (*Macaca mulatta*). Master's Thesis. Texas State University-San Marcos.

Griffis, C. M., A. L. Martin, J. E. Perlman, et al. 2013. Play caging benefits the behavior of singly housed laboratory rhesus macaques (*Macaca mulatta*). *Journal of the American Association for Laboratory Animal Science* 52 (5):534–540.

Gygax, L., N. Harley, and H. Kummer. 1997. A matrilineal overthrow with destructive aggression in *Macaca fascicularis*. *Primates* 38 (2):149–158.

Hambright, M. K. and D. A. Gust. 2003. A descriptive analysis of a spontaneous dominance overthrow in a breeding colony of rhesus macaques (*Macaca mulatta*). *Laboratory Primate Newsletter* 42 (1):8–10.

Hanya, G. 2004. Seasonal variations in the activity budget of Japanese macaques in the coniferous forest of Yakushima: Effects of food and temperature. *American Journal of Primatology* 63 (3):165–177.

Harlow, H. F., R. O. Dodsworth, and M. K. Harlow. 1965. Total social isolation in monkeys. *Proceedings of the National Academy of Sciences* 54 (1):90–97.

Harlow, H. F. and S. J. Suomi. 1971. Social recovery by isolation-reared monkeys. *Proceedings of the National Academy of Sciences* 68 (7):1534–1538.

Harris, L. D., E. J. Briand, R. Orth, et al. 1999. Assessing the value of television as environmental enrichment for individually housed rhesus monkeys: A behavioral economic approach. *Contemporary Topics in Laboratory Animal Science* 38 (2):48–53.

Hasan, M. K., M. A. Aziz, S. R. Alam, et al. 2013. Distribution of rhesus macaques (*Macaca mulatta*) in Bangladesh: Inter-population variation in group size and composition. *Primate Conservation* 26 (1):125–132.

Hill, D. A. 2004. Box 11 Intraspecific variation: Implications for interspecific comparisons. In *Macaque Societies: A Model for the Study of Social Organization*. Thierry, B., M. Singh, and W. Kaumanns (eds), pp. 262–266. Cambridge, UK: Cambridge University Press.

Hird, D. W., J. H. Anderson, and J. T. Bielitzki. 1984. Diarrhea in nonhuman-primates—A survey of primate colonies for incidence rates and clinical opinion. *Laboratory Animal Science* 34 (5):465–470.

Honess, P. 2017. Behavioral management of long-tailed macaques (*Macaca fascicularis*), Chapter 20. In *Handbook of Primate Behavioral Management*, 305–337. Schapiro, S. J. (ed.). Boca Raton, FL: CRC Press.

Johnson, J. G. 2013. Doggy door for macaque dam/infant pairs. *The Enrichment Record* 14:14–17.

Keeling, M. E., P. L. Alford, and M. A. Bloomsmith. 1991. Decision analysis for developing programs of psychological well-being in rhesus monkeys. In *Through the Looking Glass: Issues of Psychological Well-Being in Captive Nonhuman Primates*. Novak, M. A. and A. J. Petto (eds), pp. 57–65. Washington, DC: American Psychological Assocation.

Klaiber-Schuh, A. and C. Welker. 1997. Crab-eating monkeys can be trained to cooperate in non-invasive oral medication without stress. *Primate Report* 47:11–30.

Lam, K., N. Rupniak, and S. Iversen. 1991. Use of a grooming and foraging substrate to reduce cage stereotypies in macaques. *Journal of Medical Primatology* 20 (3):104–109.

Laule, G. E., M. A. Bloomsmith, and S. J. Schapiro. 2003. The use of positive reinforcement training techniques to enhance the care, management, and welfare of primates in the laboratory. *Journal of Applied Animal Welfare Science* 6 (3):163–173.

Lee, G. H., J. P. Thom, K. L. Chu, et al. 2012. Comparing the relative benefits of grooming-contact and full-contact pairing for laboratory-housed adult female *Macaca fascicularis*. *Applied Animal Behaviour Science* 137 (3):157–165.

Line, S. W., A. S. Clarke, H. Markowitz, et al. 1990. Responses of female rhesus macaques to an environmental enrichment apparatus. *Laboratory Animals* 24 (3):213–220.

Lubach, G. R., C. L. Coe, and W. B. Ershler. 1995. Effects of early rearing environment on immune-responses of infant rhesus-monkeys. *Brain Behavior and Immunity* 9 (1):31–46.

Lutz, C. K., K. Coleman, A. Maier, et al. 2011. Abnormal behavior in rhesus monkeys: Risk factors within and between animals and facilities. *American Journal of Primatology* 73:41.

Lutz, C. K., E. B. Davis, A. M. Ruggiero, et al. 2007. Early predictors of self-biting in socially-housed rhesus macaques (*Macaca mulatta*). *American Journal of Primatology* 69 (5):584–590.

Lutz, C. K. and M. A. Novak. 1995. Use of foraging racks and shavings as enrichment tools for groups of rhesus-monkeys (*Macaca mulatta*). *Zoo Biology* 14 (5):463–474.

Lutz, C. K. and M. A. Novak. 2005. Environmental enrichment for nonhuman primates: Theory and application. *ILAR Journal* 46 (2):178–191.

Lutz, C., A. Well, and M. Novak. 2003. Stereotypic and self-injurious behavior in rhesus macaques: A survey and retrospective analysis of environment and early experience. *American Journal of Primatology* 60 (1):1–15.

Magden, E. R. 2017. Positive reinforcement training and health care, Chapter 13. In *Handbook of Primate Behavioral Management*, 201–216. Schapiro, S. J. (ed.). Boca Raton, FL: CRC Press.

Markowitz, H. and S. Line. 1989. Primate research models and environmental enrichment. In *Housing, Care, and Psychological Well-Being of Captive and Laboratory Primates*. Segal, E. F. (ed.), pp. 203–212. Park Ridge, NJ: Noyes Publications.

Mason, G. J. 1991. Stereotypies—A critical-review. *Animal Behaviour* 41:1015–1037.

Mason, G. 2006. Stereotypic behaviour in captive animals: Fundamentals and implications for welfare and beyond. In *Stereotypic Animal Behaviour: Fundamentals and Applications to Welfare*. Mason, G. and J. Rushen (eds), pp. 325–356. Wallingford, CT: CABI.

Mason, G. J. and N. R. Latham. 2004. Can't stop, won't stop: Is stereotypy a reliable animal welfare indicator? *Animal Welfare* 13:S57–S69.

McCann, C., H. Buchanan-Smith, L. Jones-Engel, et al. 2007. *IPS International Guidelines for the Acquisition, Care and Breeding of Nonhuman Primates*. International Primatological Society. http://www.internationalprimatologicalsociety.org/docs/ips_international_guidelines_for_the_acquisition_care_and_breeding_of_nonhuman_primates_second_edition_2007.pdf

McCowan, B., K. Anderson, A. Heagarty, et al. 2008. Utility of social network analysis for primate behavioral management and well-being. *Applied Animal Behaviour Science* 109 (2):396–405.

McCowan, B. and B. Beisner. 2017. Utility of systems network analysis for understanding complexity in primate behavioral management, Chapter 11. In *Handbook of Primate Behavioral Management*, 157–183. Schapiro, S. J. (ed.). Boca Raton, FL: CRC Press.

McCowan, B., B. A. Beisner, J. P. Capitanio, et al. 2011. Network stability is a balancing act of personality, power, and conflict dynamics in rhesus macaque societies. *PloS One* 6 (8):e22350.

Ménard, N. 2004. Do ecological factors explain variatino in social organizaiton? In *Macaque Societies*. Thierry, B., M. Singh, and W. Kaumanns (eds), pp. 237–262. Cambridge, UK: Cambridge University Press.

Mench, J. A. and G. J. Mason. 1997. Behaviour. In *Animal Welfare*. Appleby, M. C. and B. O. Hughes (eds), pp. 127–141. Cambridge, MA: CABI.

Mineka, S., M. Gunnar, and M. Champoux. 1986. Control and early socioemotional development infant rhesus monkeys reared in controllable vs uncontrollable environments. *Child Development* 57 (5):1241–1256.

Moseley, J. and J. Davis. 1989. Psychological enrichment techniques and new world monkey restraint device reduce colony management time. *Lab Animal (USA)* 39:31–33.

Mueller, K., K. Moore, A. Maier, et al. 2008. Watching conspecifics being trained helps rhesus macaques (*Macaca mulatta*) learn faster. *Journal of the American Association for Laboratory Animal Science* 47 (5):160–161.

Nakamichi, M. and K. Asanuma. 1998. Behavioral effects of perches on group-housed adult female Japanese monkeys. *Percept Mot Skills* 87 (2):707–714.

National Research Council (Institute for Laboratory Animal Research). 2011. *Guide for the Care and Use of Laboratory Animals*, 8th edition. Washington, DC: National Academies Press.

Novak, M. A. 2003. Self-injurious behavior in rhesus monkeys: New insights into its etiology, physiology, and treatment. *American Journal of Primatology* 59 (1):3–19.

Novak, M. A. and K. H. Drewsen. 1989. Enriching the lives of captive primates: Issues and problems. In *Housing, Care and Psychological Well-Being of Captive and Laboratory Primates*. Segal, E. F. (ed.), pp. 161–182. Park Ridge, NJ: Noyes Publications.

Novak, M. A., J. H. Kinsey, M. J. Jorgensen, et al. 1998. Effects of puzzle feeders on pathological behavior in individually housed rhesus monkeys. *American Journal of Primatology* 46 (3):213–227.

Novak, M. A., J. S. Meyer, C. Lutz, et al. 2006. Deprived environments: Developmental insights from primatology. In *Stereotypic Animal Behaviour: Fundamentals and Applications to Welfare*. Mason, G. and J. Rushen (eds), pp. 153–189. Wallingford, CT: CABI.

Novak, M. A. and G. P. Sackett. 2006. The effects of rearing experiences: The early years. In *Nursery Rearing of Nonhuman Primates in the 21st Century*, Sackett, G. P., G. C. Ruppentahal, and K. Elias (eds), pp. 5–19. New York: Springer.

Oates-O'Brien, R. S., T. B. Farver, K. C. Anderson-Vicino, et al. 2010. Predictors of matrilineal overthrows in large captive breeding groups of rhesus macaques (*Macaca mulatta*). *Journal of the American Association for Laboratory Animal Science* 49 (2):196.

O'Brien, T. G. and M. F. Kinnaird. 1997. Behavior, diet, and movements of the sulawesi crested black macaque (*Macaca nigra*). *International Journal of Primatology* 18 (3):321–351.

O'Connor, J., A. Heagerty, M. Herrera, et al. 2015. Use of a tablet as enrichment for adult rhesus macaques. *American Journal of Primatology* 77 (S1):122.

Ogura, T. 2012. Use of video system and its effects on abnormal behaviour in captive Japanese macaques (*Macaca fuscata*). *Applied Animal Behaviour Science* 141 (3–4):173–183.

Ogura, T. and T. Matsuzawa. 2012. Video preference assessment and behavioral management of single-caged Japanese macaques (*Macaca fuscata*) by movie presentation. *Journal of Applied Animal Welfare Science* 15 (2):101–112.

O'Neill, P. 1989. A room with a view for captive primates: Issues, goals related research and strategies. In *Housing, Care and Psychological Well-Being of Captive and Laboratory Primates*. Segal, E. F. (ed.), pp. 135–160. Park Ridge, NJ: Noyes Publications.

Padilla, A. 2011. Training individual rhesus macaques in large social groups. *International Conference on Environmental Enrichment*, August 14–19, 2011, Portland, Oregon.

Petherick, J. C. and J. Rushen. 1997. Behavioural restriction. In *Animal Welfare*. Appleby, M. C. and B. O. Huhges (eds), pp. 89–105. Cambridge, MA: CABI.

Platt, D. M. and M. A. Novak. 1997. Videostimulation as enrichment for captive rhesus monkeys (*Macaca mulatta*). *Applied Animal Behaviour Science* 52 (1–2):139–155.

Prescott, M. 2006. *Primate Sensory Capabilities and Communication Signals: Implications for Care and Use in the Laboratory*. London, UK: National Centre for the Replacement, Refinement and Reduction of Animals in Research.

Prescott, M. J., M. E. Nixon, D. A. Farningham, et al. 2012. Laboratory macaques: When to wean? *Applied Animal Behaviour Science* 137 (3):194–207.

Reinhardt, V. 1992. Space utilization by captive rhesus macaques. *Animal Technology* 43 (1):11–17.

Reinhardt, V. 1993a. Enticing nonhuman primates to forage for their standard biscuit ration. *Zoo Biology* 12 (3):307–312.

Reinhardt, V. 1993b. Using the mesh ceiling as a food puzzle to encourage foraging behaviour in caged rhesus macaques (*Macaca mulatta*). *Animal Welfare* 2 (2):165–172.

Reinhardt, V. 2010. *Caring Hands: Discussions by the Laboratory Animal Refinement and Enrichment Forum*, Volume 2. Washington, DC: Animal Welfare Institute.

Reinhardt, V., D. Cowley, J. Scheffler, et al. 1990. Cortisol response of female rhesus monkeys to venipuncture in homecage versus venipuncture in restraint apparatus. *Journal of Medical Primatology* 19:601–606.

Reinhardt, V. and A. Roberts. 1997. Effective feeding enrichment for non-human primates: A brief review. *Animal Welfare* 6 (3):265–272.

Reichard, T., W. Shellabargar, and G. Laule. 1992. Training for husbandry and medical purposes. In *Proceedings of the American Association of Zoological Parks and Aquariums*. pp. 396–402. Wheeling, WV: AAZPA.

Robins, J. G. and C. D. Waitt. 2011. Improving the welfare of captive macaques (*Macaca* sp.) through the use of water as enrichment. *Journal of Applied Animal Welfare Science* 14 (1):75–84.

Rommeck, I., K. Anderson, A. Heagerty, et al. 2009a. Risk factors and remediation of self-injurious and self-abuse behavior in rhesus macaques. *Journal of Applied Animal Welfare Science* 12 (1):61–72.

Rommeck, I., J. P. Capitanio, S. C. Strand, et al. 2011. Early social experience affects behavioral and physiological responsiveness to stressful conditions in infant rhesus macaques (*Macaca mulatta*). *American Journal of Primatology* 73 (7):692–701.

Rommeck, I., D. H. Gottlieb, S. C. Strand, et al. 2009b. The effects of four nursery-rearing strategies on infant behavioural development in rhesus macaques (*Macaca mulatta*). *The Journal of the American Association for Laboratory Animal Science* 48 (4):395–401.

Rommeck, I., B. McCowan, D. Gottlieb, et al. 2008. The effects of intermittent play-group (surrogate-peer) rearing on infant behavioral development in rhesus macaques (*Macaca mulatta*). *American Journal of Primatology* 70:47

Ruppenthal, G. C., C. G. Walker, and G. P. Sackett. 1991. Rearing infant monkeys (*Macaca nemestrina*) in pairs produces deficient social development compared with rearing in single cages. *American Journal of Primatology* 25 (2):103–113.

Sackett, G. P., G. C. Ruppenthal, C. E. Fahrenbruch, et al. 1981. Social-isolation rearing effects in monkeys vary with genotype. *Developmental Psychology* 17 (3):313–318.

Samuels, A. and R. V. Henrickson. 1983. Brief report: Outbreak of severe aggression in captive *Macaca mulatta*. *American Journal of Primatology* 5 (3):277–281.

Schapiro, S. J. and M. A. Bloomsmith. 1994. Behavioral-effects of enrichment on pair-housed juvenile rhesus-monkeys. *American Journal of Primatology* 32 (3):159–170.

Schapiro, S. J. and M. A. Bloomsmith. 1995. Behavioral effects of enrichment on singly-housed, yearling rhesus monkeys: An analysis including three enrichment conditions and a control group. *American Journal of Primatology* 35 (2):89–101.

Schapiro, S. J., M. A. Bloomsmith, and G. E. Laule. 2003. Positive reinforcement training as a technique to alter nonhuman primate behavior: Quantitative assessments of effectiveness. *Journal of Applied Animal Welfare Science* 6 (3):175–187.

Schapiro, S. J., M. A. Bloomsmith, S. A. Suarez, et al. 1996a. Effects of social and inanimate enrichment on the behavior of yearling rhesus monkeys. *American Journal of Primatology* 40 (3):247–260.

Schapiro, S. J., P. N. Nehete, J. E. Perlman, et al. 2000. A comparison of cell-mediated immune responses in rhesus macaques housed singly, in pairs, or in groups. *Applied Animal Behaviour Science* 68 (1):67–84.

Schapiro, S. J., J. E. Perlman, and B. A. Boudreau. 2001. Manipulating the affiliative interactions of group-housed rhesus macaques using positive reinforcement training techniques. *American Journal of Primatology* 55 (3):137–149.

Schapiro, S. J., S. A. Suarez, L. M. Porter, et al. 1996b. The effects of different types of feeding enhancements on the behaviour of single-caged, yearling rhesus macaques. *Animal Welfare* 5 (2):129–138.

Shepherdson, D. J. 1998. Introduction: Tracing the path of environmental enrichment in zoos. In *Second nature: Environmental enrichment for captive animals*, ed. D. J. Shepherdson, J. D. Mellen and M. Hutchins. Smithsonial Institution Press, Washington, D.C.

Singh, M. and S. Vinathe. 1990. Inter-population differences in the time budgets of bonnet monkeys (*Macaca radiata*). *Primates* 31 (4):589–596.

Thierry, B. 2007. The macaques: A double-layered social organization In *Primates in Perspective.* Campbell, C. J., A. Fuentes, K. C. MacKinnon, et al. (eds), pp. 224–239. New York: Oxford University Press.

USDA. 1991. Animal welfare, standards, final rule (part 3, subpart d: Specifications for the humane handling, care, treatment, and transportation of nonhuman primates). United States: Federal Register

Washburn, D. A., J. P. Gulledge, and D. M. Rumbaugh. 1997. The heuristic and motivational value of video reinforcement. *Learning and Motivation* 28 (4):510–520.

Weed, J. L., P. O. Wagner, R. Byrum, et al. 2003. Treatment of persistent self-injurious behavior in rhesus monkeys through socialization: A preliminary report. *Journal of the American Association for Laboratory Animal Science* 42 (5):21–23.

Welch, J., N. D. Robertson, K. Mueller, et al. 2010. Use of operant conditioning to train non-lactating rhesus macaques as foster mothers. *American Journal of Primatology* 72:52–53.

Worlein, J. M. and G. P. Sackett. 1997. Social development in nursery-reared pigtailed macaques (*Macaca nemestrina*). *American Journal of Primatology* 41 (1):23–35.

CHAPTER 20

Behavioral Management of Long-Tailed Macaques (*Macaca fascicularis*)

Paul Honess
Bioculture (Mauritius) Ltd.

CONTENTS

INTRODUCTION

This chapter presents an overview of the key issues relating to behavioral management of the long-tailed [cynomolgus ("cyno"), crab-eating] macaque, *Macaca fascicularis*. This species is of particular importance as it is the most commonly used nonhuman primate in biomedical research and safety testing (SCHER 2009; Tasker 2012; Home Office 2014). While it may present fewer challenges for captive management than the rhesus macaque (*M. mulatta*), largely for temperamental reasons (Kling and Orbach 1963), nevertheless, the large numbers of this species in research mean that efficient and timely management of its behavior can benefit the welfare of very many individuals. Many of the issues confronting those managing long-tailed macaques are not unique to the species and are similar, if not identical, to those encountered with other macaques. Nevertheless, addressing their specific needs and management requires some nuances of approach.

The chapter begins with a brief overview of the natural history of the species. Natural history provides an essential foundation for all those wishing to work with any primate species in captivity. The chapter then discusses general and then more specific strategies relating to the behavioral management of the species, closing with important recommendations arising from what has already been presented, together with consideration of what constitutes a functionally appropriate captive environment for the species. While this chapter contributes to an increasingly political debate, it must nevertheless be considered in the context of the importance for primate welfare scientists to always ask for more for the animals than is ever likely to be delivered in a pragmatic world of limited financial resources.

NATURAL HISTORY

Taxonomy and Distribution

The long-tailed macaque occurs naturally in the wild across Southeast Asia, including southern Bangladesh, Burma, Thailand, Malaysia, Indonesia, Singapore, Laos, Cambodia, Vietnam, Brunei, the Philippines, and the Nicobar Islands (India) (Fooden 1991; Rowe 1996). Introduced populations occur in Mauritius (Stanley 2003; Tosi and Coke 2007), Angaur Island (Micronesia) (Poirier and Smith 1974a,b; Kawamoto et al. 1988), Papua New Guinea (Ong and Richardson 2008), and Hong Kong (Burton and Chan 1996).

There are nine subspecies of *M. fascicularis* (Fooden 1991, 1995), of which *M. f. fascicularis* is the most widespread and is most commonly used in biomedical research. In continental Asia, significant hybridization occurs both between subspecies (Fooden 1995; Groves 2001) and with other sympatric macaques, including rhesus (Tosi et al. 2002; Kanthaswamy et al. 2008), pig-tailed (*M. nemestrina*; Groves 2001), and potentially Assamese (*M. assamensis*) and/or Tibetan (*M. thibetana*) macaques (Tosi et al. 2003).

Habitat and Population Density

Table 20.1 indicates the key ecological features of long-tailed macaques. They are found in a range of forest types from high-altitude primary tropical forests to sea-level mangrove forests (Rowe 1996). They are also common in secondary forests and at sites where they are commensal with humans (Eudey 1994; Wheatley et al. 1996). They reach their highest densities in natural habitat in mangrove forests (Crockett and Wilson 1980), with more typical densities found in primary forests (Table 20.1; Kurland 1973; Wheatley et al. 1996). Where provisioned by humans, densities may be very high (Wheatley et al. 1996), particularly in urban environments (e.g., Bangkok, Thailand: 128/ha; Brotcorne et al. 2008). A group's home range must meet its year-round feeding

Table 20.1   Some Important Ecological Parameters of *M. fascicularis*

		Habitat	References
Elevation	Sea level—2000 m		Rowe (1996)
Density (individuals/km^2)	120	Mangrove	Crockett and Wilson (1980)
	20–90	Primary forest	Kurland (1973) and Wheatley et al. (1996)
	1111	Provisioned (temple)	Wheatley et al. (1996)
	12,800	Provisioned (urban)	Brotcorne et al. (2008)
Group size	10–48 typically	All	Melnick and Pearl (1987) and Rowe (1996)
	Max. 170	Mangrove	Son (2004)
	8–30	Undisturbed primary forest	Kurland (1973)
	22–42 (av. 30)	Undisturbed primary forest	Kurland (1973) and Wheatley et al. (1996)
	1–17 (av. 9.6)	Disturbed forest	Yanuar et al. (2009)
	11–33	Disturbed forest	van Schaik et al. (1983)
	16–28 (av. 23)	Disturbed forest	Wheatley et al. (1996)
Home range (km^2/group)	0.008–1.25		Kurland (1973) and Wheatley et al. (1996)
Day range/day	1.5 km		de Ruiter et al. (1992)
Group adult sex ratio (males:females)	1:6–1:1.8	Undisturbed primary forest	Kurland (1973) and Wheatley et al. (1996)
	1:2.5	Provisioned groups	Wheatley et al. (1996)

and shelter requirements, and therefore, the area may vary seasonally and with habitat, depending on food availability and competition (Kurland 1973; Wheatley et al. 1996). During feeding and other activities, a typical group travels ~1.5 km/day (de Ruiter et al. 1992).

Long-tailed macaques are well-known for associating with humans and human-altered habitats (Richard et al. 1989), including temple sites (Wheatley et al. 1996; Fuentes and Gamerl 2005), and commonly engage in crop raiding (Kurland 1973; Sussman and Tattersall 1986; Hadi et al. 2007; Yanuar et al. 2009). In Southeast Asia, many long-tailed macaque populations are infected with simian herpes B virus, and monkey–human contact represents a significant human health risk. Fortunately, infection rates from contact appear low (Engel et al. 2002). In contrast, the introduced Mauritian population is naturally free from herpes B and a range of other pathogens, making it a particularly important research model (Stanley 2003). The long-tailed macaques supplied for research come from both wild capture (Eudey 2008) and captive breeding programs (Stanley 2003; Carlsson et al. 2004; Honess et al. 2010).

Habitat loss and overexploitation in their native range resulted in this species being recognized as the first "widespread and rapidly declining primate" (Eudey 2008), despite its IUCN Red List status of Least Concern (Ong and Richardson 2008). However, introduced populations (e.g., in Mauritius) can present a conservation challenge where they pose a significant risk to native biodiversity (Stanley 2003).

Activity Patterns

As a result of the wide distribution and ecological flexibility of this species, its behavior in the wild is as variable as its habitats and is strongly influenced by human factors. Some representative behavioral patterns are discussed here, drawn from a limited range of field studies. Resource (food, shelter, mates) availability and distribution have a significant influence on activity patterns, contributing to the determination of the energetic and social investments required to survive and breed. Table 20.2 indicates some representative activity patterns for long-tailed macaques from a variety of ecological contexts. Of particular note is the considerable time spent resting when heavily provisioned by humans, reflecting smaller home ranges and high-calorie diets. In undisturbed habitat, feeding may focus on different items at different times of the day: fruit at ~7 am and 5 pm; more dispersed food, like insects, in the intervening hours; and increasing consumption of vegetable matter during the afternoon (van Schaik et al. 1996). From ~5 pm each evening, they begin to settle down for the night in preferred sleeping trees, often along, and extending over, rivers to gain protection from predators (Kurland 1973; van Schaik et al. 1996).

Diet and Feeding

Human encroachment leaves few long-tailed macaque populations without human influence on their diets (Wheatley et al. 1996; Son 2003). Although predominantly frugivorous (see Table 20.3), with about two-thirds of their diet made up of fruit, in undisturbed habitats they have eclectic diets, reportedly eating a range of items including seeds, bark, buds, flowers, leaves, grass, insects, frogs, worms, caterpillars, clams, crabs, roots, and even octopus and clay (Rowe 1996; Wheatley et al. 1996; Yeager 1996; Son 2003). The high sugar content of crop-raided sugarcane is implicated in glucose intolerance and natural diabetes mellitus in the Mauritius population (Tattersall et al. 1981).

Long-tailed macaques typically feed on terminal branches and are important seed dispersers, as they generally spit out, rather than chew, fruit seeds (Kurland 1973; Corlett and Lucas 1990). Food may be rubbed with hands, leaves, or on the ground (Wheatley et al. 1996) and stored in cheek pouches for later retrieval for eating (Kurland 1973). They have also been reported to use stone tools to exploit foods that are more difficult to access (Gumert and Malaivijitnond 2013).

Table 20.2 Representative Activity Patterns in Wild *M. fascicularis*

Habitat	Moving (%)	Resting (%)	Feeding/ foraging (%)	Social (%)	References
Undisturbed primary forest	45	42	13	Included in "resting"	Wheatley et al. (1996)
Human provisioned	25	65	10	Included in "resting"	Wheatley et al. (1996)
Mangrove	17.5	34.1	10.5/26.8	10.7	Son (2004)
Mauritius (introduced)	27	21.9	32.2	18.8	Sussman and Tattersall (1981)

Table 20.3 Representative Diets of Wild *M. fascicularis*

Habitat	Fruit (%)	Vegetable Matter (%)	Animal Matter (%)	Other Components	References
Undisturbed primary forest	87	4	4 (insects)	3% flowers 2% other	Wheatley et al. (1996)
Human provisioned (temple, Bali)	18	16	12	2% flowers 23% peanuts 19% sweet potatoes 1% other	Wheatley et al. (1996)
Human provisioned (Singapore)	44	8		7% flowers 14% human foods	Lucas and Corlett (1991)

Locomotion and Habitat Use

In undisturbed primary forest, long-tailed macaques are almost exclusively arboreal; spending up to 97% of their time in trees (up to 12 m high; Wheatley et al. 1996), rarely descending to the ground to forage (Ungar 1996). In mangroves, however, considerably more time is spent foraging terrestrially than arboreally (Son 2004), while in Mauritius, almost all travel is on the ground rather than in trees (Tattersall et al. 1981).

Most long-tailed macaque locomotion is quadrupedal walking (65%), with some quadrupedal climbing (6%) and leaping (11%) across gaps of 2.2 m (average) and up to 5 m (Kurland 1973; Cant 1988). Their anatomy is highly adapted for arboreal locomotion, with relatively short limbs and a long tail that acts as a balance aid during climbing and leaping (Cant 1988; Fleagle 1999). More terrestrial macaque species (e.g., pigtails) have longer limbs and shorter tails (Rodman 1979). Long-tailed macaques are good swimmers capable of covering distances of up to 100 m (Kurland 1973; Fooden 1995; van Schaik et al. 1996).

Predation

Predation on long-tailed macaques is rarely reported, but ~11% of individuals may be captured by predators each year (Cheney and Wrangham 1987). Humans and pythons are confirmed predators (Kurland 1973; van Schaik and Mitrasetia 1990), but crocodiles, Komodo dragons, tigers, leopards, sun bears, dogs, and eagles are also likely predators, depending on location (Fooden 1995; Wheatley et al. 1996). A detected predator may be approached to within 4–5 m and mobbed by the macaques, including up to 30 min of alarm calls (van Schaik and Mitrasetia 1990; Wheatley et al. 1996).

Grouping and Social Structure

Long-tailed macaques are highly gregarious and live in multi-male, multi-female groups (Melnick and Pearl 1987). Group size can be highly variable (see Table 20.1) and is largely determined by habitat quality and predation risk, but is generally around 10–48 individuals (Melnick and Pearl 1987; Rowe 1996). In undisturbed forests, the adult sex ratio within groups may also be quite varied (see Table 20.1); however, provisioned groups typically have more adult males (Kurland 1973; Wheatley et al. 1996).

Long-tailed macaque groups are female-bonded; the core of each group is made up of related, cooperating females arranged in matrilines (Melnick and Pearl 1987; de Ruiter and Geffen 1998; Dittus 2004). This social organization minimizes inbreeding by ensuring maturing males leave their natal group, while females remain with their female relatives (Wrangham 1980), as reflected in patterns of genetic relatedness within groups (de Ruiter and Geffen 1998). Social groups may be comprised of several matrilines, with dominance relationships both between and within these networks of female kin (van Noordwijk and van Schaik 1987; de Ruiter and Geffen 1998).

Males leave their birth group at between 4.5 and 7 years of age (de Ruiter et al. 1992; Fooden 1995). They often leave with related peers and integrate into new groups without significant aggression (van Noordwijk and van Schaik 1985; Wheatley et al. 1996; de Ruiter and Geffen 1998). Migrating males typically enter new groups either by gradually building alliances, or by challenging the alpha male (van Noordwijk and van Schaik 1985). Males may change social groups several times in their lives, with shorter stays (2–3 years) in larger groups and longer stays (up to 5 years) in smaller ones (de Ruiter et al. 1992). Some older males may only remain in a group for just over a year (van Noordwijk and van Schaik 1985; Fooden 1995).

Dominance status determines priority of access to resources, thereby reducing aggressive competition (Richards 1974). Hierarchies exist at different levels in long-tailed macaque groups, among males, among females, and across matrilines (Wheatley et al. 1996). Long-tailed macaques have nepotistic rank acquisition, with daughters attaining ranks similar to that of their mother, with youngest daughters ranking just below their mothers and above their older sisters (van Noordwijk and van Schaik 1999b; Chapais 2004). Sons of high-ranking females are more likely to be dominant when joining a new group (van Noordwijk and van Schaik 1999a).

Becoming an alpha male may range from a very aggressive event, with all adult males being displaced, to peaceful in-group succession (Wheatley et al. 1996). Alpha males, generally 7–9 years old, may hold tenure for 4 months to 5 years, and betas from 8 to 24 months (de Ruiter et al. 1994; Wheatley et al. 1996). Infanticide at group takeover is rare (de Ruiter et al. 1994); however, defeated alpha males may remain in the group for some time, perhaps to protect their offspring. They may then spend a period living solitarily, eventually joining a different social group, but they are unlikely to regain alpha status (van Noordwijk and van Schaik 1988; de Ruiter et al. 1992).

Large groups may permanently split (fission), typically along matrilines (de Ruiter and Geffen 1998), most likely due to the negative impacts of large group size on female energetics and life-time reproductive success (van Schaik and van Noordwijk 1988; van Noordwijk and van Schaik 1999a). Existing groups commonly temporarily split into feeding parties, reflecting the feeding requirements of different individuals and age-sex classes (van Schaik and van Noordwijk 1986; van Noordwijk and van Schaik 1988).

Reproduction

While perhaps not clear in captivity, paternity broadly correlates with dominance in wild long-tailed macaques (Shively and Smith 1985). Alpha males mate-guard receptive females for up to 5–6 weeks (van Noordwijk 1985; Engelhardt et al. 2004) and account for up to 92% of group births, while beta males account for up to 33%, and lower ranking males for 2%–8% each (Cowlishaw and Dunbar 1991; de Ruiter et al. 1994; Wheatley et al. 1996). Greater alpha male success may result from increased mounting due to female preference at peak periods of female fertility (van Noordwijk 1985; de Ruiter et al. 1994). Alpha males appear to preferentially mate with dominant females, fathering 73% of the offspring of these females, compared to only 50% of lowest ranked females (de Ruiter and Geffen 1998). Adult females exhibit sexual swellings, but swelling size may be a less reliable indicator of fertility than estrogen-related olfactory cues (Engelhardt et al. 2004, 2005). Females, even when pregnant, may mate with several males, potentially to reduce the certainty of paternity, and hence, the risk of infanticide (van Noordwijk 1985; de Ruiter et al. 1994; Engelhardt et al. 2007).

Mature females have a menstrual cycle of 28–31 days (Harvey et al. 1987; Wolfensohn and Honess 2005). Single infants, or occasionally twins, are born after 153–179 days of gestation (Harvey et al. 1987; Wolfensohn and Honess 2005). Generally, this species is viewed as a year-round, nonseasonal breeder, but may breed rather more seasonally in some areas (De Ruiter et al. 1992; Wheatley et al. 1996). In the wild, 53% of adult females (higher in good food years) breed per year, and dominant females, who are generally in better condition, have higher birth and infant survival rates than subordinates (van Schaik and van Noordwijk 1985; van Noordwijk and van Schaik 1987; van Noordwijk and van Schaik 1999a). In captivity, up to 80% of females breed per year (Timmermans et al. 1981), although stillbirth rates can be as high as 9%–12% (Sesbuppha et al. 2008; Levallois and de Marigny 2015). The ability of primiparous mothers to successfully rear their offspring is strongly linked to maternal experience; ranging from 50% for individually reared mothers (Tsuchida et al. 2008), to 93%–95% for mothers that had been peer- or harem-reared (Timmermans and Vossen 1996).

Nutritional weaning usually takes place by 6 months, but infants are not fully independent until approximately 14 months of age, still relying on their mothers for continued social learning, sup-plementary suckling, defense, and thermoregulation (Harvey et al. 1987; Wolfensohn and Honess

2005). As a result, it is currently recommended that long-tailed macaques be weaned at around 10–14 months (Prescott et al. 2012) or 14–18 months (Wolfensohn and Honess 2005) of age.

Social Behavior

In undisturbed forests, aggression between groups is rare, though frequent displays at a distance help maintain some home range exclusivity (Wheatley et al. 1996). These displays generally include branch-bouncing and shaking and "Ho!" vocalizations by dominant males (Wheatley et al. 1996). At provisioned sites, there is more aggressive intergroup competition, primarily for access to preferred feedings by humans (Wheatley et al. 1996).

In their social groups, long-tailed macaques display a more despotic than egalitarian dominance style (Aureli et al. 1997). They are more aggressive, and reconcile less, than most macaques (except rhesus) and use frequent submission displays (Thierry 1985, 2007). Wild and captive patterns of aggression and reconciliation do not differ markedly; animals self-scratch and are attacked more following aggression, and although they avoid their aggressors, victims do not leave their group (Aureli 1992). Postconflict reconciliation is vital for social stability (Cords 1992), and in long-tailed macaques, it occurs more between kin than nonkin, and less after feeding conflict than in other contexts (Aureli 1992; Aureli et al. 1997).

Wild long-tailed macaques show a range of affiliative behaviors, including grooming, playing, sitting in contact, mounting, muzzle-to-muzzle contact, genital inspection, embracing, hand touching, exchanges of lip-smacking, and eyebrow-raising (Aureli 1992). Allogrooming helps form, strengthen, and maintain social bonds, particularly between females in female-bonded groups, and reduces social tension (Terry 1970; Schino et al. 1988; Henzi and Barrett 1999). Patterns of allogrooming closely reflect kin relationships and social priorities, with females giving and receiving more than males, who receive more than they give (Thompson 1967; Mitchell and Tokunaga 1976). Female alliances, including against dominant males, are most common within matrilines, but females also give more support to those who have groomed them (Hemelrijk 1994).

GENERAL BEHAVIORAL MANAGEMENT STRATEGIES AND GOALS

The accurate interpretation of any species' captive behavior requires an understanding of its behavior in the natural environment. Understanding natural behavior helps to define and recognize the "abnormal" behaviors that are often used to assess psychological well-being (Snowdon 1994). It is also important to note that aggression is a natural behavior pattern for long-tailed macaques; we must aim to design environments that help minimize the levels and impact of aggression and abnormal behavior. It is unrealistic to expect that these macaques will never engage in aggressive interactions.

Socialization

It is clear that the best enrichment for a gregarious monkey is a compatible member of the same species (Schapiro et al. 1996; Honess and Marin 2006b; Thom and Crockett 2008). Long-tailed macaques readily invest in social behavior, like allogrooming, which is time-consuming, rewarding for both parties, and important for maintaining compatibility (Aureli et al. 1999; Shutt et al. 2007). Companions also provide social buffering and reduce stress. For example, transport can be stressful (Honess et al. 2004; Fernström et al. 2008), but when shipped in compatible pairs, rather than singly, monkeys are less stressed (Fernström et al. 2008). Social buffering should have similar beneficial effects across many captivity-related and experimental stressors. The benefits of good social housing are clear, and this section covers some of the risks of socialization in restricted captive situations, as well as successful strategies for achieving stable pairs/groups of long-tailed macaques. Detailed reviews of partner selection and introduction techniques are available elsewhere (McGrew 2017; Truelove et al. 2017). Honess and Marin (2006a,b) review the stress and aggression associated with the social enrichment of various species, including long-tailed macaques.

Dominance hierarchies are inevitable in socially housed primates; however, subordinates are not automatically more stressed (Abbott et al. 2003), and the fear of imposing species-typical social stress on potential subordinates should not deter essential social enrichment. These primates have evolved sophisticated mechanisms for managing and repairing social instability to live successfully in groups. Given a species-appropriate grouping and sufficient, well-structured space with retreat areas, social stress in captivity should not significantly exceed social stress in the wild. Species-typical temperament assessments indicate that long-tailed macaques differ temperamentally from several other macaque species, both in levels of aggression (Kling and Orbach 1963; Thierry et al. 2004) and curiosity (Clarke and Lindburg 1993). Long-tailed macaques have nevertheless been readily socialized, and accounting for their natural history suggests that, where possible, socialization strategies should be based on kin relationships and evidence of previous compatibility provided by responsible suppliers.

There are inevitable costs of social housing, including some that are financial and others that are logistical; for instance, access to animals (Line et al. 1990a). However, the *requirement* to socially house macaques is widespread (EU 2010; National Research Council 2010). Where single housing is authorized, efforts to enrich other aspects of the animal's environment must be resolute, including the use of temporary social exposure and nonsocial enrichment to stimulate a wide range of species-typical behaviors (Bayne 1991). Given the wealth of evidence on the negative effects of single housing, it is clear that in the absence of additional enrichment efforts, single-housed primates may become highly stressed, and therefore, develop behavioral, neurological, hormonal, or immunological abnormalities that are likely to undermine the validity of the primates as research models (Honess and Marin 2006a).

Pairs and Groups

While the management of animals in complex, species-typical social groups in breeding facilities is routine, it is less common in research environments. Starting with single-housed animals with no previous compatibility information, it can be challenging to form pairs or groups in space-restricted housing without risking significant injury. A basic principle that reduces risk is that, where possible, introductions should always take place in neutral caging to minimize territorial aggression (Line et al. 1990a).

Reinhardt et al. (1995b) emphasized the risks of injury from aggression during socialization of macaques. Anxiety surrounding this procedure comes from accounts of high aggression and low success in socializing adult rhesus macaques (Honess and Marin 2006b). Personal experience is that rhesus, due to differing temperament, reconciliation, and dominance styles (Thierry 1985; Honess and Marin 2006a), provide a rather poor predictor of responses of long-tailed macaques, which appear relatively easily socialized compared to many Old World monkey species in laboratories (Lee et al. 2012).

Moving animals between levels of contact as part of a socialization process is accompanied by specific behavioral changes. Baker et al. (2012) and Lee et al. (2012) both report that moving pairs of female long-tailed macaques from protected- to full-contact conditions increased aggression and allogrooming (important for tension reduction; Schino et al. 1988) and decreased abnormal and "tension-related" behavior, including autogrooming. These changes, while not always significant due to sample size, facility differences, and study design issues, erode over time, but never return to protected contact levels.

It is not always consistently easy to isosexually socialize one sex or the other; for example, Clarke et al. (1995) found that males could be peacefully grouped, but Crockett et al. (1994) found female pairs to be more affiliative and less aggressive than male pairs. Males also had higher stress levels after, rather than before, pairing: a pattern reversed in females (Crockett et al. 1994). Some, cautious of male–male aggression, have used vasectomies to enable mixed-sex socialization (Statz and Borde 2001). However, this is only an option when research is not compromised by the procedure.

Including temperament and other individual features as part of selection for socialization may help improve success (McGrew 2017; Truelove et al. 2017), even when the animal is still an infant (Capitanio et al. 2017). However, there is considerable folklore and anecdote lacking robust experimental evidence surrounding this, procedure and evidence from the literature may be frequently overlooked. Though it is likely that local conditions and breeding source significantly influence results, there is nevertheless sound biological reasoning in much of the evidence which should be integrated into socialization planning.

As for pairs, the key challenge for forming groups is the prediction of dominance between selected group mates. Body size, rather than hormonal profile, appears to be a good indicator of group stability, as body weight and reduced activity correlated significantly with subsequent dominance when single animals were formed into groups. Cortisol or testosterone levels were not predictive (Morgan et al. 2000). Dominance rank, although changeable, can remain stable despite social change; females moved among eight different groupings remained stable either as dominants or subordinates in 75% of cases (Shively and Kaplan 1991). Aggressive temperament can also be indicative of dominance, and may be especially apparent between single-housed animals in the same room (Brinkman 1996). However, given that aggressiveness can vary with context and partner, as suggested by Crockett et al. (1994), partner selection based on noncontact preference testing and aggressiveness assessment may be no more predictive of success than random pairing.

Socialization of highly institutionalized, socially impaired animals is particularly challenging (Line et al. 1990b). In many such cases, the most ethical course of action may be to minimize the life of the animal in the facility; for example, via a justified terminal research procedure. A shortage of potential social partners in small facilities need not exclude socialization when cross-species (DiVincenti et al. 2012) or adult–infant pairings (Reinhardt 2002) are possible.

Rearing

It is well established that inappropriate rearing, including poor socialization, can impact the normal development of nonhuman primates. The reproductive competence of primiparous female long-tailed macaques is an example, as the maternal competence of individually reared (Tsuchida et al. 2008) females compares negatively to the maternal competence of peer-reared females (Timmermans and Vossen 1996). Although surrogate-reared infants are similar in body weight to mother-reared infants (Timmermans et al. 1988), early removal of an infant from its mother deprives it of the opportunity to learn vital social skills, including proper mothering techniques (Prescott et al. 2012). In instances where separation cannot be avoided (e.g., maternal death or rejection), younger infants can be successfully fostered by suitable lactating females (e.g., recent loss of own infant) with rates of successful adoptions as high as 87% (Honess et al. 2013). Older infants may be successfully raised in nursery groups of peers until they are able to join normal peer groups of weaned individuals. Hand-rearing is commonly linked with inappropriate attachment to humans and should be avoided (Wolfensohn and Honess 2005).

Forming groups of weaned long-tailed macaque juveniles can be accompanied by moderate aggression as dominance relationships are established. Honess et al. (2015) found significant increases in aggression when forming mixed-sex peer groups of weaned animals when weaning age increased from 12 to 18 months. Age, proportion of males in groups, and the number of subgroups from which the groups were formed were all significant factors in increasing aggression. A refined group-formation strategy, which minimized groups comprised of 25%–75% males and involved combinations of two to four subgroups, resulted in a 50% decrease in aggression, despite a further increase in weaning age.

Environmental Enrichment

The reasons why we should enrich captive environments for primates are well-rehearsed (Novak and Suomi 1988; Poole 1997; Young 2003; Lutz and Novak 2005; Wolfensohn and Honess 2005; Honess and Marin 2006b). Enrichment programs should be well structured, goal-defined, and targeted at the specific characteristics of the animals (Bloomsmith et al. 1991; Young 2003; Honess and Marin 2006b). Important characteristics that guide the development of enrichment programs include species identity, age, sex, individual temperament, and origin (e.g., wild caught vs captive bred; Honess and Marin 2006b; Honess and Fernandez 2011; older animals; Waitt et al. 2010; temperament; Hearth-Lange et al. 1999). Any use and impact of enrichment may vary with these characteristics, and they should be carefully considered when planning and assessing enrichment. An important aspect of well-considered enrichment programs is that they not only reduce stress and aggression, but that they also improve the quality of research data (Honess and Marin 2006b; Tasker 2012). Bloomsmith et al. (1991) provides an outstanding resource for enrichment practitioners, highlighting the importance of a systematic approach when planning a coherent enrichment strategy by addressing enrichment technique, implementation, record keeping, evaluation, financial cost, and manpower implications.

Social aspects of the animal's biology, together with its physiology and anatomy, should be accounted for in the provision of a captive environment that best matches the environment to the animal's adaptive contexts and features; the "behavioral–ecological criterion" of the environment (Snowdon 1994). Environmental impoverishment results in abnormal behavior, and skewed or unrepresentative behavioral repertoires, reflecting a poor fit with the provided environment and indicative of the need to refine conditions for more appropriate and successful coping behaviors (Wechsler 1995).

A key enrichment objective should be the creation of a "naturalistic" environment (Novak et al. 1994). This requires an understanding of the ways in which the species interacts with its environment so that key functional components that stimulate species-typical behavior and development are replicated (Janson 1994; Snowdon 1994). There is an important distinction between environments designed to be naturalistic, replicating the function and appearance of features, and those based on behavioral engineering, where function alone is replicated in favor of safety and serviceability (Novak et al. 1994; Young 2003). Balancing the simulation of nature with maintaining health, practicality, and safety is necessary for the development of environmental enhancement programs that support the timely completion of research projects.

Some enrichment procedures may present hazards for the animals, and unfortunately, enrichment attempts that have gone wrong (from which others may learn) rarely appear in the literature (see Bayne et al. 1993; Bayne 2005 for two exceptions). Problems with enrichment procedures tend not to be "general," but are more likely to result from atypical use by particular animals or from simply bad luck. A single problem with an enrichment device or strategy does not mean that that device, used safely in many other circumstances, should be withdrawn from the enrichment program. Some areas of research, for instance, toxicology (Gilbert and Wrenshall 1989), require particular caution with enrichment, as its use may produce confounding effects for studies. However, many enrichment options are now well-defined and/or certified, and regulators have changed their views on feeding fruit and providing social enrichment (Bayne 2003). Even in toxicology, it is still possible to provide a varied enrichment program, including enhancements such as play pens, swimming pools, and foraging devices (Gilbert and Wrenshall 1989). All items given as enrichment, such as Kong Toys and other occupational and foraging devices, may harbor unwanted bacteria, and should therefore be regularly disinfected (Bayne et al. 1993). Enrichment can also raise logistical concerns with maintenance staff who may object to some enrichment (e.g., forage substrate, plastic balls) and its seeming incompatibility with waste-handling systems (Renquist and Judge 1985; Bayne 1989; Bennett et al. 2010). However, in most cases, simple solutions can be found that address these "problems" (Bennett et al. 2010).

Recommended Environment/Behavioral Husbandry Program

This section highlights key areas, beyond vital socialization, where aspects of a long-tailed macaque's life can be practically enriched as compensation for restricted space, reduced social complexity, and the absence of species-appropriate structural and sensory diversity. Successful enrichment can only be achieved by refining all salient aspects of the animal's captive environment. A strategy that, for example, only provides foraging enrichment, or toys, while neglecting other relevant enrichment approaches will only be partly successful and is vulnerable to failure.

Physical

The question of what defines the minimum, ethologically appropriate, cage size for nonhuman primates is the subject of recurring debate that touches on sensitive issues of resource commitment and availability (Woolverton et al. 1989; Reinhardt et al. 1996; Buchanan-Smith et al. 2004; Honess and Marin 2006b). However, both the size and configuration of caging cascades a range of other desirable enrichment options. Cages of sufficient size allow long-tailed macaques to be housed socially with compatible companions, with further enrichment designed to provide opportunities for the animals to perform species-typical behaviors, including vertical and horizontal retreat from cage mates and humans, social activities (e.g., huddling, grooming), foraging, leaping, and swinging.

Few studies that examine the effects of increased cage space report evidence of benefits; large percentage increases in cage size still produce small cages if the starting cage is small (Honess 2009). Significantly larger cages may, however, be associated with positive behavioral and reproductive benefits, while small cages that offer few opportunities for species-typical behavior, and specifically, little control for animals over their environment, are unlikely to improve animal welfare (Sambrook and Buchanan-Smith 1997).

Restrictive caging severely limits exercise, which has important cardioprotective effects in long-tailed macaques (Williams et al. 2003). Such restriction may cause skeletal problems (Faucheux et al. 1978; Rothschild and Woods 1992), obesity, and diabetes (Wolfensohn and Honess 2005). Some modern, more generously sized cages still allow few options for real exercise or extensive enrichment. Extensive socialization can help, as the social housing of several animals in one scaled-up, bigger cage does not mean that animals are limited to their portion of the cage. This assumes that compatibility and space usability are maximized, and areas where individuals can be excluded from, or trapped in, due to dominance issues, are minimized (Honess 2013). Simply reconfiguring available space can also have benefits. In groups of long-tailed macaques, halving cage length while doubling its height (2–4 m) resulted in the animals spending most of their time above the original cage height (Waitt et al. 2008). It increased positive social behavior, and decreased aggression and anxiety-related behavior by providing a better simulation of the natural habitat of this particularly arboreal macaque species.

Providing a range of perches, runners, and swings promotes species-typical locomotion, positional and social behavior, while well-placed visual barriers increase the animals' sense of control and help to reduce aggression (Honess and Marin 2006b). Furniture can be positioned to allow safe operating and cleaning spaces for veterinary and husbandry staff with swinging devices removed or fixed in position out of the way. Heavy swings should not be positioned such that their movement could hit an animal on a perch or trap it against a fixed surface.

Concerns exist about the safety of some materials used for enrichment, especially rope and wood, which may block drains, and the ingestion of which has occasionally resulted in injury or death (Eckert et al. 2000; Hahn et al. 2000; Bayne 2005). Lists of wood types considered safe for manipulanda or perches are available (Eckert et al. 2000), with manzanita (*Arctostaphylos* spp.) being considered particularly safe (Luchins et al. 2011). Given that wood is such a key element of the environment and the diet of these animals in the wild, they may reasonably be expected to cope well with it in captivity. Suitable wood that is sanitized appropriately should not present a mechanical or fomite risk to the animals. When there are concerns about the absorption and recycling of compounds through wood, it is typically inexpensive enough to be replaced between studies. While occasionally problematic, wood also has many positive properties, including being natural, thermoneutral, of novel texture, safely destructible, and inexpensive. When used for perching poles, and in combination with plastic planks or suspended fire hose, it can be used to create easily changeable structural complexity, partially simulating conditions in a forest, helping to maximize the use of cage space, and preventing the onset of stereotypic locomotion (Young 2003; Honess et al. 2012).

Visual barriers are important cage features, enabling animals to retreat from stressful visual contact with one another and with people (Bayne and Novak 1998; Honess and Marin 2006b; Honess 2013). Visual barriers can be easily constructed and positioned either on the floor or in perching areas. They need not be flat panels, as plastic barrels and similar items work equally well. When targeting enrichment strategies to specific age-sex groups, it is worth noting that young animals tend to need fewer visual barriers, but more swings and flexible structures to promote play and physical activity. Geriatric animals, on the other hand, with reduced mobility, benefit more from shorter distances between structures and from ladders on shallower angles (Waitt et al. 2010).

Long-tailed macaques should be housed in conditions that prevent discomfort, distress, or disease through extremes of temperature, humidity, lighting, or noise. While narrow ranges of environmental parameters, well within the tolerable range for the species, may assist with standardizing research conditions and are simpler for regulatory purposes, there is little evidence that they benefit the animal's well-being, given that conditions in the wild vary considerably over 24 h and across seasons. However, extreme disruptions of conditions can have dramatic effects. For example, switching to a 24-h light-only condition, from 12 light:12 dark, increased activity time during the previously dark period (6 pm–6 am) from 3% to 20% (Evans et al. 1989), disrupting sleep, which has been associated with stress. Noise levels in the animals' environment should also be monitored and controlled, as excessive noise has negative effects on both animals and the care staff.

Feeding

In the wild, long-tailed macaques have very varied diets. Providing them with a pelleted, processed diet (nutritionally balanced) in captivity is neither varied nor behaviorally enriching. Captive long-tailed macaques should receive some fresh fruit and/or vegetables every day unless sound scientific justifications suggest otherwise. The novelty associated with natural foods and their different textures, smells, colors, flavors, and processing requirements are all enriching and increase appropriate behavioral diversity. There are concerns that some fresh produce (e.g., cabbage) may cause diarrhea; however, the data suggest that a range of produce, including cabbage, can be fed without diarrhea problems (Brinkman 1996). The rotation of different fruit and vegetable types in the diet can help overcome boredom (Wolfensohn and Honess 2005). It is worth noting that most fresh produce grown for the human palate has different nutrient values (e.g., higher sugar content) compared to wild equivalents. It should not be overlooked that foliage is readily eaten by long-tailed macaques in both the wild and captivity, but toxic plants must be avoided; lists of plants that are safe to use for browse are available (Schrier 1995, 2004). While some plants are nontoxic, they may still cause physiological changes that can confound particular research results; for example, cedar may disrupt liver enzymes (Eckert et al. 2000).

Particularly high-calorie treats (e.g., peanuts) should generally be avoided, due to their impact on the maintenance of a balanced diet, though they are often used as training rewards. High-calorie treats should be fed infrequently, and only offered after veterinary consultation. High-sugar treats, such as marshmallows or chocolate, should be excluded from routine use altogether and reserved for use, in moderation, as rewards following particularly aversive procedures.

Food may be presented in ways that elicit species-typical foraging behavior, but foraging enrichment should aim to provide behavioral occupation in addition to simply feeding animals. Long-tailed macaques will invest significant time in foraging enrichment, even for very small items of low nutritional value (e.g., poppy seeds in a forage substrate; Spector et al. 1994; Wolfensohn and Honess 2005). Such occupation is natural and replicates the significant quantity of time spent foraging in the wild (see Table 20.2), and provides a valuable tool for decreasing boredom. Foraging enrichment should have at least one of the following features: (1) it extends the time and/or effort taken for animals to acquire their daily nutrient ration; (2) it requires animals to obtain, or process, their daily food ration in a way that mimics wild feeding behavior; and (3) it supplies dietary items included in, or compatible with, the diet of wild conspecifics (Honess 2013).

Foraged food must be accounted for when balancing the animals' diet and its motivation to forage. Large items in a forage substrate will be found too easily and those coated in sugar will be consumed preferentially, minimizing the desired behavior and skewing nutrition. Long-tailed macaques will sift through a forage substrate, even without food items, for prolonged periods, although the behavior will persist longer if even low-value items are to be found. Difficult food puzzles require highly motivating rewards to engage the animals beyond any initial novelty effect, and success is likely to vary with experience (Crockett et al. 1989; Lloyd et al. 2005). Natural history shows us that long-tailed macaques are extractive foragers and feeding challenges in captivity should stimulate this behavior. Foraging tasks can be rewarding in themselves, as these macaques will as readily take food from a puzzle-feeder as from an adjacent food hopper (Evans et al. 1989).

A number of reviews exist that examine the nature and effectiveness of a full range of foraging enrichment options for primates (Reinhardt 1993; Reinhardt and Roberts 1997; Honess and Marin 2006b). Devices can be diverse: some are simple containers from which food can be removed by manipulation (e.g., Kong Toys, paper bags, plastic bottles), others are much more complex and present a cognitive challenge (e.g., maze puzzles, puzzle balls, arrangements of drilled PVC pipe). Commercially available maze puzzle-feeders are relatively expensive, but do prolong foraging (Watson 1992). However, similar effects can be produced more cheaply by using PVC pipe sections with holes for food manipulation and extraction. Pipes can be presented singly or in interconnected arrangements and are readily used by long-tailed macaques, though they may be associated with a sex difference in successful use (Murchison 1991; Holmes et al. 1995). Puzzle-balls also provide a successful way of promoting foraging in this species (Lloyd et al. 2005). Kong Toys are common enrichment items in many facilities. On their own they encourage little more than species-atypical gnawing behavior, but when filled with food or frozen juice, they can be useful enrichment options that stimulate extractive foraging. However, interest in Kong Toys does not persist unless they are refilled with desirable contents (Crockett et al. 1989). Despite limitations, the ability to add different contents makes Kong Toys more complex than, for example, a nylon ball, although care should be taken in rotating such items, as both can evoke possessive behavior (Bayne 1989). Similar effects to those found with food-filled Kong Toys can be achieved with other food containers, including paper bags, cardboard boxes, and plastic bottles, all of which can also be safely destroyed by the animals in the process of extracting food.

Freezing food items, such as fruit in ice blocks (small or large), provides good enrichment, as the animals pick and chew to extract the food. Concealing forage mix in a suitable substrate (e.g., woodchips), forage box, or fleece/artificial turf simulates the behavior of sorting through leaf lit-ter or soil and represents valuable enrichment (Meunier et al. 1989; Lam et al. 1991; Honess and Marin 2006b), occupying long-tailed macaques for significant time periods (Meunier et al. 1989). Particulate forage mix may also be put into plastic bottles or holed PVC pipes for an additional chal-lenge (Brinkman 1996). Selecting a suitable forage mix is important. Commercial mixes frequently contain large items such as banana chips, dried coconut shavings, large nuts (all often coated with corn syrup), making them quite high calorie (potentially unbalancing diets), while providing limited occupational enrichment. In-house mixes can be produced, with veterinarian and study director clearance, containing lower calorie human-grade items, such as poppy or sesame seeds, pulses, raw rice, cereal grains, etc. These small items will occupy the animals for longer; also, any items left behind present fewer challenges to waste systems.

Sensory

Sensory enrichment is often the most neglected part of an animal's environment. Using varied and natural material in cage construction, furnishings and other enrichment can enhance the tactile sensory environment (Honess and Marin 2006b; Waitt et al. 2008). Music is often played in facili-ties despite a lack of evidence that it provides effective enrichment, though it may have value in screening aversive, loud, husbandry-related noise at busy times and be valued by staff (Brinkman 1996; Honess and Marin 2006b). Any sounds or video played to the animals should be screened for stressful stimuli such as predators or overly aggressively conspecifics. Smell and color have not been well-tested as enrichment, though creative enrichment in this realm would stimulate important sen-sory channels that wild long-tailed macaques use to assess food quality (Lucas et al. 1998; Dominy 2004; Osorio et al. 2004).

Occupational

Under reduced social complexity, adult long-tailed macaques may make ready use of manipulat-able objects or toys, whereas younger animals will use them even when group-housed—helping to stimulate psychomotor development. Many off-the-shelf toys are expensive, or primarily designed for non-primates, but simple, inexpensive, and effective toys that achieve species-appropriate behavioral goals can be designed in-house (Honess et al. 2012). As with all enrichment items, toys should be safe for both animals and staff, and must be regularly disinfected (Bayne et al. 1993). Items designed to be destroyed (e.g., cardboard boxes) should not contain hazards, such as staples (Honess 2013). While destruction of objects need not be hazardous, some question the enrichment value of destructible, cheap-to-replace objects (e.g., plastic crates) or cage furniture (Evans et al. 1989). Nevertheless, destruction of an enrichment object indicates occupation and potential bore-dom reduction.

There are relatively few published evaluations of the benefits of enrichment toys for normal long-tailed macaques compared to assessments of toys' therapeutic value (see below) for macaques with behavioral problems (Honess 2013). However, one study (Brinkman 1996) tested a variety of toys and found that plastic balls, rawhide bones, metal mirrors, and radios were of low inter-est compared to wall mirrors, plastic rings, phonebooks, cans with lids, ropes, fishing line reels, coconuts, cardboards, food tubes, cedar wood barbells, sections of garden hose, and plastic bottles. Mirrors, in addition to use for general enrichment, have also been used as a substitute for social housing and in socialization studies (Clarke et al. 1995), and can also be used by animals to monitor otherwise out-of-view activities (Wolfensohn and Honess 2005). However, mirrors may actually be used rather little (Brinkman 1996), and there is no conclusive evidence that long-tailed macaques can recognize themselves in a reflection (Gallup 1977; Callaway 2015).

Pools or tanks of water, with or without sinking or floating food items, make excellent enrichment for long-tailed macaques, eliciting playing, swimming, and fishing behaviors seen in the wild (Gilbert and Wrenshall 1989; Stewart et al. 2008; Waitt et al. 2008; Robins and Waitt 2010).

Positive Reinforcement Training

The management of undomesticated animals, such as long-tailed macaques, in captivity can be difficult. Reducing stress by improving human–animal relationships can have significant practical and ethical benefits (Waitt et al. 2002; Wolfensohn and Honess 2005). Familiarization of animals to humans and training them to cooperate with procedures can be conducted not just in laboratories, but also in breeding facilities. By working closely with breeders/suppliers, it is possible for researchers to identify protocols for which animals may receive preliminary, or full, training prior to supply (see Fernandez et al. 2017).

It has been well-established that primates can be trained using positive reinforcement techniques across a range of contexts (see Graham 2017; Magden 2017). Most of the published accounts of training macaques have been for rhesus, with long-tailed macaques being rather poorly represented, perhaps because they are less commonly used in areas of research (e.g., neuroscience) where complex behaviors are required and facilitated by structured animal training programs. The benefits of training are clear (Laule 1999; Prescott et al. 2005; Perlman et al. 2012; Westlund 2015) and particularly focus on stress reduction to improve welfare and data quality; however, there can also be resource benefits, including speeding up procedures and reducing staff requirements (Laule and Desmond 1998).

In the literature, the term "training" can include everything from habituation to negative and positive reinforcement. Often, the use of squeeze-backs in modular caging, where animals can be separated for individual training (Heath 1989; Perlman et al. 2012), creates very specific circumstances that facilitate training. The absence of these features in more expansive and socially complex contexts (e.g., breeding facilities) makes training incomparably more challenging. In reality, it may be necessary to use a carefully controlled mixture of positive and negative reinforcement across a training program to achieve timely results, working at all times to minimize and remove negative reinforcement techniques. In addition to housing context, other factors that influence training success include species and individual temperament, dominance status, trainer skills, and resource availability/management commitment (Perlman et al. 2012; Wergård et al. 2016). Even incorporating unfamiliar staff in training activities can set back the training progress (Heath 1989). Long-tailed macaques can successfully be trained to enter a transport box, and do so more quickly than some other macaques, but more slowly than rhesus (Clarke et al. 1988). Hence, squeeze-backs are commonly used to encourage them to enter boxes (Heath 1989). When using only positive reinforcement, transport box training of long-tailed macaques can prove challenging and is significantly less successful than target training (Fernström et al. 2009). With the availability of skilled trainers, it is possible to help resolve extraordinary challenges; for example, positive reinforcement was used to train long-tailed macaques to a cue that indicated an imminent and loud construction noise, thereby reducing their noise-associated stress (Westlund et al. 2012).

FACILITIES AND EQUIPMENT

Apart from provision of tall housing that enables this species to exhibit its species-typical highly arboreal behavior, requirements for facilities and equipment do not differ markedly from those for other macaques, and this is covered elsewhere in this volume (see Gottlieb et al. 2017).

RESEARCH-IMPOSED RESTRICTIONS/EXEMPTIONS

Single housing may be sought by some researchers for their study of animals due to fear of conspecifics tampering with implants, the need to control food or water intake, infection status, etc. However, it has been shown, including for long-tailed macaques, that careful planning can minimize or altogether remove the need for single housing (Roberts and Platt 2005; Truelove et al. 2017). Equally, minimizing nonsocial enrichment is rarely justified given the changes in regulatory perspective, careful device selection, and proper device disinfection or replacement between studies.

A not uncommon research-imposed restriction is the use of food and/or water control as a motivational tool in neuroscience research. Significant deprivation protocols to motivate task performance (particularly water deprivation) are likely to result in poor welfare through the removal of control and predisposition to abnormal behaviors, such as urine drinking and tonic immobility. In many cases, deprivation hampers performance on the task, particularly following periods of unrestricted access to water. While water deprivation may be justified to meet the challenges of, for example, obtaining sufficient recording trails, or conscious brain imaging without data artifacts associated with chewing, water should never be viewed as a treat when it is a vital life resource. Using juice or smoothies as positive rewards for performing the task may achieve the desired goals without resorting to water deprivation and experiencing its negative consequences for welfare and, potentially, data quality. Significant debate continues concerning positive reinforcement training as an alternative to such motivational paradigms (Scott et al. 2003; Prescott et al. 2010; Westlund 2012). A deep understanding of primate ethology and a high level of positive reinforcement training expertise can certainly make significant inroads in this area and reduce, or potentially eliminate, motivational techniques that can be ethically challenging.

ABNORMAL BEHAVIORS

Given the divergent nature of most captive and wild environments, it is perhaps not surprising that confinement, reduced stimulation, research interventions, and limited opportunities to express species-typical behavior result in combinations of frustration and stress that are fertile ground for the development of undesirable abnormal behavior that create ethical and compliance challenges.

Honess (2013) presents an important ethogram for long-tailed macaques, including a comprehensive list of abnormal behaviors (drawn substantially from the work of Erwin and Deni 1979). It is not uncommon for primates, including long-tailed macaques, to display combinations of these abnormal behaviors (Honess 2013).

Our aim must be to design captive environments and behavioral management programs for long-tailed macaques that minimize, or prevent, the onset of abnormal behavior. Where abnormalities occur, it is important to accurately define and quantify them, to enable the objective assessment of attempted therapies (Honess 2013). A brief selection of important abnormal behaviors is described here, followed by discussion of potentially successful intervention and prevention strategies. For a more comprehensive discussion of abnormal behavior, other sources should be consulted (Erwin and Deni 1979; Bayne and Novak 1998; Novak 2003; Honess 2013).

Self-Biting

Self-biting, which can cause serious injury, appears to be more common in rhesus than long-tailed macaques (Novak 2003; Crockett et al. 2007). Like many abnormal behaviors, self-biting can be resistant to treatment, and though successful socialization may reduce or resolve it, re-isolating may cause its reappearance (Novak 2003; Reinhardt et al. 2004). Self-biting is most common in single-housed animals, but can appear in group-housed individuals that may have experienced inappropriate rearing, early weaning, or early single housing (Lutz et al. 2003; Novak 2003; Rommeck et al. 2009; Honess 2013). The best course for these animals is to review their social compatibility and enhance nonsocial, tension-reducing enrichment (Honess 2013). Care should also be taken to monitor enrichment misuse, as noted by Bayne (1989) that inanimate items, like balls, may become incorporated into patterns of self-biting.

Hair-Plucking/Hair-Pulling

Hair-plucking or pulling, (not hair-grabbing during aggression) comprises the removal from self or a cagemate of hairs, singly or in clumps, using the mouth or hands, and is typically followed by hair ingestion (Honess 2013). Where medical, genetic, and environmental causes (Novak and Meyer 2009) are excluded, it typically represents a pathological intensification of normal grooming behavior and can result in appreciable hair loss or alopecia (Honess et al. 2005; Reinhardt 2005). Hair-plucking is rare or absent in the wild (Reinhardt 2005). Attributing alopecia to plucking should be confirmed with observational evidence, although hair-plucking may also occur outside of observation periods, for instance, at night. Examining feces for hair content helps identify hair-plucking with ingestion, and marking food (e.g., with coloring) given to each socially housed individual will help identify the animal doing the plucking (Honess 2013). Hair-loss patterns may also be informative, as self-pluckers focus on easily accessible body parts (e.g., arms, thighs, lower back), while allo-pluckers focus on body parts that are subject to normal allogrooming (e.g., mid/upper back, head, tail; Honess et al. 2005). Hair-plucking appears to be less common in long-tailed macaques than other macaques (Crockett et al. 2007; *Honess*, P., personal observation) and is more common in females and older long-tailed macaques (Honess et al. 2005; Reinhardt 2005; Crockett et al. 2007). Levels of alopecia may also be more pronounced at certain stages of the reproductive cycle, particularly during pregnancy. Mothers may cause alopecia in their infants, who typically recover well with decreased maternal contact (Honess, P., personal observation).

Scoring systems exist for quantifying alopecia in primates (Honess et al. 2005; Berg et al. 2009; Baker et al. 2017), though whole body scoring systems can be difficult to utilize for large numbers of animals or where whole body visibility is difficult. For animals in social groups, it may be best to score alopecia solely from the dorsum, as it is a common site of alopecia in social groups (Honess et al. 2005; but see Baker et al. 2017). In social housing, it may be erroneous to think the animal with alopecia is particularly stressed, as, unless the alopecia pattern indicates self-plucking, it may be an animal with a full coat that is plucking and is most in need of intervention.

Plucking can be rather resistant to treatment (Reinhardt 2005; Crockett et al. 2007), although there is some evidence of its reduction or elimination using foraging opportunities (Beisner and Isbell 2008), grooming boards (Bayne et al. 1991), or visual barriers (Honess, P., Y. Jiang, and J. McDonnell, in preparation). Some regulators focus on alopecia as a welfare indicator (Honess et al. 2005), but the possible "resistance" of hair loss to treatment may make it unproductive to invest large quantities of behavioral management effort in attempts to resolve it, particularly if more serious abnormal behaviors exist (Crockett et al. 2007).

Pacing

Stereotypic pacing may be characterized by repeated route-tracing with placing of hands and feet in the same place on each circuit, sometimes with ritualized touching points or objects (Bayne 1989; Honess 2013). This behavior may be indicated by obvious dirty marks where hands and feet are placed, which distinguish it from "patrolling"; a natural yet determined and purposeful behavior (Honess 2013). Onset of stereotypic pacing may be prevented by providing visual barriers and creating an arrangement of perches and runners that can be periodically varied (Young 2003; Honess 2013).

Depressive Posture

Animals with depressive posture generally sit in a slumped or hunched position with a downward gaze and are unresponsive to external events (Shively et al. 2006; Shively 2017). In long-tailed macaques this behavior is more common in low-ranking individuals and correlates with elevated cortisol. Individuals showing this behavior have patterns of serotonin-binding potential in the brain that are comparable to humans with major depressive order (Shively et al. 2006). Depressive posture has been noted in juvenile long-tailed macaques that were weaned into peer groups at 6 months of age, prior to being singly housed in small cages (0.34 m³) when 3 years old for 9 months prior to shipment (Camus et al. 2013). In this case, the behavior is likely a reflection of early weaning and single housing, known risk factors for the development and expression of abnormal behaviors (Novak 2003).

Intervention Strategies

Prevention is always better than cure; animals housed and managed optimally from birth are unlikely to spontaneously develop abnormal behavior patterns, making the planning of interventions unnecessary. However, many of those responsible for caring for long-tailed macaques are left with the challenge of managing animals with abnormalities that are the legacy of previous regimes. Any process of planning an intervention for behavioral abnormalities requires careful initial observation and analysis to identify, and where possible remove, responsible stressors; accounting for the importance of simple boredom in generating abnormalities. All causes may not be removable, either for practical reasons or because they are historic (e.g., early rearing experience; Bayne and Novak 1998). In such cases, behavioral therapy is the primary intervention remaining, using a systematic, analytical approach based on ethology and knowledge of the species' normal behavior. This can include assessment of the context and patterns of occurrence of the behavior and determination of ways to fill occupational voids, while redirecting behavior to species-typical, "normal" tasks. Practitioners should avoid an erratic approach that throws all available enrichment options at an animal at the same time, or in rapid succession, in the hope that something will work. Interventions may successfully address one problem, but cause others. For example, socialization may help reduce abnormal, self-directed behavior, but create problems with aggression. Honess (2013) promotes a holistic approach to intervention that considers the animal together with its environment; providing social enrichment to an animal will be more effective when combined with other improvements, such as the provision of additional space and complexity, visual barriers, and perches at different heights.

Established abnormal behavior is extremely difficult to permanently eliminate, either with behavioral or chemical therapy (Bayne and Novak 1998; Turner and Grantham 2002). Anxiolytic drugs may alleviate, but do not cure, the expression of abnormal behavior, unless the associated stressors are removed (Novak and Meyer 2009). Drug interventions may be ethically questionable, as they can affect the research model and impact the animal's quality of life if maintained on long-term drug therapy. Some important examples of successful non-drug interventions that relate to long-tailed macaques are described below.

Foraging enrichment can prove valuable for decreasing a number of abnormal behaviors. For example, foraging boards successfully reduced (up to 73%) a range of stereotypic movements in single-housed males (Lam et al. 1991). However, limited success with a similar device at a different facility (Lutz and Farrow 1996) highlights the potential impact of differences in facilities, management practices, research programs, staff, and the animals themselves (Honess 2013). More challenging devices, such as PVC foraging tubes, have also been shown to substantially (up to 85%) reduce self-directed behavior (self-grooming, scratching, hair-plucking), particularly when filled with novel, rather than familiar, food (Holmes et al. 1995). Commercial puzzles with food treats can have mixed effects; for example, decreasing some abnormalities (e.g., self-biting, floating limb, hair-pulling, and self-grasping), but increasing others (e.g., pacing and rocking) (Watson 1992).

The importance of compatible socialization for welfare has already been discussed and some studies note its success in addressing behavioral abnormalities in long-tailed macaques. For example, pairing females reduced pacing, abnormal posture, and eliminated self-injurious behavior (Line et al. 1990a), while pairing sterilized males with females dramatically reduced self-inflicted injuries and stereotypic behavior (Statz and Borde 2001).

Occasionally, studies report the benefit of a range of enrichment options in addressing abnormal behavior without being able to attribute beneficial effects to specific devices. For example, single-housed male long-tailed macaques given limited access to larger play cages enriched with forage substrate, telephone directories, a viewing panel, a swing, a rope, a ball, and a glove exhibited more species-typical behavior (e.g., locomotion, foraging and exploration), while inactivity was reduced, and abnormal behaviors (e.g., self-biting, swaying, pacing, circling, bouncing, and rocking) were almost eradicated (Bryant et al. 1988). Another study, enriching singly housed animals with toys, novel food, forage, television, and additional space, found that the enrichment reduced abnormal behavior and promoted positive behavior, particularly among males, and only unenriched controls developed alopecia or self-harming behavior (Turner and Grantham 2002).

RESPONSE TO PROCEDURES

Most scientific procedures have the potential to cause stress, or even distress, and thereby compromise animal welfare and confound scientific results (Reinhardt et al. 1995a; Russell 2002; Reinhardt 2003; Rennie and Buchanan-Smith 2006). Tasker (2012) reviews the impact on long-tailed macaques in toxicology of housing changes and restraint, but indicates that the vast array of factors involved hampers the conclusive identification of elements that have attributable positive, or negative, effects on welfare.

Not surprisingly, a whole range of scientific or veterinary procedures, including ketamine sedation, surgery, tethering, and prolonged catheterization, all produce, with some variation, physiological responses indicative of stress in long-tailed macaques (Crockett et al. 1993). Single injections of ketamine produced elevated cortisol levels for up to 36 h (Crockett et al. 1993). However, this may largely be an injection effect, as ketamine, even at different doses, introduced via a chronic venous cannula, has no effect on cortisol, plasma insulin, or blood pressure levels (Castro et al. 1981). Even multiple ketamine injections may not have a significant short-term (2 h) effect on cortisol, testosterone, or luteinizing hormone levels (Malaivijitnond et al. 1998).

Physical restraint (manual, box, board, or chair) of long-tailed macaques has been shown to significantly alter levels of cortisol (Kling and Orbach 1963), as well as key enzymes commonly monitored in toxicological and pharmacokinetic studies (Kissinger and Landi 1989; Landi et al. 1990). Physical restraint is therefore an important, potentially confounding, variable when trying to determine drug-related effects. Physical restraint also produces a fear reaction when used for venipuncture and ECG recording, and has a negative impact on sleep patterns, all indicators of stress (Tasker 2012).

While changes in certain hormones, particularly cortisol, may be indicative of a stress response, they can also be indicative of a response to non-stress-related factors (e.g., exercise) or sampling techniques (Honess and Marin 2006a; Lane 2006), and individual responses may be highly variable depending on characteristics such as sex and rank (Crockett et al. 1993; Malaivijitnond et al. 1998; Abbott et al. 2003; Honess and Marin 2006a). For example, long-tailed macaques whose blood was collected in handling cages responded differently depending on rank and whether they were conscious; dominant animals were the most stressed when bled while sedated and the least stressed when bled while conscious (Welker et al. 1992). Non-physiological and non-behavioral indicators can prove useful in assessing responses to procedures and implemented refinements. For example, alopecia was found to increase and the body condition improved among long-tailed macaques being prepared for toxicology procedures, possibly reflecting the influence of duration of stay in the facility and changes in group composition (Tasker 2012).

Husbandry and management procedures can have a dramatic impact on the animals. A good example is the effect of transport: Air transport of juvenile long-tailed macaques from breeding facilities to laboratories can result in contracted behavioral repertoires, increases in negative behaviors (Honess et al. 2004), weight loss, and reduced body condition (Tasker 2012). Even simulated transport increases stress in this species, which can be reduced by pairing the animals (Fernström et al. 2008). Behavioral changes may also accompany less dramatic relocations, though responses may vary between animals that differ in origin. Following a facility transfer, Indochinese long-tailed macaques reacted more aggressively and less affiliatively than those from the Philippines, Mauritius, or Indonesia (Brent and Veira 2002). Movements as minor as those between rooms within a facility can produce stress responses such as elevated cortisol levels (Crockett et al. 1993). Constantly seeking and implementing refinements to procedures, and general housing and husbandry are both ethically important and vital for maximizing the quality of research data. Even simple refinements, such as familiarization prior to toxicology procedures, reduces fear and improves data quality (Tasker 2012).

EXPERT RECOMMENDATIONS

The important starting point when considering a behavioral management program for long-tailed macaques is to ensure a comprehensive understanding of their behavior in the wild and of their evolutionary adaptations. For those who are in the fortunate position of being able to plan captive provision from scratch, they should always try to be innovative and expansive, providing as many natural options as is possible. Natural, or naturalistic, provision attempts to recreate as much of the wild context as can be safely and practically achieved within the constraints of research objectives. To simply copy someone else's design (and potentially mistakes!) is not *Refinement* in a global sense.

Animals should be housed in expansive cages (including providing heights over human head-height) furnished to allow the full use of available space, with compatible conspecifics in species-appropriate-sized groups. Ideally, they should have access to natural materials for playing, perching, and locomotion, and should be fed a diet that includes fresh fruit and vegetables with *ad lib* access to clean water. They should be maintained at temperature and humidity levels appropriate for the species, but not necessarily at fixed levels, and provided with sufficient ventilation. They should have access to natural daylight and be housed away from excessive noise and drafts, and have shelter from climatic excesses. Furthermore, they should be presented with appropriate, safe opportunities to express natural behaviors including, but not limited to, grooming, huddling, foraging, playing, leaping and swinging, retreat and concealment, reconciliation, exploration of their environment, and exposure to novelty. Creating a captive environment that is molded to the animal's adaptations and facilitates the expression of natural, species-typical behavior is most likely (along with sound, sympathetic management, and husbandry practices) to avoid undue stress and prevent the development of abnormal behaviors. This is my understanding of a "functionally appropriate" captive environment. On top of this basic foundation sit other vital, standard requirements, including the availability of high-quality veterinary care. Making substantial changes, such as replacing small caging with more expansive alternatives, can be daunting, particularly in large facilities where the financial costs of such changes can be prohibitive. This, however, need not prevent advancement in standards, as rooms are typically refurbished on a rotational basis, presenting ideal opportunities for gradual upgrades.

Getting an enrichment program right is as important for long-tailed macaques as for any wild animal in captivity. We are still far from being able to provide captive conditions that approach those in the wild, and with disease and predation, this is a good thing. Therefore enrichment, as we currently conceptualize it, is essentially the provision of compensation to the animals for their captive confinement. The most successful enrichment, or compensation, is designed with full understanding of the animal's natural behavior, the way in which it will interact with any device, and the enrichment's positive impact. Off-the-shelf enrichment, while being easy to source, is often expensive and designed as toys for non-primates that have very different behavioral needs than long-tailed macaques. Additionally, many of these options lack empirical evidence that demonstrates positive effects. In-house designed and constructed devices that reflect an understanding of the natural behavior of the species can be very cost-effective and have very positive effects (Schapiro et al. 1995). Enrichment as a behavioral management tool should always be well-defined in its goals (addressing behavioral deficits or abnormalities), be tuned for specific individuals and contexts, and be empirically demonstrated to be effective in the context to which it is applied (Winnicker and Honess 2014). We should not be misled into thinking that simple interaction with a device, or comments such as "the monkeys love it," indicates achievement of any therapeutic or repertoire-balancing goal (Winnicker and Honess 2014).

Even the best facilities and behavioral management schemes can fail to deliver the highest standards of welfare if historical management programs have been poor. Prenatal and early life experiences play vital roles in establishing the foundations of behavior, learning, neurological development, and stress reactivity as primate infants grow and move into adulthood (Clarke and Boinski 1995; Schneider et al. 2001; Coe et al. 2003, 2010; Novak 2003). It is therefore not surprising that one of my key recommendations is that the breeding of long-tailed macaques should be left to those who produce high–health-status research subjects in an environment that minimizes stress and maximizes welfare. The best welfare standards do not come cheaply, and those who buy predominantly based on price do little to advance global welfare standards or to ensure the quality of research models. Excessive pressure on price favors producers with the most cost-efficient, but industrialized production systems, and those that place profit at the center of their philosophy, rather than animal welfare, ethics, and scientific progress. A high-quality breeder, even one who has to ship their animals internationally by air, has to be better than a poorer, but nearer one. Transport stress is minor compared to the benefits accrued from a high-welfare source that breeds animals that are emotionally competent, neurologically and physiologically normal, and behaviorally natural.

Accounting for life-time experience is of increasing importance in meeting ethical and legal objectives (EU 2010) and experiences prior to arrival, as well as in the laboratory, should be accounted for. Russell and Burch (1959) point to the importance of accounting for "contingent" suffering in assessing the true ethical impact of the research. This includes the impact of a range of factors beyond those related directly to research procedures, including transport, housing, confinement, enrichment, loss of control, etc. An animal not "on study" is importantly, not without stressors, and continues to accumulate costs that need to be accounted for in any ethical cost–benefit analysis. While refinement of conditions and practices can reduce this cost, the only way to prevent further accumulation is to complete the program of research as quickly as possible (Honess and Wolfensohn 2010).

It is important to strive for refinements in research procedures, but it is equally important to take a balanced view that incorporates contingent costs. For example, where is the balance in bringing animals in to the lab several months early for training for cooperative blood sampling on a study that would otherwise last just a few weeks? Blood sampling is refined by the training, but only via accumulation of significant additional contingent costs. Equally, more expansive caging for group-housed animals kept for occasional catching, restraint, and blood sampling is arguably better for the animals than more spatially and socially restrictive caging where training for cooperation may be more productive, particularly given that the animals must live in the caging 24/7. Should we limit welfare improvements in housing and enrichment to refine scientific procedures? There is clearly a balance to be struck. A solution to achieving an objective, balanced view on the changing experience of the animal is to apply an individualized and holistic assessment that accounts for multiple factors that influence quality of life (Honess and Wolfensohn 2010; Wolfensohn et al. 2015).

CONCLUSIONS

Research using nonhuman primates remains a vital element of improving our understanding of both human and animal health conditions, for the prevention and treatment of disease, and to ensure the efficacy and safety of pharmaceutical products (Bateson 2011). It is hoped that this chapter, in detailing some of the key wild and captive behavioral characters of this species and reviewing the evidence base for successful enrichment and behavioral therapy, can assist those who provide for and manage captive long-tailed macaques to minimize their stress and maximize their scientific value. A guiding framework for our care for the animals is that, unless given sound scientific and ethical justification, they should be guaranteed the Five Freedoms (Farm Animal Welfare Council 2010).

It is important that we consider welfare at the level of the individual, because so much research points to the importance of individual factors in determining stress responses and the effectiveness of enrichment. It is clear, especially in socially housed animals, that different individuals have very different experiences, even under highly similar conditions, and "herd" approaches to enrichment and behavioral management may only benefit a subset of the animals. There are resource challenges for assessing individual needs, particularly in large facilities, but at least if social context, age-sex class, prior experience, and dominance status are accounted for, then enrichment and behavioral management strategies are likely to be much more successful. Finally, genuine sensitivity to the state of the animals and perception of their needs will always benefit welfare provision. Ensuring that, across an institution, the optimization of animal welfare is fostered as a philosophy, not just a policy, will ensure departure from a purely tick-box approach toward one that is more dynamic, and places the immediate and longer term needs of the individual animal at its heart.

ACKNOWLEDGMENTS

I would like to dedicate this chapter to Dr. Corri Waitt—a fellow champion of the highest animal welfare standards and a colleague of the highest scientific rigor. Taken from us much too young, she leaves a big hole in our lives and in the science of animal welfare. I must thank Steve Schapiro for asking me to contribute to this book: His relentless pursuit of primate welfare is an outstanding example to us all. Sarah Wolfensohn steered me into primate welfare and much of my understanding of the experimental and veterinary factors influencing the lives of research animals was developed under her mentorship at Oxford University. A number of others have played important roles in developing my research and understanding of primate welfare, particularly Mary-Ann and Owen Griffiths and colleagues at Bioculture, Moshe Bushmitz, my research students, particularly Carolina Marin, and the managers and staff at all the facilities where I have worked. Finally, I would like to recognize the field biologists whose hard work has taught us all we know today of the behavior of primates in the wild and the risks to their survival. Recently, we have lost two key figures that have made significant contributions in respect of macaques: Prof Charles Southwick and Dr. Ardith Eudey.

REFERENCES

Abbott, D. H., E. B. Keverne, F. B. Bercovitch, C. A. Shively, S. P. Mendoza, W. Saltzman, C. T. Snowdon, T. E. Ziegler, M. Banjevic, T. Garland, and R. M. Sapolsky. 2003. Are subordinates always stressed? A comparative analysis of rank differences in cortisol levels among primates. *Hormones and Behavior* 43(1): 67–82.

Aureli, F. 1992. Post-conflict behavior among wild long-tailed macaques (*Macaca fascicularis*). *Behavioral Ecology and Sociobiology* 31(5): 329–337.

Aureli, F., M. Das, and H. C. Veenema. 1997. Differential kinship effect on reconciliation in three species of macaques (*Macaca fascicularis*, *M. fuscata*, and *M. sylvanus*). *Journal of Comparative Psychology* 111(1): 91–99.

Aureli, F., S. D. Preston, and F. B. M. de Waal. 1999. Heart rate responses to social interactions in free-moving rhesus macaques (*Macaca mulatta*): A pilot study. *Journal of Comparative Psychology* 113(1): 59–65.

Baker, K. C., M. Bloomsmith, K. Coleman, C. M. Crockett, C. Lutz, B. McCowan, P. Pierre, J. Weed, and J. Worlein J. 2017. The behavioral management consortium: A partnership for promoting consensus and best practices, Chapter 2. In Schapiro, S. J. (ed.) *Handbook of Primate Behavioral Management*, 9–23. Boca Raton, FL: CRC Press.

Baker, K. C., C. M. Crockett, G. H. Lee, B. C. Oettinger, V. Schoof, and J. P. Thom. 2012. Pair housing for female longtailed and rhesus macaques in the laboratory: Behavior in protected contact versus full contact. *Journal of Applied Animal Welfare Science* 15(2): 126–143.

Bateson, P. 2011. *Review of Research Using Non-human Primates: Report of a Panel Chaired by Professor Sir Patrick Bateson FRS*. London: BBSRC, MRC, NC3Rs, Wellcome Trust, 51.

Bayne, K. 1989. Nylon balls revisited. *Laboratory Primate Newsletter* 28(1): 5–6.

Bayne, K. 1991. Alternatives to continuous social housing. *Laboratory Animal Science* 41(4): 355–359.

Bayne, K. 2003. Environmental enrichment of nonhuman primates, dogs and rabbits used in toxicology studies. *Toxicologic Pathology* 31(Suppl. 1): 132–137.

Bayne, K. 2005. Potential for unintended consequences of environmental enrichment for laboratory animals and research results. *ILAR Journal* 46(2): 129–139.

Bayne, K., S. Dexter, J. K. Hurst, G. M. Strange, and E. E. Hill. 1993. Kong toys for laboratory primates: Are they really an enrichment or just fomites? *Laboratory Animal Science* 43(1): 78–85.

Bayne, K., H. Mainzer, S. Dexter, G. Campbell, F. Yamada, and S. Suomi. 1991. The reduction of abnormal behaviors in individually housed rhesus monkeys (*Macaca mulatta*) with a foraging/grooming board. *American Journal of Primatology* 23: 23–35.

Bayne, K. and M. Novak. 1998. Behavioral disorders. In Bennett, B. T., C. R. Abee, and R. Hendrickson (eds) *Nonhuman Primates in Biomedical Research: Diseases*, 485–500. New York: Academic Press.

Beisner, B. A. and L. A. Isbell. 2008. Ground substrate affects activity budgets and hair loss in outdoor captive groups of rhesus macaques (*Macaca mulatta*). *American Journal of Primatology* 70(12): 1160–1168.

Bennett, A. J., C. A. Corcoran, V. A. Hardy, L. R. Miller, and P. J. Pierre. 2010. Multidimensional cost-benefit analysis to guide evidence-based environmental enrichment: Providing bedding and foraging substrate to pen-housed monkeys. *Journal of the American Association for Laboratory Animal Science* 49(5): 571–577.

Berg, W., A. Jolly, H. Rambeloarivony, V. Andrianome, and H. Rasamimanana. 2009. A scoring system for coat and tail condition in ringtailed lemurs, *Lemur catta*. *American Journal of Primatology* 71: 183–190.

Bloomsmith, M., L. Brent, and S. J. Schapiro. 1991. Guidelines for developing and managing an environmental enrichment program for nonhuman primates. *Laboratory Animal Science* 41(4): 372–377.

Brent, L. and Y. Veira. 2002. Social behavior of captive indochinese and insular long-tailed macaques (*Macaca fascicularis*) following transfer to a new facility. *International Journal of Primatology* 23(1): 147–159.

Brinkman, C. 1996. Toys for the boys: Environmental enrichment for singly housed adult male macaques (*Macaca fascicularis*). *Laboratory Primate Newsletter* 35(2): 4–9.

Brotcorne, F., M. C. Huynen, and T. Savini. 2008. Preliminary results on the behavioral ecology of a long-tailed macaque (*Macaca fascicularis*) population in disturbed urban habitats, Bangkok (Thailand). *Folia Primatologica* 79: 315.

Bryant, C. E., N. M. Rupniak, and S. D. Iversen. 1988. Effects of different environmental enrichment devices on cage stereotypies and autoaggression in captive cynomolgus monkeys. *Journal of Medical Primatology* 17: 257–269.

Buchanan-Smith, H. M., M. Prescott, and N. J. Cross. 2004. What factors should determine cage sizes for primates in the laboratory? *Animal Welfare* 13: S197–S201.

Burton, F. D. and L. Chan. 1996. Behavior of mixed species groups of macaques. In Fa, J. E. and D. G. Lindburg (eds) *Evolution and Ecology of Macaque Societies*, 389–412. Cambridge, UK: Cambridge University Press.

Callaway, E. 2015. Monkeys seem to recognize their reflections: trained macaques studied themselves in mirrors, fuelling debate over animals' capacity for self-recognition. *Nature News* doi:10.1038/nature.2015.16692.

Camus, S. M. J., C. Blois-Heulin, Q. Li, M. Hausberger, and E. Bezard. 2013. Behavioral profiles in captive-bred cynomolgus macaques: Towards monkey models of mental disorders? *PLoS One* 8(4): e62141.

Cant, J. G. H. 1988. Positional behavior of long-tailed macaques (*Macaca fascicularis*) in northern Sumatra. *American Journal of Physical Anthropology* 76(1): 29–37.

Capitanio, J. P., S. A. Blozis, J. Snarr, A. Steward, and B. J. McCowan. 2017. Do "birds of a feather flock together" or do "opposites attract"? Behavioral responses and temperament predict success in pairings of rhesus monkeys in a laboratory setting. *American Journal of Primatology* 79(1), 1–11. doi:10.1002/ajp.22464.

Carlsson, H. E., S. J. Schapiro, I. Farah, and J. Hau. 2004. Use of primates in research: a global overview. *American Journal of Primatology* 63(4): 225–237.

Castro, M. I., J. Rose, W. Green, N. Lehner, D. Peterson, and D. Taub. 1981. Ketamine-HCl as a suitable anesthetic for endocrine, metabolic, and cardiovascular studies in *Macaca fascicularis* monkeys. *Proceedings of the Society for Experimental Biology and Medicine. Society for Experimental Biology and Medicine (New York, NY)* 168(3): 389–394.

Chapais, B. 2004. How kinship generates dominance structures: A comparative perspective. In Thierry, B., M. Singh, and W. Kaumanns (eds) *Macaque Societies: A Model for the Study of Social Organization*, 186–208. Cambridge, UK: Cambridge University Press.

Cheney, D. and R. Wrangham. 1987. Predation. In Smuts, B., D. Cheney, R. Seyfarth, R. Wrangham, and T. Struhsaker (eds) *Primate Societies*, 227–239. Chicago, IL: University of Chicago Press.

Clarke, A. S. and S. Boinski. 1995. Temperament in nonhuman primates. *American Journal of Primatology* 37(2): 103–125.

Clarke, A. S., N. M. Czekala, and D. G. Lindburg. 1995. Behavioral and adrenocortical responses of male cynomolgus and lion-tailed macaques to social stimulation and group formation. *Primates* 36(1): 41–56.

Clarke, A. S. and D. G. Lindburg. 1993. Behavioral contrasts between male cynomolgus and lion-tailed macaques. *American Journal of Primatology* 29(1): 49–59.

Clarke, A. S., W. A. Mason, and G. P. Moberg. 1988. Interspecific contrasts in responses of macaques to transport cage training. *Laboratory Animal Science* 38(3): 305–309.

Coe, C. L., M. Kramer, B. Czeh, E. Gould, A. J. Reeves, C. Kirschbaum, and E. Fuchs. 2003. Prenatal stress diminishes neurogenesis in the dentate gyrus of juvenile rhesus monkeys. *Biological Psychiatry* 54(10): 1025–1034.

Coe, C. L., G. R. Lubach, H. R. Crispen, E. A. Shirtcliff, and M. L. Schneider. 2010. Challenges to maternal wellbeing during pregnancy impact temperament, attention, and neuromotor responses in the infant rhesus monkey. *Developmental Psychobiology* 52(7): 625–637.

Cords, M. 1992. Post-conflict reunions and reconciliation in long-tailed macaques. *Animal Behavior* 44(1): 57–61.

Corlett, R. T. and P. W. Lucas. 1990. Alternative seed-handling strategies in primates: Seed-spitting by long-tailed macaques (*Macaca fascicularis*). *Oecologia* 82(2): 166–171.

Cowlishaw, G. and R. I. M. Dunbar. 1991. Dominance rank and mating success in male primates. *Animal Behavior* 41: 1045–1056.

Crockett, C. M., K. L. Bentson, and R. U. Bellanca. 2007. Alopecia and overgrooming in laboratory monkeys vary by species but not sex, suggesting a different etiology than self-biting. *American Journal of Primatology* 69(S1): 87–88.

Crockett, C. M., J. Bielitzki, A. Carey, and A. Velez. 1989. Kong toys as enrichment devices for singly-caged macaques. *Laboratory Primate Newsletter* 28(2): 21–22.

Crockett, C. M., C. L. Bowers, D. M. Bowden, and G. P. Sackett. 1994. Sex differences in compatibility of pair-housed adult longtailed macaques. *American Journal of Primatology* 32(2): 73–94.

Crockett, C. M., C. L. Bowers, G. P. Sackett, and D. M. Bowden. 1993. Urinary cortisol responses of longtailed macaques to five cage sizes, tethering, sedation, and room change. *American Journal of Primatology* 30(1): 55–74.

Crockett, C. M. and W. L. Wilson. 1980. The ecological separation of *Macaca nemestrina* and *Macaca fascicularis* in Sumatra. In Lindburg, D. G. L. (ed.) *The Macaques: Studies in Ecology, Behavior and Evolution*, 148–181. New York: Van Nostrand Reinhold Company.

de Ruiter, J. R. and E. Geffen. 1998. Relatedness of matrilines, dispersing males and social groups in long-tailed macaques (*Macaca fascicularis*). *Proceedings of the Royal Society B: Biological Sciences* 265(1391): 79–87.

de Ruiter, J. R., W. Scheffrahn, G. J. J. M. Trommelen, A. G. Uitterlinden, R. D. Martin, and J. A. R. A. M. van Hooff. 1992. Male social rank and reproductive success in wild long-tailed macaques. In Martin, R. D., A. Dixson, and E. J. Wickings (eds) *Paternity in Primates: Genetic Tests and Theories*, 175–191. Basel, Switzerland: Karger.

de Ruiter, J. R., J. A. R. A. M. van Hooff, and W. Scheffrahn. 1994. Social and genetic aspects of paternity in wild long-tailed macaques (*Macaca fascicularis*). *Behavior* 129: 203–224.

Dittus, W. 2004. Demography: A window to social evolution. In Thierry, B., M. Singh, and W. Kaumanns (eds) *Macaque Societies: A Model for the Study of Social Organization*, 87–116. Cambridge, UK: Cambridge University Press.

DiVincenti, L., Jr., A. Rehrig, and J. Wyatt. 2012. Interspecies pair housing of macaques in a research facility. *Laboratory Animals* 46(2): 170–172.

Dominy, N. J. 2004. Fruits, fingers, and fermentation: The sensory cues available to foraging primates. *Integrative and Comparative Biology* 44(4): 295–303.

Eckert, K., C. Niemeyer, Anonymous, R. W. Rogers, J. Seier, B. Ingersoll, L. Barklay, C. Brinkman, S. Oliver, C. Buckmaster, L. Knowles, S. Pyle, and V. Reinhardt. 2000. Wooden objects for enrichment: A discussion. *Laboratory Primate Newsletter* 39(3): 1–4.

Engel, G. A., L. Jones-Engel, M. A. Schillaci, K. G. Suaryana, A. Putra, A. Fuentes, and R. Henkel. 2002. Human exposure to herpesvirus B-seropositive macaques, Bali, Indonesia. *Emerging Infectious Diseases* 8(8): 789–795.

Engelhardt, A., J. K. Hodges, and M. Heistermann. 2007. Post-conception mating in wild long-tailed macaques (*Macaca fascicularis*): Characterization, endocrine correlates and functional significance. *Hormones and Behavior* 51(1): 3–10.

Engelhardt, A., J. K. Hodges, C. Niemitz, and M. Heistermann. 2005. Female sexual behavior, but not sex skin swelling, reliably indicates the timing of the fertile phase in wild long-tailed macaques (*Macaca fascicularis*). *Hormones and Behavior* 47(2): 195–204.

Engelhardt, A., J.-B. Pfeifer, M. Heistermann, C. Niemitz, J. A. R. A. M. van Hooff, and J. K. Hodges. 2004. Assessment of female reproductive status by male longtailed macaques, *Macaca fascicularis*, under Natural Conditions. *Animal Behavior* 67(5): 915–924.

Erwin, J. and R. Deni. 1979. Strangers in a strange land: Abnormal behaviors or abnormal environments? In Erwin, J., T. L. Maple, and G. Mitchell (eds) *Captivity and Behavior: Primates in Breeding Colonies, Laboratories and Zoos*, 1–28. New York: Van Norstrand Reinhold.

Eudey, A. 1994. Temple and pet primates in Thailand. *Revue D'Ecologie (Terre et la Vie)* 49: 273–280.

Eudey, A. 2008. The crab-eating macaque (*Macaca fascicularis*): Widespread and rapidly declining. *Primate Conservation* 23: 129–132.

EU Directive. 2010. 63/EU of the European Parliament and of the Council of 22 September 2010 on the Protection of animals used for scientific purposes. *Official Journal of the European Union* L276: 33–79.

Evans, H. L., J. D. Taylor, and J. F. Graefe. 1989. Methods to evaluate the wellbeing of laboratory primates: Comparisons of macaques and tamarins. *Laboratory Animal Science* 39(4): 318–323.

Farm Animal Welfare Council. 2010. Five freedoms. Available at http://webarchive.nationalarchives.gov.uk/20121007104210/http:/www.fawc.org.uk/freedoms.htm. Accessed February 29, 2016.

Faucheux, B., M. Bertrand, and F. Bourliere. 1978. Some effects of living conditions upon the pattern of growth in the stumptail macaque (*Macaca arctoides*). *Folia Primatologica* 30: 220–236.

Fernandez, L., M.-A. Griffiths, and P. Honess. 2017. Providing behaviorally manageable primates for research, Chapter 28. In Schapiro, S. J. (ed.) *Handbook of Primate Behavioral Management*, 481–494. Boca Raton, FL: CRC Press.

Fernström, A. L., H. Fredlund, M. Spångberg, and K. Westlund. 2009. Positive reinforcement training in rhesus macaques: Training progress as a result of training frequency. *American Journal of Primatology* 71(5): 373–379.

Fernström, A. L., W. Sutian, F. Royo, K. Westlund, T. Nilsson, H. E. Carlsson, Y. Paramastri, J. Pamungkas, D. Sajuthi, S. J. Schapiro, and J. Hau. 2008. Stress in cynomolgus monkeys (*Macaca fascicularis*) subjected to long-distance transport and simulated transport housing conditions. *Stress* 11: 467–476.

Fleagle, J. G. 1999. Locomotor adaptations. In *Primate Adaptation and Evolution*, 297–306. London: Academic Press.

Fooden, J. 1991. Systematic review of Philippine macaques (Primates, Cercopithecidae: *Macaca fascicularis* subspp.) *Fieldiana Zoology* 64: 1–44.

Fooden, J. 1995. Systematic review of Southeast Asian longtail macaques, *Macaca fascicularis* (Raffles, 1821). *Fieldiana Zoology* 81: 1–206.

Fuentes, A. and S. Gamerl. 2005. Disproportionate participation by age/sex classes in aggressive interactions between long-tailed macaques (*Macaca fascicularis*) and human tourists at Padangtegal Monkey Forest, Bali, Indonesia. *American Journal of Primatology* 66(2): 197–204.

Gallup, G. G. 1977. Absence of self-recognition in a monkey (*Macaca fascicularis*) following prolonged exposure to a mirror. *Developmental Psychobiology* 10(3): 281–284.

Gilbert, S. G. and E. Wrenshall. 1989. Environmental enrichment for monkeys used in behavioral toxicology studies. In Segal, E. F. (ed.) *Housing, Care and Psychological Wellbeing of Captive and Laboratory Primates*, 244–254. Park Ridge, NJ: Noyes Publications.

Gottlieb, D., K. Coleman, and K. Prongay. 2017. Behavioral management of *Macaca* species (except *M. fascicularis*), Chapter 19. In Schapiro, S. J. (ed.) *Handbook of Primate Behavioral Management*, 279–303. Boca Raton, FL: CRC Press.

Graham, M. L. 2017. Positive reinforcement training and research, Chapter 12. In Schapiro, S. J (ed.) *Handbook of Primate Behavioral Management*, 187–200. Boca Raton, FL: CRC Press.

Groves, C. P. 2001. *Primate Taxonomy*. Washington, DC: Smithsonian Institution Press.

Gumert, M. D. and S. Malaivijitnond. 2013. Long-tailed macaques select mass of stone tools according to food type. *Philosophical Transactions of the Royal Society B: Biological Sciences* 368(1630): 20120413.

Hadi, I., B. Suryobroto, and D. Perwitasari-Farajallah. 2007. Food preference of semi-provisioned macaques based on feeding duration and foraging party size. *HAYATI Journal of Biosciences* 14(1): 13–17.

Hahn, N. E., D. Lau, K. Eckert, and H. Markowitz. 2000. Environmental enrichment-related injury in a macaque (*Macaca fascicularis*): Intestinal linear foreign body. *Comparative Medicine* 50(5): 556–558.

Harvey, P. H., R. D. Martin, and T. H. Clutton-Brock. 1987. Life histories in comparative perspective. In Smuts, B., D. Cheney, R. Seyfarth, R. Wrangham, and T. Struhsaker (eds) *Primate Societies*, 181–193. Chicago, IL: University of Chicago Press.

Hearth-Lange, S., J. C. Ha, and G. P. Sackett. 1999. Behavioral measurement of temperament in male nursery-raised infant macaques and baboons. *American Journal of Primatology* 47: 43–50.

Heath, M. 1989. The training of cynomolgus monkeys and how the human/animal relationship improves with environmental and mental enrichment. *Animal Technology* 40(1): 11–22.

Hemelrijk, C. K. 1994. Support for being groomed in long-tailed macaques, *Macaca fascicularis. Animal Behavior* 48: 479–481.

Henzi, S. and L. Barrett. 1999. The value of grooming to female primates. *Primates* 40(1): 47–59.

Holmes, S. N., J. M. Riley, P. Juneau, D. Pyne, and G. L. Hofing. 1995. Short-term evaluation of a foraging device for non-human primates. *Laboratory Animals* 29(4): 364–369.

Home Office. 2014. Annual statistics of scientific procedures on living animals, Great Britain 2013, 57. London: HMSO.

Honess, P. E. 2009. Beyond the final frontier: Space and its use by animals in research. Charles River Lecture on Ethics and Animal Welfare. *National Meeting of the American Association for Laboratory Animal Science (AALAS)*, Denver, CO.

Honess, P. 2013. Behavior and enrichment of long-tailed (cynomolgus) macaques (*Macaca fascicularis*). In Winnicker, C. (ed.) *A Guide to the Behavior and Enrichment of Laboratory Macaques*, 4–87. Wilmington, MA: Charles River Publications.

Honess, P., T. Andrianjazalahatra, L. Fernandez, and M.-A. Griffiths. 2012. Environmental enrichment and the behavioral needs of macaques housed in large social groups. *Enrichment Record* 10(1): 16–20.

Honess, P., T. Andrianjazalahatra, P. Matai, and S. Naiken. 2013. Primate adoption: An essential alternative to hand-rearing. *Journal of the American Association for Laboratory Animal Science* 52(3): 253.

Honess, P., T. Andrianjazalahatra, S. Naiken, and M.-A. Griffiths. 2015. Factors influencing aggression in peer groups of weaned long-tailed macaques (*Macaca fascicularis*). *Archives of Medical and Biomedical Research* 2(3): 88.

Honess, P. and L. Fernandez. 2011. Environmental enrichment for captive and wild-born macaques. *Enrichment Record* 9(4): 16–18.

Honess, P., J. Gimpel, S. Wolfensohn, and G. Mason. 2005. Alopecia scoring: The quantitative assessment of hair loss in captive macaques. *Alternatives to Laboratory Animals* 33(3): 193–206.

Honess, P. E., P. J. Johnson, and S. E. Wolfensohn. 2004. A study of behavioral responses of non-human primates to air transport and re-housing. *Laboratory Animals* 38(2): 119–132.

Honess, P. E. and C. M. Marin. 2006a. Behavioral and physiological aspects of stress and aggression in nonhuman primates. *Neuroscience and Biobehavioral Reviews* 30(3): 390–412.

Honess, P. E. and C. M. Marin. 2006b. Enrichment and aggression in primates. *Neuroscience and Biobehavioral Reviews* 30(3): 413–436.

Honess, P., M. A. Stanley-Griffiths, S. Narainapoulle, S. Naiken, and T. Andrianjazalahatra. 2010. Selective breeding of primates for use in research: Consequences and challenges. *Animal Welfare* 19: 57–65.

Honess, P. and S. Wolfensohn. 2010. A matrix for the assessment of welfare in experimental animals. *Alternatives to Laboratory Animals* 38: 205–212.

Janson, C. H. 1994. Naturalistic Environments in Captivity: A methodological bridge between field and laboratory studies of primates. In Gibbons, E. F., E. Wyers, E. Waters, and E. Menzel (eds) *Naturalistic Environments in Captivity for Animal Behavior Research*, 271–279. Albany, NY: State University of New York Press.

Kanthaswamy, S., J. Satkoski, D. George, A. Kou, B. Erickson, and D. Smith. 2008. Hybridization and stratification of nuclear genetic variation in *Macaca mulatta* and *M. fascicularis*. *International Journal of Primatology* 29(5): 1295–1311.

Kawamoto, Y., K. Nozawa, K. Matsubayashi, and S. Gotoh. 1988. A population-genetic study of crab-eating macaques (*Macaca fascicularis*) on the island of Angaur, Palau, Micronesia. *Folia Primatologica* 51(4): 169–181.

Kissinger, J. T. and M. Landi. 1989. The effect of four types of restraint on serum ALT and AST in cynomolgus monkeys. *Laboratory Animal Science* 39: 496.

Kling, A. and J. Orbach. 1963. Plasma 17-Hydroxycorticosteroid levels in the stump-tailed monkey and two other macaques. *Psychological Reports* 13(3): 863–865.

Kurland, J. A. 1973. A natural history of kra macaques (*Macaca fascicularis* Raffles, 1821) at the Kutai Reserve, Kalimantan Timur, Indonesia. *Primates* 14(2): 245–262.

Lam, K., N. M. Rupniak, and S. D. Iversen. 1991. Use of a grooming and foraging substrate to reduce cage stereotypies in macaques. *Journal of Medical Primatology* 20(3): 104–109.

Landi, M. S., J. T. Kissinger, S. A. Campbell, C. A. Kenney, and E. L. Jenkins. 1990. The effects of four types of restraint on serum alanine aminotransferase and aspartate aminotransferase in the *Macaca fascicularis*. *International Journal of Toxicology* 9(5): 517–523.

Lane, J. 2006. Can non-invasive glucocorticoid measures be used as reliable indicators of stress in animals? *Animal Welfare* 15(4): 331–342.

Laule, G. 1999. Training laboratory animals. In Poole, T. (ed.) *The UFAW Handbook on the Care and Management of Laboratory Animals*, 21–27. Oxford, UK: Wiley-Blackwell.

Laule, G. and T. Desmond. 1998. Positive reinforcement training as an enrichment strategy. In Shepherdson, D. J., J. D. Mellen, and M. Hutchins (eds) *Second Nature: Environmental Enrichment for Captive Animals*, 31–46. Washington, DC: Smithsonian Institution Press.

Lee, G. H., J. P. Thom, K. L. Chu, and C. M. Crockett. 2012. Comparing the relative benefits of grooming-contact and full-contact pairing for laboratory-housed adult female *Macaca fascicularis*. *Applied Animal Behavior Science* 137(3–4): 157–165.

Levallois, L. and S. D. de Marigny. 2015. Reproductive success of wild-caught and captive-bred cynomolgus macaques at a breeding facility. *Lab Animal* 44(10): 387–393.

Line, S., K. Morgan, H. Markowitz, J. A. Roberts, and M. Riddel. 1990a. Behavioral responses of female long-tailed macaques (*Macaca fascicularis*) to pair formation. *Laboratory Primate Newsletter* 29(4): 1–5.

Line, S., K. Morgan, J. A. Roberts, and H. Markowitz. 1990b. Preliminary comments on resocialization of aged rhesus macaques. *Laboratory Primate Newsletter* 29(1): 8–12.

Lloyd, C. R., G. H. Lee, and C. M. Crockett. 2005. Puzzle-ball foraging by laboratory monkeys improves with experience. *Laboratory Primate Newsletter* 44(1): 1–3.

Lucas, P. W. and R. T. Corlett. 1991. Relationship between the diet of *Macaca fascicularis* and forest phenology. *Folia Primatologica* 57(4): 201–215.

Lucas, P. W., B. W. Darvell, P. K. D. Lee, T. D. B. Yuen, and M. F. Choong. 1998. Colour cues for leaf food selection by long-tailed macaques (*Macaca fascicularis*) with a new suggestion for the evolution of trichromatic colour vision. *Folia Primatologica* 69(3): 139–154.

Luchins, K. R., K. C. Baker, M. H. Gilbert, J. L. Blanchard, and R. P. Bohm. 2011. Manzanita wood: A sanitizable enrichment option for nonhuman primates. *Journal of the American Association for Laboratory Animal Science* 50(6): 884–887.

Lutz, C. and R. Farrow. 1996. Foraging device for singly housed longtailed macaques does not reduce stereotypies. *Contemporary Topics in Laboratory Animal Science* 35(3): 75–78.

Lutz, C. K. and M. A. Novak. 2005. Environmental enrichment for nonhuman primates: Theory and application. *ILAR Journal* 46(2): 178–191.

Lutz, C., A. Well, and M. Novak. 2003. Stereotypic and self-injurious behavior in rhesus macaques: a survey and retrospective analysis of environment and early experience. *American Journal of Primatology* 60: 1–15.

Magden, E. R. 2017. Positive reinforcement training and health care, Chapter 13. In Schapiro, S. J. (ed.) *Handbook of Primate Behavioral Management*, 201–216. Boca Raton, FL: CRC Press.

Malaivijitnond, S., O. Takenaka, T. Sankai, T. Yoshida, F. Cho, and Y. Yoshikawa. 1998. Effects of single and multiple injections of ketamine hydrochloride in serum hormone concentrations in male cynomolgus monkeys. *Laboratory Animal Science* 48(3): 270–274.

McGrew, K. 2017. Pairing strategies for cynomolgus macaques, Chapter 17. In Schapiro, S. J (ed.) *Handbook of Primate Behavioral Management*, 255–264. Boca Raton, FL: CRC Press.

Melnick, D. and M. Pearl. 1987. Cercopithecines in multimale groups: Genetic diversity and population Structure. In Smuts, B., D. Cheney, R. Seyfarth, R. Wrangham, and T. Struhsaker (eds) *Primate Societies*, 121–134. Chicago, IL: University of Chicago Press.

Meunier, L. D., J. T. Duktig, and M. S. Landi. 1989. Modification of stereotypic behavior in rhesus monkeys using videotapes, puzzle-feeders, and foraging boxes. *Laboratory Animal Science* 39(5): 479.

Mitchell, G. and D. H. Tokunaga. 1976. Sex differences in nonhuman primate grooming. *Behavioral Processes* 1: 335–345.

Morgan, D., K. A. Grant, O. A. Prioleau, S. H. Nader, J. R. Kaplan, and M. A. Nader. 2000. Predictors of social status in cynomolgus monkeys (*Macaca fascicularis*) after group formation. *American Journal of Primatology* 52(3): 115–131.

Murchison, M. A. 1991. PVC-pipe food puzzle for singly caged primates. *Laboratory Primate Newsletter* 30(3): 12–14.

National Research Council. 2010. *Guide for the Care and Use of Laboratory Animals*. Washington DC: National Academies Press.

Novak, M. A. 2003. Self-injurious behavior in rhesus monkeys: New insights into its etiology, physiology, and treatment. *American Journal of Primatology* 59(1): 3–19.

Novak, M. A. and J. S. Meyer. 2009. Alopecia: Possible causes and treatments, particularly in captive nonhuman primates. *Comparative Medicine* 59(1): 18–26.

Novak, M. A., P. O'Neill, S. A. Beckley, and S. Suomi. 1994. Naturalistic environments for captive primates. In Gibbons, E. F., E. J. Wyers, E. Waters, and E. W. Menzel (eds) *Captive Environments for Animal Behavior Research*, 236–258. Albany, NY: State University of New York Press.

Novak, M. A. and S. J. Suomi. 1988. Psychological well-being of primates in captivity. *American Psychologist* 43(10): 765–773.

Ong, P. and M. Richardson. 2008. *Macaca fascicularis*. The IUCN Red List of Threatened Species 2008: e.T12551A3355536. Retrieved February 17, 2017. http://dx.doi.org/10.2305/IUCN.UK.2008.RLTS.T12551A3355536.en.

Osorio, D., A. C. Smith, M. Vorobyev, and H. M. Buchanan-Smith. 2004. Detection of fruit and the selection of primate visual pigments for color vision. *The American Naturalist* 164(6): 696–708.

Perlman, J. E., M. A. Bloomsmith, M. A. Whittaker, J. L. McMillan, D. E. Minier, and B. McCowan. 2012. Implementing positive reinforcement animal training programs at primate laboratories. *Applied Animal Behavior Science* 137(3–4): 114–126.

Poirier, F. E. and E. O. Smith. 1974a. The crab-eating macaques (*Macaca fascicularis*) of Angaur Island, Palau, Micronesia (Part 1 of 2). *Folia Primatologica* 22(4): 258–282.

Poirier, F. E. and E. O. Smith. 1974b. The crab-eating macaques (*Macaca fascicularis*) of Angaur Island, Palau, Micronesia (Part 2 of 2). *Folia Primatologica* 22(4): 283–306.

Poole, T. B. 1997. Happy animals make good science. *Laboratory Animals* 31: 116–124.

Prescott, M. J., V. J. Brown, P. A. Flecknell, D. Gaffan, K. Garrod, R. N. Lemon, A. J. Parker, K. Ryder, W. Schultz, L. Scott, J. Watson, and L. Whitfield. 2010. Refinement of the use of food and fluid control as motivational tools for macaques used in behavioral neuroscience research: Report of a working group of the NC3Rs. *Journal of Neuroscience Methods* 193(2): 167–188.

Prescott, M., H. M. Buchanan-Smith, and A. E. Rennie. 2005. Training of laboratory-housed non-human primates in the UK. *Anthrozoos* 18: 288–303.

Prescott, M. J., M. E. Nixon, D. A. H. Farningham, S. Naiken, and M.-A. Griffiths. 2012. Laboratory macaques: When to Wean? *Applied Animal Behavior Science* 137(3–4): 194–207.

Reinhardt, V. 1993. Foraging enrichment for caged macaques: A review. *Laboratory Primate Newsletter* 32(4): 1–5.

Reinhardt, V. 2002. Addressing the social needs of macaques used for research. *Laboratory Primate Newsletter* 41(3): 7–11.

Reinhardt, V. 2003. Working with rather than against macaques during blood collection. *Journal of Applied Animal Welfare Science* 6(3): 189–197.

Reinhardt, V. 2005. Hair pulling: A review. *Laboratory Animals* 39(4): 361–369.

Reinhardt, V., K. Baker, A. Lablans, E. Davis, J. P. Garner, C. Sherwin, S. Banjanin, L. Bell, J. Barley, and J. Schrier. 2004. Self-injurious biting in laboratory animals: A discussion. *Laboratory Primate Newsletter* 43(2): 11–13.

Reinhardt, V., C. Liss, and C. Stevens. 1995a. Restraint methods of laboratory non-human primates: A critical review. *Animal Welfare* 4(3): 221–238.

Reinhardt, V., C. Liss, and C. Stevens. 1995b. Social housing of previously single-caged macaques: What are the options and the risks? *Animal Welfare* 4: 307–328.

Reinhardt, V., C. Liss, and C. Stevens. 1996. Space requirement stipulations for caged non-human primates in the United States: A Critical Review. *Animal Welfare* 5(4): 361–372.

Reinhardt, V. and A. Roberts. 1997. Effective feeding enrichment for non-human primates: A brief review. *Animal Welfare* 6(3): 265–272.

Rennie, A. E. and H. M. Buchanan-Smith. 2006. Refinement of the use of non-human primates in scientific research. Part III: Refinement of procedures. *Animal Welfare* 15(3): 239–261.

Renquist, D. H. and F. J. Judge. 1985. Use of nylon balls as behavioral modifier for caged primates. *Laboratory Primate Newsletter* 24(4): 4.

Richards, S. M. 1974. The concept of dominance and methods of assessment. *Animal Behavior* 22(4): 914–930.

Richard, A., S. Goldstein, and R. Dewar. 1989. Weed macaques: The evolutionary implications of macaque feeding ecology. *International Journal of Primatology* 10(6): 569–594.

Roberts, S. J. and M. L. Platt. 2005. Effects of isosexual pair-housing on biomedical implants and study participation in male macaques. *Journal of the American Association for Laboratory Animal Science* 44(5): 13–18.

Robins, J. G. and C. D. Waitt. 2010. Improving the welfare of captive macaques (*Macaca* sp.) through the use of water as enrichment. *Journal of Applied Animal Welfare Science* 14(1): 75–84.

Rodman, P. S. 1979. Skeletal differentiation of *Macaca fascicularis* and *Macaca nemestrina* in relation to arboreal and terrestrial quadrupedalism. *American Journal of Physical Anthropology* 51(1): 51–62.

Rommeck, I., K. Anderson, A. Heagerty, A. Cameron, and B. McCowan. 2009. Risk factors and remediation of self-injurious and self-abuse behavior in rhesus macaques. *Journal of Applied Animal Welfare Science* 12(1): 61–72.

Rothschild, B. M. and R. J. Woods. 1992. Osteoarthritis, calcium pyrophosphate deposition disease, and osseous infection in old-world primates. *American Journal of Physical Anthropology* 87(3): 341–347.

Rowe, N. 1996. *The Pictorial Guide to the Living Primates*. New York: Pogonias Press.

Russell, W. M. S. 2002. The ill-effects of uncomfortable quarters. In Reinhardt, V. and A. Reinhardt (eds) *Comfortable Quarters for Laboratory Animals*, 1–5. Washington, DC: Animal Welfare Institute.

Russell, W. M. S. and R. L. Burch. 1959. *The Principles of Humane Experimental Technique*. London: Methuen Publishing.

Sambrook, T. D. and H. M. Buchanan-Smith. 1997. Control and complexity in novel object enrichment. *Animal Welfare* 6: 207–216.

Schapiro, S. J., M. A. Bloomsmith, S. A. Suarez, and L. M. Porter. 1996. Effects of social and inanimate enrichment on the behavior of yearling rhesus monkeys. *American Journal of Primatology* 40: 247–260.

Schapiro, S. J., G. E. Laule, M. A. Bloomsmith, and T. J. Desmond. 1995. Exploring and advancing environmental enrichment: A primate training and enrichment workshop. *Lab Animal* 24(4): 35–39.

SCHER. 2009. The need for non-human primates in biomedical research, production and testing of products and Devices. Scientific Committee on Health and Environmental Risks, European Commission, Brussells, 38.

Schino, G., S. Scucchi, D. Maestripieri, and P. G. Turillazzi. 1988. Allogrooming as a tension-reduction mechanism: A behavioral approach. *American Journal of Primatology* 16(1): 43–50.

Schneider, M. L., C. F. Moore, A. D. Roberts, and O. Dejesus. 2001. Prenatal stress alters early neurobehavior, stress reactivity and learning in non-human primates: A brief review. *Stress* 4(3): 183–193.

Schrier, J. 1995. Browse for nonhuman primates in captivity. *Laboratory Primate Newsletter* 34(4): 28.

Schrier, J. 2004. Plants for browse, Plants *not* for browse. *Laboratory Primate Newsletter* 43(1): 34.

Scott, L., P. Pearce, S. Fairhall, N. Muggleton, and J. Smith. 2003. Training nonhuman primates to cooperate with scientific procedures in applied biomedical research. *Journal of Applied Animal Welfare Science* 6(3): 199–207.

Sesbuppha, W., S. Chantip, E. J. Dick, Jr., N. E. Schlabritz-Loutsevitch, R. Guardado-Mendoza, S. D. Butler, P. A. Frost, and G. B. Hubbard. 2008. Stillbirths in *Macaca fascicularis*. *Journal of Medical Primatology* 37(4): 169–172.

Shively, C. A. 2017. Depression in captive nonhuman primates: Theoretical underpinnings, methods, and application to behavioral management, Chapter 8. In Schapiro, S. J. (ed.) *Handbook of Primate Behavioral Management*, 115–125. Boca Raton, FL: CRC Press.

Shively, C. A., D. P. Friedman, H. D. Gage, M. C. Bounds, C. Brown-Proctor, J. B. Blair, J. A. Henderson, M. A. Smith, and N. Buchheimer. 2006. Behavioral depression and positron emission tomography-determined serotonin 1a receptor binding potential in cynomolgus monkeys. *Archives of General Psychiatry* 63: 396–403.

Shively, C. A. and J. R. Kaplan. 1991. Stability of social status rankings of female cynomolgus monkeys, of varying reproductive condition, in different social groups. *American Journal of Primatology* 23(4): 239–245.

Shively, C. and D. G. Smith. 1985. Social Status and reproductive success of male *Macaca fascicularis*. *American Journal of Primatology* 9(2): 129–135.

Shutt, K., A. MacLarnon, M. Heistermann, and S. Semple. 2007. Grooming in barbary macaques: Better to Give than to Receive? *Biology Letters* 3(3): 231–233.

Snowdon, C. T. 1994. The significance of naturalistic environments for primate behavioral research. In Gibbons, E. F., E. Wyers, E. Waters, and E. W. Menzel (eds) *Naturalistic Environments in Captivity for Animal Behavioral Research.*. Albany, NY: State University of New York Press.

Son, V. D. 2003. Diet of *Macaca fascicularis* in a mangrove forest, Vietnam. *Laboratory Primate Newsletter* 42(4): 1–5.

Son, V. D. 2004. Time budgets of *Macaca fascicularis* in a mangrove forest, Vietnam. *Laboratory Primate Newsletter* 43(3): 1–4.

Spector, M., M. A. Kowalczky, J. D. Fortman, and B. T. Bennett. 1994. Design and implementation of a primate foraging tray. *Contemporary Topics in Laboratory Animal Science* 33(5): 54–55.

Stanley, M. A. 2003. The breeding of naturally occurring B virus-free cynomolgus monkeys (*Macaca fascicularis*) on the island of Mauritius. *International Perspectives: The Future of Nonhuman Primate Resources, Proceedings of the Workshop Held April 17–19, 2002*. National Academies Press, Washington, 46–48.

Statz, L. M. and M. Borde. 2001. Pairing successes with male cynomolgus macaques after vasectomy. *Contemporary Topics in Laboratory Animal Science* 40(4): 91.

Stewart, A.-M., C. Gordon, S. Wich, P. Schroor, and E. Meijaard. 2008. Fishing in *Macaca fascicularis*: A rarely observed innovative behavior. *International Journal of Primatology* 29(2): 543–548.

Sussman, R. and I. Tattersall. 1981. Behavior and ecology of *Macaca fascicularis* in Mauritius: A preliminary study. *Primates* 22(2): 192–205.

Sussman, R. W. and I. Tattersall. 1986. Distribution, abundance, and putative ecological strategy of *Macaca fascicularis* on the island of Mauritius, southwestern Indian Ocean. *Folia Primatologica* 46(1): 28–43.

Tasker, L. 2012. Linking welfare and quality of scientific output in cynomolgus macaques (*Macaca fascicularis*) used for regulatory toxicology. PhD thesis, University of Stirling.

Tattersall, I., A. Dunaif, R. W. Sussman, and R. Jamieson. 1981. Hematological and serum biochemical values in free-ranging *Macaca fascicularis* of Mauritius: Possible diabetes mellitus and correlation with nutrition. *American Journal of Primatology* 1(4): 413–419.

Terry, R. L. 1970. Primate Grooming as a tension reduction mechanism. *Journal of Psychology: Interdisciplinary and Applied* 76(1): 129–136.

Thierry, B. 1985. Patterns of agonistic interactions in three species of macaque (*Macaca mulatta, M. fascicularis, M. tonkeana*). *Aggressive Behavior* 11(3): 223–233.

Thierry, B. 2007. Unity in diversity: Lessons from macaque societies. *Evolutionary Anthropology* 16: 224–238.

Thierry, B., M. Singh, and W. Kaumanns, Eds. 2004. *Macaque Societies: A Model for the Study of Social Organization*. Cambridge, UK: Cambridge University Press.

Thom, J. P. and C. M. Crockett. 2008. Managing environmental enhancement plans for individual research projects at a national primate research center. *Journal of the American Association for Laboratory Animal Science* 47(3): 51–57.

Thompson, N. S. 1967. Some variables affecting the behavior of irus macaques in dyadic encounters. *Animal Behavior* 15: 307–311.

Timmermans, P. J. A., E. L. Röder, and A. M. L. J. Kemps. 1988. Rearing cynomolgus monkeys (*Macaca fascicularis*) on surrogate mothers with bottle feeding. *Laboratory Animals* 22(3): 229–234.

Timmermans, P. J., W. G. Schouten, and J. C. Krijnen. 1981. Reproduction of cynomolgus monkeys (*Macaca fascicularis*) in harems. *Laoratory Animals* 15(2): 119–123.

Timmermans, P. J. A. and J. M. H. Vossen. 1996. The influence of rearing conditions on maternal behavior in cynomolgus macaques (*Macaca fascicularis*). *International Journal of Primatology* 17(2): 259–276.

Tosi, A. J. and C. S. Coke. 2007. Comparative phylogenetics offer new insights into the biogeographic history of *Macaca fascicularis* and the origin of the Mauritian macaques. *Molecular Phylogenetics and Evolution* 42(2): 498–504.

Tosi, A. J., J. C. Morales, and D. J. Melnick. 2002. Y-chromosome and mitochondrial markers in *Macaca fascicularis* indicate introgression with Indochinese *M. mulatta* and a biogeographic barrier in the Isthmus of Kra. *International Journal of Primatology* 23(1): 161–178.

Tosi, A. J., J. C. Morales, and D. J. Melnick. 2003. Paternal, maternal, and biparental molecular markers provide unique windows onto the evolutionary history of macaque monkeys. *Evolution* 57: 1419–1435.

Truelove, M. A., A. L. Martin, J. E. Perlman, J. S. Wood, and M. A. Bloomsmith. 2017. Pair housing of macaques: A review of partner selection, introduction techniques, monitoring for compatibility, and methods for long-term maintenance of pairs. *American Journal of Primatology* 79(1): 1–15.

Tsuchida, J., T. Yoshida, T. Sanaki, and Y. Yasutomi. 2008. Maternal behavior of laboratory-born, individually reared long-tailed macaques (*Macaca fascicularis*). *Journal of the American Association for Laboratory Animal Science* 47(5): 29–34.

Turner, P. V. and L. E. Grantham. 2002. Short-term effects of an environmental enrichment program for adult cynomolgus monkeys. *Contemporary Topics in Laboratory Animal Science* 41(5): 13–17.

Ungar, P. S. 1996. Feeding height and niche separation in sympatric sumatran monkeys and apes. *Folia Primatologica* 67(3): 163–168.

van Noordwijk, M. A. 1985. Sexual behavior of sumatran long-tailed macaques (*Macaca fascicularis*). *Zeitschrift für Tierpsychologie* 70(4): 277–296.

van Noordwijk, M. A. and C. P. van Schaik. 1985. Male migration and rank acquisition in wild long-tailed macaques (*Macaca fascicularis*). *Animal Behavior* 33(3): 849–861.

van Noordwijk, M. A. and C. P. van Schaik. 1987. Competition among female long-tailed macaques, *Macaca fascicularis*. *Animal Behavior* 35(2): 577–589.

van Noordwijk, M. A. and C. P. van Schaik. 1988. Male careers in sumatran long-tailed macaques (*Macaca fascicularis*). *Behavior* 107(1–2): 24–43.

van Noordwijk, M. A. and C. P. van Schaik. 1999a. The effects of dominance rank and group size on female lifetime reproductive success in wild long-tailed macaques, *Macaca fascicularis. Primates* 40(1): 105–130.

van Noordwijk, M. A. and C. P. van Schaik. 1999b. The effects of dominance rank and group size on female lifetime reproductive success in wild long-tailed macaques, *Macaca fascicularis. Primates* 40(1): 105–130.

van Schaik, C. and T. Mitrasetia. 1990. Changes in the behavior of wild long-tailed macaques (*Macaca fascicularis*) after encounters with a model python. *Folia Primatologica* 55(2): 104–108.

van Schaik, C. P., A. J. J. van Amerongen, and M. A. van Noordwijk. 1996. Riverine refuging by wild Sumatran long-tailed macaques (*Macaca fascicularis*). In Fa, J. E. and D. G. Lindburg (eds) *Evolution and Ecology of Macaque Societies*, 160–181. Cambridge, MA: Cambridge University Press.

van Schaik, C. P. and M. A. van Noordwijk. 1985. Interannual variability in fruit abundance and the reproductive seasonality in Sumatran long-tailed macaques (*Macaca fascicularis*). *Journal of Zoology* 206(4): 533–549.

van Schaik, C. P. and M. A. van Noordwijk. 1986. The hidden costs of sociality: intra-group variation in feeding strategies in Sumatran long-tailed macaques (*Macaca fascicularis*). *Behavior* 99: 296–314.

van Schaik, C. P. and M. A. van Noordwijk. 1988. Scramble and contest in feeding competition among female long-tailed macaques (*Macaca fascicularis*). *Behavior* 105(1): 77–98.

van Schaik, C. P., M. A. van Noordwijk, R. J. de Boer, and I. den Tonkelaar. 1983. The effect of group size on time budgets and social behavior in wild long-tailed macaques (*Macaca fascicularis*). *Behavioral Ecology and Sociobiology* 13(3): 173–181.

Waitt, C., H. M. Buchanan-Smith, and K. Morris. 2002. The effects of caretaker-primate relationships on primates in the laboratory. *Journal of Applied Animal Welfare Science* 5(4): 309–319.

Waitt, C. D., M. Bushmitz, and P. E. Honess. 2010. Designing environments for aged primates. *Laboratory Primate Newsletter* 49(3): 5–9.

Waitt, C., P. Honess, and M. Bushmitz. 2008. Creating housing to meet the behavioral needs of long-tailed macaques. *Laboratory Primate Newsletter* 47(4): 1–5.

Watson, L. M. 1992. Effect of an enrichment device on stereotypic and self-aggressive behaviors in singly-caged macaques: A pilot study. *Laboratory Primate Newsletter* 31(3): 8–10.

Wechsler, B. 1995. Coping and coping strategies: A behavioral view. *Applied Animal Behavior Science* 43(2): 123–134.

Welker, C., C. Schäfer-Witt, and K. Voigt. 1992. Social position and personality in *Macaca fascicularis. Folia Primatologica* 58(2): 112–117.

Wergård, E.-M., K. Westlund, M. Spångberg, H. Fredlund, and B. Forkman. 2016. Training success in group-housed long-tailed macaques (*Macaca fascicularis*) is better explained by personality than by social rank. *Applied Animal Behavior Science* 177: 52–58.

Westlund, K. 2012. Can conditioned reinforcers and variable-ratio schedules make food- and fluid control redundant? A comment on the NC3Rs Working Group's report. *Journal of Neuroscience Methods* 204(1): 202–205.

Westlund, K. 2015. Training laboratory primates: Benefits and techniques. *Primate Biology* 2(1): 119–132.

Westlund, K., A. L. Fernström, E. M. Wergard, H. Fredlund, J. Hau, and M. Spangberg. 2012. Physiological and behavioral stress responses in cynomolgus macaques (*Macaca fascicularis*) to noise associated with construction work. *Laboratory Animals* 46(1): 51–58.

Wheatley, B. P., D. K. Harvya Putra, and M. K. Gonder. 1996. A comparison of wild and food provisioned long-tailed macaques (*Macaca fascicularis*). In Fa, J. E. and D. G. Lindburg (eds) *Evolution and Ecology of Macaque Societies*, 182–206. Cambridge, MA: Cambridge University Press.

Williams, J. K., J. R. Kaplan, I. H. Suparto, J. L. Fox, and S. B. Manuck. 2003. Effects of exercise on cardiovascular outcomes in monkeys with risk factors for coronary heart disease. *Arteriosclerosis, Thrombosis, and Vascular Biology* 23(5): 864–871.

Winnicker, C. and P. Honess. 2014. Evaluating the effectiveness of environmental enrichment. *Laboratory Animal Science Professional* march: 16–20.

Wolfensohn, S. E. and P. E. Honess. 2005. *Handbook of Primate Husbandry and Welfare*. Oxford, UK: Wiley-Blackwell.

Wolfensohn, S. E., S. Sharpe, I. Hall, S. Lawrence, S. Kitchen, and M. Dennis. 2015. Refinement of welfare through development of a quantitative system for assessment of lifetime experience. *Animal Welfare* 24: 139–149.

Woolverton, W. L., N. A. Ator, P. M. Beardsley, and M. E. Carroll. 1989. Effects of environmental conditions on the psychological well-being of primates: A review of the literature. *Life Sciences* 44: 901–917.

Wrangham, R. W. 1980. An ecological model of female-bonded primate groups. *Behavior* 75(3): 262–300.

Yanuar, A., D. J. Chivers, J. Sugardjito, D. J. Martyr, and J. T. Holden. 2009. The population distribution of pig-tailed macaque (*Macaca nemestrina*) and long-tailed macaque (*Macaca fascicularis*) in west central Sumatra, Indonesia. *Asian Primates Journal* 1(2): 2–11.

Yeager, C. 1996. Feeding ecology of the long-tailed macaque (*Macaca fascicularis*) in Kalimantan Tengah, Indonesia. *International Journal of Primatology* 17(1): 51–62.

Young, R. J. 2003. *Environmental Enrichment for Captive Animals.* Oxford, UK: Wiley-Blackwell.

CHAPTER 21

Behavioral Management of *Chlorocebus* spp.

Matthew J. Jorgensen
Wake Forest University

CONTENTS

INTRODUCTION

Chlorocebus aethiops, commonly referred to as vervet monkeys or African green monkeys (AGMs), have been widely used in biomedical research. However, there are very few references available describing the behavioral management of this species in captivity. Coleman et al. (2012) do not even mention vervets in their review of behavioral management and environmental enrichment of laboratory nonhuman primates, focusing on macaques and baboons when discussing Old World monkeys. While many of the techniques used for macaques can successfully be applied to vervets, there are a few important differences that should be taken into consideration. The purpose of this chapter is to provide an overview of the limited published information available pertaining to enrichment and management of vervets, and to present species-specific recommendations to help with more effective behavioral management. I will rely heavily on my 15 years of experience working at the Vervet Research Colony (VRC), an NIH-supported breeding colony and biomedical research resource.

Vervets have contributed data in a variety of research areas over the last 50 years. Field studies have included early work by Sade and Hildrech (1965), Struhsaker (1967a), Poirier (1972), and McGuire (1974; see review by Fedigan & Fedigan 1988). Perhaps the most well-known fieldwork involved groundbreaking studies by Dorothy Cheney and Robert Seyfarth on the semantics of alarm calls (Seyfarth et al. 1980) and social behavior (Seyfarth & Cheney 1984). Recent fieldwork has included studies of social learning and cultural conformity (van de Waal et al. 2013), knowledge of third-party rank relationships (Borgeaud et al. 2015), thermal benefits of social integration (McFarland et al. 2015) and population divergence (Turner et al. 2016), among others.

Laboratory work has primarily been focused on immunology (e.g., Fomsgaard et al. 1990; Zahn et al. 2008; Chahroudi et al. 2012; Pandrea et al. 2012; Briggs et al. 2014; Kim et al. 2015), but has also ranged across such varied topic areas as Alzheimer's disease (Lemere et al. 2004; Kalinin et al. 2013), brain imaging (Melega et al. 2000; Fears et al. 2009; Woods et al. 2011), pharmacology/cognition (Jentsch et al. 1997; James et al. 2007; Melega et al. 2008; Groman et al. 2013), obesity and diabetes (Kavanagh et al. 2007a,b; Cann et al. 2010), lipid biology (Parks & Rudel 1979; Rudel et al. 2002), reproduction (Kavanagh et al. 2011; Atkins et al. 2014), and behavior/temperament (Fairbanks & McGuire 1986; Raleigh et al. 1991; Fairbanks 2001; Laudenslager et al. 2011). Jasinska et al. (2013) provided an excellent overview of both the variety of biomedical research that this species has participated in and a summary of recent genetic and genomic breakthroughs (see also Freimer et al. 2007; Jasinska et al. 2012; Huang et al. 2015; Warren et al. 2015).

According to a survey by Carlsson et al. (2004), vervets are one of the most commonly employed nonhuman primate species in biomedical research. A recent analysis of PubMed citations indicated that vervets were second only to rhesus macaques in the number of citations (Jasinska et al., 2013). While it should be noted that a proportion of PubMed citations involved *in vitro* studies of commercially available cell cultures, there are still a significant number that involve live animal studies. The relatively recent growth in the participation of vervets in research projects may be due to the fact that they are safer (not carriers of Herpes B virus; Baulu et al. 2002) and less expensive than macaques (Freimer et al. 2008; Smith 2012).

NATURAL HISTORY

Taxonomy and Common Names

Jasinska et al. (2013) highlighted the often-confusing taxonomy and naming conventions used with this species (or multiple species, depending on your perspective, see the following paragraphs). She pointed out that primatologists have typically used "vervet" to refer to a whole genus (*Chlorocebus*), while immunologists and virologists have only used the term "vervet" when referring to the *pygerythrus* subspecies, preferring to use the term "African green monkey" when referring to the entire genus or to just the commonly used *sabaeus* subspecies. All of this is further confused by the fact that the genus had been formerly known as *Cercopithecus* and that all subspecies had been previously described as the single species *Cercopithecus aethiops* (Smith 2012). For this chapter, I will use the term "vervet" as the common name of the entire genus and will describe each variant as a subspecies of *Chlorocebus aethiops*. Whenever possible, I will highlight when research has focused exclusively on one or more of those subspecies (see Table 21.1).

Table 21.1   Common, Subspecies, and Species Names of *Chlorocebus*

Common Name	Subspecies Name	Alternative Species Name	Comment
Vervet monkey	*C. a. pygerythrus*	*C. pygerythrus*	Most commonly described in field studies
African green monkey (Callithrix)	*C. a. sabaeus*	*C. sabaeus*	Most commonly used in biomedical research in the U.S.
Grivet monkey	*C. a. aethiops*	*C. aethiops*	
Tantalus monkey	*C. a. tantalus*	*C. tantalus*	
Malbrouck monkey	*C. a. cynosuros*	*C. cynosuros*	

Common and alternative species names from Groves, C. P. and J. Kingdon, *The Mammals of Africa: Vol II Primates*, Bloomsbury Publishing, London, 2013, sub species names from Warren et al., *Genome Res.*, 25(12):1921–1933, 2015.

Groves and Kingdon (2013) present the alternative taxonomic viewpoint in which they refer to each variant as separate species: *Chlorocebus pygerythrus* (vervet monkey), *C. sabaeus* (green monkey, Callithrix), *C. aethiops* (grivet monkey), *C. tantalus* (Tantalus monkey), and *C. cynosuros* (Malbrouck monkey). They highlight that there is still significant controversy over the taxonomy of this genus, which would greatly benefit from extensive genetic analysis. For earlier debates regarding taxonomy, see Disotell and Raaum (2002) and Tosi et al. (2004).

Warren et al. (2015) recently published the genome of the vervet, including an analysis of the sequenced genomes of each of the main vervet subspecies (*Chlorocebus aethiops pygerythrus, C. a. sabaeus, C. a. aethiops, C. a. tantalus, and C. a. cynosuros*). Their results support the notion of a single species with multiple subspecies, and they provide a useful map of subspecies ranges across Africa and a diagram of the vervet phylogenetic tree. They conclude that *C. a. cynosuros* and *C. a. pygerythrus* are the most closely related subspecies (diverging 129 kya), with *C. a. tantalus* diverging before that (265 kya), preceded by *C. a. aethiops* (446 kya) and *C. a. sabaeus* (531 kya) (see Table 21.1).

In addition to the population in Africa, there is also a subpopulation of the *sabaeus* subspecies on the islands of St. Kitts, Nevis, and Barbados in the Caribbean (Sade & Hildrech 1965; Poirier 1972; McGuire 1974; Horrocks 1986; Denham 1987). These animals likely came to the Caribbean from West Africa during the 17th century, and have readily propagated because of the lack of predators and pathogens (Jasinska et al. 2013). Some have even advocated that the Caribbean population should be considered a separate subspecies (Palmour et al. 1997). Most of the animals used in biomedical research in the United States are derived from this Caribbean population (Smith 2012; Jasinska et al. 2013).

Social Organization

Vervets occupy a wide range of habitats, including savannahs, woodlands, and riverine forests (Pruetz 2009; Groves & Kingdon 2013). Like most Old World monkeys, vervets have a matrilineal social organization in which the females remain in their natal groups for their whole lives, while males emigrate at sexual maturity and move into neighboring groups (Cheney & Seyfarth 1983). Social groups in the wild are typically multimale and multifemale and range in size from roughly 10–40 individuals (Struhsaker 1967b; Poirier 1972; Melnick & Pearl 1987; Fedigan 1992). West African *C. a. sabaeus* group sizes observed by Dunbar (1974) were generally on the lower end of this range. Vervets are territorial, with distinct home ranges that are defended from other troops (Cheney 1981). Territorial behavior has been reported in *C. a. pygerythrus* populations (Cheney & Seyfarth 1987), as well as in *C. a. sabaeus* in Africa (Dunbar 1974) and St. Kitts (Poirier 1972).

In the wild, *C. a. pygerythrus* females reach sexual maturity between 4 and 5 years of age, while males reach full body size around 6 years of age (Cheney & Seyfarth 1990). Males are generally larger than females (4–5 kg compared to 3–4 kg), though this sexual dimorphism does not begin to emerge until 15–18 months of age (Turner et al. 1997). In general, vervets in captivity are larger than their wild counterparts (females 5.33 kg, males 7.22 kg; Kavanagh et al. 2007a). Analysis of growth patterns (Turner et al. 1997), as well as age-specific tooth eruptions and morphometric patterns (Bolter & Zihlman 2003; Bolter 2011), are also available. Patterns of growth in the VRC are presented in Figure 21.1. Vervets typically live to 11–13 years of age in the wild, but can live more than 25 years in captivity (Magden et al. 2015). The oldest animal in the VRC was 29.1 years old (see Table 21.2).

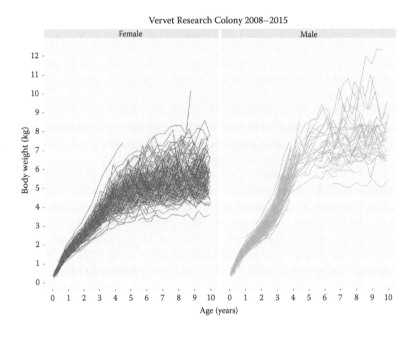

Figure 21.1   Individual growth curves for males and females between 0 and 10 years of age in the VRC. Data collected from 2008 to 2015.

Table 21.2   Vervet Life History Variables

Variable	Values
Adult female weight (kg)	2.57–3.44 (wild)[a] 5.33 (captive)[b]
Adult male weight (kg)	4.13–4.43 (wild)[a] 7.22 (captive)[b]
Life span (years)	11–13 (wild)[c] 25 (captive)[c]
Gestation length (days)	163–165[c]
Length of estrous cycle (days)	30–32[c]
Sexual maturity (years)	3–4 (females)[c] 5–6 (males)[c]

[a] From Turner, T. R., et al., *Am. J. Phys. Anthropol.*, 103:19–35, 1997.
[b] From Kavanagh, K., et al. *Obesity (Silver Spring)* 15(7):1666–1674, 2007a.
[c] From Magden, E. R., et al. *Laboratory Animal Medicine*, 3rd edition, Academic Press, San Diego, CA, 2015.

Vervets are mostly seasonal breeders in the wild, with the majority of infants born during the peak of food availability (Poirier 1972), though seasonality has been inconsistent in captivity (Hess et al. 1979; Else 1985; Seier 1986). Unlike macaques and baboons, female vervets do not develop swelling of the perineal skin and typically do not show overt signs of menstruation (Rowell 1970; Tarara et al. 1984; Eley et al. 1989; Seier et al. 1991; Carroll et al. 2007). In captivity, pregnancy is detected via palpation and/or ultrasound (Seier et al. 2000; Kavanagh et al. 2011; Redmond & Evans 2012). Females typically give birth at night. Vervet infants show a lack of facial pigmentation when young; this makes them very attractive to other members of the group (Struhsaker 1967a). Female vervets frequently engage in "aunting" or allomothering behavior, often by juveniles and adolescents, while males are generally indifferent to infants (Bloomstrand & Maple 1987; Fairbanks 1990; Fedigan 1992; though see Hector et al. 1989). Reproductive parameters including gestation periods, abortion/stillbirth rates, inter-birth intervals, and weaning have been reported for numerous populations, both captive (e.g., Rowell 1970; Bramblett et al. 1975; Hess et al. 1979; Kushner et al. 1982; Fairbanks 1984; Fairbanks & McGuire 1986; Seier 1986, 2005; Eley et al. 1989; Kavanagh et al. 2011) and in the wild (Gartlan 1969; Cheney & Seyfarth 1987; Turner et al. 1987, see Table 21.2 for a summary). Infant mortality is higher in younger females (Fairbanks & McGuire 1984) and in females with poorer metabolic health (Kavanagh et al. 2011). High-ranking females have higher fecundity and shorter inter-birth intervals than low-ranking females (Fairbanks & McGuire 1984).

Dominance hierarchies in vervets have been described as not as "clear-cut" or important/rigid as they are in macaques and baboons (Rowell 1971; Bloomstrand & Maple 1987; Kaplan 1987; Fedigan 1992). Just as dominance styles can differ across macaque species (e.g., de Waal & Luttrell 1989), the dominance style of vervets differs from that of other Old World monkeys. Dominance displays can also be more subtle (Fairbanks & McGuire 1986). As Fairbanks (1980) noted, while determination of a group's dominant female is straightforward, it can often be difficult to clearly assign the rank of the non-alpha females. Rank assignments in the VRC are typically categorized as high, medium, or low (e.g., Fairbanks 1984), or alpha and non-alpha (Fairbanks et al. 2004). Male and female dominance hierarchies also appear somewhat independent of each other, and daughters usually assume the rank below their mothers. In captive social groups, dominance relationships can be quite stable over time (Bramblett et al. 1982). One important physical difference between vervets and macaques is that adult female vervets have relatively large canines, unlike adult female macaques (Bloomstrand & Maple 1987; Fedigan 1992). Rowell (1971) speculated that this characteristic may explain some of the differences in social interactions seen in vervets compared to macaques or baboons. For example, even high-ranking adult male vervets may be chased and threatened by the lowest ranking female, something that would be very rare in a macaque group.

Behavior Patterns

While many of the behavior patterns of vervets match those of other Old World monkeys, there are a few, sometimes subtle, differences that should be noted. An overview of different behavioral patterns that are comparable to other monkeys, differing in frequency or pattern to other monkeys, or are distinct to vervets is presented. Some behavioral patterns also appear to be distinct to specific populations and/or subspecies of vervets (see Table 21.3 for a summary).

Struhsaker (1967a) described the behavior patterns of *C. a. pygerythrus* in Kenya and categorized the communicative gestures in vervets that were common to, or distinct from, those of other Cercopithecines, including rhesus macaques, baboons, and patas monkeys. He concluded that the "eyelid display" or threat gesture was common to all Cercopithecines as was head bobbing/jerking. Branch shaking was another common display during intergroup encounters in most Cercopithecines. He concluded that yawning in response to stress was rare in vervets as was homosexual mounting among males. Struhsaker then described behavior patterns seen in vervets that were rarely seen in other Cercopithecines. These included a penile display and mouthing of the lateral surface of the

Table 21.3   Partial Ethogram of Vervet/AGM Behavioral Categories Unique to This Species

Behavior	Description
Handle/ manipulate	"Touching another individual with the hands or embracing with arms including placing hands on head or shoulders, embrace, and straddle, excluding touching the genitals." Stylized behavior involving placing the hand on head or shoulders of another, embracing.
Display	"Assertion displays including partial and complete forms of the red-white-and-blue display, the stationary sideways present, and penile erections. Red-white-and-blue display behaviors consist of tail erect, encircling, rearing on hind limbs, and orienting hind quarters toward." Ritualized behavior including standing broadside to another animal with tail up, sideways prancing, circling another animal with tail up, showing the hindquarters, or showing an erection to another.
Muzzle	"Bringing muzzle to any part of another animal's body except the genitals, usually muzzle to muzzle."
Chest exposure	"…[adult] male exposes his chest by sitting upright with his arms extended at the side. The hind limbs are laterally rotated, exposing the light pigmented medial surfaces of the thighs, plus, in many cases, an erected pigmented penis. The animal leans back, lifting his head slightly. Thus the chest and thigh surfaces receive maximal exposure." Poirier (1972). Note: Presumed to be a means of intergroup avoidance. Observed in St. Kitts population, not seen in African vervet studies.

Adapted from Struhsaker, T. T., *Science*, 156(3779):1197–1203, 1967a; Poirier, F. E. *Folia Primatol. (Basel)*, 17:20–55, 1972; Fairbanks, L. A., et al., *Behav. Processes*, 3(4):335–352, 1978; Fairbanks, L. A. and M. T. McGuire, *Anim. Behav.*, 34:1710–1721, 1986.

neck, patterns of behavior performed by dominant adult males. Struhsaker also described a "red, white, and blue" display in which a "dominant male holds his tail erect and paces back and forth in front of a seated monkey, displaying his red perianus, his blue scrotum, and the white medial strip of fur extending between the perianus and the scrotum" (p. 1200). He does note that this "red, white, and blue" display was not observed in a separate population of vervets on Lolui Island in Uganda observed by J.S. Gartlan. Overall, Struhsaker concluded that 54%–63% of the communicative gestures of vervets were comparable to those of rhesus, baboons, and patas.

Poirier (1972) observed the behavior of *C. a. sabaeus* populations on St. Kitts over the course of three summers. He concluded that troop size, territoriality, and social behavior were mostly comparable to that reported for African vervets. Communication patterns were the main area of difference noted, with the St. Kitts population being generally quieter than their African counterparts. Poirier noted that the "red, white, and blue" display described by Struhsaker (1967a) was not evident in the St. Kitts *sabaeus* population, though this may have been due to seasonal differences in observation periods or the fact that the Caribbean population does not have the same scrotal coloration (Cramer et al. 2013). In contrast, Poirier described a "white chest exposure" display in which an adult male sits upright exposing his chest and thighs, often with an erect penis. Poirier suggested that this may serve as a means of intertroop avoidance, though he also noted that this pattern was not reported in most earlier studies of African vervets. Dunbar (1974) noted this same characteristic chest-displaying posture in a study of *C. a. sabaeus* in Senegal, though doubted that it was used for intertroop communication given the dense vegetation in Senegal.

In terms of behavioral patterns of captive populations of vervets, there are a number of papers that provide ethograms and/or detailed descriptions of behavioral categories. These include Rowell (1971), Fairbanks et al. (1978), Raleigh et al. (1979), Bramblett (1980), Fairbanks and McGuire (1986), Hector et al. (1989), and Erhart et al. (2005; for hybrids). There are a few behavioral patterns noted in these captive studies that are not mentioned in the field studies.

Rowell (1971) observed captive groups of Sykes monkeys and *C. a. pygerythrus* in Uganda (n = 6). She noted that these vervets failed to exhibit the "red, white, and blue" display described by Struhsaker (1967a). She described a "square face" threat in which the animal's jaw was "jutted and the corners of the closed mouth [were] pulled forward and slightly outward" (p. 629). Rowell also noted that while animals could be assigned ranks, her subjective impression was that the dominance hierarchy was of less importance than in other Old World monkeys. Rowell also noted that aggressive behavior was more frequent than threats, noting the opposite would be true in baboons or macaques. Finally, she noted that the adult male would be chased by all females, including the lowest ranking female. She suggested that this may have been due to the fact that the adult females had canines proportionally as large as those of the males. Erhart et al. (2005) echoed this notion, stating that "adult males are part of a group's dominance hierarchy, but their rank depends on their ability to form alliances and to intimidate other individuals, and they are not necessarily the most dominant animals in the group" (pp. 196–197).

Another example of a behavioral difference between vervets and macaques pertains to the somewhat confusing "hugging" behavior seen in vervets. Fairbanks et al. (1978) observed this behavior in captive *C. a. sabaeus* imported from St. Kitts and initially called this behavior "handling." They described it as "touching another individual with the hands or embracing with arms including placing hands on head or shoulders, embrace, and straddle, excluding touching the genitals" (p. 339). Fairbanks and McGuire (1986), in the same captive population, later called this behavior "manipulate" and described it as "hands-on-head, hands-on shoulders, and embrace-from-in-front" (p. 1714), citing similar patterns described by Struhsaker (1967a) and Rowell (1971). This is a relatively stylized set of behaviors that may be followed by grooming or sometimes lead to aggression. "Manipulation" is often interpreted as an assertion of dominance in situations where the recipient is then invited to stay and groom. It is commonly used by females in situations of female rank change or uncertainty, as with adolescent females or following the death of an alpha female (Fairbanks, L. A. personal communication). This behavior may often be misunderstood as an affiliative gesture. In addition to manipulation, aggressive behaviors in the VRC include "head-jerking, grabbing, slapping, chase, and attack" (Fairbanks & McGuire 1986, p. 1713).

Grooming is an important affiliative behavior in Old World monkeys, and vervets are no exception. Seyfarth (1980), Seyfarth and Cheney (1984), and Fairbanks (1980) all found that female vervets were more likely to direct grooming to dominant females rather than subordinate females, though Henzi et al. (2013) failed to find these patterns in larger groups in South Africa. Lee (1984) found that in contrast to other types of social interaction, grooming was relatively unaffected by seasonal variations in weather and food availability.

GENERAL BEHAVIORAL MANAGEMENT STRATEGIES AND GOALS

The goal of most nonhuman primate behavioral management programs is to promote psychological well-being through a combination of socialization strategies, environmental enrichment, facilities/enclosure design, and training. This should increase the expression of species-typical behavior and decrease the occurrence of abnormal behavior (Coleman et al. 2012). One historic difficulty with the use of vervets is that they have been relatively underrepresented in the national primate research centers, where most of the expertise in nonhuman primate behavior and management was developed. Therefore, smaller labs, often without dedicated behavioral management teams, are frequently the primary users of these animals. Complicating the situation is the fact that despite their widespread use in biomedical research, compared to macaques, there is relatively little published information available on the proper behavioral management of vervets (one of the goals of this chapter is to help rectify this situation). The one notable exception is a book chapter by Bloomstrand and Maple (1987) describing the management and husbandry of African monkeys.

The information that I draw from comes predominantly from three sources: (1) work by Jürgen Seier's lab in South Africa involving wild-caught and captive bred *C. a. pygerythrus* (Seier 1986, 2005), (2) work by numerous groups focused primarily on vervet breeding, using harem groups or male–female pairs, and (3) work at the VRC. The VRC is a captive breeding colony of Caribbean-origin *C. a. sabaeus* originally housed in southern California in the mid-1970s (UCLA/Sepulveda VA) that is now housed at the Wake Forest School of Medicine (WFSM; Fairbanks et al. 1978;

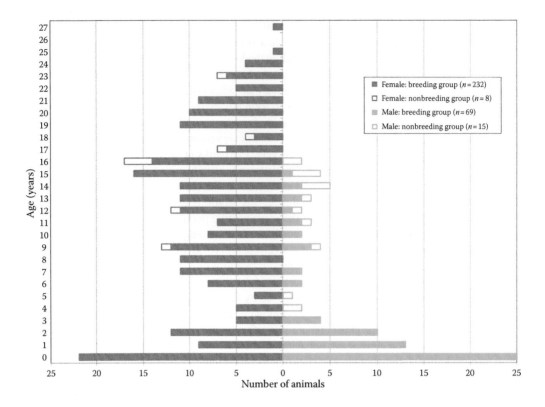

Figure 21.2 Demographic chart of VRC. Solid bars represent animals housed in the 16 matrilineal social groups. Open bars represent animals housed in small all-male groups, pairs, or individually.

Jasinska et al. 2013, see Figures 21.2 through 21.5). Throughout this chapter, I will refer extensively to these captive populations and work done describing the behavioral profiles and management practices in these *pygerythrus* and *sabaeus* populations. Given my personal experience with the VRC, and since most vervets currently used in the United States are Caribbean-origin *sabaeus*, I will tend to be biased toward the current practices at the VRC. That is not intended to ignore current captive *sabaeus* populations. As Smith (2012) noted, U.S. colonies of vervets have also been maintained at the Tulane National Primate Research Center (TNPRC) and the New Iberia Research Center (NIRC). In addition, Caribbean colonies have been maintained at the St. Kitts Biomedical Research Foundation (McGill), the Caribbean Primate Research Laboratory (Yale), and the Barbados Primate Research Center.

History and Description of the Vervet Research Colony

The VRC originated with 57 founders captured from St. Kitts and Nevis, West Indies, between the mid-1970s and the mid-1980s (Jasinska et al. 2013). In 2008, the majority of the VRC was transferred to the Wake Forest Primate Center (WFPC) at the WFSM. The VRC has been supported by NIH as an Animal and Biological Materials Resource (ABMR) since 2005 (current grant OD010965, PI: Matthew J. Jorgensen).

In the early years of the VRC, the colony consisted of groups formed from unfamiliar, wild-caught individuals (Fairbanks et al. 1978; McGuire et al. 1978). Colony management practices, maintained for the past 35 years, have been designed to reflect the typical social structure of this species seen in the wild. All animals are mother-reared, with infants and juveniles remaining in natal groups with their mothers and female kin, unless culled for experimental purposes. Males are removed at

Figure 21.3   Photos of the VRC group housing. Note the use of recycled plastic barrels, pallets, and bins that provide perches, climbing structures, and hiding places.

adolescence and temporarily held in small, all-male social groups (Fairbanks et al. 2004). These males are either transferred to other research projects or retained as future breeders. Adult males are rotated between groups at 3- to 5-year intervals or are vasectomized to control inbreeding and maintain an optimal population size. These colony management practices were developed to promote species-typical social and behavioral development, and to allow animals to express age- and sex-appropriate behavior. The strategy has also resulted in the formation of a large, complex, and inter-related pedigree (Jasinska et al. 2013). Each social group contains one to three matrilines consisting of closely related adult females and their immature male and female offspring, typically including three to four generations of females per matriline. At the time of this writing, the colony included 301 animals in the 16 social groups (see Figure 21.2). Each social group contained 10–30 individuals consisting of 1–2 adult males, 7–16 adult females, and 0–15 immature males and females. Animals are periodically removed from breeding groups for study or for clinical purposes. This necessitates the creation of pairs and/or small social groups. Currently, there are two all-male social groups (7–8 males per group) and a small cohort of females housed in indoor racks in pairs (n = 2) or individually (n = 6). Five of the females are diabetic, requiring insulin therapy twice daily (Cann et al. 2010).

Socialization

Published reports on housing techniques for vervets tend to fall into a few different categories: (1) group housing, (2) pair housing, and (3) single housing with short-term male–female pairings. Much of the published work describing vervet housing and socialization is actually focused on breeding, rather than on social housing for biomedical, behavioral, or welfare research.

Figure 21.4 Photos of concrete culverts (A) and recycled plastic barrels (B) that serve as hiding places. Photos of animal foraging for corn and seeds in river rock substrate (C) and the use of browse as enrichment items (D).

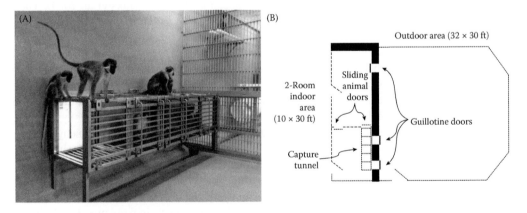

Figure 21.5 Photo of capture tunnel (A) and schematic of the housing area for a single social group, including indoor and outdoor space (B).

Group Housing

Bloomstrand and Maple (1987) provided a review of vervet group formations and a survey of zoo housing in the United States. They note that early attempts at creating captive vervet groups were either "unsuccessful" or "impossible" (Mallinson 1971). They cite captive group formations by Rowell (1971) and Else (1985) as well as a large group of 57 animals at the Johannesburg Zoo (Gartlan & Brain 1968). They suggest that successful large group formation by Gartlan and Brain (1968) may have been due to the size of the housing area, which allowed animals to avoid close contact with one another. They note that while the species is found in multi-male, multi-female groups in the wild and thrives best in complex social groupings, many zoos tend to manage vervets in small families or monogamous pairs. This tendency was reiterated by Fortman et al. (2002), who recommended that vervets be housed in harems, suggesting that the males will not tolerate the existence of other adult males, despite the fact that multi-male groups exist in the wild.

Else (1985) described the formation of 10 single-male harem groups of wild-caught vervets at the Institute of Primate Research in Kenya (described as *Cercopithecus aethiops*, but presumably *C. a. pygerythrus*). There were 200 animals initially quarantined in single cages for 3 months. After quarantine, 75 animals were randomly assigned to single-male harem groups, with 5–10 females per pen. Enclosures measured $3\,m \times 6\,m \times 2.5\,m$, with a solid partition dividing the pen in half. No details on the introduction methodology were provided. Establishment of stable groups was difficult, as considerable female fighting and/or injury required movement of females into new harem groups. They experienced 30%–75% mortality following anesthesia/ procedures, difficulties with animals adapting to commercial monkey diet, and 3 of the 10 harems failed. They concluded that "establishing breeding groups with adult vervets is difficult. Females constantly fight and it takes a minimum of 1 year for the group to stabilize" (p. 375). In hindsight, they noted that formation of captive groups from animals from the same troops may have been a better strategy.

Despite the difficulties encountered during the initial group formation, Else (1985) noted that after groups had settled down, the vervets did breed successfully. They removed male infants at 6 months of age and allowed female infants to remain in their natal groups. The "exchange of males between groups has worked well and introduction of a new male into an established female group is relatively easy" (p. 375). Morland et al. (1992), in a follow-up study at the same facility, described the changes in aggressive behavior after the introduction of new males to harem groups. Females formed coalitions against the new males and were especially aggressive when young infants were in the group. Group size, enclosure size, female density, and prior male social experience did not influence aggression patterns. They recommended avoiding the introduction of new males into groups with young infants and groups containing a single matriline.

Kushner et al. (1982) described a colony of *C. a. aethiops* and *C. a. pygerythrus* imported from Somalia and Ethiopia in 1976 to a rural facility near Philadelphia, Pennsylvania. Animals were housed in harems (1 adult male with 2–6 adult females) in 50 indoor runs that each consisted of a $4\,ft \times 5\,ft \times 7\,ft$ concrete block area and an adjacent $18\,ft \times 5\,ft \times 7\,ft$ chain-link area. The colony contained 151 feral-born adults and 161 colony-born animals. Pregnant females were removed from their groups just prior to birth, and mothers and infants were returned to their group after the infant was 1–2 weeks old. Infants were weaned at 6 months of age and housed in groups segregated by sex, age, and subspecies. Reproductive statistics were reported, but few other details were provided regarding group formation, other than noting that both subspecies adapted quickly to group housing.

Redmond and Evans (2012) described the housing arrangements at the St. Kitts Biomedical Research Foundation. Animals (n = 313 adult female *C. a. sabaeus*) were studied, 27% captive born and 73% trapped on St. Kitts. Social groups were housed in outdoor chain-link cages (no dimensions provided) and ranged from 8 to 20 females, with additional immature animals. Each group contained a single adult male who was normally kept physically separated from the group in a smaller partition of the cage. The males were released into the groups for 3- to 4- or 15- to 30-day periods for breeding. Infants were typically removed from the groups at 1 year of age. Females were periodically removed for 10-day periods for hysterectomy or sample collections and then returned. No details were provided on socialization methods.

Rowell (1970) described the formation of a group of vervets captured in Uganda (presumably *C. a. pygerythrus*). The group initially consisted of one adult male, one juvenile male, and five adult females. The group was housed in a 27 ft × 10 ft × 9 ft enclosure, separated into two sections, similar to Else (1985). The animals were unknown to each other when group formation occurred, though no other details are provided on how this was done. This group was eventually transferred to The University of Texas and work was continued by Claude Bramblett and colleagues. The group was maintained from 1968 to 1984 (Bramblett 1980; Ehardt-Seward & Bramblett 1980; Bramblett et al. 1982) and eventually was merged with a companion group of Sykes monkeys and allowed to hybridize (Erhart et al. 2005). While not entirely clear from published reports, it appears that the group was maintained as a harem for most of its existence, despite the birth of numerous male infants (Bramblett et al. 1975).

There is one paper describing the effect of decreasing cage space on affiliative and aggressive behavior in vervets. McGuire et al. (1978) described a wild-caught multi-male, multi-female group of 14 *C. a. sabaeus* that was initially observed in a large enclosure on St. Kitts (565 m²) and then observed again in a smaller enclosure (140 m²) 1.5 months after transfer to the VRC in California. The group had been formed 6 months prior to the start of the study on St. Kitts. Affiliative behavior increased in the smaller enclosure, while aggressive behavior showed no change.

As described above, the VRC is one of the few captive populations to house animals in multi-male groups. Historically, one to five adult males have been housed per breeding group in the VRC, with the modal number being two per group. After the initial formation of the VRC in the mid-1970s, in which new social groups were formed from animals imported from St. Kitts, the creation of new social groups of unfamiliar adult females was generally avoided whenever possible. Instead, most animal transfers involved the removal of adolescent males from their natal groups and the replacement of resident adult males with new, unfamiliar adult males. Either new all-male social groups were formed when these transfers occurred or males were introduced into existing all-male groups, if suitable housing was not available. Thus, the typical animal transfers within the VRC tended to mimic the types of animal movements seen in the wild (Fairbanks & McGuire 1986, 1987; Fairbanks et al. 2004).

In the early years of the VRC, males were introduced into new social groups in a gradual fashion, in a manner similar to macaque pairings (e.g., DiVincenti & Wyatt 2011). The incoming males were placed in single cages located at the edge of the new social group so that the resident females could acclimate to the new males. Unfortunately, this practice seemed to increase the anxiety of both the new males and the resident females. When the males were eventually introduced into the new social group, sometimes days later, aggressive interactions often occurred immediately after the males were released into the pens (Fairbanks, L. A. personal communication).

Fairbanks (2001) described the "Intruder Challenge Test," which was a behavioral assessment method that was developed from the early observations of the ways in which animals responded to the presence of new animals on the periphery of their home group. Some animals immediately investigated the "intruder" and engaged in threats and other aggressive behaviors. Other animals were more cautious and waited for a while before interacting with the "intruder." Still other animals kept their distance from the "intruder" and failed to interact much at all. Fairbanks (2001) showed that observations of resident animals in response to a same-sex "intruder" can yield reliable trait-like measurements of impulsivity, anxiety, and aggressiveness. The discovery of these behavioral patterns led us away from the use of a gradual introduction method when moving animals between groups.

Introduction of new adult males and the death/removal of dominant females are the two events that have the largest impact on the stability of social groups in the VRC (Fairbanks & McGuire 1986; Fairbanks et al. 2004). There is typically a concentration of aggression by females toward new males, along with a reduction in female–female aggression (Fairbanks & McGuire 1986). Fairbanks and McGuire (1987) showed that the introduction of new adult males increased maternal protectiveness of very young infants and increased maternal rejection of older infants. As also emphasized by

Morland et al. (1992), introducing new males when young infants are in the group should be avoided as females are more likely to be protective of their dependent offspring and more aggressive to new males. Raleigh and McGuire (1989) showed that the dominant female in the group can play a critical role in determining which of the new males ultimately becomes dominant. Fairbanks et al. (2004) determined that body weight, adolescent impulsivity, and measures of cerebrospinal fluid (CSF) metabolites of newly introduced males were all predictive of dominance attainment. We recently measured the sleep patterns of groups that had new male introductions and compared them with groups without male introductions. Sleep was measured via 24-h actigraphy for 2 weeks before and 2 weeks after the new male introductions. The new males showed a significant drop in nightly sleep duration for those 2 weeks, while the females in both groups, and the resident males, did not show any disruption (M. J. Jorgensen, unpublished data).

Guy and Curnoe (2011) observed a captive group of *C. a. pygerythrus* at a wildlife rehabilitation center, noting that the death of the dominant female led to prolonged aggression and instability within the group. This same pattern has been noted on numerous occasions in the VRC (Fairbanks & McGuire 1986). When adult females need to be removed from breeding groups it is important to avoid removing the dominant females whenever possible. This is not unique to vervets, as it has also been reported for other Old World monkeys (e.g., Oates-O'Brien et al. 2010).

Pair Housing

There is an established literature available on how to pair house macaques (DiVincenti & Wyatt 2011); however, there is truly a dearth of information available for isosexual pair housing of vervets. While behavioral management techniques developed in macaques can often safely be applied to vervets, I believe caution is warranted when it comes to pair housing, and would strongly recommend avoiding a "macaque-centric" approach.

Jorgensen et al. (2017) described one of the few studies of isosexual pair housing success and techniques in vervets. We described pair housing attempts in four cohorts of vervets from three different facilities: large cohorts of males and females at the VRC and the NIRC, a small cohort of males at the TNPRC, and a small cohort of imported males at WFSM. We measured the percentage of pairs that remained together 14 days after initial full contact. For the two largest cohorts (VRC and NIRC), success rates were high (96%–100%) for both males and females, while the success rates for the smaller TNPRC and WFSM male cohorts were lower (28%–50%). We found that the mean age of pair-mates, and the weight difference between pair-mates, were significant predictors of success for males, but not for females. Somewhat surprisingly, in the one cohort with available data, there was no evidence that prior familiarity of pair-mates improved success rates. We discussed differences between macaques and vervets and concluded that the gradual introduction techniques often advocated for macaque pairing may not be helpful or appropriate for vervets. We speculated that if vervets and macaques form social bonds in fundamentally different ways, then a period of protected contact may not benefit vervets. We theorized that the establishment of dominant–subordinate relationships in vervets may require physical interactions, and thus, a period of visual familiarization may simply increase anxiety and frustration and could actually constrain normal social interchange in vervets. We fully recognize that this idea is still quite speculative and that additional research is needed to more fully examine this concept.

Single Housing with Short-Term Socialization

Many of the published studies involving vervets describe animals as being housed individually. In fact, Baker et al.'s (2007) survey of housing conditions at all of the National Primate Research Centers and 11 other facilities in the United States found that all of the vervets surveyed were individually housed (note: this survey did not include the VRC). It is not clear whether these housing conditions were necessary for scientific reasons or if they merely reflected standard management practices at the time.

Seier (1986, 2005) and Seier and de Lange (1996) described their method for short-term pair housing of *C. a. pygerythrus* males and females for the primary purpose of breeding. Initially, they compared the effectiveness of breeding in pairs to breeding in harems. They ultimately concluded that harem breeding "failed due to incompatibility and rank order problems" (Seier 1986, p. 342), which included prolonged fighting and injury. They noted, however, that some of the problems they encountered in harems may have been exacerbated by the relatively small amount of space available in their facility for group housing (only $2.9\,m \times 2.0\,m \times 2.3\,m$ cages).

The housing arrangement that worked best for Seier and colleagues involved introducing females to individually housed males and then removing the females after they became pregnant. Females were then housed individually for the duration of pregnancy, parturition, and lactation. After weaning the offspring at 5–6 months of age, the mothers were remated. Pairs of weaned juveniles were housed together for 2–3 weeks and then moved to communal cages containing ~six animals of both sexes. When possible, a pregnant or lactating female was added to juvenile peer groups for socialization. Seier (1986) noted that "mother-peer sequential raising system is not optimal, but social and sexual adjustments seem to be sufficient in our colony... Our rearing environment is a compromise between practicality and optimal raising conditions" (p. 345, 347).

Cho et al. (2002) described timed mating of 36 female and 9 male Kenyan-origin, captive vervets (presumably *C. a. pygerythrus*) at the Tsukuba Primate Center in Japan. Like Seier (1986), animals were individually housed indoors with females temporarily housed together with males for 72 h for the purpose of timed mating. Under another protocol, females were housed with the same male every other day for up to 16 weeks. In this condition, animals were housed in a three-compartment cage with two "partition plates" that were moved to allow introductions. It is not clear if the animals had visual contact during periods in which the plates were in place. No other details were provided regarding housing/socialization.

Johnson et al. (1973) described an indoor colony of imported vervets (origin and subspecies not clearly identified) housed at a company in Maryland. The 29 females and 8 males were housed in "duplex cages," presumably individually housed, and females were housed with specific males for 6-day periods for the purpose of breeding. Males had canines surgically removed. Few other housing details are provided, since the focus of the paper was on reproductive outcomes. The authors do note that the animals were "excitable" after acquisition, and that they required 6–12 months of habituation before female cycling became regular. They recommended that noise and disturbances should be minimized. Hess et al. (1979) described a similar system of short-term male–female pairing for the purposes of timed mating.

Environmental Enrichment

Seier et al. (2011) published one of the few papers testing the effects of different housing and enrichment conditions on stereotypical behavior in *Chlorocebus aethiops*. They varied the cage size, the cage location (upper/lower), along with the presence/absence of a foraging log, an attached exercise cage, and a heterosexual mate/partner. Recorded stereotypies included behaviors frequent enough to be quantified, and consisted of somersaulting, head tossing, and pacing. Not surprisingly, they found that animals displayed the greatest amount of stereotypical behavior in the small, unenriched cage. They also found that females displayed more stereotypical behavior than males and that no self-injurious behavior was observed. Allowing animals access to an exercise cage attached to the home cage resulted in the lowest levels of stereotypical behavior. They further noted that, contrary to expectations, cage size alone was not responsible for the lowest amount of stereotypical behavior, nor did the addition of a social partner reduce stereotypical behavior. Animals housed in the lower cage location showed significantly greater amounts of stereotypical behavior, in contrast to the results of Schapiro et al. (2000) in cynomolgus macaques. The authors concluded that cage complexity and control may be more important factors. They also concluded that the foraging log, a blue gum tree branch drilled with holes and filled with a nut/seed paste, was an effective enrichment device for this species. Seier et al. (2005) also provide a cautionary tale about enrichment in which they described a juvenile male vervet that was housed in a communal cage with straw foraging substrate. The animal consumed the straw, obstructing the sigmoid colon, and due to necrosis had to be euthanized. While this was an isolated incident in a large colony of animals, it still highlights that there are potential risks associated with some types of enrichment.

Harris and Edwards (2004) described the use of stainless steel mirrors as environmental enrichment devices in 105 singly housed male AGMs (*Cercopithecus aethiops sabaeus*, according to Rudel et al. 2002). Fingerprint accumulation surveys over a 6-month period were used as indicators of mirror use, resulting in 97% of the animals categorized as having used the mirrors. An additional behavioral study was conducted with 25 animals that measured both contact with mirrors and looking into the mirrors without contact. Animals used the mirror 5.2% of the time (3 min/h). The authors concluded that mirrors were an effective enrichment device and that habituation did not occur, even up to a year after the initial presentation.

The *Laboratory Primate Newsletter* also contained a handful of articles describing potential enrichment devices that could be used with vervets. Watson (1997) described a study testing the utilization of enrichment devices in more than 60 wild-caught, single-caged *sabaeus* at a captive facility on Barbados. Animals were presented with bottles, PVC and bamboo feeders, and Kong toys, and were then observed to see how often they explored and interacted with the devices. Animals showed a preference for feeders, and preferred the bamboo feeder over the PVC feeder. The author concluded that the subjects may prefer natural materials over manufactured materials. Bramblett and Bramblett (1988) described the design and materials needed for building a PVC pump feeder that could be used as a time-consuming foraging device. The feeder was ~6-in. long and 2-in. in diameter and could be filled with sticky fluids (e.g., orange juice concentrate). The task for the animals was to move a central pipe up and down to gain access to small quantities of the liquid. The device was used with a colony of group-housed *pygerythrus* (see Bramblett et al. 1982).

Bloomstrand and Maple (1987), in their recommendations for enrichment in vervets and other African monkeys, emphasize the importance of (1) allowing for vertical movement; (2) maximizing useable space to "allow for cover, privacy, and social distance" (p. 228); and (3) providing for naturalistic environments when possible. To those ends, they suggested the use of tree branches, ropes, cargo nets, and other cage furniture that allow for vertical movement.

The enrichment program at WFPC, of which the VRC is a part, separates enrichment items into five categories: (1) structural, (2) foraging, (3) sensory, (4) object, and (5) social. I will now describe examples of some of these different types of enrichment used within the VRC. In our indoor–outdoor breeding groups, structural enrichment includes swings, barrels, visual barriers, pallets, and browse. We often recycle items, such as large plastic barrels that held soap for the cage washer. These barrels are cleaned, have holes cut in them, and are then placed in the pens or hung from the side of the cages. Plastic pallets are also attached to the sides of cages, or hung from the ceiling in front of shelves to allow for hiding places (see Figures 21.3 and 21.4). Recycled fire hoses have been used for climbing structures in the past; however, regular maintenance is required because younger animals can get their fingers stuck in the frayed ends of the hoses if they are not trimmed regularly. Nontoxic browse is provided periodically to allow animals to eat leaves and strip bark off the branches (Figure 21.4). Food enrichment includes the use of fruits and vegetables five times per week. Some smaller food items, such as seeds, corn, or popcorn, can be scattered throughout the outdoor cage area so that animals can forage through the river rock substrate for extended periods (Figure 21.4). Sensory enrichment includes the use of plastic tubs that are filled with water every day during the summer months. Object enrichment includes small items, such as Kong toys, mirrors, or other manipulatable objects. Animals typically are interested in these items when they are first introduced, but the interest typically diminishes rapidly. As has been stated in numerous places, social enrichment is the most stimulating kind of enrichment, though it also requires the most attention to species differences (Coleman et al. 2012). The VRC houses the majority of the animals in social groups. We currently have 16 matrilineal social groups, with 10–30 animals per group, along with 2 all-male social groups consisting of 7–8 animals per group. The combination of the rich structural and object environment, coupled with the complexity of social housing, provides for a stimulating environment.

Positive Reinforcement Training

There is relatively little published information concerning positive reinforcement training (PRT) with vervets. The available literature consists primarily of descriptions of (1) habituation/training for awake sample collections and (2) the training necessary for cognitive testing. For example, Molskness et al. (2007) described the training of 39 adult female *C. a. sabaeus* imported from St. Kitts to the Oregon National Primate Research Center. After quarantine, animals were individually housed and trained to move into a sample collection cage for daily unanesthetized vaginal swabbing and saphenous venipuncture. Stephens et al. (2013) described similar training for daily vaginal swabbing of 12 pair-housed adult female *C. a. sabaeus* at the WFPC. Those animals were not trained for awake bleeding, since a remote catheter and tether system was used instead. Reinhardt (1997), in a review of training nonhuman primates to cooperate during sample collections, cites two examples of vervets being trained for awake bleeding procedures (Wall et al. 1985; Suleman et el. 1988). Kelley and Bramblett (1981) described training animals for urine collection.

At the VRC, animal care staff working with the animals housed in large indoor–outdoor social groups are trained to shift animals inside or outside for cleaning procedures, or to lock animals inside during periods of cold temperatures (see section "Facilities and Equipment"). Animals that fail to cooperate are coerced into moving; thus, current practices do not represent full PRT, since some animals are not given choice or control in this situation (Bloomsmith 2012). Veterinary staff have periodically trained individual animals to move into a separate area of the large social group enclosure for daily oral dosing, though this has been limited to animals willing to cooperate and to groups that tolerate these procedures. Pair-housed diabetic animals have been habituated to receive twice-daily intramuscular injections of insulin and capillary blood samples from the tail.

James et al. (2007) described a spatial delayed response task used to measure working memory in adolescent male vervets (*C. a. sabaeus*) at the VRC. Animals were tested in the capture tunnel of their home cage using a modified Wisconsin General Test Apparatus. Animals were trained to move into the tunnel each day for testing and to remove food treats from boxes on the apparatus. Myers and Hamilton (2011) described a similar type of cognitive testing using individually housed adult *C. a. sabaeus* that were trained to use a touch screen testing panel attached to their home cage.

A comparable cognitive testing system was used at the VRC after the colony had been transferred to Wake Forest (between 2008 and 2010). Approximately 200 animals were trained to move into the capture tunnel to work on similar cognitive tasks. Despite extensive PRT, a subset of animals still required coercion to enter the tunnel for testing each day. Younger animals were generally more responsive to PRT than older animals.

Facilities and Equipment

Again, most of the housing facilities and equipment used with macaques and other Old World monkeys will likely be suitable for use with vervets. This section highlights the unique housing arrangement used by Jürgen Seier's laboratory, as well as the current practices at the VRC.

Seier and de Lange (1996) described the use of a mobile exercise cage or "e-cage" to allow for social contact and vertical climbing behavior in their colony of vervets (*C. a. pygerythrus*) in South Africa. The authors described introducing a small number of adult females (n = 9 and 13 in each cohort, respectively) in a serial manner to a large number of adult males (n = 54 and 20, respectively). The males were housed individually in rooms with eight individual home cages along one wall (four upper cages and four lower cages). The females were housed in the mobile C- or [-shaped e-cage that connected to the upper and lower home cages of the males. The females were introduced to individual males by connecting the e-cage to the male's home cage. No familiarization process was used prior to full contact between the two animals. In one cohort, the females were introduced to new males every 24 h. This method of moving females into the home cages of the males was frequently used at this facility for breeding purposes. Animals were fed different diets and had to be separated for feeding three times per day. The animals spent time in the e-cages running, climbing, or leaping, and often spent time observing or vocalizing to the other animals in the room. No aggressive or threatening behavior was observed and females readily entered the male's home cage (which included the only source of water). Only three cases of minor injuries were reported. The authors concluded that the e-cage was an effective way to allow for regular heterosexual social contact, while also increasing the amount of usable space for the animals. They also claimed that few females were needed to provide intermittent social contact to a large number of individually housed males. The frequent introductions to multiple males did not appear to be stressful to the females. In one cohort, 6 of the 9 females did become pregnant after 6 months. In the second cohort, 10 of 15 pairings resulted in pregnancies. The authors suggest that pregnancy rate is another reflection of social compatibility between the heterosexual pairs.

The VRC breeding groups are housed in two large buildings that were designed and built specifically for this colony in late 2007. Each building consists of eight large, indoor–outdoor enclosures with multiple perches and climbing structures. Each of the 16 total enclosures consists of a large outdoor area (111 m^2) and an indoor area comprised of two interconnected rooms (28 m^2; i.e., each group has three interconnected housing areas). Each enclosure contains a custom-designed capture tunnel that facilitates the rapid removal of animals from the group for anesthesia, examinations, and sample collections/procedures. Enclosures are equipped with shelves, hanging play structures, swings, barrels, and other enrichment devices (see Figures 21.3 and 21.4). In addition to the animal housing areas, each building contains a laboratory area for staging and procedures, and one of the buildings contains a fully-equipped clinical treatment room.

The buildings were designed around the capture tunnels (Figure 21.5) that had been used for years at the VRC to facilitate animal capture without the need for extensive training. The capture tunnels are a permanent part of each groups' housing area and the backs of the tunnels are accessible via two vertically sliding doors that provide the animals with access to the outdoor housing area. The sides of the tunnels are typically left open and animals regularly enter and exit the tunnels as they freely move between the indoor and outdoor sections of the pen. When capture is necessary, portable dividers are placed on the sides of the tunnel and the animal or animals run into the tunnel from the outside housing area via the sliding doors. Individual animals can then be (1) separated into one of five compartments using additional dividers and (2) restrained and/or anesthetized using a squeeze mechanism built into the capture tunnel. The capture process typically involves the use of nets, but only to encourage animals to enter the tunnel. Habituation and negative reinforcement training have resulted in the vast majority of animals entering the tunnel without requiring physical capture using the nets.

Special Topics

Abnormal Behavior and Stereotypy

As with many captive Old World monkeys, vervets can display stereotypies and other abnormal behavior. As previously noted, Seier et al. (2011) described the effects of enrichment, cage size, and socialization on the rates of somersaulting, head tossing, and pacing in their colony. Again, the most frequent stereotypies included somersaulting, head tossing, and pacing, with no reports of self-injury. The same lab has also described the effectiveness of administering fluoxetine (Prozac) to individually housed *C. a. pygerythrus* to diminish stereotypical behavior (including marked saluting, somersaulting, weaving, or head tossing; Hugo et al. 2003). They administered 1 mg/kg of fluoxetine daily by mixing the medication into the diet and monitoring intake. They concluded that fluoxetine significantly decreased stereotypical behavior in the treated animals compared to the five controls. Daniel et al. (2008) reported on the rates of self-directed behavior (SDB) in a group of vervets at the Lisbon Zoo. They noted that SDB rates increased after agonistic interactions, suggesting that SDB rates were indicative of anxiety.

Rates of stereotypical behavior in the *C. a. sabaeus* VRC population have historically been quite low. Typically less than one to two individuals (<0.5%) show either pacing or saluting stereotypies. Self-injurious behavior has never been observed. This is likely due to the fact that the vast majority of the animals are housed in large, indoor–outdoor pens in multigenerational species-typical social groups. Infants typically remain in their natal groups with their mothers and close kin throughout the juvenile period. Even the limited number of pair- or single-housed animals in the VRC rarely display stereotypies. Fluoxetine was used in the VRC on a single occasion, but that was to measure its impact on impulsive behavior, not on stereotypies (Fairbanks et al. 2001). Differences in stereotypical behavior between the Seier lab and the VRC are likely due to subspecies, housing, and/or rearing condition differences.

Aggression and Wounding

Aggression and wounding are always risks associated with social housing of nonhuman primates. As noted earlier, female vervets have long canines relative to other Old World monkeys (Rowell 1971; Bloomstrand & Maple 1987; Fedigan 1992). This can obviously lead to wounding in both wild and captive populations. Else (1985) reported that there was significant female fighting and injury during the formation of captive groups of *C. a. pygerythrus*. Kaplan (1987) reported that the number of deaths due to trauma in captive vervet harem groups was proportionally greater than that of comparably housed cynomolgus macaques. Fairbanks and McGuire (1986) noted that aggressive behavior was positively correlated with group size and that aggression was higher in wild-caught founders than in colony-born animals in the VRC. As noted by Coleman et al. (2012), returning animals to their social group as soon as possible after any clinical treatment is critical for long-term social stability of the group. The threshold for clinical intervention should also be adjusted during periods of initial social introductions since removals/separations can negatively impact that process. In the VRC, if animals do require separation for clinical treatment for more than 24 h, they are housed in single cages in front of their social groups whenever possible to maintain social contact. While such temporary housing in front of a group would be discouraged for introducing unfamiliar animals, it is beneficial for reintroducing familiar animals back into their social groups.

Other Clinical Issues

There are a handful of other clinical issues that may be relevant to the management of vervets. Amory et al. (2013) noted that gastrointestinal tract disease is common in vervets. Gastric dilatation (bloat) is often associated with dietary changes, antibiotic therapy, and stress and can often be fatal. Gastric ulceration can also occur in reaction to stress and is most common in wild-caught vervets. In the VRC we try to avoid bloat by restricting the feeding of preferred, denser enrichment foods (such as bananas, apples, sweet potatoes) on days when animals are anesthetized, especially after being fasted. Fincham et al. (1992) reported that the risk of torsion of the colon increased in adult male vervets that were fed a Western diet, though it did not increase in females. Neither bloat nor colon torsion has been observed during dietary studies in the VRC using Western (Jorgensen et al. 2013; Voruganti et al. 2013) or high-fiber (Fairbanks et al. 2010) diets. Isbell (1995) reported seasonal alopecia for vervets in the wild, but alopecia in the VRC is extremely rare. In fact, hair has been repeatedly used to measure cortisol in the VRC, even during periods of stress, without any noticeable hair loss (Fairbanks et al. 2011; Laudenslager et al. 2011, 2012). Tarara et al. (1995) reported that 32% of 260 necropsies of captive *C. a. pygerythrus* were found to have gastric ulcers. The rate of gastric ulcers was higher than that reported in other species and was strongly associated with individual housing. Suleman et al. (2000, 2004) reported that gastric ulcerations and other physiological markers of stress were related to capture and restraint. Normative hematology and clinical chemistries have been reported for a number of populations of vervets including the VRC (Sato et al. 2005; Chichester et al. 2015) and Caribbean populations (Liddie et al. 2010). In addition, Kagira et al. (2007) reported on how hematological values change during habituation to captivity.

EXPERT RECOMMENDATIONS

The aspects of vervet behavior and natural history that differ from other Old World monkeys typically encountered in research settings highlight the importance of developing species-typical behavioral management plans. While vervet behavior patterns show many similarities with those of macaques and other Old World monkeys, there are important differences that should be taken into consideration. For example, understanding the unique dominance displays (e.g., manipulation; see Table 21.3) of vervets can be critical when evaluating the success of different types of social housing (see Table 21.3). Knowing that vervets are highly territorial can help inform other behavioral management decisions.

Despite the fact that surveys have shown vervets to be single-housed at some facilities (Baker et al. 2007), vervets can successfully be housed in heterosexual and isosexual pairs, harems, and in multimale, multifemale social groups. Harem groups (e.g., Else 1985) and short-term heterosexual pairing (e.g., Seier 1986) have been commonly reported. Multimale, multifemale groups are the norm in the wild and certainly possible in captivity (e.g., Bramblett et al. 1982), having been employed regularly during the early years of the VRC (e.g., Fairbanks & McGuire 1986, 1987).

In the VRC, when introducing animals to new social groups, an effort is usually made to "keep everyone off balance" by making most introductions and animal transfers all at the same time. For example, in social group transfers, the following procedures will typically be done in a single day: (1) removal of natal males, (2) removal of older, resident adult males, (3) introduction of new adult males, and (4) the formation of new all-male groups or the introduction of new males into all-male groups. This results in all animals—residents and newly introduced animals—being placed in a similar, novel situation, with all animals needing to adapt to rapidly changing circumstances. This is an example of a method that is appropriate for vervets that is unlikely to be successful for macaques.

The same is true for pair housing. Ideally, animals being moved from social groups to pair housing should all be moved into a new room together, with all introductions occurring at the same time, immediately after transfer. In our experience, immediate aggressive interactions are relatively rare in vervets, as they tend to "feel things out" during the first few days or weeks following introductions using this technique. Again, the territorial nature of vervets is an important factor to keep in mind. Forming new pairs in rooms that are unfamiliar to each animal will help reduce territorial behavior, since neither animal is resident. Avoiding a "macaque-centric" technique of gradual introductions for pair-housing is recommended. While controlled comparisons of different introduction techniques have not yet been conducted, there is suggestive evidence that a gradual introduction approach may not be as beneficial for vervets as it can be for macaques (Jorgensen et al. 2017).

In group or pair-housing situations, individuals do not need to be separated at the first signs of fighting, even if wounding occurs (Coleman et al. 2012). If clinical intervention is necessary, returning animals to their group/pair as soon as possible is critical. In species-typical social groups, the removal/death of dominant females can be very disruptive to group stability. Introductions of new males should be avoided when young infants are in the groups (e.g., Morland et al. 1992).

The use of vertical space is important to vervets (Bloomstrand & Maple 1987). The need for physical contact to work things out or the need for spatial separation/avoidance is critical (Kaplan 1987). This can be particularly difficult in standard pair caging. Species-typical behavior patterns commonly observed in social groups are infrequent, nonexistent, or not physically possible within the smaller confines of an indoor, pair-housing environment. For example, in the VRC, we often fail to see some types of dominance displays in pair-housed animals that we commonly observe in animals housed in larger, social groups.

Environmental enrichment used with macaques and baboons can often be used successfully with vervets. As some have noted (Watson 1997; Seier et al. 2011), natural materials may work better with vervets. It is important that appropriate climbing and/or hiding structures are available whenever possible. There are relatively few published examples of using positive reinforcement techniques in vervets. Thus, there is clearly a need to expand the use of PRT in this species.

There are a few other important issues to be aware of when working with vervets. There is at least anecdotal evidence that wild-caught animals are qualitatively different from captive-born animals (Jorgensen et al. 2017). Not surprisingly, animals that are not raised in species-typical social groups are likely to show behavioral abnormalities as adults. Seier et al. (2011) reported higher rates of abnormal behavior and stereotypy in their population of mother–peer reared vervets when compared with the vervets raised in species-typical social groups in the VRC. Thankfully, self-injurious behavior is very rare in captive vervets, regardless of rearing condition. Alopecia, another common problem in macaques, is also relatively rare in vervets. Bloating, torsion of the colon, and gastric ulcers can be common in vervets and can be exacerbated by stress, housing, and management decisions.

CONCLUSIONS

In short: Vervets are not macaques. While many of the behavioral management techniques developed for macaques and other Old World monkeys may be perfectly appropriate for vervets, it is important to develop species-typical management plans whenever possible (Lutz & Novak 2005). As described in this chapter, there are some important behavioral and biological differences between vervets and other Old World monkeys that may not be readily apparent to animal care, veterinary, enrichment, and research staff that are new to working with this species. Vervets are an important nonhuman primate species for biomedical research and knowing how they differ from other nonhuman primates will enhance both the management and proper scientific utilization of this species.

ACKNOWLEDGMENTS

I thank Dr. Lynn A. Fairbanks and Dr. Jay R. Kaplan, the former directors of the VRC, as well as all of the past and present research, animal care, veterinary, enrichment, and data services staff at UCLA, the Sepulveda VA Medical Center, and the WFSM. Additional thanks to Lynn Fairbanks and Kelsey Lambert for comments on an earlier version of this chapter. The VRC is currently supported by NIH grant OD010965 (PI: Matthew J. Jorgensen) and the WFSM.

REFERENCES

Amory, J. T., W. M. du Plessis, A. Beierschmitt, J. Beeler-Marfisi, R. M. Palmour, and T. Beths. 2013. Abdominal ultrasonography of the normal St. Kitts vervet monkey (*d*). *J Med Primatol* 42 (1):28–38.

Atkins, H. M., C. J. Willson, M. Silverstein, M. Jorgensen, E. Floyd, J. R. Kaplan, and S. E. Appt. 2014. Characterization of ovarian aging and reproductive senescence in vervet monkeys (*Chlorocebus aethiops sabaeus*). *Comp Med* 64 (1):55–62.

Baker, K. C., J. L. Weed, C. M. Crockett, and M. A. Bloomsmith. 2007. Survey of environmental enhancement programs for laboratory primates. *Am J Primatol* 69:377–394.

Baulu, J., G. Evans, and C. Sutton. 2002. Pathogenic agents found in Barbados *Chlorocebus aethiops sabaeus* and in Old World Monkeys commonly used in biomedical research. *Lab Primate Newsl* 41:4–6.

Bloomsmith, M. A. 2012. Positive reinforcement animal training in primate laboratories. *Enrich Rec* 2012:16–18.

Bloomstrand, M. and T. L. Maple. 1987. Management and husbandry of African monkeys in captivity. In *Comparative Behavior of African Monkeys*, ed. Zucker, E. L., 197–234. New York: Alan R. Liss.

Bolter, D. R. 2011. A comparative study of growth patterns in crested langurs and vervet monkeys. *Anat Res Int* 2011:1–12.

Bolter, D. R. and A. L. Zihlman. 2003. Morphometric analysis of growth and development in wild-collected vervet monkeys (*Cercopithecus aethiops*), with implications for growth patterns in Old World monkeys, apes and humans. *J Zool Lond* 260:99–110.

Borgeaud, C., M. Alvino, K. van Leeuwen, S. W. Townsend, and R. Bshary. 2015. Age/sex differences in third-party rank relationship knowledge in wild vervet monkeys, *Chlorocebus aethiops pygerythrus*. *Anim Behav* 102:277–284.

Bramblett, C. A. 1980. A model for development of social behavior in vervet monkeys. *Dev Psychobiol* 13:205–223.

Bramblett, R. D. and C. A. Bramblett. 1988. A liquid dispenser for caged primates. *Lab Primate Newsl* 27:16.

Bramblett, C. A., S. S. Bramblett, D. A. Bishop, and A. M. Coelho. 1982. Longitudinal stability in adult status hierarchies among vervet monkeys (*Cercopithecus aethiops*). *Am J Primatol* 2 (1):43–51.

Bramblett, C. A., L. D. Pejaver, and D. J. Drickman. 1975. Reproduction in captive vervet and Sykes' monkeys. *J Mammal* 56:940–946.

Briggs, C. M., K. M. Smith, A. Piper, E. Huitt, C. J. Spears, M. Quiles, M. Ribeiro, M. E. Thomas, D. T. Brown, and R. Hernandez. 2014. Live attenuated tetravalent dengue virus host range vaccine is immunogenic in African green monkeys following a single vaccination. *J Virol* 88 (12):6729–6742.

Cann, J. A., K. Kavanagh, M. J. Jorgensen, S. Mohanan, T. D. Howard, S. B. Gray, G. A. Hawkins, L. A. Fairbanks, and J. D. Wagner. 2010. Clinicopathologic characterization of naturally occurring diabetes mellitus in vervet monkeys. *Vet Pathol* 47 (4):713–718.

Carlsson, H. E., S. J. Schapiro, I. Farah, and J. Hau. 2004. Use of primates in research: A global overview. *Am J Primatol* 63 (4):225–237.

Carroll, R. L., K. Mah, J. W. Fanton, G. N. Maginnis, R. M. Brenner, and O. D. Slayden. 2007. Assessment of menstruation in the vervet (*Cercopithecus aethiops*). *Am J Primatol* 69 (8):901–916.

Chahroudi, A., S. E. Bosinger, T. H. Vanderford, M. Paiardini, and G. Silvestri. 2012. Natural SIV hosts: Showing AIDS the door. *Science* 335 (6073):1188–1193.

Cheney, D. L. 1981. Inter-group encounters among free-ranging vervet monkeys. *Folia Primatol* 35 (2–3):124–146.

Cheney, D. L. and R. M. Seyfarth. 1983. Nonrandom dispersal in free-ranging vervet monkeys—Social and genetic consequences. *Am Nat* 122 (3):392–412.

Cheney, D. L. and R. M. Seyfarth. 1987. The influence of intergroup competition on the survival and reproduction of female vervet monkeys. *Behav Ecol Sociobiol* 21:375–386.

Cheney, D. L. and R. M. Seyfarth. 1990. *How Monkeys See the World: Inside the Mind of Another Species*. Chicago, IL: University of Chicago Press.

Chichester, L., M. K. Gee, M. J. Jorgensen, and J. R. Kaplan. 2015. Hematology and clinical chemistry measures during and after pregnancy and age- and sex-specific reference intervals in African green monkeys (*Chlorocebus aethiops sabaeus*). *J Am Assoc Lab Anim Sci* 54 (4):359–367.

Cho, F., A. Hiyaoka, M. T. Suzuki, and S. Honjo. 2002. Breeding of African green monkeys (*Cercopithecus aethiops*) under indoor individually-caged conditions. *Exp Anim* 51 (4):343–351.

Coleman, K., M. A. Bloomsmith, C. M. Crockett, J. L. Weed, and S. J. Schapiro. 2012. Behavioral management, enrichment, and psychological well-being of laboratory nonhuman primates. In *Nonhuman Primates in Biomedical Research: Biology and Management*, eds Abee C. R., K. Mansfield, S. D. Tardiff, and T. Morris, 149–176. London: Academic Press.

Cramer, J. D., T. Gaetano, J. P. Gray, P. Grobler, J. G. Lorenz, N. B. Freimer, C. A. Schmitt, and T. R. Turner. 2013. Variation in scrotal color among widely distributed vervet monkey populations (*Chlorocebus aethiops pygerythrus* and *Chlorocebus aethiops sabaeus*). *Am J Primatol* 75 (7):752–762.

Daniel, J. R., A. J. dos Santos, and L. Vicente. 2008. Correlates of self-directed behaviors in captive *Cercopithecus aethiops*. *Int J Primatol* 29:1219–1226.

Denham, W. W. 1987. *West Indian Green Monkeys: Problems in Historical Biogeography*. Basel, Switzerland: Karger.

de Waal, F. B. M. and L. M. Luttrell. 1989. Toward a comparative socioecology of the genus *Macaca*: Different dominance styles in rhesus and pigtail macaques. *Am J Primatol* 19:83–109.

Disotell, T. R. and R. L. Raaum. 2002. Molecular timescale and gene tree incongruence in the guenons. In *The Guenons: Diversity and Adaptation in African Monkeys*, eds Glenn, M. E. and M. Cords, 27–36. New York: Kluwer Academic/Plenum Publishers.

DiVincenti, L. and J. D. Wyatt. 2011. Pair housing of macaques in research facilities: A science-based review of benefits and risks. *J Am Assoc Lab Animal Sci* 50:856–863.

Dunbar, R. I. M. 1974. Observations on the ecology and social organization of the green monkey, *Cercopithecus sabaeus*, in Senegal. *Primates* 15 (4):341–350.

Ehardt-Seward, C. and C. A. Bramblett. 1980. The structure of social space among a captive group of vervet monkeys. *Folia Primatol* 34:214–238.

Eley, R. M., R. P. Tarara, C. M. Worthman, and J. G. Else. 1989. Reproduction in the vervet monkey (*Cercopithecus aethiops*): III. The menstrual cycle. *Am J Primatol* 17:1–10.

Else, J. G. 1985. Captive propagation of vervet monkeys (*Cercopithecus aethiops*) in harems. *Lab Anim Sci* 35 (4):373–375.

Erhart, E. M., C. A. Bramblett, and D. J. Overdorff. 2005. Behavioral development of captive male hybrid cercopithecine monkeys. *Folia Primatol* 76 (4):196–206.

Fairbanks, L. A. 1980. Relationships among adult females in captive vervet monkeys: Testing a model of rank-related attractiveness. *Anim Behav* 28:853–859.

Fairbanks, L. A. 1984. Determinants of fecundity and reproductive success in captive vervet monkeys. *Am J Primatol* 7:27–38.

Fairbanks, L. A. 1990. Reciprocal benefits of allomothering for female vervet monkeys. *Anim Behav* 40:553–562.

Fairbanks, L. A. 2001. Individual differences in response to a stranger: Social impulsivity as a dimension of temperament in vervet monkeys (*Cercopithecus aethiops sabaeus*). *J Comp Psychol* 115 (1):22–28.

Fairbanks, L. A., K. Blau, and M. J. Jorgensen. 2010. High-fiber diet promotes weight loss and affects maternal behavior in vervet monkeys. *Am J Primatol* 72 (3):234–241.

Fairbanks, L. A., M. J. Jorgensen, J. N. Bailey, S. E. Breidenthal, R. Grzywa, and M. L. Laudenslager. 2011. Heritability and genetic correlation of hair cortisol in vervet monkeys in low and higher stress environments. *Psychoneuroendocrinology* 36:1201–1208.

Fairbanks, L. A., M. J. Jorgensen, A. Huff, K. Blau, Y. Y. Hung, and J. J. Mann. 2004. Adolescent impulsivity predicts adult dominance attainment in male. *Am J Primatol* 64 (1):1–17.

Fairbanks, L. A. and M. T. McGuire. 1984. Determinants of fecundity and reproductive success in captive vervet monkeys. *Am J Primatol* 7:27–38.

Fairbanks, L. A. and M. T. McGuire. 1986. Age, reproductive value, and dominance-related behavior in vervet monkey females—Cross-generational influences on social relationships and reproduction. *Anim Behav* 34:1710–1721.

Fairbanks, L. A. and M. T. McGuire. 1987. Mother-infant relationships in vervet monkeys: Response to new adult males. *Int J Primatol* 8 (4):351–366.

Fairbanks, L. A., M. T. McGuire, and N. Page. 1978. Social roles in captive vervet monkeys (*Cercopithecus-aethiops sabaeus*). *Behav Processes* 3 (4):335–352.

Fairbanks, L. A., W. P. Melega, M. J. Jorgensen, J. R. Kaplan, and M. T. McGuire. 2001. Social impulsivity inversely associated with CSF5-HIAA and fluoxetine exposure in vervet monkeys. *Neuropsychopharmacology* 24 (4):370–378.

Fears, S. C., W. P. Melega, S. K. Service, C. Lee, K. Chen, Z. Tu, M. J. Jorgensen, L. A. Fairbanks, R. M. Cantor, N. B. Freimer, and R. P. Woods. 2009. Identifying heritable brain phenotypes in an extended pedigree of vervet monkeys. *J Neurosci* 29 (9):2867–2875.

Fedigan, L. M. 1992. *Primate Paradigms: Sex Roles and Social Bonds.* Chicago, IL: University of Chicago Press.

Fedigan, L. and L. M. Fedigan. 1988. *Cercopithecus pygerythrus*: A review of field studies. In *A Primate Radiation: Evolutionary Biology of the African Guenons*, eds Gautier-Hion, A., G. Bourlier, J. P. Gautier, and J. Kingdon, 389–411. Cambridge, UK: Cambridge University Press.

Fincham, J. E., J. V. Seier, and C. J. Lombard. 1992. Torsion of the colon in vervet monkeys: Association with an atherogenic Western-type of diet. *J Med Primatol* 21:44–46.

Fomsgaard, A., J. Allan, M. Gravell, W. T. London, V. M. Hirsch, and P. R. Johnson. 1990. Molecular characterization of simian lentiviruses from east African green monkeys. *J Med Primatol* 19 (3–4):295–303.

Fortman, J. D., T. A. Hewett, and B. T. Bennett. 2002. *The Laboratory Nonhuman Primate.* Boca Raton, FL: CRC Press.

Freimer, N., K. Dewar, J. Kaplan, and L. Fairbanks. 2008. The importance of the vervet (African green monkey) as a biomedical model. http://www.genome.gov/Pages/Research/Sequencing/SeqProposals/TheVervetMonkeyBiomedicalModel.pdf. Accessed October 8, 2015.

Freimer, N. B., S. K. Service, R. A. Ophoff, A. J. Jasinska, K. McKee, A. Villeneuve, A. Belisle, et al. 2007. A quantitative trait locus for variation in dopamine metabolism mapped in a primate model using reference sequences from related species. *Proc Natl Acad Sci USA* 104 (40):15811–15816.

Gartlan, J. S. 1969. Sexual and maternal behavior of the vervet monkey, *Cercopithecus aethiops. J Reprod Fertil Suppl* 6:137–150.

Gartlan, J. S. and C. K. Brain. 1968. Ecology and social variability in *C. aethiops* and *C. mitis*. In *Primates: Studies in Adaptation and Variability*, ed. Jay P., 253–292. San Francisco, CA: Holt, Rinehart and Winston.

Groman, S. M., A. M. Morales, B. Lee, E. D. London, and J. D. Jentsch. 2013. Methamphetamine-induced increases in putamen gray matter associate with inhibitory control. *Psychopharmacology (Berl)* 229 (3):527–538.

Groves, C. P. and J. Kingdon. 2013. Genus *Chlorocebus* savanna monkeys. In *The Mammals of Africa: Volume II Primates*, eds Butynski, T. M., J. S. Kingdon, and J. Kalina, 264–266. London: Bloomsbury Publishing.

Guy, A. J. and D. Curnoe. 2011. Death during parturition of a captive adult female vervet monkey (*Chlorocebus aethiops*) and its social consequences for a captive troop. *Lab Primate Newsl* 50:4–6

Harris, H. G. and A. J. Edwards. 2004. Mirrors as environmental enrichment for African green monkeys. *Am J Primatol* 64 (4):459–467.

Harvey, P. H., R. D. Martin, and T. H. Clutton-Brock. 1987. Life histories in comparative perspective. In *Primate Societies*, eds Smuts, B. B., D. L. Cheney, R. M. Seyfarth, R W. Wrangham, and T. T. Struhsaker, 181–196. Chicago, IL: University of Chicago.

Hector, A. C. K, R. M. Seyfarth, and M. J. Raleigh. 1989. Male parental care, female choice and the effect of an audience in vervet monkeys. *Anim Behav* 38:262–271.

Henzi, S. P., N. Forshaw, R. Boner, L. Barrett, and D. Lusseau. 2013. Scalar social dynamics in female vervet monkey cohorts. *Philos Trans R Soc Lond B Biol Sci* 368:1–9.

Hess, D. L., A. G. Hendricks, and G. H. Stabenfeldt. 1979. Reproductive and hormonal patterns in the African green monkey (*Cercopithecus aethiops*). *J Med Primatol* 8:273–281.

Horrocks, J. A. 1986. Life-history characteristics of a wild population of vervets (*Cercopithecus aethiops sabaeus*) in Barbados, West Indies. *Int J Primatol* 7 (1):31–47.

Huang, Y. S., V. Ramensky, S. K. Service, A. J. Jasinska, Y. Jung, O. W. Choi, R. M. Cantor, et al. 2015. Sequencing strategies and characterization of 721 vervet monkey genomes for future genetic analyses of medically relevant traits. *BMC Biol* 13:1–10.

Hugo, C., J. Seier, C. Mdhluli, W. Daniels, B.H. Harvey, D. Du Toit, S. Wolfe Coote, D. Nel, and D.J. Stein. 2003. Fluoxetine decreases stereotypic behavior in primates. *Prog Neuropsychopharmacol Biol Psychiatry* 27 (4):639–643.

Isbell, L. A. 1995. Seasonal and social correlates of changes in hair, skin, and scrotal condition in vervet monkeys (*Cercopithecus aethiops*) of Amboseli National Park, Kenya. *Am J Primatol* 36:61–70.

James, A. S., S. M. Groman, E. Seu, M. Jorgensen, L. A. Fairbanks, and J. D. Jentsch. 2007. Dimensions of impulsivity are associated with poor spatial working memory performance in monkeys. *J Neurosci* 27 (52):14358–14364.

Jasinska, A. J., M. K. Lin, S. Service, O. W. Choi, J. DeYoung, O. Grujic, S. Y. Kong, et al. 2012. A non-human primate system for large-scale genetic studies of complex traits. *Hum Mol Genet* 21 (15):3307–3316.

Jasinska, A. J., C. A. Schmitt, S. K. Service, R. M. Cantor, K. Dewar, J. D. Jentsch, J. R. Kaplan, et al. 2013. Systems biology of the vervet monkey. *ILAR J* 54 (2):122–143.

Jentsch, J. D., D. E. Redmond, J. D. Elsworth, J. R. Taylor, K. D. Youngren, and R. H. Roth. 1997. Enduring cognitive deficits and cortical dopamine dysfunction in monkeys after long-term administration of phencyclidine. *Science* 277 (5328):953–955.

Johnson, P. T., D. A. Valerio, and G. E. Thompson. 1973. Breeding the African green monkey, *Cercopithecus aethiops*, in a laboratory environment. *Lab Anim Sci* 23 (3):355–359.

Jorgensen, M. J., S. T. Aycock, T. B. Clarkson, and J. R. Kaplan. 2013. Effects of a Western-type diet on plasma lipids and other cardiometabolic risk factors in African green monkeys (*Chlorocebus aethiops sabaeus*). *J Am Assoc Lab Anim Sci* 52 (4):448–453.

Jorgensen, M. J., K. R. Lambert, S. D. Breaux, K. C. Baker, B. M. Snively, and J. L. Weed. 2017. Pair housing of vervets/African green monkeys for biomedical research. *Am J Primatol* 79 (1):1–10.

Kagira, J. M., M. Ngotho, J. K. Thuita, N. W. Maina, and J. Hau. 2007. Hematological changes in vervet monkeys (*Chlorocebus aethiops*) during eight months' adaptation to captivity. *Am J Primatol* 69 (9):1053–1063.

Kalinin, S., S. L. Willard, C. A. Shively, J. R. Kaplan, T. C. Register, M. J. Jorgensen, P. E. Polak, I. Rubinstein, and D. L. Feinstein. 2013. Development of amyloid burden in African green monkeys. *Neurobiol Aging* 34 (10):2361–2369.

Kaplan, J. R. 1987. Dominance and affiliation in the cercopithecine and papionini: A comparative examination. In *Comparative Behavior of African Monkeys*, ed. Zucker, E. L., 127–150. New York: Alan R. Liss.

Kavanagh, K., B. L. Dozier, T. J. Chavanne, L. A. Fairbanks, M. J. Jorgensen, and J. R. Kaplan. 2011. Fetal and maternal factors associated with infant mortality in vervet monkeys. *J Med Primatol* 40 (1):27–36.

Kavanagh, K., L. A. Fairbanks, J. N. Bailey, M. J. Jorgensen, M. Wilson, L. Zhang, L. L. Rudel, and J. D. Wagner. 2007a. Characterization and heritability of obesity and associated risk factors in vervet monkeys. *Obesity (Silver Spring)* 15 (7):1666–1674.

Kavanagh, K., K. L. Jones, J. Sawyer, K. Kelley, J. J. Carr, J. D. Wagner, and L. L. Rudel. 2007b. Trans fat diet induces abdominal obesity and changes in insulin sensitivity in monkeys. *Obesity (Silver Spring)* 15 (7):1675–1684.

Kelley, T.M. and C.A. Bramblett. 1981. Urine collection from vervet monkeys by instrumental conditioning. *Am J Primatol* 1:95–97.

Kim, J. R., B. C. Holbrook, S. L. Hayward, L. K. Blevins, M. J. Jorgensen, N. D. Kock, K. De Paris, et al. 2015. Inclusion of flagellin during vaccination against influenza enhances recall responses in nonhuman primate neonates. *J Virol* 14:7291–7303.

Kushner, H., N. Kraft-Schreyer, and E. T. Angelakos. 1982. Analysis of reproductive data in a breeding colony of African green monkeys. *J Med Primatol* 11:77–84.

Laudenslager, M. L., M. J. Jorgensen, and L. A. Fairbanks. 2012. Developmental patterns of hair cortisol in male and female nonhuman primates: Lower hair cortisol levels in vervet males emerge at puberty. *Psychoneuroendocrinology* 37:1736–1739.

Laudenslager, M. L., M. J. Jorgensen, R. Grzywa, and L. A. Fairbanks. 2011. A novelty seeking phenotype is related to chronic hypothalamic-pituitary-adrenal activity reflected by hair cortisol. *Physiol Behav* 104 (2):291–295.

Lee, P. C. 1984. Ecological constraints on the social development of vervet monkeys. *Behaviour* 91 (4):245–262.

Lemere, C. A., A. Beierschmitt, M. Iglesias, E. T. Spooner, J. K. Bloom, J. F. Leverone, J. B. Zheng, et al. 2004. Alzheimer's disease abeta vaccine reduces central nervous system abeta levels in a non-human primate, the Caribbean vervet. *Am J Pathol* 165 (1):283–297.

Liddie, S., R. J. Goody, R. Valles, and M. S. Lawrence. 2010. Clinical chemistry and hematology values in a Caribbean population of African green monkeys. *J Med Primatol* 39 (6):389–398.

Lutz, C. K. and M. A. Novak. 2005. Environmental enrichment for nonhuman primates: Theory and application. *ILAR J* 46:178–191.

Magden, E. R., K. G. Mansfield, J. H. Simmons, and C. R. Abee. 2015. Nonhuman primates. In *Laboratory Animal Medicine*, 3rd edition, eds Fox, J. G., L. C. Anderson, G. M. Otto, K. R. Pritchett-Corning, and M. T. Whary, 771–930. San Diego, CA: Academic Press.

Mallinson, J. J. C. 1971. Observation on the reproduction and development of vervet monkey with species reference to interspecific hybridization. *Mammalia* 35:598–609.

McFarland, R., A. Fuller, R. S. Hetem, D. Mitchell, S. K. Maloney, S. P. Henzi, and L. Barrett. 2015. Social integration confers thermal benefits in a gregarious primate. *J Anim Ecol* 84 (3):871–878.

McGuire, M. T. 1974. St. Kitts vervet (*Cercopithecus aethiops*). *J Med Primatol* 3 (5):285–297.

McGuire, M. T., S. R. Cole, and C. Crookshank. 1978. Effects of social and spatial density changes in *Cercopithecus aethiops sabaeus*. *Primates* 19 (4):615–631.

Melega, W. P., M. J. Jorgensen, G. Lacan, B. M. Way, J. Pham, G. Morton, A. K. Cho, and L. A. Fairbanks. 2008. Long-term methamphetamine administration in the vervet monkey models aspects of a human exposure: Brain neurotoxicity and behavioral profiles. *Neuropsychopharmacology* 33 (6):1441–1452.

Melega, W. P., G. Lacan, A. A. Desalles, and M. E. Phelps. 2000. Long-term methamphetamine-induced decreases of [(11)C]WIN 35,428 binding in striatum are reduced by GDNF: PET studies in the vervet monkey. *Synapse* 35 (4):243–249.

Melnick, D. J. and M. C. Pearl. 1987. Cercopithecine multimale groups. In *Primate Societies*, eds Smuts, B. B., D. L. Cheney, R. M. Seyfarth, R W. Wrangham, and T. T. Struhsaker, 121–134. Chicago, IL: University of Chicago.

Molskness, T. A., D. L. Hess, G. M. Maginnis, J. W. Wright, J. W. Fanton, and R. L. Stouffer. 2007. Characteristics and regulation of the ovarian cycle in vervet monkeys (*Chlorocebus aethiops*). *Am J Primatol* 69 (8):890–900.

Morland, H. S., M. A. Suleman, and E. B. Tarara. 1992. Changes in male-female interactions after introduction of new adult males in vervet monkey (*Cercopithecus aethiops*) groups. *Lab Primate Newsl* 31 (2):1–4.

Myers, T. M. and L. R. Hamilton. 2011. Delayed match-to-sample performance in African green monkeys (*Chlorocebus aethiops sabaeus*): Effects of benzodiazepine, cholinergic, and anticholinergic drugs. *Behav Pharmacol* 22:814–823.

Oates-O'Brien, R. S., T. B. Farver, K. C. Anderson-Vicino, B. McGowen, and N. W. Lerche. 2010. Predictors of matrilineal overthrows in large captive breeding groups of rhesus macaques (*Macaca mulatta*). *J Am Assoc Lab Anim Sci* 49:196–201.

Palmour, R. M., J. Mulligan, J. J. Howbert, and F. Ervin. 1997. Of monkeys and men: Vervets and the genetics of human-like behaviors. *Am J Hum Genet* 61 (3):481–488.

Pandrea, I., N. F. Parrish, K. Raehtz, T. Gaufin, H. J. Barbian, D. Ma, J. Kristoff, et al. 2012. Mucosal simian immunodeficiency virus transmission in African green monkeys: Susceptibility to infection is proportional to target cell availability at mucosal sites. *J Virol* 86 (8):4158–4168.

Parks, J. S. and L. L. Rudel. 1979. Isolation and characterization of high-density lipoprotein apoproteins in the non-human primate (vervet). *J Biol Chem* 254 (14):6716–6723.

Poirier, F. E. 1972. The St. Kitts green monkey (*Cercopithecus aethiops sabaeus*): Ecology, population dynamics, and selected behavioral traits. *Folia Primatol (Basel)* 17:20–55.

Pruetz, J. D. E. 2009. *The Socioecology of Adult Female Patas Monkeys and Vervets in Kenya*. Upper Saddle River, NJ: Pearson Education Inc.

Raleigh, M. J., J. W. Flannery, and F. R. Ervin. 1979. Sex differences in behavior among juvenile vervet monkeys (*Cercopithecus aethiops sabaeus*). *Behav Neural Biol* 26:455–465.

Raleigh, M. J. and M. T. McGuire. 1989. Female influences on male dominance acquisition in captive vervet monkeys, *Cercopithecus aethiops sabaeus*. *Anim Behav* 38:59–67.

Raleigh, M. J., M. T. McGuire, G. L. Brammer, D. B. Pollack, and A. Yuwiler. 1991. Serotonergic mechanisms promote dominance acquisition in adult male vervet monkeys. *Brain Res* 559 (2):181–190.

Redmond, D. E. and L. Evans. 2012. Determination of fetal age by ultrasonography in St. Kitts green monkeys. *Am J Primatol* 74:433–441.

Reinhardt, V. 1997. Training nonhuman primates to cooperate during blood collection: A review. *Lab Primate Newsl* 36 (4):1–4.

Rowell, T. E. 1970. Reproductive cycles of two *Cercopithecus* monkeys. *J Reprod Fertil* 60:500–691.

Rowell, T. E. 1971. Organization of caged groups of *Cercopithecus* monkeys. *Anim Behav* 19 (4):625–645.

Rudel, L. L., M. Davis, J. Sawyer, R. Shah, and J. Wallace. 2002. Primates highly responsive to dietary cholesterol up-regulate hepatic ACAT2, and less responsive primates do not. *J Biol Chem* 277 (35):31401–31406.

Sade, D. S. and R. W. Hildrech. 1965. Notes on the green monkey (*Cercopithecus aethiops sabaeus*) on St. Kitts, West Indies. *Carib J Sci* 5:67–81.

Sato, A., L. A. Fairbanks, T. Lawson, G. W. Lawson. 2005. Effects of age and sex on hematologic and serum biochemical values of vervet monkeys (*Chlorocebus aethiops sabaeus*). *Contemp Top Lab Anim Sci* 44 (1):29–34.

Schapiro, S. J., R. Stavisky, and M. Hook. 2000. The lower-row cage may be dark but behavior does not appear to be affected. *Lab Primate Newsl*, 39:7–11.

Seier, J. V. 1986. Breeding vervet monkeys in a closed environment. *J Med Primatol* 15 (5):339–349.

Seier, J. 2005. Vervet monkey breeding. In *The Laboratory Primate*, ed. Wolfe-Coote, S., 175–179. Amsterdam, Netherlands: Elsevier.

Seier, J. V. and P. W. de Lange. 1996. A mobile cage facilitates periodic social contact and exercise for singly caged adult Vervet monkeys. *J Med Primatol* 25 (1):64–68.

Seier, J., C. de Villiers, J. van Heerden, and R. Laubscher. 2011. The effect of housing and environmental enrichment on stereotyped behavior of adult vervet monkeys (*Chlorocebus aethiops*). *Lab Anim* 40 (7):218–224.

Seier, J.V., M.A. Dhansay, and A. Davids. 2005. Risks associated with environmental enrichment: Intestinal obstruction caused by foraging substrate. *J Med Primatol* 34 (3):154–155.

Seier, J. V., G. van der Horst, M. de Kock, and K. Chwalisz. 2000. The detection and monitoring of early pregnancy in the vervet monkey (*Cercopithecus aethiops*) with the use of ultrasound and correlation with reproductive steroid hormones. *J Med Primatol* 29:70–75.

Seier, J. V., F. S. Venter, J. E. Fincham, and J. J. Taljaard. 1991. Hormonal vaginal cytology of vervet monkeys. *J Med Primatol* 20 (1):1–5.

Seyfarth, R. M. 1980. The distribution of grooming and related behaviours among adult female vervet monkeys. *Anim Behav* 28:798–813.

Seyfarth, R. M. and D. L. Cheney. 1984. Grooming, alliances and reciprocal altruism in vervet monkeys. *Nature* 308 (5959):541–543.

Seyfarth, R. M., D. L. Cheney, and P. Marler. 1980. Monkey responses to three different alarm calls: Evidence of predator classification and semantic communication. *Science* 210 (4471):801–803.

Smith, D. G. 2012. Taxonomy of nonhuman primates used in biomedical research. In *Nonhuman Primates in Biomedical Research: Biology and Management*, eds Abee, C. R., K. Mansfield, S. D. Tardiff, and T. Morris, 57–85. London: Academic Press.

Stephens, S. M., F. K. Pau, T. M. Yalcinkaya, M. C. May, S. L. Berga, M. D. Post, S. E. Appt, and A. J. Polotsky. 2013. Assessing the pulsatility of luteinizing hormone in female vervet monkeys (*Chlorocebus aethiops sabaeus*). *Comp Med* 63 (5):432–438.

Struhsaker, T. T. 1967a. Behavior of vervet monkeys and other cercopithecines. New data show structural uniformities in the gestures of semiarboreal and terrestrial cercopithecines. *Science* 156 (3779):1197–1203.

Struhsaker, T. T. 1967b. Social structure among vervet monkeys (*Cercopithecus aethiops*). *Behaviour* 29 (2):83–121.

Suleman, M. A., J. Njugana, and J. Anderson. 1988. Training of vervet monkeys, Sykes monkeys and baboons for collection of biological samples. *Proceedings of the XIIth Congress of the International Primatological Society*, p. 12 (Abstract).

Suleman, M. A., E. Wango, I. O. Farah, and J. Hau. 2000. Adrenal cortex and stomach lesions associated with stress in wild male African green monkeys (*Cercopithecus aethiops*) in the post-capture period. *J Med Primatol* 29:338–342.

Suleman, M. A., E. Wango, R. M. Sapolsky, H. Odongo, and J. Hau. 2004. Physiologic manifestations of stress from capture and restraint of free-ranging male African green monkeys (*Cercopithecus aethiops*). *J Zoo Wildl Med* 35:20–24.

Tarara, R., J. G. Else, and R. M. Eley. 1984. The menstrual-cycle of the vervet monkey, *Cercopithecus aethiops*. *Int J Primatol* 5 (4):384–384.

Tarara, E. B., R. P. Tarara, and M. A. Suleman. 1995. Stress-induced gastric ulcers in vervet monkeys (*Cercopithecus aethiops*): The influence of life history factors. Part II. *J Zoo Wildl Med* 26:72–75.

Tosi, A. J., D. J. Melnick, and T. R. Disotell. 2004. Sex chromosome phylogenetics indicate a single transition to terrestriality in the guenons (tribe Cercopithecini). *J Hum Evol* 46 (2):223–237.

Turner, T. R., F. Anapol, and C. J. Jolly. 1997. Growth, development, and sexual dimorphism in vervet monkeys (*Cercopithecus aethiops*) at four sites in Kenya. *Am J Phys Anthropol* 103:19–35.

Turner, T. R., W. G. Coetzer, C. A. Schmitt, J. G. Lorenz, N. B. Freimer, and J. P. Grobler. 2016. Localized population divergence of vervet monkeys (*Chlorocebus* spp.) in South Africa: Evidence from mtDNA. *Am J Phys Anthropol* 159 (1):17–30.

Turner, T. R., P. L. Whitten, C. J. Jolly, and J. G. Else. 1987. Pregnancy outcome in free-ranging vervet monkeys (*Cercopithecus aethiops*). *Am J Primatol* 12 (2):197–203.

van de Waal, E., C. Borgeaud, and A. Whiten. 2013. Potent social learning and conformity shape a wild primate's foraging decisions. *Science* 340 (6131):483–485.

Voruganti, V. S., M. J. Jorgensen, J. R. Kaplan, K. Kavanagh, L. L. Rudel, R. Temel, L. A. Fairbanks, and A. G. Comuzzie. 2013. Significant genotype by diet (G×D) interaction effects on cardiometabolic responses to a pedigree-wide, dietary challenge in vervet monkeys (*Chlorocebus aethiops sabaeus*). *Am J Primatol* 75 (5):491–499.

Wall, H. S., C. Worthman, and J. Else. 1985. Effects of ketamine anaesthesia, stress and repeated bleeding on the haematology of vervet monkeys. *Lab Anim* 19:138–144.

Watson, L. M. 1997. Response of captive Barbados green monkeys (*Cercopithecus aethiops sabaeus*) to a variety of enrichment devices. *Lab Primate Newsl* 36:5–7.

Warren, W. C., A. J. Jasinska, R. Garcia-Perez, H. Svardal, C. Tomlinson, M. Rocchi, N. Archidiacono, et al. 2015. The genome of the vervet (*Chlorocebus aethiops sabaeus*). *Genome Res* 25 (12):1921–1933.

Woods, R. P., S. C. Fears, M. J. Jorgensen, L. A. Fairbanks, A. W. Toga, and N. B. Freimer. 2011. A web-based brain atlas of the vervet monkey, *Chlorocebus aethiops*. *Neuroimage* 54 (3):1872–1880.

Zahn, R. C., M. D. Rett, B. Korioth-Schmitz, Y. Sun, A. P. Buzby, S. Goldstein, C. R. Brown, et al. 2008. Simian immunodeficiency virus (SIV)-specific CD8+ T-cell responses in vervet African green monkeys chronically infected with SIVagm. *J Virol* 82 (23):11577–11588.

CHAPTER 22

Behavioral Management of *Papio* spp.

Corrine K. Lutz and C. Heath Nevill
Texas Biomedical Research Institute

CONTENTS

INTRODUCTION

The baboon shares a variety of genetic, physical, and physiological traits with humans and is therefore an important nonhuman primate model for behavioral and biomedical research. Current research areas that utilize the baboon as a model include nutrition, reproduction, genetics, cardiac disease, drug abuse, xenotransplantation, and epilepsy (VandeBerg et al. 2009). The baboon is also the most commonly used primate model for the genetic study of complex diseases, including cardiovascular disease, obesity, and hypertension (Cox et al. 2013). Although their large size makes baboons beneficial for many types of biomedical research protocols, it can also make housing and care a challenge. Fortunately, baboons are quite adaptable, allowing them to thrive in a number of housing situations. Proper behavioral management will allow for baboons to continue as a successful model in research.

NATURAL HISTORY

Baboons are part of the larger taxonomic group of Old World monkeys (family Cercopithecidae) and remain one of the most iconic and easily recognized of the African primates. There are presently five recognized species of baboons: olive (*Papio anubis*), yellow (*Papio cynocephalus*), hamadryas (*Papio hamadryas*), guinea (*Papio papio*), and chacma (*Papio ursinus*) (Groves 2005; Swedell and Leigh 2006). However, genetic analyses suggest that baboons may instead be separate subspecies under a single species *Papio hamadryas* (e.g., *Papio hamadryas anubis*; Williams-Blangero et al. 1990). Hybridization naturally occurs along the boundaries of habitats shared by species in the wild (Charpentier et al. 2012) and can also occur amongst all species in captivity.

Baboons are generally quadrupedal, but can exhibit bipedal locomotion under some circumstances (Coelho and Bramblett 1989). They have a relatively long tail that is held erect but with a downward bend. Unlike more arboreal monkeys, the baboons' tail plays little role in maintenance of balance and functions more to facilitate communication and as a means of maintaining visual contact in tall grass (Carnaby 2008). Baboons have a long, almost dog-like muzzle, hence the species name cynocephalus, or dog-headed. Each species has distinct coloration, with coat colors ranging from an olive drab agouti coat to red-brown to a brilliant silver-gray (Rowe 1996). Like other Cercopithecines, they possess cheek pouches adapted for food storage. Under conditions where the baboons' safety may be compromised, they can quickly gather food into their cheek pouches and retreat to a more secure location (Warren 2008). In general, males across the species are approximately twice the size and weight of females (Stoltz and Saayman 1970). Depending on the species, weights can vary from 21.3 to 28.1 kg for males and 11.8 to 14.5 kg for females (Rowe 1996). Not only do males weigh twice as much as females, but they also possess distinctive, large canines that are used for dominance displays and defense (Galbany et al. 2015; Rowe 1996).

Baboons inhabit a wide range of habitats across sub-Saharan Africa, from thorn scrub, grasslands, and savannahs, to montane regions, woodlands, and gallery and tropical rainforests (Rowe 1996; Whiten et al. 1991). Hamadryas are the furthest ranging species, found in far east Africa and into the Arabian Peninsula (Winney et al. 2004). Home ranges of baboons cover anywhere from 7.8 to 50 km², changing in response to resource availability (Dunbar 1988; Washburn and DeVore 1961). Troops have been observed to cover a distance of anywhere between 2.4 and 14.5 km in a given day (Stoltz and Saayman 1970).

Foraging is the predominant daytime activity, accounting for 21%–60% of the animals' time budget (Altmann 1980; Dunbar 1977; Rose 1977; Stacey 1986). A major part of feeding involves gathering foliage from herbs and shrubs on the ground, along with arboreal foraging of leaves and flowers. Subterranean foraging involves digging up fleshy bases, roots, and storage organs of a variety of herbs, grasses, and sedges (Whiten et al. 1991). In addition to foraging, baboons are active hunters. Prey includes infant ungulates, hares, birds, and sometimes other small monkeys (Rose 1977; Stoltz and Saayman 1970; Washburn and DeVore 1961). Eggs and insects are also collected (Washburn and DeVore 1961; Whiten et al. 1991). As human agriculture encroaches on their habitats, baboons have adapted by raiding crops for easy, readily available sources of food (Strum 2010), including occasional predation on livestock (Stoltz and Saayman 1970).

SOCIAL BEHAVIOR

Baboons are a social species, gathering in larger groups, or troops, which provide protection from predators and serve as the primary source for social and environmental information (Washburn and DeVore 1961). Most baboon species live in multi-male, multi-female troops ranging in size from tens to hundreds of animals, depending on species and availability of resources (Rowe 1996; Washburn and DeVore 1961). Hamadryas baboons are an exception and exist in groups of one dominant male with 1–10 subadult to adult females and their offspring. However, these one-male units can congregate in larger clans, and even larger bands, on sleeping cliffs at night (Kummer 1968; Sigg et al. 1982). Females form the core of society among most baboon species, usually remaining to breed within their natal group, while males generally emigrate at ~4 years of age (Barton 1990; Rowe 1996). An exception occurs again with the hamadryas baboon, where little kinship exists among females; females are instead acquired by the male as he establishes his one-male unit or harem group (Polo and Colmenares 2012; Sigg et al. 1982).

Dominance hierarchies are part of the social structure in baboon societies. Among most baboon species, dominance of females within a group is based on station within a matriarchal hierarchy, with a daughter's rank strongly determined by the rank of her mother. Female dominance relationships, therefore, tend to show a high degree of consistency between generations (Hausfater et al. 1982). High-ranking females enjoy preferential access to food sources, whereas low-ranking individuals tend to stay on the periphery of feeding clusters, occupying inferior feeding sites (Barton 1990). Males, on the other hand, tend to disperse from their natal group; therefore, the male hierarchy can show some degree of instability (Beehner et al. 2006). Male rank is based more on the individual's strength, body size, and/or age, rather than on kinship (Alberts et al. 2003). Although attaining a high rank in males can be costly (Gesquiere et al. 2011), it is strongly correlated with mating success (Alberts et al. 2003). Rank relationships can also be determined by approach–retreat interactions (Beehner et al. 2006) or by determining wins and losses in dyadic agonistic interactions (Galbany et al. 2015).

REPRODUCTION

Female baboons of all species have anogenital skin that changes in tumescence according to their estrous cycle. Anogenital swelling is a reliable signal of fertility status to males, occurring within a 4-day window around ovulation and lasting 2–3 days before returning to a detumescent state (Daspre et al. 2009). Males are selectively attracted to highly tumescent females. Physical changes to the anogenital skin also occur during pregnancy. For instance, in yellow baboons, paracollosal skin becomes pink, turning scarlet 2 months before parturition. No swelling occurs and there is no color change in the perineum. Contrastingly, in chacma baboons, the paracollosal skin becomes and stays scarlet, with some minor swelling within the 2 weeks leading up to parturition (Altmann 1973).

Infants are born after a 177-day gestation period with a dark black coat and bright pink skin. By 3–4 months of age, some darkening of the skin occurs along with golden hairs appearing in the coat. By 6 months of age, all natal coat and skin color is replaced by adult coloration, except for some pink on the muzzle and ears (Altmann 1980; Stoltz and Saayman 1970). At birth, infants are wholly dependent on their mothers for sustenance and transportation. Weaning begins at 4–6 months of age, when infants are refused nipple access and attempts to nurse are physically rejected. Reduction of reliance on milk alone as a form of sustenance is largely dependent on the availability of easily collected and digested food stuffs in the environment, such as fruits or flowers. This requires the development of motor skills, though, and infants must be 3–4 months old before they can obtain such food on their own, and 5–7 months old before they are capable of sufficiently supplementing their diet. By 1 year of age, an infant is quite independent and could potentially survive its mother's death (Altmann 1980).

BEHAVIOR AND TEMPERAMENT

Although baboons may appear to be aggressive due to their large size and prominent canine teeth, they tend to be less reactive than some other species of primates, such as rhesus macaques, which have a reputation for being volatile and aggressive (Beisner and Isbell 2011; Iredale et al. 2010). However, baboons are not all the same behaviorally, exhibiting a number of diverse personality traits, even within the same species. For example, in one study of chacma baboons, the females were designated as "nice," "aloof," or "loner" (Seyfarth et al. 2012), while another study of chacma baboons classified the subjects as "anxious," "calm," "bold," and "shy" (Carter et al. 2014). These personality traits remain fairly stable over time, but are not necessarily fixed (Seyfarth et al. 2014).

As with other species of nonhuman primates, baboon behavior can be organized into a number of categories. The frequency and/or duration of these behaviors may vary depending on the animal's sex, age, and aspects of the environment. A more complete ethogram can be found in Coelho and Bramblett (1989); however, a number of behavioral categories are described in the following sections.

Aggressive Behavior

Aggressive behavior can range from noncontact threats to physical attacks and wounding. The intensity, duration, and extent of threats and attacks can vary greatly based on the social situation (Hall and DeVore 1965). Threat behaviors may include branch (or cage) shake, brow raise, canine display, head bob, rub or slap the ground, yawn, and stare, whereas aggression may include charging or biting (Coelho and Bramblett 1989; Hall and DeVore 1965). In wild olive baboons, wounding frequencies from aggression are based in part on sex and age. Males received significantly more wounds than females, while females received more wounds when cycling (i.e., not pregnant or lactating). In addition, both sexes received more wounds at ages when competition for rank was the greatest (MacCormick et al. 2012). However, aggression takes up only a small percentage of the baboon's time budget (Harding 1980).

Submissive Behavior

Submissive behaviors are typically exhibited by lower ranking animals in response to a more dominant individual, and may include avoidance, flee, crouch, scratch, and a fear grimace that involves the pulling back of the lips to expose the teeth (Coelho and Bramblett 1989; Hall and DeVore 1965). Presenting (i.e., displaying of the hindquarters to another individual) may occur in a number of contexts including sexual, affiliative, and grooming, but it also occurs in a submissive context. Adult males are the primary recipients of presents, and the majority of male–male presents are directed toward the dominant animal (Hausfater and Takacs 1987; Judge et al. 2006). Similarly, "nonprime" males (i.e., young or old) presented significantly more than did "prime" males, and these presents were directed more frequently to prime males than to nonprime males (Fraser and Plowman 2007). Therefore, the act of presenting can be utilized as an indicator of submissive rank (Hall and DeVore 1965).

Reconciliation

Reconciliation is often engaged in by former opponents following conflict or aggressive interactions. Reconciliation serves to reduce arousal, as demonstrated by reductions in the self-directed behavior of the two opponents (Castles and Whiten 1998b; Judge and Bachmann 2013). Reconciliation includes behaviors such as approach and friendly grunts (Silk et al. 1996), and it reduces the likelihood of future aggression with former opponents (Castles and Whiten 1998b). Reconciliation is more likely to occur between individuals that are more closely related or that have a higher quality relationship than between those less closely related or with a lower quality relationship (Castles and Whiten 1998a; Romero et al. 2008).

Displacement Behaviors

Displacement behaviors are ambiguous types of behaviors that occur in some situations of social interaction. They appear to be irrelevant to the situation at hand and tend to occur in situations of stress (Maestripieri et al. 1992) or uncertainty (Hall and DeVore 1965). Displacement behaviors are often tension-related and may include activities such as a brow or muzzle wipe, yawn, scratch, or body shake (Castles et al. 1999; Coelho and Bramblett 1989).

Friendly Behavior

Friendly behaviors may include behaviors such as presenting, mounting, grooming, or lip-smacking (Hall and DeVore 1965). Lip-smacking is a rapid, repetitive opening and closing of the lips (Coelho and Bramblett 1989) that occurs as a greeting behavior, often when an animal approaches another animal to groom, or during grooming (Hall and DeVore 1965). Lip-smacking likely serves to facilitate positive social communication (Easley and Coelho 1991).

Grooming

Grooming has multiple functions in nonhuman primates; it aids in maintenance of coat and skin hygiene, but is also an integral form of social interaction (Figure 22.1). Grooming involves a parting of the hair with the fingers and the picking off of particulate items, such as dried skin, by hand or by mouth (Hall and DeVore 1965), and can be directed to one's own body or to another individual. Self-grooming is generally directed to easily accessible areas, such as the legs, arms, and tail, while social grooming tends to be directed to less-accessible areas, such as the head, back, and shoulders (Akinyi et al. 2013). One function of grooming is the removal of ectoparasites. For example, in a study of wild baboons, the animals that received more grooming had fewer ticks (Akinyi et al. 2013). Grooming can also be used to improve social bonds or reduce tension (Schino et al. 1988). When a group of hamadryas baboons was housed in crowded conditions, adult females groomed their harem males more than when the conditions were less crowded (Judge et al. 2006). Although female–female grooming is most common (Hall and DeVore 1965), the majority of male–female interactions also involve grooming (Harding 1980), and female hamadryas baboons groom the male more than they groom each other (Kummer 1990).

Figure 22.1   Social housing allows for grooming behavior in two female baboons.

Mating

Mating patterns are varied, both across individuals and across species. Typically, juvenile, sub-adult, and less dominant animals copulate with females during the initial stages of the female's swelling, while the dominant males form consort pairs and copulate with the female during her maximum tumescence, which typically coincides with ovulation (Hall and DeVore 1965). However, this differs in the hamadryas baboon, which lives in harem groups, and a single male has sole access to his females (Kummer and Kurt 1963). Copulation can occur in a series of mounts or a single mounting sequence followed by ejaculation (Hall and DeVore 1965). Other types of sexual behavior can include genital inspect, hip raise, and gentle biting by the male at the nape of the female's neck (Coelho and Bramblett 1989; Hall and DeVore 1965).

Vocalizations

Baboons have a number of distinctive vocalizations that are used for communication. Some examples include

- *Grunt*—A short, quiet "uh" sound (Coelho and Bramblett 1989) that is used as a contact call for short-distance communication (Ey et al. 2009) and is uttered in affiliative situations when approaching mothers with infants, while grooming and foraging, and when about to move across an open area (Gustison et al. 2012; Hall and DeVore 1965; Rendall et al. 1999).
- *Aggressive or threat vocalization*—A series of short, rapid grunts produced by the dominant animal in an aggressive encounter (Coelho and Bramblett 1989; Gustison et al. 2012; Hall and DeVore 1965).
- *Bark*—A loud, harsh call used when responding to a predator or other threat (alarm bark) (Fischer et al. 2001; Gustison et al. 2012; Hall and DeVore 1965), or a more tonal sound used when the individual is becoming separated from the group (contact bark) (Fischer et al. 2001).

- *Display call or wahoo*—Loud vocalizations typically uttered by males during displays or competitive interactions with other males (Gustison et al. 2012). This call may be used as an indicator of a male's quality or stamina (Fischer et al. 2004).
- *Scream or screech*—A loud, high-pitched call usually given by an individual that is being attacked or threatened (Coelho and Bramblett 1989; Gustison et al. 2012; Hall and DeVore 1965).
- *Ick-Ooer*—A two-phase vocalization typically performed by infants that appears to be a frustration response, such as when separated from the mother (Hall and DeVore 1965).

SPECIAL TOPIC: ABNORMAL BEHAVIOR

As with other species of captive nonhuman primates, baboons can develop a variety of abnormal behaviors in captivity (Lutz et al. 2014). Behaviors can be considered abnormal if they are qualitatively different (i.e., occur in captivity, but typically not in the wild) or are quantitatively different (i.e., occur significantly more or less often than would occur in the wild; Erwin and Deni 1979). Some examples of abnormal behavior observed in captive baboons are as follows (definitions taken from Lutz et al. 2014):

- *Abnormal mouth movements*—Repeated movement of the mouth, lips, or tongue, not associated with eating or manipulation of an object in the mouth
- *Coprophagy*—Ingesting or manipulating feces in the mouth
- *Hair-eat*—Chewing or ingestion of hair
- *Hair-pull*—Pulling out hair from own body
- *Head toss*—Repeated circular movement of the head at the neck
- *Pace*—Repeated walking in the same pattern
- *Regurgitate*—The backward flow of already swallowed food into the mouth or cheek pouches
- *Rock*—Repeated back-and-forth or side-to-side movement of the body
- *Self-bite*—Direct contact of teeth with the skin that may or may not result in wounding
- *Swing*—Repetitive back-and-forth movement when hanging from the cage side or ceiling
- *Wiggle digits*—Repeated movement of fingers or toes usually at, in, or around the mouth, often preceding regurgitation

The presence of abnormal behavior may be indicative of past or present environmental deficiencies and/or poor welfare (Mason 1991), and may be a behavioral management issue. Environmental risk factors for abnormal behavior in baboons include routine veterinary procedures, nursery rearing, and/or single housing at an early age or for an extended period of time (Brent and Hughes 1997; Lutz et al. 2014; Veira and Brent 2000). Abnormal behavior is of concern if it interferes with normal species-typical behavior or if it causes injury. Self-inflicted injury can be caused directly through behaviors such as self-biting, or indirectly through behaviors such as hair-eating, potentially resulting in trichobezoars from the ingested hair (Mejido et al. 2009). If the behavior becomes excessive, it takes up a relatively large portion of the day, significantly interferes with species-typical behaviors, and/or results in injury, behavioral assessments and interventions should be performed. Utilizing a variety of behavioral management procedures, such as food treats, inanimate enrichment, positive reinforcement training, and social housing can help to reduce abnormal behavior (Bourgeois and Brent 2005; Kessel and Brent 2001), but enrichment, such as additional forage, is not effective in all cases (Nevill and Lutz 2015). More severe self-injurious behavior has also been reduced by changing housing arrangements to allow for social interactions (De Villiers and Seier 2010) or by providing a chew toy to redirect the self-biting behavior (Crockett and Gough 2002). Drug therapy may also be helpful in treating more severe instances of self-injury (Macy Jr. et al. 2000). Interventions or therapies typically reduce, rather than eliminate, abnormal behavior. Therefore, it is important to understand the risk factors for these behaviors with a focus on developing and implementing behavioral management strategies that are likely to prevent such behavior.

TRAINING

Positive reinforcement training involves rewarding an animal for a desired behavior, thus increasing the chance that the behavior will occur again. Baboons can be trained to perform a variety of behaviors for husbandry, research, and/or clinical purposes. Through positive reinforcement, the animal cooperates voluntarily, giving it some amount of control while promoting a positive relationship with the animal care staff. Positive reinforcement training also reduces stress (Lambeth et al. 2006) and reduces the need for physical or chemical restraint.

Husbandry-related training objectives for captive baboons may include behaviors such as shifting or targeting. Shift training uses positive reinforcement to encourage the baboons to easily and readily shift between enclosures, into a chute, or into a transfer cage (Holmes et al. 1996). Training the animals to shift into different spaces increases safety for the staff and animals and improves husbandry practices. Training the animals to shift also eliminates the need for care staff to enter enclosures with the animals during cleaning, and eliminates the need to dart the animals for routine medical procedures. Alternatively, target training involves teaching the animal to hold a position or location within the cage (O'Brien et al. 2008). Targeting can be utilized to gain easier access to an individual or individuals within a group, and it can be used to reduce interference of a dominant animal when a more subordinate animal is provided with food treats or clinical care.

Training can be used in other ways to improve the efficiency of clinical or research procedures. For example, baboons have been successfully trained to voluntarily participate in noninvasive blood pressure measurements (Mitchell et al. 1980; Turkkan 1990), in the collection of saliva samples (Fagot et al. 2014; Pearson et al. 2008), and in the use of computerized testing devices (Fagot et al. 2014; Maugard et al. 2014). In cases where training is not an option, jacketing (Bentson et al. 1999) or tethering (Coelho and Carey 1990; Lukas et al. 1982) may be utilized for obtaining samples from unrestrained animals, also facilitating behavioral management.

HOUSING

Baboons are highly adaptable animals and can therefore be successfully housed in a variety of captive environments. When developing captive environments for baboons, consideration should be given to issues related to behavioral management procedures, safety (both animal and personnel), accessibility, and the welfare of the animals. However, regardless of the size and configuration of the environment, it must be sturdy enough to survive a baboon's strength and destructive nature.

Single Housing

Baboons can be housed singly indoors in stainless steel caging. Depending on the animal's size, the cages for an adult baboon can range from 0.74 to 2.33 m² or larger (Animal Welfare Act 2013) and should allow for normal postural adjustments. These cages have platforms or perches to allow the baboon to rest above the floor of the cage. This type of caging is typically used to house a baboon for limited periods for clinical or research needs that require easy access to, or separation of, the animal. A squeeze-back mechanism allows for physical restraint for treating or sedating the animal. Alternatively, transfer cages can be utilized to access the animal or to move it to another location. Although single housing is safe and efficient, it significantly diminishes animal well-being (Rennie and Buchanan-Smith 2006). Because of this, a clinical or research exemption for single housing is required and should be reviewed by the attending veterinarian every 30 days (Animal Welfare Act 2013). Individually housed animals should always have visual and auditory access to another animal; when appropriate, a companion animal can be housed in the same room.

Social Housing

Baboons can be socially housed in pairs, as well as in small- to medium-sized social groups (e.g., same-sex, harem, and juvenile groups), or large multi-male, multi-female groups consisting of 100 or more individuals (Goodwin and Coelho 1982; Figure 22.2). Social housing allows for increased species-specific behaviors, which significantly increases animal well-being. Baboons can be socially housed in indoor cages, indoor/outdoor runs or pens, or large corrals or field cages measuring an acre or more in size. Large group housing, such as in corrals, can more closely replicate social experiences in the wild; however, housing can also be tailored to species differences in social organization. For example, caging was designed to accommodate hamadryas harem groups that separate into smaller groups during the day and congregate in larger groups at night in the wild (Maclean et al. 1987). Caging for social groups should include climbing structures, perches, and visual barriers, as well as sufficient space for animals to retreat from one another. Social groups should be routinely observed for signs of aggression or food monopolization, and appropriate steps should be taken if an intervention is warranted. For example, animals could be removed from the group in cases of aggression, or the number of feedings, or feeding sites, could be increased in cases of food monopolization. As the group and enclosure sizes increase, observation of individual animals becomes more difficult. Animals housed in larger field cages or corrals may need to be viewed from observation platforms or by entering the enclosure on foot or in a vehicle. The use of binoculars may also be required. Animals of concern can be marked with a dye to more easily identify the individual.

Figure 22.2 Baboons can be housed in large social groups. Note the lack of vegetation caused by destructive foraging behavior.

Climate is one factor used to determine the extent of outdoor access. Regardless of the type of housing, adequate shelter from extreme environmental elements is necessary. For outdoor enclosures, sheltered space needs to be provided, along with supplemental heat in colder climates or during colder seasons. The shelters should be sufficiently large for all animals to enter, or alternatively, multiple shelters could be provided that allow subordinates to take refuge separately from dominants. Concrete floors can be heated by running warm water through pipes that are installed when the concrete is poured. In hot climates or seasons, an adequate amount of shade is necessary, and misters or water sprinklers may be considered to keep the animals cool. However, the animals should be given the opportunity to avoid the water if they prefer not to get wet.

To improve accessibility and safety for both the animals and the personnel, chute systems are often used in conjunction with the caging for social groups (Holmes et al. 1996). The animals can be trained to run into the chutes, allowing for the care staff to safely enter an empty cage for sanitizing or to remove animals for routine physicals or clinical care. The chutes should also contain dividing "doors" to separate the animals for easier accessibility. Animal separations using chutes can facilitate the administration of medication, the weighing of individual animals, or the feeding of individualized diets (Holmes et al. 1996; Schlabritz-Loutsevitch et al. 2004). Care should be taken to minimize the time animals spend in chutes, as the confined environment does not provide sufficient avenues for escape should fighting occur, increasing the potential for wounding.

ENVIRONMENTAL ENRICHMENT

Environmental enrichment is the process of enhancing an animal's environment in ways that promote their welfare. It provides behavioral choices to animals, promotes species-typical behavior, and promotes psychological well-being. Enrichment for captive nonhuman primates can be divided into five different categories: social, nutritional, occupational, sensory, and physical, with many enrichment items falling into multiple categories (Bloomsmith et al. 1991).

Social Enrichment

Social enrichment involves the pairing or grouping of conspecifics, but can also include visual or auditory interaction with a neighboring conspecific, or interactions with a human. Because all species of baboons live in social groups in the wild, social housing is an important feature of any environmental enrichment program. Social housing is enriching in part because it is typically varied and unpredictable. It stimulates all of the senses and is less likely to result in habituation than are other forms of enrichment. Social housing also promotes the expression of species-typical behaviors and can be used to rehabilitate animals and reduce high levels of abnormal behavior (Bourgeois and Brent 2005; Kessel and Brent 2001). The latest edition of the *Guide for the Care and Use of Laboratory Animals* acknowledges the importance of social housing by noting that all primate species should be housed socially, and that any time single housing is required, it should be limited in duration as much as possible (NRC 2011).

Nutritional Enrichment

Nutritional enrichment includes the introduction of a variety of foods into an animal's diet, but also variation in its delivery and presentation. Given the variety of the baboon's diet in the wild (DeVore and Hall 1965), a baboon's diet in captivity should be supplemented with a wide assortment of produce and other veterinarian-approved food items. Items, such as produce, can be left whole to encourage food processing behaviors and to extend the amount of time that the baboons engage in feeding. However, chopping food items into randomly shaped pieces and/or freezing them prior to distribution are options to introduce novelty. Produce should be supplied in sufficient quantity and scattered throughout the cage to ensure that all animals get access, and to prevent dominant animals from monopolizing the enrichment. Smaller pieces are easier to process and help to provide a more equitable distribution of food. Scattered grains, nuts, and legumes such as corn, soybeans, sunflower seeds, and peanuts can also be used for nutritional enrichment, serving the dual function of providing variety in the diet and encouraging foraging. For animals that are food restricted for clinical or research concerns, ice cubes may be an alternative to more highly caloric food items.

Occupational Enrichment

Occupational enrichment involves both psychological and physical tasks, and can be accomplished through the provision of diverse items, such as positive reinforcement training and foraging devices. Positive reinforcement training for simple tasks, such as stationing at a target, has proven to be an effective form of occupational enrichment, which has also been shown to reduce abnormal behavior (Bourgeois and Brent 2005) and to serve as a form of social enrichment with humans (Bloomsmith et al. 1991). Foraging devices act as another form of occupational enrichment by extending the amount of time required for retrieving food, and increasing species-typical foraging behavior to levels more closely approximating baboon time budgets in the wild. Animals have been reported to work for food in spite of the availability of another freely accessible food source (Reinhardt 1997), and foraging devices that require a baboon to engage in some manipulation to receive a food item give them this opportunity. Foraging devices also help to reduce the performance of abnormal behaviors (Brent and Long 1995; De Villiers and Seier 2010). For example, simple foraging boards, consisting of corn syrup-coated fleece sprinkled with commercially available forage bits, were shown to eliminate stereotypic behaviors in singly housed baboons (Pyle et al. 1996). When attempting to construct foraging devices, construction materials should be sturdy and consist of items such as steel, aluminum, and schedule 80 PVC. Locking nuts are necessary to keep the baboons from disassembling the device (Brent and Butler 2005). Also, similar to the distribution of produce or grain in a group setting, enough devices should be provided to avoid monopolization by a dominant animal.

Sensory Enrichment

Sensory enrichment consists of visual and auditory stimuli, but also items that stimulate the olfactory and tactile senses. Televisions are an effective way to provide both visual and auditory stimulation, and radios may act as both enrichment and as a way to mask potentially stressful ambient noises. For example, although there was no effect on the animal's behavior, baboon heart rates were shown to be significantly reduced when music was played (Brent and Weaver 1996). Aromatic and edible herbs, like basil or fennel, can also be given to baboons as a form of sensory enrichment. Additionally, popcorn can be popped within an animal room to stimulate a variety of senses, and provided as a food treat afterward. Other types of sensory enrichment include fragrant essential oils or extracts, such as vanilla, peppermint, or lavender that can be left to disseminate into the air or rubbed on a small swatch of paper or a toy that can be given to the animals.

Physical Enrichment

Physical enrichment can be further categorized into two types: structural and manipulable enrichment. Structural enrichment can include outfitting enclosures with items that increase cage complexity such as perches, culverts, climbing structures, ropes, or swings (Figure 22.3). Visual barriers, such as solid panels, are also important for providing privacy and for providing subordinate animals an opportunity to hide or avoid contact. Although baboons are more terrestrial than most monkeys in the wild, they seek trees or other elevated structures as refuge from predators and as safe resting places at night (Stoltz and Saayman 1970; Washburn and Devore 1961). Multilevel elevated platforms and perches are integral to addressing this tendency, and they give subordinates places to flee and stay out of the way of dominant animals. Providing multiple perches is also necessary to reduce competition over access to places to sit (Reinhardt 1997). Elevated perches also allow for better access to vertical space and a better view of their surroundings (Kessel and Brent 1996). Durability of construction materials should be a consideration when building perches or other physical enrichment on which baboons can climb. While natural materials like thick tree branches can be used, steel, aluminum, and PVC are more durable and less apt to be chewed. These materials are also more easily cleaned and sanitized, and are more capable of supporting the weight of an adult male baboon, which can exceed 30 kg.

Manipulable enrichment includes toys with which baboons can interact both manually and orally. Sturdy dog chew toys; pieces of hardwood, such as cherry or manzanita; and similar chewable items are excellent examples of manipulable enrichment for baboons. However, any enrichment item placed in the cage must not have small parts that can be ingested, and it must be routinely inspected and discarded if damaged (Matz-Rensing et al. 2004). Although there are many types of manipulable enrichment, baboons do show preferences (Brent and Belik 1997; Hienz et al. 1998), and preference testing prior to use may be warranted. Manipulable enrichment has also been shown to decrease abnormal, cage-directed, inactive, and self-directed behaviors (Brent and Belik 1997), and therefore has multiple benefits.

Figure 22.3 Hanging barrels can be used as both climbing structures and visual barriers.

EXPERT RECOMMENDATIONS

The behavioral management of baboons requires a multipronged approach. The baboon is a social species; therefore, social housing is critical for their well-being in captivity. Social housing can be accomplished by the formation of various types of social groups, such as harem, same-sex, and multi-male, multi-female groups. Infants should be allowed to remain with their mothers, and nursery rearing should be utilized as a last resort. Although baboons are fairly accepting of new animals into their social group, careful observation is necessary to ensure compatibility, and adjustments to the group composition should be made when necessary. Environmental enrichment, such as elevated platforms, chew toys, and foraging devices, helps to enhance the lives of captive baboons. However, the size and strength of baboons present unique challenges to their behavioral management. For example, caging needs to be big and sturdy enough to accommodate their large body size and destructive nature. Similarly, any enrichment device needs to be sturdy enough to withstand manipulation and chewing. However, such challenges can be overcome with appropriate enclosure and enrichment design and construction. Baboons also respond well to training; therefore, positive reinforcement training techniques can be utilized for both husbandry practices (e.g., shifting between enclosures), as well as enrichment, research, or clinical needs. With proper care, baboons can thrive in a captive environment.

CONCLUSIONS

The baboon is an important animal model for use in biomedical research, and its mild temperament and adaptable nature make it well suited to a captive environment. However, attention must be paid to ensure that their physical and welfare needs are met in captivity. With an understanding of the physical and behavioral needs of baboons, a suitable care and enrichment plan can be developed to ensure proper behavioral management of captive baboons with a focus on well-being.

ACKNOWLEDGMENTS

This manuscript was supported by grant P51OD011133 to Texas Biomedical Research Institute.

REFERENCES

Akinyi, M.Y., Tung, J., Jeneby, M., et al. 2013. Role of grooming in reducing tick load in wild baboons (*Papio cynocephalus*). *Anim Behav* 85:559–568.

Alberts, S.C., Watts H.E., and Altmann, J. 2003. Queuing and queue-jumping: Long-term patterns of reproductive skew in male savannah baboons, *Papio cynocephalus*. *Anim Behav* 65:821–840.

Altmann, S.A. 1973. The pregnancy sign in savannah baboons. *J Zoo Anim Med* 4(2):8–12.

Altmann, J. 1980. *Baboon Mothers and Infants*. Cambridge, MA: Harvard University Press.

Animal Welfare Act of 1966, 7 U.S.C. Sect. 2131, 2013.

Barton, R. 1990. Feeding, reproduction, and social organisation in female olive baboons (*Papio anubis*). In *Baboons: Behavior and Ecology, Use and Care*, eds Thiago de Mello, M., Whiten, A., and Byrne, R.W., 29–37. Brasilia: Selected Proceedings of the XIIth Congress of the International Primatological Society.

Beehner, J.C., Bergman, T.J., Cheney, D.L., et al. 2006. Testosterone predicts future dominance rank and mating activity among male Chacma baboons. *Behav Ecol Sociobiol* 59:469–479.

Beisner, B.A. and Isbell, L.A. 2011. Factors affecting aggression among females in captive groups of rhesus macaques (*Macaca mulatta*). *Am J Primatol* 73:1152–1159.

Bentson, K.L., Miles, F.P., Astley, C.A., et al. 1999. A remote-controlled device for long-term blood collection from freely moving, socially housed animals. *Behav Res Methods Instrum Comput* 31:455–463.

Bloomsmith, M.A., Brent, L.Y., and Schapiro, S.J. 1991. Guidelines for developing and managing an environ-mental enrichment program for nonhuman primates. *Lab Anim Sci* 41(4):372–377.

Bourgeois, S.R. and Brent, L. 2005. Modifying the behaviour of singly caged baboons: Evaluating the effec-tiveness of four enrichment techniques. *Anim Welf* 14:71–81.

Brent, L. and Belik, M. 1997. The response of group-housed baboons to three enrichment toys. *Lab Anim* 31:81–85.

Brent, L. and Butler, T.M. 2005. Baboons. In *Enrichment for Nonhuman Primates*, ed. Bayne, K., 1–10. Bethesda, MD: NIH Office of Laboratory Animal Science.

Brent, L. and Hughes, A. 1997. The occurrence of abnormal behavior in group-housed baboons. *Am J Primatol* 42:96–97.

Brent, L. and Long, K.E. 1995. The behavioral response of individually caged baboons to feeding enrichment and the standard diet: A preliminary report. *Contemp Top* 34:65–69.

Brent, L. and Weaver, D. 1996. The physiological and behavioral effects of radio music on singly housed baboons. *J Med Primatol* 25:370–374.

Carnaby, T. 2008. *Beat about the Bush: Mammals*. Johannesburg, South Africa: Jacana Media.

Carter, A.J., Marshall, H.H., Heinsohn, R., et al. 2014. Personality predicts the propensity for social learning in a wild primate. *PeerJ* 2:e283.

Castles, D.L. and Whiten, A. 1998a. Post-conflict behaviour of wild olive baboons. I. Reconciliation, redirec-tion and consolation. *Ethology* 104:126–147.

Castles, D.L. and Whiten, A. 1998b. Post-conflict behaviour of wild olive baboons. II. Stress and self-directed behaviour. *Ethology* 104:148–160.

Castles, D.L., Whiten, A., and Aureli, F. 1999. Social anxiety, relationships and self-directed behaviour among wild female olive baboons. *Anim Behav* 58:1207–1215.

Charpentier, M.J.E., Fontaine, M.C., Cherel, E., et al. 2012. Genetic structure in a dynamic baboon hybrid zone corroborates behavioural observations in a hybrid population. *Mol Ecol* 21:715–731.

Coelho, A.M., Jr. and Bramblett, C.A. 1989. Behaviour of the genus *Papio*: Ethogram, taxonomy, methods, and comparative measures. In *Perspectives in Primate Biology*, Vol. 3, eds Seth, P.K. and Seth, S., 117–140. New Delhi: Today & Tomorrow Printers and Publishers.

Coelho, A.M. and Carey, K.D. 1990. A social tethering system for nonhuman primates used in laboratory research. *Lab Anim Sci* 40:388–394.

Cox, L.A., Comuzzie, A.G., Havill, L.M., et al. 2013. Baboons as a model to study genetics and epigenetics of human disease. *ILAR J* 54:106–121.

Crockett, C.M. and Gough, G.M. 2002. Onset of aggressive toy biting by a laboratory baboon coincides with cessation of self-injurious behavior. *Am J Primatol* 57 (supplement 1):39.

Daspre, A., Heistermann, M., Hodges, J.K., et al. 2009. Signals of female reproductive quality and fertility in colony-living baboons (*Papio h. anubis*) in relation to ensuring paternal investment. *Am J Primatol* 71:529–538.

DeVore, I. and Hall, K.R.L. 1965. Baboon ecology. In *Primate Behavior: Field Studies of Monkeys and Apes*, ed. DeVore, I., 20–52. New York: Holt, Rinehart and Winston.

De Villiers, C. and Seier, J.V. 2010. Stopping self injurious behaviour of a young male Chacma baboon (*Papio ursinus*). *Anim Technol Welf* 9:77–80.

Dunbar, R.I.M. 1977. Feeding ecology of gelada baboons: A preliminary report. In *Primate Ecology: Studies of Feeding and Ranging Behaviour in Lemurs, Monkeys, and Apes*, ed. Clutton-Brock, T.H., 251–273. New York: Academic Press.

Dunbar, R.I.M. 1988. *Primate Social Systems*. New York: Springer.

Easley, S.P. and Coelho, A.M. 1991. Is lipsmacking an indicator of social status in baboons? *Folia Primatol* 56:190–201.

Erwin, J. and Deni, R. 1979. Strangers in a strange land: Abnormal behaviors or abnormal environments? In *Captivity and Behavior: Primates in Breeding Colonies, Laboratories, and Zoos*, eds Erwin, J., Maple, T.L., and Mitchell, G., 1–28. New York: Van Nostrand Reinhold.

Ey, E., Rahn, C., Hammerschmidt, K., et al. 2009. Wild female olive baboons adapt their grunt vocalizations to environmental conditions. *Ethology* 115:493–503.

Fagot, J., Gullstrand, J., Kemp, C., et al. 2014. Effects of freely accessible computerized test systems on the spontaneous behaviors and stress level of Guinea baboons (*Papio papio*). *Am J Primatol* 76:56–64.

Fischer, J., Kitchen, D.M., Seyfarth, R.M., et al. 2004. Baboon loud calls advertise male quality: Acoustic features and their relation to rank, age, and exhaustion. *Behav Ecol Sociobiol* 56:140–148.

Fischer, J., Metz, M., Cheney, D.L., et al. 2001. Baboon responses to graded bark variants. *Anim Behav* 61:925–931.

Fraser, O. and Plowman, A.B. 2007. Function of notification in *Papio hamadryas*. *Int J Primatol* 28:1439–1448.

Galbany, J., Tung, J., Altmann, J., et al. 2015. Canine length in wild baboons: Maturation, aging, and social dominance rank. *PLoS One* 10(5): e0126415. doi:10.1371/journal.pone.0126415.

Gesquiere, L.R., Learn, N.H., Simao, M.C.M., et al. 2011. Life at the top: Rank and stress in wild male baboons. *Science* 333:357–360.

Goodwin, W.J. and Coelho, A.M., Jr. 1982. Development of a large scale baboon breeding program. *Lab Anim Sci* 32:672–676.

Groves, C.P. (2005). Genus Papio. In: *Mammal Species of the World: A Taxonomic and Geographic Reference* (3rd edition), eds Wilson, D.E. and Reeder, D.M., 166–167. Baltimore, MD: Johns Hopkins University Press.

Gustison, M.L., le Roux, A., Bergman, T.J. 2012. Derived vocalizations of geladas (*Theropithecus gelada*) and the evolution of vocal complexity in primates. *Philos Trans R Soc B* 367:1847–1859.

Hall, K.R.L. and DeVore, I. 1965. Baboon social behavior. In *Primate Behavior: Field Studies of Monkeys and Apes*, ed. DeVore, I., 53–110. New York: Holt, Rinehart and Winston.

Harding, R.S.O. 1980. Agonism, ranking, and the social behavior of adult male baboons. *Am J Phys Anthropol* 53:203–216.

Hausfater, G., Altmann, J. and Altmann, S. 1982. Long-term consistency of dominance relations among female baboons (*Papio cynocephalus*). *Science* 217:752–755.

Hausfater, G. and Takacs, D. 1987. Structure and function of hindquarter presentations in yellow baboons (*Papio cynocephalus*). *Ethology* 74:297–319.

Hienz, R.D., Zarcone, T.J., Turkkan, J.S., et al. 1998. Measurement of enrichment device use and preference in singly caged baboons. *Lab Prim News* 37(3):6–10.

Holmes, K.A., Paull, M.D., Birrell, A.M., et al. 1996. A unique design for ease of access and movement of captive *Papio hamadryas*. *Lab Anim* 30:327–331.

Iredale, S.K., Nevill, C.H., and Lutz, C.K. 2010. The influence of observer presence on baboon (*Papio* spp.) and rhesus macaque (*Macaca mulatta*) behavior. *Appl Anim Behav Sci* 122:53–57.

Judge, P.G. and Bachmann, K.A. 2013. Witnessing reconciliation reduces arousal of bystanders in a baboon group (*Papio hamadryas hamadryas*). *Anim Behav* 85:881–889.

Judge, P.G., Griffaton, N.S., Fincke, A.M. 2006. Conflict management by hamadryas baboons (*Papio hamadryas hamadryas*) during crowding: A tension-reduction strategy. *Am J Primatol* 68:993–1006.

Kessel, A.L. and Brent, L. 1996. Space utilization by captive-born baboons (*Papio* sp.) before and after provision of structural enrichment. *Anim Welf* 5:37–44.

Kessel, A. and Brent, L. 2001. The rehabilitation of captive baboons. *J Med Primatol* 30:71–80.

Kummer, H. 1968. Two variations in the social organization of baboons. In *Primates: Studies in Adaptation and Variability*, ed. Jay, P.C., 293–312. New York: Holt, Rinehart and Winston.

Kummer, H. 1990. The social system of hamadryas baboons and its presumable evolution. In *Baboons: Behavior and Ecology, Use and Care*, eds Thiago de Mello, M., Whiten, A., and R.W. Byrne, 43–60. Brasilia: Selected Proceedings of the XIIth Congress of the International Primatological Society.

Kummer, H. and Kurt, F. 1963. Social units of a free-living population of hamadryas baboons. *Folia Primatol* 1:4–19.

Lambeth, S.P., Hau, J., Perlman, J.E., et al. 2006. Positive reinforcement training affects hematologic and serum chemistry values in captive chimpanzees (*Pan troglodytes*). *Am J Primatol* 68:245–256.

Lukas, S.E., Griffiths, R.R., Bradford, L.D., et al. 1982. A tethering system for intravenous and intragastric drug administration in the baboon. *Pharmacol Biochem Behav* 17:823–829.

Lutz, C.K., Williams, P.C., and Sharp, R.M. 2014. Abnormal behavior and associated risk factors in captive baboons (*Papio hamadryas* spp.). *Am J Primatol* 76:355–361.

MacCormick, H.A., MacNulty, D.R., Bosacker, A.L., et al. 2012. Male and female aggression: Lessons from sex, rank, age, and injury in olive baboons. *Behav Ecol* 23:684–691.

Maclean, J.M., Phippard, A.F., Garner, M.G., et al. 1987. Group housing of hamadryas baboons: A new cage design based upon field studies of social organization. *Lab Anim Sci* 37:89–93.

Macy, J.D., Jr., Beattie, T.A., Morgenstern, S.E., et al. 2000. Use of guanfacine to control self-injurious behavior in two rhesus macaques (*Macaca mulatta*) and one baboon (*Papio Anubis*). *Comp Med* 50:419–425.

Maestripieri, D., Schino, G., Aureli, F., et al. 1992. A modest proposal: Displacement activities as an indicator of emotions in primates. *Anim Behav* 44:967–979.

Mason, G.J. 1991. Stereotypies and suffering. *Behav Process* 25:103–115.

Matz-Rensing, K., Floto, A., and Kaup, F.J. 2004. Intraperitoneal foreign body disease in a baboon (*Papio hamadryas*). *J Med Primatol* 33:113–116.

Maugard, A., Wasserman, E.A., Castro, L., et al. 2014. Effects of training condition on the contribution of specific items to relational processing in baboons (*Papio papio*). *Anim Cogn* 17:911–924.

Mejido, D.C.P., Dick, E.J., Jr., Williams, P.C., et al. 2009. Trichobezoars in baboons. *J Med Primatol* 38:302–309.

Mitchell, D.S., Wigodsky, H.S., Peel, H.H., et al. 1980. Operant conditioning permits voluntary, noninvasive measurement of blood pressure in conscious, unrestrained baboons (*Papio cynocephalus*). *Behav Res Methods Instrum* 12:492–498.

National Research Council (NRC). 2011. *The Guide for the Care and Use of Laboratory Animals* (8th edition). Washington, DC: National Academies Press.

Nevill, C.H. and Lutz, C.K. 2015. The effect of a feeding schedule change and the provision of forage material on hair eating in a group of captive baboons (*Papio hamadryas* sp.). *J Appl Anim Welf Sci* 18:319–331.

O'Brien, J.K., Heffernan, S., Thomson, P.C., et al. 2008. Effect of positive reinforcement training on physiological and behavioural stress responses in the hamadryas baboon (*Papio hamadryas*). *Anim Welf* 17:125–138.

Pearson, B.L., Judge, P.G., and Reeder, D.M. 2008. Effectiveness of saliva collection and enzyme-immunoassay for the quantification of cortisol in socially housed baboons. *Am J Primatol* 70:1145–1151.

Polo, P. and Colmenares, F. 2012. Behavioural processes in social context: Female abductions, male herding and female grooming in hamadryas baboons. *Behav Process* 90:238–245.

Pyle, D.A., Bennet, A.L., Zarcone, T.J., et al. 1996. Use of two foraging devices by singly housed baboons. *Lab Primates Newsl* 35(2):10–15.

Reinhardt, V. 1997. Species-adequate housing and handling conditions for old world nonhuman primates kept in research institutions. In *Comfortable Quarters for Laboratory Animals* (8th edition), ed. Reinhardt, V., 85–93. Washington, DC: Animal Welfare Institute.

Rendall, D., Seyfarth, R.M., Cheney, D.L., et al. 1999. The meaning and function of grunt variants in baboons. *Anim Behav* 57:583–592.

Rennie, A.E. and Buchanan-Smith, H.M. 2006. Refinement of the use of non-human primates in scientific research. Part II: Housing, husbandry and acquisition. *Anim Welf* 15:215–238.

Romero, T., Colmenares, F., and Aureli, F. 2008. Postconflict affiliation of aggressors in *Papio hamadryas*. *Int J Primatol* 29:1591–1606.

Rose, M.D. 1977. Positional behavior of olive baboons (*Papio anubis*) and its relationship to maintenance and social activities. *Primates* 18:59–116.

Rowe, N. 1996. *The Pictorial Guide to the Living Primates*. East Hampton, NY: Patagonias Press.

Schino, G., Scucchi, S., Maestripieri, D., et al. 1988. Allogrooming as a tension-reduction mechanism: A behavioral approach. *Am J Primatol* 16:43–50.

Schlabritz-Loutsevitch, N.E., Howell, K., Rice, K., et al. 2004. Development of a system for individual feeding of baboons maintained in an outdoor group social environment. *J Med Primatol* 33:117–126.

Seyfarth, R.M., Silk, J.B., and Cheney, D.L. 2012. Variation in personality and fitness in wild female baboons. *Proc Natl Acad Sci USA* 109:16980–16985.

Seyfarth, R.M., Silk, J.B., and Cheney, D.L. 2014. Social bonds in female baboons: The interaction between personality, kinship and rank. *Anim Behav* 87:23–29.

Sigg, H., Stolba, A., Abegglen, J.-J., et al. 1982. Life history of hamadryas baboons: Physical development, infant mortality, reproductive parameters and family relationships. *Primates* 23:473–487.

Silk, J.B., Cheney, D.L., and Seyfarth, R.M. 1996. The form and function of post-conflict interactions between female baboons. *Anim Behav* 52:259–268.

Stacey, P.B. 1986. Group size and foraging efficiency in yellow baboons. *Behav Ecol and Sociobiol* 18(3):175–187.

Stoltz, L.P. and Saayman, G.S. 1970. Ecology and behaviour of baboons in the northern Transvaal. *Ann Transvaal Museum* 26(5):99–142.

Strum, S.C. 2010. The development of primate raiding: Implications for management and conservation. *Int J Primatol* 31:133–156.

Swedell, L. and Leigh, S.R. 2006. Perspectives on reproduction and life history in baboons. In *Reproduction and Fitness in Baboons: Behavioral, Ecological, and Life History Perspectives*, eds Swedell, L. and Leigh, S.R., 1–15. New York: Springer.

Turkkan, J.S. 1990. New methodology for measuring blood pressure in awake baboons with use of behavioral training techniques. *J Med Primatol* 19:455–466.

VandeBerg, J.L., Williams-Blangero, S., Tardif S.D., editors. 2009. *The Baboon in Biomedical Research.* New York: Springer.

Veira, Y. and Brent, L. 2000. Behavioral intervention program: Enriching the lives of captive nonhuman primates. *Am J Primatol* 51(S1):97.

Warren, Y. 2008. Crop-raiding baboons (*Papio anubis*) and defensive farmers: A West African perspective. *West Afr J Appl Ecol* 14:1–11.

Washburn, S.L. and DeVore, I. 1961. The social life of baboons. *Sci Am* 204(6):62–71.

Whiten, A., Byrne, R.W., Barton, R.A., et al. 1991. Dietary and foraging strategies of baboons [and discussion]. *Philos Trans R Soc B* 334:187–197.

Williams-Blangero, S., VandeBerg, J.L., Blangero, J., et al. 1990. Genetic differentiation between baboon subspecies: Relevance for biomedical research. *Am J Primatol* 20:67–81.

Winney, B.J., Hammond, R.L., Macasero, W., et al. 2004. Crossing the Red Sea: Phylogeography of the hamadryas baboon, *Papio hamadryas hamadryas. Mol Ecol* 13:2819–2827.

CHAPTER 23

Behavioral Management of *Pan* spp.

Lisa Reamer, Rachel Haller, and Susan P. Lambeth
The University of Texas MD Anderson Cancer Center

Steven J. Schapiro
The University of Texas MD Anderson Cancer Center and University of Copenhagen

CONTENTS

INTRODUCTION

Chimpanzees and bonobos share more than 98% of their DNA with humans, making them exceptionally important to study and understand. Chimpanzees and bonobos make and use tools and, in captivity, have been taught to understand human speech and to even use lexigrams or American Sign Language to communicate, underlining their intelligence and complex cognitive abilities. With such developed cognitive skills, chimpanzees and bonobos have indisputably critical needs that require appropriate behavioral management strategies that maximize their welfare. This chapter briefly covers the basic natural history of the genus; different behavioral management strategies for captive members of the genus *Pan*, including socialization, environmental enrichment, and positive reinforcement training (PRT); as well as additional issues related to the study and welfare of captive *Pan* spp.

NATURAL HISTORY

The genus *Pan* is composed of two species: *Pan troglodytes*, or the "common chimpanzee," and *Pan paniscus*, also known as the "pygmy chimpanzee" or bonobo. *Pan troglodytes* has four distinct subspecies: *P.t. verus*, found in western Africa; *P.t. vellerosus*, found along the Nigeria/Cameroon border; *P.t. troglodytes*, found in central Africa; and *P.t. schweinfurthii*, found in eastern Africa. Bonobos are only found in the Democratic Republic of the Congo. This genus inhabits a variety of habitats from primary rainforests to savannahs (see Rowe 1996). Chimpanzees and bonobos are large-bodied great apes, weighing anywhere from 31 to 60 kg for chimpanzees and 31–45 kg for bonobos (Rowe 1996; Smith and Jungers 1997). Bonobos have a more slender body type than that of chimpanzees (see Figure 23.1), and they are generally considered to be less aggressive and to reconcile more than chimpanzees. Chimpanzees and bonobos are diurnal quadrupeds, which are both terrestrial and arboreal (see Rowe 1996). While both chimpanzees and bonobos are omnivorous, fruit comprises the majority of the *Pan* diet. Compared to chimpanzees, bonobos hunt less and eat

Figure 23.1 Comparisons of male and female chimpanzees and bonobos. (a) Male bonobo. (b) Female bonobo. (c) Male chimpanzee. (d) Female chimpanzee. (Photo credits: Bonobos: Jenn Baulbit, San Diego Zoo; Chimpanzees: Kathy West, KCCMR.)

more terrestrial herbaceous vegetation (Stumpf 2007). Although it only accounts for a small portion of their diet, both chimpanzees and bonobos hunt and consume the meat of various mammals (Surbeck and Hohmann 2008; Pruetz et al. 2015). Chimpanzees in forest habitats frequently hunt in groups for arboreal monkey species; however, in addition to group hunting, savannah-dwelling chimpanzee populations have been observed to engage in tool-assisted hunting, where they probe for prey within tree cavities using a jabbing tool (Mitani and Watts 1999; Pruetz et al. 2015). Previously, it was thought that bonobo hunting focused mainly on single individuals, opportunistically catching medium-sized terrestrial prey (e.g., forest antelopes or squirrels); however, more recent studies have reported bonobos hunting arboreal monkey species as a group, much like chimpanzees do (Surbeck and Hohmann 2008).

While studies of wild chimpanzees have been continuous and abundant over the decades, studies of wild bonobos have been somewhat less visible. Jane Goodall was one of the first researchers to set up a chimpanzee field site in Africa, at Gombe Stream, Tanzania, in 1960. Field studies with chimpanzees have continued at study sites across Africa, covering multiple habitat types, from savannah to rainforest. In contrast to chimpanzee research, bonobo field sites were not established until the 1970s, and civil war disrupted the study of bonobos at the Wamba and Lomako field sites from 1996 to 2002 (Boesch 2002a; Thompson et al. 2003). Wamba and Lomako are the two most well-established field sites for studying bonobos; however, they fall within the same forest block, and although new sites are currently being established, this makes comparisons between habitat types difficult.

As social interactions are vital to chimpanzees and bonobos, group dynamics are exceptionally important to understand. With chimpanzee community sizes as large as 150 members (Watts and Mitani 2002) and bonobo communities ranging from 25 to 120 (Furuichi 1989), both live in multi-male, multi-female groups with a fluid fission–fusion social system. In general, average party size for chimpanzees is smaller than that for bonobos. Comparing 10 chimpanzee and 5 bonobo studies, Chapman et al. (1994) found that party size of chimpanzees averaged between 2.6 and 10.1 individuals, whereas average party size for bonobos ranged from 5.4 to 16.9 animals. Although females migrate out of their natal groups in both species, chimpanzees have strong male–male bonds and a linear male dominance hierarchy, whereas bonobos are female-bonded, with females considered dominant or codominant (Furuichi 1997).

Chimpanzees and bonobos display similar behavior patterns, with a few notable exceptions. Pruetz and McGrew (2001) compiled over 30 years of data, from 20 different studies across 6 field sites throughout Africa, and determined that wild chimpanzees slept approximately 12 h/day. During their waking hours, chimpanzees spent 55% of their day foraging, 18% resting, 16% traveling, 8% socializing, and 3% of their time in other behaviors. A study of bonobos at the Lomako field site showed that, of the observed waking hours, the bonobos spent 42.7% of their day foraging, 33.7% resting, 17% traveling, and 6.3% of their time socializing (White 1992). Both chimpanzees and bonobos make nests to sleep in each night, and occasionally will make day nests for resting. Particularly unique to bonobos is their propensity to exhibit high levels of sexual behavior. Bonobos copulate more than any other ape (Kano 1989) and display sexual behavior in social contexts outside of copulation (Kano 1980; de Waal 1989). While female-to-female genito-genital (g-g) rubbing has been described in chimpanzees (Anestis 2004), it is highly prevalent in bonobos. Over 8 years of data collected from the Eyengo bonobo community revealed that approximately 39% of sexual encounters were g-g rubbing and that 55% of all sexual behavior seen was homosexual (Fruth and Hohmann 2006). This high rate of sexual behavior has often been discussed in connection with the lower levels of aggression seen in bonobos compared to chimpanzees (de Waal 1995; Fruth and Hohmann 2006).

Chimpanzees and bonobos are also known for tool use and cultural transmission. One of the first significant observations that Jane Goodall made at Gombe Stream was "termite fishing," in which the chimpanzees use sticks they have designed to fit into holes in termite mounds to collect and eat termites. This showed that the chimpanzees at Gombe not only used tools, but made them as well (Goodall 1986). Chimpanzees and bonobos are very similar in terms of their cognitive and physical abilities to use tools (Takeshita and Walraven 1996; Herrmann et al. 2008) and have comparable levels of overall diversity in tool use (Gruber et al. 2010). However, Gruber et al. (2010) observed that wild chimpanzees largely use tools within the context of food acquisition (McGrew 1979; Boesch and Boesch 1990), whereas wild bonobos appear to use tools mostly during social interactions and for personal care (e.g., cleaning, protection from rain; Ingmanson 1996). While chimpanzee tool use is universal across field sites, types of tool use differ across locations. Whiten et al. (1999) found evidence for cultural transmission among seven long-term study sites across Africa, where behavioral patterns differed, even though ecological explanations were discounted. This behavioral diversity extends beyond tool use to include grooming and courtship behaviors as well (Whiten et al. 1999). Of the 65 behaviors listed in the Whiten et al. (1999) study, 14 of those behaviors were also seen in the Lomako bonobos (Hohmann and Fruth 2003). This suggests that bonobos have similar cultural tendencies to chimpanzees; however, because bonobo field sites are few in number, the collected data are not sufficient to make this claim directly (Hohmann and Fruth 2003).

The goal of captive management is to provide *functionally appropriate captive environments* for the animals in our care. In order to provide such environments, the natural history of the species, including physical traits, feeding ecology, habitat type, group dynamics, and behavior, should be taken into consideration.

GENERAL BEHAVIORAL MANAGEMENT STRATEGIES
AND GOALS FOR CAPTIVE *PAN* SPECIES

Socialization

The most important enrichment for a nonhuman primate is a conspecific companion (Schapiro et al. 1996a,b; Honess and Marin 2006); this is especially important for chimpanzees and bonobos, animals that have extremely well-developed sociocommunicative cognitive skills (Herrmann et al. 2007; Russell et al. 2011). At the Gombe field site in Tanzania, chimpanzees groomed socially for ~6%–7% of their day; however, when the group was in the provisioned feeding area of the park, grooming increased to 33% of their day (Wrangham 1977). This demonstrates the importance of social grooming to chimpanzees when they do not need to travel great distances to reach food sources, as is the case in captivity. It also illustrates that wild activity budgets may not always be appropriate yardsticks for assessing captive welfare, because wild individuals are subject to many more demands related to locating and accessing resources than are captive individuals, potentially resulting in important differences in behavior patterns.

Group Composition and Dynamics

Bloomsmith and Baker (2001) note that social housing of captive chimpanzees can be limited "by the available physical facilities, the available social partners, the behavioral and health histories of the individual chimpanzees, and the goals of the program" (p. 206). As mentioned earlier, chimpanzees and bonobos live in large fission–fusion communities. Although it may not be possible to house captive chimpanzees and bonobos in such large communities, social groups in captivity should functionally simulate the composition and complexity of wild populations. According to Boesch (1996), the average party size of chimpanzees across six study sites in Africa was 5.7 individuals. Chapman et al.'s (1994) review also reported an average chimpanzee party size of 5.7 individuals, whereas the average party size of bonobos as larger at 8.4 individuals. Pruetz and

McGrew (2001) noted that because large, stable/static groups are not normal for chimpanzees in the wild, "there is most likely a threshold of the number of individuals that can live constantly but amicably together. Higher numbers may lead to disproportionately high rates of aggression or other abnormal behavior" (p. 25). Captive groups should be mixed-age and multi-male, multi-female. Bonobos and chimpanzees have been reported in the wild as having an equal sex ratio, although some chimpanzee populations have been reported to have more females than males (Kano 1987; Nishida and Hiraiwa-Hasegawa 1987). It is important to consider social bonds when constructing groups; captive chimpanzee groups often contain one male and multiple females—a composition that is less than ideal, given that chimpanzees are a male-bonded species, and male–male grooming is the most common type of grooming observed in the wild (Wrangham 1977). Conversely, in bonobos, male–female grooming is the most common and male–male grooming is the least common form of social grooming (Kano 1980; Badrian and Badrian 1984), suggesting that all-male groups would not be particularly suitable for bonobos.

There are a number of methods that can be utilized to establish captive groups that functionally simulate wild groups. Because female chimpanzees emigrate in the wild, in captivity, sexually mature females can be moved to another group as a natural way to avoid inbreeding and to create groups with similar compositions and social bonds as in the wild. In addition, chimpanzee offspring in the wild do not start sleeping separately from their mothers until 4 or 5 years of age (Clark 1977), which underlines the importance of the mother–infant bond during the infant/toddler/juvenile phases of development. Mother–infant bonds are typically considered the strongest bonds among chimpanzees (Goodall 1986). Although the ramifications of infant socialization strategies are no longer relevant for NIH-owned chimpanzees owing to the breeding moratorium enacted in 1995, infant socialization strategies are relevant to those zoos and sanctuaries that continue to produce chimpanzees. Weaning in chimpanzees and bonobos naturally occurs at approximately 5 years of age (Goodall 1986; Johnson 1997); however, for a variety of reasons, captive-born chimpanzees are occasionally weaned earlier than this. Analyses of weaning strategies for nursery-reared chimpanzees have shown varied results. Some studies have found that rearing history had no effect on maternal competency (Toback et al. 1992; Bard 1995), whereas other studies found that early experiences did affect maternal and other behaviors later in life (King and Mellen 1994; Brent et al. 1996). Bloomsmith (1992) found that chimpanzees that were mother reared for 2 years prior to being moved to a peer group showed levels of behavior that were similar to control chimpanzees (never removed from the natal group), whereas the behaviors of subjects weaned at younger ages were significantly different from that of control (normal) individuals. Nursery-reared chimpanzees are more likely to develop abnormal behavior compared with mother-reared individuals (Maki et al. 1993; Spijikerman et al. 1994; Bloomsmith and Haberstroh 1995; Nash et al. 1999). To best simulate chimpanzee social structure in captivity, Mellen and Shepherdson (1992) suggest maintaining groups that contain a minimum of two adult males, several adult females, and immatures of various ages and both sexes.

Introductions/Forming Groups

Chimpanzees are territorial and, in the wild, will enter into intercommunity aggression when defending their home range (Goodall 1986; Boesch and Boesch-Achermann 2000). This makes the introduction of captive chimpanzees to unfamiliar conspecifics a serious endeavor—one that could result in aggression or wounding (Alford et al. 1995). The formation of new social groups should take into consideration the social experience/rearing histories, ages, and personalities (temperament) of the individuals involved. Fritz and Howell (2001) have identified certain types of chimpanzees that are likely to be successfully introduced to one another. These include juveniles/adolescents, wild-born individuals, individuals that spent a minimum of 1 year with their mother, those that were reared in a highly enriched nursery environment, and those that have lived in social groups previously and display affiliative interactions with both conspecifics and humans, while displaying a low incidence of aggressive and abnormal behaviors. Conversely, individuals that do not meet these criteria (i.e., those reared in impoverished environments or those that have spent considerable time in social isolation) are considered higher risk for introductions, and may require additional socialization efforts (Fritz and Howell 2001). It is also important to note that males may be more severely affected by periods of isolation than females, often displaying intense aggression or abnormal behavior (Fritz and Fritz 1979). Despite this, the majority of chimpanzee introductions have been successful (e.g., McDonald 1994; Alford et al. 1995; Brent et al. 1997; Howell et al. 1997). Fritz (1986) reported that 98% of captive chimpanzees were successfully integrated into a stable group, while their more recent publication (Fritz and Howell 2001) reported that only 4.3% of 853 different social introductions were unsuccessful. These introductions took anywhere from 7 days to 6 weeks to accomplish for groups of between two and eight individuals (larger groups would presumably take longer; Fritz and Howell 2001). Owing to the serious nature of introductions and the potential for injury, introductions must be done slowly, and each individual should be taken into account (their health status, social background, etc.). Introductions should never be rushed; if an individual is showing signs of stress, the introduction should be slowed or halted. Fritz and Howell (2001) also reported that groups were usually together for 3 years or less before a decline in positive interactions was observed—a change that they attributed to small social groups being housed together for extended periods of time, a social situation that is uncharacteristic of the fission–fusion society observed for wild chimpanzees. Pruetz and McGrew (2001) list reduced sexual activity, high levels of aggression, abnormal behavior, and/or inactivity as indicators of poor social group composition and stability. Alford et al. (1995), on the other hand, reported that many of the social groups at our facility were together and stable (as determined by wounding aggression) for over 14 years; indeed, some groups at the National Center for Chimpanzee Care (NCCC) contain three generations of animals and have remained stable over two to three decades (Lambeth SP personal observations).

Owing to the potential for aggression and stress, social introductions for chimpanzees should be conducted gradually and should be preceded by considerable planning, allowing for flexibility throughout (Fritz and Howell 2001; Lambeth and Schapiro 2017). Introductions require close observation, and the benefits of collecting systematic behavioral observations during introductions are well documented (e.g., Alford et al. 1995; Bloomsmith et al. 1999). Introductions should be scheduled at the beginning of a workweek to allow for the rigorous and systematic observation of the new group by behavioral management and/or caregiving staff. Even after an introduction is "completed" and the group appears stable, aggression may still escalate over time (Alford et al. 1995; Brent et al. 1997; Bloomsmith et al. 1999), emphasizing the need to closely monitor the group's compatibility, behavior, and injuries for at least several weeks following the completion of an introduction (Fritz and Howell 2001; Lambeth and Schapiro 2017). Fritz and Howell (2001) considered introductions successful when they saw a decline in fearful and abnormal behaviors and an increase in foraging and play compared to earlier stages of the introduction process. At the NCCC, we watch groups to ensure all members are thriving and are socially cohesive in the new group; if an individual is on the periphery of a group, we may implement PRT to work on cooperative interactions between group mates. It is important that all members of a group be well integrated before considering an introduction successful.

Most introductions follow a basic format in which the new individual is first allowed to have wire contact (sometimes called "groom," "protected," or "tactile" contact) with new group member(s) prior to being given full contact (McNary 1992; Sexton and Gallagher 1997; Fritz and Howell 2001). For bonobos, this usually involves the new member being adjacent to the whole group prior to being given full access to the group (assuming no aggression has been seen across the wire; Sexton and Gallagher 1997). For common chimpanzees, which are naturally more aggressive than bonobos, it is recommended that a new group member(s) be introduced one-on-one with *each* group member, before the whole group is formed (McNary 1992; Alford et al. 1995; Fritz and Howell 2001). At the NCCC, we first attempt to introduce the dominant male to the new individual. This allows us to determine whether the newcomer has been accepted by the highest-ranking member of the group, an animal that can then intervene in the event that fighting occurs between the newcomer and existing group members. After a new group member meets each new group mate one-on-one, they will then typically be introduced to strategic *subsets* of animals before putting the whole group together. This helps the animals to build bonds and helps us to see how the social dynamics change as different subgroups are put together, helping to ensure a successful introduction.

Female sexual swelling tumescence should be considered when scheduling introductions of chimpanzee and bonobo females to maximize the probability of success during group formations (McNary 1992; Sexton and Gallagher 1997). Introductions of female chimpanzees to males would be most successful if the females are at full tumescence, while female-to-female introductions would be most successful if their swellings are detumescent (McNary 1992). In the wild, when female chimpanzees immigrate into a new group, they first form bonds with male members of the group, because resident females can be aggressive toward new group members (Pusey 1980; Goodall 1986). Immigrant female bonobos, on the other hand, will first establish bonds with older female residents as a means to integrate themselves into a group (Furuichi 1989; Idani 1991). With this in mind, it makes sense to schedule introductions in captivity to introduce new female chimpanzees to the males of a group first, and to introduce new female bonobos to other females in the group first. If an introduction between two individuals is particularly difficult in either species, the use of PRT as a tool to encourage proximity and affiliative behaviors may be a useful technique (Laule and Desmond 1990; Bloomsmith and Baker 2001; Laule and Whittaker 2001).

Although fission–fusion is common in wild chimpanzees, it can be surprisingly difficult to *reintroduce* captive chimpanzees after a very short separation, even if they are from a stable group (Schapiro and Lambeth 2010). It is possible that the social hierarchy shifts during the reintroduced individual's absence, a process that can be effectively managed by reintroducing animals gradually and carefully; not proceeding to the next step in the reintroduction process until the appropriate dominance–subordinance signals have been observed.

Some introductions occur because an animal has been transported to a new facility (e.g., for a breeding loan or for a permanent relocation). One introduction technique that is often discussed, but is *not* advisable for relocated chimpanzees, is to capitalize on the novelty of the new environment and the animal's transport-related stress/disorientation to facilitate an easier/faster introduction to new group members. The behavioral indicators of transport stress significantly decrease after the first week at a new facility (Reamer et al. 2013); however, many physiological indicators of stress have still not returned to baseline levels for as long as 12 weeks after arrival at the new facility (Schapiro et al. 2012). Owing to the physiological ramifications of stress related to transport and relocation, it is *not* advisable to increase the stress experienced by the animal in order to accelerate the introduction process. Instead, introductions should always be taken at a pace that maximizes the probability of a successful introduction (Fritz and Howell 2001). If an animal arrives at a facility alone, it could be paired with a compatible individual until its physiological and behavioral parameters return to "normal"; however, whenever possible, the transport of single chimpanzees and bonobos should be avoided. Introductions and reintroductions are highly meaningful events in a captive chimpanzee's life, and everything possible should be done to ensure that the behavioral management strategies employed would enhance the animal's well-being/welfare/wellness.

Environmental Enrichment

Environmental enrichment can be broken down into five categories: social, feeding, physical, sensory, and occupational, many of which are overlapping (e.g., Bloomsmith and Else 2005; Coleman 2012). The goals of any primate enrichment program are to increase the amount of time an animal spends engaged in species-typical behaviors and to decrease the time spent in abnormal or "undesirable" behaviors (Coleman 2012). While enrichment efficacy has been studied extensively in chimpanzees, it is less well documented in bonobos. Chimpanzees and bonobos have similar behavioral repertoires and cognitive abilities; therefore, it is unlikely that they will differ greatly when discussing enrichment and the promotion of species-typical behavior patterns.

One important factor to consider is the safety of each enrichment opportunity; the safety of the animal should be the first concern of any enrichment program (Baer 1998; Coleman 2012). When providing enrichment devices, it is best if they are not easily monopolized by one, or even a few, individuals, because this can lead to undesirable competition and aggression (Honess and Marin 2006). Enrichment should be age-appropriate. Juveniles will often play with devices more than adults (Lutz and Novak 2005), and geriatric animals may require ramps and similar devices to address issues related to safe locomotion in the face of mobility impairment (Schapiro and Lambeth 2010; see section "Special Topics" below for more on geriatric care).

Overall, Pruetz and McGrew (2001) note that it is better to *stimulate* natural behavior than to attempt to *simulate* natural aesthetics. Enrichment is often required to be "natural-looking" for chimpanzees and bonobos living in zoo environments when the animals are on exhibit; however, additional, functional, non-natural-looking enrichment can be utilized off-exhibit or after visitors to the zoo are gone for the day (McMillan et al. 1991).

Social enrichment is covered fairly extensively in the section "Socialization"; however, it is important to note that social enrichment is usually the most beneficial form of enrichment for captive primates (Schapiro et al. 1996a,b; Honess and Marin 2006). When full contact with a conspecific is not possible (due to illness or for other reasons), social enrichment can include interaction with humans or non-conspecific animals. Finally, auditory, visual, and/or tactile contact (wire contact) with conspecifics is recommended, if a chimpanzee or bonobo must be housed alone (Brent 2001).

Because chimpanzees spend 55% of their waking hours foraging (Pruetz and McGrew 2001), it is exceptionally important for captive environments to provide opportunities for animals to "work" for their food (Schapiro and Lambeth 2010). Utilizing foods that required high levels of processing time (e.g., corn on the cob), high levels of searching/foraging time (e.g., popcorn, peanuts scattered into a grass substrate), and food given in puzzle feeders was found to significantly decrease aggression and abnormal behaviors in captive chimpanzees (Bloomsmith et al. 1988). Equally, providing bonobos with browse, such as bamboo, increased foraging from 3% to 17% of observation time (Damen 1990). However, when implementing enrichment programs, veterinary- and nutrition-related issues must be considered. For example, obesity is a major concern for captive chimpanzees (Lambeth et al. 2011; Reamer et al. 2014), an issue that will be discussed in greater detail in the section "Special Topics."

Whenever possible, environmental enrichment should offer choice, control, and some degree of novelty to the animals (Coleman 2012). Captive chimpanzees and bonobos should have a complex space that promotes social interaction, while allowing privacy behind visual barriers or in separate rooms (Maple and Finlay 1989; McMillan et al. 1991; Ross et al. 2011). Chimpanzees prefer upper levels, cage perimeters, and small areas over open spaces (Taylor-Holzer and Fritz 1985; also see Ross et al. 2011), and it is highly recommended that both chimpanzees and bonobos have structures for climbing and swinging, as well as elevated resting platforms (Van Puijenbroeck and De Bois 1997; Schapiro and Lambeth 2010). Such structures could include poles, hanging ropes, swings, cargo nets, fire hose, hammocks, and more. Woodchip bedding, grass, or sand substrates have been recommended for both chimpanzees and bonobos (Coe 1989; Van Puijenbroeck and De Bois 1997).

For chimpanzees, novel objects that were both portable, and/or flexible, lasted longer (Menzel 1971), and plastic toys have been shown to decrease inactivity and self-grooming (Paquette and Prescott 1988). However, manipulation of these objects decreased over time, suggesting habituation and the need to change objects regularly to maintain novelty (Paquette and Prescott 1988).

Bonobos were found to increase activity and decrease infant stealing when given novel enrichment (Csatádi et al. 2008). Destructible enrichment has been shown to maintain the animals' interest longer than indestructible objects (Bloomsmith et al. 1990a; Brent and Stone 1998). In fact, Pruetz and Bloomsmith (1992) found that chimpanzees played with kraft butcher paper approximately 27% of observed time, whereas subjects were observed playing with Kong toys only 10% of the time.

Another tool available to behavioral managers for offering choice, control, and novelty to apes is sensory enrichment. As chimpanzees and bonobos can recognize themselves in a mirror (Gallup 1970), providing mirrors is a common sensory enrichment practice. While mirrors are used regularly by chimpanzees for self-inspection, they are also used to see adjoining cages and hallways to which the chimpanzees would not otherwise have visual access (Lambeth and Bloomsmith 1992; Brent and Stone 1996). Playing the radio, as auditory enrichment, for captive chimpanzees was found to reduce aggression and increase affiliation (Howell et al. 2003); specifically, Videan et al. (2007) found that playing slow-tempo, instrumental music was the most effective at lowering aggression and increasing social interactions. However, Richardson et al. (2006) found that, regardless of music type, when given a choice, chimpanzees prefer music to be *off* rather than *on*. Similarly, playing videos for chimpanzees resulted in a decrease in abnormal behavior (Brent et al. 1989); however, no preference for the type of video watched was evident (Bloomsmith et al. 1990b; Bloomsmith and Lambeth 2000), but getting to choose the video appeared to be important (Bloomsmith et al. 2000). Essential oils (e.g., peppermint, rosemary) are another type of sensory enrichment that has been found to increase activity in chimpanzees (Struthers and Campbell 1996); however, Ostrower and Brent (1997) found that there were no significant differences in the number of times that chimpanzees touched the cloths that were either scented or not (baseline). While sensory enrichment does not appear to be as effective as other types of enrichment, and should probably not be a major component of chimpanzee or bonobo enrichment plans, in certain circumstances, it can provide valuable opportunities for choice, control, and novelty.

It is widely recommended that captive chimpanzees and bonobos be given material with which to build nests each day to functionally simulate the daily nest-building behavior of their wild counterparts (Van Puijenbroeck and De Bois 1997; Pruetz and McGrew 2001; Videan 2006; Coleman 2012; Hernandez-Aguilar et al. 2013). A variety of materials, such as blankets, burlap, hay, straw, shredded paper, and excelsior (wood wool), have been provided to chimpanzees and bonobos to promote the expression of this natural behavior. It is important to consider staff efficiency and facility design, in addition to the needs of the animals, when considering nest-building materials, because blankets require considerable staff time to wash and, depending on facility design, hay, straw, and shredded paper may block drains. At the NCCC, excelsior has proven to be the best material for nest building, both from the animals' (pulling apart the compacted flakes to form appropriately shaped nests) and the caregivers' (excelsior does not easily go down drains) perspectives (for additional discussion, see Lambeth and Schapiro 2017).

As stated earlier, chimpanzees and bonobos have exceptional social and physical cognitive abilities (Herrmann et al. 2007; Russell et al. 2011). Therefore, to prevent boredom in captivity, it is important to attempt to stimulate these abilities as much as possible. Occupational enrichment, such as puzzles, often has been implemented to provide opportunities for the animals to use their cognitive skills (Bloomsmith and Else 2005). Indeed, puzzle boards were found to decrease aggression, affiliation, inactivity, and self-directed behaviors in chimpanzees (Brent and Eichberg 1991). The use of artificial termite mounds (or "pipe feeders"), in addition to providing foraging enrichment, allows chimpanzees and bonobos opportunities to make and use tools, and simulates a unique species-typical behavior, termite fishing (Nash 1982). Pipe feeders were found to decrease abnormal behaviors and inactivity (Maki et al. 1989). Although puzzle boards and other devices are effective, behavioral or cognitive testing may be the most beneficial occupational enrichment. Such tests typically comprise voluntary tasks that promote the utilization of the animals' physical and social cognitive skills (Herrmann et al. 2007; Whiten et al. 2007; Schapiro and Lambeth 2010;

Hopper et al. 2013). Captive chimpanzees that voluntarily participated in cognitive testing throughout the day had feeding time budgets that more closely matched those of their wild counterparts than did captive chimpanzees that did not participate in behavioral experiments (Yamanashi and Hayashi 2011). Chimpanzees at the Whipsnade Zoo were found to use a "cognitive challenge device," even in the absence of food (they were rewarded with tokens), and demonstrated higher levels of problem-solving behaviors and social play when the device was present, which were clear indications of the enrichment value of such tasks (Clark and Smith 2013). Occupational enrichment and cognitive tests can also include computerized tasks involving a joystick or touch screen (Hopkins et al. 1996; Schapiro and Lambeth 2010). Ross et al. (2000) found that juveniles used computer tasks the most, and the subjects did not appear to habituate to the device.

Positive Reinforcement Training

PRT is a behavioral management technique that can be applied with captive nonhuman primates to allow them to voluntarily participate in husbandry, veterinary, and research procedures (Prescott et al. 2005; Schapiro and Lambeth 2010; Perlman et al. 2012). PRT programs provide important opportunities for the primates to exercise choice and control in captivity, while stimulating learning and offering opportunities to work for food (Laule and Desmond 1998). In addition to providing enriching opportunities for nonhuman primates (Schapiro et al. 2014), PRT can also reduce stress associated with some of the specific procedures that are required in captive environments (Laule et al. 2003; Wolfensohn and Honess 2005).

The use of PRT to encourage the voluntary participation of chimpanzees and bonobos in husbandry, veterinary, and research procedures may be particularly important, given their size, strength, and cognitive abilities. Common behaviors included in PRT programs for both chimpanzees (Laule and Whittaker 2001) and bonobos (Bell and Ballmann 1997) are examination of body parts (hands, feet, ears, mouth, etc.), measurement of rectal or tympanic temperature, presentation of a limb for an injection, and voluntary biological sample collection (e.g., urine, semen, breast milk, saliva), among others (Bell and Ballmann 1997; Lambeth et al. 2006; Urashima et al. 2009; Magden et al. 2013, 2016). Using desensitization techniques, simple body exam behaviors may be built upon to elicit voluntary participation with more complex medical behaviors, ultimately allowing programs to provide cage-side care, potentially eliminating the need for physical restraint and/or anesthesia (Laule and Whittaker 2001; Magden et al. 2013, 2016). Chimpanzees (Coleman et al. 2008) and bonobos (Bell and Ballmann 1997) can be trained for voluntary venipuncture, so that blood samples can be collected and tested without the influence, or stress, of anesthesia. Chimpanzees have also been trained to allow collection of very small "capillary" blood samples (via pricking of a finger or toe), samples that are especially useful for the management of diabetic animals whose glucose levels must be monitored frequently (Laule et al. 1996; Reamer et al. 2014). Animals have been trained to allow similar digit-pricking behaviors for the collection of slightly larger blood samples that can be used with an i-STAT handheld machine to perform more detailed bloodwork in the absence of anesthesia. This is particularly useful when an animal is sick and an anesthetic episode could increase the risks to an individual's health (Schapiro et al. 2014).

Although many behaviors can be performed voluntarily at the cage-side as part of a voluntary PRT program, sedation may be unavoidable for certain health-related reasons (e.g., annual physical examinations, wound suturing, etc.). Therefore, training primates to voluntarily present for an anesthetic injection is particularly important. Lambeth et al. (2006) found that, compared to traditional methods of sedation, the use of PRT to train chimpanzees to voluntarily present for an injection of anesthesia resulted in patterns of hematological and chemistry parameters that suggested that the animals that presented were significantly less stressed than those that did not. Furthermore, allowing the animal to choose to voluntarily participate consequently reduced the time required to perform the sedation and the stress levels experienced by the caregiving and veterinary staff. PRT can greatly improve the relationship between the animals and their human caregivers, increasing safety for both animals and humans (Bloomsmith 1992; Ross and McNary 2010).

In addition to treatments and sample collection, PRT can also facilitate the extinction of undesirable and abnormal behaviors in chimpanzees (Bourgeois et al. 2007; Martin et al. 2011) and bonobos (Bell and Ballman 1997). Pomerantz and Terkel (2009) found that abnormal and stress-related behaviors in zoo-housed chimpanzees decreased after the implementation of a PRT program, especially among lower-ranking individuals. PRT has been shown to facilitate and improve socialization in chimpanzees (Desmond et al. 1987), and to reduce self-directed behaviors and inactivity, while increasing social play (Bloomsmith 1992). PRT techniques, intended to facilitate "cooperative feeding" within groups of chimpanzees, have resulted in reduced aggression and food stealing (Bloomsmith et al. 1994). Although the behavior of the individual that stole food prior to training is altered during a PRT cooperative feeding session, the dominance hierarchy of the group is typically not affected.

Taking a personalized approach to PRT in order to meet the specific medical and/or behavioral needs of individual animals can be of great benefit. At the NCCC, we have modified multiple components of our behavioral management program to accommodate our geriatric animals, including training geriatric chimpanzees with arthritis to station for acupuncture treatments and low-level laser therapy (Magden et al. 2013, 2016). Acupuncture therapy has resulted in improved mobility scores for the animals that have received treatment (Magden et al. 2013), and the combination of acupuncture and laser therapy decreased the mean number of ventricular premature complexes (VPC/min) in a chimpanzee diagnosed with frequent VPCs (Magden et al. 2016). An implantable loop recorder, which required that the animal be trained to present its back for interrogation of the device, was used to obtain the VPC results. In addition, geriatric chimpanzees have been trained to perform specific movement behaviors as a form of physical therapy (Bridges et al. 2015). The use of PRT to address the health care requirements of individuals, especially older animals, has been a very successful component of the chimpanzee behavioral management program at the NCCC.

One of the perceived drawbacks of PRT programs is the amount of personnel time that is thought to be required to initially train the animals. In light of the PRT-related benefits briefly mentioned earlier, and the considerable number reported elsewhere in the literature (Bloomsmith et al. 1994, 1998; Lambeth et al. 2006; Coleman et al. 2008; Perlman et al. 2010; Reamer et al. 2014), the benefits of PRT as a component of a comprehensive behavioral management program are evident. Systematic implementation of PRT programs often takes less time than expected. For instance, captive chimpanzees were reliably trained to "shift" from one enclosure to another in 16 training sessions by their trainers. Once trained, the responsibility for this behavior was successfully transferred to regular caregivers, significantly increasing the efficiency of daily husbandry procedures (Bell and Ballmann 1997; Bloomsmith et al. 1998).

The training process can be expedited by understanding and taking advantage of the social learning abilities and the personality characteristics of the animals. Chimpanzees that are trained while in their social groups will learn the task by observing others perform the target behaviors (Perlman et al. 2010; Hopper 2017). Chimpanzees that rated high on the personality factor of "openness" were more successful at performing a novel behavior chain than animals that rated low on "openness" (Reamer et al. 2014). These findings suggest that training success will be enhanced when "open" chimpanzees are trained while in their social groups. Although time must be invested when implementing PRT, the benefits typically surmount the costs. PRT is an essential component of effective primate behavioral management programs, eliciting voluntary cooperation, providing control and choice, reducing stress, increasing efficiency, and enhancing well-being.

For additional discussions of the implementation of PRT programs for chimpanzees and bonobos, see Perlman et al. (2012); Colahan and Breder (2003); Bell and Ballmann (1997); and Laule and Whittaker (2001).

Special Topics

Owing to high-calorie diets and relatively low levels of physical activity, obesity is common among captive animals (Goodchild and Schwitzer 2008; D'Eath et al. 2009), including chimpanzees and bonobos (Videan et al. 2007; Lambeth et al. 2011; Reamer et al. 2014; Figure 23.2). If weight management strategies are not implemented, obese chimpanzees can be at risk of developing diabetes and/or other comorbidities (McTighe et al. 2011; Reamer et al. 2014). Lambeth et al. (2011) identified obese individuals within their captive chimpanzee population and formed a "weight management" social group, in which the number of calories available for each animal was reduced, and behavioral management techniques were employed to increase activity. Individuals were monitored closely by staff. All individuals in the weight management group reached their (reduced) weight goal while living in this special group. However, once the individuals returned to their original social groups, they regained the weight (Lambeth et al. 2011). Subsequently, Bridges et al. (2013) devised a feeding system that required obese individuals to work harder to obtain their primate biscuits (their main source of calories), while still living in their normal social groups. This modification led to a more even distribution of biscuits across group members and, importantly, overweight individuals lost weight, while non-overweight individuals maintained their weight (Bridges et al. 2013). When it comes to weight management, it is better to be proactive, rather than reactive. At the NCCC, we use healthy, unprocessed food as treats/rewards to prevent undesirable weight gain.

In addition to weight management, geriatric care is another issue that is becoming more prominent in the management of captive chimpanzees. As the animals that comprise the captive population get older and develop age-related ailments, such as arthritis, cardiovascular issues, or dementia, special accommodations must be made to maintain the well-being of these animals. At the NCCC, a comprehensive geriatric care program has been implemented, including increased

Figure 23.2 Weight management in a male bonobo. This figure shows Kanzi, a language-trained bonobo at the Ape Cognition and Conservation Initiative (ACCI) in Iowa, in (a) 2013 and (b) in 2015 after weight management strategies were implemented at the center.

monitoring and physical/behavioral modifications to accommodate the specific problems faced by each individual animal classified as geriatric. Every animal in the NCCC is observed daily by husbandry and veterinary staff; however, geriatric animals are observed more frequently to monitor mobility, cognition, and behavior (Magden et al. 2013; Bridges et al. 2015). The information gained from this additional monitoring may result in case-by-case or group-by-group modifications to physical structures, enrichment devices, and/or components of the diet. Using PRT, trainers have been able to provide physical therapy, acupuncture, and/or low-level laser therapy for animals that require such treatments (Magden et al. 2013; Schapiro et al. 2014). PRT, as one component of a comprehensive behavioral management program, is essential for providing proper care to chimpanzees with special needs, including geriatric individuals.

Beyond geriatric care, at the NCCC, we have developed a quality-of-life (QOL) program to aid in the difficult process of interpreting and quantifying pain and distress in captive chimpanzees (Lambeth et al. 2013). A QOL team is formed when an animal is diagnosed with a chronic or life-threatening condition. This team, consisting of the attending veterinarian, a veterinary technician, a primate behaviorist, the colony manager, the animal's trainer, its daily caregiver, and a pathologist, is tasked with developing specific behavioral guidelines that define the unique characteristics and traits of each individual chimpanzee. These guidelines are used, in addition to traditional clinical data, in the decision-making process related to humane euthanasia, in an attempt to minimize suffering and improve overall welfare (Lambeth et al. 2013).

These special topics and practices (weight management, geriatric care, and QOL program) could easily be adapted for use with other nonhuman primate species, including bonobos.

IMPLICATIONS FOR RESEARCH

The use of behavioral management techniques has facilitated many research projects with both captive chimpanzees and bonobos. Being our closest living relatives, chimpanzees and bonobos are extremely valuable research subjects. Research involving chimpanzees has traditionally included biomedical research, sample collection of biological specimens (obtained during routine physical examinations or voluntarily through PRT; e.g., Urashima et al. 2009), and behavioral research—all of which occur in laboratory settings. Many zoos and sanctuaries with chimpanzees also contribute to the scientific endeavor via collection of biological samples (e.g., Staes et al. 2014) and the performance of behavioral research projects (e.g., Hopper et al. 2015). In contrast to chimpanzees, bonobos have not typically participated in biomedical research, although they have provided biological specimens (e.g., Wobbera et al. 2010) and have been involved in behavioral research studies (e.g., Clay and Zuberbuhler 2009), especially studies of language (Taglialatela 2016). Research projects involving members of the genus *Pan* have resulted in insights related to the origins of communication (e.g., Taglialatela et al. 2015), an increased understanding of the mechanisms of learning (e.g., Hopper et al. 2011), an enhanced appreciation of the nutritional aspects of breast milk (e.g., Urashima et al. 2009), data relevant to identifying hormone–behavior relationships (e.g., Wobbera et al. 2010; Staes et al. 2014), and the identification of techniques to enhance the welfare of *Pan* living in captivity (Schapiro et al. 2014).

In 2011, the National Academies of Science, Institute of Medicine (IOM 2011), published a report titled "Chimpanzees in Biomedical and Behavioral Research: Assessing the Necessity." Included in this report was a recommendation that only 50 NIH-owned chimpanzees were needed for "research," a recommendation that more recently has been modified to suggest that no NIH-owned chimpanzees will be available for "research." While it is relatively clear that this prohibition on the participation of chimpanzees in research projects applies to invasive, biomedical protocols, it is less clear whether noninvasive, behavioral studies are included as well. Many behavioral studies, which involve the voluntary participation of captive chimpanzees and bonobos, not only improve our understanding of these important species, but also enhance the well-being of the animals (see Hopkins and Latzman 2017). Several groups have demonstrated that behavioral research is enriching for members of the genus *Pan* (Yamanashi and Hayashi 2011; Clark and Smith 2013), and one of the primary foci of our research program is to empirically define "functionally appropriate captive environments," including the effects of voluntary research participation on the well-being of the subjects. Since the breeding moratorium on captive NIH-owned chimpanzees in 1995, this national resource continues to dwindle, thereby increasing the motivation to gain as much information as possible (using noninvasive techniques) from the existing chimpanzees.

IMPLICATIONS FOR WELFARE

The application of behavioral management techniques for chimpanzees and bonobos clearly has positive implications for their welfare in captivity, as discussed throughout this chapter. With animals as cognitively complex as those of the genus *Pan*, those responsible for the captive management of chimpanzees and bonobos should be implementing behavioral management techniques that provide *specialized care* and opportunities for each animal, based on the specific needs and capabilities of that individual. Each chimpanzee and bonobo is unique and has its own set of needs, determined by its past experiences, personality, age, sex, and/or health status. For example, describing and quantifying the "trainability" of chimpanzees or bonobos can be a useful tool to help caregivers optimize training time. PRT can decrease stress in captive animals and increase positive interactions; it can also eliminate the need for anesthesia when collecting a diagnostic sample to assess an animal's health status. However, there can be considerable differences in the time it takes to train different individuals of the same species (chimpanzees: Coleman et al. 2008; Reamer et al. 2014). The ability to identify the factors (e.g., age, sex, rearing history, trainer, personality, and dominance) that affect training success in chimpanzees or bonobos could help identify those animals that may require extra time to learn certain behaviors (Reamer et al. 2014). In addition, geriatric animals may require specialized enrichment apparatus. For instance, if a specific individual has diminished manual dexterity, enrichment devices can be modified to accommodate these issues, while still increasing species-typical foraging behaviors. Overall, great strides in the application of behavioral management to improve animal welfare using a, more or less, "one-size-fits-all" approach have been made over the past three decades. However, attempts to take behavioral management, and captive care in general, to the next level should emphasize *specialized care*.

EXPERT RECOMMENDATIONS

Housing *Pan* in conditions that promote an extensive range of natural behaviors, such as those discussed in the section "Natural History," is one way to attempt to provide functionally appropriate captive environments for chimpanzees and bonobos. Table 23.1 contains a list of chimpanzee- and bonobo-specific behaviors accompanied by the behavioral management techniques that can be used to elicit or "promote" these behaviors.

Assessments of whether captive (physical and social) environments are functionally appropriate can be difficult and typically involve comparisons of captive time budgets to those of a species' wild counterparts (e.g., Pruetz and McGrew 2001; Yamanashi and Hayashi 2011). However, as mentioned earlier in this chapter, wild chimpanzees have different overall constraints on their time than do captive chimpanzees. For example, at the Gombe field site in Tanzania, chimpanzees spent relatively little time socially grooming (6%–7%), except when the group was in the provisioned feeding area of the park (33%; Wrangham 1977). This suggests that wild chimpanzees likely spend large portions of their day foraging/traveling out of necessity, not by choice. In captive environments, where resources are provided directly to the animals without the need to travel to obtain

Table 23.1 Chimpanzee- and Bonobo-Specific Behaviors and Suggestions on How to Promote Such Behaviors in Captivity

Wild Behavior	Citations/Examples	Captive Analog	Citations/Examples
		Sleeping/nesting	
12 h sleep cycle/rest 18% of waking hours	Pruetz and McGrew (2001)	Provide elevated sleeping platforms	Pruetz and McGrew (2001)
Sleep in new nest nightly	Pruetz and McGrew (2001)	Bedding material such as straw/hay, wood wool, shredded paper, or blankets/fleece should be provided	Pruetz and McGrew (2001) and Van Puijenbroeck and De Bois (1997)
Feeding/foraging			
Forage 55% of waking hours	Pruetz and McGrew (2001)	Use feeding enrichment and occupational enrichment to attempt to account for this time	Schapiro and Lambeth (2010)
Average of 21 different varieties of foods	Yamagiwa et al. (1996)	Offer a wide variety of foods in captivity	Bloomsmith et al. (1988)
Leaf dipping	Sousa et al. (2009)	Provide enrichment that simulates leaf dipping (e.g., wadgers with butcher paper; pond in enclosure)	Schapiro and Lambeth (2010)
Termite fishing	Goodall (1986)	Provide artificial termite mounds or "tube/pipe feeders" with sticks or bamboo	Nash (1982) and Maki (1989)
Tool making	Goodall (1986)	Provide enrichment that requires tool making (e.g., pipe feeders with leafy bamboo), or offer cognitive studies that utilize tool making and use	Nash (1982) and Maki (1989)
Hunting	Goodall (1986)	Offer cognitive studies that simulate this behavior	Bertand, D., unpublished data and Johnson (2013)
Locomotion/travel			
Travel 16% of waking hours	Pruetz and McGrew (2001)	Provide space for movement and travel	Pruetz and McGrew (2001)
Arboreal for 33%–80% of the day	Pruetz and McGrew (2001)	Provide structures for climbing and swinging/brachiation	Van Puijenbroeck and De Bois (1997)
Social/other			
Fission/fusion	Goodall (1986)	Provide space for privacy—visual barriers, culverts, access to multiple rooms	Maple and Finlay, (1989), Schapiro and Lambeth (2010), and Ross et al. (2011)
Socialize/groom 8% of waking hours (33% when provisioned)	Pruetz and McGrew (2001) and Wrangham (1977)	Provide compatible, dynamic social groups	Pruetz and McGrew (2001)
Cultural transmission/social learning	Whiten et al. (1999)	Provide compatible, dynamic social groups and offer cognitive studies that stimulate this behavior	Whiten et al. (2007)
Cooperation/behavioral economics	Boesch (2002) and Jaeggi et al. (2013)	Offer cognitive studies that stimulate this behavior	Brosnan et al. (2011) and Proctor et al. (2013)
Problem-solving	Boesch (1991)	Provide occupational enrichment and offer cognitive studies that stimulate this behavior	Bloomsmith and Else (2005) and Hopper et al. (2014)

them, comparisons to wild chimpanzee time budgets may not be the best option. Comparisons to provisioned or semicaptive populations may be more valid.

Using wild data to inform captive management decisions may also be slightly off target when defining appropriate chimpanzee group sizes in captivity. As stated already, wild chimpanzees and bonobos live in multi-male, multi-female groups (with high fission–fusion dynamics), which may include more than 100 members in some cases (Furuichi 1989; Watts and Mitani 2002). Some of the primary advantages of living in large groups in the wild are predator avoidance and enhanced foraging success (Chapman and Chapman 2000), neither of which are relevant (in a proximal sense) for captive chimpanzees or bonobos. Because these two advantages of large groups are not really advantages in captivity, and because the maintenance of such very large captive groups is quite difficult, average party sizes in the wild (5.7 for chimpanzees; 8.3 for bonobos; Chapman et al. 1994) might not be a relevant yardstick for determining optimal group sizes in captivity. There is little available empirical data that address optimal group sizes in captivity—a shortcoming that we are in the process of rectifying.

While the majority of behavioral management and welfare decisions should be based on the behavior and social structure of wild conspecifics, there are some characteristics of *Pan* for which this is not as relevant. Overall, decisions and "best practice" recommendations for functionally appropriate captive environments should be based on scientific data and the results of *relevant* comparisons.

CONCLUSIONS

Overall, a behavioral management program that appropriately integrates socialization strategies, environmental enrichment practices, and PRT can positively influence the welfare of captive chimpanzees and bonobos, stimulating a wide variety of species-typical behaviors. Table 23.1 highlights the value of cognitive behavioral testing to elicit the complex patterns of behavior that are typically observed in wild populations, such as cooperation, social learning, and problem solving. Continued empirical research to quantify the factors that comprise functionally appropriate captive environments is needed to inform regulatory bodies and to guide captive management practices.

ACKNOWLEDGMENTS

The first author would like to thank Dr. Elizabeth Magden, Jennifer Bridges, Mary Catherine Mareno, as well as Raishad Maharaj for their continual help and support throughout. All of the authors were supported in part by a cooperative agreement with NIH U42 OD-011197.

REFERENCES

Alford PL, Bloomsmith MA, Keeling ME, Beck TF. 1995. Wounding aggression during the formation and maintenance of captive, multi-male chimpanzee groups. *Zoo Biology* 14(4):347–359.
Anestis SF. 2004. Female genito-genital rubbing in a group of captive chimpanzees. *International Journal of Primatology* 25(2):477–488.
Badrian A, Badrian N. 1984. Social organization of *Pan paniscus* in the Lomako Forest, Zaire. In: Susman RL, editor. *The Pygmy Chimpanzee: Evolutionary Biology and Behavior*, 325–346. New York: Plenum Press.
Baer J. 1998. A veterinary perspective of potential risk factors in environmental enrichment. In: Shepherdson DJ, Mellen JD, Hutchins M, editors. *Second Nature: Environmental Enrichment for Captive Animals*, 277–301. Washington, DC: Smithsonian Institution Press.

Bard K. 1995. Parenting in primates. In: Bornstein MH, editor. *Handbook of Parenting, Volume 2: Biology and Ecology of Parenting*, 27–58. Mahwah, NJ: Lawrence Erlbaum.

Bell B, Ballmann S. 1997. Bonobo training. In: Mills J, Reinartz G, De Bois H, Van Elsacker L, Van Puijenbroeck B, editors. *The Care and Management of Bonobos (Pan paniscus) in Captive Environments*, Chapter 3 7.1–7.4. Milwaukee, WI: The Zoological Society of Milwaukee County.

Bloomsmith M. 1992. Chimpanzee training and behavioral research: A symbolic relationship. In: *American Association of Zoological Parks and Aquariums Annual Conference Proceedings*, Toronto, Canada, 403–410.

Bloomsmith MA, Alford PL, Maple TL. 1988. Successful feeding enrichment for captive chimpanzees. *American Journal of Primatology* 16:155–164.

Bloomsmith M, Baker K. 2001. Social management of captive chimpanzees. In: Brent L, editor. *The Care and Management of Captive Chimpanzees*, 205–242. San Antonio, TX: American Society of Primatologists.

Bloomsmith MA, Baker KC, Ross SK, Lambeth SP. 1999. Chimpanzee behavior during the process of social introductions. In: *American Zoo Aquarium Association (AZA) Annual Conference Proceedings*, Silver Springs, MD, 270–273.

Bloomsmith MA, Else JG. 2005. Behavioral management of chimpanzees in biomedical research facilities: The state of the science. *ILAR Journal* 46(2):192–201.

Bloomsmith MA, Finlay TW, Merhalski JJ, Maple TL. 1990a. Rigid plastic balls as enrichment devices for captive chimpanzees. *Laboratory Animal Science* 40(3):319–322.

Bloomsmith MA, Haberstroh MD. 1995. Effect of early social experience on the expression of abnormal behavior among juvenile chimpanzees. *American Journal of Primatology* 36:110.

Bloomsmith MA, Keeling ME, Lambeth SP. 1990b. Video: Environmental enrichment for singly housed chimpanzees. *Lab Animal* 19(1):42–46.

Bloomsmith MA, Lambeth SP. 2000. Videotapes as enrichment for captive chimpanzees *(Pan troglodytes)*. *Zoo Biology* 19(6):541–551.

Bloomsmith MA, Lambeth SP, Perlman JE, Hook MA, Schapiro S. 2000. Control over videotape enrichment for socially housed chimpanzees. *American Journal of Primatology* 51(Suppl. 1):44–45.

Bloomsmith MA, Laule GE, Alford PL, Thurston RH. 1994. Using training to moderate chimpanzee aggression during feeding. *Zoo Biology* 13:557–566.

Bloomsmith MA, Stone AM, Laule GE. 1998. Positive reinforcement training to enhance the voluntary movement of group-housed chimpanzees within their enclosures. *Zoo Biology* 17:333–341.

Boesch C. 1991. Teaching among wild chimpanzees. *Animal Behaviour* 41:530–532.

Boesch C. 1996. Social grouping in Täi chimpanzees. In: McGrew WC, Marchant LF, Nishida T, editors. *Great Ape Societies*, 101–113. Cambridge, MA: Cambridge University Press.

Boesch C. 2002a. Behavioural diversity in *Pan*. In: Boesch C, Hohmann G, Marchant LF, editors. *Behavioural Diversity in Chimpanzees and Bonobos*, 1–13. Cambridge, UK: Cambridge University Press.

Boesch C. 2002b. Cooperative hunting roles among Täi chimpanzees. *Human Nature* 13(1):27–46.

Boesch C, Boesch H. 1990. Tool use and tool making in wild chimpanzees. *Folia Primatologica* 54:86–99.

Boesch C, Boesch-Achermann H. 2000. *The Chimpanzees of the Tai Forest: Behavioural Ecology and Evolution*. Oxford: Oxford University Press.

Bourgeois SR, Vazquez M, Brasky K. 2007. Combination therapy reduces self-injurious behavior in a chimpanzee *(Pan troglodytes troglodytes)*: A case report. *Journal of Applied Animal Welfare Science* 10(2):123–140.

Brent L. 2001. Behavior and environmental enrichment of individually housed chimpanzees. In: Brent L, editor. *Care and Management of Captive Chimpanzees*, 147–171. San Antonio, TX: American Society of Primatologists.

Brent L, Eichberg JW. 1991. Primate puzzleboard: A simple environmental enrichment device for captive chimpanzees. *Zoo Biology* 10:353–360.

Brent L, Kessel AL, Barrera H. 1997. Evaluation of introduction procedures in captive chimpanzees. *Zoo Biology* 16:335–342.

Brent L, Lee DR, Eichberg JW. 1989. Evaluation of two environmental enrichment devices for singly caged chimpanzees *(Pan troglodytes)*. *American Journal of Primatology* (Suppl. 1): 65–70.

Brent L, Stone A. 1996. Long-term use of televisions, balls, and mirrors as enrichment for paired and singly caged chimpanzees. *American Journal of Primatology* 39(2):139–145.

Brent L, Stone A. 1998. Destructible toys as enrichment for captive chimpanzees. *Journal of Applied Animal Welfare Science* 1:5–14.

Brent L, Williams-Blangero S, Stone A. 1996. Evaluation of the chimpanzee breeding program at the Southwest Foundation for Biomedical Research. *Laboratory Animal Science* 46:405–409.

Bridges J, Haller R, Buchl S, Magden E, Lambeth S, Schapiro S. 2015. Establishing a behavioral management program for geriatric chimpanzees. *American Journal of Primatology* 77(Suppl. 1):111 [abstract].

Bridges J, Mocarski E, Reamer L, Lambeth S, Schapiro S. 2013. Weight management in captive chimpanzees (*Pan troglodytes*) using a modified feeding device. *American Journal of Primatology* 75(Suppl. 1):51 [abstract].

Brosnan S, Parrish A, Beran M, Flemming T, Heimbauer L, Talbot C, Lambeth S, Schapiro S, Wilson B. 2011. Responses to the assurance game in monkeys, apes, and humans using equivalent procedures. *Proceedings of the National Academy of Sciences of the United States of America* 108(8):3442–3447.

Chapman CA, Chapman LJ. 2000. Determinants of group size in primates: The importance of travel costs. In: Boinski S, Garber PA, editors. *On the Move: How and Why Animals Travel in Groups*, 24–42. Chicago, IL: University of Chicago Press.

Chapman CA, White FJ, Wrangham RW. 1994. Party size in chimpanzees and bonobos. In: Wrangham RW, McGrew WC, de Waal FBM, Heltne PG, editors. *Chimpanzee Cultures*, 41–58. Boston, MA: Harvard University Press.

Clark C. 1977. A preliminary report on weaning among chimpanzees of the Gombe National Park, Tanzania. In: Chevalier-Skolnikoff S, Poirier F, editors. *Primate Bio-Social Development: Biological, Social and Ecological Determinants*, 235–260. New York: Garland.

Clark F, Smith L. 2013. Effect of a cognitive challenge device containing food and non-food rewards on chimpanzee well-being. *American Journal of Primatology* 75:807–816.

Clay Z, Zuberbuhler K. 2009. Food-associated calling sequences in bonobos. *Animal Behaviour* 77:1387–1396.

Coe JC. 1989. Naturalizing habitats for captive primates. *Zoo Biology* (Suppl. 1):117–125.

Colahan H, Breder C. 2003. Primate training at Disney's Animal Kingdom. *Journal of Applied Animal Welfare Science* 6(3):235–246.

Coleman K. 2012. Individual differences in temperament and behavioral management practices for nonhuman primates. *Applied Animal Behavior Science* 137(3–4):106–113.

Coleman K, Pranger L, Maier A, Lambeth SP, Perlman JE, Thiele EL, Schapiro SJ. 2008. Training rhesus macaques for venipuncture using positive reinforcement techniques: A comparison with chimpanzees. *Journal of the American Association of Laboratory Animal Science* 47(1):37–41.

Csatádi K, Leus K, Pereboom J. 2008. A brief note on the effects of novel enrichment on an unwanted behaviour of captive bonobos. *Applied Animal Behaviour Science* 112:201–204.

Damen F. 1990. Effects of the addition of browse on the feeding behaviour (time budget and coprophagy) in captive bonobos. In: *Proceedings of the Scientific Session of the 45th Annual Conference of the International Union of Directors of Zoological Gardens*, Copenhagen, Denmark.

D'Eath RB, Tolkamp BJ, Kyriazakis I, Lawrence AB. 2009. "Freedom from hunger" and preventing obesity: The animal welfare implications of reducing food quantity or quality. *Animal Behaviour* 37:55–63.

Desmond T, Laule G, McNary J. 1987. Training for socialization and reproduction with drills. In: *American Association of Zoological Parks and Aquariums (AAZPA) Annual Conference Proceedings*, Portland, OR, 435–441.

de Waal FBM. 1989. Behavioral contrasts between bonobos and chimpanzees. In: Heltne PG, Marquardt LA, editors. *Understanding Chimpanzees*, 154–175. Cambridge, MA: Harvard University Press.

de Waal FMB. 1995. Sex as an alternative to aggression in the bonobo. In: Abramson PR, Pinkerton SD, editors. *Sexual Nature, Sexual Culture*, 37–56. Chicago, IL: University of Chicago Press.

Fritz J. 1986. Resocialization of asocial chimpanzees. In: Benirschke K, editor. *Primates: The Road to Self-Sustaining Populations*, 376–388. Park Ridge, NJ: Noyes Publications.

Fritz P, Fritz J. 1979. Resocialization of chimpanzees. Ten years of experience at the Primate Foundation of Arizona. *Journal of Medical Primatology* 8:202–221.

Fritz J, Howell S. 2001. Captive chimpanzee social group formation. Care and management of captive chimpanzees. In: Brent L, editor. *The Care and Management of Captive Chimpanzees*. 173–203. San Antonio, TX: American Society of Primatologists.

Fruth B, Hohmann G. 2006. Social grease for females? Same-sex genital contacts in wild bonobos. In: Sommer V, Vasey PL, editors. *Homosexual Behaviour in Animals: An Evolutionary Perspective*, 294–308. Cambridge, UK: Cambridge University Press.

Furuichi T. 1989. Social interactions and the life history of female *Pan paniscus* in Wamba, Zaire. *International Journal of Primatology* 10:173–197.

Furuichi T. 1997. Agonistic interactions and matrifocal dominance rank of wild bonobos (*Pan paniscus*). *International Journal of Primatology* 18(6):855–875.

Gallup G Jr. 1970. Chimpanzees: Self-recognition. *Science* 167:86–87.

Goodall J. 1986. *The Chimpanzees of Gombe: Patterns of Behavior.* Cambridge, MA: Harvard University Press, 673 p.

Goodchild S, Schwitzer C. 2008. The problem of obesity in captive lemurs. *International Zoo News* 55:353–357.

Gruber T, Clay Z, Zuberbühler K. 2010. A comparison of bonobo and chimpanzee tool use: Evidence for a female bias in the *Pan* lineage. *Animal Behaviour* 80:1023–1033.

Hernandez-Aguilar RA, Moore, J, Stanford CB. 2013. Chimpanzee nesting patterns in savanna habitat: Environmental influences and preferences. *American Journal of Primatology* 75:979–994.

Herrmann E, Call J, Hernandez-Lloreda MV, Hare B, Tomasello M. 2007. Humans have evolved specialized skills of social cognition: The cultural intelligence hypothesis. *Science* 317(5843):1360–1366.

Herrmann E, Wobber V, Call J. 2008. Great apes' (*Pan troglodytes, Pan paniscus, Gorilla gorilla, Pongo pygmaeus*) understanding of tool functional properties after limited experience. *Journal of Comparative Psychology* 122:220–230.

Hohmann G, Fruth B. 2003. Culture in bonobos? Between-species and within-species variation in behavior. Current Anthropology 44(4):563–571.

Honess PE, Marin CM. 2006. Behavioural and physiological aspects of stress and aggression in nonhuman primates. *Neuroscience & Biobehavioral Reviews* 30(3):390–412.

Hopkins WD, Latzman RD. 2017. Future research with captive chimpanzees in the USA: Integrating scientific programs with behavioral management, Chapter 10. In: Schapiro SJ, editor. *Handbook of Primate Behavioral Management*, 139–155. Boca Raton, FL: CRC Press.

Hopkins WD, Washburn DA, Hyatt CW. 1996. Video-task acquisition in rhesus monkeys (*Macaca mulatta*) and chimpanzees (*Pan troglodytes*): A comparative analysis. *Primates* 37(2):197–206.

Hopper LM. 2017. Social learning and decision making, Chapter 15. In: Schapiro SJ, editor. *Handbook of Primate Behavioral Management*, 225–241. Boca Raton, FL: CRC Press.

Hopper LM, Kurtycz LM, Ross SR, Bonnie KE. 2015. Captive chimpanzee foraging in a social setting: A test of problem solving, flexibility, and spatial discounting. *PeerJ* 3:e833.

Hopper LM, Lambeth SP, Schapiro SJ, Bernacky BJ, Brosnan SF. 2013. The ontogeny of social comparisons by rhesus macaques (*Macaca mulatta*). *Journal of Primatology* 2(109) doi:10.4172/2167-6801.1000109.

Hopper LM, Lambeth SP, Schapiro SJ, Brosnan SF. 2011. Chimpanzees' (*Pan troglodytes*) learning indicates both conformity and conservatism. *American Journal of Primatology* 73:77–77.

Hopper L, Price S, Freeman H, Lambeth S, Schapiro S, Kendal R. 2014. Influence of personality, age, sex, and estrous state on chimpanzee problem-solving success. *Animal Cognition* 17(4):835–847.

Howell S, Drummer L, Fritz J. 1997. Social group formation in captive chimpanzees (*Pan troglodytes*): A comparison between two institutions. *American Journal of Primatology* 42:116–117.

Howell S, Schwandt M, Fritz J, Roeder E, Nelson C. 2003. A stereo music system as environmental enrichment for captive chimpanzees. *Lab Animal* 32:31–36.

Idani C. 1991. Social relationships between immigrant and resident bonobo (*Pan paniscus*) females at Wamba. *Folia Primatologica* 57:82–95.

Ingmanson, EJ. 1996. Tool-using behavior in wild *Pan paniscus*: Social and ecological considerations. In: Russon AE, Bard KA, Parker ST, editors. *Reaching into Thought: The Minds of Great Apes*, 190–210. Cambridge, MA: Cambridge University Press.

IOM (Institute of Medicine). 2011. *Chimpanzees in Biomedical and Behavioral Research: Assessing the Necessity.* Washington, DC: The National Academies Press.

Jaeggi AV, De Groot E, Stevens JMG, van Schaik CP. 2013. Mechanisms of reciprocity in primates: Testing for short-term contingency of grooming and food sharing in bonobos and chimpanzees. *Evolution and Human Behavior* 34(2):69–77.

Johnson CM. 1997. Juvenile bonobos (ages 2–5 years). In: Mills J, Reinartz G, De Bois H, Van Elsacker L, Van Puijenbroeck B, editors. *The Care and Management of Bonobos (Pan paniscus) in Captive Environments.* Antwerp, Belgium: Royal Zoological Society of Antwerp.

Johnson ET, Lynch PA, Schapiro SJ. 2013. Hunting, cooperation and sharing in captive chimpanzees. In: *Proceedings of the 59th Annual Convention of the Southwestern Psychological Association*, Ft Worth, TX.

Kano T. 1980. Social behavior of wild pygmy chimpanzees (*Pan paniscus*) of Wamba: A preliminary report. *Journal of Human Evolution* 9:243–260.

Kano T. 1987. Social regulation for individual coexistence in pygmy chimpanzees (*Pan paniscus*). In: McGuiness D, editor. *Dominance, Aggression and War*, 105–118. New York: Paragon House.

Kano T. 1989. The sexual behavior of pygmy chimpanzees. In: Heltne PG, Marquardt LA, editors. *Understanding Chimpanzees*, 176–183. Cambridge, MA: Harvard University Press.

King NE, Mellen JD. 1994. The effects of early experience on adult copulatory behavior in chimpanzees (*Pan troglodytes*). *Zoo Biology* 13:51–59.

Lambeth S, Bernacky B, Hanley P, Schapiro S. 2011. Weight management in a captive colony of chimpanzees (*Pan troglodytes*). *American Journal of Primatology* 73(Suppl. 1):40 [abstract].

Lambeth S, Bloomsmith M. 1992. Mirrors as enrichment for captive chimpanzees (*Pan troglodytes*). *Laboratory Animal Science* 42:261–266.

Lambeth SP, Hau J, Perlman JE, Martino M, Schapiro SJ. 2006. Positive reinforcement training affects hematologic and serum chemistry values in captive chimpanzees (*Pan troglodytes*). *American Journal of Primatology* 68(3):245–256.

Lambeth SP, Schapiro SJ. 2017. Managing a behavioral management program, Chapter 18. In: Schapiro SJ, editor. *Handbook of Primate Behavioral Management*, 265–276. Boca Raton, FL: CRC Press.

Lambeth S, Schapiro S, Bernacky B, Wilkerson G. 2013. Establishing "quality of life" parameters using behavioural guidelines for humane euthanasia of captive non-human primates. *Animal Welfare* 22(4):429–435.

Laule GE, Bloomsmith MA, Schapiro SJ. 2003. The use of positive reinforcement training techniques to enhance the care, management, and welfare of primates in the laboratory. *Journal of Applied Animal Welfare Science* 6:163–173.

Laule G, Desmond T. 1998. Positive reinforcement as an enrichment strategy. In: Shepherdson D, Mellen J, Hutchins M, editors. *Second Nature: Environmental Enrichment for Captive Animals*, 302–313. Washington, DC: Smithsonian Institution Press.

Laule G, Thurston R, Alford P, Bloomsmith M. 1996. Training to reliably obtain blood and urine samples from a diabetic chimpanzee (*Pan troglodytes*). *Zoo Biology* 15:587–591.

Laule G, Whittaker M. 2001. The use of positive reinforcement techniques with chimpanzees for enhanced care and welfare. In: Brent L, editor. *Care and Management of Captive Chimpanzees*, 243–265. San Antonio, TX: American Society of Primatologists.

Lutz CK, Novak MA. 2005. Environmental enrichment for nonhuman primates: Theory and application. *ILAR Journal* 46(2):178–191.

Magden E, Haller R, Thiele E, Lambeth S, Schapiro S. 2013. Acupuncture as an adjunct therapy for osteoarthritis in chimpanzees (*Pan troglodytes*). *Journal of the American Association for Laboratory Animal Science* 52(4):475–480.

Magden E, Sleeper M, Buchl S, Jones R, Thiele E, Wilkerson G. 2016. Use of an implantable loop recorder in a chimpanzee (*Pan troglodytes*) to monitor cardiac arrhythmias and assess the effects of acupuncture and laser therapy. *Comparative Medicine* 66:1–7.

Maki S, Alford PL, Bloomsmith MA, Franklin J. 1989. Food puzzle device simulating termite fishing for captive chimpanzees (*Pan troglodytes*). *American Journal of Primatology Supplement* 1: 71–78.

Maki S, Fritz J, England N. 1993. An assessment of early differential rearing conditions on later behavioral development in captive chimpanzees. *Infant Behavioral Development* 16:373–381.

Maple TL, Finlay TW. 1989. Applied primatology in the modern zoo. *Zoo Biology* 8(Suppl. 1):101–116.

Martin AL, Bloomsmith MA, Kelley ME, Marr MJ, Maple TL. 2011. Functional analysis and treatment of human-directed undesirable behavior exhibited by a captive chimpanzee. *Journal of Applied Behavior Analysis* 44(1):139–143.

McDonald S. 1994. The Detroit Zoo chimpanzees (*Pan troglodytes*): Exhibit design, group composition and the process of group formation. *International Zoo Yearbook* 33:235–247.

McGrew WC. 1979. Evolutionary implications of sex differences in chimpanzee predation and tool use. In: Hamburg DA, McCown E, editors. *The Great Apes*, 441–463. Menlo Park, CA: Benjamin/Cummings Publishing Company.

McMillan G, Drummer L, Fouraker M. 1991. *Findings from the Chimpanzee Enclosure Design Workshop*. Knoxville, TN: Knoxville Zoological Gardens.

McNary J. 1992. Integration of chimpanzees (*Pan troglodytes*) in captivity. In: Fulk R, Garland C, editors. *The Care and Management of Chimpanzees (Pan troglodytes) in Captive Environments*, 88–103. Asheboro, NC: North Carolina Zoological Society.

McTighe MS, Hansen BC, Ely JJ, Lee DR. 2011. Determination of hemoglobin A1c and fasting blood glucose reference intervals in captive chimpanzees (*Pan troglodytes*). *Journal of the American Association for Laboratory Animal Science* 50(2):165–170.

Mellen J, Shepherson D. 1992. Environmental enrichment for captive chimpanzees. In: Fulk R, Garland C, editors. *The Care and Management of Chimpanzees (Pan troglodytes) in Captive Environments*, 64–68. Asheboro, NC: North Carolina Zoological Society.

Menzel E. 1971. Group behavior in young chimpanzees: Responsiveness to cumulative novel changes in a large outdoor enclosure. *Journal of Comparative and Physiological Psychology* 34(1):46–51.

Mitani JC, Watts DP. 1999. Demographic influences on the hunting behavior of chimpanzees. *American Journal of Physical Anthropology* 109(4):439–454.

Nash L. 1982. Tool use by captive chimpanzees at an artificial termite mound. *Zoo Biology* 1:211–221.

Nash L, Fritz J, Alford P, Brent L. 1999. Variables influencing the origins of diverse abnormal behaviors in a large sample of captive chimpanzees (*Pan troglodytes*). *American Journal of Primatology* 48:15–29.

Nishida T, Hiraiwa-Hasegawa M. 1987. Chimpanzees and bonobos: Cooperative relationships among males. In: Smuts B, Cheney D, Seyfarth R, Wrangham R, Struhsaker T, editors. *Primate societies*, 165–177. Chicago, IL: University of Chicago Press.

Ostrower S, Brent L. 1997. Olfactory enrichment for captive chimpanzees: Response to different odors. *Laboratory Primate Newsletter* 36(1):8–10.

Paquestte D, Prescott J. 1988. Use of novel objects to enhance environments of captive chimpanzees. *Zoo Biology* 7:15–23.

Perlman J, Bloomsmith M, Whittaker M, McMillan J, Minier D, McCowan B. 2012. Implementing positive reinforcement animal training programs at primate laboratories. *Applied Animal Behaviour Science* 137:114–126.

Perlman JE, Horner V, Bloomsmith MA, Lambeth SP, Schapiro SJ. 2010. Positive reinforcement training, welfare, and chimpanzee social learning. In: Lonsdorf E, Ross S, Matsuzawa T, editors. *The Mind of the Chimpanzee*, 320–331. Chicago, IL: Chicago Press.

Pomerantz O, Terkel J. 2009. Effects of positive reinforcement training techniques on the psychological welfare of zoo-housed chimpanzees (*Pan troglodytes*). *American Journal of Primatology* 71:687–695.

Prescott MJ, Bowell VA, Buchanan-Smith HM. 2005. Training of laboratory-housed non-human primates, part 2: Resources for developing and implementing training programmes. *Animal Technology and Welfare* 4(3):133–148.

Proctor D, Williamson R, de Waal F, Brosnan S. 2013. Chimpanzees play the ultimatum game. *Proceedings of the National Academy of Sciences of the United States of America* 110(6):2070–2075.

Pruetz JD, Bertolani P, Boyer-Ontl K, Lindshield S, Shelley M, Wessling EG. 2015. New evidence on the tool-assisted hunting exhibited by chimpanzees (*Pan troglodytes verus*) in a savannah habitat at Fongoli, Sénégal. *Royal Society Open Science* 2(4): 140507. doi:10.1098/rsos.140507.

Pruetz JD, Bloomsmith MA. 1992. Comparing two manipulable objects as enrichment for captive chimpanzees. *Animal Welfare* 1:127–137.

Pruetz J, McGrew W. 2001. What does a chimpanzee need? Using natural behavior to guide the care and management of captive populations. In: Brent L, editor. *The Care and Management of Captive Chimpanzees*, 17–37. San Antonio, TX: The American Society of Primatologists.

Pusey A. 1980. Inbreeding avoidance in chimpanzees. *Animal Behaviour* 28:543–552.

Reamer L, Haller R, Thiele E, Freeman H, Lambeth S, Schapiro S. 2014. Factors affecting initial training success of blood glucose testing in captive chimpanzees (*Pan troglodytes*). *Zoo Biology* 33(3):212–220.

Reamer L, Haller R, Thiele E, Schapiro S, Lambeth S. 2013. Behavioral effects of transport and relocation on captive chimpanzees (*Pan troglodytes*). *American Journal of Primatology* 75(Suppl. 1):82 [abstract].

Richardson A, Lambeth S, Schapiro S. 2006. Control over the auditory environment: A study of music preference in captive chimpanzees (*Pan troglodytes*). *International Journal of Primatology* 27:423.

Ross SK, Bloomsmith MA, Baker KC, Hopkins WD. 2000. Initiating a computer-assisted enrichment system for captive chimpanzees. *American Journal of Primatology* 51(Suppl. 1):86–87.

Ross SR, Calcutt S, Schapiro SJ, Hau J. 2011. Space use selectivity by chimpanzees and gorillas in an indoor-outdoor enclosure. *American Journal of Primatology* 73:197–208.

Ross S, McNary J. 2010. *Chimpanzee (Pan troglodytes) Care Manual.* AZA Ape TAG. Silver Spring, MD: Association of Zoos and Aquariums.

Rowe N. 1996. *The Pictorial Guide to the Living Primates.* East Hampton: Pogonias Press.

Russell JL, Lyn H, Schaeffer JA, Hopkins WD. 2011. The role of socio-communicative rearing environments in the development of social and physical cognition in apes. *Developmental Science* 14(6):1459–70.

Schapiro SJ, Bloomsmith MA, Porter LM, Suarez SA. 1996a. Enrichment effects on rhesus monkeys successively housed singly, in pairs, and in groups. *Applied Animal Behaviour Science* 48:159–171.

Schapiro SJ, Bloomsmith MA, Suarez SA, Porter LM. 1996b. Effects of social inanimate enrichment on the behavior of yearling rhesus monkeys. *American Journal of Primatology* 40:247–260.

Schapiro S, Coleman K, Akinyi M, Koenig P, Hau J, Domaingue MC. 2014. Nonhuman primate welfare in the research environment. In: Bayne K, Turner P, editors. *Laboratory Animal Welfare*, 197–209. London: Elsevier Inc.

Schapiro SJ, Lambeth SP. 2010. Chimpanzees. In: Hubrecht R, Kirkwood J, editors. *The UFAW Handbook on the Care and Management of Laboratory Animals*, 8th edition, 618–634. Oxford: Wiley-Blackwell.

Schapiro SJ, Lambeth SP, Jacobsen KR, Williams LE, Nehete BN, Nehete PN. 2012. Physiological and welfare consequences of transport, relocation, and acclimatization of chimpanzees (*Pan troglodytes*). *Applied Animal Behaviour Science* 137(3–4):183–193.

Sexton P, Gallagher B. 1997. Introduction procedures for adult animals at the San Diego Wild Animal Park. In: Mills J, Reinartz G, De Bois H, Van Elsacker L, Van Puijenbroeck B, editors. *The Care and Management of Bonobos (Pan paniscus) in Captive Environments*, Chapter 3 2.4–2.6. Milwaukee, WI: The Zoological Society of Milwaukee County.

Smith R, Jungers W. 1997. Body mass in comparative primatology. *Journal of Human Evolution* 32:523–559.

Sousa C, Biro D, Matsuzawa T. 2009. Leaf-tool use for drinking water by wild chimpanzees (*Pan troglodytes*): Acquisition patterns and handedness. *Animal Cognition* 12(1):115–125.

Spijkerman R, Dienske H, van Hoof J, Jens W. 1994. Causes of body rocking in chimpanzees (*Pan troglodytes*). *Animal Welfare* 3:193–211.

Staes N, Stevens JMG, Helsen P, Hillyer M, Korody M, Eens M. 2014. Oxytocin and vasopressin receptor gene variation as a proximate base for inter- and intraspecific behavioral differences in bonobos and chimpanzees. *PLoS One* 9(11): e113364.

Struthers E, Campbell J. 1996. Scent-specific behavioral response to olfactory enrichment in captive chimpanzees (*Pan troglodytes*). In: *XVIth Congress of the International Primatological Society and the XIXth Conference of the American Society of Primatology Proceedings*, Madison, WI.

Stumpf R. 2007. Chimpanzees and bonobos: Diversity within and between species. In: Campbell CJ, Agustin F, MacKinnon KC, Simon KB, Bearder SK, Stumpf RM, editors. *Primates in Perspective*, 321–344. Oxford: Oxford University Press.

Surbeck M, Hohmann G. 2008. Primate hunting by bonobos at LuiKotale, Salonga National Park. *Current Biology* 18(19):906–907.

Takeshita H, Walraven, V. 1996. A comparative study of the variety and complexity of object manipulation in captive chimpanzees (*Pan troglodytes*) and bonobos (*Pan paniscus*). *Primates* 37:423–441.

Taglialatela J. 2016. The neural correlates of multimodal communication in chimpanzees (*Pan troglodytes*) and bonobos (*Pan paniscus*) and their implications for the evolution of human language. *American Journal of Physical Anthropology* 159(Suppl. 62):310.

Taglialatela J, Russell JL, Pope SM, Morton T, Bogart S, Schapiro SJ, Hopkins WD. 2015. Multimodal communication in chimpanzees. *American Journal of Primatology* 77:1143–1148.

Thompson J, Hohmann G, Furuichi T. (Eds). 2003. *Behaviour, Ecology and Conservation of Wild Bonobos.* 2003 Bonobo Workshop. Inuyama, Japan: Kyoto University Primate Research Institute.

Toback E, Granholm C, McNary J. 1992. Assessment of maternal behavior in captive chimpanzees. *American Journal of Primatology* 27:61 [abstract].

Traylor-Holzer K, Fritz P. 1985. Utilization of space by adult and juvenile groups of captive chimpanzees (*Pan troglodytes*). *Zoo Biology* 4:115–127.

Urashima T, Odaka G, Asakuma S, Uemura Y, Goto K, Senda A, Saito T, Fukuda K, Messer M, Oftedal O. 2009. Chemical characterization of oligosaccharides in chimpanzee, bonobo, gorilla, orangutan, and siamang milk or colostrum. *Glycobiology* 19(5):499–508.

Van Puijenbroeck B, De Bois H. 1997. Enclosure design. In: Mills J, Reinartz G, De Bois H, Van Elsacker L, Van Puijenbroeck B, editors. *The Care and Management of Bonobos (Pan paniscus) in Captive Environments*, Chapter 5 1.1–1.5. Milwaukee, WI: The Zoological Society of Milwaukee County.

Videan EN. 2006. Bed-building in captive chimpanzees *(Pan troglodytes)*: The importance of early rearing. *American Journal of Primatology* 68:745–751.

Videan EN, Fritz J, Murphy J. 2007. Development of guidelines for assessing obesity in captive chimpanzees *(Pan troglodytes)*. *Zoo Biology* 26(2):93–104.

Watts DP, Mitani JC. 2002. Hunting behavior of chimpanzees at Ngogo, Kibale National Park, Uganda. *International Journal of Primatology* 23(1):1–28.

White FJ. 1992. Activity budgets, feeding behavior, and habitat use of pygmy chimpanzees at Lomako, Zaire. *American Journal of Primatology* 26:215–223.

Whiten A, Goodall J, McGrew WC, Nishida T, Reynolds V, Sugiyama Y, Tutin CEG, Wrangham RW, Boesch C. 1999. Cultures in chimpanzees. *Nature* 399:682–685.

Whiten A, Spiteri A, Horner V, Bonnie KE, Lambeth SP, Schapiro SJ, de Waal FBM. 2007. Transmission of multiple traditions within and between chimpanzee groups. *Current Biology* 17(12):1038–1043.

Wobbera V, Hare B, Mabotoc J, Lipsona S, Wrangham R, Ellisona PT. 2010. Differential changes in steroid hormones before competition in bonobos and chimpanzees. *Proceedings of the National Academy of Sciences of the United States of America* 107(28):12457–12462.

Wolfensohn SE, Honess PE, 2005. *Handbook of Primate Husbandry and Welfare*. Oxford, UK: Blackwell Publishing Ltd.

Wrangham R. 1977. Feeding behavior of chimpanzees in Gombe National Park, Tanzania. In: Clutton-Brock TH, editor. *Primate Ecology: Studies of Feeding and Ranging Behavior in Lemurs, Monkeys, and Apes*, 504–538. London: Academic Press.

Yamagiwa J, Maruhashi T, Yumoto T, Mwanza N. 1996. Dietary and ranging overlap in sympatric gorilla and chimpanzees in Kahuzi-Biega National Park, Zaire. In: McGrew WC, Merchant LF, Nishida T, editors. *Great Ape Societies*, 82–98. Cambridge, UK: Cambridge University Press.

Yamanashi Y, Hayashi M. 2011. Assessing the effects of cognitive experiments on the welfare of captive chimpanzees *(Pan troglodytes)* by direct comparison of activity budget between wild and captive chimpanzees. *American Journal of Primatology* 73:1231–1238.

CHAPTER 24

Behavioral Management of Neotropical Primates
Aotus, Callithrix, *and* Saimiri

Lawrence Williams
The University of Texas MD Anderson Cancer Center

Corinna N. Ross
Texas A&M University-San Antonio

CONTENTS

INTRODUCTION

Animals maintained in laboratory environments often live in conditions that are restrictive and limiting compared with their natural habitats. Although such environments meet the physical needs of the animals, they may not be sufficient to allow the animals to express the full range of their species-typical behavior. The natural environments of most animals are extremely complex, involving unpredictable elements, as well as routine tasks. Laboratory environments may lack some of the stimuli necessary to promote desirable species-specific natural behaviors. Even within a single order of animals, such as primates, there are major differences between species, both in the ways they perceive the environment and in their particular needs. Differences in ecological variables, such as social group size, food source, and mating patterns, lead to contrasts in both consummatory and appetitive patterns. These dissimilarities need to be taken into account when designing captive environments and management protocols.

It is particularly important to consider that procedures and protocols that might enhance the experiences of one species might be detrimental to another. Behavioral management plans, including husbandry protocols and enrichment, are necessary to supplement and support the captive environment. Choosing the appropriate behavioral management strategies for the species of interest, the experimental protocols, and often, the individual animal, are imperative to elicit species-specific, healthy behavioral actions. Understanding how the animals might react in times of stress or conflict is important in experimental, clinical, and diagnostic settings.

For species that are often preyed upon, individuals that exhibit atypical behaviors in a natural setting are the ones most likely to be targeted by predators. This leads to a stoicism among wild animals that makes spotting a sick animal very difficult. In experimental settings, decisions such as single housing social animals (or pair housing animals that normally live singly) can increase stress levels, affecting the behavior of the animal, steroid hormone levels, and immunological function. Any or all of these factors may be potential confounds to experimental designs. Familiarity with the natural history and behavior of the species you are working with, combined with daily observations and positive interactions with the individual nonhuman primates (NHPs) in your care, make identifying animals in need of special attention possible before it is too late to intervene. Captive environments and behavioral management routines should be designed to allow the animals to express as much of their species-typical behavior as possible.

It is beyond the scope of this chapter to discuss all Neotropical primates; so we will concentrate our discussion on the three genera most often seen in laboratory settings—*Aotus*, *Callithrix*, and *Saimiri*. Our aim is to provide a background through which the behavior of these NHPs can be better appreciated. Each species has a unique set of environmental and social needs that must be met in order for the animals to thrive in the laboratory. We will begin with general statements about the ways in which these animals behave and interact in, and with, their natural environment. Following that, we will provide specific information on selected topics, including social group formation and housing, enrichment foods and devices, and positive reinforcement training (PRT). In the end, we hope to have provided enough information to promote informed decisions concerning the behavioral management of these animals.

AOTUS

Natural History

Aotus species, commonly known as owl monkeys, occupy a wide variety of habitats, from northern Argentina to Panama. They occur in both primary and secondary forest, as well as seasonally flooded areas. Owl monkeys spend most of their time in the upper canopy, sleeping in tree holes and dense thickets of foliage or vines (Fernandez-Duque 2012). *Aotus* are territorial, with groups occupying overlapping home ranges that vary between 3 and 18 ha (Wright 1978; Robinson et al. 1987; Fernandez-Duque 2007), whereas night ranges are approximately 0.7 km. Owl monkeys are mainly frugivores that supplement their diet with flowers, insects, nectars, and leaves, depending on the season and type of habitat.

Owl monkeys are the only nocturnal genera of simian primates. Using activity monitors on animals in captivity shows a clear demarcation between active (lights off) and inactive (lights on) periods (Williams, personal observation). While most *Aotus* species are mainly active during the dark period, *Aotus azarae* can better be described as cathemeral, with active periods during both light and dark periods (Fernandez-Duque 2003). *A. azarae* occupy the extreme southern part of the *Aotus* range, and cathemerality may be an adaptation to cooler nights.

Aotus are socially monogamous, living in small groups that typically consist of an adult pair and offspring of different ages—an infant, possibly one or two juveniles, and sometimes a subadult. As with most monogamous species, both males and females disperse from their natal group. Among *A. azarae*, a significant number of adults range alone (Fernandez-Duque et al. 2008). These lone animals may be subadults that have left their natal groups or older adults that have been evicted from their groups by competitors (Fernandez-Duque and Huntington 2002; Fernandez-Duque 2004). Lone adults were observed by Villavicencio Galindo (2003) in northern Colombia. Wright (1994a) and Fernandez-Duque (2007) provide a good review of the behavior and ecology of the genus.

In the wild, male *A. azarae* reach adult weight when they are about 4 years of age, with first reproduction occurring when they are about 5 years old (Fernandez-Duque 2004). Captive male *Aotus lemurinus* reach sexual maturity when they are 2 years old, and captive female *Aotus vociferans* and *Aotus nancymaae* first breed when 3–4 years of age (Dixson 1983; Fernandez-Duque 2007). Data from our laboratory (Williams et al., in preparation) show that *A. nancymaae*, *A. vociferans*, and *A. azarae* can all be successfully paired, when they are 2 years old, with their first offspring being born approximately 1 year later.

In captivity, there does not appear to be any seasonality in reproduction, although data from the field do suggest seasonality. Wright (1978) recorded births between August and February for *Aotus nigriceps* in Peru (Manu National Park), and Aquino et al. (1990) reported a birth season between December and March for *A. nancymaae* in northeastern Peru. The differences between captive and field data may be related to seasonal differences in food availability in the wild that are not present in captivity (Fernandez-Duque and Heide 2013). Single offspring is the rule, with adult males as the primary caregivers for the infants until weaning, at about 5 months of age (Rotundo et al. 2005).

Social Housing

Owl monkeys are generally housed singly for research purposes or as monogamous pairs for breeding (Moynihan 1964; Wright et al. 1989; Fernandez-Duque 2012). Williams et al. (2015) reported that the most successful pairing strategy for adult owl monkeys is to pair a male with a female. Eighty-two percent of mixed-sex pairs were successful, whereas isosexual pairs of either two females or two males resulted in lower pairing success rates (62% for FF pairs and 40% for MM pairs). Capitanio et al. (2015) point out that most successful pairings in other commonly used laboratory primates (primarily macaques) are single-sex pairs, because mixed-sex pairs appear to have difficulty forming pair bonds. However, within *Aotus*, the primary social bond is between male and female adults, making these pairs easier to form and hence, preferred. In the field, adult owl monkeys may be expelled from a social group by intruding adults (Fernandez-Duque 2004, 2007), events which generally do not occur in captivity. However, captive pairs do seem to have a finite compatibility "life span." Pairing owl monkeys is not a "once-and-done" process. Williams et al. (2015) found a mean pair longevity of nearly 3 years for male–female pairs of *A. nancymaae*. This means that facilities with owl monkeys should be prepared to (1) separate animals that become overly aggressive to one another as their time together approaches 3 years, and (2) form new pairs as needed.

Successful pairing of owl monkeys can be accomplished with the use of protected contact panels. These panels, which may consist of a coarse mesh that allows olfactory and finger-to-finger contact, should be in place for several days prior to a physical introduction, allowing the animals some time to get used to one another. On the first day of a planned introduction, the panels separating adjacent cages can be removed and the owl monkeys allowed to engage in full-contact interactions for at least 1 h. Pairs that enter the same nest box during this first introductory period can be allowed to stay together, with monitoring, for the next several days. Pairs that do not enter the same nest box after the first hour should be separated, using the protected contact panel until the next morning. This procedure is repeated for 3 days. If, at any time, the pair enters the same nest box and rests comfortably together, the pairing can be considered successful. If, after the third day, the animals are still staying in different areas of the cage and have not engaged in agonistic interactions, the protected contact panel can be permanently removed and the pairs allowed to stay together. Animals should be separated if, at any time during the pairing protocol, they engaged in excessive agonistic interactions, defined as chases, bites, and overt fighting between potential partners.

Enrichment

Enrichment for owl monkeys should focus on those facets of the genus that contribute most to its uniqueness. Owl monkeys are nocturnal, and they should be observed during a dark or lights-out period. In most facilities, this means an altered light/dark schedule from the normal workday. Without a reversed cycle, not only will all the cleaning, feeding, and clinical observations be done during the animals' natural resting (sleep) period, but there will also be few opportunities for observation of social interaction or movement, both of which are crucial for identifying sick or injured animals. Shifting lights on/off in captivity to a schedule similar to lights on at midnight and off at noon provides the monkeys with an undisturbed period of rest from midnight until the staff arrives (at 7 am), while still providing staff with time after lights out for direct observations during the animals' active period. At least one species, *A. azarae*, shows a high degree of cathemerality in the field (Fernandez-Duque 2003; Fernandez-Duque et al. 2010), suggesting that an appropriate light cycle for this species would include times of transition from complete light to complete darkness (dusk) and vice versa (dawn).

In addition to being nocturnal, owl monkeys are territorial (Fernandez-Duque 2012), rarely interacting with nonfamily members, except during an aggressive attempt to oust an adult from a stable family unit. However, in captivity, it is not uncommon for owl monkeys to be housed in direct visual contact with other owl monkeys living in the same room. This should be avoided, and housing should be structured so that a family unit of owl monkeys has as little sensory contact with other owl monkeys as possible. *Aotus* are known to have high incidences of hypertension and cardiomyopathy in captivity (Brady et al. 2005; Smith and Astley 2007; Rajendra et al. 2010), which has been hypothesized to result from constant interactions with extragroup conspecifics. In our laboratory, we have constructed the enclosures such that it is nearly impossible for one owl monkey to see an owl monkey in another group (Figure 24.1). In addition, a waterfall provides auditory stimuli that mask the sounds of other owl monkeys, and a down-draft air recirculation system limits olfactory communication across groups. Although this degree of sensory separation may not be possible in all facilities, the use of visual barriers and the alignment of housing to prevent direct sight lines should be considered.

Owl monkeys are arboreal and, in the wild, spend a significant portion of each night traveling in search of food (Fernandez-Duque 2012). Captive enclosures should include perches, swings, and ladders to provide as much three-dimensional space as possible for the animals (Figure 24.2). Travel paths should be arranged so that the monkeys can move about their housing to get from food sources to water sources and resting platforms. "Nest" boxes, or other enclosed spaces, such as tubes and "tangles" of PVC chain, should be provided as sleeping sites and/or places of safety. Owl monkeys are neophobic (Tardif et al. 2006; Hooff and Wolovich 2008) and will retreat to an enclosed space when threatened or even presented with a novel toy or food item. Perches should be positioned at several levels within the living area to encourage climbing and jumping. Wood perches can be used, as long as they are replaced after they become excessively soiled or have otherwise deteriorated. Owl monkeys use olfactory cues to locate food within their environment (Bolen and Green 1997;

Figure 24.1 Owl monkey housing room. Social housing units along the right-hand side are large enough for family groups of up to four animals each (1 m × 1.25 m × 1.7 m). White external boxes can serve as nest boxes and transfer cages for the owl monkeys. On the left-hand side is a wall feature, with a waterfall, that separates the two sides of the room. The overhead louvres are open, providing a dusky, twilight period in which the owl monkeys awake.

Figure 24.2 An example of owl monkey social housing with multiple levels of perches, a black nest box on the upper left-hand side, and a yellow tube for sleeping and hiding.

Nevo and Heymann 2015); so care should be taken when cleaning and sanitizing their housing. To maintain some degree of olfactory consistency in *Aotus* enclosures, a specific cage device can be cleaned on an alternate sanitization schedule. We swap the nest boxes with cleaned/sanitized replacements every other week, on the weeks between the sanitization of the rest of the housing. The use of PVC and other thermoneutral materials for cage structures should be encouraged to help the animals thermoregulate.

When considering foraging enrichment, it must be remembered that owl monkeys are frugivores that supplement their diets with flowers, insects, and leaves (Wright 1994b). Beyond the standard laboratory diets, owl monkeys respond to a wide variety of food items. Fresh fruits and vegetables, rotated on a seasonal basis, provide novelty and nutritional value. Owl monkeys use a foraging board to glean seeds and soft food items, such as vegetable purees, peanut butter, or yogurt, and transfer food items both between members of the breeding pair and to their young ones (Wolovich et al. 2008). However, food that is difficult for the adult to obtain is not transferred to the young one; so providing small, easily accessible food items will encourage this positive social interaction. Unlike squirrel monkeys, owl monkeys are not destructive foragers, and so do not work particularly hard on devices that require large amounts of manipulation to obtain the reward.

Training

Owl monkeys can be trained, using PRT techniques, to participate in a number of behavioral tasks. Because owl monkeys are generally neophobic, training can take much longer than is typically seen for Old World primates, and even squirrel monkeys. Rogge et al. (2013) found that owl monkeys took twice as long to learn to target than did squirrel monkeys. However, once targeting was learned, training for additional behaviors, such as presenting a hand or foot, occurred just as quickly as for squirrel monkeys. Because *Aotus* live in small family groups or pairs, it is easy to train them in their home cage. Both clickers and verbal cues work as a bridge between the behavioral response and the reward. Primary reinforcement should consist of sweet food items, including raisins, grapes, and marshmallows. In addition, owl monkeys can be trained to jump from old to new housing by placing a jackpot reward in the new housing. Given the cardiomyopathy and hypertension (Smith and Astley 2007) seen in owl monkeys, using training either to acclimate the animals to new procedures or to reward them for actively participating in routine procedures should reduce stress-related lesions.

CALLITHRIX

Natural History

Callitrichinae, commonly known as marmosets and tamarins, are small Neotropical primates native to the rainforests of Brazil. This group of primates includes 4 genera and 50 species, and is a subfamily within Cebidae (Mittermeier 1988; Schneider et al. 2001). Callitrichines are among the smallest of the New World primates, weighing 250–600 g, and are distinguished from other New World monkeys by the presence of characteristics including the following: all digits, except the hallux, have claws rather than flattened nails; triangular upper molars; a V-shaped mandible; are typically diurnal; lack prehensile tails; and the dental formula is 2/2, 1/1, 3/3, 2/2. In these often brightly colored species, twinning is common, and they exhibit little to no sexual dimorphism (Rylands et al. 1993; Rylands 1996; Kinzey 1997; Fleagle 1999). Common marmosets (*Callithrix jacchus*) are typified by white ear tufts, a grey and brown calico body, an average size of 300 g in the wild, and have earned their name owing to their broad range throughout Northeastern Brazil. In fact, they are considered an invasive species in some southern states of Brazil (Clarke 1994).

Although marmosets were previously thought to exclusively exhibit a monogamous breeding system, field studies have revealed that populations have variable and flexible mating strategies, including polyandry and polygyny (Baker et al. 1993; Digby 1995; Ferrari and Digby 1996). Groups are typified as having singular breeding pairs, with group cooperative care; however, this has been seen to vary among social groups. Although cooperative breeding is not rare in the animal kingdom, callitrichines are considered unique among primates (Goldizen 1988; Dixson 1993; Dunbar 1995; Schaffner and French 1997; Schaffner and French 2004). Juvenile marmosets may become sexually mature by the age of 1 year, and their reproductive output is typically suppressed for more than 3 years (Abbott et al. 1993). The juveniles, rather than dispersing to form their own families, stay with their natal group and help raise their younger siblings. Males and females are found to have equal dispersion rates as adults from the natal group (Savage et al. 1996). However, if one of the breeding pair is replaced by an unrelated individual, a plural breeding system in which a son and a father, or a mother and a daughter, breeding simultaneously has been noted (Saltzman et al. 2004). In addition, a lack of inbreeding avoidance has been noted for captive and wild callitrichines (Gengozian et al. 1980; Dixson et al. 1988; Watkins et al. 1991; Dixson et al. 1992; Savage et al. 1996).

Cooperative breeding has been hypothesized to have evolved in this group of primates because of the high energetic demands of breeding on the females. Callitrichines are considered obligate fraternal twinners (Hershkovitz 1975). The average gestational length is 5 months, and at birth, the twins often weigh as much as 20% of their mother's normal body weight. This high energetic demand on the mother is further compounded by the fact that, as early as 2 weeks after giving birth, the female ovulates and may become pregnant with the next set of twins. Callitrichines do not exhibit lactational amenorrhea, or reduced fertility during lactation (Kleiman 1977; Goldizen 1988; Mittermeier 1988; Price 1992; Digby and Barreto 1993; Nievergelt and Martin 1999). This energetic demand may positively select for both cooperative breeding behavior by siblings of the new offspring, as well as the high investment by males in the young. Unlike other primates, where males generally are not involved in caring for the young (Mittermeier 1988), callitrichine males become the major caregiver for infants soon after birth. Although the mother nurses her offspring for approximately 9 weeks, males in the family group actively carry, feed, clean, and protect the infants. This behavior clearly alleviates some of the energetic demand on females and allows them to concentrate their effort on foraging and lactating (Price 1992; Nievergelt and Martin 1999; Nunes et al. 2000).

Common marmosets have been widely used as biomedical models since the 1960s in Europe, Japan, Australia, and the United States (Abbott et al. 2003). The popularity of the marmoset in the United States was limited owing to the high investment by biomedical researchers in macaque monkeys, as well as the long-held opinion that marmosets were fragile and difficult to work with (Clapp and Tardif 1985). Better understanding of the natural habits of the animals, improved husbandry techniques, and improved dietary nutrition have increased both the popularity of this animal as a model species and the health of the animals in captivity. Marmosets differ remarkably from Old World monkeys in the types of social housing that are necessary to maintain them and the ease with which they can be manipulated as a family group (Layne and Power 2003). Cost and ease of handling are often cited as reasons for choosing the marmoset as a model for research over other potential primate models (Tardif et al. 2011).

Social Housing

Marmosets in captivity prosper when maintained as a nuclear family comprising a breeding pair and several sets of offspring. While polyandrous groups have been maintained in captivity, these groups were only stable when the males were related, and tended to dissolve as a result of aggression over time (Schaffner and French 2004). Although marmosets are often described as pair bonded (Rukstalis and French 2005), they can easily be remated after the loss of one of the breeding animals, as long as care is taken with the older infants in the group. There is no indication that marmosets grieve the loss of their pair mate or avoid breeding with new individuals following a remating, either because of the loss of the mate or owing to colony maintenance/research requirements. Within a family group, the size of the group depends partly on the size of the caging, colony management-related needs, and general group temperament. It is quite typical, with standard US marmoset cages measuring $72'' \times 60'' \times 36''$, to maintain three twin litters with the parents. The oldest siblings are removed upon the weaning of the fourth litter, such that the oldest siblings are mature adults at approximately 2 years of age when removed from the family (Layne and Power 2003). In this situation, the oldest offspring should be physiologically suppressed from breeding, although of adult size and sexually mature, as long as both animals within the breeding pair are healthy (Puffer et al. 2004). It is particularly important to leave older siblings in the family group for as long as possible for two reasons. First, these helpers are necessary to ensure the viability of their parents' newest offspring, and second, the additional experience the animals gain from caring for younger siblings increases the chance that their first set of offspring will survive (Goldizen 1988; Dunbar 1995). When mature juveniles are removed from the family group, these animals can be placed directly into a breeding pair or housed temporarily with their sibling, and are expected to quickly begin cycling if female (Smith et al. 1997). Owing to limited inbreeding avoidance tendencies in this species, it is easiest to manage single-sex pairs, as sibling mating is undesirable. However, unlike many other species of primates, single-sex pairs are not a viable long-term pairing option, as they are unlikely to remain stable and amicable (Majolo et al. 2003b). If single-sex pairs are desired, it is suggested that they be maintained in rooms containing only one sex, with no mature animals of the opposite sex living in the room.

In cases of aggression or hostility between members of the family group, it is typically quite easy to identify the animals that are involved in the aggression. The animal that is being harassed or the receiver of aggression will often huddle in one area of the cage, and vigilantly visually track the harassing animal. In some cases, it is possible to alleviate the tension in the group by temporarily physically separating the animals using mesh barriers, while still maintaining visual and auditory contact among group members. If the aggression does not resolve, the aggressor is typically removed to a new cage. If, however, the aggressor is one of the breeding animals, then the recipient of the aggression and a sibling can be removed. One way to reduce the rates of aggression between family members is to ensure that the caging has visual barriers (half walls, or nest boxes) in which animals can, at least temporarily, move away from one another (Kitchen and Martin 1996; Majolo et al. 2003a). Providing alternative enrichment furniture, as discussed below in the Enrichment section, can also provide animals with additional places to hide from and/or choose not to interact with one another. The provision of multiple food dishes is necessary for marmosets, because the breeding female will often monopolize a single food dish, preventing others from accessing the food, potentially increasing aggression within the group (Yamamoto et al. 2004; Vilela et al. 2012).

The introduction process for forming new pairs is usually quite easy in marmosets and can be done with a simple "howdy" system. New pairs are typically placed in a clean cage (so that neither animal is an intruder in an established cage), and animals are visually exposed to one another for several days with protected contact panels that only allow finger–finger contact between them. When potential pair mates are released together, caregivers observe for signs of aggression, and separate the animals if it occurs. Male–female aggression is quite rare in marmosets, however. Introducing a new pair mate to a family group is a bit more complicated and is dependent on the age of the juveniles in the group. We have had good luck with introductions if the juveniles are less than 1 year of age. Animals older than a year should be removed from the family prior to adding a new pair mate in order to avoid same-sex intruder aggression. It is not possible to put unrelated male–male pairs or female–female pairs together in marmosets. It is also not possible to reintroduce sisters or brothers that have been separated for more than a few days; the animals view one another as intruders and tend to fight, potentially inflicting fatal wounds.

There remains a great deal of debate as to whether breeding family groups should be visually isolated from other breeding family groups. Some colony managers maintain their colony with visual barriers between each family group to reduce visual access. Marmosets are territorial and do react to intruders (French et al. 1995; Ross et al. 2004; Ross and French 2011); however, many colonies allow visual access, and as long as the neighbors are consistent, the families stop reacting to the neighbors. This trait does make it imperative that movement of cages within a room is minimal, and animals are housed near neighbors with whom they are familiar. Movements of neighbors and changes within a room have been associated with increased rates of miscarriage and infant death (Tardif personal communication). Care must also be taken during cage jumping, so that the order of the cages remains the same, and families are not allowed to get too close together, as fighting and stress diarrhea can result. If an animal escapes from its cage, the greatest risk of injury is due to fighting between families.

One of the toughest management decisions in marmoset care is the handling of litters that are larger than a set of twins. The family group has very low success rearing triplets with no human intervention—in the 20 years I (CR) have worked with marmosets, I know of only three successful triplet litters in the colony with no intervention (and have heard of others at other institutions, but they are rare enough that we talk about them). Each colony in the United States has a policy for handling the birth of triplets, and it is often related to the needs of the colony and to staffing. Allowing nature to take its course will result in the loss of at least one infant, and depending on the health status of the dam, it can result in the loss of the whole litter. Some colony managers assess infants at 24 h of age to determine which infant has the lowest birth weight and behavioral strength score, and will then euthanize that individual animal. Litter reduction early after birth does tend to increase the survival probability of the remaining two infants (Tardif et al. 2002). One option for triplets is to rotationally hand-rear the infants, which can be accomplished in a number of ways. Our procedure is to remove an infant from the group each day and keep it in an isolette, with staff feeding formula every 1.5 h. At the end of the day, the infant is placed back in the family group, and the next day a different infant is removed for hand rearing. This strategy increases the food intake of each infant without increasing the lactational demand on the mother and maintains the infants with typical marmoset family care and interactions. This rotational care can usually be decreased in frequency, and then stopped, when the infants are 3–4 weeks old and are beginning the weaning process. Although infants can be completely nursery reared and hand fed without interactions with the marmoset family group, this often has poor long-term outcomes in terms of future breeding and behavior.

Enrichment

Enrichment for marmosets can come in a variety of forms, and behavioral managers at each colony may have different approaches to enrichment. Marmosets are gummivores in the wild, gouging and licking the sap from tree branches, and in some species, greater than 70% of their caloric intake is from gum (Huber and Lewis 2011). Natural gums and exudates are highly variable in their chemical structure and serve as sources of dietary fiber, polysaccharides, calcium, and proteins for marmosets. Freshly cut branches offer the opportunity for captive animals to retrieve gum after gouging. Some captive colonies use gum arabic as a source of enrichment for their animals, if fresh or live branches cannot be provided. Gum can be fed in dishes or spread on enrichment devices. In addition, devices made from PVC pipes with small holes allow marmosets to retrieve gum arabic from within the tubing (Huber and Lewis 2011). Use of gum and gum feeding devices is becoming more prevalent in captive colonies and is an item that should be strongly considered when developing behavioral management and enrichment programs for marmosets (Bakker et al. 2015).

Beyond the importance of branches as a nutritional source, they often play important roles in marmoset caging as furniture, as ways to increase complexity in the cage, and to provide a substrate on which the marmosets can scent mark and gouge. Because marmosets are arboreal primates functioning in a three-dimensional space, any furniture, branches, or additions to the cage should enhance complexity and mobility in three dimensions. Marmoset behavioral diversity and activity have been found to increase with increased cage size and cage complexity related to the number of branches available (Kitchen and Martin 1996; Buchanan-Smith et al. 2009). Although increased cage complexity is known to increase the difficulty and the amount of technician time required to handle marmosets, this is a trade-off that should be assessed in all circumstances prior to cage design. Branches have also been used as a source of scent-related enrichment for the marmosets—a genus that has a variety of scent glands used to mark their territory and that is therefore particularly reliant on smell and scent marking for communication. Leaving branches in the cage that belong to the family group can provide continuity following cage changes and other cleaning protocols (Bakker et al. 2015). Alternatively, providing branches with another groups' scent for the marmosets to then cover with their own scent has the potential of increasing activity and simulating more natural conditions. However, the potential risks of disease exchange should be assessed.

As with many other captive primates, food is often offered as a form of enrichment for marmosets, although it needs to be clear in the management plan whether the addition of specific food items or delivery systems are necessary to provide basic nutrition, or are also serving as a source of enrichment. Marmosets are omnivores, and as such, their dietary tolerance can be quite broad; they show little hesitancy to consume new foods and will readily ingest foods that offer little nutritional value (e.g., human "junk" food Duarte et al. 2012).

The feeding and nutrition of marmosets is a topic of considerable controversy for which there is little standardization across colonies. Most colonies in the United States feed gel-based diets as the primary food item, including Mazuri, LabDiet, and Zupreem, all of which contain the dietary components necessary for marmoset health, including stable sources of vitamins C and D. Many colonies feed what is often referred to as the "cafeteria-style diet," in which the animals receive a base diet plus a daily mixture of other items. The added items often rotate through the week, and may include sources of additional protein, such as chicken, tuna, eggs, cheese, and cottage cheese; sources of additional vitamin C, such as oranges, tang, and other fruits; and other items to provide variety, such as cheerios, grapes, apples, bananas, raisins, potatoes, biscuits soaked in milk, peanuts, etc. The value of the additional items has been debated for years, with no general consensus as to whether this variety in feeding satisfies nutritional needs or simply serves as one (important) type of enrichment. Marmosets can be successfully maintained in captivity while being fed only a purified base diet (Harlan Teklad; Tardif et al. 2009). Additional items, such as fruits and vegetables, were provided in this colony as enrichment a few times each week, rather than as daily dietary supplements. Many investigators have suggested that the standardization of feeding practices would allow broader cross-colony comparisons of demographic and medical variables than are currently possible for macaque and rodent models.

Beyond this debate concerning whether supplemental food items should be considered nutritional and/or enrichment additions, many facilities do use special food items and foraging devices as a source of enrichment opportunities. Foraging devices, which often increase and simulate foraging, vary in complexity, style, and the amount of effort required by the animal to obtain the food item (Roberts et al. 1999; Byron and Bodri 2001; Vignes et al. 2001; Majolo et al. 2003a; Ventura and Buchanan-Smith 2003). Devices can be as simple as trays, boxes, plastic bottles, or similar receptacles filled with shredded bedding that the animals have to sift through to obtain small food items (i.e., seeds, raisins, or peanuts). More complicated feeding devices can be made or purchased that allow animals to turn wheels or lift parts to release food. However, marmosets do not tend to do well with traditional primate puzzle feeders that can require considerable strength or dexterity, and may be larger in size than the animals.

There are a number of alternatives for enrichment for marmosets in addition to foraging apparatus. It is important to keep in mind that given marmosets' status as a prey species in nature, they have higher rates of non-food-related neophobia than Old World monkeys; so the addition of any item/device to their environment needs to be closely monitored. Marmosets tend to be idiosyncratic in their responses to novel items, and in general, younger animals tend to be more exploratory than older animals. Items such as Kong toys and balls are useful when hung in the cage as swings or additional 3D objects, but they must be hung in a way that absolutely minimizes accidental entrapment and strangulation. Most marmosets do not engage with these items in the same way as larger primates, but hanging objects do add to the complexity of the space, and juveniles tend to be particularly interested in these types of devices. Mirrors and other reflective surfaces can be more stressful than enriching, especially if the marmoset has never been exposed to a similar item. Often, the response to a mirror is similar to the response to an intruding animal, (1) increasing aggression and the likelihood of displaced aggression within the family group (Ross personal observation) and/or (2) preventing animals from approaching food dishes or water sources. In general, the addition of enrichment devices needs to be monitored for marmosets to ensure that (1) neophobic responses are not preventing animals from using the cage space fully and (2) an enrichment item is not increasing the risk of injury, starvation, stress, or mortality.

European colonies have shifted to the promotion of indoor/outdoor housing to allow marmosets access to sunlight and increased interactions with natural foraging opportunities (Bakker et al. 2015). While there is some evidence that marmosets prefer to remain off the floor (Duarte et al. 2012) and may even prefer grid flooring to pine bedding (Hardy et al. 2004), the introduction of bedding and outdoor enclosures was not found to negatively affect health or social interactions (Bakker et al. 2015), facts that US colonies should consider when logistically possible.

The use of plastic hose tubing can increase flexibility in caging and allow for easy connections between cages: to increase cage space, to allow access to an enrichment cage, or to access a testing cage (Figure 24.3). Flexible clear plastic hosing can be connected between cages to allow animals to locomote between locations. We have found that juveniles are less hesitant to explore this type of novel object and to move between spaces, and early exposure for infants and juveniles makes it more likely for the animals to use it as adults (Kendal et al. 2005). Alternatively, the use of verandas

Figure 24.3 Marmoset cages connected via a flexible hosing tube, which allows the animals to access a secondary cage. Photo by Jessica Adams.

or balconies can increase visual access to other groups and serve as a form of enrichment (Ely et al. 1998). Adult marmosets can be habituated and trained to use these added cage features.

Some colonies have also instituted human interaction time as a form of enrichment for marmosets. Specifically, humans are asked to positively interact with the animals while in the room and to *not* engage in other activities, such as husbandry, feeding, or training. Animals that experienced positive interactions with humans were found to engage in more social grooming and play bouts, and decreased self-scratching and contact vocalizations (Bassett et al. 2003; Manciocco et al. 2009).

Training

Marmosets respond very well to PRT and can be habituated to participate in a number of activities that (1) can be used as enrichment activities for the animals or (2) serve a purpose in the research setting (Bassett et al. 2003, 2004; McKinley et al. 2003; Savastano et al. 2003; Scott et al. 2003; Tardif et al. 2006). Marmosets are particularly attracted to sweet items and tend to be very responsive to marshmallow-type treats. Other forms of treats that they respond to include cheerios, sugar water, syrup, banana-flavored milk shake, sweetened yogurt, and gummy fruit snacks. Training regimes should always include patient, small steps for initial exposure, and the awareness that responses to training are idiosyncratic and perhaps more variable than those of Old World monkeys. Marmosets can be trained using "bridge" devices, such as clickers; however, we recommend using low-decibel clickers. Dog clickers and whistles do not work well, because the sounds they produce are loud and often resemble marmoset alarm calls. Marmosets are easily trained to move between cages and to enter a transport box; often, no bridge is needed for this type of training, as fresh food or rewards in the new cage/box usually suffice. Marmosets can be trained to sit temporarily on a small scale placed in the cage by covering the top of the scale in marshmallow fluff. Weights can be obtained as the animal climbs on the scale to lick off the fluff. Marmosets have also been successfully trained to perform more complex behavioral tasks, such as touch screen tasks, running on treadmills, pulling weighted devices, and opening small puzzle boxes, to name a few (D'Mello et al. 1985; Roberts et al. 1999; de Rosa et al. 2003; Spinelli et al. 2004; Stevens et al. 2005; Phillips et al. 2015; Figure 24.4).

Training marmosets is typically reported to take approximately double the time it takes to train Old World monkeys, and is more effective if training sessions are conducted on a daily basis (Spinelli et al. 2004; Stevens et al. 2005). Several of the advantages associated with marmosets as subjects for training procedures relate to their small size, which allows them to be trained and tested

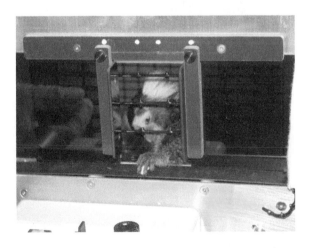

Figure 24.4 Marmoset retrieving a reward from a conveyor apparatus attached to the home cage.

in the home cage in their family group. Similarly, the marmoset social system makes it unlikely that aggression between family members will be problematic and, in fact, may facilitate training progress, as family group members will often learn target behaviors by watching others being trained, and ultimately perform the task.

Common Research Exemptions

Occasionally, research questions will require modifications to typical colony housing, feeding, and management procedures. Such questions may include diet-based studies for which exact food intake and metabolic output need to be assessed, pairing procedures in which breeding cannot occur, or studies in which instrumentation is employed and animals cannot be housed socially. In these cases, there are a variety of ways in which to alter husbandry and housing procedures in order to accommodate research needs. For feeding trials, it is possible to temporarily separate animals (for 48–72 h) with mesh dividers that allow continued visual access to the family group, but prevent food exchange. These types of separations allow for detailed assessments of food intake by individual animals (Tardif et al. 2007, 2009; Ross et al. 2013). It is also possible to implement daily restriction or overfeeding for particular animals while in group housing, without altering the social conditions for the entire day. This type of intervention requires shifting the animals of interest so that they are briefly housed separately, while the rest of the family is fed at the same time. After feeding is complete, the animals can be reunited. This protocol is staff intensive and often requires multiple feedings throughout the day. However, owing to their size and social stability, this process is much easier with marmosets than it is with many other primate models.

Many protocols and facilities require that animals be socially housed, regardless of experimental protocol. The easiest way to pair animals is in a male–female pair. There are a number of birth control methods available that minimize the risk of pregnancy in male–female pairs. Vasectomy for the male is the most successful, while injections of Estrumate can also prevent pregnancy. However, Estrumate protocols require monthly injections and close hormonal or ultrasound monitoring to ensure that conception has not taken place. Unfortunately, as mentioned earlier, it is not possible to pair unfamiliar adult animals of the same sex. It is possible to pair same-sex siblings, if they are removed from the family group at the same time and the room they live in is entirely one sex. Male–male pairs will often disintegrate due to aggression if a cycling female is present in the room. Female–female pairs may be stable but need to be monitored closely.

Although it is far from ideal for any primate to be singly housed, marmosets can be housed by themselves if they are provided with visual, olfactory, and auditory access to other marmosets (Layne and Power 2003). Marmosets are not reported to suffer from the same types of psychological ailments due to social restriction that may occur among Old World monkeys, such as hair plucking, self-injurious behavior, or depression. There have only been two cases of self-injurious behavior among marmosets at our colony, and in one case, the cause was determined to have been a stroke at the time of the behavioral shift (biting of one knee; Brasky personal communication). Care should be taken to provide supplemental enrichment for animals that are singly housed, including both food enrichment and enriched space.

SAIMIRI

Natural History

All squirrel monkeys were once considered to be a single species (*Saimiri sciureus*), with several geographically separated subspecies. However, karyotypic and phenotypic information gathered in the early 1980s led Hershkovitz (1984) to the conclusion that squirrel monkeys should be classified as a single genus with four species (*Saimiri boliviensis, Saimiri oerstedii, Saimiri sciureus, and Saimiri ustus*) and nine subspecies. Studies conducted by Assis and Barros (1987), da Silva et al. (1987), VandeBerg et al. (1990), and Cropp and Boinski (2000) support the taxonomic classification of Hershkovitz. Ayres (1985) has described a fifth species, *Saimiri vanzolinii*, which is found in central Brazil along the south bank of the Amazon River. Cropp and Boinski (2000) present evidence that *S. oerstedii*, indigenous to parts of Central America (primarily Costa Rica), is a genetically distinct population and not the result of human introduction. Lavergne et al. (2010) added a sixth species, *Saimiri collinsi*, based on cytochrome b genetic analysis.

Squirrel monkeys inhabit most types of tropical forest, including wet and dry forest, continuous and secondary forest, mangrove swamps, riparian habitat, and forest fragments (Hernandez-Camacho and Cooper 1976; Terborgh 1983; Baldwin 1985; Boinski 1987a). They appear highly flexible in their ability to adapt to different environments, and in some geographic areas, they appear to prefer disturbed habitats (Konstant and Mittermeier 1982; Boinski 1987a). They are omnivores, eating insects when they are available, but also include fruit, flowers, birds' eggs, and, occasionally, vertebrates in their diet when necessary or readily available (Thorington 1967; Baldwin and Baldwin 1972; Jones et al. 1973; Izawa 1975; Mittermeier and van Roosmalen 1981; Scollay and Judge 1981; Mitchell et al. 1991). Lima and Ferrari (2003) reported that the diet of *S. sciureus* in eastern Brazil could shift from 80% animal to 80% plant, depending on the availability of insects. Boinski (1987a) found that Costa Rican squirrel monkeys tend to forage more widely and rest less when food is scarce during the peak wet season, and travel least during the birth season.

Squirrel monkeys are typically found in large multimale, multifemale social groups of between 20 and 50 animals, with unconfirmed reports of up to 300 monkeys in a group (Thorington 1967; Baldwin and Baldwin 1971, 1985; Boinski 1987a). Group size may vary somewhat depending on habitat type (Baldwin and Baldwin 1971; Scollay and Judge 1981).

Female squirrel monkeys reach maturity and begin breeding at approximately 2.5–3 years of age. Males reach subadult status by the time they are 2.5–3 years old and generally transfer from the natal group at that time. They may then join an all-male group of juveniles and subadults until they become fully adult at about 5 years of age and are able to work their way into the male dominance hierarchy of an established group (Roder and Timmermans 2002). In the wild, the frequency of female intertroop transfer differs among groups by geographical regions (Boinski et al. 2005). A high rate of female transfer occurs in *S. oerstedii*, with males remaining in their natal groups. This is very different from the low rate of female, and high rate of male, transfer seen in *S. sciureus* and *S. boliviensis* (Boinski 1987c).

Squirrel monkeys have an annual reproductive cycle, with a roughly 3-month breeding season, followed about 5 months later by a birth season (DuMond 1967, 1968; Goss et al. 1968; Rosenblum and Cooper 1968; Baldwin 1969; Michael and Zumpe 1971; Coe and Rosenblum 1978; Boinski 1987c; Trevino 2007; Williams et al. 2010). A feature of this yearly cycle is reproductive seasonality in both breeding males and females (DuMond 1967; DuMond and Hutchinson 1967; Baldwin 1969; Kaplan 1977; Coe and Rosenblum 1978; Williams et al. 1986; Boinski 1987c; Mendoza 1987). The "fatted male" condition is a component of this reproductive cycle. Both sexes gain weight throughout the prebreeding season, attaining peak weights just prior to breeding. Weight gain in males is associated with increased spermatogenesis in preparation for breeding (DuMond and Hutchinson 1967; Coe and Rosenblum 1978; Williams et al. 1986). Nadler and Rosenblum (1972) and Chen et al. (1981) demonstrated that male fattening is closely related to circulating levels of testosterone. Estrous cycle lengths have been estimated to be approximately 8–12 days (Rosenblum and Cooper 1968; Wolf et al. 1975; Kaplan 1977; Diamond et al. 1984; Yeoman et al. 2000), and estrous cycle length can be affected by social conditions (Hutchinson 1970; Wolf et al. 1975) and day length (Rosenblum and Cooper 1968).

Dominance hierarchies within squirrel monkey groups differ across species. *S. boliviensis* has a high degree of sexual segregation and linear male dominance hierarchies, in which dominance is associated with higher testosterone levels and copulatory frequency (Wiebe et al. 1984; Boinski 1987b). A less-distinct dominance hierarchy is seen among females of this species (Mendoza et al. 1978a). The linear dominance hierarchies of *S. sciureus* spp. include both sexes and are more sexually integrated within social groups than *S. boliviensis*, with all males being dominant over all females (Mendoza et al. 1978a; Mitchell et al. 1991). *S. oerstedii* spp., in the wild, do not have dominance hierarchies within either sex; and males may cooperatively mob females to use olfactory cues to establish their state of estrus during the breeding season (Boinski 1987c; Mitchell et al. 1991). The use of visual barriers and areas within standard housing, where subordinate animals can hide or break direct eye contact with dominant individuals, will reduce the number of injuries resulting from aggression. It is particularly important for males to have access to visual barriers during the nonbreeding season.

Allomaternal care, or infant care by social group members other than the birth mother, has been documented in field studies (Ploog 1967; DuMond 1968; Baldwin 1969; Hunt et al. 1978; Morton 2000) and in the laboratory (Williams et al. 1988, 1994; Soltis et al. 2005). Infant squirrel monkeys may spend as much as 30% of their time on allomothers during the first 6 months of their lives (Baldwin 1969; Williams et al. 1994). Allomothering usually begins during the first 2 weeks of life (Williams et al. 1994). In the wild, allomothers are usually juvenile females (DuMond 1968), whereas in captivity, reports have shown that more than one-half (53%) of the allomothering is performed by young adult females, aged 4–6 years. Adult females, aged 7–9 years, provide another 20% of the allomaternal care (Williams et al. 1994). Females that had experienced a reproductive failure during the year performed almost all of the allonursing (Williams et al. 1988). This proclivity, to nurse the offspring of another female, can be used to foster orphan infants in captivity.

Social Housing

In nature, squirrel monkeys live in large social groups containing 20–50 animals (Baldwin and Baldwin 1971) with multiple males and females of different ages. Efforts should be made to house *Saimiri* in social groups in captivity whenever possible. Animals born into a social group should be maintained within their social group for as long as possible. In the South American species of squirrel monkeys, *S. boliviensis* and *S. sciureus*, females typically stay within the natal social group and males emigrate after the age of 1.5–2 years. Removing the males from their natal group at approximately this age can help reduce intragroup aggression and prevent inbreeding as the males continue to mature sexually.

Social stability in squirrel monkey societies is generally maintained by the adult females. All age-sex classes, except fully adult males, have been shown to be more attracted to adult females than to any other age-sex class (Saltzman et al. 1989). However, during the nonbreeding season, squirrel monkey social groups will separate by sex, a unique feature of primate social organization in this species. Males remain near the periphery of the group during the nonbreeding season, but the majority of social interactions between the sexes takes place during the breeding season (DuMond 1967; Baldwin 1968; Mason and Epple 1969; Baldwin and Baldwin 1972; Candland et al. 1973; Coe and Rosenblum 1974, 1978; Strayer et al. 1975; Kaplan 1977; Vaitl 1977; Hopf 1978; Mendoza et al. 1978b; Vaitl et al. 1978; Leger et al. 1981; Lyons et al. 1992; Boinski and Mitchell 1994). In captivity, where males and females typically live in the same enclosure all year, seasonal sexual segregation is generally expressed in vertical separation of the males and females. Because arboreal species typically equate height with safety, females tend to occupy the upper levels of the housing, relegating the males to perches and structures closer to the ground. Males should be provided with extra perches and places to hide in the lower reaches of their enclosure.

Squirrel monkeys will fight with unfamiliar conspecifics (Williams and Abee 1988), suggesting that frequent movements of animals between social groups should be avoided. Squirrel monkeys that must live separately from their social group for experimental or clinical purposes should be pair housed. If pair housing is not possible, animals should be provided with visual and/or auditory access to conspecifics (Schuler and Abee 2005). Forming new pairs or small groups of females is rarely a problem with squirrel monkeys. Introducing all of the females to one another simultaneously will lessen the probability of an individual animal becoming the target of aggression. Male pairs or triplets can be housed together, if care is taken when introducing the animals. Juvenile males will bond together quickly, and adult males will tolerate others, as long as there are no females housed in visual contact. When introducing new animals to an established social group, it is important to give members of the social group contact with the new animal in small subgroup situations, prior to introducing the entire group. Placing the new animal in a neutral cage with a small subset of the social group and swapping out social group members periodically will give the new animal an opportunity to establish relationships, prior to living in the whole group. This process will work for females during the entire year and for males during the breeding season only. Males can be difficult to introduce to a new group during the nonbreeding season, when they are normally segregated from the female social core. Increased levels of aggression should be expected at the beginning and end of the breeding season, when hormonal changes lead to increases in social interactions between males and females. The breeding season in squirrel monkeys occurs during short days; in the northern hemisphere, it typically runs from January through April for outdoor-housed colonies, and into May and June in indoor colonies (Schiml et al. 1996; Trevino 2007; Williams et al. 2010).

Because squirrel monkeys are arboreal, the provision of three-dimensional space and differing perch arrangements can vary the cage environment and increase the activity of the animals. Runs or large pens can be used to allow the squirrel monkeys to move over longer distances. Pens measuring 4 ft wide by 6 ft tall by 14 ft deep can house up to 15 animals. Periodically rearranging the perches and providing new materials is an easy way to provide novelty. Figure 24.5 illustrates an environment with multiple levels of perching with places for the squirrel monkeys to travel and sit.

Squirrel monkeys are social animals, and all efforts should be made to house them in social groups or, at least, in pairs. The *Guide* (National Research Council 2011) classifies squirrel monkeys as type 2 primates, requiring 3 ft^2 (0.28 m^2) for each animal. All squirrel monkeys should be

Figure 24.5 An example of squirrel monkey social housing. From left to right, a yellow tube for sleeping and huddling, a porthole style door allowing movement between multiple runs, feeding platforms outside the back of the run, and a suspended green platform and attached blue box for huddling can be seen. The squirrel monkeys at the ceiling are pulling in destructible enrichment devices consisting of paper bags filled with paper shreds and mealworms.

housed with a temperature range of approximately 75–82°F (24–27°C) and with relative humidity between 40% and 60%. Lower temperatures or humidity levels can lead to upper respiratory problems (Abee 1985).

Enrichment

Squirrel monkeys are highly social, and one of the best, and easiest, ways to enrich their environment is through housing with compatible social partners. Communication between animals within a given social group includes olfactory, visual, tactile, and vocal signals. Olfaction is most important during the breeding and birthing seasons. Males use scent to determine the receptivity of females, and females use scent to identify individual infants. When cleaning squirrel monkey enclosures, it is preferable to avoid heavily scented products. Visual cues are of primary importance in a number of displays intended to assert dominance or reduce tension between individuals. Visual cues involving facial expressions can be associated with fear or aggression. In a captive enclosure, visual barriers that a submissive monkey can use to break the stare of a more aggressive monkey should be provided. These "hide-boxes" can be made from a piece of a large diameter PVC pipe or another impervious material. However, it has to be remembered that caregivers and researchers must be able to see the animals for health checks and other observations. Vocal signals are very important for squirrel monkeys with infants possessing a wide repertoire of sounds, including grumbles, tucks, purrs, chucks, and peeps (Biben and Symmes 1986) to communicate their well-being to the mother. In adults, a number of calls are used to signal alarm, to display general disturbance or excitement, or to establish contact when group members are visually separated.

Squirrel monkeys are naturally active and will readily manipulate objects in their enclosure. They are omnivores and, in the wild, may spend up to 75% of their time eating as they move through the forest. Squirrel monkeys' captive diets can be supplemented with many fruits and vegetables, depending on seasonal availability. Additional items, such as Gatorade, multivitamins, and food pellet rewards may be added for experimental or clinical reasons.

Destructible enrichment and devices that require time to open or manipulate can be provided to occupy the animals' time. Small objects such as plastic golf balls and PVC plumbing joints can be employed, with or without food enrichment, to increase activity rates. Feeding or foraging boards can also be used, as the presence of artificial turf forage boards has been shown to increase activity (Fekete et al. 2000). Because squirrel monkeys are naturally inquisitive and active, the provision of multiple perches, swings, and other hanging devices will increase activity levels. In all caging conditions, food items and perch arrangements can be used to vary the environment of the animals and, consequently, increase activity levels and sensory stimulation. Perch and manipulanda should be rotated and replaced routinely in order to prevent habituation.

Training

Squirrel monkeys have a long history in cognitive research and can readily be trained using PRT techniques (Rogge et al. 2013). As with the marmosets, squirrel monkeys respond best to sweet food reinforcers, such as raisins, cereal, and marshmallows. When training squirrel monkeys, it is important to remember to use very small portions of food as reinforcers, because the animals can quickly become satiated and stop working if given too much food. Rogge et al. (2013) found that it only took a few sessions to train the animals to present their hands for inspection on command, even when they were trained in their social group. Squirrel monkeys can be trained to voluntarily participate in many aspects of their care, including the presentation of body parts, sitting on a scale, or moving from one cage to another on command. They can also be trained to participate in more complex tasks, such as using a joystick and computer screen (Washburn et al. 1989; Judge et al. 2005), or engaging in token exchange procedures (Talbot et al. 2011; Hopper et al. 2013).

FUNCTIONALLY APPROPRIATE CAPTIVE ENVIRONMENTS

There are a number of factors related to Neotropical NHP sociality and natural living conditions that should play a role in determining functionally appropriate captive environments and related management procedures. It should be acknowledged that the social structure of New World monkeys, especially marmosets and owl monkeys, differ significantly from the typical social structures of Old World monkeys, and housing conditions need to reflect this. Decisions related to enclosure size need to reflect the fact that New World monkeys are primarily arboreal animals, and the enclosure should include vertical space for climbing, locomoting, and jumping. Similarly, social groupings should reflect species-typical patterns, with the formation of groups containing stable breeding pairs and their immature offspring as a goal for owl monkeys and marmosets. If a breeding pair of owl monkeys must be separated, weaned infants should go with the sire rather than the dam. Infant marmosets should not be removed from the family group at weaning, nor should they be removed when the next set of infants is born. Marmosets do well in large, cooperatively breeding groups, both in nature and in captivity. Enrichment, housing, and care practices should account for the fact that these small primates are prey animals in the wild and, as such, are often hypervigilant and quite responsive to changes in the external and internal environment. Particularly for marmosets and owl monkeys, changes associated with the sounds in the building, and/or the presence of different caretakers and research staff, often have profound effects on the animals and their behaviors. The introduction of novel items, although often a form of enrichment for Old World monkeys and Great Apes, may not be enriching for some New World primates, especially owl monkeys and marmosets.

ACKNOWLEDGMENTS

CNR thanks Suzette Tardif for her continued mentorship, guidance, and dedication to the development and maintenance of the common marmoset as a biomedical model. We also thank Donna Layne Colon, Aubrey Sills, Joselyn Artavia, and Theresa Valverde for their care of the marmosets. The marmoset colony at Southwest National Primate Research Center is supported by funding from P51RR013986, and the colony at the University of Texas Health Science Center is supported by R24RR02344-01A2.

LW thanks Virginia Parks and Bethany Brock for the long-term support and care for the squirrel monkeys, and George Tustin for his care and support of the owl monkey colony at the UT MD Anderson Cancer Center's Michale E. Keeling Center for Comparative Medicine and Research. The squirrel monkeys are supported by NIH grant P40-OD010938.

Both authors would also like to thank Steve Schapiro for the opportunity to participate in developing this book and for his comments and suggestions.

REFERENCES

Abbott, D., D.K. Barnette, R.J. Colman, M. Yamamoto, and N. Schultz-Darken. 2003. Aspects of common marmoset basic biology and life history important for biomedical research. *Comparative Medicine* 53(4):339–350.

Abbott, D., J. Barrett, and L.M. George. 1993. Comparative aspects of the social suppression of reproduction in female marmosets and tamarins. In *Marmosets and Tamarins: Systematics, Behaviour, and Ecology*, edited by Rylands, A.B. New York: Oxford University Press.

Abee, C.R. 1985. Medical care and management of the squirrel monkey. In *Handbook of Squirrel Monkey Research*, 447–488. New York: Plenum Press.

Aquino, R., P. Puertas, and F. Encarnacion. 1990. Supplemental notes on population parameters of northeastern Peruvian night monkeys, Genus *Aotus* (Cebidae). *American Journal of Primatology* 21(3):215–221.

Assis, M.F.L. and R. Barros. 1987. Karyotype pattern of *Saimiri ustus*. *International Journal of Primatology* 8:552.

Ayres, J.M. 1985. On a new species of squirrel monkey, genus *Samiri*, from Brazilian Amazonia (primates, Cebidae). *Papéis Avulsos Zool. São Paulo* 36:147–164.

Baker, A.J., J.M. Dietz, and D.G. Kleiman. 1993. Behavioral evidence for monopolization of paternity in multi-male groups of golden lion tamarins. *Animal Behaviour* 46:1091–1103.

Bakker, J., B. Ouwerling, P.J. Heidt, I. Kondova, and J.A.M. Langermans. 2015. Advantages and risks of husbandry and housing changes to improve animal wellbeing in a breeding colony of common marmosets (*Callithrix jacchus*). *Journal of the American Association for Laboratory Animal Science* 54(3):273–279.

Baldwin, J.D. 1968. The social behavior of adult male squirrel monkeys (*Saimiri sciureus*) in a seminatural environment. *Folia Primatologica* 9:281–314.

Baldwin, J.D. 1969. The ontogeny of social behavior of squirrel monkeys (*Saimiri sciureus*) in a seminatural environment. *Folia Primatologica* 11:35–79.

Baldwin, J.D. 1985. The behavior of squirrel monkeys (*Saimiri*) in natural environments. In *Handbook of Squirrel Monkey Research*, 35–53. New York: Plenum Press.

Baldwin, J.D. and J.I. Baldwin. 1971. Squirrel monkeys (*Saimiri*) in natural habitats in Panama, Colombia, Brazil, and Peru. *Primates* 12(1):45–61.

Baldwin, J.D. and J. Baldwin. 1972. The ecology and behavior of squirrel monkeys (*Saimiri oerstedi*) in a natural forest in western Panama. *Folia Primatologica* 18:161–184.

Bassett, L., H.M. Buchanan-Smith, J. McKinley, and T.E. Smith. 2003. Effects of training on stress-related behavior of the common marmoset (*Callithrix jacchus*) in relation to coping with routine husbandry procedures. *Journal of Applied Animal Welfare Science* 6(3):221–33. doi:10.1207/S15327604JAWS0603_07.

Biben, M. and D. Symmes. 1986. Play vocalizations of squirrel monkeys (*Saimiri sciureus*). *Folia Primatologica* 46(3):173–182.

Boinski, S. 1987a. Habitat use by squirrel monkeys (*Saimiri oerstedi*) in Costa Rica. *Folia Primatologica* 49(3):151–167.

Boinski, S. 1987b. Mating patterns in squirrel monkeys (*Saimiri oerstedi*). *Behavioral Ecology and Sociobiology* 21(1):13–21.

Boinski, S. 1987c. Mating patterns in squirrel monkeys (*Saimiri oerstedi*): Implications for seasonal sexual dimorphism. *Behavioral Ecology and Sociobiology* 21(1):13–21.

Boinski, S., L. Kauffman, E. Ehmke, S. Schet, and A. Vreedzaam. 2005. Dispersal patterns among three species of squirrel monkeys (*Saimiri oerstedii, S. boliviensis and S. sciureus*): I. Divergent costs and benefits. *Behaviour* 142(5):525–632.

Boinski, S. and C.L. Mitchell. 1994. Male residence and association patterns in Costa Rican squirrel monkeys (*Saimiri oerstedi*). *American Journal of Primatology* 34(2):157–169.

Bolen, R.H. and S.M. Green. 1997. Use of olfactory cues in foraging by owl monkeys (*Aotus nancymai*) and capuchin monkeys (*Cebus apella*). *Journal of Comparative Psychology* 111(2):152.

Bowell, V., H. Buchanan-Smith, and K. Morris. 2004. The effect of animal age, sex and temperament on the time investment required for positive reinforcement training of common marmosets. *Folia Primatologica* 75:359–360.

Brady, A.G., V.L. Parks, S.V. Gibson, and C.R. Abee. 2005. Cardiomyopathy of owl monkeys: A study of seven cases. *Contemporary Topics in Laboratory Animal Science* 44(4):57.

Buchanan-Smith, H.M., M.R. Gamble, M. Gore, P. Hawkins, R. Hubrecht, S. Hudson, M. Jennings, et al. 2009. Refinements in husbandry, care and common procedures for non-human primates. *Laboratory Animals* 43:S1–S47. doi:10.1258/la.2008.007143.

Byron, J.K. and M.S. Bodri. 2001. Environmental enrichment for laboratory marmosets. *Lab Animal (NY)* 30(8):42–48. doi:10.1038/5000165.

Candland, D.K., L. Dresdale, J. Leiphart, D. Bryan, C. Johnson, and B. Nazar. 1973. Social structure of the squirrel monkey (*Saimiri sciureus*, Iquitos): Relationship among behavior, heart rate and physical distance. *Folia Primatologica* 20:211–240.

Capitanio, J.P., S.A. Blozis, J. Snarr, A. Steward, and B.J. McCowan. 2015. Do "birds of a feather flock together" or do "opposites attract"? Behavioral responses and temperament predict success in pairings of rhesus monkeys in a laboratory setting. *American Journal of Primatology*. doi:10.1002/ajp.22464.

Chen, J., E. Smith, G. Gray, and J. Davidson. 1981. Seasonal changes in plasma testosterone and ejaculatory capacity in squirrel monkeys (*Saimiri sciureus*). *Primates* 22(2):253–260.

Clapp, N.K. and S.D. Tardif. 1985. Marmoset husbandry and nutrition. *Digestive Diseases and Sciences* 30(12):S17–S23. doi:10.1007/bf01296967.

Clarke, J.M. 1994. The common marmoset (*Callithrix jacchus*). *ANZCCART News* 7(2):1–8.

Coe, C.L. and L.A. Rosenblum. 1974. Sexual segregation and its ontogeny in squirrel monkey social structure. *Journal of Human Evolution* 3:551–561.

Coe, C.L. and L.A. Rosenblum. 1978. Annual reproductive strategy of the squirrel monkey (*Saimiri sciureus*). *Folia Primatologica* 29:19–42.

Cropp, S. and S. Boinski. 2000. The Central American squirrel monkey (*Saimiri oerstedii*): Introduced hybrid or endemic species? *Molecular Phylogenetics and Evolution* 16(3):350–365.

D'Mello, G.D., E.A.M. Duffy, and S.S. Miles. 1985. A conveyor belt task for assessing visuo-motor coordination in the marmoset: Effects of diazepam, chlorpromazine, pentobarbital, and D-amphetamine. *Psychopharmacology* 86:125–131.

Da Silva, B.T.F., M.I.C. Sampico, M.P.C. Scheider, and H. Schneider. 1987. Preliminary analysis of genetic distance between squirrel monkeys. *International Journal of Primatology* 8:828.

de Rosa, C., A. Vitale, and M. Puopolo. 2003. The puzzle-feeder as feeding enrichment for common marmosets (*Callithrix jacchus*): A pilot study. *Laboratory Animals* 37(2):100–107. doi:10.1258/00236770360563732.

Diamond, E.J., S. Aksel, J.M. Hazelton, R.A. Jennings, and C.R. Abee. 1984. Seasonal changes of serum concentrations of estradiol and progesterone in Bolivian squirrel monkeys (*Saimiri sciureus*). *American Journal of Primatology* 6(2):103–113.

Digby, L.J. 1995. Infant care, infanticide, and female reproductive strategies in polygynous groups of common marmosets (*Callithrix jacchus*). *Behavioral Ecology and Sociobiology* 37:51–61.

Digby, L.J. and C.E. Barreto. 1993. Social organization in a wild population of *Callithrix jacchus*. I. Group composition and dynamics. *Folia Primatologica* 61:123–134.

Dixson, A.F. 1983. The owl monkey (*Aotus trivirgatus*). In *Reproduction in New World Primates*, 69–113. New York: Springer.

Dixson, A.F. 1993. Callitrichid mating systems: laboratory and field approaches to studies of monogamy and polyandry. In *Marmosets and Tamarins: Systematics, Behaviour, and Ecology*, edited by Rylands, A.B. New York: Oxford University Press.

Dixson, A.F., M.G. Anzenberger, M.A.O. Monteiro Da Cruz, I. Patel, and A.J. Jeffreys. 1992. DNA fingerprinting of free ranging groups of common marmosets in NE Brazil. In *Paternity in Primates: Genetic Tests and Theories*, edited by Martin, R.D., A.F. Dixson, and E.J. Wickings. Switzerland: Karger.

Dixson, A.F., N. Hastie, I. Patel, and A.J. Jeffreys. 1988. DNA fingerprinting of captive family groups of common marmosets (*Callithrix jacchus*). *Folia Primatologica* 51:52–55.

Duarte, M.H.L., V.D.L.R. Goulart, and R.J. Young. 2012. Designing laboratory marmoset housing: What can we learn from urban marmosets? *Applied Animal Behaviour Science* 137(3–4):127–136. doi:10.1016/j.applanim.2011.11.013.

DuMond, F.V. 1967. Semi-free-ranging colonies of monkeys at Goulds Monkey Jungle. *International Zoo Yearbook* 7:202–207.

DuMond, F.V. 1968. The squirrel monkey in a seminatural environment. In *The Squirrel Monkey*, edited by L. Rosenblum and R. Cooper. 87–145. New York: Academic Press.

DuMond, F.V. and T.C. Hutchinson. 1967. Squirrel monkey reproduction: The "fatted" male phenomenon and seasonal spermatogenesis. *Science* 158(3804):1067–1070. doi:10.1126/science.158.3804.1067.

Dunbar, R.I.M. 1995. The mating system of callitrichid primates. 2. The impact of helpers. *Animal Behaviour* 50:1071–1089. doi:10.1016/0003-3472(95)80107-3.

Ely, A., A. Freer, C. Windle, and R.M. Ridley. 1998. Assessment of cage use by laboratory-bred common marmosets (*Callithrix jacchus*). *Lab Animal* 32(4):427–33.

Fekete, J.M., J.L. Norcross, and J.D. Newman. 2000. Artificial turf foraging boards as environmental enrichment for pair-housed female squirrel monkeys. *Journal of the American Association for Laboratory Animal Science* 39(2):22–26.

Fernandez-Duque, E. 2003. Influences of moonlight, ambient temperature, and food availability on the diurnal and nocturnal activity of owl monkeys (*Aotus azarae*). *Behavioral Ecology and Sociobiology* 54(5):431–440. doi:10.1007/s00265-003-0637-9.

Fernandez-Duque, E. 2004. High levels of intra-sexual competition in sexually monomorphic owl monkeys. *Folia Primatologica* 75:60.

Fernandez-Duque, E. 2007. Social monogamy in the only nocturnal haplorhines. In *Primates in Perspective*, edited by Bearder, S., C.J. Campbell, A. Fuentes, K.C. MacKinnon, and M. Panger, 139–185. Oxford, UK: Oxford University Press.

Fernandez-Duque, E. 2012. Owl monkeys *Aotus* spp in the wild and in captivity. *International Zoo Yearbook* 46(1):80–94. doi:10.1111/j.1748-1090.2011.00156.x.

Fernandez-Duque, E., H.O. de la Iglesia, and H.G. Erkert. 2010. Moonstruck primates: owl monkeys (*Aotus*) need moonlight for nocturnal activity in their natural environment. *PLoS One* 5(9):e12572. doi:10.1371/journal.pone.0012572.

Fernandez-Duque, E. and G. Heide. 2013. Dry season resources and their relationship with owl monkey (*Aotus azarae*) feeding behavior, demography, and life history. *International Journal of Primatology* 34(4):752–769. doi:10.1007/s10764-013-9689-5.

Fernandez-Duque, E. and C. Huntington. 2002. Disappearances of individuals from social groups have implications for understanding natal dispersal in monogamous owl monkeys (*Aotus azarai*). *American Journal of Primatology* 57(4):219–225.

Fernandez-Duque, E., C. Juárez, and A. Di Fiore. 2008. Adult male replacement and subsequent infant care by male and siblings in socially monogamous owl monkeys (*Aotus azarai*). *Primates* 49(1):81–84.

Ferrari, S.F. and L.J. Digby. 1996. Wild *Callithrix* groups: Stable extended families? *American Journal of Primatology* 38(1):19–27.

Fleagle, J.G. 1999. *Primate Adaptation and Evolution*. San Diego, CA: Academic Press.

French, J.A., C.M. Schaffner, R.E. Shepherd, and M.E. Miller. 1995. Familiarity with intruders modulates antagonism toward outgroup conspecifics in Wied's black tufted-ear marmoset. *Ethology* 99:24–38.

Gengozian, N., J.G. Brewen, R.J. Preston, and J.S. Batson. 1980. Presumptive evidence for the absence of functional germ cell chimerism in the marmoset. *Journal of Medical Primatology* 9:9–27.

Goldizen, A.W. 1988. Tamarin and marmoset mating systems: Unusual flexibility. *Trends in Ecology & Evolution* 3(2):36–40. doi:10.1016/0169-5347(88)90045-6.

Goss, C.M., L.T. Popejoy II, J.L. Fusiler, and T.M. Smith. 1968. Observations on the relationship between embryological development, time of conception, and gestation. In *The Squirrel Monkey*, edited by L. Rosenblum and R. Cooper, 171–191. New York: Academic Press.

Hardy, A., C.P. Windle, H.F. Baker, and R.M. Ridley. 2004. Assessment of preference for grid-flooring and sawdust-flooring by captive-bred marmosets in free-standing cages. *Applied Animal Behaviour Science* 85(1–2):167–172. doi:10.1016/j.applanim.2003.09.004.

Hernandez-Camacho, J. and R.W. Cooper. 1976. The nonhuman primates of Colombia. In *Neotropical Primates: Field Studies and Conservation*, edited by R. W. Thorington Jr. and P. G. Heltne, 35–69. Washington, DC: National Academy of Sciences.

Hershkovitz, P. 1975. Living new world monkeys (*Platyrrhini*): With an introduction to primates. Volume 1. Chicago, IL: University of Chicago Press.

Hershkovitz, P. 1984. Taxonomy of squirrel monkeys genus *Saimiri* (Cebidae, Platyrrhini): A preliminary report with description of a hitherto unnamed form. *American Journal of Primatology* 7(2):155–210.

Hooff, S.B. and C.K. Wolovich. 2008. Captive owl monkeys respond to novel flavors with neophobia, discrimination, and food sharing. *American Journal of Primatology* 70:48.

Hopf, S. 1978. Huddling subgroups in captive squirrel monkeys and their changes in relation to ontogeny. *Biology of Behavior* 3:147–162.

Hopper, L.M, A.N. Holmes, L.E. Williams, and S.F. Brosnan. 2013. Dissecting the mechanisms of squirrel monkey (*Saimiri boliviensis*) social learning. *PeerJ* 1:e13.

Huber, H.F. and K.P. Lewis. 2011. An assessment of gum-based environmental enrichment for captive gummivorous primates. *Zoo Biology* 30(1):71–78. doi:10.1002/zoo.20321.

Hunt, S.M., K.M. Gamache, and J.S. Lockard. 1978. Babysitting behavior by age/sex classification in squirrel monkeys (*Saimiri sciureus*). *Primates* 19:179–186.

Hutchinson, T.C. 1970. Vaginal cytology and reproduction in the squirrel monkey (*Saimiri sciureus*). *Folia Primatologica* 12:212–223.

Izawa, K. 1975. Foods and feeding behavior of monkeys in the upper Amazon Basin. *Primates* 16(3):295–316.

Jones, T.C., R.W. Thorington, M.M. Hu, E. Adams, and R.W. Cooper. 1973. Karyotypes of squirrel monkeys (*Saimiri sciureus*) from different geographic regions. *American Journal of Physical Anthropology* 38:269–277.

Judge, P.G., T.A. Evans, and D.K. Vyas. 2005. Ordinal representation of numeric quantities by brown capuchin monkeys (*Cebus apella*). *Journal of Experimental Psychology: Animal Behavior Processes* 31(1):79.

Kaplan, J.N. 1977. Breeding and rearing squirrel monkeys (*Saimiri sciureus*) in captivity. *Laboratory Animal Science* 27:557–567.

Kendal, R.L., R.L. Coe, and K.N. Laland. 2005. Age differences in neophilia, exploration, and innovation in family groups of callitrichid monkeys. *American Journal of Primatology* 66(2):167–88. doi:10.1002/ajp.20136.

Kinzey, W.G. 1997. Synopsis of new world primates (16 Genera). In *New World Primates: Ecology, Evolution and Behavior*, edited by Kinzey, W.G. New York: Aldine de Gruyter.

Kitchen, A.M. and A.A. Martin. 1996. The effects of cage size and complexity on the behaviour of captive common marmosets, *Callithrix jacchus jacchus*. *Laboratory Animals* 30(4):317–326. doi:10.1258/002367796780739853.

Kleiman, D.J. 1977. Monogamy in mammals. *Quarterly Review in Biology* 52:39–69.

Konstant, W.R. and R.A. Mittermeier. 1982. Introduction, reintroduction and translocation of neotropical primates: Past experiences and future possibilities. *International Zoo Yearbook* 22:69–77.

Lavergne, A., M. Ruiz-Garcia, F. Catzeflis, S. Lacote, H. Contamin, O. Mercereau-Puijalon, V. Lacoste, and B. de Thoisy. 2010. Phylogeny and phylogeography of squirrel monkeys (genus *Saimiri*) based on cytochrome B genetic analysis. *American Journal of Primatology* 72(3):242–53. doi:10.1002/ajp.20773.

Layne, D.G. and R.A. Power. 2003. Husbandry, handling, and nutrition for marmosets. *Comparative Medicine* 53(4):351–9.

Leger, D.W., W.A. Mason, and D.M. Fragaszy. 1981. Sexual segregation, cliques, and social power in squirrel monkey (*Saimiri*) groups. *Behaviour* 76:163–181.

Lima, E.M. and S.F. Ferrari. 2003. Diet of a free-ranging group of squirrel monkeys (*Saimiri sciureus*) in eastern Brazilian Amazonia. *Folia Primatologica* 74(3):150–158.

Lyons, D.M., S.P. Mendoza, and W.A. Mason. 1992. Sexual segregation in squirrel monkeys (*Saimiri sciureus*): A transactional analysis of adult social dynamics. *Journal of Comparative Psychology* 106(4):323–330.

Majolo, B., H.M. Buchanan-Smith, and J. Bell. 2003a. Response to novel objects and foraging tasks by common marmoset (*Callithrix jacchus*) female pairs. *Lab Animal (NY)* 32(3):32–38. doi:10.1038/laban0303-32.

Majolo, B., H.M. Buchanan-Smith, and K. Morris. 2003b. Factors affecting the successful pairing of unfamiliar common marmoset (*Callithrix jacchus*) females: Preliminary results. *Animal Welfare* 12(3):327–337.

Manciocco, A., F. Chiarotti, and A. Vitale. 2009. Effects of positive interaction with caretakers on the behaviour of socially housed common marmosets (*Callithrix jacchus*). *Applied Animal Behaviour Science* 120(1–2):100–107. doi:10.1016/j.applanim.2009.05.007.

Mason, W.A. and G. Epple. 1969. Social organization in experimental groups of *Saimiri* and *Callicebus*. In *Proceedings of the Second International Congress of Primatology, Volume 1: Behavior*, 59–65. Basel, Switzerland: Karger.

McKinley, J., H.M. Buchanan-Smith, L. Bassett, and K. Morris. 2003. Training common marmosets (*Callithrix jacchus*) to cooperate during routine laboratory procedures: Ease of training and time investment. *Journal of Applied Animal Welfare Science* 6(3):209–20. doi:10.1207/S15327604JAWS0603_06.

Mendoza, S.P. 1987. Breeding in Groups: The influence of social context on the reproductive potential of squirrel monkeys. *International Journal of Primatology* 8(5):459.

Mendoza, S.P., E.L. Lowe, and S. Levine. 1978a. Social organization and social behavior in two subspecies of squirrel monkeys (*Saimiri sciureus*). *Folia Primatologica* 30:126–144.

Mendoza, S.P., E.L. Lowe, J.A. Resko, and S. Levine. 1978b. Seasonal variations in gonadal hormones and social behavior in squirrel monkeys. *Physiology and Behavior* 20:515–522.

Michael, R.P., and D. Zumpe. 1971. Patterns of reproductive behavior. In *Comparative Reproduction of Nonhuman Primates*, edited by S.E.S Hafez, 205–242. Springfield, MO: Charles C Thomas.

Mitchell, C.L., S. Boinski, and C.P. van Schaik. 1991. Competitive regimes and female bonding in two species of squirrel monkeys (*Saimiri oerstedi* and *S. sciureus*). *Behavioral Ecology and Sociobiology* 28(1):55–60.

Mittermeier, R.A., A.B. Rylands, A.F. Coimbra-Filho, G.A.B. da Fonseca. 1988. *Ecology and Behavior of Neotropical Primates*. Washington, DC: World Wildlife Fund.

Mittermeier, R.A., and M.G. van Roosmalen. 1981. Preliminary observations on habitat utilization and diet in eight surinam monkeys. *Folia Primatologica* 36(1–2):1–39.

Morton, L.S. 2000. Maternal and non-maternal contributions to infant caregiving in wild and captive Peruvian squirrel monkeys (*Saimiri spp.*). PhD Dissertation, Psychology, University of California, Davis.

Moynihan, M. 1964. *Some Behavior Patterns of Platyrrhine Monkeys: I. The Night Monkey (Aotus trivirgatus)*. Washington, DC: Smithsonian Institution.

Nadler, R.D., and L.A. Rosenblum. 1972. Hormonal regulation of the fatted male phenomenon in squirrel monkeys. *The Anatomical Record* 173(2):181–187.

National Research Council. 2011. *Guide for the Care and Use of Laboratory Animals*. 8th ed. Washington, DC: National Academy Press.

Nevo, O., and E.W. Heymann. 2015. Led by the nose: Olfaction in primate feeding ecology. *Evolutionary Anthropology* 24(4):137–48. doi: 10.1002/evan.21458.

Nievergelt, C.M., and R.D. Martin. 1999. Energy intake during reproduction in captive common marmosets (*Callithrix jacchus*). *Physiology and Behavior* 65:849–854.

Nunes, S., J.E. Fite, and J.A. French. 2000. Variation in steroid hormones associated with infant care behaviour and experience in male marmosets (*Callithrix kuhlii*). *Animal Behaviour* 60:857–865.

Phillips, K.A., M.K. Hambright, B.M. Schilder, C. Ross, and S. Tardif. 2015. Take the monkey and run. *Journal of Neuroscience Methods* 248:27–31.

Ploog, D.W. 1967. The behavior of squirrel monkeys (*Saimiri sciureus*) as revealed by sociometry, bioacoustics, and brain stimulation. In *Social Communication among Primates*, edited by S. Altmann, 149–184. Chicago: University of Chicago Press.

Price, E.C. 1992. The costs of infant carrying in captive cotton-top tamarins. *American Journal of Primatology* 26:23–33.

Puffer, A.M., J.E. Fite, J.A. French, M. Rukstalis, E.C. Hopkins, and K.J. Patera. 2004. Influence of the mother's reproductive state on the hormonal status of daughters in marmosets (*Callithrix kuhlii*). *American Journal of Primatology* 64(1):29–37. doi: 10.1002/ajp.20059.

Rajendra, R.S., A.G. Brady, V.L. Parks, C.V. Massey, S.V. Gibson, and C.R. Abee. 2010. The normal and abnormal owl monkey (*Aotus sp.*) heart: Looking at cardiomyopathy changes with echocardiography and electrocardiography. *Journal of Medical Primatology* 39(3):143–50. doi: 10.1111/j.1600-0684.2010.00403.x.

Roberts, R.L., L.A. Roytburd, and J.D. Newman. 1999. Puzzle feeders and gum feeders as environmental enrichment for common marmosets. *Contemporary Topics in Laboratory Animal Science* 38(5):27–31.

Robinson, J.G., P.C. Wright, and W. G. Kinzey. 1987. Monogamous cebids and their relatives: Intergroup calls and spacing. In *Primate Societies*, edited by B.B. Smuts, D.L. Cheney, R.M. Seyfarth, R.W. Wrangham and T.T Struhsaker, 44–53. Chicago: University of Chicago Press.

Roder, E.L., and P.J.A. Timmermans. 2002. Housing and care of monkeys and apes in laboratories: Adaptations allowing essential species-specific behaviour. *Laboratory Animals* 36(3):221–242. doi: 10.1258/002367702320162360.

Rogge, J., K. Sherenco, R. Malling, E. Thiele, S. Lambeth, S. Schapiro, and L. Williams. 2013. A comparison of positive reinforcement training techniques in owl and squirrel monkeys: Time required to train to reliability. *Journal of Applied Animal Welfare Science* 16(3):211–20. doi: 10.1080/10888705.2013.798223.

Rosenblum, L.A., and R.W. Cooper (eds). 1968. *The Squirrel Monkey*. New York: Academic Press.

Ross, C.N., and J.A. French. 2011. Female marmosets' behavioral and hormonal responses to unfamiliar Intruders. *American Journal of Primatology* 73:1072–1081.

Ross, C.N., J.A. French, and K.J. Patera. 2004. Intensity of aggressive interactions modulates testosterone in male marmosets. *Physiology and Behavior* 83:437–445.

Ross, C.N., M. Power, J.M. Artavia, and S. Tardif. 2013. Relation of food intake behaviors and obesity development in young common marmoset monkeys. *Obesity* 21:1891–1899.

Rotundo, M.A., E. Fernandez-Duque, and A.F. Dixson. 2005. Infant development and parental care in free-ranging *Aotus azarae azarae* in Argentina. *International Journal of Primatology* 26(6):1459–1473.

Rukstalis, M., and J.A. French. 2005. Vocal buffering of the stress response: Exposure to conspecific vocalizations moderates urinary cortisol excretion in isolated marmosets. *Hormones and Behavior* 47(1):1–7. doi: 10.1016/j.yhbeh.2004.09.004.

Rylands, A.B. 1996. Habitat and the evolution of social and reproductive behavior in Callitrichidae. *American Journal of Primatology* 38:5–18.

Rylands, A.B., A.F. Coimbra-Filho, and R. Mittermeier. 1993. Systematics, geographic distribution, and some notes on the conservation status of the Callitrichidae. In *Marmosets and Tamarins: Systematics, Behaviour and Ecology*, edited by Rylands, A.B. New York: Oxford University Press.

Saltzman, W., S.P. Mendoza, and W.A. Mason. 1989. Social dynamics of sequential group formation among female squirrel monkeys. *American Journal of Primatology* 18(2):164.

Saltzman, W., R.P. Pick, O.J. Salper, K.J. Liedl, and D. Abbott. 2004. Onset of plural cooperative breeding in common marmoset families following replacement of the breeding male. *Animal Behaviour* 68:59–73.

Savage, A., L.H. Giraldo, L.H. Sotto, and C.T. Snowdon. 1996. Demography, group composition and dispersal in wild cotton-top tamarin (*Saguinus oedipus*) groups. *American Journal of Primatology* 38:85–100.

Savastano, G., A. Hanson, and C. McCann. 2003. The development of an operant conditioning training program for new world primates at the Bronx Zoo. *Journal of Applied Animal Welfare Science* 6(3):247–261.

Schaffner, C.M., and J.A. French. 1997. Group size and aggression: Recruitment incentives in a cooperatively breeding primate. *Animal Behaviour* 54:171–180.

Schaffner, C.M., and J.A. French. 2004. Behavioral and endocrine responses in male marmosets to the establishment of multimale breeding groups: Evidence for non-monopolizing facultative polyandry. *International Journal of Primatology* 25(3):709–732.

Schiml, P.A., S.P. Mendoza, W. Saltzman, D.M. Lyons, and W.A. Mason. 1996. Seasonality in squirrel monkeys (*Saimiri sciureus*): Social facilitation by females. *Physiology and Behavior* 60(4):1105–1113.

Schneider, H., F.C. Canavez, I. Sampaio, M.Â.M. Moreira, C.H. Tagliaro, and H.N. Seuánez. 2001. Can molecular data place each neotropical monkey in its own branch? *Chromosoma* 109(8):515–523. doi: 10.1007/s004120000106.

Schuler, A.M., and C.R. Abee. 2005. Squirrel monkeys (*Saimiri*). In *Enrichment of Nonhuman Primates*, 73–88. Bethesda, MA: NIH Office of Laboratory Animal Welfare.

Scollay, P.A., and P. Judge. 1981. The dynamics of social organization in a population of squirrel monkeys (*Saimiri sciureus*) in a seminatural environment. *Primates* 22:60–69.

Scott, L., P. Pearce, S. Fairhall, N. Muggleton, and J. Smith. 2003. Training nonhuman primates to cooperate with scientific procedures in applied biomedical research. *Journal of Applied Animal Welfare Science* 6(3):199–207. doi: 10.1207/S15327604JAWS0603_05.

Smith, O.A., and C.A. Astley. 2007. Naturally occurring hypertension in new world nonhuman primates: Potential role of the perifornical hypothalamus. *American Journal of Physiology-Regulatory, Integrative and Comparative Physiology* 292(2):R937–45. doi: 10.1152/ajpregu.00400.2006.

Smith, T.E., C.M. Schaffner, and J.A. French. 1997. Social and Developmental influences on reproductive function in female Wied's black tufted-ear marmosets (*Callithrix kuhlii*). *Hormones and Behavior* 31:159–168.

Soltis, J., F.H. Wegner, and J.D. Newman. 2005. Urinary prolactin is correlated with mothering and allo-mothering in squirrel monkeys. *Physiology & Behavior* 84(2):295–301.

Spinelli, S., L. Pennannen, A. Dettling, J. Feldon, G.A. Higgins, and C.R. Pryce. 2004. Performance of the marmoset monkey on computerized tasks of attention and working memory. *Cognitive Brain Research* 19:123–137.

Stevens, D.J., R.J. Hornsby, D.L. Cook, G.D. Griffiths, E.A.M. Scott, and P.C. Pearce. 2005. A simple method for assessing muscle function. *Laboratory Animals* 39:162–168.

Strayer, F.F., M. Taylor, and P. Yanciw. 1975. Group composition effects on social behavior of captive squirrel monkeys (*Saimiri sciureus*). *Primates* 16:253–260.

Talbot, C.F., H.D. Freeman, L.E. Williams, and S.F. Brosnan. 2011. Squirrel monkeys' response to inequitable outcomes indicates a behavioural convergence within the primates. *Biology Letters* 7(5):680–2. doi: 10.1098/rsbl.2011.0211.

Tardif, S., K. Bales, L. Williams, E.L. Moeller, D. Abbott, N. Schultz-Darken, S. Mendoza, W. Mason, S. Bourgeois, and J. Ruiz. 2006. Preparing new world monkeys for laboratory research. *ILAR Journal* 47(4):307–315.

Tardif, S., D.G. Layne, and D. Smucny. 2002. Can marmoset mothers count to three? Effect of litter size on mother-infant interactions. *Ethology* 108:825–836.

Tardif, S., K. Mansfield, R. Ratman, C. Ross, and T.E. Ziegler. 2011. The marmoset as a model of aging and age related disease. *ILAR Journal* 52:54–65.

Tardif, S., M. Power, D. Layne, D. Smucny, and T. Ziegler. 2007. Energy restriction initiated at different gestational ages has varying effects on maternal weight gain and pregnancy outcome in common marmoset monkeys (*Callithrix jacchus*). *British Journal of Nutrition* 92(05):841. doi: 10.1079/bjn20041269.

Tardif, S., M. Power, C. Ross, J. Rutherford, D. Layne-Colon, and M. Paulik. 2009. Characterization of obese phenotypes in a small nonhuman primate, the common marmoset. *Obesity* 17:1499–1505.

Terborgh, J. 1983. *Five New World Primates: A Study in Comparative Ecology.* Princeton, NJ: Princeton University Press.

Thorington, R.W., Jr. 1967. Feeding and activity of *Cebus* and *Saimiri* in a Columbian forest. In *Neue Ergebnisse Der Primatologie*, 180–184. Stuttgart: Gustav Fischer Verlag.

Trevino, H.S. 2007. Seasonality of reproduction in captive squirrel monkeys (*Saimiri sciureus*). *American Journal of Primatology* 69(9):1001–1012.

Vaitl, E.A. 1977. Social context as a structuring mechanism in captive groups of squirrel monkeys (*Saimiri sciureus*). *Primates* 18:861–874.

Vaitl, E.A., W.A. Mason, D.M. Taub, and C.C. Anderson. 1978. Contrasting effects of living in heterosexual pairs and mixed groups on the structure of social attraction in squirrel monkeys (*Saimiri*). *Animal Behavior* 26:358–367.

VandeBerg, J.L., S. Williams-Blangero, C.M. Moore, M.L. Cheng, and C.R. Abee. 1990. Genetic relationships among three squirrel monkey types: Implications for taxonomy, biomedical research, and captive breeding. *American Journal of Primatology* 22(2):101–111.

Ventura, R., and H.M. Buchanan-Smith. 2003. Physical environmental effects on infant care and development in captive *Callithrix jacchus*. *International Journal of Primatology* 24(2):399–413. doi: 10.1023/a:1023061502876.

Vignes, S., J.D. Newman, and R.L. Roberts. 2001. Mealworm feeders as environmental enrichment for common marmosets. *Contemporary Topics in Laboratory Animal Science* 40(3):26–9.

Vilela, J.D.V., A.L. Miranda-Vilela, E.V. Stasieniuk, G.M. Alves, F.N. Machado, W.M. Ferreira, F.M.O.B. Saad, P.A.R. Machado, C.C.G.M. Coelho, and N.A.M. da Silva. 2012. The influence of behavioral enrichment on dry food consumption by the black tufted-ear marmoset, *Callithrix penicillata* (Mammalia: Callithricidae): A pilot study. *Zoologia* 29(1):1–6. doi: 10.1590/s1984-46702012000100001.

Villavicencio Galindo, J.M. 2003. Distribución geográfica de los primates del género *Aotus* en el Departamento Norte De Santander, Colombia. In *Primatología del Nuevo Mundo*, edited by Pereira-Bengoa, V., F. Nassar-Montoya, and A. Savage, 264–271. Bogotá, Colombia: Centro de Primatología Araguatos.

Washburn, D.A., W.D. Hopkins, and D.M. Rumbaugh. 1989. Automation of learning-set testing: The video-task paradigm. *Behavior Research Methods, Instruments, & Computers* 21(2):281–284.

Watkins, D.J., T.L. Garber, Z.W. Chen, A.L. Hughes, and N.L. Letvin. 1991. Evolution of new world primate MHC class I genes. In *Molecular Evolution of the Major Histocompatibility Complex*, edited by Klein, J. and D. Klein. Berlin; Heidelberg: Springer-Verlag.

Wiebe, R.H., E. Diamond, S. Akesel, P. Liu, L.E. Williams, and C.R. Abee. 1984. Diurnal variations of androgens in sexually mature male bolivian squirrel monkeys (*Saimiri sciureus*) during the breeding season. *American Journal of Primatology* 7(3):291–297.

Williams, L.E., and C.R. Abee. 1988. Aggression with mixed age-sex groups of bolivian squirrel monkeys following single animal introductions and new group formations. *Zoo Biology* 7(2):139–145.

Williams, L.E., C.R. Abee, and S. Barnes. 1988. Allo-maternal behavior in *Saimiri boliviensis*. *American Journal of Primatology* 14(4):452.

Williams, L.E., A.G. Brady, and C.R. Abee. 2010. Squirrel monkeys. In *The UFAW Handbook on the Care and Management of Laboratory and Other Research Animals*, edited by Hubrecht, R. and J. Kirkwood. 8th ed. Oxford, UK: Wiley-Blackwell.

Williams, L.E., C.S. Coke, and J.L. Weed. 2015. Socialization of adult owl monkeys (*Aotus sp.*) in captivity. *American Journal of Primatology* 79:1–7. doi: 10.1002/ajp.22521.

Williams, L., S. Gibson, M. McDaniel, J. Bazzel, S. Barnes, and C. Abee. 1994. Allomaternal interactions in the Bolivian squirrel monkey (*Saimiri boliviensis boliviensis*). *American Journal of Primatology* 34(2):145–156.

Williams, L.E., R.R. Yeoman, and C.R. Abee. 1986. Estrus cycle influences on mating behavior in *Saimiri boliviensis*. *Primate Report* 14:112–113.

Wolf, R.H., R.M. Harrison, and T.W. Martin. 1975. A review of reproductive patterns in new world monkeys. *Laboratory Animal Science* 25:814–821.

Wolovich, C.K., J.P. Perea-Rodriguez, and E. Fernandez-Duque. 2008. Food transfers to young and mates in wild owl monkeys (*Aotus azarai*). *American Journal of Primatology* 70 (3):211–221. doi: 10.1002/ajp.20477.

Wright, P.C. 1978. Home range, activity pattern, and agonistic encounters of a group of night monkeys (*Aotus trivirgatus*) in Peru. *Folia Primatologica* 29:43–55.

Wright, P.C. 1994a. The behavior and ecology of the owl monkey. In *Aotus: The Owl Monkey*, edited by Baer, J.F., R.E. Weller, and I. Kakoma, 97–112. San Diego, CA: Academic Press.

Wright, P.C. 1994b. The behavior and ecology of the owl monkey. In *Aotus: The Owl Monkey*, edited by Baer, J., R.E. Weller, and I. Kakoma, 97–112. New York: Academic Press.

Wright, P.C., D.M. Haring, M.K. Izard, and E.L. Simons. 1989. Psychological well-being of nocturnal primates in captivity. In *Housing, Care and Psychological Well-Being of Captive and Laboratory Primates*, edited by Segal, E.F, 61–74. Park Ridge, NJ: Noyes Publications.

Yamamoto, M.E., C. Domeniconi, and H. Box. 2004. Sex differences in common marmosets (*Callithrix jacchus*) in Response to an unfamiliar food task. *Primates* 45(4):249–54. doi: 10.1007/s10329-004-0088-6.

Yeoman, R.R., F.H. Wegner, S.V. Gibson, L.E. Williams, D H. Abbott, and C.R. Abee. 2000. Midcycle and luteal elevations of follicle stimulating hormone in squirrel monkeys (*Saimiri boliviensis*) during the estrous cycle. *American Journal of Primatology* 52(4):207–211.

CHAPTER 25

Behavioral Management of Prosimians

Meg H. Dye
Duke Lemur Center

CONTENTS

INTRODUCTION

Behavioral management is a comprehensive strategy that includes using enrichment, positive reinforcement training, facilities and enclosure design, positive staff–animal interactions, and behavioral monitoring to promote psychological well-being (Coleman et al. 2012). This multifaceted approach allows for a holistic view of an animal's experiences within the context of its environment and daily experiences. Behavioral management programs (1) strive to create an environment and a staff culture that encourage species-specific behavior and (2) are proactive in avoiding the occurrence of abnormal behavior. With the use of nonhuman primates (NHPs) in biomedical research and the high profile of great apes, the components of behavioral management for many primates, including gorillas, chimpanzees, and macaques, have been well documented (Bloomsmith and Else 2005; Carrasco et al. 2009). However, the behavioral management of prosimians is relatively new in the literature, as these primates have not been used in biomedical research to an extensive degree. While the majority of information related to the care and behavioral management of prosimians has been refined in zoos and noninvasive research settings, the principles and practices followed here are highly relevant to prosimians in a more traditional research setting.

With 103 species within the suborder of Prosimian, it would be unrealistic to thoroughly discuss each species within the context of this chapter. Instead, the focus will be on species that are currently under human care and thus part of a behavioral management program. Physiological, behavioral, and social differences of prosimians from other nonhuman primates will be highlighted. These differences create the behavioral management framework on which facility design, socialization, enrichment, and positive reinforcement training are based, and enhance the overall well-being of prosimians.

NATURAL HISTORY

Prosimian (meaning "pre-monkey") primates comprise lemurs, lorises, pottos, and galagos. The classification of primates remains somewhat controversial, and taxonomic structure continues to be revised as the increasing quantity of genetic information is reconciled with earlier methods of classifications based on morphology and fossil records. Most authorities now follow a systematic arrangement in which the primates are divided into two suborders: (1) Strepsirhini (i.e., tooth-combed primates) and (2) Haplorhini, which includes tarsiers, monkeys, apes, and humans. The Strepsirhine group is further divided into two infraorders: Lorisiformes and Lemuriformes. The infraorder Lorisiformes includes all the extant African and Asian species of lorises, pottos, and galagos, which are represented by 9 genera and 18 species of small-bodied, nocturnal primates (Nowak 1999). The infraorder Lemuriformes comprises five families endemic to Madagascar: (1) Lemuridae (bamboo lemurs, ring-tailed lemurs, true lemurs, and ruffed lemurs), (2) Indriidae (indri, sifaka, and woolly lemurs), (3) Cheirogaleidae (mouse lemurs, dwarf lemurs, and fork marked lemurs), (4) Lepilemuridae (sportive lemurs), and (5) Daubentoniidae (aye-aye) (Williams 2015).

Lemurs are found in a wide range of ecologic niches in Madagascar, including the low- to high-altitude tropical rain forest on the east coast, the dry deciduous forests of the west, and the spiny deserts of the south (Williams 2015). Lorises are native to Southeast Asia and the tropical forests of India and Sri Lanka, and galagos (bushbabies) and pottos are distributed throughout Africa, south of the Sahara (Nowak 1999). Lemurs are considered to be the most endangered group of mammals on the planet due to habitat loss and environmental degradation in Madagascar, and loris populations continue to suffer from ongoing illegal wildlife trade in Asia (Moore et al. 2014).

Many prosimian species are not represented in captivity or are present only in very small numbers. Of the lemur species housed in zoos, ring-tailed lemurs are most numerous, followed by ruffed lemurs (*Varecia*). Increasingly, Coquerel's sifaka (*Propithecus coquereli*) and aye-aye (*Daubentonia madagascariensis*) are housed in North American and European zoos, as better husbandry and feeding programs for these species are developed (Williams 2015).

Prosimian activity patterns can be nocturnal (predominantly active at night), diurnal (predominantly active during the day), or cathemeral (mixture of day and night activity) (Curtis 2006). Although most primates are presumed to fall within the strict diurnal versus nocturnal dichotomy, recent research shows that cathemerality is more common among prosimians than originally thought (Lafleur et al. 2014; Santini et al. 2015; see Table 1).

Diurnal and cathemeral lemurs tend to live in larger, female-dominant social groups and are generally more gregarious than the nocturnal lorisids and cheirogaleids (Izard 2006). Ring-tailed lemurs (*Lemur catta*) are the most social species, with large family groups of sizes up to 30 members living in multimale, multifemale groups (Jolly 1967). The nocturnal aye-aye is likely the most solitary of the lemurs; nonetheless, evidence suggests the regular proximity of satellite males on the perimeter of the female's territory (Ancrenaz et al. 1994).

In all diurnal and cathemeral species, grooming is regarded as one of the best indicators of social bonds between individuals, and serves multiple functions, including reconciliation, alliance formation, stress reduction, and hygiene (Barton 1985). Vocalizations can also serve as a way to maintain social bonds when members are not in close proximity to one another. Studies have shown that the vocal responses of ring-tailed lemurs to conspecific contact calls can be highly selective (Kulahci et al. 2015). Instead of exchanging vocalizations with all group members, ring-tailed lemurs have been shown to reserve their vocal responses to contact calls mainly from the group members whom they frequently groom.

Strepsirrhine primates are considered the more macrosmatic (or "keen-scented") suborder of primates and display the most complex array of scent-marking behavior within the Order Primate (Drea and Scordato 2008). Scent marking is very important in many contexts, including territoriality, reproduction, and social hierarchy. In species that tend to be solitary foragers, scent marking is often the principal means of communication by which breeding activities are coordinated (Dröscher and Kappeler 2014).

With the exception of the aye-aye, lemurs are highly seasonal breeders. There is evidence that lorises (*Nycticebus*) are seasonal breeders (Ratajszczak 1998), while galagos (*Otolemur*) and pottos (*Perodicticus potto*) are not. Photoperiod is the single most important environmental cue controlling estrus cycles in females and testicular change in males (Van Horn 1975; Perret and Aujard 2001). This is evident as institutions housing lemurs in the northern hemisphere will experience breeding and birth seasons opposite to those in Madagascar. The breeding season for most species of lemurs housed in regions of the northern hemisphere occurs during the fall, when day length is short.

SOCIALIZATION

Establishing and maintaining species-specific social groups is critical to the successful housing of prosimians when considering their behavioral well-being. Social management of group composition should ideally occur throughout a colony's evolution as animals are added through breeding and/or the introduction of new individuals. Recognizing and understanding species-typical social behaviors is a key component of maintaining stable social groupings.

Pregnancy and Parturition

Birthing suites provide a protected area in the animal's normal living space where a dam and her new infant can be separated with visual and olfactory access to the rest of her family, while ensuring a safe zone for uninterrupted nursing and bonding time. The first few days following parturition are the most critical for a new infant, with mortality the highest in the first 72h following delivery.

Uninterrupted bonding and nursing time with the dam provide the best scenario for an infant's survival (Figure 25.1; Williams, C. personal communication). In an analysis of birth records, Katz (1980) reported that separating a dam with her new infant significantly decreased mortality in the first 48 h. The benefit of separating a dam and her infant was reinforced in the following breeding season when 51 lemurs were born at the Duke Lemur Center and 100% survived.

Brief and highly seasonal breeding in diurnal and cathemeral lemurs allows for the estimation of parturition dates based on observed behavioral and physical indicators. Estimated due dates can assist with preparation for an infant and for facilities that separate females prior to parturition, by maximizing the amount of time pregnant females can remain with their social group. At the earliest date of expected birth, pregnant females are given access to their family groups during the day and separated at night. Because of the potential for rapid changes in social hierarchies, pregnant ring-tailed lemurs are not separated until the female gives birth (Katz 2014).

Of the nocturnal prosimians, mouse lemurs (*Microcebus murinus*) and fat-tailed dwarf lemurs (*Cheirogaleus medius*) are seasonal breeders. While generally singly housed, compatibly housed pregnant mouse lemurs or fat-tailed dwarf lemurs can be separated at the onset of the earliest parturition date. Early separation of pregnant females can decrease infant injuries and infanticide (Katz 2014). Similarly, an aye-aye dam that is not already singly housed will generally be separated from her mate or previous offspring prior to parturition.

Prosimians show two distinct styles of maternal care. Ruffed lemurs, aye-ayes, mouse lemurs, and dwarf lemurs build nests for their infants, while lorises park their infants, but do not build nests. Dams will leave infants to forage and return to care for the infants. The remaining species of lemurs carry their infants on their bodies. The ability of newborn infants to cling to the mother's lower abdomen is an important behavior to assess when determining whether an infant is thriving. The infant's hands and feet should be grasping tightly with no space between the infant's and the dam's bodies. Evidence of head lolling, inability to grasp with one or more limbs, or visible space between the infant's body and the dam's abdomen are causes for concern.

Mouse lemur dams that do not come out of the nest for food after the earliest estimated parturition date are presumed to have newborn infants. Management of the environment is crucial to keep noise levels and disturbances at a minimum. Luring or disturbing a dam out of her nest box too early, or too often, can lead to infant abandonment or infanticide (DLC, unpublished data). In species that do not carry their infants, but instead keep them in a nest, a strong indicator of a nervous dam is if she does not settle in the nest immediately after birth, and instead carries infants from site to site in the first 24 h.

Figure 25.1 Sifaka newborn and dam.

Introducing Infants

Introducing a thriving infant and dam to other members of the social group is a strategic management practice and is best done gradually. Generally, the sire is the first member of the family to have physical access to the infant. Positive behaviors performed by the sire during a reunion with the dam and new infant may include singing (in sifaka), grooming the dam and the infant, or sitting quietly close to the dam. The process of allowing additional family members access to the dam continues through selective introductions of siblings, often starting with the youngest and progressing to older siblings, with unrelated individuals introduced last. Introductions can occur over a period of days or weeks, depending on the group's social dynamics. The immersion of an infant into the family unit, for all species, allows for natural socialization and communal behaviors, such as grooming, play, and shifts in dominance hierarchies, to occur.

Reintroducing Family Members

The temporary removal of an individual animal from an established group for research or medical reasons can create a change in the social hierarchy in some species. This is particularly true for ring-tailed lemurs, in which a short absence during times of flux within the social hierarchy, particularly during the months leading up to and during the breeding season, can promote social reshuffling (Jolly 1967). Social reorganization can be avoided by simultaneously separating all members of the troop in cases where the dominant female is temporarily removed or other factors, such as social dynamics, age, or breeding, warrant proactive separations. When all individuals are separated for the duration of the absence, members of the troop, including the returning animal, can be simultaneously reunited with one another. In situations where only one member of the family is separated for periods greater than a day, reintegration into the group needs to be evaluated on a case-by-case basis, while taking into account the scenario that led to the separation, and the personalities and dominance status of the other family members.

New Introductions

Introduction of a new individual is most successful when animals have visual and olfactory access to each other through a "howdy" cage or barrier prior to introduction. While housed side-by-side, visual, auditory and olfactory cues can be freely exchanged between individuals. Affiliative behaviors displayed on either side of the barrier, such as sitting together, calling, and efforts to solicit grooming, are desired behavioral indicators that should precede a physical introduction. Aggressive behaviors, including attacking the barrier or aggressive calls, warrant a longer visual introductory phase.

Species-specific behavior during physical introductions varies greatly. What is socially acceptable with one species may not be appropriate with other species. The behaviors of gray mouse lemurs may include a dominant animal pinning another on its back for periods of time, a behavior that typically does not result in injury. However, similar behavior between diurnal lemurs can lead to bite wounds and injury. Positive parameters of a successful diurnal introduction may include allogrooming, sharing food, or resting and sleeping together with tails curled over the back. Sifaka do not curl their tails over their back, but do display relaxed mutual grooming and resting. Compatibility of nocturnal species is often shown when the animals sleep in the same nest box and antagonistic behaviors, such as chasing or displacing, are absent.

Aggression

An important aspect of a behavioral management program is the recognition of precursors to aggression. Precursors to potentially injurious situations between lemurs include overly exaggerated displacement at feeding time, chasing, keeping an individual away from core group members, hair pulls, and submissive vocalizations. In many species, aggression may be directed from sires toward sexually mature sons at the onset of the breeding season, young females to their siblings (usually other females), or by offspring who may challenge their aging or medically compromised parents. The incidence and severity of injury resulting from intraspecies aggression varies by species, with ring-tailed lemurs, brown lemurs, black lemurs, and sifaka more often biting or injuring one another, and ruffed lemurs and aye-ayes infrequently sustaining injuries (Williams, C. personal communication).

Separations with visual access can be an effective management tool to de-escalate aggression. Physical separation of the aggressor from the group for a few days or even up to a week is often followed with a successful reintroduction into the group. Hostile behavior, for example, lunging at the caging or reaching through the caging, that continues during physical separation with visual access, is an indicator of poor compatibility and may require permanent separation of the aggressor from the group (PTAG 2014).

Brief bouts of displacement and chasing are often a normal part of social interaction as juveniles reach sexual maturity and leave, or are forced to leave, their natal group (PTAG 2014). When younger animals of a group begin to assert their dominance, several factors are considered prior to permanently removing a juvenile from the group. The age of the juvenile is important to consider in order to avoid interfering with natural socialization and the learning of key behaviors from parents. In addition, preparations for integration of the younger animal into a new social group are important to avoid long-term housing of a single animal.

ENVIRONMENTAL ENRICHMENT

The primary goals of an environmental enrichment program are applicable to all ex situ species—to promote physical and psychological well-being by providing a functionally appropriate environment that gives the animals choice and control. The benefits of social and sensory enrichment with primates have been well documented (Lutz and Novak 2005; Wolfensohn and Honess 2005; Coleman et al. 2012). Among prosimians, the wide diversification of physiology, behavior, locomotion, and habitat use requires an understanding of species-specific behavior in order to provide enrichment that is effective for each individual. Not all prosimian enrichment is equally effective. An enrichment program designed for a mouse lemur will not fulfil the psychological and physiological needs of a sifaka. Similarly, some enrichment items or practices designed for monkeys and apes may not be effective with prosimians. Prosimians differ from other primate taxa in possessing few object-oriented behaviors, such as tool use, as prosimians lack the highly developed manual dexterity of monkeys and apes. As a result, techniques commonly used for other primate species, such as puzzle-solving and tool-driven enrichment, may not be as effective with lemurs (Dishman et al. 2009). Lemurs have never been observed using tools in the wild; however, research suggests that lemurs can learn to use pulling tools as quickly as other more dexterous primates (Santos et al. 2005b).

While species-specific differences should be considered for providing effective prosimian enrichment, the fundamentals are the same. Enrichment is not a single, isolated event, but an ongoing process. Enriching experiences that elicit species-specific behavior, increase activity, and promote mental simulation are intertwined in the animal's entire captive milieu (Mellen and MacPhee 2001). Enrichment can take many forms, including social enrichment, sensory enrichment, cognitive challenges, changes to the structure of the enclosure, and changes in feeding practices (Quirke and O'Riordan 2011; Dye 2015). A holistic look at stimulating experiences over the course of days and weeks should reveal enriching opportunities that enhance an animal's overall well-being.

Social Enrichment

In the wild, diurnal and cathemeral lemurs are social animals even though the degree of physical and auditory contact with conspecifics may vary by species and by group. As such, the well-being of these species in a controlled environment depends on social housing, rather than single housing. Breeding groups of red-ruffed lemurs (*Varecia rubra)*, black-and-white ruffed lemurs (*V. variegata)*, crowned lemurs (*Eulemur coronatus)*, blue-eyed black lemurs (*E. flavifrons)*, collared lemurs (*E. collaris)*, and sifaka (*Propithecus coquereli)* are most often housed as family groups consisting of a dam, sire, and one to three offspring. Ring-tailed lemurs are often housed in larger social groups, with multiple adult males and females. Bachelor groups of ring-tailed lemurs can also be maintained successfully. For individuals in which a companion of the same species is not available, compatible inter-species pairings can provide social interaction.

A wide variety of diurnal and cathemeral lemurs have been successfully housed together (AZA 2011). Ruffed lemur and ring-tailed lemur mixed-species groups are the most common and are generally very successful (Franck-McCauley 2015). Mixed-species groups work best when they consist of different genera and, thus, do not occupy the same ecological niche (Manna et al. 2007). With any combination of species, individual personalities and group dynamics need to be evaluated when attempting to establish compatible groups. While there are many benefits to a mixed-species exhibit, particular attention should be paid to inter-group dynamics during birth seasons and months in which young infants could be vulnerable to inter-species aggression or infanticide (Jolly et al. 2000; AZA 2011).

Large and small nocturnal prosimians have also been successfully housed together (AZA 2011). Aye-ayes have been successfully housed with loris species, fat-tailed dwarf lemurs, and greater bush-babies (*Otolemur crassicaudatus)* (Gibson 2015). While the success of mixed pairs will depend on individual personalities and habitat characteristics, increased complexity of perching, and multiple sleeping and feeding spots encourage natural behavior from both species. Access to species-specific diets and quantities of food should be regularly assessed to prevent obesity in the smaller species.

Foraging Enrichment

In the wild, lemurs spend a large amount of time locating and acquiring food, most often in arboreal settings. The one exception to this pattern is the semi-terrestrial behavior of ring-tailed lemurs (Maloney et al. 2006). Using feeding devices and strategically changing the placement of food (Figure 25.2) in captive environments has been shown to significantly increase foraging time

Figure 25.2 Fresh browse as food enrichment for diurnal and cathemeral species.

and movement, while simultaneously decreasing inactivity and stereotypic behavior (Maloney et al. 2006). For example, by suspending whole, unpeeled fruits, Kerridge (2005) noted a significant increase in feeding and foraging time that approached those of wild ruffed lemurs. In addition, a physiological benefit of increased bipedal and tripedal suspension was noted as the animals actively changed physical positions to reach the food. Captive white-fronted lemurs that spent most of their time near the ground and resting significantly increased their foraging and locomotion levels with the addition of self-operated feeder boxes (Sommerfeld et al. 2006). Techniques, such as scatter feeding, can have different effects when used with different species. Britt (1998) found no significant effect of scatter feeding on activity levels of black-and-white ruffed lemurs. In contrast, Dishman et al. (2009) found manipulating food presentation on the ground significantly increased the foraging time and movement of ring-tailed lemurs and brought activity levels closer to those of wild conspecifics. The difference in outcomes reflects the importance of understanding the nature of each species, as ring-tailed lemurs are more terrestrial than the highly arboreal ruffed lemurs.

Species-specific foraging behaviors of lorises and pottos can be elicited when gum arabic is provided in an enrichment device rather than simply presented in a dish. Foraging behaviors include extractive foraging, vertical or horizontal clinging, and gouging (Huber and Lewis 2011). Craig and Reed (2003) presented puzzle feeders that necessitated gouging to pygmy slow lorises that resulted in increased activity. Fitch-Snyder et al. (2001) used gum arabic placed in a treat log to stimulate activity. This resulted in both the northern slow loris (*Nycticebus bengalensis*) and the pygmy slow loris (*N. pygmaeus*) gouging into the log far beyond the original diameter of the holes initially drilled into the device. In addition to enhancing psychological well-being, increased foraging activity can have physiological benefits for lorises, including the maintenance of optimal body weights and improvements in dental health (Gray et al. 2015).

The unique characteristics of aye-ayes, including acoustic foraging and ever-growing front incisors, can present challenges in creating effective and safe enrichment. Food presentation of daily dietary items, such as worms or nuts, enclosed in bamboo or wood feeders can reliably elicit natural foraging behavior (Figure 25.3; Gibson and McKinney 2015). Additional food enrichment, such as peanut butter sandwiched between two wood boards sewn together with fleece, can be a very effective form of enrichment, both in duration of interest and in eliciting the performance of natural behaviors, such as tapping and gouging.

Figure 25.3 Aye-aye foraging from a bamboo feeder.

Olfactory Enrichment

When compared to other primates, prosimians rely more on their sense of smell than on vision due to their retention of extensive neuroanatomical structures associated with olfactory discrimination (Valenta et al. 2015). For this reason, scent enrichment may be more effective with prosimians than other primate species for eliciting olfactory-related, species-specific behaviors. Literature on the use and effectiveness of scent enrichment with prosimians is sparse. There is anecdotal evidence that smells of extracts and natural oils can increase levels of activity. The provision of garlic and/or onions can be beneficial, as animals have been observed to rub the garlic on their bodies and carry it around for extended periods of time (Keith, B. personal communication).

A suite of studies that focused on olfactory communication in ring-tailed lemurs investigated whether this species can discriminate the complex chemical information encoded in conspecific scent marks (Greene and Drea 2014). To accomplish this, researchers used wooden dowels rubbed with swabs of scent gland material from familiar and unfamiliar ring-tailed lemurs (Scordato and Drea 2007). The dowels were presented to different ring-tailed lemurs and elicited responses such as sniffing, licking, scent marking, and even stink fighting (rubbing their tail with their wrist glands and wafting scent at opponents). Based on the level of activity the scented dowels elicited during the research trials, the animal care staff continues to provide scented dowels as an effective form of olfactory enrichment for ring-tailed lemurs.

Auditory and Visual Enrichment

Auditory enrichment has been shown to effectively decrease stereotypic behavior in a variety of primates, including baboons, macaques, chimpanzees, and gorillas (Robbins and Margulis 2014). Currently, only a few studies have been conducted on the effect of auditory enrichment with prosimians. Hanbury (2009) reported that bushbabies exposed to Mozart did not show reduced levels of stereotypy when compared to baseline levels. Clark and Melfi (2012) studied the effect of playing rainforest sounds in a mixed-species nocturnal habitat that included bushbabies. While rainforest sounds were playing, bushbabies spent over 90% of the time in two nest boxes, which the authors interpreted as hiding. Fistarol et al. (2008) reported that ring-tailed lemurs showed changes in behavior when music was provided as enrichment. In addition, different types of music affected behavior differently. Further research is needed to determine if auditory enrichment can assist in decreasing stereotypic behavior during daily routines or periods of increased noise levels (e.g., during construction).

Similar to auditory enrichment, little is published detailing the effectiveness of visual enrichment with prosimians. Unlike the great apes that will often display self-directed behaviors when looking in a mirror, prosimians have only shown exploratory or social behaviors (Figure 25.4; Inoue-Nakamura 1997). While the interactions with a mirror may vary between species, ruffed lemurs, sifaka, and ring-tailed lemurs have been observed actively looking at and behind a mirror and reaching toward a mirror. Upon seeing their reflection, ring-tailed lemurs have been observed to actively stink fight the perceived intruder in the mirror.

Research as Enrichment

Noninvasive research tasks can provide an array of novel enrichment opportunities for prosimians. Lemurs have learned to use touch screens to indicate serial ordering (Drucker et al. 2016), numerical quantities (Santos et al. 2005a; Merritt et al. 2011), visual discrimination (Joly et al. 2014), and color discrimination (Vagell, R. personal communication). Research projects, including

Figure 25.4   Ring-tailed lemur looking in a mirror.

studies of locomotion (Granatosky et al. 2014), facial recognition (Marechal et al. 2010), personality with novel objects (Verdolin 2013), spatial foraging (Rosati et al. 2014), and social cognition (Bray et al. 2014; Reddy et al. 2015), among others, have been conducted successfully, and have provided mental and psychological enrichment for prosimians.

Spatial Complexity

Prosimians are housed in a variety of habitats, ranging from modest-sized enclosures to large, free-ranging environments encompassing several acres. The physical dimensions of the space are not the only significant factor for nonhuman primates. The complexity of the space and the availability of opportunities to make use of three-dimensional space are also critical elements for creating a functionally appropriate captive environment for prosimians. Several primate studies have shown that the complexity of the space is more important than its size (Hosey 2005). In addition, complexity of space may be especially beneficial for highly arboreal species, such as lemurs (Tarou et al. 2005).

Branches and perches can be added to maximize the use of three-dimensional space (Figure 25.5). Additional elevated resting and perching areas can be created with hammocks, bread trays, fire hose, milk crates, and PVC pipes (Figure 25.6). For sifaka, multiple vertical structures are required to accommodate the species' vertical clinging and leaping form of locomotion. Studies on the leaping behavior of the genus *Eulemur* suggest variation in landing style, with some species landing hind limbs first, others with fore limbs first, and still others landing on all four limbs (Terranova 1996; Demes et al. 2005). Artificial and natural branches, platforms, and other structures situated in both horizontal and sloping positions can accommodate the variety of landing styles of these species (Ferrie et al. 2013). In lorises, a fast, almost trot-like locomotion is almost exclusively seen on longer horizontal branches and gaps between branches that promote bridging (Schulze 1998). Reporting on the effects of enrichment added to a potto habitat, Frederick (1996) noted an increase in activity and a 10% increase in social behaviors after vegetative covering and hiding spots were added. The addition of branches for locomotion, feeding sites, and sleeping sites decreased levels of aggression and increased exploratory behavior in mouse lemurs. Altering the complexity of branching effectively encouraged mouse lemurs to "map" their surroundings when a portion of branches were purposely made unstable or led to dead ends (Faber 2012).

Figure 25.5 Branches and perches are available to maximize the use of three-dimensional space.

Figure 25.6 Elevated resting and perching areas can be created with hammocks, bread trays, fire hose, milk crates, and PVC pipes.

Effective Enrichment

Comparing activity budgets of captive species with activity budgets of wild conspecifics can assist animal care staff in identifying behavioral goals for the animals in their care. An understanding of activity budgets allows staff to create effective enrichment plans to significantly modify the behavior of captive animals to more closely resemble the behavior of wild prosimians.

Kerridge (2005) compared behaviors of wild black-and-white ruffed lemurs with captive black-and-white ruffed lemurs. Results showed that captive ruffed lemurs spent significantly more time self-grooming, scratching, scent marking, and engaging in solitary play than wild ruffed lemurs (Kerridge 2005). Similarly, Sommerfeld et al. (2006) compared captive and wild white-fronted lemurs (*E. albifrons*) and found activity budgets for resting, foraging, and locomotion to be significantly lower in the captive animals. Data from wild observations can reinforce the efforts of animal care staff to increase natural species-typical behaviors of prosimians in their care by including opportunities to forage and move within all dimensions of their habitat. By creating a stimulating and complex environment that elicits natural behavior, enrichment can effectively decrease abnormal frequencies of natural behaviors, such as self-grooming and scent marking, as well as proactively prevent or decrease behaviors not seen in the wild, such as extensive pacing or head rolling.

POSITIVE REINFORCEMENT TRAINING

Historically, good animal care consisted of three pillars: veterinary care, good nutrition, and proper environment. In the 1980s, positive reinforcement training (PRT) became recognized as an essential fourth pillar of care (Ramirez 2013). Today, PRT is a refinement in animal handling methods with applicable benefits to animal welfare, animal husbandry, veterinary care, and research methodology. The primary reasons for providing an animal training program directly relate to the individual welfare of an animal: physical exercise, mental stimulation, and cooperative (husbandry) behaviors. Assessments of PRT on husbandry, behavior problems, veterinary care, and research procedures show that animals trained with PRT have enhanced welfare (Perlman et al. 2012). Thus, the primary reasons for training directly align with the primary goals of a behavioral management strategy and create an environment in which animals physically and psychologically thrive.

The successful use of PRT to refine husbandry and research methods has been well documented in gorillas, chimpanzees, orangutans, and several species of monkeys (Laule et al. 2003; Bloomsmith et al. 2007; Coleman and Maier 2010). Peer-reviewed articles and documentation of prosimian training programs are less abundant. However, in the last 10 years, PRT has increasingly been incorporated into the behavioral management of prosimians with experiences and methods shared through professional conferences and publications (Dye 2011b; Flanagan 2012; Doss and Dye 2015; Ellison and Laverick 2015; Stierhof 2015; Wall 2015; Whipple 2015). An inaugural Prosimian Keeper Workshop in 2008 was the first venue dedicated to creating a network for prosimian caregivers and keepers to exchange information, including PRT and enrichment data, and remains a valued meeting for advancing prosimian behavioral management.

PRT and Animal Management

Daily management of prosimians is greatly enhanced with the use of PRT. By conditioning animals to calmly move from one part of their enclosure to another, separate temporarily from conspecifics or enter a transport kennel, a system of voluntary participation is created that emphasizes relaxed behavior and decreases animal stress (Schapiro et al. 2003). Having the flexibility to move animals quickly and easily greatly enhances management of prosimians for cage cleaning, annual physical exams, research, and to rearrange or separate individuals as needed (Ferrie et al. 2013).

Separation training is particularly helpful when working with a large social group. Separating animals from their groups can be a time-consuming process and can be stressful for both the animals removed and those remaining behind (Schapiro et al. 2003). In the case of a troop of seven ring-tailed lemurs that were not part of a PRT program, temporary separations prompted stress-related behavior in individuals, including diarrhea, hyperactivity, and contact calling. The process took an extended amount of time, as separations were dependent on opportunistic closing of shift doors. With regular training sessions and successive approximations, the troop was successfully conditioned to calmly separate with no signs of stress within 1 min of the start of the session (Dye, M. H. personal observation).

Training a prosimian to voluntarily enter a transport kennel has benefits for multiple aspects of a behavioral management program, including animal management. Regardless of whether an animal is moving within a facility, participating in research, visiting the veterinary staff, or transferring to another facility, a reinforcement history of calmly entering a kennel reduces stress and is safer for both the animal and staff (Coleman 2012). This is especially true when using protected-contact techniques with aggressive animals or animals with a history of biting. Multiple species, including sifaka, blue-eyed black lemurs (*E. flavifrons*), and ring-tailed lemurs have been conditioned to enter a kennel on cue, allowing the staff to safely enter the enclosure. Free-contact kennel training has also been accomplished with individuals from the majority of captive diurnal and cathemeral species, as well as nocturnal prosimians, including slow lorises, mouse lemurs, and aye-ayes (Dye, M. H. personal observation).

At least two North American facilities house lemurs in a "free ranging" system in which the animals have seasonal access to multiple acres of forest. PRT has been utilized for mandatory weekly lockup of free-ranging animals to meet Institutional Animal Care and Use Committee (IACUC) requirements for consistent recall of animals into a secure building. Training sessions have been invaluable for the rapid recovery of individuals when, in rare incidences, a lemur has breached a perimeter fence (Dye 2011a). The use of training has also assisted with providing medical care to an injured lemur. At one facility, an injured sifaka quickly descended from a 40-foot-tall tree when it heard the cue for the start of a training session. The trainer was then able to move the sifaka to a kennel for immediate medical attention for its broken ankle (Dye 2013).

Husbandry Training

Conditioning prosimians to participate in their own health care has steadily increased over the last decade. Behaviors range from simple targeting and scale training to more advanced behaviors, such as voluntarily accepting injections, eye drops or physical therapy; allowing auscultation of the heart and lungs with a stethoscope; presenting for visual examination and collecting vaginal swabs for estrus detection; voluntary restraint; prenatal exams, including abdominal palpation and lactation confirmation; ultrasounds (Figure 25.7); infant removal for exams; and postpartum exams (Kellerman and Dye 2015). PRT has been successfully used to remove a chronically sick sifaka infant from its dam to receive regular vet care, including supplemental feeding, antibiotic administration, and physical examinations. Due to the consistent success of removing the infant during training sessions, the infant could remain housed with the dam and did not need to be housed separately to accommodate medical interventions (Dye 2011b). Today, many yearlings and young adults experience training within the first days or weeks of life as they cling to their dam or observe training sessions from the safety of their nest.

Early observation and participation in training greatly benefits the individual animal, as well as staff, as behaviors the infants experience in the presence of their dam become familiar and reliable behaviors as adults. Behaviors such as entering a kennel, walking onto a scale for weighing, or separating from other family members (while still with their dam) are done easily as a result of observational learning from an early age.

Figure 25.7   Aye-aye trained to voluntarily allow an ultrasound examination.

Figure 25.8   A veterinarian participating in the positive reinforcement training program.

Veterinarian Participation

Animals trained with PRT consistently and voluntarily participate in veterinary and research procedures and are less stressed while doing so (Lambeth 2006; Perlman et al. 2012). Veterinarians can have a tremendous impact on the successful outcome of conditioned behaviors by regularly participating in training sessions (Figure 25.8). Incorporating consistent veterinary participation in the early approximations of a medical procedure is critical to behavioral success for both the staff and the animal. Creating positive associations, such as delivering positive reinforcement or enrichment to an animal, establishes a valuable veterinarian–animal relationship that serves as the basis for successful voluntary medical procedures (see section "Husbandry Training").

Research Training

Using PRT sessions to desensitize primates to research areas, equipment, and tasks prior to data collection is a proven technique for reducing anxiety and stress (Laule et al. 2003; Melfi 2005;

Perlman et al. 2012; McMillan et al. 2014; Spiezio et al. 2015). PRT also works well to prepare lemurs for events they may experience from the time they are separated for research until they return back to their social group. Conditioning voluntary separation and kenneling allows a lemur to arrive at a data collection session in a calm state. To avoid handling stress associated with capture and transport for data collection, Trouche et al. (2010) conditioned mouse lemurs to enter a wooden box in which they were carried to an apparatus for data collection then returned to their home cage. Similarly, a variety of species, including ruffed lemurs, ring-tailed lemurs, sifaka, and aye-aye have been conditioned to voluntarily kennel before and after data collection (Dye, M. H. personal observation).

Prior to data collection, lemurs can be desensitized to the research environment and conditioned for specific research tasks. In a suspensory locomotion study, aye-ayes were desensitized to an unfamiliar room in which data collection would take place. Through short training sessions and small approximations in the new environment, the aye-aye were conditioned to complete the desired locomotion in a relaxed and reliable manner during data collection (Granatosky et al. 2016). Simple behaviors, such as following a target or following the point of a trainer's finger are valuable tools when data collection depends on a subject moving in a specific manner, for example, across a force plate or to a specific location in front of a video camera (Doss and Dye 2015).

In addition to reducing stress and increasing data quality (see section "Implications for Research"), noninvasive research projects can present cognitive challenges that clearly fit with the definition and goals of enrichment (see section "Research as Enrichment"). With an increase in the use of PRT with prosimians, there are exciting opportunities for increased voluntary participation in a wide variety of noninvasive research projects (Figure 25.9).

Compared to mice, the gray mouse lemur is of interest to researchers as a more practical model for noninvasive Alzheimer's disease research. The mouse lemur has a relatively short life expectancy, with animals over 5 years of age considered to be elderly (Bons et al. 2006). Learning and memory capacities of mouse lemurs have been tested to identify age-related deficits (Picq et al. 1999; Trouche et al. 2010; Joly et al. 2014). In these studies, mouse lemurs were conditioned to participate in a series of visual discrimination tasks. Results revealed that aged mouse lemurs showed cognitive impairments that have, so far, only been found in aged humans and rhesus monkeys (*Macaca mulatta*; Joly et al. 2014).

Figure 25.9 Lemur participating in a locomotion research trial.

FACILITY DESIGN

Facility design can greatly benefit the overall well-being of prosimians, as well as assist with the management of the animals. Arboreal by nature, accessible vertical space is an important feature for prosimian housing (see section "Space Complexity"). Habitats designed with the flexibility to attach branches and enrichment to walls, ceilings, and heavy gauge mesh directly impact the success of creating a complex and species-specific environment.

Sunning is a thermoregulatory behavior most commonly used by ring-tailed lemurs, ruffed lemurs, and sifaka (Figure 25.10). As such, indoor windows and outdoor access are important environmental variables for eliciting natural behaviors. Sunning behaviors are generally observed in early morning, late afternoon, or on cool days and include hanging, lying, or sitting "Buddha-style" facing the sun. Lemurs housed indoors may sun less than those housed outdoors, despite having exposure to natural light through a glass roof or windows (Shire 2012).

To encourage species-appropriate behaviors, behind-the-scenes enclosures should be similar in height and complexity to public viewing areas. Multiple enclosures or stalls are essential in off-exhibit areas for separating individuals, or species in mixed-species groupings, when necessary (Ferrie et al. 2013).

The timing of estrus can be unintentionally altered when artificial photoperiods, such as extended viewing hours at a zoo, result in lemurs living under longer or shorter than normal astronomical photoperiods (Van Horn 1975). In addition, studies have shown evidence of decreased activity and melatonin suppression when nocturnal prosimians are housed under blue light as opposed to red light (Fuller 2013).

Enclosures for social groups and multispecies groups should contain multiple feeding stations, multiple locations for sleeping, and multiple outside access doors to prevent dominant animals from monopolizing food, perches, or cornering a subordinate animal. The addition of visual barriers allows subordinate animals the option of moving out of sight of dominant individuals and decreases tension and fighting (Williams and Bernstein 2012). Flexibility with design, including on- and off-exhibit areas, as well as indoor and outdoor access, is essential to accommodate groups of various sizes and to perform introductions and separations as needed.

Figure 25.10 Ring-tailed lemurs "sunning."

STEREOTYPIC BEHAVIOR

In a 2005 survey of 440 zoo-housed prosimians (representing *Lemur, Eulemur, Varecia, Otolemur, Microcebus, Cheirogaleus, Galago, Perodicticus, Nycticebus,* and *Loris*) in facilities accredited by the Association of Zoos and Aquariums (AZA), pacing was found to be the most prevalent form of stereotypic behavior. In the survey, 63% of the individuals were reported to engage in stereotypical behaviors (pacing, somersaulting), overgrooming, and self-injurious behavior while indoors only, as compared to 18.3%, who performed such behaviors outdoors only (Tarou et al. 2005). Analyses showed that individuals of the genus *Lemur* were significantly less likely to show stereotypic behavior than individuals of the genus *Varecia*, which the authors suggested was the result of the more arboreal and territorial nature of the latter.

Management and husbandry practices can inadvertently contribute to stereotypic behavior. In an attempt to provide zoo guests with a natural, aesthetically pleasing habitat, facilities may restrict the use of non-natural items, including enrichment and provisioned food, while animals are on exhibit (Shire 2012). The restrictive use of synthetic materials can inadvertently create an unstimulating environment for the animals and thus create a habitat in which increased incidences of stereotypic behavior are observed (Shire 2012). Anticipatory behaviors in primates, including increased vocalizations and pacing, have been correlated with delays in routine feeding (Gottlieb et al. 2013). While not documented with prosimians, further research would be of interest to assess the behavioral effects of variation in daily routines.

Similar to other nonhuman primates, there is a grave concern for the increasing numbers of lemurs that are in personal possession as pets. A recent survey on the species of primates accepted into AZA institutions from personal possession sources showed lemurs as being the third most common taxa of primates taken in by AZA facilities (Fenn 2015). Stereotypic behaviors, social skill deficits, self-injurious behaviors, and hyper-aggression toward conspecifics and/or caretakers are some of the abnormal behaviors that are commonly observed in lemurs that are former pets (AZA 2015). When incorporated into a professional behavioral management program, many of these behaviors can be reduced, given a high level of staff dedication, time, and patience.

IMPLICATIONS FOR RESEARCH

Understanding the behavioral implications of social separation as well as the presence of novel observers in novel environments is critical in assessing a prosimian's psychological well-being during participation in research. Data from an animal that is calm and confident differs from that of an animal that is anxious, pacing, or vocalizing (Poole 1997). There is evidence that good welfare is strongly linked to good science (Herrelko et al. 2012).

Application of the 3Rs (refinement, reduction, and replacement) is an underlying goal as advancements in research methodology take place that strive to improve animal welfare (Guhad 2005). Positive reinforcement training (PRT) is a valuable tool for research refinement and reduction. Prior to the start of a research project, lemurs can be conditioned to specific tasks that are required for data collection. Through successive approximations, PRT sessions teach the animal the criteria for specific research behaviors. In addition, PRT sessions establish a reinforcement history of positive associations with multiple aspects of data collection, including the process of being separated and entering a kennel, as well as exposure to novel equipment, environments, and people. Frustration, confusion, and anxiety can be successfully avoided by teaching the lemur what is expected prior to data collection. PRT can also serve as a tool to reduce the number of data collection sessions. Creating a comfortable and familiar environment for the animal has a direct impact on the quality, and quantity, of data that are obtained (Bloomsmith and Else 2005). An increase in consistently reliable data points translates to productive and valuable sessions, and may result in a reduction in the number of animals that are used. In addition, lemurs that are trained prior to data collection greatly reduce the amount of time that researchers need to devote to data collection (Ehmke, E. personal communication).

Observational data collection works best when prosimians are relaxed and not overly interested in the presence of observers. Ex situ free-ranging lemurs that are habituated to the presence of observers allow researchers to collect data from subjects that behave similarly in many ways to wild conspecifics. Success of home enclosure observations is species-specific. In general, diurnal and cathemeral lemurs may initially be curious in the presence of an observer, but will lose interest in a relatively short amount of time. Observations of nocturnal animals are undoubtedly more difficult, both logistically and because the species tend to be more elusive. Aye-aye in particular have been noted to increase locomotion and repetitive behaviors when an observer is in the room. Video recordings of animals offer a mechanism for obtaining accurate assessments of activity levels without any human presence-related influences on the data (Gibson and McKinney 2015).

IMPLICATIONS FOR WELFARE

Inclusion of animal-based behavioral indicators is becoming more prevalent in comprehensive assessments of animal welfare (Whitham and Wielebnowski 2013). In contrast to measuring what the animal is provided (resource-based assessments), animal-based assessments allow animal care staff and researchers to quantify and qualify the behavior of the animal as part of an overall welfare assessment. The inclusion of animal-based measures complements what animal care staff instinctively know: Animals are sentient, not all animals thrive in the same environment, and animal welfare must be assessed at the level of the individual.

A well-developed behavioral management program utilizes staff from multiple departments/ sections/groups to provide an integrated and continuous welfare assessment of the animals in their care. The individual components that fall under the behavioral management umbrella, including enrichment, positive reinforcement training, facilities and enclosure design, positive staff–animal interactions, and behavioral monitoring, are interlinked in their contribution to animal welfare and are not always defined by a single department/section/group. For example, animal care staff can teach the animals to calmly participate in their own health care, veterinary technicians understand the behavioral effects and consequences of separations and reintroductions, research serves as a form of enrichment when animals are given access to cognitively and physically challenging tasks, and veterinarians work to create positive animal associations to decrease avoidance and stress. The result of teamwork and collaboration is the creation of a culture that prioritizes the individual animal's experience and understands the holistic approach a behavioral management program contributes to the achievement of optimal animal welfare.

There is a growing understanding of the relationship between behavioral management and animal welfare through all life stages of an individual. From the time an infant is born, through the adult years, and then as a geriatric animal, a continuous assessment of the physical and psychological needs of the animals in our care is required to ensure an appropriate environment in which the individual can thrive. Behavioral management plays a role in proactively managing the increasing population of geriatric prosimians as accommodations are made to provide a high quality of life for elderly animals (Williams, C. personal communication). For example, arthritis can make it difficult for a geriatric lemur to jump to high ledges or through shift doors. Strategically placed ramps, ladders, and shelving allow for movement-at-will by a geriatric individual. Temporary separations may be necessary at feeding time if an elderly animal is being displaced or is eating slower than the rest of the group. Proactive quality of life discussions provide a reference point to assist decisions regarding the nutritional, physical, and psychological needs of an aging animal.

EXPERT RECOMMENDATIONS

Environmental enrichment and behavioral management are sometimes used interchangeably, as are psychological well-being and welfare (Coleman et al. 2012), with relatively little confusion. It is not as clear that "functionally appropriate captive environment" is interchangeable with "behaviorally appropriate environment." Specifically, do functionally appropriate captive environments only include those enclosures with expansive, natural habitats large enough to provide sufficient opportunities to range, form subgroups, and travel (NIH 2013)? The opinion of this author is that the term "functionally appropriate captive environment" must be used in the context of species-specific behavior, as additional resources, such as expansive space, do not guarantee improved welfare (Hosey 2005). At the same time, synthetic materials, such as plastic crates and PVC poles, can complement the complexity and functionality of an enclosure. For these reasons, it is believed that the premise for using the term "functionally appropriate" is reflected in current practices in the captive housing community by creating environments that promote species-specific social organization and behavior by utilizing all available resources, natural and synthetic. Animal welfare can then be measured with continuous assessments of habitat effectiveness using a variety of qualitative and quantitative behavioral measures.

CONCLUSIONS

The evolution of behavioral management strategies for nonhuman primates has been well documented for species that participate in "high-profile" research projects. While the documentation of prosimian behavioral management programs is relatively sparse in the scientific literature, there is an increasing number of publications that document the benefits of prosimian enrichment, positive reinforcement training, and staff–animal interactions. Within the last 5 years, this author has seen an increase, not only in the quantity, but also in the complexity, of voluntary behaviors that have been conditioned with prosimians to assist with husbandry and research behaviors. The advancement of prosimian behavioral management depends on continued opportunities for the exchange of information, as well as publications of quantitative and qualitative research.

ACKNOWLEDGMENTS

I express much gratitude for detailed editing and content advice to Dr. Cathy Williams, Dr. Erin Ehmke, Britt Keith, and Andrea Katz. I thank Christie Eddie, Gina Ferrie, Alison Grand, and Dawn Neptune for graciously reviewing and commenting on the chapter. I also thank Tracy Fenn and Dr. Ken Glander for additional help and comments. All photos should be credited to David Haring, Duke Lemur Center. For endless and unlimited support, I lovingly thank my husband, Greg.

REFERENCES

Ancrenaz, M., I. Lackman-Ancrenaz, and N. Mundy. 1994. Field observations of aye-aye (*Daubentonia madagascariensis*) in Madagascar. *Folia Primatologica* 62:22–36.

AZA. 2011. *Prosimian TAG Mixed-Species Exhibit Manual*. Silver Spring, MD: Association of Zoos and Aquariums.

AZA. 2015. Personal possession of non-human primates. https://www.aza.org/assets/2332/personal_possession_of_non-human_primates_7212015.pdf.

Barton, R. 1985. Grooming site preferences in primates and their functional implications. *International Journal of Primatology* 6 (5):519–532.

Bloomsmith, M. A. and J. G. Else. 2005. Behavioral management of chimpanzees in biomedical research facilities: The state of the science. *ILAR Journal* 46 (2):192–201.

Bloomsmith, M. A., M. J. Marr, and T. L. Maple. 2007. Addressing nonhuman primate behavioral problems through the application of operant conditioning: Is the human treatment approach a useful model? *Applied Animal Behaviour Science* 102 (3–4):205–222. doi:10.1016/j.applanim.2006.05.028.

Bons, N., F. Rieger, D. Prudhomme, et al. 2006. *Microcebus murinus*: A useful primate model for human cerebral aging and Alzheimer's disease? *Genes, Brain and Behavior* 5 (2):120–30. doi:10.1111/j.1601-183X.2005.00149.x.

Bray, J., C. Krupenye, and B. Hare. 2014. Ring-tailed lemurs (*Lemur catta*) exploit information about what others can see but not what they can hear. *Animal Cognition* 17 (3):735–744. doi:10.1007/s10071-013-0705-0.

Britt, A. 1998. Encouraging natural feeding behavior in captive-bred black and white ruffed lemurs (*Varecia variegata variegata*). *Zoo Biology* 17(5):379–392.

Carrasco, L., M. Colell, M. Calvo, et al. 2009. Benefits of training/playing therapy in a group of captive lowland gorillas (*Gorilla gorilla gorilla*). *Animal Welfare* 18:8–19.

Clark, F. E. and V. A. Melfi. 2012. Environmental enrichment for a mixed-species nocturnal mammal exhibit. *Zoo Biology* 31 (4):397–413.

Coleman, K. 2012. Individual differences in temperament and behavioral management practices for nonhuman primates. *Applied Animal Behaviour Science* 137 (3–4):106–113. doi:10.1016/j.applanim.2011.08.002.

Coleman, K., M. A. Bloomsmith, C. M. Crockett, et al. 2012. Behavioral management, enrichment, and psychological well-being of laboratory nonhuman primates, Chapter 6. In *Nonhuman Primates in Biomedical Research: Biology and Management*, edited by Abee, C. R., K. Mansfield, S. Tardif, et al., 149–176. London: Elsevier Inc. doi:10.1016/b978-0-12-381365-7.00006-6.

Coleman, K. and A. Maier. 2010. The use of positive reinforcement training to reduce stereotypic behavior in rhesus macaques. *Applied Animal Behaviour Science* 124 (3–4):142–148. doi:10.1016/j.applanim.2010.02.008.

Craig, J. and C. Reed. 2003. Diet-based enrichment ideas for small primates. *International Zoo News* 16–20.

Curtis, D. J. 2006. Cathemerality in lemurs. In *Lemurs: Ecology and Adaptation*, edited by Gould, L. and M. L. Sauther, 133–157. New York: Springer.

Demes, B., T. M. Franz, and K. J. Carlson. 2005. External forces on the limbs of jumping lemurs at takeoff and landing. *American Journal of Physical Anthropology* 128 (2):348–358. doi:10.1002/ajpa.20043.

Dishman, D. L., D. M. Thomson, and N. J. Karnovsky. 2009. Does simple feeding enrichment raise activity levels of captive ring-tailed lemurs (*Lemur catta*)? *Applied Animal Behaviour Science* 116 (1):88–95. doi:10.1016/j.applanim.2008.06.012.

Doss, S. and M. H. Dye. 2015. Positive reinforcement training in lemur species. *Laboratory Animal Science Professional* 3 (2):16–20.

Drea, C. M. and E. S. Scordato. 2008. Olfactory communication in the ringtailed lemur (*Lemur catta*): Form and function of multimodal signals. In *Chemical Signals in Vertebrates 11*, edited by Hurst, J. L., R. J. Beynon, S. C. Roberts et al., 91–102. New York: Springer.

Dröscher, I. and P. M. Kappeler. 2014. Maintenance of familiarity and social bonding via communal latrine use in a solitary primate (*Lepilemur leucopus*). *Behavioral Ecology and Sociobiology* 68 (12):2043–2058. doi:10.1007/s00265-014-1810-z.

Drucker, C. B., T. Baghdoyan, and E. M. Brannon. 2016. Implicit sequence learning in ring-tailed lemurs (*Lemur catta*). *Journal of the Experimental Analysis of Behavior* 105(1):123–136. doi:10.1002/jeab.180.

Dye, M. H. 2011a. DLC External Scientific Advisory Board Presentation.

Dye, M. H. 2011b. *Voluntary Infant Removal of Coquerel's Sifaka (Propithecus coquereli)*. Denver, CO: Animal Behavior Management Alliance.

Dye, M. H. 2013. DLC External Scientific Advisory Board Presentation.

Dye, M. H. 2015. Opportunities for expanding the enrichment process. *Animal Keepers' Forum* 42 (11&12):391–392.

Ellison, E. and L. Laverick. 2015. Looking at lemurs: Confirming pregnancies in ring-tailed lemurs (*Lemur catta*) via voluntary ultrasound. *Animal Keepers' Forum* 42 (11&12):383–384.

Faber, F. 2012. Multi-layered enrichment for mouse lemurs. *The Shape of Enrichment* 21 (4):1.

Ferrie, G. M., T. Bettinger, and C. Kuhar. 2013. *Eulemur Care Manual.* Silver Spring, MD: Association of Zoos and Aquariums.

Fistarol, L., D. Grassi, and C. Spiezio, 2008. The influence of music on behavior: A comparative study of two species of non-human primates in captivity. *Folia Primatologica* 79(5):328–329.

Fitch-Snyder, H. and H. Schulze. 2001. *Management of Lorises in Captivity. A Husbandry Manual for Asian Lorisines.* San Diego, CA: Zoological Society of San Diego, Center for Reproduction of Endangered Species (CRES).

Flanagan, C. 2012. You want to stick what? where?: Redirecting focus to facilitate injection training of mongoose lemurs. *Animal Keepers' Forum* 39 (9):428.

Franck-McCauley, J. 2015. Creation of a mixed-species group of red ruffed (*Varecia rubra*) and ring-tailed lemurs (*Lemur catta*). *Animal Keepers' Forum* 42 (11&12):332–333.

Frederick, C. and D. Fernandes. 1996. Behavioral changes in pottos (*Perodicticus potto*): Effects of naturalizing an exhibit. *International Journal of Primatology* 17(3):389–399.

Fuller, G. 2013. The night shift: Lighting and nocturnal strepsirrhine care in zoos. Doctor of Philosophy, Department of Biology, Case Western University.

Gibson, D. 2015. The night is alive with nocturnals. How about at your zoo? *Animal Keepers' Forum* 42 (11&12):357–360.

Gibson, D. and J. McKinney. 2015. Managing aye-aye (*Daubentonia madagascariensis*) on a natural photoperiod (NPP). *Animal Keepers' Forum* 42 (11&12):370–374.

Gottlieb, D. H., K. Coleman, and B. McCowan. 2013. The effects of predictability in daily husbandry routines on captive rhesus macaques (*Macaca mulatta*). *Applied Animal Behaviour Science* 143 (2–4):117–127. doi:10.1016/j.applanim.2012.10.010.

Granatosky, M. C., C. E. Miller, D. M. Boyer, et al. 2014. Lumbar vertebral morphology of flying, gliding, and suspensory mammals: Implications for the locomotor behavior of the subfossil lemurs *Palaeopropithecus* and *Babakotia*. *Journal of Human Evolution* 75:40–52. doi:10.1016/j.jhevol.2014.06.011.

Granatosky, M. C., C. H. Tripp, and D. Schmitt. 2016. Gait kinetics of above- and below-branch quadrupedal locomotion in lemurid primates. *Journal of Experimental Biology* 219 (Pt 1):53–63. doi:10.1242/jeb.120840.

Gray, A. E., Wirdateti, and K. A. I. Nekaris. 2015. Trialling exudate-based enrichment efforts to improve the welfare of rescued slow lorises *Nycticebus* spp. *Endangered Species Research* 27 (1):21–29. doi:10.3354/esr00654.

Greene, L. K. and C. M. Drea. 2014. Love is in the air: Sociality and pair bondedness influence sifaka reproductive signalling. *Animal Behaviour* 88:147–156. doi:10.1016/j.anbehav.2013.11.019.

Guhad, F. 2005. Introduction to the 3R's (refinement, reduction and replacement). *Contemporary Topics* 44 (2):58–59.

Hanbury, D. B., M. B. Fontenot, L. E. Highfill, et al. 2009. Efficacy of auditory enrichment in a prosimian primate (*Otolemur garnettii*). *Lab Animal* 38(4):122.

Herrelko, E. S., S. J. Vick, and H. M. Buchanan-Smith. 2012. Cognitive research in zoo-housed chimpanzees: Influence of personality and impact on welfare. *Journal of the American Association for Laboratory Animal* 74 (9):828–840. doi:10.1002/ajp.22036.

Hosey, G. R. 2005. How does the zoo environment affect the behaviour of captive primates? *Applied Animal Behaviour Science* 90 (2):107–129. doi:10.1016/j.applanim.2004.08.015.

Huber, H. F. and K. P. Lewis. 2011. An assessment of gum-based environmental enrichment for captive gummivorous primates. *Zoo Biology* 30 (1):71–78. doi:10.1002/zoo.20321.

Inoue-Nakamura, N. 1997. Mirror self-recognition in nonhuman primates: A phylogenetic approach. *Japenese Psychological Research* 39 (3):266–275.

Izard, M. K. 2006. Nursery-reared prosimian primates. In *Nursery Rearing of Nonhuman Primates in the 21st Century*, edited by Sackett, G. P., G. Ruppenthal, and K. Elias, 111–119. New York: Springer.

Jolly, A. 1967. *Lemur Behavior: A Madagascar Field Study.* Chicago, IL: University of Chicago Press.

Jolly, A., S. Caless, S. Cavigelli, et al. 2000. Infant killing, wounding and predation in eulemur and lemur. *International Journal of Primatology* 21 (1):21–40. doi:10.1023/a:1005467411880.

Joly, M., S. Ammersdorfer, D. Schmidtke, et al. 2014. Touchscreen-based cognitive tasks reveal age-related impairment in a primate aging model, the grey mouse lemur (*Microcebus murinus*). *PLoS One* 9 (10):e109393. doi:10.1371/journal.pone.0109393.

Katz, A. S. 1980. Management techniques to reduce perinatal loss in a lemur colony. *American Association of Zoological Parks and Aquariums Regional Proceedings* 137–140.

Katz, A. S. 2014. Housing and temperature management for pregnant lemurs and dams with new infants (*Lemur, Eulemur, Varecia, Propithecus*). Duke Lemur Center.

Kellerman, L. and M. H. Dye. 2015. PTAG behavioral husbandry advisory committee: How can we help? *Animal Keepers' Forum* 42 (11&12):388–390.

Kerridge, F. J. 2005. Environmental enrichment to address behavioral differences between wild and captive black-and-white ruffed lemurs (*Varecia variegata*). *American Journal of Primatology* 66 (1):71–84. doi:10.1002/ajp.20128.

Kulahci, I. G., D. I. Rubenstein, and A. A. Ghazanfar. 2015. Lemurs groom-at-a-distance through vocal networks. *Animal Behaviour* 110:179–186. doi:10.1016/j.anbehav.2015.09.016.

Lafleur, M., M. Sauther, F. Cuozzo, et al. 2014. Cathemerality in wild ring-tailed lemurs (*Lemur catta*) in the spiny forest of Tsimanampetsotsa National Park: Camera trap data and preliminary behavioral observations. *Primates* 55(2): 207–217. doi:10.1007/s10329-013-0391-1.

Lambeth, S. P. 2006. Positive reinforcement training affects hematologic and serum chemistry values in captive chimpanzees (*Pan troglodytes*). *American Journal of Primatology* 68:245–256. doi:10.1002/2014810.1002/ajp.

Laule, G. E., M. A. Bloomsmith, and S. J. Schapiro. 2003. The use of positive reinforcement training techniques to enhance the care, management, and welfare of primates in the laboratory. *Journal of Applied Animal Welfare Science* 6 (3):163–73. doi:10.1207/S15327604JAWS0603_02.

Lutz, C. K. and M. A. Novak. 2005. Environmental enrichment for nonhuman primates: Theory and application. *ILAR Journal* 46 (2):178–191.

Maloney, M. A., S. T. Meiers, J. White, et al. 2006. Effects of three food enrichment items on the behavior of black lemurs (*Eulemur macaco macaco*) and ringtail lemurs (*Lemur catta*) at the Henson Robinson Zoo, Springfield, Illinois. *Journal of Applied Animal Welfare Science* 9 (2):111–27. doi:10.1207/s15327604jaws0902_2.

Manna D., M. Rodeano, and E. Ferrero. 2007. A lemur mixed exhibit at Parco Zoo Punta Verde, Italy. *International Zoo News* 54 (8):452–457.

Marechal, L., E. Genty, and J. J. Roeder. 2010. Recognition of faces of known individuals in two lemur species (*Eulemur fulvus and E. macaco*). *Animal Behaviour* 79 (5):1157–1163. doi:10.1016/j.anbehav.2010.02.022.

McMillan, J. L., J. E. Perlman, A. Galvan, et al. 2014. Refining the pole-and-collar method of restraint: Emphasizing the use of positive training techniques with rhesus macaques (*Macaca mulatta*). *Journal of the American Association for Laboratory Animal Science* 53 (1):61–68.

Melfi, V. 2005. The appliance of science to zoo-housed primates. *Applied Animal Behaviour Science* 90 (2):97–106. doi:10.1016/j.applanim.2004.08.017.

Mellen, J. and M. MacPhee. 2001. Philosophy of environmental enrichment: Past, present and future. *Zoo Biology* 20:211–226.

Merritt, D. J., E. L. Maclean, J. C. Crawford, et al. 2011. Numerical rule-learning in ring-tailed lemurs (*Lemur catta*). *Frontiers in Psychology* 2:23. doi:10.3389/fpsyg.2011.00023.

Moore, R. S., Wihermanto, and K. A. I. Nekaris. 2014. Compassionate conservation, rehabilitation and translocation of Indonesian slow lorises. *Endangered Species Research* 26 (2):93–102. doi:10.3354/esr00620.

NIH. 2013. Council of Councils Working Group on the Use of Chimpanzees in NIH-Supported Research. http://dpcpsi.nih.gov/council/pdf/FNL_Report_WG_Chimpanzees.pdf.

Nowak, R. M. 1999. *Walker's Mammals of the World*, 6th edition, Vol. 1. Baltimore, MD: John Hopkins University Press.

Perlman, J. E., M. A. Bloomsmith, M. A. Whittaker, et al. 2012. Implementing positive reinforcement animal training programs at primate laboratories. *Applied Animal Behaviour Science* 137 (3–4):114–126. doi:10.1016/j.applanim.2011.11.003.

Perret, M. and F. Aujard. 2001. Regulation by photoperiod of seasonal changes in body mass and reproductive function in gray mouse lemurs (*Microcebus murinus*): Differential responses by sex. *International Journal of Primatology* 22 (1):5–24. doi:10.1023/A:1026457813626.

Picq, J. L., M. Dhenain, J.-L. Michot, et al. 1999. A comparison of old and young adult mouse lemurs, *Microcebus murinus*, on two memory tasks. *Folia Primatologica* 70 (4):218.

Poole, T. 1997. Happy animals make good science. *Lab Animal* 31:116–124.

PTAG. 2014. Ring-tailed Lemur (*Lemur catta*) Seasonal Aggression and Introduction Management: Tips and Suggestions.

Quirke, T. and R. M. O' Riordan. 2011. The effect of different types of enrichment on the behaviour of cheetahs (*Acinonyx jubatus*) in captivity. *Applied Animal Behaviour Science* 133 (1–2):87–94. doi:10.1016/j.applanim.2011.05.004.

Ramirez, K. 2013. *Cognitive Connections: Dolphins, Dogs, and Us!*. Chicago, IL: Chicago Humanities Festival.

Ratajszczak, R. 1998. Taxonomy, distribution and status of the lesser slow loris *Nycticebus pygmaeus* and their implications for captive management. *Folia Primatologica* 69 (1):171–174.

Reddy, R. B., E. L. MacLean, A. A. Sandel, et al. 2015. Social inhibitory control in five lemur species. *Primates* 56 (3):241–52. doi:10.1007/s10329-015-0467-1.

Robbins, L. and S. W. Margulis. 2014. The effects of auditory enrichment on gorillas. *Zoo Biology* 33 (3): 197–203. doi:10.1002/zoo.21127.

Rosati, A. G., K. Rodriguez, and B. Hare. 2014. The ecology of spatial memory in four lemur species. *Animal Cognition* 17 (4):947–961. doi:10.1007/s10071-014-0727-2.

Santini, L., D. Rojas, and G. Donati. 2015. Evolving through day and night: Origin and diversification of activity pattern in modern primates. *Behavioral Ecology* 26 (3):789–796. doi:10.1093/beheco/arv012.

Santos, L. R., J. L. Barnes, and N. Mahajan. 2005a. Expectations about numerical events in four lemur species (*Eulemur fulvus, Eulemur mongoz, Lemur catta and Varecia rubra*). *Animal Cognition* 8 (4):253–262. doi:10.1007/s10071-005-0252-4.

Santos, L. R., N. Mahajan, and J. L. Barnes. 2005b. How prosimian primates represent tools: Experiments with two lemur species (*Eulemur fulvus* and *Lemur catta*). *Journal of Comparative Psychology* 119 (4):394–403. doi:10.1037/0735-7036.119.4.394.

Schapiro, S. J., M. A. Bloomsmith, and G. E. Laule. 2003. Positive reinforcement training as a technique to alter nonhuman primate behavior: Quantitative assessments of effectiveness. *Journal of Applied Animal Welfare Science* 6 (3):175–87. doi:10.1207/S15327604JAWS0603_03.

Schulze, H. 1998. A "checklist" of possible items for prosimian husbandry manuals and research. *Folia Primatologica* 69:152–170.

Scordato, E. S. and C. M. Drea. 2007. Scents and sensibility: Information content of olfactory signals in the ringtailed lemur, *Lemur catta*. *Animal Behaviour* 73 (2):301–314. doi:10.1016/j.anbehav.2006.08.006.

Shire, T. 2012. Differences in behavior between captive and wild ring-tailed lemur (*Lemur catta*) populations: Implications for reintroduction and captive management. Master of Arts, Anthropology, Iowa State University.

Sommerfeld, R., M. Bauert, E. Hillmann, et al. 2006. Feeding enrichment by self-operated food boxes for white-fronted lemurs (*Eulemur fulvus albifrons*) in the Masoala exhibit of the Zurich Zoo. *Zoo Biology* 25 (2):145–154. doi:10.1002/zoo.20082.

Spiezio, C., F. Piva, B. Regaiolli, et al. 2015. Positive reinforcement training: A tool for care and management of captive vervet monkeys (*Chlorocebus aethiops*). *Animal Welfare* 24 (3):283–290. doi:10.7120/09627286.24.3.283.

Stierhof, M. 2015. Husbandry challenges associated with managing an individual with diabetes in a ring-tailed lemur (*Lemur catta*) troop. *Animal Keepers' Forum* 42 (11&12):394–394.

Tarou, L. R., M. A. Bloomsmith, and T. L. Maple. 2005. Survey of stereotypic behavior in prosimians. *American Journal of Primatology* 65 (2):181–96. doi:10.1002/ajp.20107.

Terranova, C. J. 1996. Variation in the leaping of lemurs. *American Journal of Primatology* 40 (2):145–165. doi:10.1002/(SICI)1098-2345(1996)40:2<145:AID-AJP3>3.0.CO;2-Y.

Trouche, S. G., T. Maurice, S. Rouland, et al. 2010. The three-panel runway maze adapted to Microcebus murinus reveals age-related differences in memory and perseverance performances. *Neurobiol Learn Mem* 94 (1):100–6. doi:10.1016/j.nlm.2010.04.006.

Valenta, K., K. A. Brown, R. R. Rafaliarison, et al. 2015. Sensory integration during foraging: The importance of fruit hardness, colour, and odour to brown lemurs. *Behavioral Ecology and Sociobiology* 69 (11):1855–1865. doi:10.1007/s00265-015-1998-6.

Van Horn, R. N. 1975. Primate breeding season: Photoperiodic regulation in captive *Lemur catta*. *Folia Primatologica* 24 (2–3):203–220.

Verdolin, JL. 2013. Are shy individuals less behaviorally variable? Insights from a captive population of mouse lemurs. *Primates* 54:309–314. doi:10.5061/dryad.5m083.

Wall, B. 2015. When tactile training comes in handy: Helping a red-ruffed lemur (*Varecia rubra*) recover from four metacarpal fractures. *Animal Keepers' Forum* 42 (11 & 12):378–382.

Whipple, M. A. 2015. Using operant conditioning to manage reproduction in Coquerel's sifaka (*Propithecus coquereli*). *Animal Keepers' Forum* 42 (11&12):385–387.

Whitham, J. C. and N. Wielebnowski. 2013. New directions for zoo animal welfare science. *Applied Animal Behaviour Science* 147 (3–4):247–260. doi:10.1016/j.applanim.2013.02.004.

Williams, C. V. 2015. Prosimians. In *Fowler's Zoo and Wild Animal Medicine*, edited by Eric Miller, R. and Fowler, M. E. 291–300. St. Louis, MO: Elsevier Saunders.

Williams, L. E. and I. S. Bernstein. 2012. Study of nonhuman primate social behavior, Chapter 5. In *Nonhuman Primates in Biomedical Research: Biology and Management*, edited by Abee, C. R., K. Mansfield, S. Tardif, et al., 131–147. London: Elsevier Inc. doi:10.1016/b978-0-12-381365-7.00005-4.

Wolfensohn, S. and P. Honess. 2005. Psychological well-being. In *Handbook of Primate Husbandry and Welfare*, edited by Wolfensohn, S. and P. Honess, 99–114. Oxford, UK: Blackwell Publishing Ltd.

Products, Equipment, Techniques, and Services

PART V

Resources, Equipment, Techniques, and Services

CHAPTER 26

Behavioral Management, Primate Jackets, and Related Equipment

Teresa Woodger
Lomir Biomedical Inc.

CONTENTS

Driven by early space exploration (NASA; Burgess and Dubbs 2007), fabric jackets have been used with laboratory nonhuman primates (NHPs) since the 1950s. Early fabric jackets were often part of the physical restraint system, although in their day, these fabric jackets were considered refinements compared to earlier models, since they significantly reduced the degree of physical restraint and the associated stress. Almost 60 years later, jackets for NHPs have evolved considerably, in terms of materials and fabrication, and can now be used with numerous biological models and scientific applications. The use and design of jackets for NHPs has adapted to, and kept pace with, the ongoing changes in research with NHPs, in relation to housing, conduct, and scope of studies, and appreciation of the need for behavioral management. With the development and adoption of modern behavioral management practices, especially considerations related to acclimation to the jackets, they are no longer simply restraints, but are now protective garments that allow animals to rapidly return to normal physiological states following their application. This helps promote normal behavior patterns, which should be associated with reductions in stress, allowing for more consistent qualitative and quantitative data, and therefore more reliable and valid animal models.

During the 1960s and 1970s, Alice King Chatham continued the development and commercialization of jackets and associated swivel–tether systems, primarily for the administration of test materials via intravenous infusion, with the jacket and tether serving as protection for surgical sites and exteriorized indwelling catheters (Chatham 1985; see Figures 26.1 and 26.2). Alternatives to cloth jackets, at the time, included individual plaster casts, and leather or metal harnesses used with swivel–tether systems (Pearce et al. 1989). Jacket and swivel–tether systems are still in widespread use today; however, the restrictions associated with tethering means that many of these applications are associated with singly housed animals (Geisbert et al. 2010). Current tether jackets for NHPs include features that address the animal's comfort, and adaptations that facilitate the ease of use and maintenance of the system by staff (Kelly et al. 2014; see Figure 26.3).

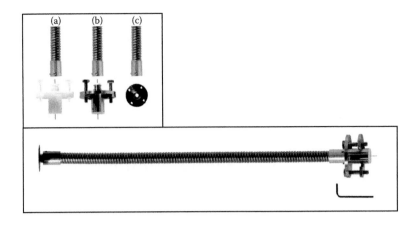

Figure 26.1   Large animal tether set. Swivel (a), large animal tether (b), animal tether end plate (c).

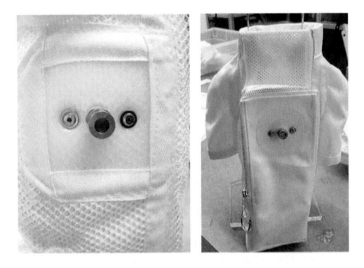

Figure 26.2   Jackets with Animal Tether Endplate 1 (ATE1) in place. Patch on left and patch on back on pocket.

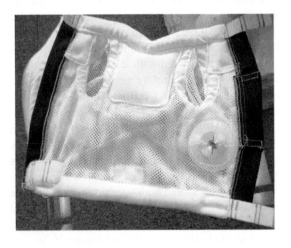

Figure 26.3   Tether jacket with comfort features.

Application of jackets for NHPs began to diversify during the 1990s, driven, in part, by the availability of ever smaller, but much more powerful, biologgers (telemetry devices and data-loggers). The development of biologging systems for the collection and processing of physiological parameters goes back to early space exploration, and some equipment, like Holter monitors, go back as far as the late 1940s. Early systems required data collection via hardwires, which for NHPs were generally passed through a protective jacket, tether, and electronic swivel set-up (Vogel 1991). The transfer of recorded data required physical downloading from the device with associated interventions and restraint of the animal. The recent development of smaller, more powerful biologgers occurred in concert with the use of (1) complex mathematical programs that were capable of collecting and analyzing vast amounts of physiological data and (2) near-field communication systems, such as Bluetooth, Wi-Fi, and similar applications. These applications enable data collection (transmission to a remote site) directly "from" the NHPs, minimizing the need for intervention or additional handling of the animals. Adoption of these rapidly evolving technologies has been facilitated by the emergence and acceptance of behavioral management programs with NHPs, in particular, those that involve the application of positive reinforcement training (PRT) techniques (Laule et al. 2013; Graham et al. 2012; Graham 2017). Current training protocols have resulted in refined acclimation programs for jacketed NHPs that enhance animal welfare; jacketed primates can remain in their social housing situations, moving freely while rapidly returning to their prejacketed behavioral and physiological states (Derakhchan et al. 2014; Field et al. 2015).

During the 1990s, portable intravenous infusion pumps were adopted for research applications, enabling the development of an ambulatory model for intravenous infusion. This model allowed for completely freely moving animals, with opportunities for group and social housing, particularly important refinements for NHP models. The ambulatory infusion pumps that were selected were small and lightweight enough to be carried by an NHP and were secured in a pocket placed like a backpack attached to the protective jacket (Turner et al. 2011; see Figure 26.4). Access from the pump to the indwelling catheter is contained inside the pocket and jacket, reducing the possibility of interference. Additional refinements to the ambulatory infusion model include the use of vascular access ports (VAPs), with the connection to the pump made via a Huber needle. Many infusion pumps in use today feature an interface with computer programs and (1) provide confirmation of accurate dosing, (2) allow calculation of dosage based on body weights and preset volumes, and (3) include a range of failsafe features (some systems will send notifications of any changes outside normal or preset parameters). These systems comply with the requirements of good laboratory practice (GLP) and Code of Federal Regulations (CFR) part 11, confirming dosing regimens and further refining the technique.

Figure 26.4 Ambulatory jacket with comfort features.

The types of biologgers that are particularly applicable for use with NHPs include external telemetry systems primarily for cardiovascular and related data (Ingram-Ross et al. 2012); actigraph systems to collect and transmit data relating to activity monitoring, diurnal changes, and event logging; and equipment to monitor and record movement via a static video camera. NHP jackets have been developed for these different biologgers to provide a protective garment that secures devices, attachments, and sensors in place, while protecting the subject and the integrity of the data. These uses led to the creation of undershirts for NHPs; the tight-fitting undershirt holds the leads and attachments, reduces requirements for handling and bandaging, and allows the animals to behave naturally while wearing the garments (Figure 26.5).

The most recent applications for jackets are in areas related to video tracking technology. Video tracking, as one might expect, records animals' movements and events via a static camera. Data are then sent to a remote device. Recordings can be used to assess movement, gait, posture, distance traveled, velocity, and other parameters and behaviors. Innovative strategies to optimize the use of jackets with video tracking systems include using different colored jackets for each NHP, enabling the tracking software to identify individual animals within a group (Figure 26.6). Other tracking software systems can utilize Pearl Markers attached to the jacket at specific points

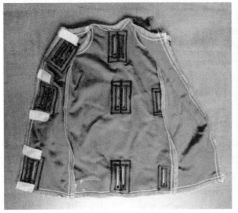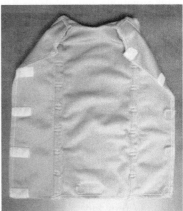

Figure 26.5 Data Sciences International Jacketed External Telemetry & EMKA Technologies undershirts.

Figure 26.6 Colored jackets for individual identification. Color primate jackets.

on the body to track and analyze preset motion parameters (Figures 26.7 and 26.8). Again, these types of tracking systems and jackets allow recordings to be made in the animal's normal social and physical environments.

Many of these biologging devices are regarded as noninvasive, with no requirement for major surgery; however, a few systems involve minimal surgery to record additional physical parameters. There are certain biomodels where NHP jackets can be designed to incorporate more than one of the available technologies, maximizing the scientific data that can be gained at any one time. The majority of biologgers currently in use capture data in real time, even though protocols and regulatory requirements may only require reporting at set time points. Additional data are thus retained and available for further study, potentially reducing the need to involve additional animals. Applications utilizing biologgers will continue to evolve, as circuitry is further

Figure 26.7   Pearl Markers.

Figure 26.8   A full suit with long sleeves and a reinforced chest area.

miniaturized and microprocessors continue to shift toward lower energy consumption, extending battery life in portable devices.

The emergence in the early 2000s of the field of safety pharmacology and the regulatory requirements that effects on vital organ systems of new pharmaceuticals be assessed as part of preclinical development (ICH 2000; Safety Pharmacology Society 2001) necessitated the adoption of new biomodels, including external telemetry, among other techniques. A fundamental factor in the success of these models is the acceptance of wearing the jacket and equipment by the NHP. This promotes normal behavioral and physiological states, which allow for the collection of robust, reliable, and valid data. These technologies are especially valuable when they are accompanied by successful behavioral management and training programs focused on issues relevant to animals in jackets.

In addition to serving as protection for apparatus and equipment, specialized jackets have also been designed to be used on their own as protective covering for surgical or treatment sites. Such applications can help to reduce unnecessary interventions and handling. Individual designs allow options, such as "pants" and long sleeves (Figure 26.8) to protect the limbs and to address variations in the morphology and preferences of individual NHPs.

The benefits of acclimating NHPs to equipment and training them to voluntarily participate in research procedures have long been recognized. Behavioral management and, especially, cooperative PRT programs are now widely utilized, enabling researchers to develop a wide range of reliable and valid biomodels involving jacketed NHPs. In the revised *Guide for the Care and Use of Laboratory Animals*, considerable emphasis has been placed on training and behavioral management of laboratory animals. While there are no specific timelines for training and acclimation to jackets, the Association of Primate Veterinarians (APV) recognizes that acclimation periods should be provided and suggests 1–2 weeks as a target for investigators to consider for acclimation. There are currently few studies available that discuss strategies or protocols to acclimate NHPs to wearing jackets and/or equipment. It is recommended that acclimation protocols be established in collaboration with veterinarians, technical staff, behaviorists, and trainers. Once acclimation protocols have been established, detailed records must be maintained. Several chapters in this handbook deal with the processes and techniques involved with acclimation, desensitization, and positive reinforcement training (Graham 2017; Magden 2017).

DESIGN

Design of jackets for NHPs must take into consideration the species, the scientific application, housing conditions, and the duration of jacket use. Jackets must protect the wearer from interfering with surgical sites, devices, and equipment without harm to the subject. They should include features for comfort, such as soft fabric lining at key points and protective panels under zippers. Key points to consider when selecting jackets include ensuring that the jacket is the correct size and fit for each individual primate and establishing a strategy for acclimation to the jacket in advance of any protocol procedures. Jacket manufacturers supply them in various sizes that are matched to species and to specific weight ranges, and can confirm measurements to facilitate proper sizing (Figure 26.9). Investigators must also consider any other equipment or instruments that may be required for certain procedures or handling systems, such as the pole-and-collar, where the jackets need to have a short "V" neck to prevent the restraint collar from sitting too high on the neck (Figure 26.10). Primate jackets have been created that meet the needs of the individual species of primates that are commonly used in biomedical research, taking into account species-specific body shapes and morphology.

Figure 26.9 Primate jackets for different species (catalogue photo).

Figure 26.10 V-neck full body suit.

GENERAL FITTING AND MAINTENANCE OF JACKETS

Selecting jackets that are suitable for the species of NHP subjects, and that are the correct size for each individual animal, facilitates the acclimation process, as poorly fitting jackets can cause a number of problems during the adaptation phase and beyond. A jacket that fits correctly allows one human finger to fit comfortably inside the neck and two fingers to fit comfortably inside the sleeve, while ensuring that there are no other areas that are tight or that result in rubbing or chafing. Animals routinely have 1–3 jackets each to make certain that a clean jacket is always available. Jackets are routinely rotated and removed for cleaning. They are typically disinfected and then laundered and are checked for wear and tear prior to re-use. For training and acclimation to jackets that contain apparatus, researchers first use either a plain training jacket or a previously used jacket. As the training continues, the NHP is gradually acclimated to jackets containing sham apparatus of approximately the same size and weight of the device to be placed in the protective garment.

JACKET DESIGN AND FEATURES

The essential design of a basic jacket for laboratory NHPs has changed little over the years. However, the use of modern materials and fasteners, combined with adaptations for species and specific scientific models, provide researchers with refined protective garments that contribute to animal well-being and welfare. Modern fabrics include spandex and various elastane-based materials with outstanding stretch capabilities and smooth finishes that enhance comfort. Currently used fasteners include Velcro, which is available in multiple forms, including stretch, back-to-back, and One-Strap. Other available materials include ballistic canvas for equipment pockets, and fabrics used for sports uniforms that are selected for their strength and moisture-wicking properties.

The body of most primate jackets is sewn from a double layer of polyester mesh fabric, chosen for its durability and the fact that if it is ripped or torn, it does not come apart. Specialized jackets that include outer layers in less robust fabrics still retain the inner double layer of mesh to ensure the integrity of the finished product.

Jackets are available with a variety of adaptations that are based on the scientific application and experimental protocol for which they are needed. The following section highlights some of the major design features of primate jackets.

Jackets for swivel–tether systems feature a mounting point where the tether-end plate of the swivel–tether is fixed into the jacket so that it does not turn. The location of the mounting point may vary according to the surgical technique used and the location of the externalized indwelling catheter or vascular access port.

For models using ambulatory infusion pumps, jackets have a pump-specific pocket on the back, which is matched to the volume of the test article to be delivered, either in cassettes supplied with the pump, or for larger volumes, drug bags. There is a small opening on the inside of the pocket that accommodates the tubes between the animal and the pump. Some pump pockets may also include room for a control device. The pump and accessories can be secured inside the pocket with elastic or Velcro, or the entire device can be held in a protective padded pouch or sleeve within the pocket. In social housing situations, NHP jackets need extra features so that the animals are protected from interfering with their own jackets or those of their group mates. This is achieved, in part, by hiding pocket zippers under a fabric flap with Velcro fasteners, making it difficult for the NHPs to reach the zipper. Jacket zippers are typically secured using hooks on the zipper tabs that are clipped into solid "D" rings that are stitched adjacent to the ends of the zippers (Figure 26.11). If required, additional security can be achieved by using small padlocks at the zipper tabs. Extra padding and a soft lining can be added at various potential "hot spots" on the jackets, such as the underarm and upper chest area.

Jackets for use with telemetry systems are similar to those used for ambulatory pumps. Pocket dimensions are matched to the device and a somewhat larger opening is needed from inside the pocket to the animal to accommodate the leads (Figure 26.12). Depending on the telemetry device, there may be pouches inside the jacket for a blood pressure transponder or for a cuff-like device. Telemetry systems typically utilize a tight-fitting undershirt, in addition to the jacket, to secure ECG leads and other connectors. Undershirts are made from spandex and are fastened with Velcro, permitting maximum adjustability, while replacing the need for bandaging and/or tape to hold connectors. This considerably reduces the stress experienced by the animal.

Jackets for video and motion tracking have an outer layer of individually colored robust matte fabric derived from sports uniforms, so that software systems can read the different wavelengths. Individual NHPs can be individually identified and their data separated from the rest of the group. Jackets for motion tracking use Pearl Markers on a black matte base jacket to absorb light and create high contrast with the marker. Motion tracking jackets are typically made to order for specific investigations. NHPs are acclimated to the base jacket, after which the markers are attached at the required locations on the body. Oversized anchor points at specific locations on the jacket allow the

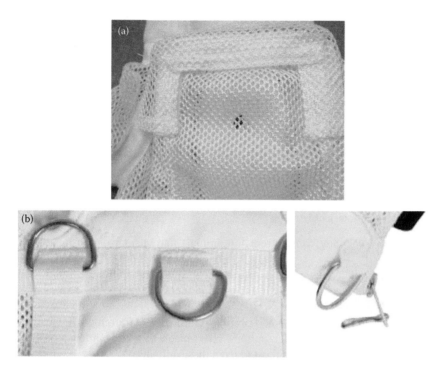

Figure 26.11 (a) Jacket with pocket and zipper protectors (b) "D" rings.

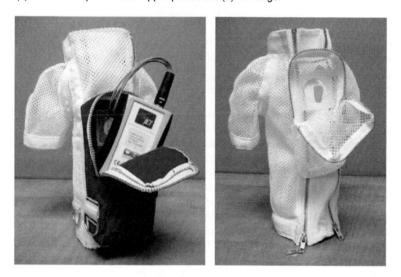

Figure 26.12 Data Sciences International Jacketed External Telemetry & EMKA Technologies jackets.

markers to be positioned in the exact places that facilitate optimal measurement of the location of the animal (Figure 26.10). As mentioned earlier, the use of a plain training jacket for the acclimation period is recommended, with Pearl Markers only added once the acclimation process is complete.

Protective and other custom jackets can be used in conjunction with specific equipment, protect surgical and application sites, in addition to addressing many other research-related needs. Long pants and sleeves provide a protective layer and can be made removable to be added to a base jacket as required (Figure 26.13). Further customizations available with jackets include special

Figure 26.13 Extras on pocket jackets.

reinforcements to prevent interference from the animals and custom dimensions for individual animals.

Jackets and their applications with NHPs are likely to undergo continued rapid evolution. As new textiles and wearable technologies are incorporated into biologging and related systems, further refinements in NHP biomodels are expected. The continued application of behavioral management techniques, especially acclimation and positive reinforcement training, will ensure that many of the available welfare benefits will accrue for NHPs in jackets, while simultaneously increasing the reliability and validity of the data collected from these animals.

REFERENCES

Association of Primate Veterinarians. 2013. Guidelines for jacket use for nonhuman primates. http://www.primatevets.org/Content/files/Public/education/NHP_Jacket_Use_Guidelines.pdf (Accessed February 1, 2016).

Burgess, C. and C. Dubbs. 2007. *Animals in Space: From Research Rockets to the Space Shuttle*. New York: Springer.

Chatham, A.K. 1985. Jacket- and-swivel tethering system. *Laboratory Animal* 14:29–33.

Derakhchan, K., R.W. Chui, D. Stevens, W. Gu, and H.M. Vargus. 2014. Detection of QTc interval prolongation using jacket telemetry in conscious non-human primates: Comparison with implanted telemetry. *British Journal of Pharmacology* 171(2):509–522.

Field, A.E., C.L. Jones, R. Kelly, S.T. Marko, S.J. Kern, and P.J. Rico. 2015. Measurement of fecal corticosterone metabolites as a predictor of the habituation of rhesus macaques (*Macaca mulatta*) to jacketing. *Journal of the American Association for Laboratory Animal Science* 54:59–65.

Geisbert, T.W., A.C.H. Lee, M. Robbins, J.B. Geisbert, A.N. Honko, V. Sood, J.C. Johnson, et al. 2010. Postexposure protection of non-human primates against a lethal Ebola virus challenge with RNA interference: A proof-of-concept study. *Lancet* 375:1896–1905.

Graham, M.L. 2017. Positive reinforcement training and research, Chapter 12. In Schapiro, S.J. (ed.) *Handbook of Primate Behavioral Management*. Boca Raton, FL: CRC Press. 187–200.

Graham, M.L., E.F. Rieke, L.A. Mutch, E.K. Zolondek, A.W. Faig, T.A. DuFour, J.W. Munson, J.A. Kittredge, and H.-J. Schuurman. 2012. Successful implementation of cooperative handling eliminates the need for restraint in a complex nonhuman primate disease model. *Journal of Medical Primatology* 41:89–106.

ICH. 2000. International Conference on Harmonisation of Technical Requirements for Registration of Pharmaceuticals for Human Use (ICH S7A guidance) http://www.ich.org/products/guidelines/safety/article/safety-guidelines.html (Accessed February 1, 2016).

Ingram-Ross, J.L., A.K. Curran, M. Miyamoto, J. Sheehan, G. Thomas, J. Verbeeck, E.J. de Waal, B. Verstynen, and M.K. Pugsley. 2012. Cardiorespiratory safety evaluation in non-human primates. *Journal of Pharmacological and Toxicological Methods* 66:114–124.

Kelly, R., A. Carlson, S.J. Kern, A. Field, S. Marko, E. Bailey, S. Norris, A. Honko, and P. Rico. 2014. Evaluation of the use of primate undershirts as a refinement practice for jacketed rhesus macaques (*Macaca mulatta*). *Journal of the American Association for Laboratory Animal Science* 53:267–272.

Laule, G.E., M.A. Bloomsmith, and S.J. Schapiro. 2013. The use of positive reinforcement training techniques to enhance the care, management, and welfare of primates in the laboratory. *Journal of Applied Animal Welfare Science*, 6(3):163–173.

Magden, E.R. 2017. Positive reinforcement training and health care, Chapter 13. In Schapiro, S.J. (ed.) *Handbook of Primate Behavioral Management*. Boca Raton, FL: CRC Press. 201–216.

NASA. 2009. http://history.nasa.gov/animals.html (Accessed February 1, 2016).

Pearce, P.C., M.J. Halsey, J.A. Ross, N.P. Luff, R.A. Bevilacqua, and C.J. Maclean. 1989. A method of remote physiological monitoring of a fully mobile primate in a single animal cage. *Laboratory Animals* 23:180–187.

Safety Pharmacology Society. 2001. https://www.safetypharmacology.org/index.asp (Accessed February 1, 2016)

Turner, P.V., C. Pekow, M.A. Vasbinder, and T. Brabb. 2011. Administration of substances to laboratory animals: Equipment considerations, vehicle selection, and solute preparation. *Journal of the American Association for Laboratory Animal Science* 50:614–627.

Vogel, A.P., G.P. Jaax, T.M. Tezak-Reid, S.I. Baskin, and J.L. Bartholomew. 1991. Ambulatory electrocardiography (Holter monitoring) in caged monkeys. *Laboratory Animals* 25:16–20.

CHAPTER 27

Nutrition, Feeding, and Behavioral Management

Carrie L. Schultz
LabDiet

CONTENTS

INTRODUCTION

Nonhuman primates (NHPs) play an important role in biomedical research. They serve as an animal model for the cause, prevention, and treatment of many different types of maladies, including cardiovascular disease, cancer, AIDS, Parkinson's disease, and Alzheimer's disease. The involvement of NHPs in laboratory investigations has demanded that close attention be paid to the diet of these animals in order to help maintain health, improve longevity, and to minimize the potential of diet as a confounding variable in scientific research.

While nutritional requirements for most production species are well established, comparatively few data exist to establish nutritional requirements for many laboratory species, including NHPs. The nutritional requirements for some species of NHPs are more comprehensive than for others. The majority of studies that focus on the nutrient requirements of NHPs have been done with macaques (*Macaca mulatta* and *M. fascicularis*). Limited information is available on the nutritional requirements of growing, pregnant, or lactating NHPs, and nutritional assumptions are typically based on information from the dietary requirements of humans. Care should be taken when feeding animals in these life stages to ensure proper nutrition during these times.

This chapter will summarize the known nutritional information concerning NHPs, as well as the use of dietary (feeding) enrichment to enhance the well-being of the animals. Examples of common, commercially available, natural ingredient diets will be provided, in addition to examples of dietary enrichment items that are commonly used as components of comprehensive behavioral management programs.

INGREDIENTS AND DIET FORMULATION

The feeding ecology of NHPs in the wild has been recorded by many, with a review of these studies referenced in the National Research Council's 2003 publication (NRC 2003). To promote the foraging and feeding behavior of NHPs, it would be ideal to provide the animals with a species-typical diet comprising appropriate proportions of fruits, gums, leaves, roots, seeds, and insects and other small prey items. However, in a laboratory setting, the provision of such a diverse array of foods (including live prey) can be problematic. It is difficult to identify the optimal variety and proportions of food that meet the animals' nutritional requirements. Feeding NHPs a "natural diet" requires temperature- and humidity-controlled facilities to prevent spoilage and is typically more expensive than feeding them a processed commercial diet. Therefore, to satisfy effectively the nutrient requirements of captive NHPs, while also accounting for many of the complexities associated with feeding an entirely fresh diet, processed commercial diets typically serve as the primary foodstuff provided to the animals. Virtually all captive primates are provided with fresh fruits and vegetables, seeds, and other "treats" as a regular supplement to their daily ration of chow/biscuits/pellets and/or as foraging enrichment.

Commercial diets typically comprise cereal grains, such as corn, wheat, soybean meal, and oats. Dietary sources of protein typically include soybean meal, corn gluten meal, and fish meal. Energy sources commonly used, in the form of lipids, are soy and corn oils and pork animal fat. Tables 27.1 and 27.2 illustrate the ingredient composition of regularly used diets for Old World and New World nonhuman primates, respectively.

Commercial diets for laboratory NHPs are typically closed, proprietary formulas, in which the inclusion rates of each raw material are not disclosed to the end user. However, within the industry, most feed suppliers are willing to work with individuals on components of the diet so that they can design the most appropriate control and experimental diets for their research-specific studies. Open-formula diets can also be used, but are less common with nonhuman primates than with rat and mouse models. With an open formula, the inclusion level of each raw material is provided, as

Table 27.1 Ingredients of Commonly Used Diets for Old World Nonhuman Primates

Diet	Manufacturer	Ingredients (Descending Order of Inclusion by as-is Weight)
5049	LabDiet[a]	Ground corn, dehulled soybean meal, ground soybean hulls, ground oats, corn gluten meal, ground wheat, porcine animal fat, dehydrated alfalfa meal, sucrose, dicalcium phosphate, monocalcium phosphate, dried whey, fish meal, calcium carbonate, brewers dried yeast, salt, L-ascorbyl-2-polyphosphate, pyridoxine hydrochloride, menadione, dimethylpyrimidinol bisulfite, cholecalciferol, DL-methionine, choline chloride, vitamin A acetate, folic acid, calcium pantothenate, ferrous sulfate, dl-α-tocopheryl acetate, vitamin B12 supplement, biotin, thiamin mononitrate, nicotinic acid, riboflavin, zinc oxide, copper sulfate, L-lysine, manganous oxide, ferrous carbonate, zinc sulfate, calcium iodate, cobalt carbonate, sodium selenite.
2050	Teklad[b]	Ground corn, soybean hulls, wheat middlings, dehulled soybean meal, corn gluten meal, corn gluten feed, ground wheat, dehydrated alfalfa meal, dried whey, fish meal, sucrose, dicalcium phosphate, soybean oil, porcine animal fat (preserved with BHA), calcium carbonate, iodized salt, brewers dried yeast, L-ascorbyl-2-polyphosphate, choline chloride, calcium propionate, kaolin, ferrous sulfate, vitamin E acetate, manganous oxide, niacin, menadione sodium bisulfite complex (source of vitamin K activity), zinc oxide, copper sulfate, calcium pantothenate, vitamin A acetate, folic acid, pyridoxine hydrochloride, thiamine mononitrate, riboflavin, vitamin D_3 supplement, cobalt carbonate, vitamin B_{12} supplement, ethylenediamine dihydriodide, biotin
OWM	SDS[c]	Wheat, wheatfeed, dehulled extracted toasted soya, maize, soya oil, whey powder, macrominerals, yeast, vitamins, micro minerals, amino acids.

[a] Purina LabDiet [subsidiary of Land O'Lakes Purina Feed LLC (St. Louis, MO)].
[b] Envigo (Madison, WI).
[c] Special Diet Services (England, UK).

Table 27.2 Ingredients of Commonly Used Diets for New World Nonhuman Primates

Diet	Manufacturer	Ingredients (Descending Order of Inclusion by as-is Weight)
5040	LabDiet[a]	Ground corn, wheat middlings, sucrose, acid casein, dried egg product, wheat germ, dehulled soybean meal, dried whey, corn gluten meal, ground wheat, soybean oil, dried beet pulp, brewers dried yeast, porcine animal fat, wheat bran, dehydrated alfalfa meal, calcium carbonate, fish meal, dicalcium phosphate, salt, folic acid, L-ascorbyl-2-polyphosphate, choline chloride, pyridoxine hydrochloride, menadione, dimethylpyrimidinol bisulfite, cholecalciferol, ferrous sulfate, calcium pantothenate, vitamin A acetate, *dl*-α-tocopheryl acetate, DL-methionine, biotin, thiamin mononitrate, nicotinic acid, vitamin B_{12} supplement, riboflavin, zinc oxide, copper sulfate, manganous oxide, ferrous carbonate, zinc sulfate, calcium iodate, cobalt carbonate, sodium selenite.
8794	Teklad[b]	Ground corn, ground wheat, wheat middlings, wheat germ, dehulled soybean meal, corn gluten meal, sucrose, soybean oil, dried beet pulp, porcine fat, egg product, dried whey, calcium carbonate, dehydrated alfalfa meal, fish meal, casein, dicalcium phosphate, iodized salt, L-ascorbyl-2-polyphosphate, choline chloride, calcium propionate, taurine, kaolin, vitamin E acetate, manganous oxide, zinc oxide, ferrous sulfate, niacin, menadione sodium bisulfite complex (source of vitamin K activity), copper sulfate, calcium pantothenate, folic acid, vitamin A acetate, thiamin mononitrate, pyridoxine hydrochloride, riboflavin, vitamin D_3 supplement, cobalt carbonate, ethylenediamine dihydriodide, vitamin B_{12} supplement, biotin.

[a] Purina LabDiet [subsidiary of Land O'Lakes Purina Feed LLC (St. Louis, MO)].
[b] Envigo (Madison, WI).

well as the identity of each ingredient. This allows for enhanced interpretation of the results within a study, but also supports the ability to replicate a diet based on a published study.

Diets are formulated using only one of two formulation methods within the laboratory animal industry—fixed formulations or managed formulations. Least cost formulation, a method of formulation that allows raw materials to be added or removed from the formula based on their market value, is not used by suppliers that focus on the laboratory animal industry. Although this method may result in a lower overall price of the diet, it is not an ideal method for those conducting research, because there can be considerable variability in the raw materials. In fixed formulation diets, the amount of each ingredient does not change over time. Managed formulation techniques allow for ingredient composition to change slightly, although nutrient composition remains constant. Both methods are acceptable practices for diets fed to laboratory NHPs, as well as other research species.

PROCESSING OF NONHUMAN PRIMATE DIETS

Commercial diets for NHPs in the biomedical industry are typically extruded; however, other preparations such as meal, pelleted or gel diets can be used. In the extrusion process, ingredients (raw materials) are first ground to a specific particle size. This mixture is passed through a conditioner where steam is injected to start the cooking process. The extrudate is then passed through the extruder, which consists of a large rotating screw within a stationary barrel. The extruder's rotating screw then forces the extrudate in the direction of a die, which essentially determines the size of the biscuit. After it passes through the die, the biscuit is then cut by knives to a specified length. The extruded product usually puffs and changes texture as it is extruded. Depending on the ingredients used, their bushel weights, and moisture levels, the amount of puff can vary from product to product, but may also vary within a given product. Thus, when individual portions are based on a specified number of biscuits or on a volume basis, it is important to understand that such variability can lead to unintended changes in energy and nutrient intake (NRC, 2003). Finally, the product is cooled and dried to create a rigid biscuit, as well as to reduce the overall

Jumbo   Standard   Peanut   Mini

Figure 27.1   Commonly used biscuit sizes and shapes for nonhuman primates.

moisture levels (typically <14% moisture) in the diet prior to packaging. The extrusion process can decrease some vitamin concentrations because of the high temperatures used during manufacturing, and therefore, natural ingredient diets are typically fortified with vitamins to account for losses during processing.

Feed processing is multifunctional. One reason for processing ingredients into pellets or biscuits is to prevent waste by the animals. If animals were allowed to select only those ingredients that they found most palatable, then it is likely that they would consume a diet that is nutritionally incomplete. A processed diet prevents this by allowing for the homogeneous delivery of ingredients and nutrients to the animals. Commercial diets can also be manufactured into biscuits of varying sizes and shapes (Figure 27.1). Larger biscuits are often intended for larger primates, such as macaques and baboons, while the smaller biscuits are intended for primates with smaller hands, such as marmosets, and squirrel and owl monkeys. Cage design, especially flooring materials/substrates, can also influence the utility of biscuits of particular sizes and shapes. Fewer biscuits are wasted when they do not easily fit through the mesh/slats on the cage floor.

Processing may also aid in the animals' ability to digest starches by making the starch more readily available for enzymatic hydrolysis. Much of this research has been done in vitro. Snow and O'Dea (1981) showed that cooking wheat and oats greatly increased the percentage of starch hydrolyzed in vitro, compared to raw oats and wheat. The percentage starch hydrolyzed in cooked rolled oats was almost 50% greater than for raw oats and nearly 45% greater for cooked rolled wheat versus raw rolled wheat. In vitro, Sun et al. (2006) also found that starch digestion was higher for extruded grains than for their raw counterparts. The coefficients of ileal tract and total tract apparent digestibility were increased significantly in pigs when raw materials were extruded; however, the extent of digestibility also varied depending on the raw materials themselves, when evaluated both in vitro and in vivo (Sun et al., 2006).

DIETARY (FORAGING) ENRICHMENT

The welfare of NHPs in captivity is not only dependent on meeting the nutritional requirements of the animals, but also on providing them with environmental and/or dietary enrichment that facilitates their natural foraging and feeding behaviors. To be consistent with earlier chapters in this handbook, I will refer to this type of enrichment as "foraging enrichment." Because nonhuman primates are typically motivated by food, foraging enrichment is a primary component of many NHP enrichment programs. Foraging enrichment plays a much smaller role for many other common laboratory species, such as mice and rats.

Forms of Foraging Enrichment

A key component to foraging enrichment is variety. The variety of foraging enrichment items (and devices) that facilities provide to their animals is quite extensive, as many use the local grocery store to obtain "treats." This section will specifically focus on the somewhat less diverse array of treats that can be supplied by feed manufacturers within the biomedical industry. Figure 27.2 provides some examples of foraging enrichment products that are available from feed manufacturers.

Most of the foraging enrichment items provided to NHPs are not considered nutritionally complete for the animals. Some laboratory animal feed manufacturers can supply a few types of treats that are considered to be nutritionally complete since they provide the same level of nutrients as the available processed diets. Feedstuffs frequently used in foraging enrichment programs that are not considered nutritionally complete include, but are not limited to, grains, fruits, vegetables, peanut butter, dehydrated fruit–nut blends, some tableted treats, and cereals. Feed manufacturers provide a number of foraging items, including extruded feeds, fruit–nut–seed blends, and sunflower seeds. The average nutrient content of various seeds, fruits, and vegetables in comparison to the minimum recommended nutrient requirements for macaques is presented in Table 27.3. Tablets (pellets), which have a precise weight, size, and shape, are often used as enrichment and/or as a reward in behavioral studies. Pellets are designed to work within a specialized feeding chamber, in which the animal can press a trigger to release the reward. Other enrichment treats supplied by laboratory animal feed suppliers include banana chips, live insects, dehydrated yogurt pieces and fruits. Many of the treats supplied by diet manufacturers are intended to be used within devices that require the animal to work for the food; to search for, or solve a puzzle or task, to access the food. Such devices/tasks provide opportunities to stimulate the animal's cognitive and motor skills. Table 27.4 illustrates the macro nutrient content of common treats supplied by laboratory animal feed manufacturers for NHPs in research settings, distinguishing those treats that are nutritionally complete from those that are not.

The foraging enrichment items provided by feed manufacturers only account for a small portion of effective foraging enrichment strategies, and an even smaller portion of a comprehensive behavioral management program. It is important to note that special dies and manufacturing equipment are required to make treats of different sizes, shapes, and colors, contributing to the complexity of manufacturing and distributing such treats. Even fresh produce treats present challenges; it can be difficult to maintain fruits and vegetable at optimal temperatures and humidity levels to prevent the growth of molds and minimize bioburdens.

Impact of Enrichment

Environmental enrichment, in general, has been shown to enhance the welfare of captive NHPs by reducing atypical and agonistic behaviors (Chamove et al. 1982; Line at al. 1990; Byrne and Suomi 1991). In some species, enrichment has even been associated with improvements in the

Figure 27.2 Examples of dietary enrichment feeds.

Table 27.3   Nutrient Content of Commonly Used Dietary Enrichment Items

Nutrient[a]	NRC Requirement[b]	Seeds[c]	Fruits[d]	Vegetables[e]
Protein (%)	8.0	19.0–21.0	0.0–1.0	1.0–3.0
Essential n3 FA (%)	0.5	0.07–0.15	0.009–0.03	0.0–0.052
Essential n6 FA (%)	2.0	0.02–23.0	0.05–0.079	0.02–0.12
Carbohydrates (%)	NR	0.0–20.0	14.0–23.0	3.0–10.0
Fiber (%)	NR	0.0–9.0	1.0–3.0	1.0–3.0
Calcium (%)	0.55	~0.8	0.005–0.014	0.02–0.05
Phosphorous (%)	0.33	0.2–0.7	0.01–0.022	0.02–0.07
Magnesium (%)	0.04	~0.3	0.005–0.027	0.007–0.02
Iron (ppm)	100.0	12.0–52.0	3.0	2.0–7.0
Zinc (ppm)	13–20	15.0–50.0	0.0–2.0	1.0–4.0
Vitamin A (IU/g)	5.0	~5.0	1.0–6.4	0.98–167.1
Vitamin E (mg/kg)	68.0	~332.0	1.0–2.0	2.0–8.0
Vitamin C (ppm)	110.0	1.0–14.0	40.0–87.0	28.0–892.0
Vitamin K (ppm)	>0.06	0.0	0.01–0.15	0.24–1.02
Thiamin (ppm)	1.1	1.0–15	0.0–1.0	0.0–1.0
Riboflavin (ppm)	1.7	1.0–4	0.0–1.0	0.0–1.0
Niacin (ppm)	16.0	12.0–83.0	1.0–7.0	1.0–6.0
Pyridoxine (ppm)	4.4	1.0–13	0.0–4.0	0.0–2.0
Folate (ppm)	1.5	0.2–2.3	0.04–0.2	0.19–0.63
Vitamin B12	0.01	0–0.02	0.0	0.0
Pantothenic acid (ppm)	20.0	7.0–11.0	0.0–3.0	1.0–6.0

[a] USDA, ARS or www.nutritiondata.self.com.
[b] Macaques—NRC (2003).
[c] Pumpkin and sunflower seeds (deshelled).
[d] Bananas, apples, grapes.
[e] Raw iceberg lettuce, carrots, celery, cabbage, broccoli.

Table 27.4   Macro Nutrient Content of Dietary Enrichment Supplied by Laboratory Animal Feed Suppliers

Nutrient	Sucrose Tablets[a,b]	Foraging Bits[a,c]	Monkey Morsels[a]	Grain-Based Tablet[a,b,c]	Yogurt Drops[b]	Veggie Relish[a]
Protein (%)	0.0	15.2	13.0	20–21	5.5	10.5
Carbohydrates (%)	100.0	49.0	66.0	54–55	66.0	60–65
Fiber (%)	0.0	15.0	10.0	4.0	0.6	5.0
Calories (kcal/g)	4.0	3.02	4.52	3.36	4.89	3.81

[a] Purina LabDiet [subsidiary of Land O'Lakes Purina Feed LLC (St. Louis, MO)].
[b] BioServ (Flemington, NJ).
[c] Nutritionally complete—meets the minimum nutrient requirements.

overall health of research animals. Vitalo et al. (2009) found that burn-injury healing was more rapid in those Sprague–Dawley rats that had been group reared than it was for similar rats reared in isolation. Rats reared in isolation and provided with nesting materials as a form of enrichment also healed more quickly than those reared in isolation, but not as quickly as those that had been group reared. Mice that were injected with B16 melanoma cells and reared in an enriched environment (cages with more living space, nesting materials, and apparatus) for 3 weeks prior to inoculation exhibited a delay in tumor development compared to mice reared in an unenriched environment

(Cao et al. 2010). Mice exposed to an enriched environment (EE) for 3 weeks prior to inoculation and for an additional 3 weeks after inoculation exhibited tumors that were smaller in volume and weight, compared to control mice. The occurrence of tumors was also delayed for EE mice over the 17-day period compared to control mice, further demonstrating that enrichment for laboratory animals can improve their overall health.

Alternatively, environmental enrichment can be viewed to be in conflict with the overriding goal of standardizing experimental treatments, because enriched animals exhibit more diverse behavior, and therefore increased variation in their responses to experimental treatments (Van de Weerd and Aarsen 2002). Standardization should enhance the replicability of experiments, thereby reducing unwanted variation. Van de Weerd and Aarsen (2002) evaluated the effects of an enriched environment (Kleenex and a nest box of perforated sheet metal with a metal climbing grid) or a super-enriched environment (Kleenex, a nest box of perforated sheet metal with a metal climbing grid, and two Aspen wood gnawing blocks) on weight gain, food intake, locomotion, interaction with the enrichment items, climbing, and rearing in mice. Body weight and food intake were greater in enriched animals compared to control animals (Kleenex only). Frequencies of locomotion and interaction with objects were also significantly higher for mice in the super-enriched condition compared to the standard condition. Eskola et al. (1999), however, only saw slight, nonsignificant, trends in physiological parameters measured in mice that were exposed to either an enriched or a nonenriched environment. A review of the literature indicates that the findings, conclusions, and interpretations of studies evaluating the effects of environmental enrichment on behavior or physiology are mixed, particularly with regard to species, gender, age, and strain (Van de Weerd et. al. 2002; Toth et al. 2011).

One potential negative effect of foraging enrichment that is typically considered, but has not been extensively evaluated, is the excessive provision of enrichment foods to animals. Like humans, animals can become obese and develop metabolic disorders due to the consumption of excess calories from inappropriate sources. When using foraging enrichment, it is important to account for the number of calories that are provided to the animals in addition to those included in their daily nutritionally complete diet. Calories from fat and refined carbohydrates should be kept to a minimum. Animals may also choose to consume their highly palatable enrichment items at the expense of their regular diet, further reducing their micronutrient intake and potentially resulting in significant wasting of the diet.

Although there may be the potential for minor adverse effects of both foraging enrichment specifically, and behavioral management in general, on research animals or the standardization of experimental models, most agree that the benefits of appropriate foraging enrichment strategies and comprehensive behavioral management programs far outweigh any of these potential detriments. The welfare of laboratory animals, especially nonhuman primates, has been greatly improved as feed manufacturers have contributed to the enhanced definition and standardization of foraging enrichment programs and behavioral management practices.

CONCLUSIONS

Nonhuman primates play an important role in biomedical research and it is important that we continue to find ways to improve the well-being of our animal subjects. Foraging enrichment improves welfare, thereby contributing to the availability of high-quality, well-defined research subjects. When considering the general diet and foraging enrichment opportunities for laboratory animals, it is essential to make certain that nutritional requirements are met for the animals at all stages in their life cycles, and that dietary treats are not fed at the expense of required nutrients. All supplements should be fed in moderation to allow our animal models to live long, healthy lives and to contribute reliable and valid data to advance the biomedical research endeavor.

REFERENCES

Byrne, G. D. and S. J. Suomi. 1991. Effects of woodchips and buried food on behavior patterns and physiological well-being of captive rhesus monkeys. *Am. J. Primatol.* 23: 141–151.

Cao, L., X. Liu, E. J. D. Lin, C. Wang, E. Y. Choi, V. Riban, B. Lin, and M. J. During. 2010. Environmental and genetic activation of a brain-adipocyte BDNF/leptin axis causes cancer remission and inhibition. *Cell* 142(1): 52–64.

Chamove, A. S., J. R. Anderson, S. C. Morgan-Jones, and S. P. Jones. 1982. Deep woodchip litter: Hygiene, feeding, and behavioral enhancements in eight primate species. *Int. J. Study Anim. Probl.* 3(4): 308–18.

Eskola, S., M. Lauhikari, H. M. Voipio, M. Laitinen, and T. Nevalainen. 1999. Environmental enrichment may alter the number of rats needed to achieve statistical significance. *Scand. J. Lab. Anim. Sci.* 26: 134–44.

Line, S. W., A. S. Clarke, H. Markowitz, and G. Ellman. 1990. Responses of female rhesus macaques to an environmental enrichment apparatus. *Lab. Anim.* 24: 213–20.

National Research Council (NRC). 2003. *Nutrient Requirements of Nonhuman Primates*, 2nd revised edition. National Academies Press, Washington, DC, 184.

SELFNutritionData. 2014. SELFNutritionData: Know what you eat. http://nutritiondata.self.com (Accessed December 30, 2016).

Snow, P. and K. O'Dea. 1981. Factors affecting the rate of hydrolysis of starch in food. *Am. J. Clin. Nutr.* 34: 2721–27.

Sun, T., H. N. Lærke, H. Jørgensen, and K. E. B. Knudsen. 2006. The effect of extrusion cooking of different starch sources on the *in vitro* and *in vivo* digestibility in growing pigs. *Anim. Feed Sci. Technol.* 131:66–85.

Toth, L. A., K. Kregel, L. Leon, and T. I. Musch. 2011. Environmental enrichment of laboratory rodents: The answer depends on the question. *Comp. Med.* 61(4): 314–21.

USDA-ARS. 2015. National nutrient database for standard reference. Release 28. The National Agricultural Library. http://ndb.nal.usda.gov (Accessed December 30, 2016).

Van de Weerd, H. A., E. L. Aarsen, A. Mulder, C. L. J. J. Kruitwagen, C. F. M. Hendriksen, and V. Baumans. 2002. Effects of environmental enrichment for mice: Variation in experimental results. *J. Appl. Anim. Welf. Sci.* 5(2): 87–109.

Vitalo, A., J. Fricchione, M. Casali, Y. Berdichevsky, E. A. Hoge, S. L. Rauch, F. Berthiaume, M. L. Yarmush, H. Benson, G. L. Fricchione, and J. B. Levine. 2009. Nest making and oxytocin comparably promote wound healing in isolation reared rats. *PLoS One* 4(5): e5523.

CHAPTER 28

Providing Behaviorally Manageable Primates for Research

Luis Fernandez, Mary-Ann Griffiths, and Paul Honess
Bioculture (Mauritius) Ltd.

CONTENTS

Nonhuman primates (NHPs) have played a critical role in the evolution of scientific and biomedical understanding for many years. Without the important contributions of NHPs to medical and scientific understanding, progress toward the elimination of morbidity and mortality afflicting animals and humans may not have occurred or would have been delayed [1].

Since 1994, when fewer than 10,000 animals were imported, demand for NHPs for research has nearly tripled. Starting in fiscal year (FY) 2003 (October 1, 2003 to September 30, 2004), the United States began importing over 20,000 NHPs per year—primarily, *Macaca fascicularis* (cynomolgus macaques) from Asia and Mauritius. With the exception of FYs 2011–2013, which saw reductions due to challenges associated with access to air transportation and the global financial crisis, the number of NHPs imported into the United States has exceeded 20,000 animals each year through FY 2016 [2; see Figure 28.1].

This large increase in demand has necessitated an increased understanding of the techniques required for the proper care and rearing of *Macaca* species in captivity, particularly those related to psychological well-being. As regulatory and societal changes have increasingly focused on the psychological well-being of laboratory animals, managers and researchers have had to modify care paradigms to more explicitly consider the impact of research-related factors on the psychological well-being of the animals in their care. The supply chain has had to adapt to regulatory and client expectations as well, caring for, and raising, NHPs under increasingly enriched and psychologically beneficial conditions [3–9].

Studies have shown that animals that are highly enriched can better adapt to new and changing environmental conditions [10–18]. As has been reported, physiologic and hematologic parameters of the research subject not only reflect the well-being of the animals, but can impact the quality of the science being produced [19–27]. In addition to continuously evaluating and modifying the conditions in breeding facilities, third-party audits, reciprocal site visits, and professional continuing education opportunities have increased animal behavior-related expectations at the end user and at

Number of NHPs

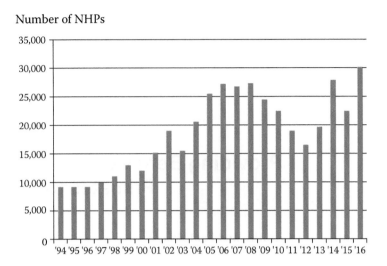

Figure 28.1   Imports of NHPs increased dramatically from 1994 to 2008. (From Mullan R, et al. Nonhuman Primate Importation and Quarantine: United States, Fiscal Year 2016. Paper presented at: Association of Primate Veterinarians Annual Workshop 2016. Charlotte, North Carolina.)

the farm of origin. As an example, McGrew [28] has discussed (1) origin-specific differences in the percentage of animals that approach the front of the cage and accept hand-feeding upon arrival in the United States, and (2) the desire to have all animals approach the front of the cage and accept hand-feeding after the Centers for Disease Control and Prevention quarantine period.

Breeders of macaques invest varying amounts of operating capital in facilities, people, and husbandry procedures aimed at raising the animals in appropriate environments that promote optimal levels of production, low disease incidences, and maximal levels of psychological well-being for both the animals and their caregivers. With so much emphasis on producing the highest quality model for studies of human and animal diseases, breeders want to make certain that they are preparing their animals to have the best possible experience during the transition period from the country of origin to life at a research facility.

As will be discussed in the following sections, differences exist in the behavioral management practices experienced by macaques between, and within, geographic regions.

CHINA AND SURROUNDING COUNTRIES

At the same time that the increase in demand for NHPs, particularly cynomolgus macaques, was rising dramatically, the participation of NHPs in drug development programs in China was also rapidly expanding. The large increase in demand for animals, primarily, cynos from Asia, created a rapid expansion in demand for wild-caught animals in Cambodia and Laos, as well as construction and expansion of large holding, breeding, and export facilities in China, Vietnam, and Cambodia. In some cases, rapid expansion came with little end-user input concerning the necessary adaptations required to prepare the animals for their eventual use in research settings. Instead, emphasis appeared to be placed on issues related to animal production. The establishment of the housing and husbandry practices necessary to maximize the welfare of the animals in these new facilities was often underemphasized, and sufficient numbers of veterinarians, trained behavioral scientists, and animal care personnel may not have been available to work with the large quantity of animals. Cultural and language barriers relative to animal welfare also may have created impediments to progress during this period. While "Sharing Conferences" [29] and exchanges between US and EU sponsors and nonclinical Asian Contract Research Organizations (CROs) became more frequent, these conferences primarily emphasized improving animal production, and the reduction or elimination of viral and bacterial diseases (e.g., B-virus, SRV, STLV, and tuberculosis), with relatively little attention devoted to psychological well-being in captivity or the reduction of stress during nonclinical studies conducted in a research setting.

At the breeder level, some companies have made substantial modifications to the configurations and housing conditions used to hold their animals, adding considerable variation and complexity to the animals' environment-related experiences. The typical square pen of the past, with no perches or hanging structures, has been replaced by more complex pens with perches, swings, and concrete tunnels or barrels (see Figure 28.2).

In the past, enhancements that were provided may have been of relatively low quality or the density of animals in the pen may have diminished the enrichment value of the enhancements. In addition, many enrichment attempts did not add sufficient complexity or variety to functionally simulate natural environments. Some key features, such as vertical and horizontal escape zones, visual barriers, and comfortable resting areas, may have been absent in the past. Fortunately, many breeders continue to retrofit and improve the overall design of their pens in an attempt to minimize

Figure 28.2 (a, b) Rhesus enclosure with swings, perches, and concrete tunnel for hiding.

Figure 28.3 (a, b) Typical Chinese pens.

the potential negative consequences that barren housing can have on captive cynos, including hierarchy-related aggression within groups (see Figure 28.3a and b).

Because herpes B virus is a human health hazard and NHPs infected with retroviruses can be detrimental to many types of research studies, breeders have undertaken a number of additional changes to their husbandry practices in an attempt to minimize the transfer of viruses from older animals to younger animals. One of these changes has been a shift to weaning animals at, or prior to, 6 months of age, creating additional behavioral and husbandry-related challenges. Housing for recently weaned juveniles may be too crowded to promote normal play and foraging behaviors, and some enclosures may not provide optimal temperature and lighting conditions. This period of intensive housing and testing, lasting several months, could be used for focused behavioral management procedures, including enhanced socialization and/or training efforts.

One aspect of behavioral management that could use improvement among the animals at these facilities is the implementation of techniques that will reduce the macaques' fear of humans. Many of these cynos frequently huddle in the back of the pen and will not approach the cage front when humans are nearby. Additionally, hair coat quality may be less than optimal while these animals are living in groups.

The facilities housing these animals tend to comprise a single ~2 m tall cage, with limited or no complexity to the physical environment. Visual barriers, vertical escape areas, manipulanda, ladders, and swings may not be present in sufficient quantities to yield a positive psychological environment for the macaques. Food is usually scattered on the limited floor area, a technique which may make it difficult for subordinate animals to obtain an ample quantity of food.

In spite of the challenges mentioned earlier, these facilities continue to make significant progress, improving husbandry practices, especially with respect to stocking densities, health monitoring, nutrition, winter heating, and veterinary care. The best facilities have often enjoyed longer term relationships with US, UK, and EU institutions and have instituted some of the improvements in behavioral management necessary to develop appropriate human–animal relationships. Although noticeable levels of fear and aggression persist, during visits to the breeding farms, it is obvious that hand-feeding of the animals does occur. However, the behavior of the animals suggests that hand-feeding may not be a prioritized component of the routine husbandry program. It is clear that many Asian suppliers are eager to learn and implement best practices (see Figure 28.4) for their NHPs, to improve the macaques' well-being and in turn, expand their market share in response to the strong demand for Asian-sourced cynos.

Figure 28.4 Improved focus on hand-feeding and interacting with humans in a positive, non-threatening way will continue to improve in Asia.

VIETNAM

NHP suppliers from Mauritius and Vietnam have had long-term relationships with US and European clients; have established programs that incorporate health and welfare recommendations from the United States, United Kingdom, and the European Union; and have been awarded accreditation by AAALAC-International. As a result, animal care and welfare standards continue to evolve and align themselves with the needs and expectations of the end user. In the Asian market, an AAALAC-International-accredited site in Vietnam (VN) is the premier provider of cynos from the many suppliers in China, Vietnam, and Cambodia. VN has made investments in housing and processes to regularly supply high-quality cynos for researchers around the world. An example of VN's commitment to the highest level of care is the neonatal care unit. VN's well-trained and well-equipped veterinary staff maintains incubators from which orphans can be provided intensive care and later returned to a surrogate mother or, less frequently, to peer groups at the normal weaning age. The typical detrimental consequences of nursery rearing on the psychological well-being of infant macaques do not appear to be exhibited by the cynos VN provides to clients (see Figures 28.5 and 28.6).

Figure 28.5   A stable social group in Vietnam sharing a resting platform for grooming and sleeping.

Figure 28.6   Intensive neonatal housing demonstrates the commitment to care for each individual macaque.

The overall high quality and adequate psychological well-being of the cynos produced by the Vietnamese breeders are likely due, in part, to the comprehensive two-tiered behavioral management program that they employ. The two tiers of the program, *routine* and *purposed*, aim to produce a cyno that is a willing participant in human–animal interactions and that voluntarily comes to the front of the cage, demonstrating an amelioration of its natural aversion to humans. The *routine* program involves regular hand-feeding of novelty food items, such as peanuts or raisins. During periods of stress, such as preexport quarantine and hospitalizations, routine hand-feeding is intensified. The *purposed* portion of the behavioral management program is client-driven training of animals for specific tasks, such as presentation for blood sampling. As further evidence of VN's commitment to continuously improving quality, VN has added animal training to its program at the production site, thereby enhancing the animal's ability to acclimate to similar experiences at the study site. By training and rewarding animals for voluntarily presenting an arm for blood collection in Vietnam, the need for restraint and handling at the study site, and hence, the level of adaptation stress, are reduced (see Figures 28.7 and 28.8).

Figure 28.7 Vietnamese infant approaching visitor freely. Note the tire swing and multi-level perches available for resting.

Figure 28.8 Training animals to present voluntarily for blood sampling offered by VN reflects the commitment to continuous improvement and reducing animal stress through positive reinforcement training.

VN has focused on providing consistently healthy and uniform animals, while investing in facilities (caging) and behavioral management programs (environmental enrichment, socialization practices, and training) to try to meet EU, UK, and US client demands. VN also routinely utilizes hand-feeding, as evidenced by the high number of juveniles that approach the front of the enclosures and willingly interact with caregivers and visitors. Manipulanda and other apparatus are provided during the preexport period to offset the potential impact of indoor, single housing during this period. More recently, VN has been utilizing pair housing in quarantine caging, which is intended to minimize the impact of isolation and prevent the development of stereotypic behavior during this period.

MAURITIUS

Due to the unique, specific, viral-free, and genetic characteristics of *M. fascicularis* from Mauritius, a close working association between several US and EU research companies has been ongoing since 1985. As a result of this close partnership, animal care and well-being has been the centerpiece of long-time breeders, such as Bioculture, which employs full-time animal welfare officers and behaviorists to oversee their management programs. The animal welfare officers and behaviorists evaluate *all* management practices in the context of their impact on animal care and welfare. While breeding and housing conditions in Mauritius have been carefully designed and adapted over time to allow animals to express species-typical social behaviors and social structures in captivity, more recently, the emphasis has expanded to include better understanding of the conditions that the animals will experience at the client's facility [5,9,30–32; Figure 28.9a and b].

Given their concern for the well-being of the animals, one of the most important goals for some of the Mauritian breeders is to familiarize the animals with caregivers to reduce the macaques' innate aversion to human intruders [9,13,15,17]. It is often during this normal aversive response to the presence of a human that animals get injured, as they attempt to run and hide from the human, or receive redirected aggression from a more dominant animal. Not only is there adult-to-adult aggression during these situations, but infants may be injured, or even attacked and killed by adults. Unfortunately, if attempts are not made to curtail this dangerous escape response, a vicious cycle may develop, whereby fear of humans results in aggression among monkeys, which leads to continued fear and aggression. Fear and flight become the standard response, as infants simply learn from their parents that the normal response to caregivers is to be fearful and run away.

While breeders do not want their animals to seek human attention and interfere with day-to-day husbandry tasks by being "friendly" or seeking excessive human contact, considerable efforts are made to minimize the flight response and redirect aggression in the captive animals. In essence, the goal is for the animals to comfortably approach the caregiver in appropriate circumstances.

In order to tackle the challenge of minimizing the negative impact of the normal fear and flight response in a captive situation, a breeder must adopt a holistic management approach that includes housing, feeding, routine husbandry, and animal handling practices, while building a practical human–animal relationship. As has been discussed by Honess [30] and others, the design of group caging facilities is a critical component for achieving normal social group interactions and breeding success.

If animal housing situations are not properly managed, it may be difficult to build useful human–animal relationships within the social housing situation at the breeding site, and subsequently, when the animals are living in more confined research spaces. Housing situations must allow dominant animals to feel comfortable and unthreatened by humans and the other members of the social group, in part as a function of the vertical and horizontal distribution of the monkeys in the group. When dominant animals "sanction" interactions between subordinate monkeys and humans, positive reinforcement training and the construction of a cooperative human–animal relationship can begin. It is helpful to have positive reinforcement training (e.g., hand-feeding), performed by multiple people, so that the alpha or other dominant animals are occupied, and subordinates can be interacted with

(a) (schematic)

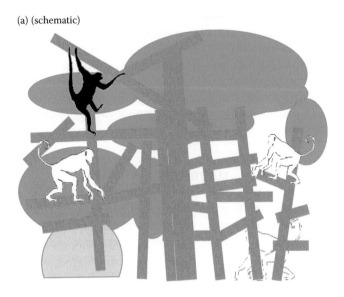

(b) (actual)

Figure 28.9 (a, b)   Bioculture cages designed to mimic the normal arboreal environment of the cynomolgus macaque, including multiple levels for hierarchical spread, climbing, foraging, and visual barriers.

successfully. Cooperative feeding is a training technique that is used to work with subordinate macaques in the presence of dominants. Honess has shown that although subordinates may not directly interact with caregivers, when dominant animals are removed from the social group or the subordinates are relocated to another group, the desired behavior has been learned through observation of the interactions between the caregiver and the dominant animals [30].

In order to accomplish this in a systematic manner across all of its sites and animal age categories, Bioculture Mauritius refined its existing culture of emphasizing primate well-being by forming a specialty department, the Familiarization and Training Team (F&T). The F&T was created to focus a group of caregivers and behaviorists on maximizing the care and well-being of not only the animals, but also the animal caregivers and husbandry staff across all components of the company. The F&T team members work collaboratively with the primary caregivers, veterinarians, and members of management, and in concert with the oversight of the Ethical Review Committee (an analog to the Institutional Animal Care and Use Committee (IACUC)), to continuously drive best practices and ensure animal well-being is maximized. Bioculture commits approximately 10% of the workforce to F&T activities on a full-time basis to familiarize and train approximately 14,000 macaques, while all the rest of the members of the husbandry and technical teams join in hand-feeding activities during the afternoon feeding period. The efforts of the F&T toward familiarization begin with unweaned infants observing the alpha males approaching and accepting food directly from caregivers. Soon thereafter, groups of infants are commonly observed sitting alongside alpha males as they are fed by hand, while the infants' mothers watch. Next, weaned monkeys are hand fed, while simultaneously learning more advanced cooperative tasks, such as clicker and target training, box training, drinking from a syringe, or presenting for swabbing. The goal of Bioculture's program is to familiarize cynos throughout their life cycles, so that macaques that are accustomed to a culture of positive caregiver–animal interactions are provided to the end user (see Figures 28.10 and 28.11a and b).

Ultimately, Bioculture animals are expected to approach caregivers and to come to the front of the cage when appropriate, rather than cowering at the back. Husbandry processes are designed to minimize

Figure 28.10 Multigenerational learning. Hand-feeding and positive interactions with the dam and her infant allow the infant to accept human–animal interactions, while her older sibling participates as well.

Figure 28.11 (a) Using target training to adapt cynos at Bioculture to enter transport boxes for reduction of handling stress. (b) Target training used in conjunction with drinking from a syringe as the initial stages of training to accept medication without restraint at Bioculture.

negative impacts of social stress and the potential for injury to caregivers and animals, while rewarding macaques for cooperation. With very low levels of turnover in the caregiver ranks, refining husbandry practices to decrease "hard labor," and thereby accommodate employees with diminishing physical abilities, is critical to retaining employees with high levels of expertise and commitment to animal care and well-being. In recent years, husbandry practices have evolved from catching macaques by hand or by net, to the use of procedure boxes and chutes/tunnels for capture and/or restraint. While these procedural refinements require some additional animal and caregiver training, the reduction in the amount of time necessary to complete routine husbandry procedures, as well as the diminished stress experienced by the monkeys, justifies the investment. In addition, as more animals are provided for clients that use social housing systems, including procedure tunnels, prefamiliarization with this management system becomes increasingly beneficial. The procedure tunnel is now an integral part of the housing furniture in Bioculture pens, and the design of the tunnel and the protocols for their use are continuously refined to maximize beneficial outcomes for the animals and their caregivers. In the future, suppliers may be working even more closely with their clients to acclimate animals to the specific procedure tunnels they will encounter during their research experience. The overall goal is to harmonize the animals' experiences at its facility of origin with its experiences at the facility of the client (see Figure 28.12a and b).

While all Bioculture animals are familiarized and desensitized to the people they encounter in their environment, calm, interactive, and interested cynos are increasingly important as the animals transition to new situations. Clients increasingly expect their animals to willingly participate with their caregivers and to (1) be clicker trained, (2) be target trained, (3) enter a transport

Figure 28.12 (a, b)   Procedure tunnels combined with positive reinforcement reduce the stress of everyday husbandry procedures and strengthen the human–animal bond at Bioculture.

box, or (4) have received specialized training based on client-specified protocols to ease the transition to the new housing situation. Building on these basic levels, animals have been acclimated to company-specific Personal Protective Equipment (PPE) or preferred targets. Association of the client's PPE with the positive reinforcement training process was felt to be beneficial to the animals once they arrived at the client facility. Similarly, the use of company-specific targets at the farm of origin helps to minimize the stress of the animals after arrival at the research facility. In addition to training macaques with company-specific targets, animals have been trained to present their chests for easy reading of the tattoo or microchip and for palpation and/or auscultation (see Figure 28.13).

More advanced familiarization techniques involve training animals to novel procedures, such as ingesting unpalatable solutions without using nasogastric gavage techniques. Honess [30] and others reported the use of a bitter gourd to approximate the taste of a pharmaceutical preparation that would need to be administered to macaques during drug development studies. Using positive reinforcement techniques, cynos were trained to voluntarily ingest (orally) a bitter gourd solution, eliminating the need to restrain and gavage the animals. Not only was the stress associated with handling and restraining the monkeys avoided, but the potential risks associated with gavage could be avoided as well (see Figure 28.14).

Figure 28.13 Client-specified target and PPE training aimed at acclimating animals to PPE and devices that will be encountered at the study site.

Figure 28.14 Training cynos to take bitter medications through positive reinforcement training to reduce the need for restraint and gavage tubes to improve cooperation and reduce stress at the study site.

Table 28.1   FY 2013 Shipments by Port of Entry

Station	Number	Percent (%)	Number of Shipments	Average per Shipment	% of Shipments
New York	11,937	40	51	234	43
Houston	8,265	27	8	1033	7
Chicago	7,189	24	48	150	40
Dulles	1,876	6	10	188	8
Newark	660	2	4	165	3
Others	168	1	11	15	9
Totals	30,095	100	120	251	100

Note: The increase in average shipment size and the growing percentage of large shipments necessitates specialized care and staffing. FY October 1 through 30 September; CDC data

Other important behavioral management-related practices that can be carried out at the farm of origin include working closely with clients during the animal selection and transportation segments of the supply chain processes. For example, in order to ease the transition from farm to study facility, animals are routinely grouped in compatible pairs on the farm prior to pre-export processing. Upon boxing/crating for shipments, pairmates are housed individually, but next to one other, and are then unloaded into pair housing upon arrival at the CDC quarantine facility. Compatible pairs can then be maintained throughout quarantine, and in most cases, they can be maintained together on study as well. Recently, further refinements include some suppliers pairing animals based on specific genetic characteristics, to prevent the need to separate pairs after arrival at the study site. Some clients, especially in the United Kingdom, request that Bioculture group animals in cohorts of 3 or 4 pre-export, to better match study facility caging designs.

Adding to the many challenges associated with preparing nonhuman primates for research, activist pressure on transportation of research animals has changed the dynamics of animal movement in the last several years. Not only has the United States seen a large increase in the number of imported NHPs, the average size of primate shipments has increased from relatively small cohorts of 60–120 animals, to charters of up to 1700 animals (see Table 28.1), as the number of commercial carriers willing to transport research animals, especially primates, has diminished.

The increase in the size of the typical shipment will further challenge breeders and recipients to creatively address the psychological well-being of these larger groups. Given the commitment of all of the stakeholders, this challenge should be addressable, and the care and well-being of research animals will continue to improve.

CONCLUSIONS

There are many other processes that are incorporated into the rearing of high-quality cynomolgus macaques, which individually may not have a huge impact on the psychological well-being of the cynos used in studies, but collectively may have a major impact. While processes and procedures in countries like Mauritius and Vietnam have led the way to better prepare animals to cope with the stress of transportation and relocation to a study site, clients and study sponsors will continue to drive the development of improved and refined behavioral management programs at the breeders and at research facilities. Regardless of the country of origin of the farm, only suppliers that continue to invest in understanding and delivering a well-adapted, psychologically fit research primate will be successful.

REFERENCES

1. Foundation for Biomedical Research. *Critical Role of Nonhuman Primates in Medical Research* https://fbresearch.org/wp-content/uploads/2016/08/NHP-White-Paper-Print-08-22-16.pdf.

2. Mullan R, Galland G, and Murphy G. Nonhuman primate importation and quarantine: United States, fiscal year 2016. *Paper presented at: Association of Primate Veterinarians Annual Workshop 2016.* Charlotte, NC.

3. National Research council. *Guide for the Care and Use of Laboratory Animals*, 8th edition. Washington DC: National Academies Press, 2011. http://www.nap.edu/catalog/12910.html.

4. Animal Welfare Act and Regulations. United States Department of Agriculture Animal Plant Health Inspection Service, November 2013. https://www.aphis.usda.gov/animal_welfare/downloads/Animal%20Care%20Blue%20Book%20-%202013%20-%20FINAL.pdf.

5. Honess P, Andrianjazalahatra T, Fernandez L, and Griffiths M-A. Environmental enrichment and the behavioural needs of macaques housed in large social groups. *Enrichment Rec* 10(1): 16–20, 2012.

6. Laule G. Positive reinforcement training for laboratory animals. In: Hubrecht R, Kirkwood J (eds) *The UFAW Handbook on the Care and Management of Laboratory and Other Research Animals.* Chichester, UK: Wiley-Blackwell, 206–218, 2010.

7. Schapiro SJ, Perlman JE, Thiele E, and Lambeth S. Training nonhuman primates to perform behaviors useful in biomedical research. *Lab Anim (NY)* 34(5): 37–42, 2005.

8. Scott L, Pearce P, Fairhall S, Muggleton N, and Smith J. Training nonhuman primates to cooperate with scientific procedures in applied biomedical research. *J Appl Anim Welf Sci* 6(3): 199–207, 2003.

9. Winnicker C, Honess P, Schapiro SJ, Bloomsmith MA, Lee DR, McCowan B, et al. *A Guide to the Behavior and Enrichment of Laboratory Macaques.* Shrewsbury, MA: Charles River Publications, 2013.

10. Bloomsmith MA, Marr MJ, and Maple TL. Addressing nonhuman primate behavioral problems through the application of operant conditioning: Is the human treatment approach a useful model? *Appl Anim Behav Sci* 102(3–4): 205–222, 2007.

11. Buynitsky T and Mostofsky DI. Restraint stress in biobehavioral research: Recent developments. *Neurosci Biobehav Rev* 33(7): 1089–1098, 2009.

12. Coleman K, Pranger L, Maier A, Lambeth SP, Perlman JE, Thiele E, et al. Training rhesus macaques for venipuncture using positive reinforcement techniques: A comparison with chimpanzees. *J Am Assoc Lab Anim Sci* 47(1): 37–41, 2008.

13. Coleman K, Bloomsmith MA, Crockett C, Weed, JL, and Schapiro, SJ. Behavioral management, enrichment and psychological well-being of laboratory nonhuman primates. In: Abee C, Mansfield K, Tardif S, and Morris T (eds) *Nonhuman Primates in Biomedical Research*, 2nd edition. New York: Elsevier, 149–176, 2012.

14. Gilbert SG and Wrenshall E. Environmental enrichment for monkeys used in behavioral toxicology studies. In: Segal EF (ed.) *Housing, Care and Psychological Wellbeing of Captive and Laboratory Primates.* Park Ridge, NJ: Noyes Publications, 244–254, 1989.

15. Heath M. The training of cynomolgus monkeys and how the human/animal relationship improves with environmental and mental enrichment. *Anim Technol* 40(1): 11–22, 1989.

16. Laule GE, Bloomsmith MA, and Schapiro SJ. The use of positive reinforcement training techniques to enhance the care, management, and welfare of primates in the laboratory. *J Appl Anim Welf Sci* 6(3): 163–73, 2003.

17. Novak MA and Suomi SJ. Psychological well-being of primates in captivity. *Am Psychol* 43(10): 765–773, 1988.

18. Schapiro SJ and Lambeth SP. Control, choice, and assessments of the value of behavioral management to nonhuman primates in captivity. *J Appl Anim Welf Sci* 10(1): 39–47, 2007.

19. Graham ML, Rieke EF, Mutch LA, Zolondek EK, Faig AW, DuFour TA, et al. Successful implementation of cooperative handling eliminates the need for restraint in a complex non-human primate disease model. *J Med Primatol* 41(2): 89–106, 2012.

20. Landi MS, Kissinger JT, Campbell SA, Kenney CA, and Jenkins EL. The effects of four types of restraint on serum alanine aminotransferase and aspartate aminotransferase in the *Macaca fascicularis*. *Int J Toxicol* 9(5): 517–523, 1990.

21. Lane J. Can non-invasive glucocorticoid measures be used as reliable indicators of stress in animals? *Anim Welf* 15(4): 331–342, 2006.
22. Lapin B, Gvozdik T, and Klots I. Blood glucose levels in rhesus monkeys (*Macaca mulatta*) and cynomolgus macaques (*Macaca fascicularis*) under moderate stress and after recovery. *Bull Exp Biol Med* 154(4): 497–500, 2013.
23. Lee J, Shin J, Lee J, Jung W, Lee G, Kim M, et al. Changes of N/L ratio and cortisol levels associated with experimental training in untrained rhesus macaques. *J Med Primatol*. 42(1): 10–4, 2012.
24. Schapiro SJ, Bloomsmith MA, and Laule GE. Positive reinforcement training as a technique to alter nonhuman primate behavior: Quantitative assessments of effectiveness. *J Appl Anim Welf Sci* 6(3): 175–87, 2003.
25. Schapiro SJ, Nehete PN, Perlman JE, and Sastry, KJ. A comparison of cell-mediated immune responses in rhesus macaques housed singly, in pairs, or in groups. *Appl Anim Behav Sci* 68(1): 67–84, 2000.
26. Shirasaki Y, Yoshioka N, Kanazawa K, Maekawa T, Horikawa T, and Hayashi T. Effect of physical restraint on glucose tolerance in cynomolgus monkeys. *J Med Primatol* 42(3): 165–168, 2013.
27. Viswanathan K. Stress-induced enhancement of leukocyte trafficking into sites of surgery or immune activation. *Proc Natl Acad Sci USA* 102(16): 5808–5813, 2005.
28. McGrew K. Pairing strategies for cynomolgus macaques, Chapter 17. In: Schapiro SJ (ed.) *Handbook of Primate Behavioral Management*. Boca Raton, FL: CRC Press, 255–264, 2017.
29. Global Research Education and Training (GR8). Office of Shanghai Administrative Committee for Laboratory Animals, AAALAC INTERNATIONAL. *The 1st Annual Shanghai Laboratory Animal Welfare Sharing Conference*, September 26–28, 2009, Shanghai, PRC. http://gr8tt.com/2009/10/06/2009-sharing-conference/.
30. Honess P. Behavior and enrichment of long-tailed (Cynomolgus) macaques (*Macaca fascicularis*). In: Winnicker C (ed.) *A Guide to the Behavior and Enrichment of Laboratory Macaques*. Wilmington, MA: Charles River Publications, 4–87, 2013.
31. Honess P and Wolfensohn S. The extended welfare assessment grid: A matrix for the assessment of welfare and cumulative suffering in experimental animals. *Altern Lab Anim* 38: 205–212, 2010.
32. Honess P. Behavioral management of long-tailed macaques (*Macaca fascicularis*), Chapter 20. In: Schapiro SJ (ed.) *Handbook of Primate Behavioral Management*. Boca Raton, FL: CRC Press, 305–337, 2017.

Conclusion

Behavioral Management of Laboratory Primates
Principles and Projections

Mollie A. Bloomsmith
Emory University

CONTENTS

INTRODUCTION

In 1979, Joe Erwin, Terry Maple, and Gary Mitchell published an edited volume titled *Captivity and Behavior: Primates in Breeding Colonies, Laboratories and Zoos*. The book included chapters on behavioral information and research, focusing on the impact of the captive environment on the behavior and development of nonhuman primates. The book was different from others in that the findings were presented with the intention that they would be put to use by those "concerned with the design, evaluation, or maintenance of animals in captive environments" including colony managers, animal scientists, veterinarians, zoo keepers, and scientists (p. vii). The book seems to be the first with such a focus for primates involved in research. There have been other landmark books published on the well-being of primates living in research laboratories including the 1989 *Housing, Care and Psychological Wellbeing of Captive and Laboratory Primates*, edited by Eve Segal. Segal's volume was followed in 1991 by *Through the Looking Glass: Issues of Psychological Well-being in Captive Nonhuman Primates*, edited by Melinda Novak and Andrew Petto, and was based on their successful conference "The Psychological Well-being of Captive Primates" held in 1988. In 1998, *The Psychological Well-Being of Nonhuman Primates* was published by the National Research Council's Committee on Well-being of Nonhuman Primates chaired by Irwin Bernstein. Since that time, the field has grown remarkably, with many important publications and books focusing on various aspects of captive primate well-being and care, but there has not been a comprehensive behavioral management book published recently, so the current volume is greatly needed.

The primate research community has made remarkable progress in addressing the psychological well-being of their primates. Behavioral research documenting which potential improvements are effective and which are not, has grown exponentially. Grant funding has been focused on the topic, and there are now many hundreds of published peer-reviewed articles that evaluate a wide variety of approaches to improving the welfare of many species of laboratory primates. Scientists publishing this work are leading the way toward meaningful change in the animals' lives. Books and newsletter-type publications are widely available to those working with laboratory primates. Many professional conferences have multiple sessions devoted to primate welfare, and there are entire conferences that are devoted just to environmental enrichment. Workshops are teaching methods to improve laboratory primate welfare. Commercial companies that provide enrichment supplies, enhanced caging, and other equipment designed to improve the welfare of research primates have grown. Even language has evolved, as terms like "behavioral management," "psychological well-being," "positive reinforcement training," "cognitive enrichment," and many others have become common parlance.

Alongside jobs in animal care and veterinary care, there are now positions that specialize in animal behavior, environmental enrichment, training of primates using positive techniques, socialization methods, and behavioral management. These types of jobs are ubiquitous in primate laboratories, although 30 years ago, virtually none of them existed. A recent survey indicated that full-time technicians dedicated to behavioral management programs are employed at 78% of the primate research facilities surveyed (Baker 2016).

The development of formal behavioral management programs in primate research facilities represents a maturation of this field as a recognized specialty. Research facilities for primates now include more innovative housing that provides better social opportunities and more complex and interesting environments for laboratory primates. Facilities have environmental enrichment programs with the budgets and resources needed to carry out the daily tasks of increasing complexity and stimulation in species-appropriate ways. Behavior of the primates is routinely assessed to identify problems, and treatments are implemented. Many laboratory primates are now being trained to cooperate with husbandry, veterinary, and research procedures using positive reinforcement training techniques, garnering the animals' cooperation, rather than relying on coercive methods. And, some laboratory primates are now retired to sanctuaries following their time in biomedical research. These changes are now considered the modern way of caring for primates—the "new normal"—and the progress is undeniable. This current *Handbook of Primate Behavioral Management* summarizes these advances. It will be a valuable resource to those caring for laboratory primates and will help set the course for future research needed to guide behavioral management in the most meaningful directions. In this final chapter of the book, I will reiterate some of its major themes, highlight some especially noteworthy contributions in the book, and make some suggestions regarding the focus of future work in behavioral management.

SCIENTIFIC UNDERSTANDING OF BEHAVIOR MUST BE INCORPORATED INTO MODERN PRIMATE CARE

One major theme of this book is the value of applying a deep and scientific understanding of primate behavior to their care. Many of the prominent issues in managing primates are fundamentally behavioral—addressing breeding and production problems, managing severe aggression, eliminating abnormal behavior, and dealing with manifestations of stress. As a field, we are recognizing that behavioral specialists can begin to understand and address these issues, and successful efforts will not only benefit the animals under our care, but will also improve the quality of the resulting scientific research.

Natural History of Species Is Critical to Behavioral Management

Basing the behavioral care of a species on a thorough understanding of that species' natural history is a long-standing tenet of behavioral management. This volume includes many wonderful chapters that provide this type of information in the "Genera-specific Behavioral Management" section of the book (Part IV). This information should be used to design species-specific behavioral management programs and should be part of personnel training for all of those working with non-human primates (NHPs). Understanding the natural behavior of each primate species with whom one works is required for their optimal care.

An understanding of species-typical behaviors can guide optimum environmental design. For example, it is important to attend to the olfactory component of the environment for New World species (providing olfactory-based enrichment, for example) because they rely heavily on olfaction for gaining information about their physical and social environments (Williams and Ross, 2017). Similarly, climbing structures and the availability of vertical space may be more meaningful for arboreal primates than for more terrestrial ones, and so those opportunities should be emphasized in their behavioral management plans. Providing nesting opportunities for prosimians, such as ruffed lemurs and aye-ayes that build nests for their infants as described by Dye (2017), is critical for them, but might not be appropriate for many other laboratory primates.

Awareness of species differences can also optimize behavioral management practices. Facilitating NHP learning in a positive reinforcement training (PRT) program by providing them with opportunities to observe others that have learned the same task will only be effective if that particular species is capable of social learning. The well-known allomothering behavior of vervet monkeys, including extensive handling of infants by unrelated group members (McKenna 1979; Jorgensen 2017), might be disconcerting for those unfamiliar with this natural behavior. Further, Jorgensen describes the vervet monkeys' "red-white-and-blue" genital display, their sideways prancing, and the behavior of encircling another monkey with their raised tail. Managers unfamiliar with vervet behaviors, such as these, will not be equipped to manage aggression in this species. Jorgensen (2017) makes this point concisely: "In short: Vervets are not macaques."

Why Behavior Matters

As Bettinger et al. (2017) point out in their chapter, focusing on behavior when evaluating welfare is essential, as "behavior is the external manifestation of everything occurring inside of the animal, ranging from the physiological to the psychological to the social" (p. 2). In this way, behavior is a readily accessible indicator of how the animal is doing. Having staff members who know each individual monkey's history, preferences, common reactions to events, and behavior is important, as is having staff members with a scientific understanding of behavior. The ability to interpret animal behavior is arguably one of the most important skills of a primate caregiver or manager.

However, the correct interpretation of behavior may not be obvious. Gottlieb et al. (2017) provide an important example (stereotypic behavior) of how cursory interpretations of behavior may be inadequate. While the presence of stereotypic behavior is generally seen as undesirable and as evidence of poor welfare, scientific evaluation has established that when animals are faced with the same suboptimal environment, those individuals that develop stereotypic behavior (such as pacing or repeated flipping) are often coping better than those animals that do not express stereotyped behavior, as assessed by measures such as fear-related behavior, depression-related behavior, alarm calling, and elevated levels of corticosteroids (Mason and Latham 2004). In other words, the stereotyped behavior may be a way of coping with an environment that is lacking some essential features, and those animals that show this coping mechanism are better off than those that do not. With this understanding, one may look very differently at individuals with particular behavioral profiles, since those that pace may be more effectively coping with their inadequate environment than those that do not.

Another behavioral topic that illustrates the necessity of understanding complex behavior is aggression. While aggression is a natural part of the lives of primates, severe and damaging aggression is also a significant animal welfare problem for some species that live in certain types of social groups. The answer to the question "Why are those rhesus macaques fighting?" is actually quite complicated. The chapter by McCowan and Beisner (2017) on understanding how to better manage large groups of macaques through the application of network analysis techniques exemplifies how a relatively superficial understanding of behavior was not adequate to address certain management problems. For many years, there have been reports of "cage wars" or "over throws" in large groups of macaques, and a few empirical evaluations of the factors that could account for this have been published (Bernstein and Ehardt 1985, 1986; Ehardt and Bernstein 1986; Westergaard et al. 1999), but the aggression continues and often seems "unpredictable."

McCowan's body of research now illustrates the rigor and depth of the behavioral data collection and statistical analysis that is required to come to meaningful conclusions concerning the management and prevention of such aggression. As McCowan and Beisner (2017) describe, the network analysis approach provides tools to assess less obvious patterns in the relationships among individuals, by statistically accounting for indirect relationships (i.e., "friends of friends") within complex social groups. This focus on indirect relationships requires the collection and analysis of data at multiple levels within social groups, information that may not necessarily be readily observed by those working with the animals. In fact, McCowan and Beisner state that "Little or no predictive power results from monitoring simple rates of aggression or other behaviors in these groups." Based on their findings, they have recommended a number of improvements to current practices. They also indicate that joint network modeling can reveal aberrant patterns in the social dynamics of groups and is useful in *detecting social instability before the collapse* of a group. Ultimately, being able to predict, and intervene to prevent, social collapse would be a major improvement in the care and management of large social groups of macaques.

ROLE OF BASIC BEHAVIORAL SCIENCE IN BEHAVIORAL MANAGEMENT

Applied behavioral research is directly focused on finding solutions to practical problems related to the management of primates, but basic research findings can also be very useful in applied settings. Several chapters in this book are great examples of how basic behavioral science can or might be applied to enhance the care of primates.

Animal Temperament

The study of animal temperament is one topic area currently undergoing this evolution in application, as described in the chapters by Capitanio (2017) and by Coleman (2017). Capitanio argues that by understanding and measuring individual variation in NHPs, we can then use that information to improve our ability to care for the individual primates. For example, the results of temperament testing can predict those primates that are more likely to develop certain behavioral problems, such as locomotor stereotypies, or certain health problems, such as diarrhea. Additionally, we can use individual differences to improve the quality of the science that is conducted with primates, since we know (based on the findings of Capitanio's work) that enduring behavioral characteristics of animals is often related to their physiological functioning, which may influence dependent measures in the biomedical sciences. Coleman points out that homogenous caging and enrichment are typically provided to laboratory primates, despite growing evidence that individual differences, based on temperament, might be better accommodated through offering different types of environments. She describes how social housing issues (such as weaning age or choosing social partners), selecting the best type of enrichment for an individual, or determining the steps in training for a particular procedure can all be influenced by temperament. Animals with a more exploratory temperament, for example, may progress more rapidly through training than those with an inhibited temperament. Similarly, temperament type could predict the best enrichment strategies for an individual, and help to guide social housing choices.

There is a growing interest from those involved with behavioral management of primates in learning to assess temperament, and to determine how this information can be used within their own programs [see McGrew (2017) for an excellent example of applying this to pairing macaques]. Recent conference presentations and instructional workshops on measuring temperament have been well-attended. In some cases, the results of temperament testing are now included in animal records, requested when monkeys are being purchased from vendors, and used to select social partners for introduction or to determine appropriate training plans for individuals. I doubt that such an application was envisioned by those first studying human temperament, or even initial, basic science work on animal temperament, but NHPs are now reaping the benefits, as this knowledge spreads within the behavioral management community.

Social Learning

The field of social learning as described by Hopper (2017) has not received as much attention as temperament from those responsible for behavioral management programs, but social learning may be one of the next fields of basic science to make this transition into application within behavioral management programs. Hopper details how a primate's learning can be affected by his/her social environment, with factors such as the individual's relationship to the "expert" (who exhibits a certain behavior) and the expert's own identity influencing the aspects of social learning that may be manifested. This type of approach might be used to encourage desirable forms of learning, including a monkey learning to cooperate with a research procedure. Hopper points out that, based on the present state of the literature, it seems that primates may learn "better" when observing live experts than when observing video recordings of experts. However, there is still much to be learned before firm recommendations can be made concerning the ways in which animal training programs can best be facilitated by our understanding of the social learning abilities of primates. The fascinating possibility that we could encourage prosocial behaviors in primate social groups (such as sharing), based on designing social learning opportunities for some group members, is an exciting area for future research and application.

Anti-predator Behavior

Caine (2017) described some basic science findings regarding the anti-predator behavior of primates, and this area of research seems to have some additional, untapped value related to the care of primates. Caine explained that primates involved in biomedical research have not lost their fundamental anti-predator instincts and that behavioral management programs should take these instincts into account when planning for the care of the animals. Three aspects of care that she contends could be modified are

1. Providing preferred sleeping locations for primates, such as sleeping boxes that are optimally designed and located for each species. Caine argues that the provision of these appropriate sleeping spaces would not only lead to a decrease in fear and an increase in sleep quality, but that this, in turn, could affect research results designed to assess measures of metabolic function, cognition, and immunology.
2. Examining routine management procedures and the effects of human presence on the behavior of the primates. Caine says that some situations that seem benign from a human perspective, may, in fact, lead to either defensive or fearful behaviors from the primates, which could lead to variability in experimental measures and a negative effect on welfare. It may be possible to alter these situations to reduce the animals' fear.
3. She suggests that designing captive experiences for some primate species to elicit species-typical anti-predator behavior may benefit the animals by promoting cooperative social behaviors, and perhaps even by reducing stress.

These recommendations concerning the ways in which our basic understanding of anti-predator behavior could be applied to the care of primates are insightful and novel, and will be interesting to apply.

Depression

Research on depression in primates was described by Shively (2017), and these findings have also not yet been fully incorporated into behavioral management programs. Shively provides a long list of physiological (e.g., heart rate, cholesterol, body weight, skeletal health, atherosclerosis) and neurobiological (e.g., 5-HT1A receptor binding potential, global reductions in hippocampus volume) measures that vary between monkeys that exhibit depressive behavior and those that do not. Since these are measures used in many types of research, they may be confounded in a wide array of studies. The techniques described by Shively to detect primates with symptoms of depression should be incorporated into the monitoring systems already in place for primate welfare assessment. For example, for macaques, behaviors such as slumped or collapsed body posture; immobility; lack of responsiveness to environmental events; time devoted to species-typical behaviors, such as locomotion; and time in contact with another monkey should all be included in behavioral data collection programs to assess welfare. Shively describes different sampling methods that successfully detect depressive behavior, and those criteria should be kept in mind when designing a behavioral system to detect depression. The prevention strategies, such as avoiding separation from long-term companions and avoiding single housing, and the treatment strategies she described, such as dietary manipulations, increasing exercise and reducing obesity, should be considered for monkeys expressing depressive behaviors. Although it is not yet known if these treatments will be successful in nonhuman primates, they should be evaluated. Successful treatment of depression would improve the welfare of the individual animal, as well as the quality of the research projects involving these monkeys, by reducing confounding influences.

Using the findings of this basic research, we can better detect and treat behavioral problems, such as fear and depression, and by reducing unexplained variability across subjects, simultaneously improve the quality of the research in which primates participate. To continue to advance the field of primate behavioral management, we must recruit individuals to the field who are not only committed to improving the welfare of the animals, but who are also trained in the natural history of the animal, and in interpreting and analyzing behavior. We must also recruit scientists with the theoretical background and skills needed to turn basic research in a variety of areas into practical and effective solutions for the management of captive primates.

STAFFING FOR BEHAVIORAL MANAGEMENT PROGRAMS

Because of the complicated, multifaceted nature of behavior, the myriad factors that can influence behavior, and the importance of proper evaluation of behavior to assess animal welfare, it is imperative that behavior and behavior measurement are well understood by those working with primates (also suggested by McCowan and Beisner, 2017). Animal care staff, veterinary staff, and research staff members all need to have a basic understanding of behavior. Those working within the behavioral management field need to know much more.

The American Society of Primatologists commented on these jobs in one of their recent public statements (2015; www.asp.org), stating that:

> There are roles for both Applied Primate Behavioral Scientists and for Primate Behavior Managers in the research laboratory environment. Applied Primate Behavioral Scientists study captive primate behavior in research laboratory conditions with the goal of understanding and improving their welfare. They typically hold a professional degree (PhD, MS) in a relevant discipline (e.g., Psychology, Anthropology, Zoology, Ethology, Behavioral Ecology, Animal Welfare, or Biology), have expertise in the science of animal behavior, welfare science, and in managing the behavior of captive nonhuman primates…[T]hey frequently work closely with a Primate Behavior Manager…[who] are often responsible for implementing behavioral management programs (e.g., social housing, enrichment, and animal training). They typically hold bachelor's degrees in a relevant field (e.g., Psychology, Anthropology, Zoology, Ethology, Behavioral Ecology, Animal Welfare, or Biology), have experience working directly with primates and are knowledgeable about the behavior of individual primates held at their institutions. … [Clearly] an advanced level of behavioral expertise is required to meet the spirit of the Animal Welfare Act and to guide any changes in standards for primate behavioral management.

A recent survey provides information concerning progress being made on this front. In U.S. research institutions with more than 100 primates, 41% of the primate behavioral management programs are overseen by individuals with PhDs, 37% by individuals with Doctorates in Veterinary Medicine, 7% by those with Master's degrees, and the remaining 15% by individuals with a Bachelor's or Associate's degree (Baker 2016). This advanced level of education indicates that many leaders do have the desired expertise in the science of behavior. As mentioned earlier, full-time technicians dedicated to implementing behavioral management programs are employed at 78% of the facilities surveyed, and the other 22% of facilities have individuals with responsibilities in behavioral management in addition to other responsibilities. Among the people working at the technical level (working directly with the primates), the highest level of educational experience is a Bachelor's or Associate's degree at 40% of facilities, a Master's degree at 24%, and a PhD at 15%, demonstrating that many of these individuals are also highly trained. When results were compared between an earlier survey in 2003 and the 2014 survey, it was revealed that the proportion of facilities employing technicians without college degrees has fallen from 25% to 8% during that period. Facilities also appear to be increasing their staffing of behavioral management employees, with primate-to-personnel ratios falling from 643:1 (2003) to 310:1 (2014) in facilities that employ full-time technicians and house most of their primates indoors. So, by multiple measures, staffing within behavioral management programs has made real advancements.

Behavioral management specialists are needed at the levels of PhD scientists, practitioners who manage the behavioral programs, and those working directly with the animals. The roles of conducting science as opposed to implementing daily programs are quite different but in some cases can be performed by the same person. In other situations, multiple people will be responsible for the various tasks. Continual interaction and mutual respect among the individuals carrying out these differing duties is essential to keep the field moving forward. Behavioral experts should be serving as a resource for others in the research setting (e.g., animal care, veterinary, and research staff). As Hutchinson (2017) stated, "It is important for veterinarians to note that behavioral management is a complicated field unto its own, not necessarily a sub-field of veterinary medicine, and that even in areas where veterinary medicine and behavior overlap, behavior scientists bring unique experience and expertise to bear." Behavioral management specialists should have formal and informal means of communicating with, and teaching, others about: the natural history and natural behavior of primates, identifying behavior, interpreting behavior, and treating and preventing behavioral problems. Formal classes, certification programs and workshops, libraries with relevant materials, group discussions, and daily interactions for sharing information are all needed. The information provided in this volume can serve as the basis for a good deal of the curriculum for some of this education. It will also be useful to have behavioral staff participate in veterinary rounds and clinical meetings, and participate in the training of any laboratory animal medicine residents or visiting veterinary students, as suggested by Hutchinson (2017).

To continue the growth of the behavioral management field, it will be necessary to develop formalized training programs for people working at the various levels described earlier. Training programs should include college-based programs that grant undergraduate and graduate degrees in animal welfare (there are a number of these in Europe already), and technical-level educational programs in primate behavioral management. Fostering partnerships between college programs in animal behavior, psychology or animal welfare, and primate research facilities, could be very helpful in this process. Incorporating more individuals with backgrounds in applied behavior analysis (see the following discussion) will also be good for the field of behavioral management. The conferences and meetings of organizations such as the American Society of Primatologists, the Association of Primate Veterinarians, and the American Association for Laboratory Animal Science are already important gathering places for the exchange of information and recent findings related to behavioral management.

THE FUTURE OF BEHAVIORAL MANAGEMENT PROGRAMS

We will surely continue to evolve and improve both the implementation of behavioral management programs and the science that underlies them. In the coming years, daily programs should be improved by a wider implementation of proven techniques, and of practices that are more customized for each individual primate.

Wider Implementation of Techniques to Promote Welfare

Chapters in this book have portrayed many of the significant steps forward in caring for laboratory primates and in promoting their welfare. New techniques (e.g., releasing monkeys into larger "activity cages" for periods of time, using positive reinforcement training for blood sample collection, providing hay as a substrate) have been developed and evaluated, and those evaluations are published in peer-reviewed journals. While these academic publications are an essential aspect of driving forward behavioral management practices, they are generally not sufficient to ensure that these new techniques are widely applied. For example, the value of periodically releasing a monkey into an activity cage has been repeatedly demonstrated (Kessel and Brent 1995a,b; Tustin et al.

1996; Griffis et al. 2013), but how many indoor, cage-housed monkeys are actually getting access to such an enriched experience? Although it has been established that primates can be trained for blood sampling (Coleman et al. 2008) and that this approach is less stressful (Lambeth et al. 2006), I estimate that only a small percentage of the blood samples collected have utilized that approach. I believe that a major goal for the future needs to be the fostering of the *broad implementation* of these advances in behavioral management.

For new and effective techniques to be applied/implemented by others, they must be further refined, endorsed, and encouraged. To facilitate putting new techniques into practice, those who develop them should try to promote them by teaching others about them in conference presentations, webinars, and workshops; writing about them in more informal ways to supplement more formal publications; and consulting with facilities considering the new approaches. To facilitate implementation, communication will be needed among staff in different roles in the organizations, including behavioral scientists, veterinarians, behavioral management technicians, colony managers, animal care supervisors and staff, shop staff, principal investigators, and administrators. People in these roles will have different types of questions concerning the new techniques. For example, if a new technique is to use hay as a substrate for monkeys living in indoor/outdoor runs, a behavioral scientist may ask about which behaviors changed significantly with the use of hay and about habituation over time; a veterinarian may ask about handling animals with possible allergies to the hay or possible gut impactions from consuming it; an animal care supervisor may want to know about clean up and disposal of the hay; and administrators may ask about the associated costs and savings. This type of communication is important before a new technique is applied for many primates across different facilities. The creation of mechanisms to facilitate this type of practical discussion would be a useful enhancement to the field of behavioral management.

Individually Customized Behavioral Management Practices

Tailoring behavioral management practices to individual animals is another direction of growth for the implementation of behavioral management programs in the years ahead. As Capitanio (2017) succinctly states, "It is my thesis that understanding this variation, measuring it, and using that information, can improve both the science that is done with nonhuman primates and the ability to care for our animals." (p. 1 of his chapter). The chapters by Capitanio (2017) and Coleman (2017) lay the foundation for this approach to understanding individual differences, and to planning optimal behavioral management strategies for those individuals. The studies by McCowan and Beisner (2017) also incorporate an individual differences approach to selecting adult male macaques who may be best suited to serve as breeding males in large groups, such as choosing those with the best ability to "police" the behavior of other group members and thereby help maintain group stability.

Capitanio (2017) and Coleman (2017) describe many domains in which individual variation could be accommodated to improve the welfare of our research primates. These include varying the social conditions for primates (e.g., their age when weaned from mother, choosing the most appropriate partners for pairing), based on some of their individual characteristics; developing training routines for individuals that emphasize the needs of that animal (e.g., more time devoted to desensitizing them); and providing enrichment items on a schedule that will maximize use by individuals (since more inhibited animals might respond to novelty with some hesitation, a program with more time for habituation could be helpful). An individualized approach may also help predict those animals that are more likely to develop motor stereotypies or alopecia, and the individualized treatment that may best promote their welfare. For example, since macaques with exploratory temperaments have been shown to be more likely to have alopecia, as described by Coleman (2017), they may benefit from additional enrichment that may effectively reduce hair plucking behavior.

Another possible avenue for modifying behavioral management approaches based on the characteristics of individual primates is to emphasize the role of individual choice. Choice is recognized as an important factor in promoting well-being (Schapiro and Lambeth 2007; National Research Council 2011; Coleman et al. 2012), and many of our behavioral management programs could benefit from providing more agency to individual animals in determining various aspects of their environments. Could environments be designed so that primates could choose their own social group members? (See the CHOICE primate cage http://www.britzco.com/CatalogItem.aspx?ProductId=102300 as an example.) Could primates have spaces within their enclosures in which they could choose to be alone for some period? Could they choose between options for fresh produce during an enrichment feeding, or choose between different videos to watch? Using a preference testing method prior to training sessions, so that a monkey's most preferred foods can be used as reinforcers (Clay et al. 2009), is another way of allowing the animals more choice. Considerable choice is provided to the animal when PRT methods are used exclusively (with no negative reinforcement employed), and the training progresses at the pace set by the individual primate being trained (Laule et al. 2003).

Some of the individual differences described by Capitanio (2017) and by Coleman (2017) can also be used as tools to select the most appropriate subjects for certain research studies. We may be able to pick monkeys who will be less disturbed by the procedures for a particular study. For example, monkeys with "bold" temperaments can be selected for studies that involve chair restraint, since those animals may have an easier time being trained and working in close proximity to humans (Coleman 2012). Temperament characteristics are also associated with some of the physiological responses that may be important to consider for research projects. For example, temperament may influence the dependent variables in studies that measure antibody responses (as cited in Coleman (2017)); therefore, the temperament of subjects should be considered to reduce variability in those types of studies.

To fully embrace this customized approach to behavioral management will require additional resources. Behavioral management specialists will need to work with investigators and veterinarians at their facilities to determine which characteristics would be the most relevant to measure at their institutions; depending on the animal population, the research being conducted, and the cost. These characteristics might include behavioral, physiological, emotional, and other measures. Robust methods to systematically carry out the measurements will then need to be determined, in addition to decisions related to the ways in which the data will be maintained, stored, accessed, and used. Staff may need training to accomplish these tasks.

THE FUTURE OF BEHAVIORAL MANAGEMENT RESEARCH

In addition to enhancements on the implementation side of behavioral management programs (described previously), there are many other areas in need of additional research.

Evaluating the Impact of Behavioral Management Issues on Biomedical Research

Some chapters in this book have made reference to the impact of behavioral management practices on various dependent measures used in biomedical research studies, but considerably more research effort should be devoted to this issue. Since primates are involved in a huge variety of biomedical research studies, there is a very long list of measures that might be impacted by behavioral management practices, either positively (reducing variability) or negatively (increasing variability), and these should be detailed. Since we know that effective behavioral management practices are defined as those that change behavior, and often also change associated physiological measures (such as reducing measures associated with stress), it is certainly possible that measures relevant to studies are also altered. As Honess (2017) points out, research procedures can have a confounding impact on certain study measures, and behavioral management efforts, such as training primates to cooperate with procedures, can reduce confounds by normalizing cortisol levels and enzyme values used in certain studies, and by reducing fear. There are many additional examples of behavioral management techniques enhancing the quality of data collected for research studies. Evaluating the impact of manipulations, such as social housing, training methods, and living in larger and more enriched enclosures must be completed to improve or sustain the integrity of the research in which our primates are involved. There will undoubtedly be situations where behavioral management practices have negative impacts, but these must be identified and managed, rather than left unrecognized. This type of research will be best accomplished by behavioral scientists taking the opportunity to work with veterinarians and biomedical investigators to test these ideas, and to design solutions to discovered problems that balance animal welfare with research quality.

Another benefit of conducting such studies is that we will be broadening the assessments of animal welfare beyond behavioral measures. Even though behavior is important, comprehensive assessments of welfare should also include physiological, health, and emotional measures. The fine science that has developed around the study of domestic animal welfare has incorporated a large number of health and behavioral measures, in addition to evaluating human/animal relationships. The biobehavioral assessment program described by Capitanio (2017) is probably the very best example of such a comprehensive approach being taken with NHPs, and should serve as a model for many others to emulate. I propose that such an approach with NHPs would help to drive our science forward and would lead to a more thorough understanding of NHP welfare. Again, this would also benefit from behavioral scientists and veterinarians working together on these welfare topics.

Addressing More Challenging Situations through Research

In the future, behavioral management research needs to address some of the more difficult animal welfare problems that remain. When enrichment was first getting started, the most common practices were easy to implement; providing simple toys to monkeys, for example. These practices did not interfere with research for the most part, could be accommodated by animal care procedures, and were safe and fairly inexpensive. As the field matured, more difficult practices have been initiated; practices that slow down research, that require a great deal more time to maintain, that are more expensive, that require new skills in staff members, and that may even lead to health problems for some animals (e.g., social housing that may involve wounding and other stress), and yet, also allow for meaningful improvements in animal welfare. There are other aspects of caring for laboratory primates that should still be addressed by behavioral management research. I will give three examples of such areas—advancing animal training, the issue of minimum cage size, and the retirement of research primates. These issues will not be easy to study, and will require the commitment of more human and monetary resources than have been invested in the past. However, if built upon a firm foundation of behavioral science, this work could lead the way to improved animal welfare.

Animal Training Advancement and Assessment

First, animal training techniques should be further refined to more fully address difficult situations. I agree with Honess (2017) that skilled trainers can help resolve extraordinary challenges to welfare. He describes the creative approach by Westlund et al. (2012) who taught long-tailed macaques a cue to precede loud construction noise, which increased predictability of the noise and therefore reduced the stress they experienced. Another area that needs more attention is training primates to comply with veterinary care procedures. Magden (2017) described great success with the use of acupuncture, laser therapy, and physical therapy, all based on PRT, to gain access to the animals and to elicit cooperation in their care. Surely primates can be trained for other, complex medical treatments that require long periods of cooperation, and even some that involve some discomfort for the animals. There is an important opportunity to advance the methods used for restraint (e.g., manual restraint, use of sliding squeeze-back mechanisms in cages, primate restraint chairs). There has been some work done on refinement (Graham et al. 2012; Bliss-Moreau et al. 2013; McMillan et al. 2014), but much more is needed to determine the most effective training techniques, the necessary timeframes for preparation of the animals, and the selection of the most appropriate animals for studies requiring restraint (Coleman 2012). Animal training expertise should also be focused to address the use of food or fluid control in research studies. I anticipate that expert trainers with a thorough understanding of operant and classical conditioning procedures, who understand: the use of various schedules of reinforcement and their effects on animal responses, the impact of varying the magnitude of reinforcement, the use of conditioned reinforcers, and many other intricacies of training methods, would be able to reduce the need for food and fluid control to motivate monkeys to respond (Westlund 2011). The use of training methods to minimize fear should also be more fully exploited. Fear, including fear of people caring for them, can have a major negative impact on welfare in domestic species (Waiblinger et al. 2006), and is likely to be even more detrimental in nondomestic species, such as nonhuman primates. PRT methods have not been fully realized in the care and management of large groups of primates. There is opportunity to refine these techniques to facilitate shifting, accessing, and collecting biological samples from primates in outdoor breeding groups. PRT could also be used to facilitate social introductions and to help manage social dynamics in these groups. Finally, I think training can be used to better address abnormal and other problematic behaviors in research primates (Bloomsmith et al. 2007; Martin 2017). The psychology specialty area of applied behavioral analysis uses operant conditioning to increase adaptive behaviors and decrease unwanted behaviors among humans in clinical settings, and has been highly successful in treating self-injurious behaviors and stereotypies (Kahng et al. 2002; Kurtz et al. 2003). If we model NHP analytical and treatment methods after their successful application in humans, similar success might be achieved, as has already been found in some early studies (Dorey et al. 2009; Martin et al. 2011; Farmer-Dougan 2014). All of these advancements in animal training will require expert trainers, including those with academic backgrounds in applied behavioral analysis, suggesting that the field needs to invest in employing such individuals.

Scientific Assessment of Cage Size

Identifying the appropriate minimum cage size for research primates is a thorny issue. Many research primates live in enclosures much larger than the required minimum space, and there are recent innovative attempts to develop other types of larger caging such as indoor pens attached to smaller caging to facilitate animal access for research. These types of modernizations should continue and should be carefully evaluated. But for those primates living in minimum-sized cages, the important question remains as to whether their behavioral needs are met. Some studies have shown that modest changes in cage size do not improve their welfare. A study by Crockett et al. (2000) found that cage sizes ranging from 20% to 148% of (United States Department of Agriculture) USDA regulation floor area did not lead to changes in abnormal behavior, self-grooming, manipulating the environment, eating/drinking, activity cycle, cortisol excretion, or biscuit consumption. Locomotion and the frequency of behavior change were significantly reduced in the smallest cage, but were unchanged in cages ranging from 77% to 148% of regulation size. Therefore, it appears that moderate (up to 48%) increases in cage size have not currently been shown to have a meaningful impact on the animals. Whether larger increases in cage size would be beneficial will need to be the subject of future studies; however, significant increases may not be feasible in the near future. We also know that larger but empty spaces are not likely to improve behavior; instead, the space must function to encourage activity and to make interesting activities available to the primates.

Increasing cage size requirements will demand a tremendous outlay of resources for new cages, possibly larger rooms, extensive facility modifications and buildings to hold the cages, new animal management procedures to care for the animals, new research equipment to adapt to different caging, and simultaneously may reduce the animal holding capacity of existing facilities, therefore having a detrimental impact on research. Despite these many difficulties, this issue should be addressed with practical short- and long-term solutions in mind. Effective solutions should be launched as soon as solid science can begin to lead the way. Questions should be experimentally addressed that might define effective compromises between space, cost, and animal welfare. Could living in larger caging for intermittent periods of time substantially improve behavior? For example, could the routine use of activity cages on a daily or weekly basis be sufficient to avoid the development of behavioral problems? Could arrangements with traditional caging that are linked to larger "playground" spaces be beneficial? Is age of the primate an issue? For example, should younger and more active animals be prioritized to be kept in the largest caging that is available? Does temperament play a role in the cage environment needed to support the welfare of individual primates? I am advocating for additional scientific research to support alternative approaches that are practical to implement in the research environment (such as the suggestions described earlier), and that will make the optimal use of limited funding.

Evaluating Retirement of Research Primates

There is a growing movement of relocating research primates into retirement or sanctuary facilities at the end of their research careers (Seelig and Truitt 1999). These sanctuaries should provide the animals with an opportunity to live the remainder of their lives in socially and environmentally enriched surroundings, in some cases freeing up space for additional primates at research facilities. Retirement of a large number of research chimpanzees was established first (Brent et al. 1997; Noon 1999), and there are now multiple accredited sanctuaries that care for chimpanzees that formerly lived in laboratories (www.chimpcare.org). There is a growing interest in retiring other primates after the conclusion of their involvement in research (Kerwin 2006; https://asp.org/welfare/retirement.cfm; http://www.junglefriends.org). In a recent survey of primate veterinarians working in research facilities, 55% of them reported having been involved in the retirement of at least one monkey (Turner, P., in preparation).

Practices related to retiring research primates to sanctuary facilities should be thoroughly studied by those with expertise in behavioral management. The purpose of sanctuaries for primates is to improve their welfare, and this should be demonstrated. While the positive value of primates living in large social groups, and in large and more complex outdoor enclosures, is well established, it does not mean that every primate will benefit from moving to every sanctuary. Transfer to a sanctuary may involve the primate being removed from compatible social partners and from familiar people; moving to a different climate; living in very different types of enclosures with more physical, environmental, and social demands on the animal (e.g., moving to a large outdoor enclosure with varied terrain and a large social group); being introduced to unfamiliar conspecifics; dealing with complex social dynamics within a new group; being cared for with different methods; and changes in diet, enrichment, and animal training procedures. These changes are undoubtedly challenging for the primates, but there are very few evaluations published to indicate the degree of challenge, the ability of the animals to adjust to those challenges, and the long-term benefits they may experience.

There are many questions to be empirically addressed including: how to select the best candidates for retirement (e.g., should aged animals be retired?), the effects of the transportation to the sanctuary on health and welfare, and the animals' adjustment to new facilities and (possibly) new social groups. Animal welfare scientists should carefully evaluate the pros and cons of such moves, so that the short-term and long-term costs and benefits can be weighed. An individualized approach should be studied to assure that the welfare of each unique primate can be carefully considered when deciding whether and how that individual should be retired. Primates with certain early rearing experiences, of certain ages, with some health problems, or with certain temperamental characteristics may not be as well suited to transfer to a sanctuary as others. All of these issues can, and should, be addressed with rigorous, quantitative assessment methods that require collaboration between qualified behavioral scientists and veterinarians at both research and sanctuary facilities.

CONCLUSIONS

The primate research community has made remarkable progress in addressing the psychological well-being of their primates. The development of formal behavioral management programs has been a pivotal means for achieving this, and behavioral management has matured to be recognized as a specialty within primate care. This current *Handbook of Primate Behavioral Management* summarizes these advances, will be a valuable resource to those caring for laboratory primates, and will help set the course for future research needed to guide behavioral management in the most meaningful directions.

One major theme of this book is the value of applying a scientific understanding of primate behavior to the animals' care. Many of the prominent issues in managing primates are fundamentally behavioral, and we are recognizing that behavioral specialists can begin to understand and address these issues to benefit the animals under our care, and to improve the quality of the resulting scientific research. Basing the behavioral care of a species on a thorough understanding of that species' natural history is a long-standing tenet of behavioral management. Having staff members who know each individual monkey's behavior is important, as is having staff members with a scientific understanding of behavior. These dual aspects of behavioral management are both important. In the coming years, the daily implementation of programs should be improved by a wider implementation of proven techniques, and by developing behavioral management practices that are more customized to the individual primate. In the future, behavioral management research should focus on evaluating the effects of behavioral management practices on dependent measures used in biomedical research studies, on advancing animal training approaches, on determining minimum cage sizes, and on issues related to the retirement of research primates.

As we continue to nurture both the daily implementation of behavioral management and the research that serves as its foundation, we should keep in mind the "bias for action" that Michale Keeling championed at his own facility, advocated to the laboratory animal science community, and wrote about (Keeling et al. 1991). Keeling believed that we should complement theoretical discussions and slow-moving science with a more aggressive approach to doing things to address the psychological needs of our primates. He wrote that the absence of complete data should not preclude our immediate action, even though that action may be based on less information than we might like. This continuing back-and-forth of doing things we believe will benefit the animals, and carefully evaluating the outcomes, will serve us best as we progress in caring for research primates.

ACKNOWLEDGMENTS

I thank Steve Schapiro for the opportunity to write this chapter and for his continuing efforts to bring people together on the topic of primate behavioral management. Thanks also to Allison Martin and Susan Haverly for their assistance with the chapter. I acknowledge support of the Yerkes National Primate Research Center by the Office of Research Infrastructure Programs through grant P51OD011132.

REFERENCES

Baker, K. C. 2016. Survey of 2014 behavioral management programs for laboratory primates in the United States. *American Journal of Primatology* 78:780–96.

Bernstein, I. S. and C. L. Ehardt. 1985. Age-sex differences in the expression of agonistic behavior in rhesus monkey (*Macaca mulatta*) groups. *Journal of Comparative Psychology* 99:115–32.

Bernstein, I. S. and C. L. Ehardt. 1986. The influence of kinship and socialization on aggressive behaviour in rhesus monkeys (*Macaca mulatta*). *Animal Behaviour* 34:739–47.

Bettinger, T. L., Leighty, K. A., Daneault, R. B., et al. 2017. Behavioral management: The environment and animal welfare, Chapter 4. In *Handbook of Primate Behavioral Management*, 37–51, ed. Schapiro, S. J. Boca Raton, FL: CRC Press.

Bliss-Moreau, E., Theil, J. H., and G. Moadab. 2013. Efficient cooperative restraint training with rhesus macaques. *Journal of Applied Animal Welfare Science* 16:98–117.

Bloomsmith, M. A., Marr, M. J., and T. L. Maple. 2007. Addressing nonhuman primate behavioral problems through the application of operant conditioning: Is the human treatment approach a useful model? *Applied Animal Behaviour Science* 102:205–22.

Brent, L., Butler, T. M., and J. Haberstroh. 1997. Surplus chimpanzee crisis: Planning for the long-term needs of research chimpanzees. *Lab Animal* 26:36–9.

Caine, N. G. 2017. Anti-predator behavior: Its expression and consequences in captive primates, Chapter 9. In *Handbook of Primate Behavioral Management*, 127–138, ed. Schapiro, S. J. Boca Raton, FL: CRC Press.

Capitanio, J. P. 2017. Variation in biobehavioral organization, Chapter 5. In *Handbook of Primate Behavioral Management*, 55–73, ed. Schapiro, S. J. Boca Raton, FL: CRC Press.

Clay, A. W., Bloomsmith, M. A., Marr, M. J., et al. 2009. Systematic investigation of food preferences in captive orangutans. *Journal of Applied Animal Welfare Science* 12:306–13.

Coleman, K. 2012. Individual differences in temperament and behavioral management practices for nonhuman primates. *Applied Animal Behaviour Science* 137:106–13.

Coleman, K. 2017. Individual differences in temperment and behavioral management, Chapter 7. In *Handbook of Primate Behavioral Management*, 255–264, ed. Schapiro, S. J. Boca Raton, FL: CRC Press.

Coleman, K., Bloomsmith, M. A., Crockett, C. M., et al. 2012. Behavioral management, enrichment and psychological well-being of laboratory nonhuman primates. In *Nonhuman Primates in Biomedical Research*, eds Abee, C. R., Mansfield, K., Tardiff, S., et al., 149–76. New York: Academic Press.

Coleman, K., Pranger, L., Maier, A., et al. 2008. Training rhesus macaques for venipuncture using positive reinforcement techniques: A comparison with chimpanzees. *Journal of the American Association for Laboratory Animal Science* 47:37–41.

Crockett, C. M., Shimoji, M., and D. M. Bowden. 2000. Behavior, appetite, and urinary cortisol responses by adult female pigtailed macaques to cage size, cage level, room change, and ketamine sedation. *American Journal of Primatology* 52:63–80.

Dorey, N. R., Rosales-Ruiz, J., Smith, R., et al. 2009. Functional analysis and treatment of self-injury in a captive olive baboon. *Journal of Applied Behavior Analysis* 42:785–94.

Dye, M. H. 2017. Behavioral management of prosimians, Chapter 25. In *Handbook of Primate Behavioral Management*, 435–457, ed. Schapiro, S. J. Boca Raton, FL: CRC Press.

Ehardt, C. L. and I. S. Bernstein. 1986. Matrilineal overthrows in rhesus monkey groups. *International Journal of Primatology* 7:157–81.

Erwin, J., Maple, T. L., and G. Mitchell. 1979. *Captivity and Behavior: Primates in Breeding Colonies, Laboratories, and Zoos.* New York: Van Nostrand Reinhold.

Farmer-Dougan, V. A. 2014. Functional analysis of aggression in a black and white ruffed lemur (*Varecia variegata variegata*). *Journal of Applied Animal Welfare* 17:283–93.

Gottlieb, D., Coleman, K., and K. Prongay. 2017. Behavioral management of *Macaca* species (except *M. fascicularis*), Chapter 19. In *Handbook of Primate Behavioral Management*, 279–303, ed. Schapiro, S. J. Boca Raton, FL: CRC Press.

Graham, M. L., Rieke, E. F., Mutch, L. A., et al. 2012. Successful implementation of cooperative handling eliminates the need for restraint in a complex nonhuman primate disease model. *Journal of Medical Primatology* 41:89–106.

Griffis, C. M., Martin, A. L., Perlman, J. E., et al. 2013. Play caging benefits the behavior and activity of singly-housed rhesus macaques (*Macaca mulatta*) in the laboratory. *Journal of the American Association for Laboratory Animal Science* 52:534–40.

Honess, P. 2017. Behavioral management of long-tailed macaques (*Macaca fascicularis*), Chapter 20. In *Handbook of Primate Behavioral Management*, ed. Schapiro, S. J. Boca Raton, FL: CRC Press.

Hopper, L. M. 2017. Social learning and decision making, Chapter 15. In *Handbook of Primate Behavioral Management*, 225–241, ed. Schapiro, S. J. Boca Raton, FL: CRC Press.

Hutchinson, E. 2017. The veterinarian—Behavioral management interface, Chapter 14. In *Handbook of Primate Behavioral Management*, ed. Schapiro, S. J. Boca Raton, FL: CRC Press.

Institute for Laboratory Animal Research (U.S.). 1998. *The Psychological Well-Being of Nonhuman Primates*. Washington, DC: National Academies Press.

Jorgensen, M. J. 2017. Behavioral management of *Chlorocebus* spp., Chapter 21. In *Handbook of Primate Behavioral Management,* 339–365, ed. Schapiro, S. J. Boca Raton, FL: CRC Press.

Kahng, S., Iwata, B. A., and A. B. Lewin. 2002. Behavioral treatment of self-injury, 1964 to 2000. *American Journal on Mental Retardation* 107:212–21.

Keeling, M. E., Alford, P. L., and M. A. Bloomsmith. 1991. Decision analysis for developing programs of psychological well-being: A bias-for-action approach. In *Through the Looking Glass*, eds Novak, M. A. and A. J. Petto, 57–65. Washington, DC: American Psychological Association.

Kerwin, A. M. 2006. Overcoming the barriers to the retirement of old and new world monkeys from research facilities. *Journal of Applied Animal Welfare Science* 9:337–47.

Kessel, A. L. and L. Brent. 1995a. An activity cage for baboons, part I. *Contemporary Topics in Laboratory Animal Science* 34:74–9.

Kessel, A. L. and L. Brent. 1995b. An activity cage for baboons, part II: Longterm effects and management issues. *Contemporary Topics in Laboratory Animal Science* 34:80–3.

Kurtz, P. F., Chin, M. D., Huete, J. M., et al. 2003. Functional analysis and treatment of self-injurious behavior in young children a summary of 30 cases. *Journal of Applied Behavior Analysis* 36:205–19.

Lambeth, S. P., Hau, J., Perlman, J. E., et al. 2006. Positive reinforcement training affects hematologic and serum chemistry values in captive chimpanzees (*Pan troglodytes*). *American Journal of Primatology* 68:245–56.

Laule, G. E., Bloomsmith, M. A., and S. J. Schapiro. 2003. The use of positive reinforcement training techniques to enhance the care, management, and welfare of primates in the laboratory. *Journal of Applied Animal Welfare Science* 6:163–74.

Magden, E. R. 2017. Positive reinforcement training and health care, Chapter 13. In *Handbook of Primate Behavioral Management*, ed. Schapiro, S. J. Boca Raton, FL: CRC Press.

Martin, A. L. 2017. The primatologist as a behavioral engineer. *American Journal of Primatology* 79(1):1–10. doi:10.1002/ajp.22500.

Martin, A. L., Bloomsmith, M. A., Kelley, M. E., et al. 2011. Functional analysis and treatment of human-directed undesirable behavior exhibited by a captive chimpanzee. *Journal of Applied Behavior Analysis* 44:139–43.

Mason, G. J. and N. R. Latham. 2004. Can't stop, won't stop: Is stereotypy a reliable animal welfare indicator? *Animal Welfare* 13:S57–S69.

McCowan, B. and B. Beisner. 2017. Utility of systems network analysis for understanding complexity in primate behavioral management, Chapter 11. In *Handbook of Primate Behavioral Management*, 157–183, ed. Schapiro, S. J. Boca Raton, FL: CRC Press.

McGrew, K. 2017. Pairing strategies for cynomolgus macaques, Chapter 17. In *Handbook of Primate Behavioral Management*, 255–264, ed. Schapiro, S. J. Boca Raton, FL: CRC Press.

McKenna, J. J. 1979. The evolution of allomothering behavior among colobine monkeys: Function and opportunism in evolution. *American Anthropologist* 81:818–40.

McMillan, J., Perlman, J. E., Galvan, A., et al. 2014. Refining the pole-and-collar method of restraint: Emphasizing the use of positive training techniques with rhesus macaques (*Macaca mulatta*). *Journal of the American Association for Laboratory Animal Science* 53:61–8.

National Research Council. 2011. *Guide for the Care and Use of Laboratory Animals*. Washington, DC: National Academic Press.

Noon, C. 1999. Chimpanzees and retirement. *Journal of Applied Animal Welfare Science* 2:141–6.

Novak, M. A. and A. J. Petto. 1991. *Through the Looking Glass: Issues of Psychological Well-Being in Captive Nonhuman Primates*. Washington, DC: American Psychological Association.

Schapiro, S. J. and S. P. Lambeth. 2007. Control, choice, and assessments of the value of behavioral management to nonhuman primates in captivity. *Journal of Applied Animal Welfare Science* 10:39–47.

Seelig, D. and A. Truitt. 1999. Postresearch retirement of monkeys and other nonhuman primates. *Laboratory Primate Newsletter* 38:1–4.

Segal, E. 1989. *Housing, Care and Psychological Wellbeing of Captive and Laboratory Primates*. Park Ridge, NJ: Noyes Publications.

Shively, C. A. 2017. Depression in captive nonhuman primates: Theoretical underpinnings, methods, and application to behavioral management, Chapter 8. In *Handbook of Primate Behavioral Management*, 115–125, ed. Schapiro, S. J. Boca Raton, FL: CRC Press.

Tustin, G. W., Williams, L. E., and A. G. Brady. 1996. Rotational use of a recreational cage for the environmental enrichment of Japanese macaques (*Macaca fuscata*). *Laboratory Primate Newsletter* 35:5–8.

Waiblinger, S., Boivin, X., Pedersen, V., et al. 2006. Assessing the human–animal relationship in farmed species: A critical review. *Applied Animal Behaviour Science* 101:185–242.

Westergaard, G. C., Izard, M. K., Drake, J. H., et al. 1999. Rhesus macaque (*Macaca mulatta*) group formation and housing: wounding and reproduction in a specific pathogen free (SPF) colony. *American Journal of Primatology* 49:339–47.

Westlund, K. 2011. Can conditioned reinforcers and variable-ratio schedules make food- and fluid control redundant? A comment on the NC3Rs Working Group's report. *Journal of Neuroscience Methods* 204:202–5.

Westlund, K., Fernström, A. L., Wergård, E. M., et al. 2012. Physiological and behavioural stress responses in cynomolgus macaques (*Macaca fascicularis*) to noise associated with construction work. *Laboratory Animals* 46:51–8.

Williams, L. and C. N. Ross. 2017. Behavioral management of neotropical primates: *Aotus*, *Callithrix*, and *Saimiri*, Chapter 24. In *Handbook of Primate Behavioral Management*, 409–434, ed. Schapiro, S. J. Boca Raton, FL: CRC Press.

Index

A

Abnormal Behavior Ethogram, 14–16
Abnormal behaviors, 222, 236
    in baboons, 373–374
    in long-tailed macaques, 320–323
    in macaques, 290–292, 291*t*
    prevention of, 294–295
    stress in, 75–89
    in vervets, 355–356
Abnormal mouth movements, 373
Actigraph systems, 464
Activity, 62, 63, 66, 307–308
Acupuncture, 209–211, 249
Adaptation, 58
Adaptiveness of depressive behavior, 121–122
Addison's disease, 81
Adrenal insufficiency (Addison's disease), 81
Adrenocortical activity, 80
Adrenocorticotrophic hormone (ACTH), 59, 77, 85
African crowned eagles (*Stephanoaetus coronatus*), 128
African green monkeys (AGMs), *see Chlorocebus* spp.
Africa, NHP research in, 32
Aggression, 63, 416, 439–440, 500
    problem of, 172
    and wounding, 356
Aggressive behaviors, 82, 439, *see also* Abnormal
        behaviors
    in baboons, 270
    in vervets, 345
Aggressive/threat vocalization, 372
Airway hyperresponsiveness, 97
Alarm bark, 372
Allomaternal care, 423
Allport, Gordon, 58
Alopecia, 18, 83–84, 88, 108, 324, 357, 358, 505
Alopecia Scoring Scale, 18–20
Ambulatory jacket, 463, 468
American Association for Laboratory Animal Science,
    504
American College of Laboratory Animal Medicine
    (ACLAM), 218
American Journal of Primatology, The, 218
American Sign Language, 385
American Society of Primatologists, 503, 504
American Veterinary Medical Association (AVMA), 270
Amygdala, 129
Anesthesia, 202, 212
Anesthetization introductions, 259, 259*t*
Animal and Biological Materials Resource (ABMR), 346
"Animal as client" model, 38
"Animal-as-interconnected-organ-systems," 220
Animal Behaviour (journal), 218
Animal Care Policy Manual, 30
Animal management, and positive reinforcement training,
    446–447

Animal planning model
    enclosure design, 47–48
    individual history, 47
    natural history, 46–47
"Animal Rights and Human Morality," 220
Animal temperament, 501
Animal training advancement and assessment, 507–508
Animal Transfer Form, 20–21
Animal welfare
    environment and, 37–49
    infrastructure, 38
Animal Welfare Act (AWA), 13, 217, 271, 374
    exemptions, 14*t*
    regulations, 271, 281
Animal Welfare Information Center, 218
Anointing behavior, 40
Anticipatory behaviors, in primates, 451
Antipredator behavior
    and captive management
        humans as predators, 132–133
        sleeping, 130–132
        using antipredator reactions, 133–134
    in captive primates, 129–130, 502
    fear module, 129
    predators of primates, 128–129
    visual evolution, 129
Anxiety, 133, 230
Anxiolytic drugs, 133
Anxious behavior, 98
Aotus azarae, 411, 412
Aotus lemurinus, 411
Aotus nancymaae, 411
Aotus nigriceps, 411
Aotus species (owl monkeys)
    enrichment, 412–414
    natural history, 410–411
    social housing, 411–412
    temperament and behavioral management, 101
    training, 213, 414
Aotus vociferans, 411
Ape Cognition and Conservation Initiative (ACCI), 396*f*
Applied Primate Behavioral Scientists, 503
Arboreal locomotion, 309
"Artificial fruit" device, 247, 248*f*
Asio madagascariensis, 128
Assamese (*Macaca assamensis*) macaques, 306
Association for Assessment and Accreditation of
        Laboratory Care (AAALAC), 28
Association of Primate Veterinarians (APV), 29, 466, 504
Association of Zoos and Aquariums (AZA), 451
Assortativity, 167
Auditory enrichment, 287–288, 393, 443
Auscultation, 202, 203*f*
Australia, NHP research in, 32–33
Aye-aye (*Daubentonia madagascariensis*), 436–437, 441,
        442, 448*f*, 449, 452

Printed and bound by CPI Group (UK) Ltd, Croydon, CR0 4YY

24/10/2024

01778290-0012